T0135315

Book of Abstracts of the 68th Annual Meeting of the European Federation of Animal Science

EAAP
European Federation of Animal Science

The European Federation of Animal Science wishes to express its appreciation to the
Ministero delle Politiche Agricole Alimentari e Forestali (Italy) and the
Associazione Italiana Allevatori (Italy)
for their valuable support of its activities.

Book of Abstracts of the 68th Annual Meeting of the European Federation of Animal Science

Tallinn, Estonia, 28 August – 1 September 2017

EAAP Scientific Committee:

E. Strandberg
G. Savoini
H.A.M. Spoolder
H. Sauerwein
M. Tichit
M. Klopčič
J. Conington
E.F. Knol
A.S. Santos
T. Veldkamp
I. Halachmi
G. Pollott

EAN: 9789086863129
e-EAN: 9789086868599
ISBN: 978-90-8686-312-9
e-ISBN: 978-90-8686-859-9
DOI: 10.3920/978-90-8686-859-9

ISSN 1382-6077

First published, 2017

© Wageningen Academic Publishers
The Netherlands, 2017

Wageningen Academic
P u b l i s h e r s

Welcome to Tallinn, Estonia

On behalf of the Estonian Organising Committee, it is my pleasure to welcome you to the 68th Annual Meeting of the European Federation of Animal Science at the Solaris Tallinn. It is our first opportunity to host the EAAP Annual Meeting – the Europe's largest animal scientific conference.

The main theme of this years' meeting is Patterns of Livestock Production in the Development of Bioeconomy, which is a hot topic worldwide and a very appropriate subject in view of the current challenges for both human society and livestock industry. Knowledge-based innovation in the livestock sector is needed to integrate animal production into a viable bioeconomy value chain and ensure food security along with improving animal health and reducing environmental impacts. The programme will cover the latest findings and views on developments in animal genetics, health and welfare, nutrition, physiology, livestock farming systems, precision livestock farming, insects and cattle, horse, pig, sheep and goat production, as well as their allied industries.

The participants will have the opportunity to attend a very interesting scientific programme, to meet scientists working with a wide range of animal species and in various disciplines, to make new contacts and discuss the latest developments in animal sciences. The social events will offer the participants a unique occasion to get a glimpse of Estonian culture and the beautiful city of Tallinn.

We hope that all of you will have a very productive meeting and that you will enjoy the social events and our warm and friendly atmosphere.

Toomas Kevvai

Chairman of the Estonian Organising Committee
Deputy Secretary General for Food Safety, Research and Development
Ministry of Rural Affairs of the Republic of Estonia

National organisers of the 68th EAAP Annual Meeting

Estonian National Organising Committee

Ministry of Rural Affairs of the Republic of Estonia
- **Mr Toomas Kevvai,** chairman
- **Mr Martin Minjajev,** deputy chairman
- **Ms Pille Tammemägi**
- **Ms Maria Liisa Luur**
- **Ms Moonika Kuningas**
- **Ms Janika Salev**
- **Ms Kai Kasenurm**
- **Ms Inda Vaht**
- **Ms Eva Lehtla**
- **Ms Kaisa Väärtnõu**
- **Ms Sirje Jalakas**

The Veterinary and Food Board
- **Ms Katrin Reili,** deputy chairman
- **Ms Anneli Härmson**

Estonian University of Life Sciences
- **Prof Haldja Viinalass,** deputy chairman
- **Prof Mait Klaassen**
- **Prof Ülle Jaakma**
- **Dr Andres Aland**
- **Ms Krista Rooni**

EAAP 2017

Ministry of Rural Affairs of the Republic of Estonia
Lai 39 // Lai 41
15056 Tallinn
Estonia
eaap2017@agri.ee
+372 731 3420
Conference website: www.eaap2017.org

Estonian Scientific Committee

Animal Genetics
- **Dr Tanel Kaart,** Estonian University of Life Sciences
- **Dr Sirje Värv,** Estonian University of Life Sciences

Animal Health and Welfare
- **Dr Andres Aland,** Estonian University of Life Sciences

Animal Nutrition
- **Prof Meelis Ots,** Estonian University of Life Sciences

Animal Physiology
- **Prof Ülle Jaakma,** Estonian University of Life Sciences

Livestock Farming Systems
- **Dr Ragnar Leming,** Estonian University of Life Sciences

Cattle Production
- **Prof Haldja Viinalass,** Estonian University of Life Sciences

Horse Production
- **MSc Krista Rooni,** Estonian University of Life Sciences

Pig Production
- **Dr Alo Tänavots,** Estonian University of Life Sciences

Sheep and Goat Production
- **Dr Peep Piirsalu,** Estonian University of Life Sciences

Friends of EAAP

By creating the 'Friends of EAAP', EAAP offers the opportunity to industries to receive services from EAAP in change of a fixed sponsoring amount of support every year.
- The group of supporting industries are layered in three categories: 'silver', 'gold' and 'diamond' level.
- It is offered an important discount (one year free of charge) if the sponsoring industry will agree for a four years period.
- EAAP will offer the service to create a scientific network (with Research Institutes and Scientists) around Europe.
- Creation of a permanent Board of Industries within EAAP with the objective to inform, influence the scientific and organizational actions of EAAP, like proposing choices of the scientific sessions and invited speakers and to propose industry representatives for the Study Commissions.
- Organization of targeted workshops, proposed by industries.
- EAAP can represent and facilitate activities of the supporting industries toward international legislative and regulatory organizations.
- EAAP can facilitate the supporting industries to enter in consortia dealing with internationally supported research projects.

Furthermore EAAP offers, depending to the level of support (details on our website: www.eaap.org):
- Free entrances to the EAAP annual meeting and Gala dinner invitation.
- Free registration to journal *animal*.
- Inclusion of industry advertisement in the EAAP Newsletter, in the banner of the EAAP website, in the Book of Abstract and in the Programme Booklet of the EAAP annual meeting.
- Inclusion of industry leaflets in the annual meeting package.
- Presence of industry advertisements on the slides between presentations at selected standard sessions.
- Presence of industry logos and advertisements on the slides between presentations at the Plenary Sessions.
- Public Recognition by the EAAP President at the Plenary Opening Session of the annual meeting.
- Discounted stands at the EAAP annual meeting.
- Invitation to meetings (at every annual meeting) to discuss joint strategy EAAP/Industries with the EAAP President, Vice-President for Scientific affair, Secretary General and other selected members of the Council and of the Scientific Committee.

Contact and further information

If the industry you represent is interested to become 'Friend of EAAP' or want to have further information please contact jean-marc.perez0000@orange.fr or EAAP secretariat (eaap@eaap.org, phone : +39 06 44202639).

The Association

EAAP (The European Federation of Animal Science) organises every year an international meeting which attracts between 900 and 1500 people. The main aims of EAAP are to promote, by means of active co-operation between its members and other relevant international and national organisations, the advancement of scientific research, sustainable development and systems of production; experimentation, application and extension; to improve the technical and economic conditions of the livestock sector; to promote the welfare of farm animals and the conservation of the rural environment; to control and optimise the use of natural resources in general and animal genetic resources in particular; to encourage the involvement of young scientists and technicians. More information on the organisation and its activities can be found at www.eaap.org

Acknowledgements

Thank you
to the 68th EAAP Annual Congress Sponsors and Friends

Gold

Silver

Bronze

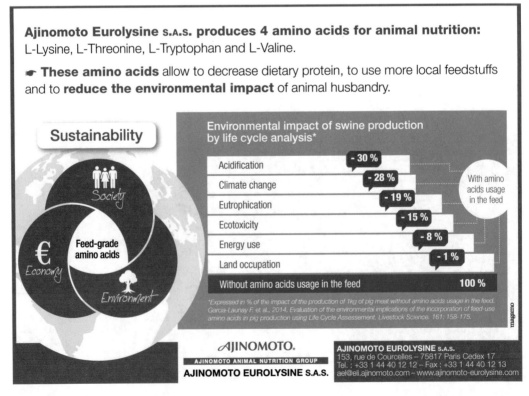

LIVESTOCK, MEAT & SOCIETY

CIV Meat, Science & Society

 ENVIRONMENT

 SOCIETY

 NUTRITION

 ANIMAL HEALTH

Impact environnemental de la production de viande : comprendre et évaluer

Animal, viande et société : des liens qui s'effilochent

La consommation de viande en France

Zoonoses and Livestock

To meet CIV's studies and bibliographic ressources, just connect to www.civ-viande.org

European Federation of Animal Science (EAAP)

President:	M. Gauly
Secretary General:	A. Rosati
Address:	Via G. Tomassetti 3, A/I
	I-00161 Rome, Italy
Phone/Fax:	+39 06 4420 2639
E-mail:	eaap@eaap.org
Web:	www.eaap.org

69th Annual Meeting of the European Federation of Animal Science

Dubrovnik, Croatia, 27 – 31 August 2018

The EAAP 2018 organising team can be reached at:
Croatian Agricultural Agency
Ilica 101, 10 000 Zagreb
Croatia
eaap2018@hpa.hr
+385 1 3903 176

Organized by:

Ministry of Agriculture
- Mr Tomislav Tolušić
- Mr Dalibor Janda

Croatian Agricultural Agency
- Dr Zdravko Barać
- Dr Marija Špehar

In cooperation with:

Faculty of Agriculture – University of Zagreb
- Prof Boro Mioč
- Prof Zlatko Janječić
- Prof Josip Leto
- Dr Nikica Šprem

Faculty of Agriculture – University J.J. Strossmayer in Osijek
- Prof Vesna Gantner
- Prof Pero Mijić
- Prof Anđelko Opačak
- Prof Zvonko Antunović

Faculty of Veterinary Medicine – University of Zagreb
- Prof Velimir Sušić
- Prof Anamaria Ekert Kabalin

University of Zadar
- Prof Slaven Zjalić
- Dr Bosiljka Mustać

Agricultural Advisory Service
- Mr Zdravko Tušek

Croatian Chamber of Agriculture
- Dr Danko Dežđek

Conference website: www.eaap2018.org

Scientific Programme EAAP 2017

Monday 28 August 8.30 – 12.30	Monday 28 August 14.00 – 18.00	Tuesday 29 August 8.30 – 12.30	Tuesday 30 August 14.30 – 18.00
ONE-DAY INTERBULL SEMINAR			**Session 14** New traits for new breeding goals Chair: S. König/ M. Klopčič
Session 01 Managing genetic diversity in cattle in the era of genomic selection (with Interbull) Chair: R. Reents/ A.W.M. Roozen	**Session 07** Genetic defects in cattle – identification, finding the mutation and managing it in breeding plans (with Interbull) Chair: H. Jorjani		**Session 15** Non-additivity and predicting crossbred performance in the era of genomics Chair: H.A. Mulder
Session 02 Innovation in cattle husbandry and land use Chair: R. Keatinge/ P. Galama	**Session 08** Young Train – Dairy innovation in research and extension Chair: P. Aad/ A. Kuipers		**Session 16** Communication of science to the general public Chair: J.A.M. Van Arendonk
Session 03 Effective population size as a tool for the management of animal genetic resources Chair: I. Curik	**Session 09** Genomics and genetic evaluation in small ruminants Chair: J. Conington/ A. Cecchinato		**Session 17** Emission and feed management of cattle Chair: Y. Montanholi
Session 04 Mult-discplinary approaches for pig breeders for a sustainable production Chair: P. Trevisi	**Session 10** Food trust in the food chain: how can the animal production sector contribute? Chair: J.L. Peyraud/ M. Gauly	**Session 13** **Welcome Ceremony and Plenary Session** Development of automatized methods and their applications to improve livestock production Chair: M. Gauly/ I. Halachmi	**Session 18** Animal behaviour and genetics Chair: L.A. Boyle
			Session 19 Bioeconomy, industrial ecology and territorial metabolism: how to manage flows and cycles in LFS and at what scale? Chair: R. Ripoll-Bosch
ONE-DAY SEMINAR ON PRODUCTION EFFICIENCY			
Session 05 Physiological limits for production efficiency Chair: C.H. Knight	**Session 11** LFS efficiency: indicators and scales? Chair: M. Zehetmeier		**Session 20** Alternatives to surgical castration of male piglets without pain relief Chair: S. Millet
Session 06 Welfare free communications Chair: H.A.M. Spoolder	**Session 12** Improving health and welfare through precision farming Chair: T. Norton/ H.A.M. Spoolder		**Session 21** PLF and its generated animal-data usage on farm and along the entire chain Chair: M. Guarino/ I. Halachmi
			Session 22 SheepNet and iSAGE networking to improve efficiency in sheep and goats Chair: J. Conington
11.30 – 12.30 **Commission meeting** • Physiology	19.00 – 21.00 **Welcome reception and Poster session**	13.30 – 14.30 **Poster session**	17.00 – 18.00 **Commission meeting** • PLF

Wednesday 30 August 8.30 – 12.00	Wednesday 30 August 14.00 – 18.00	Thursday 31 August 8.30 – 11.30	Thursday 31 August 14.00 – 18.00
Session 23 Genomic selection Chair: A. Cecchinato	**Session 32** The nature of inbreeding and genomic approaches to assess inbreeding rate and inbreeding depression Chair: H. Simianer	**Session 40** Genetics free communications Chair: G.F. Difford	**Session 49** Impact of selection methods and reproduction technologies on breeding programs Chair: E. Wall
ONE-DAY FEED-A-GENE SEMINAR		**Session 41** New traits for new breeding goals Chair: S. König	**Session 50** Human health and animal products Chair: M. De Marchi/ A. Kuipers
Session 24	**Session 33**		
Feed-a-gene Chair: J. Van Milgen		**Session 42** Livestock transport and slaughter Chair: S. Messori	
Session 25 Health challenges for modern farming: farming intensity and emerging animal diseases Chair: E. Sossidou	**Session 34** Rules and guidelines for successful funding and research: Opportunities for young scientists Chair: C. Lambertz	**Session 43** Nutrition Free Communications Chair: G. Savoini	**Session 51** Nutrition, intestinal health and immunity Chair: E. Tsiplakou
ONE-DAY CATTLE NETWORK SEMINAR		**Session 44** Non-invasive monitoring of physiological state (animal biology) (with EU FP7 GplusE and COST DairyCare) Chair: C.H. Knight	**Session 52** Alternative versus traditional feed protein sources Chair: S. De Campeneere
Session 26 Better management for better economics in dairy Chair: J.F. Hocquette/ G. Thaller	**Session 35** Better management for better economics in beef Chair: J.F. Hocquette/ M. De Marchi		**Session 53** Resilience: immunology, breeding and stress reduction Chair: G. Das
ONE DAY NUTRITION SEMINAR		**ONE-DAY INSECT SEMINAR**	
Session 27 Innovation in animal nutrition through speciality feed ingredients (with Fefana) Chair: E. Apper	**Session 36**	**Session 45** Safety, regulatory, environmental issues and consumer acceptance of insects Chair: J. Eilenberg/ T. Veldkamp	**Session 54** Insects in a circular economy Chair: T. Veldkamp/ J. Eilenberg
Session 28 Local solutions, global answer: facing global challenges with local productions systems? Chair: M.R.F. Lee	**Session 37** Impact of the design of smart housing systems on climate, health, welfare and the individual variation between animals (with OptiBarn and FreeWalk projects) Chair: T. Amon/ I. Halachmi	**Session 46** Pig free communications Chair: E.F. Knol	**Session 55** Science supporting the health and welfare of horses Chair: M.L. Cox
Session 29 Weaning pigs in a healthy way Chair: G. Bee		**Session 47** The global farm platform initiative: towards sustainable livestock systems Chair: M.R.F. Lee	**Session 56** Free communication in cattle Chair: M. Klopčič/ Y. Montanholi
Session 30 The future role of native horse breeds Chair: J. Kantanen	**Session 38** The role of advanced genetics in the equine sector Chair: K.F. Stock	**Session 48** Health and welfare of small ruminants Chair: H.A.M. Spoolder	
Session 31 Biology of adipose tissue and lipid metabolism Chair: I. Louveau	**Session 39** More from less-enhancing resilience and productivity from sheep and goats Chair: O. Tzamaloukas		
12.00 – 13.00 **Poster session**	17.00 – 18.00 **Commission meeting** · Horse	11.30 – 12.30 **Commission meeting** · Genetics · Cattle · Health and Welfare · Nutrition	· Insect · Pig · LFS · Sheep and Goat

Commission on Animal Genetics

Dr Strandberg	President	Swedish University of Agricultural Sciences
	Sweden	erling.strandberg@slu.se
Dr Wall	Vice-President	SRUC
	United Kingdom	eileen.wall@sruc.ac.uk
Dr Lassen	Vice-President	Aarhus University
	Denmark	jan.lassen@mbg.au.dk
Dr Mulder	Vice-President	Wageningen University
	The Netherlands	han.mulder@wur.nl
Dr Cecchinato	Secretary	Padova University
	Italy	alessio.cecchinato@unipd.it
Dr Gredler	Industry rep.	Qualitas AG
	Switzerland	birgit.gredler@qualitasag.ch

Commission on Animal Nutrition

Dr Savoini	President	University of Milan
	Italy	giovanni.savoini@unimi.it
Dr Tsiplakou	Vice-President	Agricultural University of Athens
	Greece	eltsiplakou@aua.gr
Dr De Campeneere	Secretary	ILVO
	Belgium	sam.decampeneere@ilvo.vlaanderen.be
Dr Goselink	Secretary	Wageningen Livestock Research
	The Netherlands	roselinde.goselink@wur.nl
Dr Apper	Industry rep.	Tereos Syral
	France	emmanuelle.apper-bossard@tereos.com

Commission on Health and Welfare

Dr Spoolder	President	Wageningen Livestock Research
	The Netherlands	hans.spoolder@wur.nl
Dr Das	Vice-President	Leibniz Institute for Farm Animal Biology
	Germany	gdas@fbn-dummerstorf.de
Dr Sossidou	Vice President	Hellenic Agricultural Organization
	Greece	sossidou.arig@nagref.gr
Dr Messori	Secretary	SIRCAH
	France	stefano.messori@yahoo.it
Dr Ferrante	Secretary	Milan University
	Italy	valentina.ferrante@unimi.it
Dr Rosner	Industry rep.	CIV-Viande
	France	pm.rosner@civ-viande.org

Commission on Animal Physiology

Dr Sauerwein	President	University of Bonn
	Germany	sauerwein@uni-bonn.de
Dr Louveau	Vice president	INRA
	France	isabelle.louveau@inra.fr
Dr Silanikove	Vice-President	Agricultural Research Organization (ARO)
	Israel	nsilaniks@volcani.agri.gov.il
Dr Knight	Secretary	University of Copenhagen
	Denmark	chkn@sund.ku.dk
Dr Troescher	Industry rep.	BASF
	Germany	arnulf.troescher@basf.com

Commission on Livestock Farming Systems

Dr Tichit	President	INRA
	France	muriel.tichit@agroparistech.fr
Dr Lee	Vice-President	University of Bristol
	United Kingdom	michael.lee@rothamsted.ac.uk
Dr Ripoll Bosch	Secretary	Wageningen University
	The Netherlands	raimon.ripollbosch@wur.nl
Mrs Zehetmeier	Secretary	Institute Agricultural Economics and Farm Management
	Germany	monika.zehetmeier@tum.de

Commission on Cattle Production

Dr Klopčič	President	University of Ljublijana
	Slovenia	marija.klopcic@bf.uni-lj.si
Dr König	Vice-President	Justus-Liebig-University Gießen
	Germany	sven.koenig@agrar.uni-giessen.de
Dr Montanholi	Vice-President	Dalhousie University
	Canada	yuri.montanholi@dal.ca
Dr Fürst-Waltl	Secretary	University of Natural Resources and Life Sciences
	Austria	birgit.fuerst-waltl@boku.ac.at
Dr De Marchi	Secretary	Padova University
	Italy	massimo.demarchi@unipd.it
Dr Keatinge	Industry rep.	Agriculture & Horticulture Development Board
	United Kingdom	ray.keatinge@ahdb.org.uk

Commission on Sheep and Goat Production

Dr Conington	President	SRUC
	United Kingdom	joanne.conington@sac.ac.uk
Dr Ådnøy	Vice-President	NMBU
	Norway	tormod.adnoy@nmbu.no
Dr Tzamaloukas	Secretary	Cyprus University of Technology
	Cyprus	ouranios.tzamaloukas@cut.ac.cy
Dr Ligda	Secretary	Hellenic Agricoltural Organisation
	Greece	chligda@otenet.gr
Dr Yates	Industry rep.	Texel
	United Kingdom	johnyates@texel.co.uk

Commission on Pig Production

Dr Knol	President	TOPIGS
	The Netherlands	egbert.knol@topigs.com
Dr Millet	Vice President	ILVO
	Belgium	sam.millet@ilvo.vlaanderen.be
Dr Bee	Secretary	Agroscope Liebefeld-Posieux ALP
	Switzerland	giuseppe.bee@alp.admin.ch
Dr Velarde	Secretary	IRTA
	Spain	antonio.velarde@irta.es
Dr Trevisi	Secretary	Bologna University
	Italy	paolo.trevisi@unibo.it

Commission on Horse Production

Dr Santos	President	CECAV - UTAD - EUVG
	Portugal	assantos@utad.pt
Dr Evans	Vice president	Norwegian University College of Agriculture and Rural Development
	Norway	rhys@hlb.no
Dr Stock	Vice president	VIT-Vereinigte Informationssysteme Tierhaltung w.V.
	Germany	friederike.katharina.stock@vit.de
Dr Potočnik	Vice president	University of Ljubljana
	Slovenija	klemen.potocnik@bf.uni-lj.si
Dr Holgersson	Secretary	Swedish University of Agriculture
	Sweden	anna-lena.holgersson@slu.se
Dr Cervantes Navarro	Secretary	University of Madrid
	Spain	icervantes@vet.ucm.es
Dr Cox	Industry rep.	CAG GmbH – Center for Animal Genetics
	Germany	melissa.cox@centerforanimalgenetics.de

Commission on Insects

Dr Veldkamp	President	Wageningen Livestock Research
	The Netherlands	teun.veldkamp@wur.nl
Dr Epstein	Vice president	Medical University of Vienna
	Austria	michelle.epstein@meduniwien.ac.at
Dr Eilenberg	Secretary	University of Copenhagen
	Denmark	jei@plen.ku.dk
Dr Agazzi	Secretary	University of Milan
	Italy	alessandro.agazzi@unimi.it
Dr Peters	Industry rep.	IIC (International Insect Centre)
	The Netherlands	marianpeters@ngn.co.nl
Dr Angot	Industry rep.	Ynsect
	France	aan@ynsect.com
Dr Boersma	Industry rep.	Protix
	The Netherlands	roel.boersma@protix.eu

Commission on Precision Livestock Farming (PLF)

Dr Halachmi	President	Agriculture research organization (ARO)
	Israel	halachmi@volcani.agri.gov.il
Dr Guarino	Vice president	University of Milan
	Italy	guarino@unimi.it
Dr Norton	Vice president	M3-BIORES
	Belgium	tomas.norton@kuleuven.be
Dr Lokhorst	Secretary	Wageningen UR Livestock Research
	The Netherlands	kees.lokhorst@wur.nl
Dr Umstaetter	Industry rep.	Agroscope
	Switzerland	christina.umstaetter@agroscope.admin.ch

Scientific programme

Session 01. Managing genetic diversity in cattle in the era of genomic selection (with Interbull)

Date: Monday 28 August 2017; 8.30 – 12.30
Chair: R. Reents / A.W.M. Roozen

Theatre Session 01

Poster Session 01

Session 02. Innovation in cattle husbandry and land use

Date: Monday 28 August 2017; 8.30 – 12.30
Chair: R. Keatinge / P. Galama

Theatre Session 02

Poster Session 02

Session 03. Effective population size as a tool for the management of animal genetic resources

Date: Monday 28 August 2017; 8.30 – 12.30
Chair: I. Curik

Theatre Session 03

Poster Session 03

Session 04. Mult-discplinary approaches for pig breeders for a sustainable production

Date: Monday 28 August 2017; 8.30 – 12.30
Chair: P. Trevisi

Theatre Session 04

Poster Session 04

Session 05. Physiological limits for production efficiency

Date: Monday 28 August 2017; 8.30 – 12.30
Chair: C.H. Knight

Theatre Session 05

Poster Session 05

Session 06. Welfare free communications

Date: Monday 28 August 2017; 8.30 – 12.30
Chair: H.A.M. Spoolder

Theatre Session 06

Poster Session 06

Session 07. Genetic defects in cattle – identification, finding the mutation and managing it in breeding plans (with Interbull)

Date: Monday 28 August 2017; 14.00 – 18.00
Chair: H. Jorjani

Theatre Session 07

Poster Session 07

Session 08. Young Train - Dairy innovation in research and extension

Date: Monday 28 August 2017; 14.00 – 18.00
Chair: P. Aad / A. Kuipers

Theatre Session 08

Poster Session 08

Session 09. Genomics and genetic evaluation in small ruminants

Date: Monday 28 August 2017; 14.00 – 18.00
Chair: J. Conington / A. Cecchinato

Theatre Session 09

Poster Session 09

Session 10. Food trust in the food chain: how can the animal production sector contribute?

Date: Monday 28 August 2017; 14.00 – 18.00
Chair: J.L. Peyraud / M. Gauly

Theatre Session 10

Session 11. LFS efficiency: indicators and scales?

Date: Monday 28 August 2017; 14.00 – 18.00
Chair: M. Zehetmeier

Theatre Session 11

Session 12. Improving health and welfare through precision farming

Date: Monday 28 August 2017; 14.00 – 18.00
Chair: T. Norton / H.A.M. Spoolder

Theatre Session 12

Poster Session 12

Session 13. Welcome Ceremony and Plenary session: Development of automatized methods and their applications to improve livestock production

Date: Tuesday 29 August 2017; 8.30 – 12.30
Chair: I. Halachmi

Theatre Session 13

Session 14. New traits for new breeding goals

Date: Tuesday 29 August 2017; 14.30 – 18.00
Chair: S. König / M. Klopčič

Theatre Session 14

Poster Session 14

Session 15. Non-additivity and predicting crossbred performance in the era of genomics

Date: Tuesday 29 August 2017; 14.30 – 18.00
Chair: H.A. Mulder

Theatre Session 15

Poster Session 15

Session 16. Communication of science to the general public

Date: Tuesday 29 August 2017; 14.30 – 18.00
Chair: J.A.M. Van Arendonk

Theatre Session 16

Session 17. Emission and feed management of cattle

Date: Tuesday 29 August 2017; 14.30 – 18.00
Chair: Y. Montanholi

Theatre Session 17

Poster Session 17

Session 18. Animal behaviour and genetics

Date: Tuesday 29 August 2017; 14.30 – 18.00
Chair: L.A. Boyle

Theatre Session 18

Poster Session 18

Session 19. Bioeconomy, industrial ecology and territorial metabolism: how to manage flows and cycles in LFS and at what scale?

Date: Tuesday 29 August 2017; 14.30 – 18.00
Chair: R. Ripoll-Bosch

Theatre Session 19

Session 20. Alternatives to surgical castration of male piglets without pain relief

Date: Tuesday 29 August 2017; 14.30 – 18.00
Chair: S. Millet

Theatre Session 20

Session 21. PLF and its generated animal-data usage on farm and along the entire chain

Date: Tuesday 29 August 2017; 14.30 – 18.00
Chair: M. Guarino / I. Halachmi

Theatre Session 21

Poster Session 21

Session 22. SheepNet and iSAGE networking to improve efficiency in sheep and goats

Date: Tuesday 29 August 2017; 14.30 – 18.00
Chair: J. Conington

Theatre Session 22

Session 23. Genomic selection

Date: Wednesday 30 August 2017; 8.30 – 12.00
Chair: A. Cecchinato

Theatre Session 23

Poster Session 23

Session 24. Feed-a-gene

Date: Wednesday 30 August 2017; 8.30 – 12.00
Chair: J. Van Milgen

Theatre Session 24

Poster Session 24

Session 25. Health challenges for modern farming: farming intensity and emerging animal diseases

Date: Wednesday 30 August 2017; 8.30 – 12.00
Chair: E. Sossidou

Theatre Session 25

Poster Session 25

Session 26. Better management for better economics in dairy

Date: Wednesday 30 August 2017; 8.30 – 12.00
Chair: J.F. Hocquette / G. Thaller

Theatre Session 26

Poster Session 26

Session 27. Innovation in animal nutrition through speciality feed ingredients (with Fefana)

Date: Wednesday 30 August 2017; 8.30 – 12.00
Chair: E. Apper

Theatre Session 27

Poster Session 27

Session 28. Local solutions, global answer: facing global challenges with local productions systems?

Date: Wednesday 30 August 2017; 8.30 – 12.00
Chair: M.R.F. Lee

Theatre Session 28

Poster Session 28

Session 29. Weaning pigs in a healthy way

Date: Wednesday 30 August 2017; 8.30 – 12.00
Chair: G. Bee

Theatre Session 29

Poster Session 29

Session 30. The future role of native horse breeds

Date: Wednesday 30 August 2017; 8.30 – 12.00
Chair: J. Kantanen

Theatre Session 30

Poster Session 30

Session 31. Biology of adipose tissue and lipid metabolism

Date: Wednesday 30 August 2017; 8.30 – 12.00
Chair: I. Louveau

Theatre Session 31

Poster Session 31

Development of a mature adipocyte culture system 310
P.Y. Aad and S. Bou Karroum

Session 32. The nature of inbreeding and genomic approaches to assess inbreeding rate and inbreeding depression

Date: Wednesday 30 August 2017; 14.00 – 18.00
Chair: H. Simianer

Theatre Session 32

Poster Session 32

Session 33. Feed-a-gene

Date: Wednesday 30 August 2017; 14.00 – 18.00
Chair: J. Van Milgen

Theatre Session 33

Session 34. Rules and guidelines for successful funding and research: Opportunities for young scientists

Date: Wednesday 30 August 2017; 14.00 – 18.00
Chair: C. Lambertz

Theatre Session 34

Session 35. Better management for better economics in beef

Date: Wednesday 30 August 2017; 14.00 – 18.00
Chair: J.F. Hocquette / M. De Marchi

Theatre Session 35

Session 36. Innovation in animal nutrition through speciality feed ingredients (with Fefana)

Date: Wednesday 30 August 2017; 14.00 – 18.00
Chair: E. Apper

Theatre Session 36

Poster Session 36

Session 37. Impact of the design of smart housing systems on climate, health, welfare and the individual variation between animals (together with OptiBarn and FreeWalk projects)

Date: Wednesday 30 August 2017; 14.00 – 18.00
Chair: T. Amon / I. Halachmi

Theatre Session 37

Poster Session 37

Session 38. The role of advanced genetics in the equine sector

Date: Wednesday 30 August 2017; 14.00 – 18.00
Chair: K.F. Stock

Theatre Session 38

Poster Session 38

Session 39. More from less-enhancing resilience and productivity from sheep and goats

Date: Wednesday 30 August 2017; 14.00 – 18.00
Chair: O. Tzamaloukas

Theatre Session 39

Poster Session 39

Session 40. Genetics free communications

Date: Thursday 31 August 2017; 8.30 – 11.30
Chair: G.F. Difford

Theatre Session 40

Poster Session 40

Session 41. New traits for new breeding goals

Date: Thursday 31 August 2017; 8.30 – 11.30
Chair: S. König

Theatre Session 41

Session 42. Livestock transport and slaughter

Date: Thursday 31 August 2017; 8.30 – 11.30
Chair: S. Messori

Theatre Session 42

Poster Session 42

Session 43. Nutrition Free Communications

Date: Thursday 31 August 2017; 8.30 – 11.30
Chair: G. Savoini

Theatre Session 43

Poster Session 43

Session 44. Non-invasive monitoring of physiological state (animal biology) (with EU FP7 GplusE and COST DairyCare)

Date: Thursday 31 August 2017; 8.30 – 11.30
Chair: C.H. Knight

Theatre Session 44

Poster Session 44

Session 45. Safety, regulatory, environmental issues and consumer acceptance of insects

Date: Thursday 31 August 2017; 8.30 – 11.30
Chair: J. Eilenberg / T. Veldkamp

Theatre Session 45

Session 46. Pig free communications

Date: Thursday 31 August 2017; 8.30 – 11.30
Chair: E.F. Knol

Theatre Session 46

At-line carcass quality. NIRS determination of fat composition at Swiss commercial slaughterhouses 416
S. Ampuero Kragten, M. Scheeder, M. Müller, P. Stoll and G. Bee

Poster Session 46

Session 47. The global farm platform initiative: towards sustainable livestock systems

Date: Thursday 31 August 2017; 8.30 – 11.30
Chair: M.R.F. Lee

Theatre Session 47

Session 48. Health and welfare of small ruminants

Date: Thursday 31 August 2017; 8.30 – 11.30
Chair: H.A.M. Spoolder

Theatre Session 48

Poster Session 48

Session 49. Impact of selection methods and reproduction technologies on breeding programs

Date: Thursday 31 August 2017; 14.00 – 18.00
Chair: E. Wall

Theatre Session 49

Session 50. Human health and animal products

Date: Thursday 31 August 2017; 14.00 – 18.00
Chair: M. De Marchi / A. Kuipers

Theatre Session 50

Poster Session 50

Session 51. Nutrition, intestinal health and immunity

Date: Thursday 31 August 2017; 14.00 – 18.00
Chair: E. Tsiplakou

Theatre Session 51

Poster Session 51

Session 52. Alternative versus traditional feed protein sources

Date: Thursday 31 August 2017; 14.00 – 18.00
Chair: S. De Campeneere

Theatre Session 52

Poster Session 52

Session 53. Resilience: immunology, breeding and stress reduction

Date: Thursday 31 August 2017; 14.00 – 18.00
Chair: G. Das

Theatre Session 53

Poster Session 53

Session 54. Insects in a circular economy

Date: Thursday 31 August 2017; 14.00 – 18.00
Chair: T. Veldkamp / J. Eilenberg

Theatre Session 54

Poster Session 54

Session 55. Science supporting the health and welfare of horses

Date: Thursday 31 August 2017; 14.00 – 18.00
Chair: M.L. Cox

Theatre Session 55

Poster Session 55

Session 56. Free communication in cattle

Date: Thursday 31 August 2017; 14.00 – 18.00
Chair: M. Klopčič / Y. Montanholi

Theatre Session 56

Genomic tools to improve progress and preserve variation for future generations

P.M. Vanraden

USDA-ARS, Animal Genomics and Improvement Lab, Building 5 BARC-West, Beltsville, MD 20705, USA; paul.vanraden@ars.usda.gov

Use of genomic tools has greatly decreased generation intervals and increased genetic progress in dairy cattle, but faster selection cycles can also increase rates of inbreeding per unit of time. Average pedigree inbreeding of Holstein cows increased from 4.6% in 2000 to 5.6% in 2009 to 6.6% in 2015. The 0.17% per year recent rate of inbreeding is larger than in the previous decade but similar to the rate of 0.16% from 1985 to 2000. Recent rates from 2009 to 2015 were only 0.09% per year for Jersey and 0.13% for Brown Swiss cows. Breeders have reduced the effects of inbreeding depression in the short term by focusing on genomic rather than pedigree inbreeding in mating programs, avoiding carrier matings, selecting against lethal defects, or crossbreeding. Genomic relationships of >1 million genotyped females with >5,000 marketed males have been used in industry mating programs since 2014 and are updated monthly for new females. Haplotype or laboratory genotype tests for 22 recessive defects, red, and polled are provided for all genotyped animals. Pedigree breed composition has been provided since 2007 and genomic breed base representation since 2016, allowing breeders to examine ancestry of each animal even if pedigrees are incomplete or missing. Genomic evaluations on all-breed scale were computed but not released yet for 25 traits of 44,023 crossbreds including 20,367 with no official evaluation because of breed check edits. The new evaluations were for 1,822 Jersey × Holstein crossbreds with >40% of both breeds (F1 crosses), 7,237 Holstein backcrosses with >67% and <94% Holstein, 7,820 Jersey backcrosses, 388 Brown Swiss crosses, plus other mixtures. In the long term, selecting for rare favourable alleles within the same breed, introgressing new alleles from another breed, and gene editing or transgenic tools could help maintain or increase genetic variation. Many genomic tools are available to help breeders make progress and manage diversity in the current generation while preserving variation for future selection.

Genetic diversity of sires used in the Irish dairy industry

F. Hely[1], F. Kearney[2], A. Cromie[2], D. Matthews[2] and P. Amer[1]

[1]AbacusBio Ltd, P.O. Box 5585, Dunedin 9058, New Zealand, [2]Irish Cattle Breeding Federation, Shinagh, Bandon, Co. Cork, Ireland; fhely@abacusbio.co.nz

Maintenance of genetic diversity in the Irish dairy industry is one of the driving motivations for having a nationally co-ordinated progeny test structure. This paper reports on analysis of co-ancestry and inbreeding trends between and among cohorts of bulls with progeny in Ireland over the past 15 years. The cohorts include bulls used widely and not included in the progeny test, bulls sourced from Ireland included in the progeny test, foreign sourced bulls included in the progeny test, and those foreign bulls not included in the progeny test. Average co-ancestry for these bulls with subsets of the cow population are also presented. The cow population is divided into those herds who have had a contract mating organised with an artificial insemination breeding company at least once in their history, versus those that have not.

Pedigree inbreeding underestimate true inbreeding in genomic dairy cattle breeding schemes

J.R. Thomasen[1,2], H. Liu[2] and A.C. Sørensen[2]
[1]*VikingGenetics, Ebeltoftvej 16, 8960 Randers SØ, Denmark,* [2]*Aarhus University, Department of Molecular Biology and Genetics, Blichers Allé 20, 8830 Tjele, Denmark; jorn.rthomasen@mbg.au.dk*

In this study, we compare the rate of true inbreeding (ΔF_{true}) and inbreeding calculated from pedigree (ΔF_{ped}) in different dairy cattle breeding schemes with different reference population structures. We hypothesize, that pedigree inbreeding underestimates the true inbreeding in a fully genomic breeding scheme. The reasoning is, that pedigree inbreeding assumes that all changes in allele frequencies is attributed to genetic drift, meaning all loci are expected to be neutral. This assumption is violated in all breeding schemes and in genomic breeding schemes in particular. To test our hypothesis, we evaluated true inbreeding across the genome using simulated populations and genome structures that mimic previously reported LD in dairy cattle. Two population-structures were considered; a high and a low LD (LDH and LDL). We simulated two daughter proven bull schemes (PB) mimicking a small breeding program and a large breeding program with 5 and 20 yearly selected PB (PB-POPS and PB-POPL). For the LDH structure, a large and a small breeding program (LDH-POPL and LDH-POPS) were studied. For the POPL, a reference population was built, consisting of 10,000 PB. Each year, 2,000 genotyped cows and 100 genotyped PB were included in the reference population. For POPS, the reference population consisted of 1000 PB and each year 2,000 genotyped cows and 50 genotyped PB were included. For the LDL structure, only POPL was analyzed. All breeding schemes were investigated over a timeframe of ten years. The average reliabilities of the genomic estimated breeding values for the three genomic schemes were 0.42 (LDH-POPL), 0.38 (LDL-POPL) and 0.34 (LDH-POPS). For the PB schemes (PB-POPS and PB-POPL), ΔF_{ped} was approximately 4% lower than ΔF_{true}. For the POPS-LDH scheme, ΔF_{ped} was 50% lower than ΔFtrue and for POPL-LDH, it was 56% lower. In POPL-LDL, which had more heterogeneity, ΔF_{ped} was estimated to be 37% lower. The result confirmed our hypothesis as we can conclude that ΔF_{ped} underestimates true inbreeding in a genomic breeding scheme where breeding candidates are selected on genomic estimated breeding values.

Genetic diversity measures of local European cattle breeds for conservation purposes

V. Kukučková[1], N. Moravčíková[1], I. Curik[2], M. Simčič[3], G. Mészáros[4] and R. Kasarda[1]
[1]*Slovak University of Agriculture in Nitra, Department of Animal Genetics and Breeding Biology, Tr. A. Hlinku 2, 94976 Nitra, Slovak Republic,* [2]*University of Zagreb, Department of Animal Sciences, Svetošimunska 25, 10000 Zagreb, Croatia,* [3]*University of Ljubljana, Department of Animal Science, Jamnikarjeva 101, 1000 Ljubljana, Slovenia,* [4]*University of Natural Resources and Life Sciences, Division of Livestock Sciences, Gregor-Mendel Str. 33, 1180 Vienna, Austria; veron.sidlova@gmail.com*

The Slovak Pinzgau cattle faces the bottleneck effect and the loss of diversity due to unequal use of founders and significant population decline. Further population size reduction can lead to serious problems. High-throughput molecular information of 179 genotyped individuals was used to characterise genetic diversity and differentiation of Slovak Pinzgau, Austrian Pinzgau, Cika and Piedmontese cattle by Bayesian clustering algorithm. A gene flow network for the clusters estimated from admixture results was produced. The low estimate of genetic differentiation (F_{ST}) in Pinzgau cattle populations indicated that differentiation among these populations is low, particularly owing to a common historical origin and high gene flow. The changes in the log marginal likelihood indicated Austrian Pinzgau as the most similar breed to Slovak Pinzgau. All population except Piedmontese showed two ways of gene flow among populations which means that Piedmontese was involved in formation of analysed breeds while these breeds were not involved in creation of Piedmontese. Genetic evaluation represents an important tool in breeding and cattle selection. It is more strategically important than ever to preserve as much of the livestock diversity as possible, to ensure a prompt and proper response to the needs of future generations. The information provided by the fine-scale genetic characterization of this study clearly shows that there is a difference in genetic composition of Slovak and Austrian populations as well as Cika and Piedmontese. Despite economic pressures and intensive action of dairy breeds, due to Pinzgau population size, Slovakia has the potential to serve as a basic gene reserve with the European and World importance.

Long-term impact of optimum contribution selection on breeds with historical introgression

Y. Wang, J. Bennewitz and R. Wellmann

University of Hohenheim, Institute of Animal Science, Garbenstr. 17, 70599 Stuttgart, Germany; yu.wang@uni-hohenheim.de

In the breeding schemes of small local breeds, two conflicts have to be addressed: the conflict between increasing genetic gain while managing the inbreeding level, and the conflict between maintaining genetic diversity while controlling the loss of genetic uniqueness. Advanced optimum contribution selection (OCS) approaches were compared for balancing these conflicting objectives via constraining migrant contributions (MCs) and kinships at native alleles in the offspring. The aim of the study was to evaluate the performance of different OCS strategies in subsequent simulated generations obtained from genomic data of Angler cattle, a local cattle breed in northern Germany. Genotypes were available for 131 bulls and 137 cows born between 1986 and 2014. A total of 23,448 SNPs were used in the analysis and were phased with Beagle. The R package *optiSel* was used to estimate foreign breed contributions and kinships from genotype data and to perform the optimization. A base population with a size of 1000 animals (500 males and 500 females) was generated from the Angler genotypes. Pedigree and genotypes of each generation were simulated according to the optimum contributions of selection candidates in the previous generation. Three scenarios were compared with truncation selection. In all scenarios the aim was to maximize genetic gain, and segment based kinship was constrained to achieve an effective population size of 100. The first scenario was traditional OCS with only the constraint on segment based kinship. In the second scenario there was an additional constraint on segment based kinship at native alleles, and MCs were maintained at the current level. In the third scenario segment based kinship at native alleles was constrained as well, but MCs were reduced each generation. Results showed that traditional OCS procedure has slight advantages in increasing genetic gain whilst controlling relatedness compared to truncation selection. However, the introgression of foreign genetic material by traditional OCS is not desirable in the local breed conservation. In the long run, constraining MC and kinship at native alleles in the OCS procedure is a promising approach to increase genetic gain whilst maintaining genetic uniqueness and diversity.

Using recombination to maintain genetic diversity

D. Segelke[1], J. Heise[2] and G. Thaller[3]

[1]IT solutions for animal production (vit), Heinrich-Schroeder-Weg 1, 27283 Verden, Germany, [2]University of Goettingen, Animal Breeding and Genetics, Albrecht-Thaer-Weg 3, 37075 Göttingen, Germany, [3]Christian-Albrechts-University of Kiel, Institute of Animal Breeding and Husbandry, Hermann-Rodewald-Straße 6, 24098 Kiel, Germany; dierck.segelke@vit.de

The rate of genetic gain increased due to genomic selection. Consequently, inbreeding rate increased and a loss of genetic diversity might occur. Recombination breaks existing haplotypes and leads to new haplotypes, which is necessary to maintain genetic diversity. To investigate the nature of recombination, genotypes of over 200,000 German-Holstein were analyzed. The average global number of crossovers ranged from 23.3 in males to 24.3 in females. A univariate animal model was used to estimate variance components and breeding values for the number of global recombinations in males and females separately. The chiptype was modeled as a fixed effect. Heritability of the global number of recombinations was higher in females than in males. The 50k-based genome-wide association study showed previously reported additive QTL for genome-wide number of recombinations on chromosome 6 and 10 in males and on chromosome 1, 3, 6, 10 in females. We identified additional QTL on chromosome 3 for males and on chromosome 20 for females. Furthermore, we investigated dominance effects on the global number of recombinations and found significant effects for the mentioned QTL. Additionally, heritabilities and genetic correlations between chromosome-wise numbers of recombination were investigated, using a multivariate model for chromosomes 1 to 12. Different genetic correlation patterns between chromosomes were observed. For example, the number of recombinations on chromosomes 2 and 11 were genetically highly correlated with chromosomes 3 to 5. In contrast, chromosome 10 showed low genetic correlations to all other chromosomes except chromosome 8. The global number of recombinations dropped over the last decades in both sexes. This decrease accelerated with the implementation of genomic selection. To stop this loss of genetic variation, new approaches are needed, e.g. to the inclusion of the global number of recombinations in breeding schemes.

Using genomics to manage progress and diversity: an industry perspective

S. Van Der Beek and H. Geertsema
CRV, P.O. Box 454, 6800 AL, Arnhem, the Netherlands; sijne.van.der.beek@crv4all.com

In the last decade, genomics has become the routine tool to breed AI-bulls. In most countries most dairy farmers now use genomic bulls ad no longer daughter proven bulls to breed the next generation. The advent of genomics coincides with a further professionalization of commercial breeding programs. Because we now have the better tools, it also pays off to make maximum use of it. Consequently, breeding programs not only make best use of genomic prediction, but also focus on reducing the generation interval and increasing the use of reproduction technology. In addition, one can see that breeding goals has become more alike across countries. The net effect of all these developments is a huge risk of inbreeding levels climbing faster and higher than ever before. The only way to mitigate this risk is by managing genetic diversity within breeding programs with the same rigour as we manage genetic progress. Genetic diversity management consists of the following elements: (1) Segmentation. Customers desire choice. Creating a segmented portfolio of bulls gives a geat opportunity to also create genetic diversity (when executed smartly); (2) Implementing genetic contribution theory; (3) Limiting the use of individual animals; (4) Sourcing of new animals for the breeding nucleus via extensive genomic screening; (5) Strict monitoring Although both the genetic and inbreeding trend in the population are driven by what happens within the breeding programs of AI companies, the same companies can also help farmers to overcome the inbreeding challenge. First by offering a wide portfolio of bulls. Second by advocating mating programs that include (genomic) inbreeding management and account for inbreeding depression. Third by providing tools that allow farmers to assess the relatedness of potential bulls to the herd of a farmer during the semen purchasing process. Fourth by creating insights via monitoring tools.

Biodiversity within and between European Red dairy breeds – conservation through utilization

D. Hinrichs[1], M. Calus[2], D.J. De Koning[3], J. Bennewitz[4], T. Meuwissen[5], G. Thaller[6], J. Szyda[7], J. Tetens[8], V. Juskiene[9] and B. Guldbrandtsen[10]
[1]Humboldt-University, Unter den Linden 6, 10099 Berlin, Germany, [2]Wageningen Livestock Research, P.O. Box 338, 6700 AH Wageningen, the Netherlands, [3]Swedish University of Agricultural Sciences, P.O. Box 7070, 750 07 Uppsala, Sweden, [4]University Hohenheim, Garbenstr. 17, 70599 Stuttgart, Germany, [5]Norwegian University of Life Sciences, Ås, Norway, [6]Christian-Albrechts-University, Hermann-Rodewald-Straße 6, 24098 Kiel, Kiel, Germany, [7]Wroclaw University of Environmental and Life Sciences, Norwida 25, Wrocław, Poland, [8]Georg-August-University, Albrecht-Thaer-Weg 3, 37075 Göttingen, Germany, [9]Lithuanian University of Health Sciences, R. Zebenkos 12, 82317 Baisogala, Lithuania, [10]Arhus University, Blichers Allé 20, 8830 Tjele, Denmark; hinricdi@agrar.hu-berlin.de

Red dairy breeds across Europe represent a unique source of genetic diversity and are partly organized in transnational breeding programs but are also well adapted to local conditions. ReDiverse's objective is to develop and to set in place collaborative and integrated novel breeding and management concepts to achieve a resilient and competitive use of these resources and to strengthen best practices for small farm holders to improve product quality and to supply ecosystem services according to their specific circumstances. The challenge of establishing appropriate breeding and maintenance strategies for diverse farm systems and regional markets is met by multi-actor operations considering economic, structural and social diversity in participating countries to offer tailored solutions. The holistic approach relies on integrative research of scientists in the fields of animal genetics, proteomics, economy and social sciences. Cutting edge technology such as large scale genomic and proteomic tools will be implemented to enhance genetic progress and to characterize specific properties. Innovative survey approaches will assess the impact of the sector on social acceptance and the needs of farmers. The project will generate novel knowledge and concepts that will be timely disseminated to lead-users in the breeding and dairy industries, food sector, farmers' cooperatives and among farmers.

The effect of using whole genome sequence instead of a lower density SNP chip on GWAS

S. Van Den Berg[1,2], R.F. Veerkamp[2], F.A. Van Eeuwijk[1], A.C. Bouwman[2], M.S. Lopes[3] and J. Vandenplas[2]
[1]*Wageningen University and Research, Biometris, Droevendaalsesteeg 1, 6700 AH Wageningen, the Netherlands,* [2]*Wageningen University and Research, Animal Breeding and Genomics, Droevendaalsesteeg 1, 6700 AH Wageningen, the Netherlands,* [3]*Topigs Norsvin Research Center, Schoenaker 6, 6641 SZ Beuningen, the Netherlands; sanne.vandenberg@wur.nl*

The interest in using whole genome sequence (WGS) data for genomic prediction is increasing with the increasing availability of WGS data. WGS data can be seen as the ultimate resolution for DNA genotyping, since it contains all variants, including the causal mutations underlying traits. Therefore, our assumption of using WGS data for Genome Wide Association Study (GWAS) was to find SNP closer to the QTLs, that explain a higher proportion of phenotypic variance and that will hold stronger across line. To investigate this assumption, GWAS was performed for the number of teats using 80K, and 660K SNP chip data, and imputed WGS data from 12,184 Large White and 4,993 Dutch Landrace pigs. A SNP was defined as significant if it had a -log10(P>5); and a QTL region was defined as a region of 1 MB around the most significant SNP. GWAS results were validated in an independent set of animals within and across line by estimating the phenotypic variance explained by the significant SNP. Results show that the number of QTL regions increased from low SNP density chips to WGS data. In addition to newly found QTL regions, some QTL regions were linked to known from previous studies QTL regions. Within line, we found that increasing the SNP density from 80K to 660K increased the proportion of phenotypic variance explained by the significant SNP. However, increasing the SNP density from 660K to WGS did not increase of the proportion of phenotypic variance explained. Across line, the proportion of phenotypic variance explained increased with increasing the SNP density, even from 660K to WGS. In conclusion, increasing the SNP density to WGS data increased the number of observed QTL and the proportion of phenotypic variance explained by the significant SNP across line, suggesting that the observed significant SNP were closer to the QTLs when using WGS.

Genetic trends from single-step GBLUP and traditional BLUP for production traits in US Holstein

I. Misztal[1], Y. Masuda[1], P.M. Vanraden[2] and T.J. Lawlor[3]
[1]*University of Georgia, Animal and Dairy Science, Athens GA, 30605, USA,* [2]*AGIL USDA, Beltsville, MD 20705, USA,* [3]*Holstein Association, Bratteleboro, VT 05301, USA; ignacy@uga.edu*

The objective of this study was to compare genetic trends from a single-step genomic BLUP (ssGBLUP) and the traditional BLUP models for milk production traits in US Holstein. Phenotypes were 305-day milk, fat, and protein yield from 21,527,040 cows recorded between January, 1990 and August, 2015. The pedigree file included 29,651,623 animals limited to 3 generations back from recorded or genotyped animals. Genotypes for 764,029 animals were utilized, and analyzes were by a three-trait repeatability model as used in the US official genetic evaluation. Unknown parent groups were incorporated into the inverse of a relationship matrix (H^{-1} in ssGBLUP and A^{-1} in BLUP) with the QP-transformation. In ssGBLUP, 18,359 genotyped animals were randomly chosen as core animals to calculate the inverse of genomic relationship matrix with the APY algorithm. Computations with BLUP took 6.5 h and 1.4 GB of memory, and computations with ssGBLUP took 13 h and 115 GB of memory. For genotyped sires with at least 50 daughters, the trends for PTA and GPTA were similar up to 2008, with a higher trend for ssGBLUP later (approx. by 82 kg milk, 5 kg fat and 3 kg protein for bulls born in 2010). For genotyped cows, the trends were similar up to 2006, with a higher trend for ssGBLUP later (approx. by 245 kg milk, 10 kg fat and 7 kg protein for cows born in 2013). For all cows, the trends were slightly higher for ssGBLUP, with much smaller differences than for genotyped cows, and with a decline in 2013 for milk and protein. Trends for BLUP indicate bias due to genomic preselection for genotyped sires and cows. ssGBLUP seems to at least partially account for that bias and is computationally suitable for national evaluations.

Development of genomic selection in dairy cattle in two emerging countries: South Africa and India

V. Ducrocq[1], L. Chavinskaia[2], M. Swaminathan[3], A. Pande[3], M. Van Niekerk[4] and F.W.C. Neser[4]
[1]GABI, INRA, AgroParisTech, Université Paris-Saclay, 78350 Jouy-en-Josas, France, [2]Lisis, INRA, Université Paris-Est, Marne la Vallée, France, [3]BAIF, Central Research Station, Uruli Kanchan, Maharashtra, India, [4]Department of Animal, Wildlife and Grassland Sciences, University of the Free State, Bloemfontein, South Africa; vincent.ducrocq@inra.fr

In large developed countries, genomic selection has radically transformed dairy cattle breeding, with collateral impact on the international semen market. Emerging countries are often strong semen importers. Few of them are Interbull members. However, if they are a member, MACE evaluations of progeny tested bulls give them access to EBV on their own scale. At least in theory, this allows for a rational choice of semen to import. This is no longer the case with genomic evaluations, for which the constitution of a proper reference population is complex. As a consequence, the gap between countries with expertise in genomic selection and the others is widening fast. However, genomic selection also offers new opportunities in emerging countries, because it does not require sophisticated national performance recording systems and it more easily leads to selection adapted to local environments. The project 'GENOSOUTH2' of the INRA SELGEN metaprogramme aims at promoting the local implementation of genomic selection of dairy cattle, with 2 countries as case study: in South Africa, the objective is to develop a prototype for national genomic evaluation in Holstein cattle, exploiting the participation of the country to Interbull MACE evaluations, in order to enlarge the reference population while accommodating two very contrasted feeding systems. In India, BAIF, the largest Indian NGO in agriculture, is developing a female reference population of crossbred *Bos taurus* × *Bos Indicus* animals, facing a number of challenges such as very small average herd sizes, scarce identification and limited performance recording. We will describe the motivations, the current constraints and the expected benefits of these two initiatives. Local genomic evaluations are fundamental for sustainable genetic progress of dairy cattle in harsh environments.

Genetic diversity of the Mexican and Spanish Lidia populations by using a subset of non-linked SNPs

P.G. Eusebi, O. Cortés, S. Dunner and J. Cañón
Universidad Complutense de Madrid, Departamento de Produccion Animal, Avenida Puerta de Hierro S/N, 28040 Madrid, Spain; paulinaeusebi@gmail.com

Retaining features of the auroch (*Bos taurus primigenius*), the Lidia bovine is a primitive breed that has its origin ~250 years ago in the Iberian Peninsula, where is still distributed along with France and several American countries. Selected upon a behavior that enhances their aggressiveness, these bovines were raised to take part in popular festivities that nowadays reinforce the identity of regional cultures. Different festivities demanded diverse behavior patterns, prompting a fragmentation of the breed into small lineages. In Mexico, mainly two families of breeders imported Lidia bovines from Spain in the early XXth century specializing their production either reproducing the new arrivals among them or realizing systematic crosses with local populations. Genetic diversity and structure of the Mexican and Spanish Lidia populations has been assessed with microsatellite data, but nowadays SNP molecular markers allows higher resolution level. Genetic diversity of the Mexican and Spanish Lidia populations and their relationship were assessed by using the 50K BeadChip on 468 individuals, who were genotyped, and after applying strict filtering criteria, 573 non-linked SNPs were selected. In both populations Similar gene diversity values were observed: remarkable for the Mexican population. Significant FIS values mean a strong subdivision within and higher F_{ST} genetic distances were observed in the Spanish than in the Mexican population. Genetic structure analysis showed similarity of three Spanish lineages with González family and some Llaguno breeders, but most Llaguno family clustered separated: genetic differentiation along with high gene diversity suggest an introgression of creole cattle in the constitution of the Mexican population.

Inbreeding trend and pedigree evaluation in Polish Holstein-Friesian cattle

E. Sell-Kubiak[1,2], Ł. Czarniecki[2] and T. Strabel[1,2]
[1]Department of Genetics and Animal Breeding, Poznan University of Life Sciences, Wolynska 33, 60-637 Poznan, Poland,
[2]Centre for Genetics, Polish Federation of Cattle Breeders and Dairy Farmers, Kłosowa 17, 61-625 Poznan, Poland;
sell@up.poznan.pl

The aim of this study was to evaluate observed and future inbreeding level in Polish Holstein-Friesian cattle population. In total over 9.8 mln animals were used in the analysis coming from the pedigree of Polish Federation of Cattle Breeders and Dairy Farmers. Inbreeding level, as an average per birth year, was estimated with the method accounting for missing parents information with the assumption of year 1950 as the base year. If an animal had no ancestral records, an average inbreeding level from its birth year was assigned. The future inbreeding was estimated as an average inbreeding of possible offspring of bulls and cows available for mating in a certain year. It was observed that 30-50% of animals born between 1985 and 2015 had no relevant ancestral information, which was caused by a high number of new farms entering the national milk recordings. For the last 20 years the observed inbreeding has been clearly higher than future inbreeding. In year 2015 the observed inbreeding level was 3.30%, while future inbreeding was 2.89%. The average increase of inbreeding in years 2001-2015 was 0.10%, which was similar to other countries monitored by World Holstein Friesian Federation. The estimates of future inbreeding suggested that observed inbreeding could be even lower than currently observed. This may be achieved with computerized assisted matings, however, still deep pedigrees are essential. Thus, the importance of keeping pedigree entries complete should be highlight. The obtained inbreeding levels indicated that population of Holstein-Friesian in Poland has one of the lowest values in the world and that its increase over the past decades remained on a safe level similar to other countries.

Genotyping females improves genomic breeding values for new traits

L. Plieschke[1], C. Edel[1], E.C.G. Pimentel[1], R. Emmerling[1], J. Bennewitz[2] and K.-U. Götz[1]
[1]Bavarian State Research Center for Agriculture, Institute of Animal Breeding, Prof.-Dürrwaechter-Platz 1, 85586 Poing-Grub, Germany, [2]University Hohenheim, Institute of Animal Science, Garbenstraße 17, 70599 Stuttgart, Germany;
laura.plieschke@lfl.bayern.de

Genotyping females is considered to provide gains in reliabilities of estimated breeding values of young selection candidates for new traits for which there is no reference set yet. By the use of a simulation we investigated two different strategies to establish a reference set for so called new traits where performance recording has begun only recently and phenotypes of daughters were available for one or two generations of bulls only. In design 'NTsires' cows are phenotyped but only their sires are genotyped and used in the reference set. In design 'NTcows' genotypes of phenotyped cows are available and cows are used directly as the reference set. Simulated heritability of the trait was 0.05. We studied the effects on validation reliabilities and unbiasedness of predicted values for selection candidates. We additionally illustrate and discuss the effects of a selected daughter sample and an unbalanced sampling of daughters. If the number of phenotypes is limited, as it is in the case of a new trait, it is always better to collect and use genotypes and phenotypes on the same animals instead of aggregating phenotypes of daughters and using them by relating them to their sires' genotypes. We found that the benefits that can be achieved are also sensitive to the sampling strategy used when selecting females for genotyping.

Genetic patterns predicting fertility in Estonian dairy cattle

S. Värv[1,2], T. Kaart[1,2], E. Sild[2] and H. Viinalass[1,2]
[1]*Bio-Competence Centre of Healthy Dairy Products LLC, Kreutzwaldi St. 1, 51014, Estonia,* [2]*Estonian University of Life Sciences, Institute of Veterinary Medicine and Animal Sciences, Kreutzwaldi St. 1, 51014 Tartu, Estonia; sirje.varv@emu.ee*

Aim of the study was to assess the genetic variability in connection with fertility of Estonian Holstein cows. Firstly, 2619 cows with estimated production and fertility breeding values (EBV) were randomly selected from Estonian Livestock Performance Recording Ltd database and their genotypes in genes relevant to production and fertility (ABCG2, DGAT1, CSN gene cluster, CYP11B1, LEP, LGB and SPP1) were determined. Secondly, ~5% of cows with the most extreme fertility EBV were selected from the both sides of fertility EBV distribution, resulting in 141 cows with low fertility EBV (average 81.6 points) and 147 cows with high fertility EBV (average 122.4 points). These 288 cows were genotyped using the BovineLD SNP chip (Illumina, Inc.). The analysis of variance revealed strong statistically significant effects of DGAT1, CSN2, CSN3 and CSN1S2 on fertility. Also, DGAT1 and LGB were statistically significantly related to milk fat content EBV, and DGAT1 and CSN2 to milk protein content EBV (for all listed relationships $P<10^{-5}$). The Fisher exact test revealed more than 10% of the chip SNPs across the whole genome associating with fertility ($P<10^{-5}$) in Estonian Holstein. The genome regions with the highest proportion of significant SNPs were found at chromosomes X and 7. The strongest relationships between SNP and fertility ($P<10^{-15}$) were discovered in BTA 6, 8, 9, 10, 17, 24, 29 and in X chromosome. From the adjacent SNPs of the previously studied genes only one marker in casein gene region was highly significantly ($P<10^{-5}$) related with the fertility EBV groups. We also detected heterozygosity excess in the group of cows with low fertility that potentially could indicate the deleterious polymorphisms causing decreased fertility. Finally, the assignment test performed chromosome-wise indicated clear patterns in SNPs distinguishing the fertility groups. Estonian Ministry of Education and Research (grants IUT8-1, IUT8-2) and Bio-Competence Centre of Healthy Dairy Products LLC (project EU48686).

Linkage disequilibrium and imputation power in Gyr beef cattle

J.A.I.I.V. Silva[1], A.M.T. Ospina[2], A.M. Maiorano[2], R.A. Curi[1], J.N.S.G. Cyrillo[3] and M.E.Z. Mercadante[3]
[1]*FMVZ/Unesp, Melhoramento e Nutrição Animal, C.P. 560, Botucatu, SP, 18618-970, Brazil,* [2]*FCAV/Unesp, Jaboticabal, SP, 14884-900, Brazil,* [3]*Instituto de Zootecnia, C.P. 63, Sertãozinho, SP, 14160-900, Brazil; jaugusto@fmvz.unesp.br*

The aim of this study was to evaluate imputation accuracy and to estimate linkage disequilibrium (LD) at 25-50 kb, 50-100 kb and 100-500 kb distances in a Gyr small closed population selected for post weaning weight. Genotypes of 155 and 18 animals were obtained through Illumina Bovine LDv4 (33K) and BovineHD BeadChip panels, respectively. Missing genotypes were imputed using FImpute software considering HD panel as reference. After imputation step, quality control was performed for minor allele frequency (MAF) less than 0.02 and call rate less than 0.10, remaining 173 animals and 418.086 SNPs for LD analyzes. Accuracy of imputation was evaluated by three methodology: (1) concordance rate (CR), (2) imputation quality score (IQS) and (3) simple correlation (ρ). The LD correlation coefficient (r^2) was obtained by Plink and RStudio software. Values of 96, 97 and 97% were obtained for CR, IQS and ρ, respectively, showing high imputation quality. FImpute was appropriated for low MAF SNPs imputation. The r^2 at 25-50 kb, 50-100 kb and 100-500 kb distances were 0.35, 0.29 and 0.18, respectively. LD persistence was observed at distances up to 100 kb, decreasing as the distance length increases. The r^2 values were higher than those described in the literature for zebu cattle, showing that the animals shared long segments of haplotypes in this closed herd.

Optimum genetic contribution in the Swiss original Braunvieh cattle population

H. Signer-Hasler[1], A. Burren[1], B. Bapst[2], M. Frischknecht[1,2], B. Gredler[2], F.R. Seefried[2], D. Garrick[3], C. Stricker[4], F. Schmitz-Hsu[5] and C. Flury[1]
[1]*Bern University of Applied Sciences, School of Agricultural, Forest and Food Sciences HAFL, Länggasse 85, 3052 Zollikofen, Switzerland,* [2]*Qualitas AG, Chamerstrasse 56, 6300 Zug, Switzerland,* [3]*Iowa State University, Department of Animal Science, 1221 Kildee Hall, Ames, IA 50011-3150, USA,* [4]*agn Genetics, Börtjistrasse 8b, 7260 Davos, Switzerland,* [5]*Swissgenetics, Meilenfeldweg 12, 3052 Zollikofen, Switzerland; mirjam.frischknecht@qualitasag.ch*

The Original Braunvieh (OB) cattle population is a local breed with about 10,000 cows in the herd-book. The breed is well-known for its dual-purpose characteristics and is characterized by a high proportion of matings through natural service (around 50%). Genomic breeding values are available since 2014. In a retrospective study we investigated the use of optimum genetic contributions (OGC) in OB using the conventional total merit index, milk index and pedigree-information for breeding animals having offspring born in 2014. We compared the realized genetic parameters (selection response and average inbreeding coefficients) with the results from optimized matings that would have resulted by following recommendations of EVA software. Computing capacity was a limiting factor if all cows were to be considered in the analysis. Based on our data a weight of 1 on merit combined with weights ranging from 200-400 on relationship lead to an improvement of the actual selection scheme. The results support, that selection response and average inbreeding coefficients could have been optimized by using EVA. For practical breeding programs the application of OGC on the two paths sires to breed bulls and dams to breed bulls seems promising.

First genetic analysis of four Algerian cattle populations from dense SNP data

N. Boushaba[1], N. Tabet-Aoul[1,2], D. Laloë[3], K. Moazami-Goudarzi[3] and N. Saïdi-Mehtar[1]
[1]*Laboratoire Génétique moléculaire et cellulaire LGMC, Département de Génétique Moléculaire Appliquée, B.P 1505 El M'Naouar, USTOMB, 31025, Algeria,* [2]*Département de Biotechnologie, Université d'Oran Es-Sénia, Oran, 31000, Algeria,* [3]*Génétique Animale et Biologie Intégrative, UMR 1313, INRA, Jouy-en-Josas, 78352, France; nacerat@yahoo.com*

This work is realized in the framework of biodiversity of animal genetic resources in general and cattle in particular. In Algeria, we were interested in the constitution of a first unrelated DNA library (121) of four (04) local cattle populations: Cheurfa, Guelmoise, 'Chélifienne' and Biskra. We genotyped these individuals DNA using the Illumina Bovine SNP50. A statistical analysis of these four bovine populations was carried out using Principal Component Analysis (PCA) based on all molecular data of SNPs genotyped using software R. However, in order to better clarify the respective roles of geographic isolation, gene flow and the historical origin of the populations studied, a comparison with a larger set of 19 breeds corresponding to African, European taurines and zebu was needed. The results of the ACP show a clear distribution of the four Algerian bovine populations as a representation of triangular gradient with peaks corresponding to African taurines, European taurines and zebu breeds. We also detected the influence of zebu breeds in Cheurfa and Guelmoise populations. Moreover, our results were consistent with previous historical and archaeological reports on the gene flow that existed between North Africa and South European breeds. The aims of for this work would be, in the medium term, to safeguard biodiversity by setting up a strategy for the conservation and preservation of our bovine genetic resources and, in the long term, a possible selection for an improvement of this species for economic purposes.

Vision systems aspects of dairy farming towards 2030

P. Galama[1], K. De Koning[1] and A. Kuipers[2]
[1]Wageningen University and Research (WUR), Livestock Sciences, De Elst 1, 6708 WD Wageningen, the Netherlands,
[2]Wageningen University and Research (WUR), Economic Research, Alexanderveld 5, 2585 DB Den Haag, the Netherlands;
paul.galama@wur.nl

As a case, the vision of the Dutch dairy chain towards 2020 is presented on basis of the so called 'Sustainable Dairy Chain' program: a sector in which we work safely and with pleasure; with fair income and high quality food; we handle our animals and environment respectfully; a sector which is appreciated by society. Objectives: Climate neutral development (20% reduction in GHG versus 1990; 16% sustainable energy; energy-efficiency +2%/year); maintain grazing (80% of farms should apply grazing); continuous improvement of animal welfare and health (70% reduction antibiotic use; longevity + 6 mo. versus 2011; improve score welfare index, in operation by 2017); maintain biodiversity and environment (100% use of responsible soya; national kg phosphate at level 2002; ammonia emission – 5 kton versus 2011; biodiversity tool available in 2017). Goals are annually monitored by Wageningen Economic Institute. For each theme a sector team is active to realize goals. In 2017 new goals will be formulated towards 2030, intended to be in line with the ever increasing societal influence. Current examples of societal participation are: existence of the 'Animal party' in the parliament; phosphate plafond; discussions about obligatory grazing and calf with cow rearing in region; dispute about no. of animals/farm (Industrial); less meat consumption in relation to climate change. We will outline the development directions for the dairy sector: high tech farming (incl. milk robot of which 1st prototype was at our station) versus low cost; resource efficiency including labor; specialization versus diversification and a climate neutral dairy chain. Specific practices to be explained: cow transition period/pregnancy gymnastics; smart grazing techniques; sensor technology in- and outside barn; mobile milking and feeding systems (also in combination with cows as nomads in nature areas); innovative housing concepts, like cow-garden (slatted floors and boxes under scrutiny). Future images about energy saving, manure handling and soil quality, and the circular economy will be addressed, as well as opportunities of new communication techniques.

Dairy farmers' attitudes towards farm animal welfare

H. Heise and L. Theuvsen
University of Goettingen, Agricultural Economics & Rural Development, Platz der Goettinger Sieben 5, 37073 Goettingen, Germany; hheise@gwdg.de

The demand for products with higher farm animal welfare (FAW) standards is growing. Nevertheless, there are hardly any dairy products from pure animal welfare programs (AWPs) on the market. Although dairy farmers are an important stakeholder group for the successful implementation of AWPs, little is known about their attitudes. For this study, 258 dairy farmers in Germany were questioned via an online survey. We found five clusters that significantly differ with regard to their attitudes toward AWPs, FAW, and their willingness to part in AWPs. Cluster A consists of farmers who strongly oppose AWPs; these farmers will probably not take part in AWPs. Farmers in cluster B view AWPs and the associated market effects with some skepticism; however, they are willing to improve their level of FAW and, thus, may someday become willing to participate in AWPs. Cluster C farmers have diverse attitudes toward AWPs; since they are comparatively most optimistic concerning the market effects of higher FAW standards, these farmers could also become AWP participants in the future. Farmers in cluster D have positive attitudes toward AWPs and show the highest willingness among the five clusters to improve FAW on their farms. However, they are the most skeptical towards market effects of higher national FAW standards and the market potential for more animal-friendly products. If the economic security of AWPs were guaranteed, Cluster D farmers would probably constitute an important target group. Farmers in cluster E have positive attitudes toward AWPs, show a high willingness to improve the own FAW, and tend to be less skeptical about the market effects of higher animal welfare standards; these farmers constitute the most important potential target group for AWPs. Our results have important managerial implications and provide a starting point for the design of tailor-made AWPs that fulfil the requirements of both dairy farmers and the broader public Participating in AWPs could create an opportunity for dairy farmers to escape from the pressure to produce at the level of world market prices and, instead, to take advantage of a more differentiated market segment for milk produced with higher FAW standards which is accepted and financially rewarded by society.

Analysis of dairy farmers' expectations and challenges in four European countries

M. Klopčič[1], A. Malak-Rawlikowska[2], A. Stalgiene[3] and A. Kuipers[4]

[1]University of Ljubljana, Biotechnical Faculty, Dept. of Animal Science, Groblje 3, 1230 Domžale, Slovenia, [2]Warsaw University of Life Sciences, Faculty of Economic Sciences, ul Nowoursynowska 166, 02-787 Warsaw, Poland, [3]Lithuanian Institute of Agrarian Economics, V. Kudirkos g. 18-2, 03105 Vilnius, Lithuania, [4]Wageningen UR, Expertise Centre for Farm Management and Knowledge Transfer, P.O. Box 35, 6700 AA Wageningen, the Netherlands; marija.klopcic@bf.uni-lj.si

On the base of survey, data from dairy farmers were collected from 2011 to 2016 concerning their future expectations, as part of a Leonardo da Vinci and an ERASMUS+ project. The goal was to examine if a coherent outlook on future developments and challenges can be found within and between dairy farmers in four European countries. Topics dealt with were choice of development paths, availability of resources, perceived opportunities & threats and farmers' abilities, and economic expectations. In 2011/2012, 1,039 questionnaires from farmers were received, spread over Poland, Lithuania and Slovenia. In 2013, 352 of these farmers filled in the questionnaire again, as well as 122 farmers from the Netherlands. Those farmers were again questioned in 2016. PCA, cluster analysis and ANOVA methods were used to analyse the dataset. On base of 10 strategies, seven farmer segments were identified. Differences in opinions between these segments were rather limited, while significant differences between countries were found. For instance, farmers in the Netherlands show much more confidence in dealing with the free quota situation and the dairy market than in the other countries. Farmers in Lithuania and Slovenia are more diversified, while the latter farmers are also more consumers oriented. The trend in opinions over years will be discussed. However, the outlook on the future is quite different between countries. Results of this study provide insights, which can support the development of the dairy sector in the various regions of Europe.

Stakeholder opinions on development paths, opportunities and barriers in dairy sector

A. Kuipers[1], A. Malak-Rawlikowska[2], A. Stalgiene[3], P. Kristof[4] and M. Klopčič[5]

[1]Expertise Centre for Farm Management and Knowledge Transfer WUR, P.O. Box 35, 6700 AA Wageningen, the Netherlands, [2]SGGW, Economic faculty, Nowoursynowska 166, Warsaw, Poland, [3]Lithuanian Agrarian Institite, V. Kudirkos Str. 18-2, Vilnius, Lithuania, [4]Corinthian Agricultural Chamber, Museumgasse 5, Klagenfurt, Austria, [5]University of Ljubljana, Biotechnical faculty, Groblje 3, Domzale, Slovenia; abele.kuipers@wur.nl

An analysis was performed on how stakeholders, i.e. leading persons in the dairy chain, envisage the future of dairy farming in a period of radical policy change and what opportunities and barriers they foresee to their objectives. Perceptions on development strategies, availability of resources, opportunities & threats (O&T), farmer skills and future expectations were examined in 5 EU countries. In 2015 and early 2016, a total of 161 completed questionnaires were collected from stakeholders in the Netherlands, Slovenia, Lithuania and Poland. Data from Austrian stakeholders were added during 2016 and are being processed. Data were analysed by PCA, cluster and stepwise regression methods. Results for the first 4 listed countries are presented here. Eight strategic groups of stakeholders were defined. Farm expansion and specialisation was the most expected development strategy (57% of stakeholders). Almost 1/4 of the stakeholders took a wait and see approach, of which 15% looked for opportunities to activate at a particular moment in time, while 8% were generally pessimistic about the future. Diversification in combination with organic farming was chosen by 5% of the stakeholders, 10% of stakeholders focused on cooperation, service and high tech, and another 5% placed their trust in skills, subsidies and labour. The opinions were highly affected by the country of origin, while only minor variations in opinions were observed between different categories of stakeholders. Polish stakeholders showed the most specialised view on the dairy chain, but they scored relatively low on cooperation. Development towards diversification and organic agriculture received higher scores in Slovenia and Lithuania. Netherlands' stakeholders were the most positive about the future e.g. they foresee expansion and market opportunities. It was shown that strategies, resources and O&T each directly affect future expectations, which was in agreement with the hypothetical model used. The Austrian results will be integrated in the presentation.

Model-based analysis of fertility components in dairy cows

M. Derks[1,2], P. Blavy[2,3], N.C. Friggens[2,3], F. Blanc[1,4] and O. Martin[2,3]
[1]INRA, UMR 1213 Herbivores, 63122, Saint Genès Champanelle, France, [2]INRA, UMR 791 MoSAR, 75005, Paris, France, [3]AgroParisTech, UMR 791 MoSAR, 75005, Paris, France, [4]Clermont Université, VetAgro Sup UMR 1213 Herbivores, 63000, Clermont-Ferrand, France; marjoleinderks@hotmail.com

A simulation model that is able to investigate hypotheses about dairy cow fertility is presented here. Over the last decades, in parallel with changes in genetics, productivity and management, dairy cattle have faced a decline in fertility. A number of possible physiological mechanisms underlying the decline in reproductive efficiency have been proposed, with a major focus on the interaction between lactation, nutrition and reproduction. However, given that fertility can be impaired on different levels (e.g. oocyte quality, oestradiol levels, uterine health), there is no clear consensus on how mechanisms like negative energy balance or increased milk yield in itself impair fertility in dairy cows: everything is connected to everything. The model presented here combines a simple reproductive physiology model, where each step in the fertility cycle is represented separately, with an existing animal model of lifetime performance based on a conceptual representation of energy partitioning. The reproductive physiology model incorporates genetic, endocrine, metabolic and stochastic regulatory components, meaning that at each step in the fertility cycle the individual effects of animal factors on that step can be adjusted. The resulting whole model allowed us to test hypotheses on the animal factors modulating reproduction, like milk potential, parity or energy balance. The model can easily be adjusted when new insights are achieved, based on the compartmental nature and the different regulatory components. Model outputs are in good agreement with findings reported in literature, suggesting that it adequately represents the major effects of cow performance on reproduction. This model provides a way to quantify the effects of impaired fertility on long term productive performance.

Heritability of methane emission from dairy cows measured over long time period

M. Pszczola, K. Rzewuska, S. Mucha and T. Strabel
Poznan University of Life Sciences, Department of Genetics and Animal Breeding, Wolynska 33, 60-637 Poznan, Poland; strabel@man.poznan.pl

Methane (CH_4) emission from dairy cattle is important due to its environmental and economic impact. CH_4 emission may change across lactation. Current reports on heritability of CH_4 emission are based on short measuring periods. In this study, we estimated heritability of CH_4 emission measured over a long period of time. Measurements were taken at two commercial Polish farms on 485 Polish Holstein-Friesian cows being in up to eight lactation from 5 to 305 DIM. CH_4 was measured during milking in automated milking systems (AMS) in the following periods: from 2014/12/02 to 2016/02/03, and from 2016/06/01 to 2016/09/17 on Farm1; and from 2016/02/05 to 2016/03/14 on Farm2 (in total 442,813 milkings). CH_4 emission was quantified in liters per day after accounting for diurnal variation and AMS unit. Variance components were estimated using random regression model that included: fixed parity effect (1 and 2+), overall fixed lactation curve modeled with third order Legendre polynomials, fixed farm by year week effect, regression on milk yield and daily body weight, and individual genetic and permanent environmental lactation curves modeled with second order Legendre polynomials. Average daily CH_4 emission was 417.56 l/d (S.D. 95.40). Average genetic and permanent environment variances were 930 $(l/d)^2$ and 760 $(l/d)^2$, respectively. Average heritability equaled 0.18 (average SE 0.06). Heritability was the highest at the beginning of lactation 0.32 (SE 0.11) and dropped considerably towards the end of lactation to 0.06 (SE 0.06). Genetic correlations were high between neighboring DIM and decreased when the distance between DIM increased. The lowest genetic correlation (0.58) was reached after the distance between DIM reached 302 days. Phenotypic correlations were much lower than genetic ones (0.15 to 0.51). This study showed an importance of collecting CH_4 phenotypes repeatedly on the cows in the course of lactation and that it is important to include DIM in modeling of individual variation and estimation of the genetic parameters. Low genetic correlation between distant DIM suggests that CH_4 emission at the beginning and at the end of lactation may have a different genetic background.

Relative Importance of factors affecting GHG emissions and profitability of dairy farms

M. Zehetmeier[1], B. Zerhusen[1], A. Meyer-Aurich[2], H. Hoffmann[3] and U.K. Müller[4]
[1]Bavarian State Research Center for Agriculture, Freising, 85354, Germany, [2]ATB, Potsdam, 14469, Germany, [3]TUM, Freising, 85354, Germany, [4]California State University Fresno, Fresno, 93740, USA; monika.zehetmeier@lfl.bayern.de

Agricultural farms increasingly need to contribute to greenhouse gas (GHG) mitigation actions so governments will be able to fulfill reduction commitments. To identify factors that are suitable targets for GHG mitigation on farms, many studies use normative models. Normative models assume 'best practices' or 'typical farming systems' to predict which factors contribute most strongly to GHG emissions. By ignoring actual farm practices, however, normative models cannot predict inter-farm variability. Yet determining inter-farm variability is essential to identify policy targets: factors with high inter-farm variability are more likely to be effective targets for GHG mitigation, as long as they do not also strongly affect profitability. The objectives of our study were to (1) identify factors that strongly affect GHG emissions and profitability of actual dairy farms and (2) determine their inter-farm variability, which indicates potential to reduce GHG emissions or increase profitability. To assess inter-farm variability, our model requires in-depth and site- or farm-specific input variables for a high number of farms. To this end, we combined data and models for 100 dairy farms for the year 2013 in Bavaria, Germany, from the following sources: administrative databases available for individual farms, farm data from discussion groups, and site specific models (e.g. site specific N_2O emissions from a biophysical model). Our global sensitivity analysis identified five factors affecting GHG emissions in the following order of relative importance (proportion of inter-farm variability explained): nitrogen use efficiency (28%), feed use efficiency (26%), site specific nitrogen emission factor (21%), milk yield (15%), and replacement rate (10%). Of these five factors, feed use efficiency, replacement, and milk yield also affected profitability. This leaves two main determinants for inter-farm variability of GHG emissions that did not affect variability of profitability and hence need to be addressed by additional incentives for farmers, advisory service, or stricter rules.

Milk production and grazing behavior of dairy cows grazing two timing of pasture and herbage mass

I. Beltran[1], A. Morales[1], P. Gregorini[2], O. Balocchi[1] and R.G. Pulido[1]
[1]Universidad Austral de Chile, Independencia 647, 5090000, Valdivia, Chile, [2]Lincoln University, P.O. Box 7647, 7647, Lincoln, New Zealand; rpulido@uach.cl

We evaluated the herbage mass and time of pasture allocation effects on herbage chemical composition, dry matter intake and grazing behavior of dairy cows during autumn. Forty-four Holstein-Friesian cows (24.7 kg milk/d, 581 kg BW, 74 DIM) were allocated to one of four treatments resulting from the combination of two time of pasture allocation (TA) [access to new pasture in the morning (MTA) or afternoon (ATA)] and two herbage mass (HM) [2,000 kg DM/ha (LHM) and 3,000 kg DM/ha (HHM)]. Treatment groups were strip-grazed with an herbage allowance of 21 kg DM/day. All cows were supplemented with concentrate and grass silage (3.5 and 3.0 kg DM/d, respectively) which were fed in equal amounts during the morning and afternoon. Herbage was evaluate for dry matter (DM), crude protein (CP), and soluble carbohydrates (WSC). Herbage DM intake was back-calculated using productive data. Grazing, rumination and idling time were recorded for each cow during 24 h, twice in the experiment. Herbage DM was affected by TA, being 20% greater for ATA than MTA. Herbage CP was 15% greater for MTA than ATA. In addition, herbage CP was 18% greater for HHM than LHM. Herbage WSC content was affected by TA, being 32% greater for ATA than MTA. Herbage WSC/CP ratio was affected for by herbage mass and TA, being the highest for HHM and ATA. Total and herbage DMI were similar among treatments, averaging 14.8 and 8.3 kg DM, respectively. There was significant TA×HM interaction for total grazing time, being the lowest for LHA×MTA. Total rumination time was not affected by treatment, averaging 411 min/day. A significant TA×HM interaction was found for total idling time, being longer for HHM×MTA than HHM×MTA and LHM×ATA. Grazing time (07:00 to 14:00 h) during the morning was 86% higher for MTA than ATA. Afternoon grazing time (15:00 to 06:45 h) was 56% higher for ATA than MTA. In addition, pm grazing time was 15% longer LHM than HHM. In conclusion, herbage mass and time of pasture allocation did not modify DM intake, however, TA and HM changed the grazing patterns, motivating for longer grazing time when cows were allocated pasture in the afternoon, at the time herbage had a greater nutritive value.

A method to estimate cow potential and subsequent milk responses to dietary changes

J.B. Daniel[1,2], N.C. Friggens[1], H. Van Laar[2] and D. Sauvant[1]
[1]INRA, AgroParisTech, Université Paris-Saclay, 75005 Paris, France, [2]Trouw Nutrition, R&D, P.O. Box 220, 5830 AE Boxmeer, the Netherlands; jean-baptiste.daniel@trouwnutrition.com

The application of equation that predict production responses to dietary changes requires an estimation of the production potential of the dairy cows in question. This is because the further away the animal is relative to its potential, the greater the expected response to extra nutrient will be. However, definitions of potential in nutrition model are rather abstract which render its estimation difficult. Therefore the aim of this study is to propose a method usable by nutrition models to estimate cow production potential. The observed efficiencies in net energy for lactation (NEL) and metabolizable protein (MP) are proposed as a basis to estimate the production potential and how far the cow is from this potential. The rationale for using NEL and MP efficiency (ratios of milk energy yield/NEL above maintenance supply and milk protein yield/MP above maintenance supply) builds on the uniformity of the observed relationships between size of the milk responses and extra NEL supply and MP supply, when centred on a given efficiency. From there, a pivot nutritional situation where NEL and MP efficiency are 1.00 and 0.67, respectively, was defined, from which milk responses could be derived across animals varying in production potential. An implicit assumption of using response equations centred on reference efficiency pivots is that the size of the response to a fixed change in nutrient supply, relative to the pivot, is identical for animals with different potentials. Further, by using a fixed and clearly defined reference efficiency pivots the effects of stage of lactation on NEL and MP efficiency would have consequences on the expected size of the response. Typically, NEL and MP efficiency decreases with stage of lactation. This implies that the predicted responses to a dietary change would be greater in early- than in late-lactation, in agreement with the literature. This method was evaluated against two independent datasets across stages of lactation. Overall, milk component yields were predicted with RMSPE<13.5% and CCC>0.784, which indicates that the method has the potential to be used in practice for prediction of responses.

Extracellular vesicles: a new paradigm in cell communication controlling the success of reproduction

P. Mermillod[1], A. Alcantara-Neto[1], E. Corbin[1], G. Tsikis[1], V. Labas[1], S. Bauersachs[2] and C. Alminana[1]
[1]Institut National de la Recherche Agronomique (INRA), UMR7247 PRC, 37380 Nouzilly, France, [2]ETH Zurich, Animal Physiology, Institute of Agricultural Sciences, Zurich, Switzerland; pascal.mermillod@inra.fr

The success of early reproductive events relies on intensive communication between gametes – embryos and maternal tissues, particularly in the oviduct. Besides traditional pathways involving ligand-receptor binding, extracellular vesicles (EV), including exosomes, are emerging as another way of cell communication. We identified the presence of EV in the bovine oviduct fluid (boEV) and showed that the boEV composition changes across the female estrous cycle, in terms of proteins and miRNA. From a total of 336 proteins identified in boEV by mass spectrometry, 170 were differentially expressed between the stages, including proteins with suspected function in the regulation of gametes function or embryo development (OVGP1, HSPA8, MYH9 and HSP90). Moreover, a total of 168 miRNA were identified in boEV, with 35 of them differentially expressed along the estrous cycle. Some of these miRNAs are known to be involved in cell proliferation (miR-10b; miR-34b) and embryo development (miR-34b; miR-125b). Furthermore, we demonstrated that after boEV fluorescent labelling, these EVs are able to cross the zona pellucida and be uptaken by bovine developing embryos in vitro. When bovine in vitro produced (IVP) embryos where developed in the presence of boEV, the development rate and quality were increased, mimicking the positive effect of coculture of IVP embryos with bovine oviduct epithelial cells (BOEC). Indeed, we observed that BOEC in culture were also able to produce boEV with size distribution similar to in vivo ones but with a different protein contents. These differences may reflect the partial dedifferentiation of BOEC in culture, translating in a modified transcriptomic profile, observed previously in our laboratory. These findings will lead to a better understanding of the regulation of fertilization and early embryo development in mammals. Furthermore, it may open new directions for the optimization of in vitro embryo production and provide new tools for the diagnosis of very early pregnancy as well as new targets for selection on female fertility.

Genotype by environment interactions for Norwegian Red crossbreds in Ireland, Canada and USA

E. Rinell and B. Heringstad

Norwegian University of Life Sciences, Universitetstunet 3, 1430, Norway; ellen.rinell@nmbu.no

Genetic material from Norwegian Red (NR) has been exported to over 20 countries and used for crossbreeding with Holstein (HO) dairy cattle. Improvement has been observed in fertility and health of the offspring, while a slight decrease in milk production has been observed. Although NR crossbreds have been successful, NR bulls were selected using a breeding goal developed in Norway without taking other countries' environments into consideration and are some of the same bulls used for breeding in Norway. The aim of this study was to compare performance of F1 HO × NR crossbreds (NRX) with their HO herdmates in different climatic environments and production systems, as well as to investigate possible GxE interactions. After edits, the data consisted of 3,631 NRX and 75,573 HO parities from USA, 1,713 NRX and 9,422 HO parities from Ireland, and 910 NRX and 23,824 parities from Canada. Available traits that were similar across countries were 305-day milk yield, calculated from test day observations using a polynomial equation developed from lactation curves, and lactation-mean somatic cell score (mSCS), calculated from test day records. Fixed effects were parity, herd-year (Ireland and USA only), age at calving (Ireland and USA only) and region-age-season of calving (Canada only). The American data included parities 1-6, Irish 1-8, and Canadian 1-3. There were 299 and 797 herd-year levels in the USA and Ireland, respectively. The effect of region-age-season in the Canadian data included 60 levels. Least-squares means in kg for 305-day milk yield (NRX and HO, respectively) were 9,957 and 11,084 in the USA, 5,980 and 6,125 in Ireland, and 8,662 and 9,481 in Canada. Least-squares means for mSCS (NRX and HO, respectively) were 1.96 and 2.44 in the USA, 2.75 and 2.93 in Ireland, and 2.37 and 2.23 in Canada. We will compare the level of NRX performance between the countries, using their HO counterparts as a baseline. Genetic correlations will be calculated to find GxE interactions between countries for each trait and examine the effect of breed group. Additional traits such as 305-day fat- and protein yields and calving difficulty will be included. The final results of the study will be presented at the EAAP Annual Meeting.

Automatic milking rotary AMR™ in practical use – investigations on the success rate of attachment, c

H. Scholz, A. Harzke, M. Moik, P. Kühne and B. Füllner

Anhalt University of Applied Sciences, LOEL, Strenzfelder Allee 28, 06406 Bernburg, Germany; heiko.scholz@hs-anhalt.de

The use of automatic milking systems now holds even in very large dairy herds with up to 1000 cows sustainable catchment. Thus, the proportion of automatic milking systems, all by the year 2011, realized and planned milking project in the farms is around 50%. At farm 1 the success rate by the robot was evaluated at regular intervals starting up in August 2013. With the change of the visual-system to TOF-cameras (Time of Flight) in October 2013 the success rate of attaching teats could be increased to more than 95% (February 2017). The remaining 5% of not attached teats could either be attached manually by the supervisor or those cows are sorted back to the rotary for an additional turn, but this may reduce the real throughput somewhat. A significant effect of the linear evaluation of the udder (DHV) scheme on the success rate could not be observed. The success rate of clearing the teats showed against the success rate of attachment significant lower values (April 2013: 65%; February 2017: 81%). In farm 1 the score of hyperkeratosis were recorded in the course of lactation in cows since the start of automatic milking. The classification was based on the scale of My et al., with score 1 'no keratin' and score 4 'rough, fissured hyperkeratosis'. It was recorded that the two front teats with an average of 81% compared to the rear teats with an average of 88% had a slightly lower proportion of the score 1 + 2. However during the course of lactation no significant differences between the three recording periods have been observed, which indicates a very good and gentle milking done by the system.

Changes in SCC and their origin depending on milking phase in organic and conventional dairy cattle

P. Wójcik[1], P. Dudko[2] and A. Mucha[1]
[1]1National Research Institute of Animal Production, Sarego 2, 31-047 Krakow, Poland, [2]Poznań University of Life Sciences, Wojska Polskiego 28, 60-637 Poznań, Poland; aurelia.mucha@izoo.krakow.pl

The aim of the study was to determine SCC changes in three milking phases of organic and conventional cattle. Two groups of 15 cows per group were investigated. The highest SCC values (above 1 million cells/ml) were observed during the final milking phase in both herds, and the lowest during the first milking phase. In the organic farm, the release of lymphocytes and macrophages was the highest in the second milking phase, and that of granulocytes and epithelial cells in the third phase. In the conventional farm, the highest lymphocyte release occurred in the third phase. In healthy udders, the first and second milking phase was characterized by the lowest level of SCC, including lymphocytes, granulocytes and epithelial cells. In cows with subclinical mastitis, granulocytes were most abundant in the first milking phase. The highest SCC in the third milking phase was associated with epithelial cells. In acute mastitis, the third milking phase was characterized by the highest SCC, including the number of epithelial cells and granulocytes. No statistically significant differences in terms of quality and species were found between bacterial flora from the healthy and mastitic milk samples. Natural flora of milk was predominant, but the analysed samples also contained staphylococci, numerous Gram-negative bacilli (*Pseudomonas* sp.), bacilli (*Bacillus* sp.) and corynebacteria (*Corynebacterium* sp.).

Using motion sensors fitted in collars to monitor behaviour of grazing cattle

J. Maxa, S. Thurner, M. Kaess and G. Wendl
Bavarian State Research Centre for Agriculture, Voettingerstr. 36, 85354 Freising, Germany; jan.maxa@lfl.bayern.de

Global Navigation Satellite Systems together with other motion sensors have been recently used to monitor behaviour of wildlife as well as domesticated animals. Such monitoring techniques are predestined for livestock on rangeland where accurate prediction of its behaviour can provide important information about health status of single animals as well as improve pasture management. Therefore, the main aim of this study was to use motion sensor data in order to develop classification algorithms for behavioural monitoring of grazing cattle. Two trials were conducted in 2015 and 2016 on paddocks located in Bavaria, Germany. In total 11 heifers were fitted with a tracking collar prototype collecting Global Position System (GPS) data and motion sensors data (3-axis accelerometer and magnetometer) at 1 and 3 Hz, respectively. Seven behavioural activities were manually recorded based on continuous sampling observations of random animals in trial 1 and video-based observations in trial 2. Transformed tracking collar data were aggregated into 10 second intervals with calculated mean values and standard deviations (SD) giving information about speed, position of the neck and activity level of the animal. Mixed effects regression analyses were conducted in order to detect the significances among sensor data and observed behaviours. Furthermore, probability density functions were fitted to data with mixture distributions and thereof threshold values for a classification algorithm obtained. The results of regression analyses showed significant effects ($P<0.05$) for observed behaviours on SD of speed, X-axis of accelerometer and Y-axis of magnetometer data. Based on the data from motion sensors it was possible to distinguish among heifers' grazing, ruminating and lying behaviour. The resulting classification algorithm showed the highest accuracy for grazing (87%) followed by ruminating (81%) and lying (78%). Possible development of more complex classification algorithms for health monitoring of livestock on pasture by sensor fusion and thereby an increase in prediction accuracies will be further investigated.

Prediction of calving from prepartum intravaginal temperature and electric resistance of cows

H. Kamada

Institute of Livestock and Grassland Science NARO, Animal Reproduction, Ikenodai-2, Tsukuba, Ibaraki, 305-0901, Japan; kama8@affrc.go.jp

Possibility to predict the time or day of calving from intravaginal temperature or electric resistance was investigated. Sensor probe for temperature and electric resistance measurement was inserted into the vagina of pregnant cows (n=15) two weeks before they were due to calve. Sensor data was sent wirelessly from transmitter on the back of cows to the receiver connected to personal computer. In order to observe the changes relating calving clearly, circadian changes of intravaginal temperature data was removed by subtraction of average daily change for first three days from raw data. It is known that rectal temperature decreases before calf discharge at delivery. In this study, minimum intravaginal temperature was detected 1,290±93 min before calving on average (n=7). And it was obtained as a new knowledge that intravaginal temperature increased once about 0.4 degrees Celsius over the average temperature three days before calving. We can predict the calving day from increased intravaginal temperature earlier than before. When the calving was induced by prostaglandin (PG) injection seven days before they were due to calve (n=8), the intervals between PG injection and minimum intravaginal temperature varied (332±134 min); however, the intervals between minimum value and calving were relatively stable (2,011±98 min). We can predict the time of induced calving from the monitoring of prepartum intravaginal temperature. Intravaginal electronic resistance gradually decreased several hours before calving. We can notice the imminent calving from the change of electric resistance earlier than the discharge of sensor probe. This work was supported by Cabinet Office, Government of Japan, Cross-ministerial Strategic Innovation Promotion Program (SIP), 'Technologies for creating next-generation agriculture, forestry and fisheries'.

Effect of ketone bodies on milk freezing point of Polish Holstein-Friesian cows

A. Otwinowska-Mindur[1], E. Ptak[1] and A. Zarnecki[2]

[1]University of Agriculture in Krakow, Department of Genetics and Animal Breeding, al. Mickiewicza 24/28, 30-059 Krakow, Poland, [2]National Research Institute of Animal Production, ul. Krakowska 1, 32-083 Balice, Poland; rzmindur@cyf-kr.edu.pl

Milk freezing point is an important indicator of milk quality, and β-hydroxybutyrate and acetone content are important indicators of subclinical ketosis. Ketosis often occurs in high-producing dairy cows during the first two months after calving as a result of negative energy balance. The objective of this study was to determine the effect of β-hydroxybutyrate (BHB) and acetone in milk on the milk freezing point (MFP) of Polish Holstein-Friesian cows. The data, collected in 2014 and made available by the Polish Federation of Cattle Breeders and Dairy Farmers, comprised 3,077,229 test day milk samples from 866,167 lactations of 714,440 cows. Four classes of lactations were created: first, second, third, and fourth to seventh lactations (≥4 class). Both BHB and acetone in milk were grouped in five classes: (1) ≤0.05 mmol/l, (2) 0.06-0.10 mmol/l, (3) 0.11-0.20 mmol/l, (4) 0.21-0.50 mmol/l and (5) >0.50 mmol/l. Milk freezing point was analyzed using the MIXED procedure in SAS and a linear model in which effects of lactation, BHB and acetone classes were included, with interactions between lactation and BHB and between lactation and acetone classes. MFP increased with lactation number, from -0.5354 °C in first to -0.5289 °C in ≥4 lactations. MFP was lowest (-0.5330 °C) in the fourth class of BHB (0.21-0.50 mmol/l), and highest (-0.5312 °C) when BHB in milk was lower than 0.05 mmol/l (first class). Milk with acetone content from 0.11 mmol/l to 0.20 mmol/l (third class) showed the highest MFP (-0.5308 °C); the lowest MFP (-0.5333 °C) was found for milk with acetone exceeding 0.50 mmol/l (fifth class). There were highly significant differences in MFP between all pairs of acetone classes, and in most cases between BHB classes (P<0.001). There was no significant difference in two pairs of BHB classes: first and fifth, and second and fifth. Highly significant interactions between lactation and BHB classes and between lactation and acetone classes were also observed (P<0.001). This means that the effects of BHB as well as acetone class were different in different lactations.

Body and milk traits as indicator of cow's energy status

P. Mäntysaari[1], T. Kokkonen[2], C. Grelet[3], E.A. Mäntysaari[1] and M.H. Lidauer[1]
[1]Natural Resources Institute Finland (Luke), Green technology, Jokioinen, Finland, [2] Department of Agricultural Sciences, University of Helsinki, Helsinki, Finland, [3]Walloon Agricultural Research Center, (CRA-W), Gembloux, Belgium; paivi.mantysaari@luke.fi

In early lactation the feed intake (FI) of high producing cow seldom fulfils her energy demands, forcing the cow to mobilize body reserves. A long lasting and deep energy deficiency can cause health and reproduction problems. Also, if FI is considered in the breeding goal, the postpartum energy status (ES) has to be monitored. ES can be estimated by calculating the energy balance (EB) from cow's energy intake and output. Alternatively, EB can be predicted by indicator traits like changes of body weight (ΔBW) and body condition score (ΔBCS), milk fat-protein ratio (FP) or milk fatty acid (FA) composition. The precision of these predictions has often been low. This may be related to a lack of precision in estimated EB itself, when standard energy requirements are used in EB. We used plasma non-esterified fatty acids (NEFA) concentration as a biomarker of ES, and addressed associations between NEFA concentration and ES indicators. Data included daily BW, milk and FI and monthly BCS of 102 and 43 cows on the 1st and 2nd lactation. Plasma samples for NEFA were collected twice on lactation weeks 2 and 3 and once on week 20. Milk samples for fat and protein concentration and FA composition (using MIR) were taken on the days of NEFA sampling. Milk FA contents were predicted by calibration equations from University of Liège/CRA-W, Belgium. On lactation weeks 2, 3 and 20 NEFA concentrations were on average 0.60 ($\pm$0.32), 0.46 ($\pm$0.23) and 0.14 ($\pm$0.06) mmol/l. First, a multiple linear regression model to predict NEFA was developed without milk FA. The best fit model (M1) included ΔBW, FP, ΔBCS, BCS$\times\Delta$BCS, parity and days in milk (AIC -153.2). Five milk FA or FA groups (C10, C16, C18:1, monounsaturated, and saturated FAs) were chosen by step-wise regression on NEFA. With FAs included, the best model (M2) contained ΔBW, ΔBCS, BCS$\times\Delta$BCS and FAs (AIC -366.5). The correlations between predicted and observed NEFA were 0.73 (M1) and 0.79 (M2), which were clearly higher than the correlation -0.47 between NEFA and EB. Body and milk indicators predict ES better than the calculated EB.

Impact of body condition at calving on milk production and reproduction

A. Ule and M. Klopčič
University of Ljubljana, Biotechnical Faculty, Dept. of Animal Science, Groblje 3, 1230 Domžale, Slovenia; Anita.Ule@bf.uni-lj.si

Body condition score (BCS) is a reflection of the body fat reserves, which change significantly during lactation. It is estimated by observation and palpation of specific areas of the body. The aim of this paper was to evaluate effects of BCS at calving and during lactation on the milk production and reproduction of dairy cows. Body condition was evaluated by 9-point scale in two dairy herds. During one year observations, we collected 931 records for BCS, milk production and reproduction data at each milk recording. BCS ranged between 3.0 and 8.0 and the average score was 5.64 and average daily milk yield was 33.05 kg. We used multiple trait model for BSC and milk yield with fixed effect of number of lactation, year-month interaction as season, breeder and breed. Ali-Schaeffer equations was used to describe the lactation curves. The model explained 61.2% variability in milk yield and 41.6% variability in BCS. In early lactation milk yield increased, the milk fat and protein decreased and BCS declined. After the peak of lactation milk yield started to decrease while milk fat, protein and BCS increased. Primiparous cows had a higher BCS, lower milk yield and better milk persistency than multiparous cows. Multiparous cows showed higher BCS loss and higher milk yield. BCS was positively correlated with milk fat (0.19) and protein (0.37) and negatively with milk yield (-0.36). Milk yield was significantly influenced by breeder, parity, season and breed, while BCS was influenced by number of lactation, season and breed. BCS was not influenced by breeder.

Comparison of definitions for lactation persistency in Polish Holstein-Friesian cattle

M. Graczyk[1], S. Mucha[1,2], J. Jamrozik[3,4] and T. Strabel[1,2]
[1]*Poznan University of Life Sciences, Departament of Genetics and Animal Breeding, 33 Wolynska, 60-637 Poznan, Poland,* [2]*Polish Federation of Cattle Breeders and Dairy Farmers, Centre for Genetics, 17 Klosowa, 61-625 Poznan, Poland,* [3]*University of Guelph, Animal Biosciences, ON, N1G 2W1 Guelph, Canada,* [4]*Canadian Dairy Network, ON, N1K 1E5 Guelph, Canada; s.mucha@cgen.pl*

Persistency of lactation is defined as an ability to maintain a stable milk production during lactation. It is not only related to milk yield but also to health and reproduction of cows and the profitability of production. The aim of this study was to analyse the properties of three definitions of lactation persistency in the Polish Holstein-Friesian cows. The following definitions were used: second principal component of additive genetic covariance matrix for a particular lactation (Pers1), second regression coefficient of random regression sub-model used to model the additive genetic effect (Pers2), and the difference in breeding values at DIM 280 and at the peak of lactation (DIM 30 to 50) (Pers3). The data consisted of test-day milk yield records from lactations 1 to 3 collected on 1,759,084 Polish Holstein-Friesian cows. Variance components were estimated using the Gibbs-sampling approach based on a three-lactation random regression model with the third order of Legendre polynomials. Estimates of heritability for persistency ranged from 0.07 (Pers3 in lactation 1) to 0.20 (Pers2 in lactation 2). The lowest correlation between EBVs for 305-day milk yield and persistency was found for Pers1 (0.10, 0.03 and 0.06 for first, second and third lactation, respectively). Lactation curves of high and low Pers1 EBV bull daughters differed in shape. The genetic curves of sires with high Pers1 EBVs had a clearly increasing trend with DIM. The daily EBVs of sires with high lactation persistency were about +4 kg of milk higher between the beginning and the end of lactation, compared to sires with low lactation persistency (the difference of -3 kg of milk). Pers1 seems to be the most suitable definition of lactation persistency for Polish Holstein Friesians and it is therefore recommended for routine genetic evaluations.

Quantile regression mixed model at different milk production levels of Iranian Holsteins

H. Naeemipour Younesi[1], M. Shariati[1], S. Zerehdaran[1], M. Jabbari Noghabi[1] and P. Lovendahl[2]
[1]*Ferdowsi University of Mashhad, Azadi Square, 9177948974 Mashhad, Iran,* [2]*Aarhus University, Blichers Allé 20, 8830 Tjele, Denmark; peter.lovendahl@mbg.au.dk*

The objective of this study was to estimate the effects of environmental factors such as herd, year and season of calving, Holstein gene percentage, age at first calving as fixed effects, and random animal effects on milk yield using quantile regression (QR) mixed model. An animal model was used to compare and discuss the results. Data comprised of 64,530 primiparous cows from 754 herds calved between 1996 and 2010. Mixed QR model was carried out using LQMM package of R software. The results showed that the effects of age at first calving and Holstein gene percentage across different quantiles were not similar. The spearman correlation between predicted breeding values from QR analyses at different quantiles with ones from animal model were high (0.88). It shows the ranking of animals based on EBV from different models is almost identical, while, the fixed effects differ across quantiles. The genetic trends estimated with QR at different quantiles were different, such that the genetic progress was highest among high producing cows (0.75[th] quantile).

Situation and prospects of livestock farms in the Umbrian areas, affected by the earthquake in 2016

L. Morbidini, M. Pauselli, G. Luciano and D. Grohmann

University of Perugia, DSA3, Borgo XX Giugno, 74, 06100 Perugia, Italy; luciano.morbidini@unipg.it

The ruinous consequences of the earthquake in Central Italy (Marche, Abruzzo and Umbria) in 2016 are forcing the agricultural sector to evaluate the possible strategies to be undertaken. All the involved stakeholders share the consciousness that new production models are now needed. A recent study of the features of the livestock production in 124 Umbrian farms affected by the earthquake, to estimate the needs of interventions, delivered a clear picture. The survey confirmed the primary importance of livestock production and highlighted its diversified traits. In details: sheep production is the leading segment (35% of total farms), followed by dairy (14.4%) and beef (14.4%) cattle and by horse, pigs and donkeys (4, 1.6 and 0.8%, respectively). An interesting peculiarity is the importance of multispecies combinations, with more than 14 different combinations being found in almost 30% of the total farms. Another clear trait is the predominance of farms with a limited number of animals: <30 heads for 30% and 67% dairy and beef herds, respectively, while sheep flocks <50 animals represent almost 40%. Therefore, this mapping highlighted the vocation of the damaged areas for sheep and the increasing importance of beef cattle, both representing the traditional productions over centuries in Central Italy. This description is linked with other considerations such as the high fragmentation of the territory and the difficult morphology of these mountain areas, which contribute to connote these farms as strategic for environmental protection. Additionally several farms have diversified the offer, opening to rural tourism. Whereby, possible directions to optimize the interventions are being proposed, as promoting the modern development of traditional extensive systems or strengthening an agro-forestry activity, which could most probably be the target production system. These systems would deliver products with perceived superior value linked to both unique objective quality traits and to the link with the 'terroir'. Additionally, these production models would serve to maintain the environment. Proposed models may, for example, include raising local breeds in the forest and promoting the use of mountain pastures.

Microbial protein and milk production of dairy cows grazing two timing of pasture and herbage mass

I. Beltran[1], A. Morales[1], P. Gregorini[2], O. Balocchi[1] and R.G. Pulido[1]

[1]Universidad Austral de Chile, Independencia 641, 5090000, Chile, [2]Lincoln University, P.O. Box 7647, 7647, Lincoln, New Zealand; ignacio.beltran.gonzalez@gmail.com

We evaluated the herbage mass and time of pasture allocation effects on milk production, milk composition and microbial protein synthesis of grazing dairy cows during autumn. Forty-four Holstein-Friesian cows (24.7 kg milk/d, 581 kg BW, 74 DIM) were allocated to one of four treatments resulting from the combination of two time of pasture allocation (TA) [access to new pasture in the morning (MTA) or afternoon (ATA)] and two herbage mass (HM) [2,000 kg DM/ha (LHM) and 3,000 kg DM/ha (HHM)]. Treatment groups were strip-grazed with an herbage allowance of 21 kg DM/cow per day. All cows were supplemented with concentrate and grass silage (3.5 and 3.0 kg DM/d, respectively) which were fed in equal amounts during the morning and afternoon. Milk production was recorded at each milking (07:00 and 14:00 h). Milk samples were collected 4 times during the experiment, at morning and afternoon milking to estimate the protein, fat and urea in milk. Spot of urine samples were collected on days 19 and 47 after each milking. Urine samples were analyzed to estimate purines derivate (PD; allantoin, uric acid) and creatinine. The microbial nitrogen production (MN, g/d) was calculated from PD/Creatinine ratio using equations derived by Chen and Ørskov. Milk production was affected by TA and HM. Cows receiving a MTA had a greater milk production than ATA. In addition, cows receiving a HHM had a greater milk production than LHM. Fat and protein in milk were not affected by treatment, averaging, 3.9 and 3.2%, respectively. Urea in milk was affected by HM, being greater for LHM than HHM. Urinary excretion of PD and creatinine were not affected by treatment. The DP/creatinine was not affected by treatment, averaging 2.8. Ruminal microbial nitrogen was similar among treatments, averaging 254 g N/d. In conclusion, the milk production was greater when cows receive a morning pasture allocation or high herbage mass. On the other hand, urea in milk was just higher for cows receiving a low instead of high herbage mass, however, these results were not reflected on synthesis of rumen microbial protein.

The effective population size: difficulties of practical interpretation

J.P. Gutiérrez
Universidad Complutense de Madrid, Departamento de Producción Animal, Facultad de Veterinaria, Avda. Puerta de Hierro s/n, 28040 Madrid, Spain; gutgar@vet.ucm.es

The idealized population was conceptually defined as that evolving under random mating, and it was developed as a reference in order to compare with any other population. Therefore, effective population size (Ne) can be computed in any population as the equivalent size of the idealized population having the same genetic drift as the population under study. Ne has become a key parameter in population and quantitative genetics, and many approaches have been developed in order to find its value as a single number summarizing consequences of past or future, artificial or natural breeding schemes in a wide range of real scenarios. Although Ne can have several utilities, its probably main and most controversial use has been the definition of populations risks, which, in turn, would decide the distribution of grants for conservation. Thus, reliably assessing Ne has become a highly relevant issue. A first global classification of the methodologies conducting to obtain Ne can be based on how it was conceptually understood, either in the context of the changes in the gene frequency or in the context of the increase in inbreeding, leading to methods based on identity by descent and identity by state, respectively, identified also respectively with those using pedigree/demographic information or those arisen from molecular information. The broad variety of available methodologies and sources of information have then provided a wide range of computed values in the literature, which has promoted a big debate about it. Here the most and recently used methodologies to be used regarding the information proceeding from survey, pedigree or molecular data are reviewed, jointly with the corresponding usual range values and guides to understand their difficulties in interpretation.

The effective population size and its relevance in the conservation of farm animal genetic resources

M.A. Toro
Universidad Politécnica de Madrid, Departamento de Producción Agraria, ETSIAAB, Avda. Puerta de Hierro 2-4, 28040 Madrid, Spain; miguel.toro@upm.es

The effective inbreeding (coancestry) population size N_{eF} (N_{ef}) is defined as $1/(2\Delta F)$ ($1/(2\Delta f)$) were ΔF (Δf) are the rates of increase in F (f) calculated using either genealogical or molecular data. These concepts are illustrated in a simulated population and in a real conserved population of Iberian pigs. Based on genomic data, recent studies has revealed heterogeneity in N_e throughout the genome that is shown in an analysis of a Spanish Holstein population. The causes of such heterogeneity and its relevance for conservation of farm animal genetic resources is discussed. Finally, the N_e has been used to establish the degree of endangerment (or risk category of a breed). The approaches of FAO, EAAP or RBST are discussed in the context of setting priorities for conservation. Another issue is the concept of minimum viable size that it should be discussed accounting for purging upon the inbreeding load.

Long term survival and management of populations with small effective population size

J.J. Windig
Wageningen University & Research, Animal Breeding and Genomics, P.O. Box 338, 6700 AH Wageningen, the Netherlands; jack.windig@wur.nl

Lethal genetic defects caused by recessive deleterious alleles, such as CVM in cattle and PRA in dogs, frequently occur in breeds with small effective population (Ne) sizes. Much effort is put in the discovery and elimination of such defects. We investigated long term survival of populations with genetic defects by computer simulation of existing breeds. In breeds with small Ne lethal recessive alleles may occasionally reach allele frequencies up to 50% but such alleles are invariably eliminated within 100 years by natural selection. In breeds with larger Ne some lethal alleles remained for more than 100 years in the population at low frequencies, but high frequencies were never reached. Mildly deleterious alleles, for example with 10% mortality in the homozygous state, were less effectively removed from populations by natural selection. In populations with small Ne they even could get fixed and sometimes caused extinction of the entire population. Thus the smaller the lethal effect of an allele, the more difficult its identification and elimination and the lower the long term survival. Therefore, genetic management of small populations is best done by increasing Ne, for example by avoiding the use of a small number of related sires, rather than eliminating all carriers of all lethal defects, which may decrease the Ne of populations.

Changes in genetic diversity over time in the Dutch Holstein Friesian genome

H.P. Doekes[1,2], R.F. Veerkamp[2], S.J. Hiemstra[1,2], P. Bijma[2] and J.J. Windig[1,2]
[1]Centre for Genetic Resources the Netherlands, Wageningen University & Research, P.O. Box 16, 6700 AA Wageningen, the Netherlands, [2]Wageningen University & Research Animal Breeding and Genomics, P.O. Box 338, 6700 AH Wageningen, the Netherlands; harmen.doekes@wur.nl

The objective of this study was to obtain estimates of changes in genome-wide and region-specific genetic diversity in the Dutch Holstein Friesian (DHF) population over tme. Traditionally, genetic diversity has been quantified and managed with genealogical coefficients of inbreeding (F_{PED}) and kinship (f_{PED}). A drawback of this approach is that F_{PED} and f_{PED} only provide genome-wide expectations based on neutral and selection-free loci. In practice, only few (or even no) neutral unlinked loci exist and it is anticipated that there are substantial differences in diversity across the genome due to selection, drift and variation in recombination rate. With the increasing availability of SNP-data, it is possible to quantify these differences. Approximately 5,400 DHF bulls, born between 1996 and 2014 and included in the Dutch gene bank, were genotyped with the Illumina BovineSNP50 BeadChip and imputed to 76 k. Genome-wide levels and rates of F_{PED} and f_{PED}, marker-by-marker homozygosity (H) and similarity (S) and segment-based genomic inbreeding (F_{ROH}) and kinship (f_{SEG}) were estimated and compared. Region-specific levels and rates of H, S, F_{ROH} and f_{SEG} were also mapped across the genome. As expected, substantial differences were observed in (changes in) diversity across the genome. Combined with the increasing knowledge on functional significance of genomic regions, the insight in region-specific diversity will allow for a better and more customised control of genetic diversity in breeding and conservation programmes.

Use of genealogical information in the assessment of genetic diversity in native cattle breeds

S. Addo[1], J. Schäler[1], D. Hinrichs[2] and G. Thaller[1]

[1]Institute of Animal Breeding and Husbandry, Christian-Albrechts-University of Kiel, Hermann-Rodewald-Straße 6, 24118, Kiel, Germany, [2]Albrechts Daniel Thaer Institute of Agricultural and Horticultural Sciences, Humboldt University, Invalidenstr. 42, 10115, Berlin, Germany; saddo@tierzucht.uni-kiel.de

Herdbook numbers of native cattle breeds are on a decline in Germany and this is partly due to the use of high performing breeds in advanced production systems where genetic materials of elite animals are widely spread. The aim of the study was to assess the within breed genetic diversity of the Angler (ANG) and Red and White dual purpose (RDN) cattle breeds. Genealogical data (ANG, n=93,078 and RDN, n=184,358) were analyzed employing ENDOG v4.8 to calculate parameters including pedigree completeness index (PCI), inbreeding coefficient (f), effective population size (N_e), effective number of founders (f_e) and effective number of ancestors (f_a). Average f was higher for ANG (1.39) than for RDN (0.41) and the corresponding N_e values were 156 and 170 respectively. Pedigree completeness was high in the 1st parental generation (ANG=88%, RDN=64%) but decreased steadily to below 10% in the 13th parental generation for both breeds. Parameters f_e and f_a which are more robust to pedigree errors were respectively, 310 and 90 for ANG and 519 and 189 for RDN. Compared to the actual number of founders, our recorded f_e values suggest an unbalanced genetic representation in the founder populations while only a few ancestors explained the complete genetic variability of the current gene pool. Measures that minimize relatedness between breeding animals are highly recommended to achieve long term genetic diversity within native breeds. Furthermore, the estimation of population parameters based on the Bovine 54k chip is being investigated.

Impact of conservation schemes on genetic variability

G. Leroy[1], R. Baumung[1], E. Gicquel[1], C. Danchin-Burge[2], S. Furre[3], M. Sabbagh[4] and J. Fernandez Martin[5]

[1]Food and Agriculture Organization of the United Nations, Viale delle Terme di Caracalla, 00153, Roma, Italy, [2]Institut de l'Elevage, 149 Rue de Bercy, 75012 Paris, France, [3]Norwegian Horse Center, Starumsvegen 71, 2850 Lena, Norway, [4]Institut francais du cheval et de l'equitation, La Jumenterie du Pin, 61310 Exmes, France, [5]Instituto Nacional de Investigación y Tecnología Agraria y Alimentaria, Crta. A Coruña Km. 7,5, 28040 Madrid, Spain; gregoire.leroy@fao.org

If literature is abundant on the theoretical approaches to manage genetic variability within a population, there is only limited information available on the actual efficiency of those conservation schemes in livestock breeds. In practice, breed associations consider a diversity of approaches to manage genetic variability, from simple measures to raise farmer's awareness and improve their exchanges of reproducers, to more elaborate ones, such as elaboration of mating plans minimizing loss of diversity. Using various case studies from monogastric and ruminant breeds raised in France, Norway and Spain, we investigate how the genetic variability within those breeds is impacted by the implementation of management schemes. Based on pedigree data, the evolution of demographic parameters and genetic variability indicators is assessed over time, considering especially effective population size metrics (individual IBD rate, population IBD rate between successive generations, …). Those indicators are interpreted and discussed in relation to the effect of the implementation of different strategies and change in management schemes.

Current effective population size estimated from genomic data in a turbot aquaculture population

M. Saura[1], A. Fernández[1], B. Villanueva[1], M.A. Toro[2], S. Cabaleiro[3], P. Martínez[4], A. Millán[5], M. Hermida[4], A. Blanco[4] and J. Fernández[1]

[1]Instituto Nacional de Investigacion y Tecnologia Agraria y Alimentaria, Ctra. Coruña Km 7,5, 28040 Madrid, Spain, [2]ETSIA, UPM, Producción Agraria, Av. Puerta de Hierro 2-4, 28040 Madrid, Spain, [3]CETGA, Cluster de Acuicultura de Galicia, Punta do Couso s/n, 15695 Aguiño-Ribeira, Spain, [4]Facultade de Veterinaria, USC, Xenética, Av. Carballo Calero, 27002 Lugo, Spain, [5]Geneaqua SL, Av. Coruña 500, 27002 Lugo, Spain; jmj@inia.es

Turbot aquaculture production almost doubles wild capture fisheries in Europe and has a high potential worldwide. Genetic breeding programmes for this species started in the early 1990s and are now well established. The success of these programmes critically depends on the sustainable use of genetic variability and thus on the effective population size (Ne). In this study we have used genomic data to estimate the current Ne of a commercial Spanish turbot population. RAD-sequencing was used to identify and genotype 18,824 SNPs in 1,393 fish coming from 36 families (23 sires and 23 dams). Linkage disequilibrium (LD) was estimated as the squared correlation (r^2) between non syntenic SNP pairs, assuming a linkage distance between them of 0.5 Morgan (equivalent to a recombination frequency of 0.5). Then, an estimate of the current Ne was obtained from LD. Average r^2 between non syntenic SNPs was low (<0.01) and leaded to a Ne estimate >500. These preliminary results suggest that the stock used as breeders in this population comes initially from a large natural population and that the current population harbours enough genetic variation as to guarantee response to selection in a long term horizon.

Insights on the genetic determinism and evolution of recombination rates in Sheep

M. Petit, S. Fabre, J. Sarry, C. Moreno and B. Servin

INRA Toulouse, 24 Chemin de Borde Rouge, 31326 Castanet-Tolosan, France; morgane.petit@inra.fr

Patterns of recombination can be inferred from genome-wide polymorphisms data using different approaches: meiotic recombination rates can be estimated by identifying crossovers within individual meioses from pedigree data; more precise estimates can be obtained using population genetic approaches in densely genotyped samples of unrelated individuals. However these population-based estimates can be biased as they are affected by evolutionary pressures, most notably, selection. In this work, we present a strategy to establish fine scale recombination maps by combining these two approaches. We exploiting two different datasets from one Sheep breed, the Lacaune: a large pedigree genotyped with a 50K SNP array and a sample of unrelated individuals genotyped with a 600K SNP array. Our analyses of these two datasets show that recombination patterns in the Sheep resemble of those observed in other mammals, in particular we demonstrate the presence of small intervals of a few kilobases exhibiting very large recombination rates (i. e. harbouring crossover hotspots). Then, we developped a statistical model aimed at formally combining the two datasets. This model allows first to scale the population-based estimates from the meiotic ones while accounting for their respective uncertainties and to produce high-resolution recombination maps. Second, we show that the combination of meiotic and population-based inference highlights the effect of selection on the genome, revealing already known selection signatures, but also new ones. Finally, we compared our recombination maps to those obtained from an analysis of pedigree data in Soay Sheep. This comparison allows to demonstrate that Soay Sheep exhibit about 20% more crossovers per meiosis compared to Lacaune Sheep and that this inflation is most likely due to selection in the Soay. Despite this strong past selection, we were able to show that the genetic determinism of recombination rate in the two populations remain very similar, two major loci (RNF212 and HEI10) contributiong greatly to its inter-individual variation in both populations. This suggest that recombination rate is genetically determined by multiple independant biological pathways.

Effective population size in four Lithuanian horse breeds

R. Sveistiene and A. Rackauskaite
Lithuanian University of Health Sciences, Institute of Animal Science, R. Zebenkos 12, 82317, Lithuania;
ruta.sveistiene@lsmuni.lt

Management of animal genetic resources in order to minimize loss of genetic diversity within breed has recently received attention at conservation programs, sustainable improvement and accurate estimates of population parameters, such as rate of inbreeding (F) and effective population size (Ne). The aim of this study was to research the effective population size and inbreeding of three Lithuanian native horse breeds - Zemaitukai (Z), large type Zemaitukai (LTZ), Lithuanian Heavy Drought (LHD) breeds and one traditionally bred - Trakehenen breed (T). Data consisted of all registration history for all breeds, including 5,408 Z, 1,989 LTZ, 11,590 LHD and 6,462 T individuals. The data were analysed with the software system POPREP using data of live populations in 2015. The beginning of substantial pedigree recording in the datasets varies from 1,857 for the LHD and T to 1,894 for the Z and LTZ populations. The average pedigree completeness for all breeds horses born within the last 10 years were from 100% in 1 generation deep and 83.7% - 6 generations. The average inbreeding coefficient for horses born within the last 10 years increased for Zemaitukai population 5,4% (F=0.263), LTZ – 3.4% (F=0.192), LHD – 1.7%, (F= 0.186), except T population in which average inbreeding coefficient were stabile (F=0.150). Since the breeding populations two horse breeds are relatively small. The total population size of Zemaitukai horses is 708, LTZ -698, LHD-1,678, T – 1,083. The effective population size (Ne) were 81 in the Z, 45 - LTZ, 169 - LHD and 48 - T. All three native horse populations have very narrow genealogy structure but breeder's association's strongly keeps breeding schemes to monitor inbreeding coefficient and use different stallions in breeding to increase Ne.

Population structure analysis of the New World Creole cattle

G. Mészáros[1], M. Naves[2], R. Martínez[3], C. Lucero[3] and J. Sölkner[1]
[1]University of Natural Resources and Life Sciences, Vienna, BOKU, Gregor-Mendel Str.33, 1180 Vienna, Austria, [2]Institut National De La Recherche Agronomique, INRA, UR 143, 97170 Petit-Bourg, Guadeloupe, France, [3]Colombian Corporation of Agricultural Research, CORPOICA, Km 14 Via Bogotá- Mosquer, CP 250047, Bogotá, Colombia; gabor.meszaros@boku.ac.at

The Creole cattle are a group of *Bos Taurus* breeds, originating from Spain, first imported to the New World in the 16th century. There were later influenced by local selection processes, herd management, and complex admixtures, during the historical period. The genotype data of 200 Creole cattle from Colombia (Blanco Orejinegro (BON, n=50), Costeño con cuernos (CCC, n=50), Romosinuano (ROM, n=50), Sanmartinero (SAM, n=50)) and 223 Creole cattle from Guadeloupe was analyzed to assess within and between breed genetic diversity. Total of 41,011 autosomal single nucleotide polymorphism (SNP) markers, after standard quality control, were used for the analysis. The population structure was analyzed via logistic factor (LFA) and principal component analysis (PCA). Both confirmed well separated, non-overlapping clusters for all breeds. The first principal component explained 10% and the second 7.7% of the variance. The F_{st} values between breeds were in the range of 0.09 to 0.12. While most of the animals between breeds were unrelated, some non-zero IBD coefficients were found between animals from different Colombian breeds. The within breed genetic distances were fairly similar in all cases, with slightly higher values for ROM, confirming the LFA and PCA results. The genomic inbreeding coefficients assessed via runs of homozygosity (ROH) were computed with restrictions of minimal ROH lengths of 2, 4 8 and 16 Mb. The longest ROH segments denoted recent inbreeding up to 3 generations ago, with the highest values in the GUA population (F_{ROH}=0.04, sd=0.05). When considering ancestors up to 25 generations ago via minimum ROH length of 2 Mb the ROM population was the most inbred (F_{ROH}=0.12, sd=0.04). The ROH segments were distributed across the whole genome, with notable exceptions of several breed specific ROH islands. These regions of excessive homozygosity indicate differentiation in selection pressure in various Creole populations.

Population structure and genetic diversity of Bovec sheep from Slovenia – preliminary results

M. Simčič[1], M. Žan Lotrič[1], D. Bojkovski[1], E. Ciani[2] and I. Medugorac[3]
[1]University of Ljubljana, Biotechnical Faculty, Jamnikarjeva 101, 1000 Ljubljana, Slovenia, [2]University of Bari, Via Amendola 165a, 70126 Bari, Italy, [3]LMU Munich, Faculty of Veterinary Medicine, Veterinaerstr. 13, 80539 Munich, Germany; mojca.simcic@bf.uni-lj.si

The aim of the study was to obtain unbiased estimates of the population structure, genetic diversity parameters, and the degree of admixture in Bovec sheep. The breed is known under the name Krainer Steinschaff in Austria and Plezzana in Italy. According to the literature, the breed is the direct descendant of the Small White sheep, widespread in the Alps centuries ago. In 80's, the part of the Bovec sheep population was improved with the White East Friesian sheep. A new breed named Improved Bovec sheep was developed. Genetic analyses were performed on the genome-wide Single Nucleotide Polymorphism (SNP) Illumina Ovine SNP50 array data of 44 animals of Bovec sheep, 9 animals of Improved Bovec sheep and 255 animals from 12 reference populations representing neighbor breeds and possible sources of admixture. To obtain unbiased estimates we used short haplotypes spanning four markers instead of single SNPs to avoid an ascertainment bias of the OvineSNP50 array. Genome-wide haplotypes confirmed typical type traits of purebred Bovec animals. Unfortunately, in some purebred Bovec animals according to the pedigree data, an introgression of the White East Friesian sheep was found. Phylogenetic analyses demonstrated unique genetic identity of Bovec sheep with higher number of private alleles. Genetic distance matrix presented by Neighbour-Net showed independent origin of the breed. Unsupervised clustering performed by the Admixture analysis between individuals also suggested Bovec sheep is a distinct breed. Animals identified as the most purebred represent a nucleus for the conservation of the native genetic background. Phenotypic selection performed over several centuries resulted in the well-adapted dairy breed for the Alpine grazing. Despite the part of the Bovec sheep population was admixed in the past, the implemented procedure is capable to detect animals representing old local Bovec sheep and could be used to prevent the future uncontrolled introgression in this unique breed preserved in the Alps.

Genetic resources conservation of two local population of Central European honeybee in Poland

G.M. Polak and A. Chelminska
National Research Institute of AnimalProduction, Department of Animal Genetic Resources Conservation, Krakowska 1, 32-083 Balice, Poland; grazyna.polak@izoo.krakow.pl

Central European honeybee (*Apis mellifera mellifera* L.) is the native subspecies in North and Central Europe. Historically, these bees are adapted to the harsh conditions, more resistant to disease and lack of food than other breeds. From 2,000 two local line of *A. mellifera mellifera*: Augustowska (MA) and Northern (MP) are covered by genetic resources conservation programs. Beekeepers are located in wooded areas of north-eastern Poland, where environmental conditions are extremely difficult. The aim of work was to analyze the variability of number of bee families covered by conservation programs in climatic conditions of north-west Poland. Studies conducted from 1999 to 2016 indicate a permanent slight increase in both population MA and MP. It was found that in 16 years from 2000 to 2016, the population of MA increased from 100 to 775 families (225 in breeding herds and 550 in the protected zones) of the Augustow Forest. MP Line increased from 22 to 200 families in 4 breeding herds located in Mazury region.

Estimating the genetic diversity of Rhodeus amarus populations in caspian sea`s river by PCR-RFLP

S. Jafari Kenari[1], G.H. Rahimi Mianji[2], H. Rahmani[2] and A. Farhadi[2]
[1]Gorgan University of Agricultural Sciences & Natural Resources, fisheri and enviroment, iran, mazandaran,sari, taleghani Blv, amirafshari alley, 4817894431, Iran, [2]Sari University of Agricultural Sciences & Natural Resources, animal science and fisheri, iran,mazandaran,sari,Km 9 Farah Abad Road, 4818168984, Iran; ss.jk.sara@gmail.com

So far, no studies have been done about genetic differences of *Rhodeus amarus* species in the rivers of the southern Caspian Sea. In this study the genetic diversity of this species in different rivers were examined using polymorphic cytochrome b and morphomeristic characteristics. In total, 90 samples (30 per river) were collected from Babolrood, Tajan and Siahrod rivers with electrofishing. To estimate the genetic diversity of populations of DNA extraction from abdominal and pectoral fins was carried out by salting out method. A fragment with the length of 1,141 bp from cytochrome b gene was amplified using specific primer pairs. PCR products were digested with *Alu*I, *Bsl*I and *Taq*I restriction enzymes. The samples were genotyped on digested products by electrophoresis on agarose gel and stained with ethidium bromide. No different banding pattern (polymorphism) were observed for *Alu*I enzyme and all the samples showed monomorph genotype. But *Bsl*I and *Taq*I enzymes showed different banding pattern. The highest allele and genotype frequency was related to A allele and AA genotype which was observed in cty *b*-I marker site in Siahrud population. The most frequent haplotype was related to AAA which was observed in Siahrud population. The comparison of allele frequency for each enzyme marker site and haplotype frequency between populations was carried out by SAS program using Fisher's exact test. The results showed that there was a significant difference in allele frequency between Tajan-Siahrod and Babolrod-Siahrod populations in cyt- *Taq*I marker site.

A molecular analysis of the patterns of genetic diversity in local chickens from western Algeria

F.Z. Mahammi[1], S.B.S. Gaouar[1,2], N. Tabet Aoul[1,3], D. Laloe[4], R. Faugeras[5], X. Rognon[4,6], M. Tixier-Boichard[4] and N. Saidi-Mehtar[1]
[1]Département de Génétique Moléculaire Appliquée, Laboratoire de Génétique Moléculaire et Cellulaire, Université des Sciences et de la Technologie d'Oran, 31025, Oran, Algeria, [2]Département de biologie, Université de Tlemcen, 13000 Tlemcen, Algeria, [3] Département de Biotechnologie, Université d'Oran 1, 31000 Oran, Algeria, [4]UMR1313 Génétique Animale et Biologie Intégrative, INRA, 78352 Jouy-en-Josas, France, [5]Laboratoire d'analyses génétiques pour les espèces animales (LABOGENA), INRA, 78352 Jouy-en-Josas, France, [6]UMR1313 Génétique Animale et Biologie Intégrative, Agro-Paris-Tech, Paris, 75000, France; nacerat@yahoo.com

The objectives of this study were to characterize the genetic variability of village chickens from three agro-ecological regions of western Algeria: coastal (CT), inland plains (IP) and highlands (HL), to reveal any underlying population structure, and to evaluate potential genetic introgression from commercial lines into local populations. A set of 233 chickens was genotyped with a panel of 23 microsatellite markers. Geographical coordinates were individually recorded. A genetic diversity analysis was conducted both within and between populations. Multivariate redundancy analyses were performed to assess the relative influence of geographical location among Algerian ecotypes. The results showed a high genetic variability within the Algerian population, with 184 alleles and a mean number of 8.09 alleles per locus. The values of heterozygosity (He and Ho) ranged from 0.55 to 0.62 in Algerian ecotypes. Although the structuring analysis of genotypes did not reveal clear subpopulations within Algerian ecotypes, the supervised approach using geographical data showed a significant ($P<0.01$) differentiation between the three ecotypes which was mainly due to altitude. Thus, the genetic diversity of Algerian ecotypes may be under the influence of two factors with contradictory effects: the geographical location and climatic conditions may induce some differentiation, whereas the high level of exchanges and gene flow may suppress it.

The present and future role of native breeds of cattle in Poland

E.M. Sosin-Bzducha and A. Chełmińska

National Research Institute of Animal Production, Department of Animal Genetic Resurces Conservation, ul. Krakowska 1, 32-083 Balice, Poland; ewa.sosin@izoo.krakow.pl

In Poland, conservation programs of genetic resources cover 4 cattle breeds: Polish Red, Polish Whitebacked, Polish Red and White as well as Polish Black-and-White. The protection is carried out mainly by *in situ* method supported by cryopreservation. In 2016, a total of 7,930 cows in 797 herds were subject to the protection. Implementation of conservation programs relying, i.a., on a gradual reduction of the share of foreign genes in highly overcrossed earlier populations leads to a reduction in animal growth and a decline in their productivity, which is especially evident in breeds where for the purpose of the ennobling crossing in the past the Holstein-Friesian breed was used. Attributes of conservative breeds such as longevity, high resistance, good health and high quality of products can still help to maintain these breeds in the landscape of the Polish countryside. At present, studies aim to determine the suitability of conservative breeds of cattle for food production on generally understood high quality. Also initiatives aimed at supporting the production and promotion of product lines derived from individual breeds, which is connected, in turn, with linking them to local tradition and the original place of occurrence or produce, are undertaken. The increased use of conservative breeds in activities aimed at the active protection of the landscape are pointed at. An additional issue is the use of conservative breeds in the long-term and future policy of the breeding. Implemented program of cryopreservation of genetic resources of animals presupposes the ability to use accumulated biological material not only for the restoration of conservative breeds, but also use for the occasion of the rapid changes in the direction of breeding or improvement of the allelic variability of population, but also preventing the negative effects of inbreeding or reducing the frequency of genes determining the occurrence of genetic defects. The programs of conservative breeding provide the possibility of using part of cryopreserved genetic material outside the protected herds.

Multi-disciplinary approaches for improving complex traits in pig: a case study in feed efficiency

R. Quintanilla[1], D. Torrallardona[2], I. Badiola[3], E. Fàbrega[4], Y. Ramayo-Caldas[1] and J.P. Sánchez[1]

[1]Institute for Food and Agriculture Research and Technology (IRTA), Animal Breeding and Genetics, Torre Marimon, Caldes de Montbui, Barcelona, 08140, Spain, [2]Institute for Food and Agriculture Research and Technology (IRTA), Animal Nutrition, Mas de Bover, Constantí, 43120, Spain, [3]Institute for Food and Agriculture Research and Technology (IRTA), Animal Health, Center for Research in Animal Health (CReSA), Autonomous University of Barcelone (UAB), Bellaterra, 08193, Spain, [4]Institute for Food and Agriculture Research and Technology (IRTA), IRTA-Animal Welfare, Finca Camps i Armet, Monells, 17121, Spain; raquel.quintanilla@irta.cat

Under the societal challenge regarding sustainable food security, improving feed efficiency is essential to increase the economic sustainability while reducing environmental impact of pig production systems. Genetics and nutrition are recognized as key disciplines to optimise livestock production efficiency through new selection strategies and feed technologies, but animal feeding behaviour and social interactions cannot be disregarded. Furthermore, gut microbiota has emerged as a relevant phenotype due to its well-recognized contribution to host physiology. New high-throughput technologies have revolutionized biological research providing vast amounts of 'omics' data, but the real challenge is to bridge the knowledge gap between heterogeneous data and complex phenotypes. Multidisciplinary and collaborative approaches are needed to have a better understanding of the interactions within the animal biological system allowing optimising pigs feed use.

'pigFit' – molecular genetic analysis of immune traits associated with piglet survival

E. Heuß[1], M.J. Pröll[1], C. Große-Brinkhaus[1], H. Henne[2], A.K. Appel[2], K. Schellander[1] and E. Tholen[1]
[1]Institute of Animal Science, University of Bonn, Endenicher Allee 15, 53115 Bonn, Germany, [2]BHZP GmbH, An der Wassermühle 8, 21368 Dahlenburg-Ellringen, Germany; esther.heuss@uni-bonn.de

Piglet mortality has a negative impact on animal welfare, public acceptance and decreases the subsequent viability of pig performance. Moreover, the profitability of piglet producers is mainly determined by the number of weaned piglets per sow. This situation is intensified by an increasing litter size which influences negatively birth weight and piglet survival. Furthermore, the control of diseases and infections is a crucial challenge in pig production, because the survival rate and performance are decreased. For that reason and to enhance animal welfare, improved resistance to pathogens by breeding is a major trend in the selection of pigs. However, the phenotypic recording of resistance traits is hardly possible. Therefore, the analysis of immune traits of innate and adaptive immunity for an evaluation of the immunocompetence is promising to gain a generally improved immune response. In this context, the 'pigFit'-project was initiated, aiming to improve the survivability, health and immune status of piglets and growing pigs in the maternal lines Landrace (LR) and Large White (LW). In a first step (WP 1), variance components and estimated breeding values (EBV) were predicted for stillbirth (SB) and birth weight (BW) in piglets (n=33,000 LW, 42,000 LR) as well as reproductive traits in their respective dams (2,300 LW, 2,400 LR). In the following work package (WP 2), blood samples for the determination of complete blood counts and cytokine levels were taken from sows (n=500) and piglets born alive (n=1000) after farrowing and post weaning, respectively. In addition, all animals including stillborn siblings (n=500) were weighted and genotyped. The objective of WP 2 was to evaluate a breeding-based improvement of health traits and survival of piglets and growing pigs through immune profiling and genomic selection. In a final step (WP 3), we intend to verify the findings of WP 1 and 2 by an exact trial with genetically divergent families for piglet survival. Based on these insights, genomic methods will be applied, aiming to improve the vitality and the robustness of piglets by breeding.

Genetics as a considerable factor affecting gut microbiota and gene expression of healthy piglets

D. Luise[1], V. Motta[1], M. Bertocchi[2], P. Bosi[1] and P. Trevisi[1]
[1]University of Bologna, V. G. Fanin 46, 40127, Italy, [2]University of Molise, Via F. De Sanctis, 86100, Italy; paolo.trevisi@unnibo.it

Recently the interest of genetics as factor playing important interaction with the microbiota and immune response is increasing. In pigs, some polymorphisms have been associated with susceptibility and resistance to *Escherichia coli* infection and immune response. This work aims to investigate on *MUC4, FUT1, ABO, BPI* and *TLR4* genotypes distribution and their effect on the gut microbiota, on intestinal expression of genes related to inflammation and on performances parameters of weaned healthy pigs. For this study, 71 post weaned piglets (35 day of age, 7,063±936 g) were reared for six weeks, weekly weighed and then scarified to collect the jejunum content and mucosa. The DNA and RNA were extracted from jejunum and animals were genotyped for *MUC4* g.8227C>G, g.307 FUT1 G>A, g. *BPI* exon 10, g.962 TLR4, g. 611 *TLR4* and *ABO* group. Semi-quantitative PCR of *IL8, GPX2, REG3G, TFF3, CCL20, ST3GAL, LBPI* were performed on cDNA. Bacteria DNA was isolated from jejunum content and V3-V4 regions of 16S rRNA gene were sequenced on MiSeq-Illumina® platform. Generated sequences were analyzed using subsampled open-reference OTU strategy in QIIME (v1.9.1). Expression and performance data were analyzed using a GLM model, including genes and batch as fixed effects. MUC4 genotype induced different expression (P<0.05) on *REG3G, TFF3* and *CCL20* genes. Piglets with susceptible genotype (*MUC4* GG AG) had higher values compared to the *MUC4* CC pigs. *FUT1* gene affected the microbiota composition. Evenness and the Shannon index (P<0.01) and Bray-Curtis distance (P=0.02) were significantly different between groups. The abundance of Fusobacteia was lower on resistant animals (*FUT1* AA) and Actinobacteria level was higher on AG compared to GG pigs. The present study highlights that single SNPs for *FUT1* and *MUC4* genes significantly influence the microbiota variability and the expression of genes related to inflammation in the jejunum of healthy piglets respectively and encourage to consider them as relevant factors to be included in the experimental design of future piglet's studies.

Genetic correlation between the residual energy intake of purebred and crossbred pigs

R.M. Godinho[1,2], R. Bergsma[3], C.A. Sevillano[2,3], S.E.F. Guimarães[1], F.F. Silva[1] and J.W.M. Bastiaansen[2]
[1]Universidade Federal de Viçosa, Animal Science, Avenida Peter Henry Rolfs, s/n, 36570-900 Viçosa, Brazil, [2]Wageningen University and Research, Animal Breeding and Genomics, Droevendaalsesteeg 1, 6708 PB Wageningen, the Netherlands, [3]Topigs Norsvin Research Center, Schoenaker 6, 6641 SZ Beuningen, the Netherlands; rodrigo.mezenciogodinho@wur.nl

Selection for improved feed efficiency is a strategy to minimize the production costs per unit of animal product, therefore, one of the major objectives of current animal breeding programs. While pork is typically produced from crossbred pigs (CB), data recording for breeding programs is traditionally organized at the nucleus level on purebred pigs (PB). The genetic correlation between the performance of PB and CB (r_{pc}) is therefore an important parameter when breeding for CB performance. The current study estimated the genetic parameters for the trait residual energy intake (REI), which is a possible measure of feed efficiency. REI was calculated based on the average daily metabolizable energy intake, the energy required for maintenance, and the calculated lipid and protein deposition. Individual feed intake records were available on 22,984 PB and 8,657 CB. The PB population consisted of five sire and four dam lines, and the CB population consisted of three-way crosses between those PB lines. Heritability estimates were 0.16 in both PB and CB. REI presented moderate genetic correlations with energy intake, lipid and protein deposition, and was not genetically associated with growth. The estimated r_{pc} was 0.67, indicating that part of the genetic progress being realized in PB is expected to be seen at the production level. The difference from unity may be attributed to genetic factors and also to genotype by environment interaction (GxE) given the usual differences between the nucleus environment and commercial farms, where CB were raised. This estimate of r_{pc} is relevant because in practical pork production the environments typically are different for PB and CB. Including phenotypic records from CB on commercial farms is expected to lead to higher genetic progress for this trait when selecting PB for CB performance. Disentangling the effects of r_{pc} and G×E is important to allow further optimization of breeding programs.

Copy number variations in two Finnish pig breeds

T. Iso-Touru[1], M.-L. Sevón-Aimonen[1], D. Fischer[1], T. Serenius[2], P. Uimari[3] and A. Sironen[1]
[1]Natural Resources Institute Finland, Green technology, Myllytie 1, 31600 Jokioinen, Finland, [2]Figen Oy, PL 319, 60101 Seinäjoki, Finland, [3]University of Helsinki, Department of Agricultural Sciences, Koetilantie 5, 00014 Helsingin yliopisto, Finland; terhi.iso-touru@luke.fi

The purpose of this study was to identify copy number variations (CNVs) from the Finnish Yorkshire and Finnish Landrace pig breeds and to identify putative markers for breeding purposes. In addition, genome-wide association study (GWAS) was used to identify genomic regions associated with reproduction traits. The CNV identification was done within 605 Finnish Yorkshire and 386 Finnish Landrace progeny tested boars genotyped with the Illumina PorcineSNP60 BeadChip. The CNV calling was done using the log R-ratio -values that measure signal intensities of the markers in a specific location in the genome for each breed separately using the univariate CNAM methodology implemented to the SVS software package (Golden Helix). The GWAS was done using single-locus mixed model (EMMAX) method implemented in the SVS with 816 Finnish Yorkshire and 405 Finnish Landrace boars using deregressed estimated breeding values (Mix99) from nine reproduction traits (obtained from Figen Oy). We identified altogether 46 copy number variation regions (CNVRs) encompassing 28 genes. Among the 46 CNVRs, 11 were shared between the breeds, 20 were unique to the Finnish Yorkshire and 15 unique to the Finnish Landrace. The GWAS analysis identified zero to five reproduction associated genomic regions per trait. The identified CNVRs were compared against associated genomic regions observed in this study but no overlaps were detected. To extend the analysis, QTLs were downloaded from the Animal Genome Database and overlaps between the identified CNVRs were searched. The CNVRs were enriched for regions containing meat and carcass related QTLs for both of the breeds (Fisher's Exact Test $P < 2.39×10^{-10}$ and 1.15 fold enrichment). This indicates that CNVs may have a role in control of production related traits thus making them as potential markers for the breeding programs.

Organic enrichment material in pig husbandry: any risk for animal health?

K.M. Wagner, J. Schulz and N. Kemper

University of Veterinary Medicine Hannover, Foundation, Institute for Animal Hygiene, Animal Welfare and Farm Animal Behaviour, Bischofsholer Damm 15, 30173 Hannover, Germany; nicole.kemper@tiho-hannover.de

In the EU, pigs have to have permanent access to enrichment material. Because of the new Commission Recommendation (EU) 2016/336, demanding these materials to be edible, only organic materials should be used. However, an EFSA Scientific Opinion reminds on possible hygienic risks and states that this has to be investigated further. Therefore, the aim of this study is to analyse the hygienic status of different, mainly commercially available, organic enrichment materials and the possible risks of pathogen transfer. In total, 21 organic materials were examined. While most of them were commercially available in Germany, only three were produced on-farm. These materials consisted of wood and partly compressed straw and hay. Additionally, six miscellaneous materials were analysed. Parameters tested were the total viable count (TVC) and the coliform count, the presence of *Escherichia coli*, *Klebsiella* spp., *Yersinia* spp., *Salmonella* spp., fungi, methicillin resistant *Staphylococcus aureus* (MRSA), and *Mycobacterium* spp. In addition, a high-performance liquid chromatography-mass spectrometry based multi-mycotoxin analysis was performed. Out of the analysed bacteria, only *Mycobacterium* spp. (*M. smegmatis*) was isolated from hemp litter. The TVC ranged from no detection of colony forming units per g dry matter (cfu/g DM) to 7.7×10^7 cfu/g DM. Wood and compressed straw and hay products showed a lower microbial load than loose straw and hay. The analysis of mycotoxins revealed a high mycotoxin load in some materials, especially products made of maize. In their hygienic status, the tested materials differed widely. Important pathogens such as *E. coli*, or zoonotic agents such as MRSA or *Salmonella* spp. were not found in any material. Some materials contained high amounts of mycotoxins which might pose a health risk for pigs. In conclusion, with regard to the microbiological status, most tested materials are suitable as enrichment material for pigs, the maize products with restrictions concerning mycotoxins.

Quality of PDO Noir de Bigorre pork products according to pig feeding and season in extensive system

B. Lebret[1], H. Lenoir[2], A. Fonseca[3], J. Faure[1] and M.J. Mercat[2]

[1]INRA, Pegase, Agrocampus-Ouest, 35590 Saint-Gilles, France, [2]IFIP, La Motte au Vicomte, 35651 Le Rheu, France, [3]Consortium Noir de Bigorre, Route de Lourdes, 65290 Louey, France; benedicte.lebret@inra.fr

The Noir de Bigorre (NB) pork chain relying on pure Gascon autochtonous breed obtained in 2015 the French AOC label required for further European Protected Designation of Origin (PDO) registration, for both fresh meat and dry-cured hams. Pigs produced in extensive conditions (pasture, 20 pigs/ha) consume large quantities of grass and, depending on the season, acorns and chestnuts in addition to a conventional diet. Variations in feed resources and climatic conditions along the fattening period can modulate animal growth, muscle and fat tissue traits and the intrinsic properties of pork products. Within the European H2020 TREASURE project, we aimed at better characterizing the variability of various phenotypic traits of pork and processed products from NB chain according to the pig finishing season. Gascon castrated male pigs were produced in compliance with PDO specifications in two farms of the NB chain with similar husbandry practices and slaughtered at the end of winter (W, n=25) or spring (S, n=23). Season influenced carcass and meat traits. At similar slaughter weight and age (166.5 kg and 414 d on average), S pigs exhibited slightly lower carcass fatness (P<0.10) but higher lipid content in the Gluteus (ham) muscle (3.44 vs 2.79%, P<0.001) than S pigs. In loin and ham muscles, S pigs showed higher ultimate pH (+0.06 to 0.15 units, P<0.01) and lower meat lightness (P<0.05) and hue angle (P<0.10), indicating redder meat, than W pigs. Season associated to feed resources affected tissue lipid profile with higher (P<0.05) n-3 polyunsaturated fatty acid content and lower (P<0.001) n6-/n-3 ratio in Gluteus muscle and external fat of green hams from S pigs, underlying the positive effect of pig grass consumption during spring on the nutritional value of pork. Present and future results on quality of fresh pork and dry-cured hams from pigs slaughtered at the end of autumn following acorns consumption will be useful for actors of the local NB chain to adapt pig husbandry practices, in order to further improve the eating quality and differentiation of pork products within PDO specifications.

Influence of rearing environment on growth rate and meat characteristics of highly muscled pigs

V. Juskiene, R. Juska, R. Juodka and R. Leikus
LUHS Institute of Animal Science, Department of Ecology, R. Zebenkos 12, Baisogala, 82317, Baisogala, Lithuania;
violeta.juskiene@lsmuni.lt

The aim of this study was to compare the influence of indoor and outdoor environmental conditions on the growth rate, carcass and meat quality of intensively raised highly muscled pigs. The study with Lithuanian White × Norwegian Landrace × Pietrain crossbred pigs was carried out at the Lithuanian University of Health Sciences, Institute of Animal Science. Two groups of analogous pigs of 12 animals each were used in the study. The pigs were housed in either indoor pens (6 pigs per 9.37 m^2 area) or an outdoor enclosure (12 pigs per 2,000 m^2 area). The data of investigation indicated that the Outdoor pigs at the end of the growing and finishing periods weighed, respectively, on average 14.5 and 7.4% more than the Indoor pigs. Also, Outdoor pigs had by 22.1% higher average daily gain during the growing period and by 9.0% higher during the growing-finishing periods. During the experiment Outdoor pigs consumed daily 5.2% more compound feed, however, feed consumption per kg gain was 3.5% lower than that of the Indoor pigs. Rearing environment had no significant influence on the physicochemical indicators of meat, excluding the crude protein level which was higher by 0.7% for the Outdoor pigs. It can be concluded from the results of our study that outdoor rearing of pigs has a positive effect on the growth rate of pigs and higher protein content in the meat but did not affect the carcass traits and other physicochemical indicators of meat.

Fitting the 'parity curve': application of random regression to litter size in pigs

E. Sell-Kubiak[1], E.F. Knol[2] and H.A. Mulder[3]
[1]Poznan University of Life Sciences, Department of Genetics and Animal Breeding, Wolynska 33, 60-637 Poznan, Poland, [2]Topigs Norsvin Research Centre B.V., P.O. Box 43, 6640 AA Beuningen, the Netherlands, [3]Wageningen University & Research, Animal Breeding and Genomics, P.O. Box 338, 6700 AH Wageningen, the Netherlands; sell@up.poznan.pl

The objective of this study is to analyze the genetic background of variation in reproductive performance between parities of a sow with application of random regression. Performance of a sow (e.g. litter size) changes with the parity. Usually, the 1st and 2nd parity sows have smaller litters than sows in their 3rd to 5th parity. After parity 6, the performance declines again. The shape of litter size across parities closely resembles the lactation curve observed in dairy cattle. Fitting the 'parity curve' could help reducing differences between parities of a sow, preferably, keeping only the first parity with the lowest performance. Litter size data of Large White sows from commercial farms were provided by Topigs Norsvin. The records of 246,799 litters (total number born) came from 53,794 sows with at least two observations in parities 1-10. The analysis was performed in ASReml 4.1. Firstly, the 3rd order polynomials were selected as the best fit for the fixed parity curve. Secondly, the comparison of 3 models with respect to order of polynomials for random effects, indicated that the best fit was the model with 3rd order polynomials for both additive genetic and permanent environmental effects with simultaneous accounting for heterogeneity of residual variance per parity (10 levels). The heritability estimates were increasing with the lowest value in parity 1 with 0.11 and the highest value in parity 9 with 0.21. Genetically, parity 1 is the most different from parities 7-10 (r_g from 0.56 to 0.60), whereas it is most similar to parities 2 (r_g=0.96) and 3 (r_g=0.85). Already the 2nd parity has correlations between 0.72 to 0.96 with other parities, whereas parities 3 to 10 are genetically almost identical traits with r_g from 0.84 to 0.99. The non-unity genetic correlations between parities 1 and 2 with parities 3-10 show existence of genetic variation to genetically change the parity curve, e.g. to make litter size more equal among parities.

Precision feeding reduces N and P excretion in growing pigs reared in high ambient temperatures

L.S. Santos[1], P.H.R.F. Campos[2], L. Hauschild[1], W.C. Silva[1], J.P. Gobi[1], A.M. Veira[1] and C. Pomar[3]
[1]UNESP, FCAV, Paulo Donato Castellane,w/n, 14884900, Brazil, [2]UFVJM, Rd MG 367, 5000, 39100000, Brazil, [3]AAFC QC, 2000 College St, J1M0C8, Canada; luanddos@gmail.com

Previous studies have demonstrated that precision feeding can significantly reduce nutrient intake and excretion in growing pigs reared in thermoneutral (TN) conditions. However, feeding pigs with low protein diets has been shown to attenuate the negative effects of heat stress in pigs reared in hot climate areas. The aim of this study was to evaluate the N and P excretion of pigs fed in conventional group-phase feeding systems (CONV) or feed individually with daily tailored diets (PREC: precision feeding) under 2 thermal conditions (TN: 23 °C; and high temperature, 30 °C). Forty-eight barrows (41.0±0.98 kg BW) were assigned to treatments in a 2×2 factorial arrangement and fed during 2 28-days consecutive feeding phases. CONV fed pigs received within each phase a constant blend of diets A (high nutrient density) and B (low nutrient density) supplying the estimated nutrient requirements of the group, while PREC pigs received daily a personalized blend providing the estimated required nutrients. Total body minerals, fat and lean content were assessed by dual-energy X-ray absorptiometry at beginning and at the end of each phase. Data were analyzed using the MIXED procedure of SAS including the fixed effects of feeding systems (FS), ambient temperature and their interaction. The initial BW was included as a covariate. During phase 1, FS did not affect N and P intake and excretion in pigs kept at 23 °C, whereas lower ($P<0.01$) N (31.4 vs 37.8 g/d), and P intake (6.2 vs 7.4 g/d) and lower N (12.7 vs 17.5 g/d) and P excretion (4.5 vs 5.3 g/d) was observed, respectively, in PREC and CONV pigs reared at 30 °C. During phase 2, N excretion was lower ($P<0.01$) for PREC pigs at 23 °C (26.5 vs 31.8 g/d) and 30 °C (17.9 vs 22.0 g/d) than for CONV pigs. The lower intake and excretion of nutrients in PREC compared to CONV pigs did not affect P and N retention efficiency. According to our results, precision feeding was an efficient method of reducing the impacts of N and P intake by the lower excretion of these nutrients without impairing your efficiency in growing pigs reared in thermoneutral and high ambient temperatures.

Effect of slaughter weight and sex on carcass composition, N- and P- efficiency of pigs

A. Van Den Broeke, F. Leen, M. Aluwé, J. Van Meensel and S. Millet
ILVO (Institute for Agricultural and Fisheries Research), Scheldeweg 68, 9090 Melle, Belgium; sam.millet@ilvo.vlaanderen.be

This study aimed at assessing the effect of sex and slaughter weight and their interaction on carcass composition and N- and P- efficiency of pigs between 25 and 130 kg. In 2 rounds, 24 pens of 4 pigs per sex (entire males (EM), barrows (BA), immunocastrates (IC) and gilts (GI)) were raised. They were fed *ad libitum* on a multiphase feeding regime (NE: 9.7; 9.5; 9.5; 9.3 MJ/kg and SID LYS: 9.0; 8.2; 7.7; 7.0 g/kg for phase 1, 2, 3, and phase 3 of the BA respectively). Pens were randomly divided into 3 groups of different slaughter weights: 105, 117 and 130 kg. Four piglets of each sex (21±0.6 kg) and one pig per pen at slaughter weight were euthanized and used to analyze the carcass composition at start and at slaughter. Nutrient accretion (dry matter, crude protein, crude fat, crude ash, and P) in the carcass was calculated as [mean bodyweight of the pen at slaughter, multiplied by nutrient contents of the carcass] minus [mean bodyweight of the pen at start multiplied by nutrient content of the piglets]. Nutrient intake was calculated as the sum of feed ingested per feeding phase multiplied by the nutrient content of the particular feed for the three phases. The efficiency of nutrient use was calculated as the nutrient accretion divided by the nutrient intake. Slaughter weight did not affect the protein content per kg carcass. BA had a lower crude protein content (16.9%) compared to EM (17.7%) and GI (17.8%) with IC (17.5%) as intermediates. Slaughter weight and sex had an interacting effect on N- efficiency: only in BA, N-efficiency decreased with higher slaughter weights (49%, 45%, 42% for 130, 117 and 105, respectively). In EM, GI and IC, no effect of slaughter weight was observed. Overall, BA (46%) had the lowest N- efficiency compared to EM (54%) and IC (52%), with GI (49%) as intermediates. Carcass phosphorus content (0.47% on average) or P- efficiency (45% on average) was not affected by sex or slaughter weight. In conclusion, we can state that EM were fed the most efficient in this trial and although BA received an adapted diet in phase 3, the ideal protein content can still be lowered to increase N-efficiency.

Multiobjective formulation, a method to formulate eco-friendly and economic feed for monogastrics

L. Dusart[1], B. Méda[2], S. Espagnol[3], P. Ponchant[1], D. Gaudré[3], A. Wilfart[4] and F. Garcia-Launay[5]
[1]ITAVI, UMT BIRD, 75009 Paris, France, [2]INRA, URA, 37380 Nouzilly, France, [3]IFIP, Pig Institute, 35650 Le Rheu, France, [4]INRA, SAS, 35000 Rennes, France, [5]INRA, PEGASE, 35590 Saint-Gilles, France; bertrand.meda@inra.fr

Feed production represents more than 70% of several environmental impacts estimated by Life Cycle Assessment (LCA) in pig (P) and broiler (B) productions. Yet, least cost feed formulation (LCF) is not relevant to reduce the environmental impacts of P and B feeds. Optimising feed composition while taking into account environmental impacts of feedstuffs (FS) is a way to do so. This study describes a new method, multiobjective formulation (MOF), based on the simultaneous use of an economic index (based on feed cost) and an environmental index (using LCA impacts) in a single objective function to minimize. A weighting coefficient (α) allows giving more or less influence to these indexes. The best trade-off was considered to be reached for an optimal weighting coefficient (α_{opt}) set to be the coefficient beyond which the marginal increase of the economic index exceeds the marginal decrease of the environmental index. In the actual French context of FS availability, important reductions in impacts of P and B feeds were achieved with MOF in comparison with LCF. In B, at α_{opt}, reductions by 12, 18, 7, 4, and 12% were achieved for 'climate change' (CC), 'non-renewable energy use', 'eutrophication', 'acidification' and 'phosphorus consumption' impacts, respectively, for an extra cost of 3%. In P, similar reductions of these impacts were achieved at αopt: -14, -13, -11, -7, and -6%, respectively, for an extra cost of 1%. For 'land occupation' impact, MOF had little effect in B (+4%) whereas in P, it decreased by 13%. LCA impacts at farm gate of one kg of live weight of B and P fed with these eco-feeds were also assessed. They confirmed that MOF reduces the environmental impacts of P and B productions (e.g. for CC, -7 and -10%, respectively) without pollution swapping and with moderate extra costs (e.g. +1% and +2% in P and B, respectively). Greater reductions of impacts are even possible when FS availability is less limiting (e.g. -19% for CC in B), questioning the actual availability of low-impact FS such as pea or sorghum.

Pig genotypes associated with ETEC susceptibility and the fucosylation of the intestinal mucosa

P. Trevisi[1], G. Pelagalli[1], D. Luise[1], V. Motta[1], M. Mazzoni[2] and P. Bosi[1]
[1]University of Bologna, Department of Agricultural and Food Science, V. le Fanin 46, 42123 Bologna, Italy, [2]University of Bologna, Department of Veterinary Medical Science, Via Tolara di Sopra, 50, 40064 Ozzano dell'Emilia, Italy; paolo.trevisi@unibo.it

The pattern of host mucus carbohydrates may affect the gut microbiota, in a genetically depended way. Colibacilloses by enterotoxigenic *Escherichia coli* F4 and F18 were respectively associated with variations of Mucin 4 marker gene (MUC4), and with two potentially causative genes, fucosyltransferase 1 (FUT1) and transferase A (α 1-3-N-acetylgalactosaminyltransferase), the presence or absence of the latter characterizing the porcine A0 blood group system. However, it is not in general explored if these differences affect the degree of fucosylation in the mucine-secreting cells and/or in the brush border, that represent the membrane anchored glycoproteins. For the study 21 PIC pigs were selected to be almost balanced for the FUT1 polymorphism (307A>G) (24% AA, resistant; 33% AG; 43% GG) and MUC4 (8227C>G) (33.3% each genotype, with CC being the resistant); in addition A0 gene was also checked (24% A0; 76% 00 – favoring resistance). The subjects were weaned at 24 days of age and reared for 7 weeks, fed a standard balanced diet and individually penned. At the end (19.9±2.7 kg live weight) they were slaughtered and from each subject a sample of jejunum was isolated (at 75% of its length). The fucose profile was evaluated by microscopic screening of Ulex europaeus agglutinin I (UEA) lectin binding. The intensity of staining was scored for presence in the brush border (from 0 to 2) and in the goblet cells (from 1 to 4), separately for cells in 20 villus and 20 crypts per pig. The intensity score for the fucose presence was lower in the brush border (P=0.0001), and in the Goblet cells in the villus (P=0.003) and crypt (P=0.0002) of FUT1 AA pigs, vs AG and GG, in the crypt Goblet cell of FUT1 AG pigs, vs GG (P=0.033) and in the in the villus Goblet cell of MUC4 CC pigs, vsCG and GG (P=0.040). Intensity scores did not differ for AB0. It is confirmed that the polymorphism at FUT1 gene affects the degree of fucosylation in the porcine jejunal mucosa. The brush borders are particularly affected with implications for the intestinal microbiota multiplication.

Effect of GPAT1 gene polymorphisms on pork quality

I. Mitka[1], M. Tyra[1] and K. Ropka-Molik[2]
[1]National Research Institute of Animal Production, Department of Genetics and Animal Breeding, Sarego 2, 31-047 Kraków, Poland, [2]National Research Institute of Animal Production, Department of Animal Genomics and Molecular Biology, Sarego 2, 31-047 Kraków, Poland; miroslaw.tyra@izoo.krakow.pl

Glycerol-3-phosphate acyltransferase mitochondrial enzyme encoded by GPAT1 gene, plays an essential role in triacylglycerols (TG) biosynthesis. TG in adipose tissue serves as the major energy storage. The aim of this research was to determine the effect of three selected polymorphisms within porcine GPAT1 locus on pork quality traits, with particular emphasis on intramuscular fat (IMF) content. A total of 709 pigs raised in Poland (represented by five breeds: Polish Landrace, Polish Large White, Puławska, Pietrain, Duroc) were used in this study. The frequency of selected SNPs localized in the first intron of GPAT1: c.37+55C>T and in 3'UTR region: GPAT1: c.*116T>C and GPAT1: c.*186C>T were estimated using PCR-RFLP method (with the use of RsaI, SfaNI and BbvI endonuclease; respectively). The association of meat quality traits and different SNPs was evaluated using GLM procedure (SASv. 8.02). The present research showed the highly significant association of GPAT1: c.37+55C>T polymorphism and IMF content. Pigs with CC genotype were characterized by the highest IMF content (2.35%), while these with TT genotype by the lowest (1.34%) (P<0.01). Furthermore, CC pigs had the highest value for meat colour (L*) (55.12) (P<0.05). The GPAT1: c.*116T>C SNP affected also IMF content as well as meat colour parameters. The highest IMF content was observed in homozygotes TT (2.34%) (P<0.01) and these animal showed also the highest value according to L* parameter (55.53)(P<0.05). The GPAT1: c.*186C>T polymorphism had no significant effect on any of examined meat quality traits. GPAT1 gene may be considered as one of the candidate genes related to some meat quality traits. Regarding the low frequency of favorable genotype according to GPAT1: c.37+55C>T and GPAT1: c.*116T>C especially with respect to IMF content, there can exist theoretical possibility of its practical use. The obtained results can be the basis for further studies.

Association of selected GPAT1 gene polymorphisms with slaughter traits in pigs

I. Mitka[1], K. Ropka-Molik[2] and M. Tyra[1]
[1]National Research Institute of Animal Production, Department of Genetics and Animal Breeding, Sarego 2, 31-047 Kraków, Poland, [2]National Research Institute of Animal Production, Department of Animal Genomics and Molecular Biology, Sarego 2, 31-047 Kraków, Poland; miroslaw.tyra@izoo.krakow.pl

GPAT1 gene encodes a glycerol-3-phosphate acyltransferase mitochondrial enzyme, which plays an essential role in triacylglycerols biosynthesis. The aim of this study was to analyse three selected polymorphisms of the ENSSSCT00000011627 and to estimate theirs associations with slaughter traits in pigs. Analysis was performed on five pig breeds (Polish Landrace, Polish Large White, Puławska, Pietrain, Duroc) maintained in the Pig Performance Testing Stations, in total on 709 pig. The analysis of ENSSSCT00000011627.2: c.37+55C>T; c.*116T>C; c.*186C>T genotypes was performed using PCR-RFLP method (with the use of RsaI, SfaNI and BbvI endonuclease; respectively). The association between slaughter traits and different GPAT1 genotypes was performed using GLM procedure (SASv. 8.02). The RsaI c.37+55C>T polymorphism affected only loin eye area (P<0.05) while SfanI c.*116T>C SNP was related to loin eye area as well as to loin eye height (P<0.01). The highest number of associations was noted according to the third polymorphism (BbvI c.*186C>T). Apart from the traits mentioned above on which highly significant effect was shown, there was also association of the BbvI c.*186C>T SNP on weight of ham without skin and bone (P<0.05), average backfat thickness (P<0.01) and lean meat percentage (P<0.05). The present study confirmed significant associations of all selected SNPs in GPAT1 locus and slaughter traits in pigs. However, regarding the high frequency of favorable genotype in pig population, there is no possibility of using these result in breeding practice. The possibility may exist only for the BbvI c.*186C>T SNP.

Identifying genes associated with feed efficiency in pigs by integrative analysis of transcriptome

Y. Ramayo-Caldas[1], M. Ballester[1], O. González-Rodríguez[1], J.P. Sánchez[1], M. Revilla[2], D. Torrallardona[3] and R. Quintanilla[1]
[1]IRTA, Genètica i Millora Animal, Torre Marimon, 08140, Spain, [2]Center for Research in Agricultural Genomics (CRAG), Department of Animal Genetics, UAB, Bellaterra, 08193, Spain, [3]IRTA, Animal Nutrition Program, IRTA, Mas de Bover, Constantí, 43120, Spain; yuliaxis.ramayo@irta.cat

Improving feed efficiency (FE) contributes to increase economic sustainability and reduce ecological footprint of pig production systems. Carried out within the ECO-FCE project, the main goal of this study was to identify candidate genes associated with feed efficiency in pigs through an integrative approach. A population of 288 pigs of both sexes were raised under control conditions at IRTA control station, and 80 individuals (40 males and 40 females) with divergent residual feed intake (RFI) were selected for RNA-Seq analyses. Transcriptome from liver and duodenum samples was sequenced on an Illumina Hiseq2000. Reads were mapped to Sscrofa10.2 with STAR, and counted by htseq-count. Gene expression data was explored through a combination of weighted gene co-expression network analyses and multivariable methods integrating transcriptomic and phenotypic information. Co-expression analyses revealed highly connected networks with similar topological parameters in both liver and duodenum tissues. Genes were clustered according to their connectivity, and a total of 23 (liver) and 25 (duodenum) modules were identified. Four of these co-expressed gene modules (one in liver and three in duodenum) were significantly associated with FE traits, including RFI, and residual intake and body weight gain (RIG). In both tissues, RFI and RIG showed opposite correlation patterns. Intra-module analyses revealed interesting tissue-specific candidate genes, as well as 12 genes that were common to both tissues. Pathway analyses indicated that these genes take part in a wide variety of physiological and biological events, such as inflammation, immune response and heat shock protein binding. Our results provide a basis for future studies to gain novel insights into the molecular mechanisms related to FE in pigs.

Effects of feed restriction on pig skeletal muscle transcriptome

M. Ballester[1], M. Amills[2], O. González-Rodríguez[1], T. Figueiredo-Cardoso[2], M. Pascual[1], E. Mármol-Sánchez[2], J. Tibau[1] and R. Quintanilla[1]
[1]Institute for Research and Technology in Food and Agriculture (IRTA), Animal Breeding and Genetics, IRTA-Torre Marimon, 08140 Caldes de Montbui, Spain, [2]Center for Research in Agricultural Genomics (CSIC-IRTA-UAB-UB), Department of Animal Genetics, Campus UAB, 08193 Bellaterra, Spain; raquel.quintanilla@irta.cat

This study aims to determine changes in the pig skeletal muscle transcriptome profile derived from feed restriction by RNA-Seq. A total of 24 females belonging to a Duroc commercial line received either a restricted (RE) or *ad libitum* (AL) diet during the first fattening period (60-120 days of age). Animals were slaughtered at ~150 days of age. Transcriptome from gluteus medius muscle samples was sequenced on an Illumina Hiseq2000. Quality control of RNA sequencing data was assessed with FASTQC, and STAR was used to map reads to the porcine reference genome (*Sscrofa10.2*). A total of 2,430 million of 75 paired-end reads were generated. About 89.6% of reads mapped to the porcine reference genome, and 70.6% of them corresponded to annotated genes. Besides, 80.06% were located in exons, 8.22% in introns and 11.72% mapped to intergenic regions. EdgeR and Deseq2 programs were used to perform differential expression analyses with the counting reads mapped to each gene by htseq-count. The comparison of AL and RE expression profiles made possible to identify 101 ($|logFC|>1.2$, FDR<0.05, EdgeR) and 30 ($|logFC|>1.2$, $p_{adj}<0.05$, Deseq2) differentially expressed (DE) genes. A total of 26 DE genes, 23 over-expressed in AL and three over-expressed in RE, were consistently detected with both edgeR and Deseq2. A functional categorization of the overlapped list of DE genes was performed with IPA. The most significantly enriched canonical pathway was AMPK Signaling, which plays a key role in the maintenance of energy homeostasis. Among the top annotated biological functions, we found glycolysis, oxidation of fatty acids, transport and metabolism of carbohydrate, and biogenesis of mitochondria whose genes were down-regulated in the RE group. Further studies are being conducted to associate these transcriptomic changes with the modifications observed in the intramuscular fat composition of the animals under feed restriction.

Evaluation of protein level and lysine:net energy ratio in finishing pigs

W. Yun[1], J.H. Lee[1], C.H. Lee[1], W.G. Kwak[1], S.D. Liu[1], I.H. Kim[2] and J.H. Cho[1]
[1]Chungbuk National University, Animal Science, 1, Chungdae, Cheongju-si, Chungcheongbuk-do, 28644, Korea, South, [2]Dankook University, 119, Dandae-ro, Dongnam-gu, Cheonan-si, Chungcheongnam-do, Republic of Korea, 31116, Korea, South; adidats@naver.com

A total of 72 finishing pigs (54.16±0.42 kg) were used in 6-wk and 11-wk experiments to evaluate the effects of reduced crude protein (CP), Lysine (Lys):NE ratio on growth performance, nutrient digestibility, blood profiles, and meat quality in finishing pigs. The diets were formulated to contain 10.38 MJ/kg NE, 169.5 g/kg, 169.5 g/kg, 158 g/kg CP, and Lys:NE ratios of 0.79, 0.75, 0.75 g/MJ NE. No differences (P>0.05) were found on average daily gain (ADG), average daily feed intak (ADFI) and G:F during wk 0-6, wk 6-11 and the over-all period. The apparent total tract digestibility (ATTD) of N in 11 wk was greater in pigs fed a diet contained a Lys:NE ratio of 0.79 g/MJ (P<0.05) than those fed a diet containing a Lys:NE ratio of 0.75 g/MJ. No difference was observed on the ATTD of dry matter (DM) and energy as Lys:NE ratio decreased. The Minolta color parameter lightness (L*) of pigs fed CON diet was similar to that of pigs fed T2 diet, however, it was higher (P<0.05) than pigs fed T1 diet. Redness (a*) was decreased (P<0.05) and the marbling, firmness score increased (P<0.05) when the Lys:NE ratio decreased. The CP level and Lys:NE ratio had no effect on the pH value, TBARS content, drip loss, water holding capacity, longissimus muscle area, and backfat thickness. In conclusion, growth performance was not influenced when dietary protein level and Lys:NE ratio decreased in this experiment.

Characteristics of high and low prolific sows on high-performing Spanish farms

S. Tani[1], C. Piñeiro[2] and Y. Koketsu[1]
[1]Meiji University, Higashi-mita 1-1-1 Tama-ku, Kawasaki, Kanagawa, 214-8571, Japan, [2]PigCHAMP Pro Europa S.L., c/Santa Catalina 10, Segovia, 40003, Spain; composition.013@gmail.com

The objectives of this retrospective cohort study were to compare high and low prolific sows' performances between high- and low-performing farms. Reproductive performance in consecutive parity and lifetime performance were analyzed in a three sow groups × three farm groups factorial arrangement. Data from herd entry to removal included 444,429 first service records of 91,971 females in 98 Spanish herds, served between 2008 and 2013. Farms were categorized into three groups based on the upper and lower 25th percentiles of the farm means of annualized lifetime pigs weaned per sow: high- (HP), intermediate- (IP) and low-performing (LP) farms. Also, sows were categorized into three groups based on the upper and lower 10th percentiles of pigs born alive (PBA) at parity 1 as follows: 15 pigs or more (H-PBA), 8 to 14 pigs, and 7 pigs or fewer (L-PBA). Linear regression models were applied to examine the three × three factorial arrangement data for either by-parity reproductive performance or lifetime performance of sows. Also, repeated measures models were applied to the by-parity data. From parity 2 to 6, H-PBA sows on HP farms had 0.8-1.1 more PBA than those on LP farms (P<0.05). Also, from parity 2 to 6, L-PBA sows on HP farms had 1.4-1.7 more PBA than those on LP farms (P<0.05). Farrowing rate was higher in parity 1 to 4 H-PBA sows than L-PBA sows in all farm groups. Also, from parity 0 to 6, HP farms had higher farrowing rate than LP farms. With regard to lifetime performance, there was no difference in parity at removal between the three farm groups (P>0.05). H-PBA sows on HP farms had 6.7 more lifetime PBA than similar sows on LP farms (P<0.05), but there was no difference between farm groups for L-PBA sows (P>0.05). Also, H-PBA and L-PBA sows on HP farms had 29.7 and 30.7 fewer lifetime non-productive days than respective sows on LP farms. Meanwhile, H-PBA sows on HP farms and LP farms had only 9.0 and 8.0 days more lifetime non-productive days than L-PBA sows on the respective farms. In conclusion, HP farms achieved more pigs weaned in lifetime for both H-PBA and L-PBA sows, by having high PBA and high farrowing rate, and also reducing non-productive days.

Culling risks for served females and farrowed sows at consecutive parities in commercial herds

S. Tani[1], C. Piñeiro[2] and Y. Koketsu[1]

[1]Meiji University, Higashi-mita 1-1-1 Tama-ku, Kawasaki, Kanagawa, 214-8571, Japan, [2]PigCHAMP Pro Europa S.L., c/Santa Catalina 10, Segovia, 40003, Spain; composition.013@gmail.com

The objectives of the present study were: (1) to characterize culling and retention patterns of parity 0 to 6 served females and farrowed sows in two herd productivity groups; and (2) to quantify the factors associated with by-parity culling risks of these pigs. The lifetime data from first service of gilts to removal included 465,974 service records of 94,691 females in 98 Spanish herds, served between 2008 and 2013. Herds were categorized into two groups based on the upper 25th percentiles of the herd means of annualized lifetime pigs weaned per sow: high-performing (>24.7 pigs) and ordinary herds (<24.7 pigs). Two-level log-binomial regression models were used to examine risk factors and risk ratios associated with by-parity culling risks for served females and farrowed sows. Mean by-parity culling risks (± SE) for the served females and farrowed sows were 5.9 ± 0.03 and $12.4\pm0.05\%$, respectively. Increased culling risks for pregnant females were associated with an increased age of gilts at first-mating, and sows having a weaning-to-first-mating interval (WMI) of 7 days or more (P<0.05). The relative risk ratios of sows having WMI of 7 days or more were 1.56-1.84 across parity, compared to the same parity sows having WMI of 0-6 days. In contrast, a decreased culling risk for served females was associated with sows being fed in high-performing herds. For farrowed sows, increased culling risks were associated with being fed in high-performing herds, sows having farrowed 8 or fewer pigs born alive, and sows having farrowed 3 or more stillborn piglets (P<0.05). However, herd size was not associated with any culling risk (P>0.88).The relative risk ratios of sows having farrowed 8 or fewer pigs born alive were 3.52-4.11 across parity, compared to the respective reference sows that had farrowed 16 pigs or more. Our study indicates that high-performing herds cull more farrowed sows in mid-parity than ordinary herds, and also cull less served sows from mid-parity to late parity. It is recommended that producers provide extra attention to sows farrowing stillborn piglets and having prolonged WMI.

Strategies to achieve a sustainable dairy cow production under consideration of physiological limits

J.J. Gross and R.M. Bruckmaier

Veterinary Physiology, Vetsuisse Faculty, University of Bern, Bremgartenstrasse 109a, 3012 Bern, Switzerland; josef.gross@vetsuisse.unibe.ch

Health disorders and poor reproductive performance account distinctly for culling and economical costs in dairy production. Sustainability is currently not achieved when considering the negative impact of selection for high genetic potential on immune competence, fertility, lifetime performance, animal welfare and meeting nutritional requirements. Due to limitations in nutrient quality of forages a high dependency on importing protein-rich feed and competition for human food developed. Furthermore, the use of antibiotics, fate of male calves, detrimental ecological and economic effects represent relevant issues in the social awareness. A routine identification of biomarkers accurately characterizing the physiological status of an animal is crucial for decisive strategies. Recent research revealed that concentrations of free fatty acids and ketone bodies (i.e. β-hydroxybutyrate) due to extended periods of insufficient energy supply as present in high-yielding dairy cows further affect the limited availability of glucose required for milk production, fertility and immune system. The present contribution will elucidate open potentials and their sustainability, but also restrictions of alternative production strategies under consideration of existing physiological limitations. Such strategies might include for instance the use of dual purpose breeds, shortening the dry-period length, extend lactation by later insemination, and individualized feeding. Feed additives have only a limited potential to alleviate short-term deficiencies. Breaking the peak of the lactation curve by either omitting the dry period or manipulating the endocrine regulation of early galactopoesis imply economic losses due to only partial use of the available genetic potential, but reduce the metabolic load, risk of health disorders and culling. Within each individual local production system efficiency must be maximized. Intensifications in pasture-based milk production systems with suitable genotypes may result in higher methane emission, but allow welfare orientated animal husbandry and efficient use of nitrogen and grassland sources without competition for human food and costly concentrate supplementation.

Metabolic profiling of calves reared on whole milk versus milk replacer in a metabolomics approach

A. Kenez[1], M. Korst[2,3], C. Koch[2], F.J. Romberg[2], K. Eder[4], H. Sauerwein[3] and K. Huber[1]
[1]University of Hohenheim, Fruwirthstr. 35, 70599 Stuttgart, Germany, [2]Hofgut Neumuehle, Neumuehle 1, 67728 Muenchweiler a.d. Alsenz, Germany, [3]University of Bonn, Katzenburgweg 7-9, 53115 Bonn, Germany, [4]Justus-Liebig University, Heinrich-Buff-Ring 26-32, 35392 Giessen, Germany; akos.kenez@uni-hohenheim.de

Optimized nutrition of calves is a fundamental requirement for an efficient production in adult age. An early metabolic dysregulation due to suboptimal nutrition of calves could have detrimental long-term effects on metabolic health and production efficiency. This study performed a targeted metabolomics analysis for comparative metabolic profiling of calves reared on whole milk vs milk replacer, in order to generate novel hypotheses regarding the modulatory effect of whole milk vs milk replacer rearing on metabolic maturation. German Holstein calves were fed either with whole milk (WM; n=10) or with milk replacer (MR; n=9) ad libitum until d 25 post natum. On d 22 one blood plasma sample was collected from every calf, which was then analyzed by a targeted metabolomics analysis using the AbsoluteIDQ p180 Kit of Biocrates Life Science AG (Innsbruck, Austria). Profiling of metabolomics data was performed by heatmap visualization and multivariate data analysis techniques such as principal component analysis. A quantitative identification of 187 plasma metabolites was possible, belonging to the metabolite classes of acyl-carnitines, amino acids, biogenic amines, phosphatidylcholines (PC), lyso-phosphatidylcholines (lyso-PC), sphingomyelines (SM) and hexoses. The plasma metabolome profile of WM and MR calves was found to be different due to distinct concentration patterns of specific metabolite groups. The metabolites contributing the most to the differences between groups were identified to be SM, lyso-PC and PC of specific chain lengths. Dissimilar metabolic profiles were associated with particular lipid species, the most of which are responsible for membrane function, cellular turnover and integrity, and signaling cascades such as inflammatory pathways. The characterization of these pathways in a follow-up study can help us improve our understanding of modulatory links between nutrition and metabolic maturation, with the goal of ensuring a healthy production efficiency.

New insights into the physiology of mineral homeostasis in lactation from beef cows

S.T. Anderson[1], M. Benvenutti[2], J. Spiers[1], K. Goodwin[2], L. Kidd[3], M.T. Fletcher[1] and R.M. Dixon[1]
[1]The University of Queensland, Brisbane, 4072, Australia, [2]Department of Agriculture and Fisheries, Gayndah, 4625, Australia, [3]The University of Queensland, Gatton, 4343, Australia; stephen.anderson@uq.edu.au

At parturition a key physiological mechanism in mineral homeostasis in cows is up-regulation of 1,25-dihydroxy vitamin D (1,25-diOH Vit D) to promote intestinal absorption of Ca and P and enhance bone resorption. This activation of Vit D has been extensively studied in dairy cows and appears to be critical in milk fever. However only a transient increase in Vit D activation occurs within the first week of lactation. Here we examined 1,25-diOH Vit D in beef cows fed diets adequate or deficient in P. Droughtmaster cows (n=32; 6-11 years) at calving were housed in individual pens. For 14 weeks the cows were fed ad libitum diets of high (2.16 g P/kg DM) or low (0.54 g P/kg DM) P or Ca: HP-HCa, LP-HCa and LP-LCa. A fourth diet comprised LP-LCa, with inclusion of ammonium chloride to provide a negative DCAD diet (LP-LCa-DCAD). The diet Ca/P ratios were ca. 2:1, 5:1, 3:1 and 3:1. Plasma concentrations of Vit D were measured using commercial kits (IDSystems, UK). Results are mean ± SE. At calving cows (n=32) were in adequate P and Ca status (PiP 1.93±0.10 mM, plasma Ca 2.13±0.04 mM), with 25-OH Vit D (42±3 ng/ml) and 1,25-diOH Vit D (115±15 pg/ml) concentrations. 1,25-diOH Vit D remained high in early lactation in HP-HCa cows (128±30, 88±12, 75±41 pg/ml at 2, 4 and 8 weeks), and declined thereafter (46±5 pg/ml at 14 weeks). All LP diet cows had higher (P<0.05) 1,25-diOH Vit D during later lactation (81±5 LP-HCa, 114±13 LP-LCa, and 127±9 LP-LCa-DCAD pg/ml at 14 weeks). These results suggest that beef cows in early lactation have a greater physiological capacity for activation of 1,25-diOH Vit D than that reported in dairy cows (<50 pg/ml after 4 weeks). This intriguing difference suggests a physiological limitation in dairy cows that likely reflects their genetic selection for high milk production, together with adaptations to improved nutritional regimes. Research supported by Meat and Livestock Australia and Qld Department of Agriculture and Fisheries.

Prediction of variability in pig growth rate using immune and metabolism blood indicators

A. Chatelet[1], N. Le Floc'h[1], F. Gondret[1] and N.C. Friggens[2]
[1]*INRA, Agrocampus Ouest, PEGASE, Saint-Gilles, 35590, France,* [2]*INRA, AgroParisTech, Université Paris-Saclay, UMR MoSAR, Paris, 75005, France; alexandra.chatelet@inra.fr*

Pig growth rate is influenced by metabolism and immune status. This study aimed to assess whether immune and metabolic blood indicators can predict variability in growth rate of two pig lines differing in their Residual Feed Intake or RFI (a measure of feed efficiency) and submitted to an immune challenge. A total of 160 growing pigs were allotted in a 2×2 factorial design comparing the two lines (low RFI or LRFI is more feed efficient vs high RFI or HRFI is less efficient) housed in good or poor hygiene conditions (clean vs dirty). This design permitted us to evaluate whether predictors of variability in growth rate applied across the different combinations of hygiene and line. Average Daily Gain (ADG, g/d) was calculated for 6 weeks (W0 to W6) and fasting blood samples were taken at W0, W3 and W6 to measure 47 variables related to immune system or metabolism. A preliminary descriptive Principal Component Analysis showed a great variability as well as strong multicollinearity within the 47 blood indicators measured at W3. In addition, the experimental design had introduced a great variability in growth rate, with ADG of the HRFI line housed in dirty conditions exhibiting the largest range. Thus, Partial Least Square (PLS) analyses were performed within each RFI line using the 47 variables measured at W3 to predict ADG. Several PLS models were fitted to estimate the proportion of explained ADG and the Root Mean Square Error of Prediction (RMSEP). Finally, a model reduction step was also performed to select the most relevant ADG predictors taking into account their biological meaning. Using the 47 variables and irrespective of the hygiene conditions, it was possible to explain 42% and 73% of ADG variations for the LRFI and HRFI, respectively. Within the 47 predictors, immune traits such as haptoglobin concentration and some amino acids (taurine, hydroxyproline) were associated with variation in growth rate. This study may provide useful information on biological pathways explaining the ability of pigs to cope with an immune challenge. Research has received funding from the EU FP7 Prohealth project (no. 613574)

Roles of size, behavior, metabolites and body composition in the variation of cattle feed efficiency

A.B.P. Fontoura[1], Y.R. Montanholi[2], A.K. Ward[1] and K.C. Swanson[1]
[1]*North Dakota State University, 1300 Albrecht Blvd, Fargo, ND 58102, USA,* [2]*Dalhousie University, 58 Sipu Awti, B2N 5E3, Truro, NS, Canada; ananda.fontoura@ndsu.edu*

The biological variation in feed efficiency (FE) is regulated by multiple physiological mechanisms relevant to energy use. Thus, objectives were to evaluate the associations between animal size (AS), feeding behavior (FB), blood plasma metabolites (PM), body composition (BC) with different measures of FE. Steers (n=61; BW=401±54 kg) were fed a corn-based diet and tested over a 142-d performance evaluation and housed in pens equipped with an automated feeding system that allowed the measurement of FB traits. Measurements of AS were performed on days 1 and 142. Blood and BW were collected every 28 days. Steers were slaughtered at the end of feeding test and BC traits were further collected on the carcasses. Residual feed intake (RFI), residual gain (RG) and gain to feed (G:F) were calculated as measures of FE. Data were analyzed using the SAS software. Correlations between AS, FB, BM, BC traits and FE were measured through the MANOVA/PRINTE statement within the general linear model procedure. Relative contribution to the variation in RFI, RG and G:F were performed using partial regression analysis. Gain to feed was strongly correlated with AS traits (range: -0.38 to -0.58, $P \leq 0.004$). Visits to the feeder and feed intake per meal (kg) were associated with decreased FE (RFI, 0.46, P=0.0004; G:F, -0.30, P=0.02). The lack of correlations between PM (plasma urea N, glucose, and non-esterified fatty acids) with FE, combined with the low contribution estimates of these traits with RFI, RG, and G:F (1%, 6%, and 4%, respectively) suggest that the measured PM play minor roles in the variation of FE. Similarly, BC played minor roles in FE variation (3% of RFI, 2% of RG and G:F). Overall, AS and FB traits may constitute important physiological factors influencing FE (14% of RFI and RG, 39% of G:F; 50% of RFI, 10% of RG, 16% of G:F, respectively). This data indicate that selection criteria and performance evaluation based on FE measures should account for the role of AS and FB, animal's stage of production and system's outputs of interest. Ongoing analysis will investigate the association between these traits in cattle at different stages of production.

Immune function, colostrum and blood metabolomics in performance tested beef heifers at peripartum

V. Peripolli[1], A. Kenez[2], A. Fontoura[3], N. Karrow[4], S. Miller[4,5] and Y. Montanholi[6]

[1]Instituto Federal Catarinense, Abelardo Luz, 89830-000, Brazil, [2]Hohenheim University, Stuttgart, 70599, Germany, [3]North Dakota State University, Fargo, ND 58102, USA, [4]University of Guelph, Guelph, N1G 2W1, Canada, [5]Angus Genetics Inc., Saint Joseph, MO 64506, USA, [6]Dalhousie University, Truro, B2N 5E3, Canada; yuri.r.montanholi@gmail.com

The fine balance between production and efficiency in livestock is related to the immune system and metabolites. The objective was to identify relationships of feed efficiency with immune function and metabolome of blood and colostrum from beef heifers. Thirty-six heifers were evaluated for feed efficiency using residual feed intake (RFI). Heifers were immunized with ovalbumin at 33 ± 16 and 25 ± 15 d prior to parturition; with blood collected prior to these events, at calving and at 47 ± 0 d during nursing. Colostrum was sampled within 4 h post-calving. Blood serum and colostrum whey were analyzed for total immunoglobulin G (IgG) and A (IgA) and specific IgG to ovalbumin (IgG-OVA). Additionally, these samples were submitted to NMR spectroscopy. Least square means comparison of metabolic state (baseline, booster, calving or nursing), RFI (efficient or inefficient) and responsiveness to IgG-OVA (low and high) were calculated. Regression analysis were conducted to define the relationship of immune variables and RFI. Principal component analysis was used on the metabolome data based on the classes above. The RFI classification of heifers suggested increased IgG-OVA in the colostrum (44.5 vs 32.7 1000 OD; $P<0.10$) and decreased IgA in the blood during nursing (193 vs 283 mg/ml) in efficient vs inefficient. In regards to metabolome no distinction between feed efficiency, IgG-OVA classes were observed; however clear distinctions were noted for metabolic state. The regression analysis, with IgG-OVA as dependent variable, suggested an effect of RFI ($P<0.10$; $R^2=0.29$), supporting a synergism with feed efficiency. The evaluation of immunoglobulins and metabolome over metabolic states highlights shifts to be considered in the development of biomarkers. While analysis of RFI and immunoglobulins had discriminatory potential for feed efficiency, further analysis of the metabolome may reveal key analytes related to feed efficiency and immune response.

Relationship between milk fat and urea, cow productivity and reproduction indices in the dairy herd

I. Sematovica and L. Liepa

Latvia University of Agriculture, Faculty of Veterinary Medicine, Clinical Institute, Kr. Helmana street 8, 3004 Jelgava, Latvia; isem@inbox.lv

The study was a part of the State Research Project (AgroBioRes) No. 2014.10-4/VPP-7/5; subproject AP4. Nowadays, in the dairy cow ration, a large amount of easy digestible carbohydrates and proteins are included in order to achieve maximum productivity. That kind of feeding boarders on the ruminants' physiological ability to maintain homeostasis in the rumen regarding to microflora, pH, and the health in all. To provide the adequate content of energy, protein and other ingredients in cows' ration, milk fat (MF), milk fat/protein ratio (F/P) and urea in milk (MU) must be evaluated. Low MF content could be one of the signal variable to suspect subacute rumen acidosis (SARA) in the dairy herd. The aim of the study was to evaluate the effect of the low MF content on the indices of reproduction (times of artificial insemination (AI) for pregnancy) and productivity in the dairy herd with milk yield of 8,500 kg/cow per year. The data of cows productivity, milk composition and reproduction performance were collected and analysed in the early and late period of lactation. Results. MF content was too variable in the first 60 days of lactation (1.3-7.41%), exceeding an optimal average values (3.0-4.5%). Cows with MF below 2.5% had a higher productivity (46.4 ± 7.65 kg) compared to cows with MF 3-4.5% (41.5 ± 8.36 kg). Results of calculating energy corrected milk (ECM) were opposite (35.0 ± 6.08 kg; 43.5 ± 8.87 kg, respectively) ($P<0.05$). In cows which had the lowest F/P ratio (0.75 ± 0.11 mg d/l), the MU level was higher (33.7 ± 9.6 mg d/l) than in cows with optimal (1.2-1.5) and high (>1.5) F/P ratio. There was a poor agreement between the low MF content and AI times and the length of service period as a whole; however cows with MF below 3% had 1.9 ± 1.11 AI times per pregnancy, cows with MF above 3% had 2.2 ± 1.42 AI times per pregnancy ($P<0.05$). Conclusion. Cows with a lower MF are more productive regarding to milk yield per cow a day, but they are less productive regarding to ECM. Low MF does not influence the cow to become pregnant. MF and F/P ratio signal the necessity to make corrections with regard to the composition of the feed and management.

Performance of Dorper × Santa Inês lambs with divergent efficiency for residual body weight gain

E.M. Nascimento, M. Figueiredo, L.M. Alves Filho, J.C.S. Lourenço, C.I.S. Bach, W.G. Nascimento, S.R. Fernandes and A.F. Garcez Neto
Federal University of Paraná, Animal Science, Rua Pioneiro, 2153, Jd. Dallas, Palotina, Paraná, 85950000 Palotina, Brazil; americo.garcez@ufpr.br

Residual body weight gain (RG) is an efficiency index calculated by regressing average daily gain (ADG) against dry matter intake (DMI) and mean metabolic weight (MMW). The difference between observed ADG and expected ADG obtained for each animal from regression equation is defined as RG. The trial was carried out to evaluate the performance of lambs with divergent efficiency for RG during the growing and finishing phases. Twenty four crossbred ½ Dorper × ½ Santa Inês non-castrated male lambs with four months of age and 24.1±3.2 kg of body weight (BW) were used. Lambs were fed *ad libitum* with diets composed of 64% of *Cynodon* spp. and 36% of concentrate feed, with 15.8% of crude protein (CP) and 67.0% of total digestible nutrients (TDN). DMI was measured based on the amount of feed provided to lambs and their orts, which were quantified daily. A completely randomized design was used for analysis of variance with two treatments based on two classes of RG, one efficient (RG from 0.3 to 21.1 g BW/d) and other inefficient (RG from -1.1 to -32.7 g BW/d). Pearson correlation analysis was run between the RG and the other performance and feed efficiency traits. There was correlation (P<0.05) between RG and ADG (r=0.54). It was found a higher (P<0.05) ADG for the efficient group (141.9 vs 109.3 g/d) which is explained by the better feed conversion ratio (FCR) (P<0.05) of this group (6.89 vs 8.48 g DM/g gain). Although not significant, the DMI for the efficient group was 60 g higher than inefficient group (975 vs 915 g DM/d). It may also have contributed to the better performance of efficient lambs. There were correlations (P<0.05) between RG and FCR (r=-0.93), gross feed efficiency (r=0.94), relative growth rate (r=0.74) and Kleiber ratio (r=0.71). These traits are influenced by ADG, which serves for their calculations. Based on the significant correlations, it may be concluded that the RG allows the selection of lambs with better FCR, demonstrating a high productive potential of these animals during the growth and finishing phases.

Effects of feed offer length on commercial traits, heart and liver weight of broiler chickens

A. Karimi[1], R. Hosseini[2] and S. Razaghzadeh[3]
[1]University of Tabriz, Ahar Faculty of Agriculture and Natural Resources, Tabriz, 5166616471, Iran, [2]Orum Gohar Daneh Co., Product management, Miandoab, 59716-56711, Iran, [3]West Azerbaijan Agricultural Research, Education & Extension center(AREEO), Animal Science, Orumieh, 5716963963, Iran; akkg2000@yahoo.com

Any shortage in feed stuff can negatively affect broiler chick's growth and also, body weight gain is more rapid than organs such as liver and heart that are vital to survive. The aim of this trial was to determine the effect of different feed offering programs on commercial traits of broiler chicks in special condition such as feed stuff shortage and internal organs weight gain. A total of 800 unsexed day-old boiler chicks were randomly distributed in 20 floor pens of 40 birds each. Five replicates were allocated to one of four experimental program feedings include: (1) control group was fed *ad libitum* throughout the experiment (Con), (2) 8 h restricted feeding per day from 9 to 24 day of age (8 h), (3) 16 h restricted feeding per day from 9 to 21 day of age (16 h) and (4) every other day feed restriction from 9 to 18 day of age (24 h). Diets of treatments were iso-energetic and iso-nitrogenous base on Ross 308 guide and the experiment lasted for 42 days. The birds on Con and 8 h groups consumed significantly more feed than other groups in week and 4 of age (P<0.05); but there was no difference among experimental groups in other weeks (P>0.05). There was a significant difference in body weight between control group and other groups in 28 day of age (P<0.05). But restricted groups could compensate body weight in next weeks without significant difference (P>0.05). Also there was no significant difference in feed conversion ratio (FCR) and mortality rate among experimental groups in different weeks of age (P>0.05). Heart and liver rational weights were higher in group 24 h vs control group (P<0.05) and total productivity efficiency factor (PEF) in end of rearing in group 24 h was higher than others (P<0.05). Overall, it seems birds with every other day feed restriction (24 h group) show the best commercial performance because in spite lower total feed consumption, body weight was as same as control and also with higher PEF.

Bad news about production efficiency

K. Huber[1] and H. Sauerwein[2]
[1]University of Hohenheim, Faculty of Agricultural Sciences, Institute of Animal Science, Fruwirthstrasse 35, 70599 Stuttgart, Germany, [2]Institute for Animal Science Physiology & Hygiene, University of Bonn, Katzenburgweg 7-9, 53115 Bonn, Germany; sauerwein@uni-bonn.de

Production efficiency can be described as an animal's ability to adapt properly to physiological changes related to performance (milk, eggs, growth) and to cope adequately with inverse environmental conditions, at least to certain degree. For that, morphological and functional features of the organism have to be fully developed. However, in modern intensive production systems not all of the animals are able to be efficient in this way. This inability causes severe health and welfare issues in livestock due to resource allocation mismatch and due to overrunning physiological limitations. Aim of this presentation is to overview pathophysiological conditions in farm animals indicating physiological limitations. The underlying causes may be based on inadequate feeding and management and on selective breeding for single production traits. In Holstein dairy cows we are facing an increasing proinflammatory condition and immune deficiency which could be derived from accelerated development of calves and heifers. In chicken, selective breeding for breast muscle growth led to cardio-vascular and respiratory insufficiency and increasingly, to myopathies, especially in the breast muscle. In sows and ewes, selection for high litter size led to intrauterine growth retardation, colostrum deficiency and higher mortality of the offspring. Although this presentation can only cover a few pathophysiologies observed in the field of animal production, their relation to animal's production efficiency will be clear. It is suggested to re-think about 'animal's production efficiency' in terms of optimization of production, not maximizing production. The former includes health and welfare of farm animals which most likely increases production efficiency significantly.

Towards defining efficiency: a systems and an animal oriented dialogue

K. Huber[1], M. Zehetmeier[2], H. Sauerwein[3] and M. Tichit[4]
[1]University of Hohenheim, Institute of Animal Science, Fruwirthstrasse 35, 70593 Stuttgart, Germany, [2]Bavarian State Reseach Center for Agriculture, Institute of Agricultural Economics, Menzinger Straße 54, 80638 München, Germany, [3]University of Bonn, Institute for Animal Science, Katzenburgweg 7, 53115 Bonn, Germany, [4]INRA, UMR SAD-APT INRA/AgroParisTech, 16 Rue Claude Bernard, 75231 Paris Cedex 05, France; sauerwein@uni-bonn.de

This part of the session is aiming to bridge the current session of the EAAP Physiology Commission with the following session of the EAAP Livestock Farming Systems (LFS) Commission. As evident from the sessions' titles, i.e. 'Physiological limits of production efficiency' and 'LFS efficiency: indicators and scales?', efficiency is a topic that is of central interest for both commissions. Moreover, for all animal science and animal production worldwide, efficiency is of outstanding importance, but the definition and the scale of efficiency is erratic and may vary between different disciplines and stakeholders. The animal oriented view dominating in Physiology will be addressed in the preceding presentations, whereas the systems oriented view will be in focus of the afternoon LFS session. The dialogue planned herein aims at a mutual exchange about efficiency not only between the speakers and members of the two commissions but also with the audience to start a general discussion that may eventually yield common definition (-s) of efficiency in context of livestock farming.

Development of an ELISA for the acute-phase protein alpha 1-acid glycoprotein (AGP) in cattle

T. Blees[1], C. Catozzi[2], C. Urh[1], S. Dänicke[3], S. Häussler[1], H. Sauerwein[1] and F. Ceciliani[2]
[1]University of Bonn, Institute of Animal Science, Katzenburgweg 7, 53115 Bonn, Germany, [2]Università di Milano, Department of Veterinary Medicine, Via Celoria 10, 20133 Milano, Italy, [3]Friedrich-Loeffler Institute, Institute of Animal Nutrition, Bundesallee 50, 38116 Braunschweig, Germany; sauerwein@uni-bonn.de

Being part of the non-specific immune system, alpha 1-acid glycoprotein (AGP) increases in response to systemic tissue injury, inflammation, or infection, where it exerts its anti-inflammatory functions. In cattle, AGP is considered as minor or moderate acute phase protein but quantitative assays were limited to immuno diffusion. To accurately measure bovine AGP concentrations, we developed an ELISA. As a first application, we aimed to characterize the effects of a rapid weight gain due to increasing the energy content of the diet. The competitive ELISA was developed using a Protein G purified, polyclonal rabbit antibody against AGP purified from bovine serum. Microtiterplates were coated with serum with assumingly high AGP concentrations (mastitis). The anti-AGP antibody was detected via a monoclonal mouse anti-rabbit peroxidase-labelled antibody; the signal obtained in the competitive ELISA was inversely related to the AGP concentration in the sample. A serum standard, calibrated against purified bovine AGP, was used to determine AGP concentrations. The intra- and interassay variation was 8.77 and 8.15%, respectively. Serum AGP concentrations in non-pregnant, non-lactating Holstein cows (n=8) under extensive conditions was 2.52±0.17 mg/ml (mean ± SEM). After 15 weeks, when the cows were fed increasing portions of concentrate until 60% were reached within 6 weeks and maintained for further 9 weeks, the AGP concentrations increased up to 3.64±0.30 mg/ml (P=0.002; paired t-Test, SPSS 24). In view of the increased body weights (+243±33 kg), the increase of AGP might be related to an increase in body fat mass that may either directly affect the serum concentration by secretion of AGP from adipose tissue and/or indirectly by metabolic alterations that stimulate hepatic AGP synthesis. Using the ELISA developed herein allows for assessing even subtle changes of AGP concentrations that are below the 2-3-fold increase expected for a minor acute phase protein.

Performance of Dorper × Santa Inês lambs with divergent efficiency for residual feed intake

M. Figueiredo, E.M. Nascimento, T. Rodrigues, C.I.S. Bach, H. Maggioni, W.G. Nascimento, S.R. Fernandes and A.F. Garcez Neto
Federal University of Paraná, Animal Science, Rua Pioneiro, 2153, Jd. Dallas, Palotina, Paraná, 85950000 Palotina, Brazil; americo.garcez@ufpr.br

Different from feed conversion ratio (FCR), the residual feed intake (RFI) is based on intake adjusted to the mean metabolic weight (MMW) and average daily gain (ADG) of each animal, improving the selection of animals with lower intake to the same level of production of the herd. The trial was carried out to evaluate the performance of lambs with divergent efficiency for RFI during the growing and finishing phases. Twenty four crossbred ½ Dorper × ½ Santa Inês non-castrated male lambs with four months of age and 24.1±3.2 kg of body weight (BW) were used. Lambs were fed *ad libitum* with diets composed of 64% of *Cynodon* spp. hay and 36% of concentrate, with 15.8% of crude protein (CP) and 67.0% of total digestible nutrients (TDN) on dry matter (DM) basis. A completely randomized design was used for analysis of variance with two treatments based on two classes of RFI, one efficient (RFI from -0.9 to -89.4 g DM/d) and other inefficient (RFI from 22.7 to 81.2 g DM/d). Pearson correlation analysis was run between the RFI and the other performance and feed efficiency traits. There was correlation (P<0.05) between dry matter intake on body weight basis (DMI_{BW}) and RFI (r=0.78). The efficient group had lower (P<0.05) DMI_{BW} (3.18% vs 3.46% BW/d). This result may be due to higher (P<0.05) gross feed efficiency (GFE) to the first group, with lower feed intake to BW maintenance and growth. Although not significant, the dry matter intake (DMI) for the efficient group was 72 g less than the inefficient group (919 vs 991 g/d). There were not correlations between RFI and FCR and GFE. However, the efficient group had better (P<0.05) FCR and GFE (7.16 vs 8.11 g DM/g of gain; and 141.1 vs 125.0 g gain/kg DM, respectively). These values represent the better capacity of the efficient lambs in using the feed. The RFI is indicated to sheep selection allowing choosing animals with better FCR and GFE without affect their productive performance, which reflects positively on feedlot profitability for lambs in growing and finishing phases.

Evaluation of the suitability of a marker for oxidative stress in dairy cows of a dual purpose breed

C. Urh[1], T. Ettle[2], E. Gerster[3], U. Mohr[4] and H. Sauerwein[1]

[1]University of Bonn, Institute for Animal Science, Katzenburgweg 7-9, 53115 Bonn, Germany, [2]Bavarian State Research Center for Agriculture, Institute for Animal Nutrition, Prof.-Duerrwaechter-Platz 3, 85586 Poing, Germany, [3]Landwirtschaftliches Zentrum Baden-Württemberg (LAZBW), Atzenberger Weg 99, 88326 Aulendorf, Germany, [4]Center for Agricultural Learning, Markgrafenstraße 1, 91746 Weidenbach, Germany; urh@uni-bonn.de

To evaluate the suitability of derivates of reactive oxygen metabolites (dROM) as universal marker or predictor for distressed cows, we characterized its variability as affected by time, different farms, and feeding regimens. 118 multiparous lactating Simmental cows from three different experimental farms (A: n=26; B: n=43; C: n=49) were allocated in equal shares per farm to receive either low or high amounts of concentrate relative to their milk yield, i.e. 150 or 250 g/kg of energy corrected milk. All cows were fed *ad libitum* with a diet in which the roughage portion had 6.5 MJ NEL/kg of DM. The dROM concentration was measured in serum samples from d-50 (dry off), -14, +8, +28, and +100 days relative to calving. The results are given as H_2O_2 equivalents (means ± SEM). The linear mixed model (dependent variable: dROM; fixed effects: time, feeding, farm, and their interactions; random effect: cow; Post-Hoc: Bonferroni) was used for statistical analyses (SPSS 24). Significance was declared at P<0.05. Time and farm had a significant effect with lowest values on day -14 (74.0±3.4 µg H_2O_2/ml) and lower values in B (76.6±3.7 µg H_2O_2/ml) than in C (91.0±3.6 µg H_2O_2/ml). Neither feeding nor the interactions tested were significant. The oxidative status in this study showed a nadir close to parturition as previously described and was not influenced by the different feeding regimens. Animals showed similar time courses but different levels in the different farms. Therefore it could be problematic to set general thresholds for dROM as a biomarker but assessing its potential association with other variables assessed in blood (e.g. fatty acids and ketone bodies) is needed for a final conclusion. The study is part of the optiKuh project funded by the German Federal Ministry of Food and Agriculture (BMEL).

Plasma homocysteine concentrations vary due to group in commercially raised Australian pigs

K. Di Giacomo, J.J. Cottrell, F.R. Dunshea and B.J. Leury

The University of Melbourne, Agriculture and Food Systems, Parkville, Vic., 3010, Australia; kristyd@unimelb.edu.au

Homocysteine (HCY) is an intermediary product of the methionine cycle that provides methyl groups to DNA, protein and methylation reactions, thus influencing many cellular processes. HCY is not incorporated into protein but is recycled to methionine, undergoes tissue specific transulfuration or is excreted by the kidney. If methyl donor supply is inadequate there is an increase in remethylation of HCY. This experiment quantified the methyl donor balance of commercial pigs of various growth stages from the Australian pig herd. Blood samples were obtained via jugular venipuncture on one occasion from four piggeries (n=30 per site, per group). Plasma HCY concentrations were quantified via LC-MS and creatinine via HPLC; with plasma HCY >20 µM considered abnormal. Statistical analysis was undertaken using the REML function in GenStat. The majority (71%) of pigs sampled had HCY concentrations above 20 µM. Plasma HCY was greatest in weaners, lowest in sows and intermediate in growers and finishers (129 vs 46, 56 and 21±9.1 µM for weaners, growers, finishers and sows respectively, P<0.001). Plasma creatinine was lowest in weaners, greatest in lactating sows and intermediate in growers and finishers (300 vs 376, 402 and 525±42.0 µM for weaners, growers, finishers and sows respectively P<0.001). Weaning pigs had the highest HCY concentration which is likely not due to reduced renal clearance as weaning pigs had the lowest creatinine concentration. This may reflect reduced recycling of methionine or transulfuration to cysteine with consequences for oxidative stress. Plasma HCY was lowest in lactating sows who also had the highest creatinine concentration, perhaps due to increased muscle catabolism. That hyperhomocysteinemia occurs in weaners and not sows may reflect that weaning causes significant disruption of the gastrointestinal tract, which is a primary site of methionine cycling and transulfuration. Conversely, lactation is marked by substantial mobilisation of tissue stores, which does not appear to influence HCY metabolism. Thus, HCY concentrations in Australian pigs appear to be abnormally high and elevated HCY in weaner pigs may reflect deficiencies in methyl donors.

Binding and distribution of cortisol in bovine and equine saliva and plasma during ACTH challenge

A.-C. Schwinn[1], F. Sauer[2], V. Gerber[2], J.J. Gross[1] and R.M. Bruckmaier[1]
[1]Veterinary Physiology, Vetsuisse Faculty, University of Bern, Bremgartenstrasse 109a, 3012 Bern, Switzerland, [2]Swiss Institute for Equine Medicine (ISME), Vetsuisse Faculty, University of Bern, Länggassstrasse 124, 3012 Bern, Switzerland; ann-catherine.schwinn@vetsuisse.unibern.ch

Plasma cortisol represents the hypothalamic-pituitary-adrenocortical (HPA) axis activity. While most plasma cortisol is supposed to be bound, only free cortisol (FC) is involved in the metabolic and immunological regulation. Its proportion relative to total cortisol (TC) is affected by various circumstances such as stress. We have established a method to assess FC in cows and horses to interpret proportional changes of cortisol in saliva and blood following controlled HPA axis activation. Synthetic ACTH was i.v. administered to 8 dairy cows (0.16 µg/kg BW) and 5 horss (1 µg/kg BW) through a jugular catheter. Blood and saliva was collected every 30 min for 3 h. Plasma TC was measured by RIA, and saliva cortisol (SC) by ELISA. An ultrafiltration (UF) technique was used to determine the proportion of unbound cortisol (pUC) and hence to calculate FC in plasma (cFC). Pearson's correlation coefficients were calculated and differences between species were evaluated with the t-test. Data are means ± SD. Throughout the ACTH test, TC was paralleled by cFC, pUC, and SC in both cows and horses. Basal TC concentrations were 8.3±5.2 ng/ml in cows and 21.8±7.0 ng/ml in horses. Peak TC concentration reached 63.2±9.6 and 73.2±11.8 ng/ml in bovine and equine plasma, resp. Basal SC and pUC were similar in both species (1.0±0.7 vs 0.8±0.1 ng/ml; 9.0±0.6 vs 10.2±2.3%, resp.). In conclusion, the short-term activation of the HPA axis causes not only an elevation of TC, but also increases, similar in both species, the proportion of free cortisol. UF is a suitable method for assessing proportional changes of free and total cortisol following HPA axis activation.

Optimisation of claw traits evaluation in accordance to claw and metabolic disorder of dairy cows

M. Vlček and R. Kasarda
Slovak University of Agriculture in Nitra, Department of Animal Genetics and Breeding Biology, Tr. A. Hlinku 2, 949 76 Nitra, Slovak Republic; xvlcekm@uniag.sk

The Aim of the study was to optimize use of digital image analysis in evaluation of claw formation after regular claw trimming and association of observed claw parameters to production, reproduction traits and metabolic disorders. Digital images of 127 animals were studied. Images were taken during animals' fixation in vertical trimming box. Two images were obtained per one animal on right rear leg: from the bottom and outer lateral side. Production and reproduction data were taken from official milk recording provided by Breeding Services of Slovak Republic s.e. Bratislava. Digital images were analysed using NIS software (Nikon). Highest significant positive correlations were observed between claw height and length as well as total claw area to functional area. Highest negative correlation was observed in contrary between claw angle and width. Effect of fat to protein ratio (F/P ratio) obtain from test-day records affected prevalence of interdigital dermatitis and heel erosion (IDHE), digital dermatitis (DD) and sole ulcer (SU). With use of animal model, was based on observed claw measures estimated heritability of claw angle by 0.1, claw length by 0.2, claw height by 0.03 and claw diagonal by 0.02. Claw parameters and F/P ratio might be useful to achieve the correct gait and decrease the occurrence of claw disorders. Future strategies for selection of animals fitting better into the present production system, with higher importance of type traits and fitness to increase longevity and production life length are expected, because decreased herd replacement will increase total profitability of dairy farms.

Genetic evaluation of claw health – challenges and recommendations

B. Heringstad[1], C. Egger-Danner[2], K.F. Stock[3], J.E. Pryce[4], N. Gengler[5], N. Charfeddine[6] and J.B. Cole[7]
[1]Norwegian University of Life Sciences, Department of Animal and Aquacultural Sciences, Ås, 1432, Norway, [2]ZuchtData, EDV-Dienstleistungen GmbH, Vienna, Austria, [3]IT Solutions for Animal Production, (vit), Verden, Germany, [4]La Trobe University, Agribio, Bundoora, VIC, Australia, [5]Gembloux Agro-Bio Tech, University of Liège, Gembloux, Belgium, [6]CONAFE, ctra. Andalucia, Valdemoro, Spain, [7]Animal Genomics and Improvement Laboratory, United States Department of Agriculture, Beltsville MD, USA; bjorg.heringstad@nmbu.no

Routine recording of claw health status at claw trimming of dairy cattle have been established in several countries, providing a valuable database for genetic evaluation. In this review, issues related to genetic evaluation of claw health are examined; data sources, trait definitions, and data validation procedures are discussed; and a review of genetic parameters, possible indicator traits, and status of genetic and genomic evaluations for claw disorders are presented. Different sources of data and traits can be used to describe claw health, including veterinary diagnoses, data from lameness and locomotion scoring, activity-related information from sensors, and foot and leg conformation traits. The most reliable and comprehensive information is probably data from claw trimming. Heritability of the most commonly analyzed claw disorders based on data from routine claw trimming were in general low, with linear (threshold) model estimates ranging from 0.01 to 0.14 (0.06-0.39). Estimated genetic correlations among claw disorders varied from -0.40 to 0.98. The strongest genetic correlations were found among sole hemorrhage (SH), sole ulcer (SU), and white line disease (WL), and between digital/interdigital dermatitis (DD/ID) and heel horn erosion (HHE). Genetic correlations between DD/ID and HHE on the one hand and SH, SU, or WL on the other hand were low in most cases. Although some studies were based on relatively few records and the estimated genetic parameters had large standard errors, results were generally consistent across studies. Routine genetic evaluations of direct claw health have been implemented in The Netherlands (2010), Denmark, Finland and Sweden (joint Nordic evaluation; 2011), and Norway (2014), and other countries plan to implement this in the near future.

Evaluation of different data sources for genetic improvement of claw health

C. Egger-Danner[1], A. Koeck[1], J. Kofler[2], J. Burgstaller[2], F. Steininger[1], C. Fuerst[1] and B. Fuerst-Waltl[3]
[1]ZuchtData EDV-Dienstleistungen GmbH, Dresdner Straße 89/19, 1200 Vienna, Austria, [2]University of Veterinary Medicine, Austria, Veterinärplatz 1, 1210 Vienna, Austria, [3]University of Natural Resources and Life Sciences Vienna, Gregor Mendel-Straße 33, 1190 Vienna, Austria; astridkoeck@gmx.net

Various studies worldwide have shown the need to work on genetic improvement of claw health. The challenge is the availability of phenotypes. Within the Austrian project 'Efficient Cow' various different phenotypes related to claw health have been recorded in 167 farms and about 3.500 Fleckvieh (Simmental), 1.000 Brown Swiss and 1.000 Holstein cows during the observation period of 1 year. Within the project claw trimming was documented and recorded. Additionally lameness was assessed at each time of milk recording by the trained staff of the milk recording organizations using the scoring system of ZINPRO. Within the health monitoring in Austria veterinarian diagnoses are recorded on routine bases. The same is true for culling data. Heritabilities and genetic correlations between the traits have been estimated for veterinarian diagnoses, lameness scores, claw health based on claw trimming data and culling information with a linear animal model. The heritability based on veterinarian diagnoses is 0.025 for Fleckvieh (Simmental) and 0.013 for Brown Swiss. The respective values based on claw trimming data are 0.042 for Fleckvieh (Simmental) and 0.075 for Brown Swiss. The trait was defined as 0/1 or including the number of events as the second value of the results. Based on the lameness score a lactation lameness value was calculated taking the frequency of different severity cases into account. The heritability for this value is 0.092 for Fleckvieh and 0.109 for Brown Swiss. The genetic correlations between veterinarian diagnoses, claw trimming data and lactation lameness value are in the range of 0.3-0.9. The results confirm the usability of various data sources for genetic improvement of claw health.

Assessment of the footbaths contamination by dairy cattle manures under field conditions
J.M. Ariza[1,2], N. Bareille[2], K. Oberle[1] and R. Guatteo[2]
[1]*Qalian, Neovia group., Rue Jean Monet, Segré, 49500, France, [2]BIOEPAR, INRA, Oniris., La Chantrerie, 44307 Nantes, France; juan-manuel.ariza@oniris-nantes.fr*

Among the strategies for the control of foot infectious diseases, footbaths represent a useful alternative to treat concomitantly an important numbers of animals against conditions associated to lameness in ruminants. However, the active compounds used in footbaths can be challenged against incrasing number of cow passages and therefore, to different volume losses and levels of manure contamination which could alter their activity. The objective of this study was to explore different physico-chemical variables of footbaths under field conditions The study was carried out in 6 dairy cattle farms from western France. The hygienic status of the farms was determined by the overall feet cleanliness score of the herd. A footbath filled with water was placed at the usual location of each farm. Footbath samples were taken after 0, 50, 100, 150 and 200 cow passages to analyze the evolution of the microbial loads and the OM concentration. The number and moment of cow defecations were recorded. During the sampling, the ambient and inside temperature, the pH and the approximate residual volume of the footbath were recorded. The results indicate that the microbial loads and the OM concentration increased mainly as the number of passages of animals rose, and to a lesser extent, with the number of defecations. This increase was independent of the hygienic status of the farms, indicating that the number of cow passages influenced the OM concentration in footbaths and not the dirtiness of the feet. On average 6% of cows defecated in footbaths. Only slight variations on the temperature and the pH measures across the different numbers of cow passages were recorded. However, the volume decreased drastically in the order of 40 to 50% after 200 cow passages. The OM concentration after 150 and 200 cow passages reaches the maximal concentrations in which the bactericidal efficacies of footbath products are tested (20 g/l), indicating the relevance of this renewal rates. Nevertheless, the renewal rates must be mainly adapted according to the footbath remaining volume, provided that the entire foot should be cover by the footbath solution.

Gender and seasonal differences in bone mineral density in swine
M. Bernau, J. Schrott, S. Schwanitz, L.S. Kreuzer and A.M. Scholz
Ludwig-Maximilians-Universität München, Livestock Center Oberschleissheim, St. Hubertusstrasse 12, 85764 Oberschleissheim, Germany; Maren.Bernau@lmu.de

Bone mineral density (BMD) and bone mineral content (BMC) are used as indicators for osteoporosis or osteochondrosis in humans. Both diseases affect health due to lameness and pain. In pigs lameness and pain are an important issue in pig fattening and animal welfare. The present study aimed at evaluating BMD/BMC during growth in fattening pigs of three male gender types in three seasons. A total number of 101 male pigs were used. The data consisted of 34 entire boars (EB), 34 immunocastrated boars (IB), and 33 barrows (CB); all piglets of a Piétrain sire mated with German Landrace sows. They were examined in 3 experimental groups representing different seasons, with gender type equally distributed. All pigs were scanned via Dual Energy X-ray Absorptiometry (DXA) to evaluate BMD at an age of 30, 60 and 90 kg body weight. Data were analysed by creating 9 body parts: total body, head, left & right front leg, loin area, left & right femur and left & right lower hind leg. A mixed model was used to calculate LSM + SEE by using a REML procedure. Gender and group were used as fixed effect and age and weight as random effect. The results showed significant differences between the gender types. EB and IB showed the significant lowest BMD with 60 and 90 kg (90 kg; EB: 0.958 ± 0.008 g/cm^2, IB: 0.974 ± 0.007 g/cm^2, CB: 0.991 ± 0.007 g/cm^2). Additionally, significant differences were found between the experimental groups already at 30 kg, showing seasonal differences. Animals raised in a season with increasing daylight showed the highest BMD. Seasonal and gender differences could be detected regarding BMD. It is necessary to evaluate whether a low BMD affects leg health, as it is published in human medicine where declining bone density is surrogate for declining bone strength. If this is the case, the differences could result in more lame animals in a season or especially in finishing a special gender type. Further studies including lameness scorings are necessary. Acknowledgements: This project was financially supported by the Lehre@LMU project of the Ludwig-Maximilians-Universität Munich (LMU), funded by the German Ministry of Education and Research (grant 01PL12016).

Verification of the sample size of the Welfare Quality® protocol applied to sows and piglets

L. Friedrich[1], I. Czycholl[1], N. Kemper[2] and J. Krieter[1]
[1]Institute of Animal Breeding and Husbandry, Christian-Albrechts-University, Olshausenstr. 40, 24098 Kiel, Germany,
[2]Institute for Animal Hygiene, Animal Welfare and Farm Animal Behaviour, University of Veterinary Medicine Hannover,
Foundation, Bischofsholer Damm 15, 30173 Hannover, Germany; lfriedrich@tierzucht.uni-kiel.de

The Welfare Quality® (WQ) protocols are animal-based on-farm assessment tools for welfare. Aim of the study was to verify the size of the random samples recommended in the protocol. Data collection took place on 12 farms in Northern Germany, on which in total 40 protocol assessments were performed. This study focused on scans for individual parameters (IP) in gestating sows. According to the protocol a random sample of 30 gestating sows is selected. Following the instructions of the WQ protocol, IP are assessed on only one side (the side better visible at that moment) of the sow. In the present study, not only a random sample, but all gestating sows were assessed on both sides of the body for all IP. These data were used to calculate the overall frequency of each IP for each of the 40 visits. The visits were ranked from 1 to 40 for each IP regarding their frequency. Then, random samples were drawn from the totality of the animals of each visit, whereby left and right side was also chosen randomly. For each sample, the frequency of each IP was calculated. Within these samples, again the farms were ranked. The ranks of the visits (RV) were compared with the ranks of the samples (RS). The agreement in percent between RV and RS was calculated, whereby exact agreement between RV and RS was assessed as well as agreement within a tolerance of ±4 ranks (equivalent to ±10%). Samples with the recommended sample size of WQ (30 sows) showed insufficient exact agreement of RV and RS in all IP (e.g. bursitis: 24.2% agreement). Applying a tolerance of ±4 ranks, agreement of RV and RS improved (e.g. bursitis: 66.5% agreement). It was determined that agreement with applied tolerance must be ≥90% in order to be acceptable. If this was not the case, sample size was increased in steps of 1 until appropriate agreement was reached (e.g. bursitis: 90% agreement at sample size of 63). In conclusion, regarding the assessment of IP in gestating sows, a larger sample size should be considered in the WQ protocol.

Validity of behavioural scans included in the Welfare Quality® protocol for growing pigs

I. Czycholl, K. Büttner and J. Krieter
Institute of Animal Breeding and Husbandry, Christian-Albrechts-University Kiel, Olshausenstraße 40, 24118 Kiel,
Germany; iczycholl@tierzucht.uni-kiel.de

For the evaluation of social and investigative behaviour, behavioural observations (BO) are included in the Welfare Quality® protocols. Thereby, positive and negative social behaviour, use of enrichment material, pen investigation, resting and other behaviours are scored. However, it remains questionable if there is an effect of the observer on the behaviour of the pigs and if the analysed time frame is too short to capture the real behaviour. Thus, on 9 farms in Germany, 6 protocol assessments each were carried out by one trained observer, resulting in overall 54 assessments. Additionally the BO were carried out on video sequences one and two days after the protocol assessments. In the present preliminary study, 18 assessments are included. For sufficient validity, BO on-farm with a potential observer effect should reflect the real behaviour of the pigs. Moreover, a certain consistency over time is expected. Thus, two comparisons were carried out: (1) on-farm assessment compared to the two days of video analysis; (2) comparison between video analyses. Therefore, Spearman rank correlation coefficients (RS), intraclass correlation coefficients (ICC), smallest detectable changes (SDC) and limits of agreement (LoA) were calculated. In the first comparison, none of the behaviours suggested agreement (e.g. negative social behaviour: RS: 0.17, ICC: 0.04, SDC: 0.80, LoA: -0.77-0.90, use of enrichment material: RS: 0.10, ICC: 0.12, SDC: 0.13, LoA: -0.12-0.16). Regarding the comparison between the video analyses, sufficient agreement was only detected for use of enrichment material (RS: 0.50, ICC: 0.40, SDC: 0.09, LoA: -0.10-0.08). This suggests that BO might only be valid for the evaluation of the use of enrichment material. Regarding pen investigation, a clearer distinction to other active behaviours might enhance reliability and validity. Regarding social behaviours which usually last for a rather short time period, probably longer observation intervals are necessary. In the ongoing study, sample size will be enlarged and whole days of each farm will be observed continuously from video data in order to detect the true occurrences of the behaviours.

Effects of farrowing and rearing systems on welfare indicators of growing and finishing pigs

A. Szulc[1], C. Lambertz[2], M. Gauly[2] and I. Traulsen[1]
[1]Georg-August-University, Livestock Production Systems, Albrecht-Thaer-Weg 3, 37075 Göttingen, Germany,
[2]Free University of Bozen-Bolzano, Faculty of Science and Technology, Piazza Università 5, 39100 Bolzano, Italy;
aszulc@agr.uni-goettingen.de

While group housing of gestating sows is mandatory in the EU, lactating sows are still commonly kept in farrowing crates, which alter welfare of sows but prevent piglet crushing. It is questionable to which extent the housing system during lactation influences piglet's behavior during their later life. In two runs, effects of three farrowing systems, namely farrowing crates (FC), free farrowing pens (FF) and group housing of lactating sows (GH), on welfare indicators of pigs during the rearing and finishing period were studied. Tails were docked in the first, but not in the second run. Two rearing systems were compared: single-phase (wean-to-finish) and conventional two-phase (regrouping at the start of the finishing period) rearing. Finishing pigs were only observed in the 1st run. Skin lesions (SL) and presence of blood, swellings and necrosis on the tail were scored on a scale from 0 to 2 (SL) and 0 to 3 (tail) during the rearing and finishing period. A total of 390 piglets were scored in the 1st and 424 in the 2nd run. In the 1st run, 49.0% of the FC piglets were scored with moderate and severe SL, while 11.4% of the FF and 6.7% of GH piglets showed SL on day 6 of the rearing period. In the 2nd run, 64.3% of the FC and 47.2% of the FF piglets showed moderate and severe SL, while only 3.5% of the GH piglets showed moderate SL. In two-phase rearing, 83.4% of the FC animals showed moderate and severe SL, whereas 65.5% of FF and 67.8% of GH piglets were scored with lesions on day 3 of the finishing period. Only moderate SL (FF: 47.6%, FC: 11.1%, GH: 7.1%) were found in single-phase housed fattening pigs, but no severe. Single-phase reared pigs showed less blood (1.7%), swelling (1.2%) and necrosis (2.2%) during rearing than two-phase reared piglets (4.5%, 4.4% and 4.5%). In conclusion, GH lowered the incidence of SL during the rearing period, especially when tails were not docked. Single-phase reared pigs were scored with less SL on first days of the fattening period and showed less blood, swelling and necrosis on the piglets' tails, than two-phase reared animals.

Prevalence of health and welfare issues in the weaner and finisher stages on 31 Irish pig farms

N. Van Staaveren[1], A. Hanlon[2], J.A. Calderón Díaz[3,4] and L. Boyle[4]
[1]University of Guelph, Department of Animal Science, ON, Canada, [2]University College Dublin, School of Veterinary Medicine, Dublin, Ireland, [3]Institute of Genetics and Animal Breeding, Polish Academy of Sciences, Magdalenka, Poland, [4]Teagasc, Pig Development Department, Moorepark, Fermoy, Ireland; laura.boyle@teagasc.ie

This work describes a large-scale welfare assessment of pigs in different production stages on 31 Irish farms. Ultimately such information could be used to inform herd health and management plans. Farrow-to-finish pig farms were visited between July and Nov 2015 to assess pig welfare using an adapted version of the Welfare Quality protocol. On each farm 6 randomly selected pens in first weaner (S1, 4-8 wks), second weaner (S2, 8-13 wks) and finisher stage (S3, 13-23 wks) were observed for 10 min during which time the number of pigs affected by different welfare outcomes was recorded. The percentage (median [IQR]) of pigs affected was calculated to identify the most prevalent outcomes within each stage. Differences between stages were analyzed using generalized linear mixed models with pen within stage and farm as a random effect. Large variation was observed for the recorded welfare outcomes, reflecting different challenges pigs experience as they grow. In S1 welfare outcomes with higher prevalence were those related to post-weaning stress i.e. poor body condition (4.4% [3.5-7.1%] of pigs affected), sickness (1.5% [0.9-2.5%]), and huddling (3.7% [0.0-7.3%]). In S2 and S3 outcomes associated with injurious behavior i.e. tail (S2: 5.9% [4.1-7.7%]; S3: 10.5% [8.4-13.2%]), ear (S2: 9.1% [2.6-26.4%]; S3: 3.3% [1.8-14.3%]) and flank (S2: 0.4% [0.0-0.9%]; S3: 1.3% [0.0-3.5%]) lesions were more common. Additionally, lameness (S2: 0.8% [0.0-1.1%]; S3: 1.1% [0.0-1.9%]), bursitis (S2: 3.9% [2.7-6.2%]; S3: 7.5% [4.9-11.6%]), and hernias (S2: 1.6% [0.8-2.4%]; S3: 1.8% [1.0-3.5%]) were more prevalent in these stages reflecting heavier weights and higher growth rates of pigs in these stages. These findings indicate where changes in management are needed. For example, identification of sick pigs in pens suggests that practices relating to euthanasia and/or the use of hospital pens need to be reconsidered. Furthermore, better prevention and intervention strategies are required to address problems with injurious behavior.

An ethogram of biter and bitten weaner pigs during ear biting events by automatic video recordings

A. Diana[1,2,3], L. Carpentier[3], D. Piette[3], L. Boyle[1], E. Tullo[4], M. Guarino[4], T. Norton[3] and D. Berckmans[3]
[1]Teagasc, Moorepark, Co. Cork, Ireland, [2]UCD, School of Veterinary Medicine, Dublin, Ireland, [3]KU Leuven, M3-BIORES, Leuven, Belgium, [4]UniMi, Dept. of Veterinary and Technological Sciences for Food Safety, Milan, Italy; laura.boyle@teagasc.ie

There are reports that ear biting (EB) is increasing on intensive pig farms. The lack of research on this behaviour and on the resulting ear lesions (EL) poses a challenge in trying to address this welfare issue. Furthermore, performance of pigs affected by EL does not appear to be compromised making it difficult to convince producers of its importance. So far on-farm assessments of damaging behaviour are made by direct observation. This procedure is expensive, time-consuming and potentially inaccurate. PLF technologies could help in the development of intervention strategies for EB. One of the first steps in developing an automatic monitoring system for the detection of EB events is to identify reliable behavioural indicators. Hence the objective of this study was to identify and describe specific behaviours performed by biter (BR) and bitten (BT) pigs during an EB event on a commercial farm. Six pens of 35 grower pigs were video recorded for 13.2 h (2.2 h/pen). Videos were observed by one person and following review of EB events an ethogram was developed for BR and BT pigs. BR behaviours were identified and described as quick biting (short duration bites directed towards penmate's ears), shaking head (lateral movement of the head with penmate's ear in mouth), pulling the ear (taking hold of penmate's ear in the mouth and moving it towards itself) and chewing (prolonged mastication of the penmate's ear). All were accompanied by a reaction by the bitten pig. These were described as head knocking (forceful and rapid vertical action of the head against the body of the biter), moving head away (lateral/vertical movement of the head away from the biter), moving/walking away and vocal response. 'Gentle manipulation' was also recognized (received by BT and performed by BR). Understanding the behaviours is a first step in the development of an automatic monitoring system. In fact, these ethograms will be used for the labelling procedure to identify reliable features of the behaviours for the development of an algorithm.

Is tail biting in growing pigs influenced by the amount of crude fibre in feed ration?

A. Honeck[1], I. Czycholl[1], O. Burfeind[2] and J. Krieter[1]
[1]Institute of Animal Breeding and Husbandry, Christian-Albrechts-University, Olshausenstr. 40, 24098 Kiel, Germany, [2]Futterkamp, agricultural research farm of the Chamber of Agriculture of Schleswig-Holstein, Gutshof 1, 24327 Blekendorf, Germany; ahoneck@tierzucht.uni-kiel.de

A higher crude fibre content in the feed ration offers better conditions for the gut microflora and prevents the genesis of gastric ulcers and thus can lead to an improved animal health. Moreover, due to its longer time for digestion a longer feeling of satiety is given. As tail biting has multifactorial causes such as health issues and dissatisfaction, the influence of different crude fibre contents in the feed ration on the occurrence of tail biting was analysed. The study comprises five batches with overall 480 pigs kept in conventional pens. Until weaning all pigs were fed the same prestarter. During the rearing period four experimental groups were investigated: A control group (CR) with a conventional feed (up to 3.4% crude fibre), two groups with an increased crude fibre content up to 5% (G5) resp. 6% (G6) and a group with conventional feed and crude fibre provision *ad libitum* (AL). All feed rations contained the same energy content (12.1-14.0 MJ) and crude protein (14.2-17.3%). Tail lesions (unharmed; scratches; small lesions; large lesions) and tail losses (intact; tip loss; partial loss with more than one third lost; partial loss with more than two thirds lost; complete loss) were scored weekly. Directly after weaning, in all four groups, 22-33% of the tails showed scratches or more severe lesions which occurred again in the 5th week after weaning. Only in G6, large tail lesions occurred already in the 4th week after weaning. Tail losses appeared mostly in the 4th or 5th week after weaning except for G5 which showed no tail losses during the rearing period. Furthermore, tail losses appeared only within the 3rd to 5th batch. A possible explanation for the result of the 1st and 2nd batch is that these pigs were marked twice a week for video analysis and were thus more handled. In summary, G6 had the highest level of tail losses at the end of rearing (8.6%) and G5 had the lowest level (0%). Thus, this preliminary analysis suggests that the level of crude fibre in the feed ration had only a small effect on tail biting and that the batch effect was most influential.

The impact of straw ration on tail biting and other pig directed behaviours in undocked growing pigs

T. Wallgren, A. Larsen and S. Gunnarsson
Swedish University of Agricultural Sciences (SLU), Department of Animal Environment and Health, P.O. Box 234, 53233
Skara, Sweden; torun.wallgren@slu.se

According to the Council Directive 2008/120/EC tail docking may not be carried out routinely. Still, the majority of the pigs in within EU remain docked. There is a gap in knowledge about how to rear pigs with intact tails without risking tail biting outbreaks, which makes the transition from docking elongated. It has previously been determined that straw provision to pigs reduces tail biting also in pigs with intact tails. The objective of the study was to evaluate the impact of an increased straw ration on tail biting and pig directed behaviours in pig farms in Sweden, where tail docking is prohibited. Previous studies in Sweden showed that tail biting is kept at a low level (~2-3%) even though the straw rations were considered low (8-85 g/pig for growers and 9-225 g/pig for finishers). The experiment was conducted on five commercial farms; three units with growers (10-30 kg) and four units with finishing pigs (30-115 kg). On two farms, the pigs were followed from the grower stable to the finishing pig stable. All pigs were undocked. Each unit was divided into control and experiment group. The control groups were given the farm normal amount of straw (on average 12 g/pig for growers and 9 g/pig for finishing pigs). The experimental group were given a doubled amount of straw (~27 vs 18 g). The aim was to see if the increased straw ration had an impact on the behaviour and tail lesions of the pigs. The hypotheses were that an increased amount of straw would lead to a decrease in unwanted behaviours (e.g. pig directed behaviour) and an increase in exploratory behaviours and a reduced amount of lesions on the ears and tails. The results of this study are currently being processed but the preliminary results show that pig in the experimental group has a numerically higher proportion of active pigs manipulating straw, but also higher percentage of pigs exploring the pen fittings.

Behavioural adaptation of growing pigs to illuminated and dark pen compartments

S. Ebschke, P. Koch and E. Von Borell
Martin-Luther-University Halle-Wittenberg, Theodor-Lieser-Str. 11, 06120 Halle, Germany;
stephan.ebschke@landw.uni-halle.de

The effects of light intensity on the rhythmicity and behavioural activity of growing pigs are not well documented. Results from our previous studies indicate that a 8-h daylight intensity of 600 lux under laboratory conditions affected behavioural activities that are associated with increased skin lesion scores. As a follow up study we gave 2 groups of 10 pigs a choice between two identical pen compartments (2.5×1.7 m) that were either kept under a 12L:12D-h standard light regimen of 80 lux or permanent darkness (<1 lux) in a climatically controlled room. Four wk old pigs were first adapted to uniform 12L:12D conditions in all compartments for the first wk. Behavioural states (lying, locomotion) as well as agonistic activities, play and compartment use were analyzed on 2 days in each of wks 1, 2 and 5 between 6:30-8:30 and 14:00-17:00. Data were analyzed with SAS 4.9 using a linear mixed model considering compartment, wk, the interaction between the two and random individual effects. Behavioural activities and preferences for compartments in which these are performed were consistently determined by the initial communal resting behaviour during the first week under uniform 12L:12D conditions. These preferences were consistent ($P<0.001$) for all behaviours, ranging from <50 to >250 min of total time pigs spent in individual compartments during the first wk of observation. Pigs did not significantly change resting behaviour from their initially preferred compartment to another when they had a choice between illuminated and dark compartments. However, a significant change in preference was observed for locomotion, play and agonistic activities with time (wk1 vs wks 2 and 5, $P=0.0019/P=0.045$, $P<0.0001$, $P=0.705/P=0.0061$). Locomotion and play was most frequently observed in illuminated pen compartments during wk 5. It can be concluded that the preferences of pigs for resting and other activities in illuminated or dark pen compartments seems to be strongly predetermined by their initial selection of a communal lying area. In further studies we will assess the threshold level of light intensities at which pigs avoid illuminated areas for resting as our previous studies with light intensities of 600 lux indicated that these are perceived as aversive.

Relationships among mortality, performance, and disorder traits in broiler chickens

D.A.L. Lourenco[1], X. Zhang[1], S. Tsuruta[1], S. Andonov[2], R.L. Sapp[3], C. Wang[3] and I. Misztal[1]
[1]University of Georgia, 425 River Rd, Athens, GA 30602, USA, [2]University Ss Cyril and Methodius, 9 Goce Delcev, 1000 Skopje, Macedonia, [3]Cobb-Vantress, Inc., US-412, Siloam Springs, AR 72761, USA; danilino@uga.edu

Four performance related traits [growth trait (GROW), feed efficiency trait 1 (FE1) and trait 2 (FE2), and dissection trait (DT)] and 4 categorical traits [mortality (MORT) and 3 disorder traits (DIS1, DIS2 and DIS3)] were analyzed using linear and threshold single- and multiple-trait models. Field data included 186,596 records for commercial broilers from Cobb-Vantress, Inc. Genotypes for 60k SNP were available for 18,047 birds. Average-information REML and Gibbs sampling-based methods were used to obtain estimates of the (co)variance components, heritabilities, and genetic correlations in a traditional approach without using genomics. The ability to predict future breeding values was checked for genotyped birds in the last generation, when traditional BLUP and single-step genomic BLUP (ssGBLUP) were used. Use of genomic information increased realized accuracy for all 8 traits, especially performance traits. A gain of 18 points over BLUP was observed for performance, whereas the gain for MORT and disorders was 7 points. Heritabilities for GROW, FE1, and FE2 in single- and multiple-trait models were similar and moderate (0.22 to 0.26) but high for DT (0.48 to 0.50). For MORT, DIS1, and DIS2, heritabilities were 0.13, 0.24, and 0.34, respectively; estimates were similar for single- and multiple-trait models. However, heritability for DIS3 was higher from the single-trait threshold model than for the multiple-trait linear-threshold model (0.29 vs 0.19). Genetic correlations between growth traits and MORT were weak, except for maternal GROW, which had a moderate negative correlation (-0.50) with MORT. The genetic correlation between MORT and DIS1 was strong and positive (0.77). Feed efficiency 1, which was moderately heritable (0.25) and is highly selected for, was not genetically related to MORT of broilers and other disorders. Broiler MORT also had low to moderate heritability (0.13), which suggests that MORT and FE1 can be improved through selection without negatively impacting other important traits. Selection of heavier maternal GROW also may decrease offspring MORT.

A pipeline to annotate PDF documents with different ontology concepts related to farm animal welfare

N. Melzer[1], S. Trißl[1], M. Zebunke[1], M. Joppich[2] and R. Zimmer[2]
[1]Leibniz Institute of Farm Animal Biology (FBN), Institute of Genetics and Biometry, Wilhelm-Stahl-Allee 2, 18196 Dummerstorf, Germany, [2]Ludwig-Maximilians-Universität München (LMU), Institute of Informatics, Amalienstr. 17, 80333 München, Germany; melzer@fbn-dummerstorf.de

Scientific knowledge is usually published in text form and made available as PDF documents. This way, researchers are able to describe and classify their work and share it with the community. The disadvantage of text publication is that it is an unstructured way to share knowledge and thus makes it difficult to automatically extract knowledge such as behavior test methods to evaluate farm animal welfare. In contrast, an ontology contains highly structured knowledge of a certain domain with hierarchically structured concepts (definitions and keywords). The 'Animal Trait Ontology for Livestock' (INRA, France) contains concepts for traits in farm animals, including concepts for animal welfare. We use also an ontology for behavioral test methods, which we started to develop. Our aim is to investigate how precisely knowledge can be revealed using more than 2,500 publications in PDF format and both ontologies. To achieve this goal the following has to be done: (1) transform PDF documents into structured text using PDFEx (FBN Dummerstorf). PDFex, in contrast to other freely available tools, produces structured text in JATS format, reflecting the usual structure of a scientific article. (2) Annotate the structured text with synonyms corresponding to ontology concepts using syngrep (LMU München). (3) Perform a co-occurrence analysis to identify commonly occurring concepts of the behavior test method and animal welfare ontology. Finally, we evaluate the outcome by manually checking the annotations. In general, the success depends on the goodness of the applied tools, the structure of the PDF and the completeness of the ontologies. Hence, we will present the advantages and disadvantages of the applied tools. This study is helpful to create or enhance ontologies by using one's own PDF library.

Responses of bone ash content and bone strength in pigs and poultry on dietary mineral concentration

P. Bikker[1], W. Spek[1], C.V. Nicolaiciuc[1], S. Millet[2] and M.M. Van Krimpen[1]
[1]Wageningen University & Research, Wageningen Livestock Research, P.O. Box 338, 6700 AH Wageningen, the Netherlands, [2]Institute for Agricultural and Fisheries Research (ILVO), Animal Sciences Unit, Scheldeweg 68, 9090 Melle, Belgium; paul.bikker@wur.nl

Phosphorus (P) and calcium (Ca) are essential minerals for adequate bone development in pigs and broilers. Bone minerals provide strength and mechanical rigidity to the bone whereas the organic matrix (mainly collagen) provides flexibility, elasticity and tensile strength. We conducted two dose-response studies, with dietary P content (constant Ca:P ratio) increasing from 50 to 130% of current recommendations in 5 equidistant steps in individually housed growing male and female pigs (25-125 kg BW) and from 75 to 137.5% of current recommendations in 6 equidistant steps in group-housed male Ross 308 broilers (0-39 d of age). Bone characteristics, i.e. dimensions, weight, mineral content in fat-free-dry matter, and breaking strength (with Instron 5,564 materials tester), were determined in metacarpus 3 and 4 of pigs at 50, 80 and 125 kg BW and in tibia of broilers of 10, 21, 30, and 39 d of age. In pigs, metacarpus ash content increased linearly ($P<0.001$) and quadratically ($P<0.05$), while total ash mass, bone breaking strength and stiffness increased linearly ($P<0.001$) with increasing dietary P content. At the same dietary P supply, metacarpus breaking strength and stiffness were substantially higher ($P<0.01$) in female pigs than in entire male pigs. This effect was partly related to a higher bone ash content in the female pigs. In broilers, tibia ash content and total ash mass increased linearly ($P<0.001$) with increasing dietary P content, whereas an increasingly significant linear effect of dietary P content on tibia breaking strength was observed with increasing age of the birds. This may suggest that in young birds breaking strength was less suited to determine bone development. In general, dietary P content required for maximum growth performance was below P intake for maximum bone strength. Hence, minimum or optimum bone characteristics need to be defined. The relevance and value of different characteristics and similarities between pigs and birds will be discussed.

Effects of stocking density on feather pecking and aggressive behaviors in Thai crossbred chickens

P. Na-Lampang[1] and X. Hou[2]
[1]Suranaree University of Technology, School of Animal Production Technology, 111 University Avenue, T. Suranaree, A. Mueang, Nakhon Ratchasima 30000, Thailand, [2]Nakhon Ratchasima Rajabhat University, Veterinary Technology Program, Faculty of Science and Technology, A. Mueang, Nakhon Ratchasima 30000, Thailand; pongchan@sut.ac.th

The purpose of this study was to investigate the influence of stocking density on feather pecking and aggressive behaviors of Thai crossbred chickens from 4 to 12 weeks of age. A total of 900 day-old mixed sex Thai crossbred chickens were assigned as 3 replicates of 100 birds per pen, to stocking densities of 8, 12 and 16 birds/m^2, respectively. Frequency occurrences of feather pecking, number of pecks per bout, pecking intensity, and frequency occurrences of aggressive behaviors were recorded once a week by scanning all the birds in the pen. It was found that stocking density had no effect on the frequencies of feather pecking on body areas except on wings area ($P<0.05$). Stocking density had no effect on occurrence of 1 to 4 pecks per bout or 5 to 9 pecks per bout. Stocking density had no significant influence on pecking, pinching and plucking intensity, except on intensity of pulling. The different types of aggressive behavior such as stand off, fight, threat, leap, chase, avoidance and peck were not affected by stocking density. In conclusion, from our study, stocking density did not affect the feather pecking activities and aggressive behaviors of Thai crossbred chickens.

Effect of housing system on the behaviour of laying hens during elevated temperatures

E. Sosnówka-Czajka, E. Herbut and I. Skomorucha

National Research Institute of Animal Production, Department of Technology, Ecology and Economics of Animal Production, Krakowska Street 1, 32-083 Balice n. Kraków, Poland; ewa.sosnowka@izoo.krakow.pl

The factors that influence avian behaviour include the housing system and climatic conditions. The aim of the study was to compare the effect of housing laying hens in three systems (enriched cage, litter-based system with or without outdoor access) on their behaviour at elevated temperatures during the summer heat. The experiment was performed in summer (June, July, August). Hy-Line commercial layers were assigned to experimental groups according to the housing system: enriched cages (750 cm^2/bird), litter-based system without outdoor access (9 birds/m^2) and litter-based system (9 birds/m^2) with outdoor access (area of 2.5 m^2/bird). During the period of elevated outdoor temperatures and for two days after the end of the heat, behavioural observations were made on: number of birds drinking, eating, sitting, standing, and preening; aggression; feather pecking; cannibalism; perch use; nesting; litter pecking; pecking at other inanimate objects (except for food); panting; standing with wings spread; lying with wings spread. The results were statistically analysed by variance analysis and estimated by chi-square test. The elevated ambient temperatures had an effect on the birds which drank more often, rested more often, and stood individually with raised wings, were less active, and pecked and scratched litter less often. The housing system also determined the behaviour of layers. During the summer heat, hens with outdoor access performed dustbathing more often, were more active, with a greater proportion of litter pecking and scratching birds compared to the other two production systems. Despite the high temperatures, hens from enriched cages were characterized by more frequent feed intake and perch use compared to the other two systems. The results indicate that under elevated rearing temperatures hens activate thermoregulatory behaviour, thus contributing to maintaining thermal homeostasis.

Effect of flooring type in the farrowing crate on reasons for piglet mortality

N. Van Staaveren[1], J. Marchewka[2], M. Bertram[3] and J.A. Calderón Díaz[2,4]

[1]University of Guelph, Department of Animal Biosciences, ON, Canada, [2]Institute of Genetics and Animal Breeding, Polish Academy of Science, Magdalenka, Poland, [3]First Choice Livestock, Des Moines, IA, USA, [4]Teagasc – Pig Development Department, Moorepark, Fermoy, Ireland; j.marchewka@ighz.pl

The aim of this study was to investigate the effects of floor type in the farrowing crate on piglet mortality. Mortality records including number of and reasons for mortality during lactation were available for 3,469 piglets housed on either woven wire (WW, n=1,991 piglets) or slatted steel (SS, n=1,478 piglets) floor during lactation. Piglets originated from 235 sows parity 1 to 13. Six-hundred-and-one piglets died during lactation due to crushing (n=301), piglet unviability (n=153), scour (n=43), starvation (n=32), splay legs (n=2), Streptococcus infection (n=6), savaging (n=14), rupture (n=5) or unknown reasons (n=45). These reasons were categorized as 'crushing', 'unviability' and 'other' due to the low number of dead piglets in the rest of categories. Logistic regressions models were created to examine the effect of floor type on the probability of the different piglet mortality categories. SS piglets were 1.3 times more likely to die during lactation than WW piglets (P<0.01) irrespective of cause of death. When examining mortality reasons, SS piglets were 0.4 times less likely to die due to crushing (P<0.0001) than WW piglets, while no differences were found for the likelihood of unviable piglet mortality (P>0.05). However, SS piglets were 7.15 times more likely to die because of 'other' reasons than WW piglets during lactation (P<0.001). Piglets born in larger litters were at greater risk of mortality during lactation (P<0.001). Additionally, there was a greater risk of dead in early lactation (P<0.001). Results suggest that the use of slatted steel in farrowing crates had a detrimental impact on piglet survival during the lactation period and that it should not be recommended. Other studies have reported that SS piglets sustain more injuries feet and leg lesions that could lead to secondary infections that could affect piglet health.

Implications of mosaicism for deleterious de novo mutations in artificial insemination bulls

H. Pausch[1,2], C. Wurmser[2], S. Ammermueller[2], D. Segelke[3], A. Capitan[4] and R. Fries[2]
[1]*Animal Genomics, ETH Zurich, Tannenstr. 1, 8092 Zurich, Switzerland,* [2]*Animal Breeding, TU Muenchen, Liesel-Beckmann Str. 1, 85354 Freising, Germany,* [3]*vit w.V., Heinrich-Schroeder-Weg 1, 27283 Verden, Germany,* [4]*INRA, Allée de Vilvert, 78352 Jouy-en-Josas, France; hubert.pausch@usys.ethz.ch*

Animals that carry known disease-causing alleles can be detected using haplotype information or direct gene tests from custom genotyping arrays. However, the early identification of carriers of deleterious *de novo* mutations is barely possible because whole-genome sequencing of breeding animals is not routinely performed and the prediction of phenotypes associated with new sequence variants is difficult. Within a short time, breeding consultants noticed a high proportion of calves with lethal congenital malformations among the descendants of two young sires from the Fleckvieh (FV) and Holstein (HOL) breeds. The analysis of 442 and 275 calving records revealed that 140 (32%) and 57 (21%) calves sired by the FV and HOL bull, respectively, were stillborn or perished within 48 h of birth. Clinical and pathological findings of affected calves were similar to those described for osteogenesis imperfecta and lethal chondrodysplasia (bulldog). Because the disorders were detected in such a high proportion of the paternal half-sibs and both sexes were affected, autosomal dominant inheritance of deleterious mutations with mosaicism in the sires was likely. The whole-genome sequencing of both bulls, affected calves and healthy half-sibs revealed that a frameshift and a missense mutation in the *COL1A1* and *COL2A1* genes segregated with osteogenesis imperfecta and chondrodysplasia, respectively. The mutations were heterozygous in affected calves while healthy half-sibs and 1577 animals from the 1000 bull genomes project were homozygous for the reference alleles. Sanger- and pyrosequencing of DNA extracted from blood and semen samples confirmed mosaicism in both sires indicating that the mutations occurred early in development. Our findings show that mosaicism for deleterious mutations in artificial insemination bulls may result in a high number of paternal half-sibs with congenital malformations, particularly when mosaic bulls are used for thousands of inseminations before the birth of their first progeny.

Genotype prediction for a structural variant in Brown Swiss cattle using Illumina Beadchip data

F.R. Seefried[1], P. Von Rohr[1] and C. Drögemüller[2]
[1]*Qualitas AG, Chamerstr. 56, 6300 Zug, Switzerland,* [2]*University of Bern, Institute of Genetics, Bremgartenstr. 109a, 3001 Bern, Switzerland; franz.seefried@qualitasag.ch*

Generally Brown Swiss (BSW) cattle are characterised by solid brown coloured coat, which is devoid of white spotting. However, two hypopigmented coat colour phenotype variants exist at low frequencies. One of these is colour sidedness (Cs), where animals have pigmented sectors on the flanks, ears and snout combined with unpigmented areas on head, legs and spine. In breeding history, colour sided animals have been excluded from the national herdbook since complete pigmentation was declared as a major breed characteristic. Nonetheless, due to dominant inheritance colour sided animals still occur. Homozygous (Cs/Cs) and heterozygous (Cs/+) animals differ by the extent of pigmentation, explaining why colour sidedness represents a semi-dominant trait. Colour sidedness in BSW cattle is caused by a complex structural copy number variant (*Cs6*), namely by a serial translocation of hundreds of kb-sized chromosome segments between BTA6 encompassing the *KIT* gene and BTA29. In total the *Cs6* allele encompasses two duplicated fragments on BTA6 and BTA29 of 3.87 Mb. Illumina Infinium assay provide genotypes and signal intensities for each assayed SNP. Structural variants are easily seen in log R ratio data. Log R ratio values are calculated by the binary logarithm of the ratio between normalized vs expected intensity values. Increased log R ratios relative to the base value represent higher signal intensity due to increased copy number variants in the genomic sequence. A Support Vector Machine (SVM) classification algorithm on Log R ratios from Illumina BovineHD SNPs located within the 3.87 Mb fragment was used to predict BTA6/BTA29 *Cs6*-genotypes in BSW. Confirmed *Cs6*-genotypes were available for a total of 2,003 animals, including 37 heterozygous and 7 homozygous colour-sided animals. Leave-One-Out cross validation with 2,000 replicates resulted in an estimated error-rate of 0.22%. Therefore, Illumina Beadchip data together with a SVM-based classification algorithm seem to be applicable for genotype prediction of larger structural variants. Limiting factors may be SNP-density and size of the structural variant.

Chasing deleterious recessives in Italian Holstein dairy cattle

P. Ajmone Marsan[1], M. Milanesi[1], S. Capomaccio[2], L. Colli[1], S. Biffani[3], J.T. Van Kaam[4], R. Finocchiaro[4], R. Negrini[1], C.J. Rubin[5], A. Nardone[6], N.P.P. Maciotta[7] and J.L. Williams[8]

[1]Università Cattolica del Sacro Cuore, Zootechnics, Via Emilia Parmense 84, 29122 Piacenza, Italy, [2]Università degli Studi di Perugia, Dipartimento Medicina Veterinaria, Via San Costanzo, 06123 Perugia, Italy, [3]Associazione Italiana Allevatori, Via Molini 36, Roma, Italy, [4]Anafi – Associazione Nazionale Allevatori Frisona Italiana, Via Bergamo, 26100 Cremona, Italy, [5]Science for Life Laboratory Uppsala, Uppsala University, Husargtan, Uppsala, Sweden, [6]Universita della Tuscia, Department for innovation in biological, agro-food and forest systems (DIBAF), Via de Lellis, Viterbo, Italy, [7]Università degli studi di Sassari, Dipartimento di Agraria, Via Enrico de Nicola, Sassari, Italy, [8]University of Adelaide, School of Animal and Veterinary Sciences, Faculty of Sciences, Roseworthy, Roseworthy, Australia; paolo.ajmone@unicatt.it

Deleterious recessive variants have been searched in Italian Holstein dairy cattle combining High Density SNP genotypes from more than 1000 progeny tested bulls and exome sequence data from 18 animals sampled from the extremes of the male and female fertility effective daughter performance deregressed proof (EDP) distribution. SNP data were used to identify high linkage disequilibrium haplotype blocks. Haplotypes significantly lacking one class of homozygotes, compared to HWE expectation were classified as Homozygous Haplotype Deficient (HHD). In parallel, variants with a putative deleterious effect were identified from exome data. Candidate deleterious variants mapping within HHD were identified and further investigated in silico to assess their conservation across vertebrates and gene function. Thirteen candidate deleterious variants resided in nucleotides or genomic regions highly conserved across vertebrates and occurred in genes functionally linked to fertility or defects in mouse and human. These are pursued as candidates to influence fertility and fitness in Italian Holstein animals and are being validated to evaluate their use in genome-informed breeding programs.

Reverse genetics to describe a recessive defect in different breeds

C. Grohs[1], P. Michot[2], S. Chahory[3], M.C. Deloche[2], S. Barbey[4], M. Boussaha[1], C. Danchin-Burge[5], S. Fritz[2], D. Boichard[1] and A. Capitan[2]

[1]GABI, INRA, AgroParisTech, Université Paris Saclay, 78350 Jouy-en-Josas, France, [2]Allice, 149 rue de Bercy, 75595 Paris, France, [3]ENVA, Université Paris-Est, Ophtalmology, 7 Avenue du Général de Gaulle, 94704 Maisons-Alfort, France, [4]INRA, Domaine Expérimental du Pin, 61310 Exmes, France, [5]Idele, 149 rue de Bercy, 75595 Paris, France; cecile.grohs@inra.fr

Cattle breeds have a narrow genetic basis favoring emergences of recessive defects. In recent years, detection of cases by dedicated observatories (e.g. ONAB in France) combined with homozygosity mapping based on high density SNP genotyping data have proven to be efficient tools, leading to the characterization of more than 130 genetic defects in cattle. However, considering that more than 4,900 defects have been identified in human, this might be only the tip of the iceberg. Indeed, this approach relies on the observation of affected animals with distinctive symptoms, and mutations resulting in non-specific symptoms or symptoms with little economic importance are likely to be missed. The increasing availability of whole-genome sequences has opened new research avenues such as reverse genetics for the identification of mutations impacting animal's health. Recently, using this top-down strategy, we described 2,489 putative deleterious mutations in 1,923 genes, segregating at a minimal frequency of 5% in at least one of 15 breeds studied. The genes observed to be enriched in this study were mainly associated with nervous, visual and auditory systems, suggesting that those genes have not been taken into account for production purposes. Among them, we identified an ancestral deleterious variant in retinitis pigmentosa-1 (RP1) gene causing progressive retinal degeneration in several breeds. ONAB used the well-established and motivated network of partners to investigate the phenotypic consequences of the frameshift candidate variant. Clinical and functional analysis were performed and permit us to validate the causal mutation. Large scale genotyping showed that the mutated variant is very frequent in Normande breed (27%) but also segregates in other breeds at lower frequency (1.5% in Holstein). The conserved sequence haplotype suggests that the mutation is around 3,800 years old.

Identification and management of recessive genetic defects in Belgian Blue beef cattle

T. Druet[1], A. Sartelet[2], X. Hubin[3], N. Tamma[1], M. Georges[1] and C. Charlier[1]
[1]Unit of Animal Genomics, GIGA-R, 11 avenue de l'Hôpital B34, 4000 Liège, Belgium, [2]Clinical Department of Production Animals, FARAH, Avenue de Cureghem 3, 4000 Liège, Belgium, [3]Association Wallonne de l'Elevage, Rue des Champs Elysées 4, 5590 Ciney, Belgium; tom.druet@ulg.ac.be

The Belgian Blue cattle breed has been intensively selected for extreme muscular development, causing a reduction of its effective population size. Several outbursts of recessive genetic defects have been observed in recent years. With the availability of high-density SNP panels, the Unit of Animal Genomics developed highly effective mapping and identification methods. Causative variants for two recessive defects (congenital muscular dystonia 1 & 2) were identified in a early pilot study (2006) and five more subsequently (crooked tail syndrome, dwarfism, gingival hamartome, prolonged gestation, lethal arthrogryposis syndrome). To improve genome-wide association studies (GWAS), we developed a haplotype-based GLMM suited for binary trait, effective for different scenarii (recessive, dominance, heterogeneity) and robust to isolated misclassifications. Surprisingly, some of the defects segregated at high frequency in the population. Based on segregation analysis and GWAS for selected traits, we provided strong evidence that at least two recessive defects were under balancing selection. More recently, with the reduction of costs for whole genome sequencing, an alternative strategy based on reverse genetic screen was developed to identify defects and embryonic lethal variants. That strategy is particularly important for small populations where screens for depletion in homozygous haplotypes lack of power. Five putative embryonic lethal variants were validated by carrier by carrier crosses. Genetic tests were rapidly implemented and massively used by the breeders (only bulls free of all defects were selected). As a result, a desirable drastic reduction of calf peri-natal mortality was observed but this imposes new constraints affecting genetic diversity. Currently, tools are tested to reduce the risk of mating two carriers of the same recessive defect, either based on genotypes from bulls (sire, maternal grand-sires, etc.) or based on segregation analysis for cows. First results indicate that this can be done effectively.

A review of bioinformatic methods to locate a new recessive mutation on the genome

G.E. Pollott
Royal Veterinary College, Royal College Street, London, NW1 0TU, United Kingdom; gpollott@rvc.ac.uk

The advent of a new recessive Mendelian condition can be critical if it is associated with a lethal or disease phenotype. Such a scenario has several challenges such as little information, few cases and a general fear amongst breeders that their breed will get a bad reputation if word gets out, causing a drop in prices for breeding stock. One might expect that the introduction of molecular genetic technologies would make the situation more tractable but, if the literature reports are anything to go by, this appears not to be the case. Two major contrasting bioinformatic approaches have been suggested to find the location of such novel recessive genetic diseases using molecular genetic markers. The first is based on the use of chi-squared tests (CST) at each marker location and the second utilises runs of homozygosity (ROH). Other candidate gene methods have been used in certain circumstances but this relies heavily on a well-annotated genome and some preliminary information on the likely candidate genes. Methods based on CST may be useful under certain circumstances but suffer from significant locations only being found when one allele segregates with the new mutation and is found at a relatively low allele frequency. Mutations in a ROH in both cases and controls will never be found by this method. There has been much debate about the efficacy of certain ROH methods. The most reliable seem to be those based on measuring ROH in cases using bp lengths and taking into account any similar ROH in controls. Phenotypic permutation allows some degree of probability to be attached to identified regions. Such methods are extendable from SNP to WGS approaches and only require ~10 cases and ~10 controls to detect the region containing the mutation. If this ROH method is used with NGS data there should not be any need for further resequencing to locate the exact position of the mutation. Apart from DNA quality control issues and reliance on a good reference genome these methods also critically depend on good phenotyping at the field level, which is not necessarily as accurate as geneticists would like. This presentation will use real examples to illustrate these methods.

Managing lethal alleles using genomic optimum contribution selection

L. Hjortø[1], J.R. Thomasen[1,2], P. Berg[1,3], M. Kargo[1,4], M. Henryon[5,6], H. Liu[1] and A.C. Sørensen[1]
[1]Aarhus University, Department of Molecular Biology and Genetics, Blichers Allé 20, 8830, Denmark, [2]Viking Genetics, Ebeltoftvej 16, 8960 Randers SØ, Denmark, [3]NordGen Husdyr, Postboks 115, 1431 Ås, Norway, [4]Seges, Agro Food Park 15, 8200 Aarhus N, Denmark, [5]The University of Western Australia, School of Animal Biology, 35 Stirling Highway, CRAWLEY WA 6009, Australia, [6]Seges, Axeltorv 3, 1609 København V, Denmark; achristian.sorensen@mbg.au.dk

We tested the hypotheses that (1) culling animals that carry a recessive lethal allele reduces genetic gain when animals are truncation selected on genetic merit, and (2) optimum-contribution selection (OCS) reduces the frequency of lethal alleles without reducing genetic gain. We tested these hypotheses by simulating rates of genetic gain realized by three selection strategies that reduce the frequency of a single lethal allele: truncation selection, OCS penalising average-genetic relationship based on pedigree information, and OCS penalising average relationship based on genomic information. In each strategy, carriers of the lethal allele were either culled or not culled prior to selection. The strategies were simulated at 1% rate of increase in identity-by-descent averaged across a 30 M genome. We simulated breeding schemes with 10 discrete generations of selection for a single trait with a heritability of 0.2. Only females were phenotyped, all animals were genotyped prior to selection, and genetic merit was predicted using GBLUP. We found that when selection was by truncation selection, culling carriers reduced initial genetic gain by up to 15% compared to no culling of carriers. Genetic gain in subsequent generations was unaffected by culling strategy. On the other hand, when selection was by OCS, culling carriers reduced the frequency of lethal alleles without reducing genetic gain. The reason was that culling carriers distorted the desired distribution of parental contributions in the remaining selection candidates. OCS was able to restore the desired genetic contributions in the selected animals. Truncation selection did not have this tidying function and allowed the distorted distribution to persist. This implies that OCS should be used in breeding schemes, where carriers of lethal alleles are culled prior to selection.

Using a MQTL matrix to test for pleiotropic effects of Mendelian trait loci on quantitative traits

C. Scheper and S. König
Justus-Liebig-University Gießen, Ludwigstr. 21b, 35390 Gießen, Germany; carsten.scheper@agrar.uni-giessen.de

Genotypes for Mendelian trait loci can be used to set up a MQTL relationship matrix to enhance variance component estimations. The approach allows to test for pleiotropic QTL-effects of favorable Mendelian traits (e.g. polledness in cattle) or Mendelian inherited detrimental genetic disorders on quantitative traits. The present study aimed on (1) the validation of a MQTL matrix approach to test for pleiotropic QTL-effects of Mendelian trait loci via simulation, and (2) the application of this approach to real data, i.e. considering pleiotropic QTL-effects of the polled locus on quantitative production and reproduction traits. Via stochastic simulations, a Mendelian trait locus with a pleiotropic QTL-effect on a simulated quantitative trait was generated. The MQTL matrix approach successfully detected the simulated pleiotropic QTL-effects using bivariate models, but with the tendency for slight overestimation of QTL-effects. The real data considered 48,046 test day and 3,834 calving and insemination records from 1,746 German Simmental cows kept in 12 herds. The pedigree traced back to five generations, and included 8,624 animals with known polled pheno- or genotypes. The MQTL matrix was constructed considering reliable reconstructed polled marker genotypes. Results for model comparisons based on likelihood values indicated generally better model fits for bivariate models including the QTL-effects compared to base models without QTL-effects. For milk yield, fat percentage, somatic cell score, non-return-rate 56, days to first service and days open, the proportion of additive genetic variance explained by a QTL-effect of the polled locus was negligible. However, the QTL-effect of the polled locus explained 2.7% of the additive genetic variance for protein percentage, whereas the corresponding genetic correlation of -0.005 for MQTL matrix based random effects indicated no genetic antagonism. We conclude that the reported inferiority of polled animals in quantitative traits is not due to inevitable detrimental effects of the polled locus. Furthermore, the presented approach allows for further integration of polledness in present breeding goals using estimated (co)variance components in selection indexes.

Bovine genetic disease and trait frequencies in Ireland: >85 causative alleles in >1M animals

M.C. McClure[1], P. Flynn[2], R. Weld[2], T. Pabiou[1], M. Mullen[3], J.F. Kearney[1] and J. McClure[1]
[1]Irish Cattle Breeding Federation, Bandon, Cork, Ireland, [2] Weatherbys Ireland, Johnstown, Kildare, Ireland, [3]Athlone Institute of Technology, Athlone, Westmeath, Ireland; mmcclure@icbf.com

On the International Dairy and Beef (IDB) custom Illumina bovine SNP chip we have included diagnostic probes for as many disease and trait causative alleles available. The current build, IDBv3, includes 238 diagnostic probes, of which 86 have been validated. Via multiple national Department of Agriculture, Food, and the Marine schemes over 1 million Irish cattle have now been genotyped and have had their data deposited in the ICBF database. These include AI sires, pedigree cattle, and a large number of commercial and crossbred animals. Most studies that report the allele frequencies in a breed or national population are only able to include data from AI or pedigree animals, a few are able to include a small amount of commercial animals. Given the size of our dataset we are able to show that the frequency of undesirable alleles is often lowest in the AI population, highest in the commercial population, and that the frequencies between them to be highly variable. This trend is understandable as traditionally only animals with a high economic value were tested for genetic diseases as their value, or sale price, offset the cost of multiple genetic diagnostic tests. By incorporating all of the causative alleles on the IDBv3 we are able to provide carrier status information for all animals tested, although some traits do require an additional royalty fee for individual reporting. By having the animal's genomic status known Ireland will be able to reduce its genetic disease risk by advising against carrier × carrier matings. To aid the farmers and advisors we also developed a booklet that describes each trait in plain language. The booklet lists known carrier ancestors, for example top AI sires, and pictures of affected animals when possible. For scientists we have developed a second booklet which expands on the first and include flanking DNA sequence, HSPC information, a more in depth trait description, and references. Both booklets are available via the ICBF website at https://www.icbf.com/wp/?page_id=2170.

Management of Mendelian traits in breeding programs by gene editing

J.B. Cole
Agricultural Research Service, USDA, Animal Genomics and Improvement Laboratory, 10300 Baltimore Avenue, Beltsville, MD 20705-2350, USA; john.cole@ars.usda.gov

SNP genotypes have been used to identify several new recessive mutations that adversely affect fertility in dairy cattle, and to track conditions such as polled. Recent findings suggest that the use of sequential mate allocation strategies that account for increases in genomic inbreeding and the economic impact of affected matings result in faster allele frequency changes than those which do not. However, the effect of gene editing on selection programs also should be considered because it has the potential to dramatically change allele frequencies in livestock populations. Computer simulation was used to study the effect of clustered regularly interspaced short palindromic repeat, transcription activator-like effector nuclease, and zinc finger nuclease technologies for gene editing on dairy cattle breeding programs. A hypothetical technology with a perfect success rate was used to establish an upper limit on attainable progress, and a case with no editing served as a baseline for comparison. Technologies differed in the rate of success of gene editing, as well as the success rate of the embryo transfer step, based on literature estimates. Number of alleles edited was assumed to have no effect on success rate. The two scenarios evaluated considered only the horned locus, or 12 recessive alleles segregating in the US Holstein population. The top 1, 5, or 10% of bulls were edited each generation, and either no cows or the top 1% of cows were edited. Inefficient editing technologies produced less cumulative genetic gain and lower level of inbreeding than efficient ones. Gene editing was very effective at reducing the frequency of the horned haplotype (increasing the frequency of polled animals in the population), and allele frequencies of the 12 recessives segregating in the US Holstein population decreased faster with editing than without. These results suggest that gene editing can be an effective tool for reducing the frequency of harmful alleles or increasing the frequency of desirable alleles in a dairy cattle population even if only a small proportion of elite animals are modified. The source code for the simulation and scripts used to analyze the data are available on GitHub: https://github.com/wintermind/gene-editing.

International registration and management of genetic defects: general discussion

H. Jorjani

Department of Animal Breeding & Genetics, Box 7023, 75007 Uppsala, Sweden; hossein.jorjani@slu.se

Discussing national and intrernational initiatives to standardize the nomenclature, register and manage the genetic defects.

APOB associated cholesterol deficiency in Holstein cattle is not a simple recessive disease

U. Schuler[1], M. Berweger[1], B. Gredler-Grandl[1], S. Kunz[1], S. Hofstetter[2], T. Mock[3], T. Mehinagic[4], M. Stokar-Regenscheit[4], M. Meylan[3], F. Schmitz-Hsu[5], F.R. Seefried[1] and C. Drögemüller[2]

[1]Qualitas AG, Chamerstr. 56, 6300 Zug, Switzerland, [2]University of Bern, Institute of Genetics, Bremgartenstr. 109a, 3001 Bern, Switzerland, [3]University of Bern, Clinic for Ruminants, Bremgartenstr. 109a, 3001 Bern, Switzerland, [4]University of Bern, Institute of Animal Pathology, Längasstr. 122, 3012 Bern, Switzerland, [5]Swissgenetics, Meielenfeldweg 12, 3052 Zollikofen, Switzerland; urs.schuler@qualitasag.ch

Cholesterol deficiency (CD) has been reported for the first time in 2015 as a new recessive inherited genetic defect in Holstein cattle. It was initially mapped on BTA11 and subsequently a causative loss of function mutation in *APOB* was identified by whole genome sequencing. Affected homozygous mutant calves showed poor development, intermittent diarrhea and hypocholesterolemia. Usually heterozygous carriers did not show any clinical signs of maldigestion but had in general lower cholesterol and lipoprotein concentrations, suggesting a codominant effect of the *APOB* mutation on lipid homeostasis. In the meantime we collected 15 CD affected animals heterozygous for the *APOB* mutation indicating a more complex inheritance compatible with a dominant disease with incomplete penetrance. We have also analysed possible effects on additional traits and used routine phenotypes and official models from Swiss Holstein genetic evaluation for fertility, birth, conformation and beef traits. A fixed effect decoding the risk status for CD-homozygosity was fitted in genetic evaluation models. Since dams are usually not genotyped, nine subclasses were defined for three possible conditions of each sire / maternal grandsire status. Interestingly, effects were found on non-return rate and on interval from first to last insemination in both, heifers and cows. Furthermore, negative effects on birth weight and stillbirth could be identified whereas effects on beef traits have not been detected. In conclusion, as beyond malabsorption of dietary lipids, deleterious effects of *APOB* deficiency can be expected on hepatic lipid metabolism, steroid biosynthesis, and cell membrane function. Therefore the *APOB* mutation may also explain unspecific symptoms of reduced fertility.

Associations between a causal mutation for Mulefoot and production traits in Holstein Friesian cows

M.P. Mullen[1], J. McClure[2], F. Kearney[2] and M. McClure[2]

[1]Bioscience Research Institute, Athlone Institute of Technology, Athlone, Co. Westmeath, Ireland, [2]Irish Cattle Breeding Federation, Highfield road, Bandon, Co. Cork, Ireland; mmullen@ait.ie

The impact of causal mutations with undesirable effects on animal growth and development are of particular interest to the livestock breeding industry. In dairy cattle, a doublet substitution mutation (NG1621KC) in *LRP4* is attributed to the fusion or non-division of the functional digits of the hoof, termed syndactyly or Mulefoot. The identification and exclusion of Mulefoot carriers from breeding is a desirable goal, however, strategic mating may be considered in the case of otherwise high genetic merit animals. Estimation of the possible pleiotropic effects of such mutations would enable more informed strategic mating decisions. The objective therefore of this study was to estimate the effects of the NG1621KC causal mutation for Mulefoot in *LRP4* on milk, fertility, carcass and health traits (n=16) in dairy cows in Ireland. Genotypes and phenotypes on 10,707 dairy cows were obtained through the Irish cattle breeding federation (ICBF). Phenotypes were expressed as predicted transmitting abilities (PTAs). PTAs were deregressed following the removal of parental contributions. Only animals with an adjusted reliability of >10% were included in the analysis which included n=6,876, 1,198, 264, 4,566, 8,564, 152, 2,280, 3,194, 518, 360, 1,374, 5,747 cows for milk traits (n=5), calving interval, survival, gestation length, calf mortality, maternal calving difficulty, carcass weight, carcass conformation, carcass fat, cull cow weight, and somatic cell score, respectively. The association between NG1621KC and deregressed PTAs were analysed in ASREML using a weighted mixed animal model. The NG1621KC mutation was associated with increased calving difficulty (3.41, s.e. 1.56, P<0.05) and decreased maternal calving difficulty (-4.24, s.e. 1.85, P<0.05), however, there was no association (P>0.05) identified with any of the other milk, carcass or health related traits examined. Assuming average genetic merit the results of this study provide no evidence to support the maintenance of carriers of the NG1621KC mutation on farm or in the national herd in relation to the production traits analysed.

Relationships between mutations responsible for Holstein haplotype 1, 3 and 4 and bovine leukocyte A

M. McClure[1], J. McClure[1], L. Ratcliffe[2], J.F. Kearney[1] and M. Mullen[2]

[1]Irish Cattle Breeding Federation, bandon, cork, Ireland, [2]Athlone Institute of Technology, Bioscience Research Institute, Athlone, Westmeath, Ireland; mmcclure@icbf.com

Identification of carriers of mutations with lethal effects in cattle populations enables more informed decision making by the farmer be it elimination from breeding stock or management through strategic mating schemes for high genetic merit carriers. In order to best advise farmers on the use of this information, estimation of the effects of these mutations on routinely recorded production traits in carrier animals is needed. Therefore, the objective of this study was to estimate if the mutations associated with HH1, 3, 4 or BLAD showed any evidence of effects across any production traits (milk, fertility, carcass and health traits (n=16)) in dairy cows. Genotypes and phenotypes (expressed as predicted transmitting abilities (PTAs)) on 10,707 dairy cows were obtained from the Irish Cattle Breeding Federation (ICBF) database. Only animals with an adjusted reliability of >10% were included in the analysis which included n=6,876, 1,198, 264, 4,566, 8,564, 152, 2,280, 3,194, 518, 360, 1,374, 5,747 cows for milk traits(n=5), calving interval, survival, gestation length, calf mortality, maternal calving difficulty, carcass weight, carcass conformation, carcass fat, cull cow weight, and somatic cell score, respectively. The association between each SNP and PTA (deregressed) was analysed in ASREML using weighted mixed animal models. BLAD carriers were associated with increased somatic cell score (P<0.05) and calf mortality (P<0.05), however, there was no association (P>0.05) with any of the other milk, fertility or carcass traits analysed in this study. No association (P>0.05) was observed between HH1 and any of the traits examined. Cows with a HH2 allele were associated (P<0.05) with decreased gestation length with no other effects identified. Cows with a HH3 allele were associated (P<0.05) with increased calving interval with no other effects observed. Unless carriers of either BLAD, HH1, HH3 or HH4 are of otherwise high genetic merit these results provide no evidence to support the maintenance of carriers on farm or in the national herd.

A region on BTA5 is significantly associated with Brachygnathia inferior in dairy cattle

C. Flury[1], H. Signer-Hasler[1], M. Frischknecht[1,2], A. Lussi[1], F.R. Seefried[2] and C. Drögemüller[3]
[1]Bern University of Applied Sciences, School of Agricultural, Forest and Food Sciences HAFL, Länggasse 85, 3052 Zollikofen, Switzerland, [2]Qualitas AG, Chamerstrasse 56, 6300 Zug, Switzerland, [3]University of Bern, Institute of Genetics, Bremgartenstrasse 109a, 3001 Bern, Switzerland; christine.flury@bfh.ch

Brachygnathia inferior (BI) in cattle is considered as a heritable condition. However the mode of inheritance and underlying genes still remain unclear. Braunvieh cows being affected from BI are recognized during linear type classification in the first lactation and penalized in their overall type note. At 1.1.2014, 104 living cows with BI were registered by Braunvieh Schweiz. Out of these, hair samples of 81 cows were collected and genotyped for the 80k-GGHDP. These genotypes were combined with 440 HD-genotypes from unaffected cows. After filtering data from 81 cases and 440 controls and 50,864 SNP were used in a genome wide association study. 17 SNP located in a 4.2 Mb interval on BTA5 were significantly associated with BI. The interval is positioned between 29.1 and 33.3 Mb and contains several genes. Subsequently runs of homozygosity (ROH) were calculated and their distribution was compared between cases and controls. More than 70% of the cases had SNP between 29.9 and 33.5 Mb on BTA5 in a ROH, while the same holds only for less than 16% of the controls. For SNP on all other autosomes the deviations between the two groups were much less pronounced.

Screening for missing homozygosity in a local Swiss dual purpose breed

F.R. Seefried[1], M. Berweger[1], B. Gredler-Grandl[1], S. Kunz[1] and C. Drögemüller[2]
[1]Qualitas AG, Chamerstr. 56, 6300 Zug, Switzerland, [2]University of Bern, Institute of Genetics, Bremgartenstr. 109a, 3001 Bern, Switzerland; franz.seefried@qualitasag.ch

However the population size of the autochthonous Swiss dual-purpose breed Original Braunvieh (OB) is rather small, genomic selection has been implemented. Therefore large-scale SNP genotype data became available and enables an identification of haplotypes with reduced or missing homozygosity that may harbour deleterious recessive mutations. Genome-wide scans for missing homozygosity were applied based on 1,379 animals genotyped at of at least 50K density level. Sliding window approach using windows sizes between 0.5 and 10 Mb was applied for inferring haplotypes. FImpute software was used for phasing, and expected numbers of homozygous animals were calculated assuming random mating. Phenotypes were taken from routine genetic evaluations for fertility-, birth- and beef traits. Official models were adjusted by adding a fixed effect decoding the risk of homozygosity. Since dams are usually ungenotyped, nine subclasses were defined for three possible conditions of each sire / maternal grandsire status. In summary, haplotype analyses detected a single genome region at significant level: a 2 Mb segment on BTA11. Additional seven genome regions located on BTA1, BTA5, BTA13, BTA14, BTA17 and BTA20 were identified slightly below significance threshold. Subsequent phenotypic analyses were performed for all identified genome regions revealing that an increased risk of homozygosity for the BTA11 region is associated with a reduced birth weight. Effects on fertility traits, in detail reduced non-return rate and extended interval from first to last insemination, were detected for regions on BTA1, BTA5, BTA17 and BTA20. Interestingly, regions on BTA1 and BTA20 exhibited effects on beef traits either. Based on further analyses, one may assume a co-segregation of the growth-related and recessive FH2-disorder originally reported in Simmental behind the observations made for the haplotype region at the telomeric end of BTA1 in OB.

Searching for, finding, and fixing genetic diseases: we can't afford not to

J. McClure[1], P. Flynn[2], S. Waters[3], F. Kearney[1], M. Mullen[4], T. Pabiou[1], R. Schnabel[5], J. Taylor[5], R. Weld[2] and M. McClure[1]

[1]Irish Cattle Breeding Federation, Highfield House, Shinagh, Bandon, County Cork, Ireland, [2]Weatherby's DNA Laboratory, Johnstown, Co. Kildare, Ireland, [3]AGRIC, Teagasc, Grange, Co. Meath, Ireland, [4]AIT, Department of Life and Physical Sciences, Faculty of Health and Science, Athlone, Co. Westmeath, Ireland, [5]University of Missouri Columbia, Animal Sciences, UMC, Columbia, MO 65211, USA; jmcclure@icbf.com

Genetic diseases cost the livestock industry millions every year. It is estimated that every animal carries 20-100 genetic disease causing alleles and that every animal born carries 50 spontaneous mutations, though not all are detrimental to the viability or productivity of the animal, in fact, some may be beneficial. Since 2013, 650,000 Irish cattle have been genotyped and allele frequencies were analysed to calculate carrier frequency on 33 different diseases using the International Dairy and Beef (IDB) SNP Chip. Economic losses and gains will be calculated for disease and trait genes tested on the IDB SNP chip. Genotyping costs have shrunk dramatically and made developing a 'Breed Smarter Strategy' realistic for the Irish population. Implementation of this strategy, will allow the industry to retain high value animals even if they carry known adverse traits through mating them with animals that do not carry those genes. The idea behind the program has incentivised ICBF to seek out new disease SNP in the population. A survey was developed allowing farmers and veterinarians to report genetic defects and provide samples to identify diseases causing economic loss. The limitations in a program like this are removing the stigma linked with having an atypical animal, and identifying diseases that are genetic. While implementation of this program has been slow, articles online and in popular press have increased the frequency of reports every year. After a few key diseases were identified, DNA from affected animals was sent off for whole genome sequencing and analysis of the genomes will commence to identify candidate causal mutations. This is a constant process as new genetic diseases appear every year. Success of the program will save the industry millions of euros annually once new SNP are identified and added to the next version of the IDB chip.

Influence of live weight and genotype on efficiency of dairy cows

M. Ledinek[1], L. Gruber[2], F. Steininger[3], B. Fuerst-Waltl[1], M. Royer[2], K. Krimberger[2], K. Zottl[4] and C. Egger-Danner[3]
[1]University of Natural Resources and Life Sciences, Gregor-Mendel-Str. 33, Vienna, Austria, [2]AREC Raumberg-Gumpenstein, Raumberg 38, Irdning, Austria, [3]ZuchtData, Dresdner Str. 89/19, Vienna, Austria, [4]LKV NOE, Pater Werner Deibl-Str. 4, Zwettl, Austria; maria.ledinek@boku.ac.at

The aim was firstly to evaluate the influence of live weight on feed and energy efficiency of dairy cows and secondly to analyse the current state of population within the Efficient Cow Project of the Austrian Cattle Breeding Association. Data of Fleckvieh (FV), Holstein (HF) and Brown Swiss (BS) dairy cows were recorded on-farm (161 farms) at each performance testing during a whole year. Feed intake was estimated (6,480 cows) using cow individual (diet) information. The relationship of milk yield to live weight was shown to be non-linear. Milk yield decreased after the live weight class of 750 kg for HF and BS but less dramatically and later for FV at 800 kg. This resulted in an optimum live weight for feed and energy efficiency. BS and HF had highest efficiency in a narrower and lighter live weight range (550 to 700 kg) due to a stronger curvature of the parabolic curve. Contrary to this efficiency of FV did not change as much as of the dairy breeds with increasing live weight. So FV had a similar efficiency in a range of 500 to 750 kg. Therefore live weight seems to influence efficiency of dairy breeds more than of dual purpose breeds. The difference of breeds vanished when live weight ranged between 750 and 800 kg. The average live weights of the breeds studied (FV 728 kg, BS and HF 656 kg) are in the optimum range. FV is located at the upper end of the decreasing part. Furthermore a cow which is heavier has to produce more milk based on a higher concentrate level to be as efficient as a lighter cow. In conclusion an optimum range of live weight for efficiency exists due to the non-linear relationship of milk yield and live weight. Milk yield and efficiency of cows with a high genetic potential of milk production depend more on live weight than of dual purpose cows. Cows with medium weights within population are the most efficient ones. Heavy cows (>750 kg) produce even less milk. A further increase of dairy cows' live weights should hence be avoided.

Utility of whole genome sequence data for across breed genomic prediction

B. Raymond[1,2], A.C. Bouwman[2], C. Schrooten[3], J. Houwing-Duistermaat[4] and R.F. Veerkamp[2]

[1]Biometris, Wageningen University and Research, P.O Box 6700 AA Wageningen, the Netherlands, [2]Animal Breeding and Genomics, Wageningen Livestock Research, P.O. Box 338, 6700 AH Wageningen, the Netherlands, [3]CRV, BV, P.O. Box 454, 6800 AL Arnhem, the Netherlands, [4]Medical Statistics and Bioinformatics, Leiden University Medical Centre, Leiden, the Netherlands; biaty.raymond@wur.nl

Genomic prediction (GP) across breeds have so far resulted in low accuracies of the estimated genomic breeding values. Our objective was to evaluate if using the whole genome sequence (WGS) instead of low density markers can improve across breeds GP especially when markers are pre-selected from a GWAS, and to test our hypothesis that many non-causal markers in WGS have a diluting effect on across breed prediction accuracy. Estimated breeding values (EBVs) for stature and BovineHD genotypes were available for 595 Jersey bulls from New Zealand, 957 Holstein bulls from New Zealand and 5,553 Holstein bulls from the Netherlands. BovineHD genotypes for all bulls were imputed to WGS. Genomic prediction across these breeds and countries was performed using genomic restricted maximum likelihood. Besides the 50k, HD and WGS, markers significantly associated with stature in a large meta-GWAS analysis were selected and used for prediction. Furthermore, we estimated the proportion of genetic variance explained by markers in each scenario as a form of validation. Across breeds, 50k, HD and WGS resulted in very low accuracies of prediction ranging from -0.04 to 0.13. Accuracies were better in scenarios with pre-selected markers from the previously conducted meta-GWAS. For example, using only 133 most significant markers from 133 QTL regions form the meta-GWAS yielded accuracies ranging from 0.08 to 0.23, while 23,125 markers with -log10(p)>7 resulted in accuracies of up 0.35. Using WGS did not significantly improve the proportion of genetic variance captured across breeds compared to scenarios with few but selected markers. Our results demonstrated that the accuracy of across breed GP can be improved using WGS data instead of lower density marker panels. However, such improvement becomes apparent only when markers in WGS are pre-selected based on their potential causal effect.

Effect of semen pharbitidis on ruminal methane production and bacterial abundance – *in situ* approach

B.D. Rajaraman[1], C.H. Lee[2], Y.W. Woo[1] and K.H. Kim[1,2]

[1]Seoul National University, Graduate School of International Agricultural Technology, Pyeongchang, Korea, South, [2]Seoul National University, Institute of Green Bio Science and Technology, Pyeongchang, Korea, South; bharanitharshan76@gmail.com

Besides *in vitro* and respiratory chamber techniques, a useful, simple *in vivo* method is needed to confirm the effectiveness of numerous additives for mitigating methane (CH_4) emissions. In this study, we performed live continuous culture with an *in situ* bag containing 60 g of *Semen pharbitidis* seed (ISS) in the rumen and analyzed the CH_4 percentage in ruminal gas sampled via a cannula stopper using syringe. Four rumen-cannulated Holstein steers fed on a basal diet given in equal amounts twice daily at 09:00 h and 18:00 h (control for 15 days) were exposed to two phases during the trial: first, three ISS were placed in the rumen for 12 days (treatment); then,10 days after the ISS had been removed (recovery). The results revealed 76% dry matter (DM) degradability of the seeds during ruminal fermentation. The mean CH_4 (%) was decreased with the ISS treatment compared with the control (1.89 vs 4.41%; P=0.07) and increased in the recovery phase compared with the treatment (2.63 vs 1.89%; P=0.57). The ISS treatment decreased the pH, NH_3-N (mg/l), and isobutyrate (%) significantly (P<0.01) compared with the control and increased (P<0.05) them in the recovery phase compared with the treatment. The total volatile fatty acid production increased (P<0.005) with the ISS and decreased relatively (P=0.05) after removing the ISS. The significantly (P<0.05) lower acetate:propionate ratio with the treatment reflected the shift in hydrogen transfer towards the formation of more propionate. This study demonstrated the effect of *Semen pharbitidis* on ruminal methane production using an *in situ* approach; however, the relative abundances of the bacterial sp. *Ruminococcus albus*, *Ruminococcus flavefaciens*, *Prevotella*, *Butyrivibrio*, *Succinivibrio* and *Fibrobacter succinogenes*, and total methanogens in the control and treatment do not provide a plausible explanation for this.

Temporal changes in gene expression during CLA induced milk fat depression in lactating dairy cow

A.M. Abdelatty[1], M.E. Iwaniuk[2], M. Garcia[2], K.C. Moyes[2], B.B. Teter[2], P. Delmonte[3], M.A. Tony[1], F.F. Mohamed[1] and R.A. Erdman[2]
[1]Cairo University, Giza, 12211, Egypt, [2]University of Maryland, Maryland, 20742, USA, [3]FDA, Maryland, 20740, USA; raramehmet@gmail.com

In order to explore the mechanistic pathways associated with milk fat depression, a detailed investigation of mammary gene expression over time needs to be conducted. We hypothesized that the RNA present in the cytosolic crescent in the milk fat globule could represent mammary epithelial cell RNA. The objective of this study was to investigate the effects of conjugated linoleic acid (CLA) induced milk fat depression on temporal changes in gene expression using RNA extracted from milk fat. Ten multiparous Holstein dairy cows averaging 100 (±11) days in milk were used. All cows were *ad libitum* fed the same basal diet during an 18 d covariate (COV) period. From days 19 to 24 (EXP), 5 cows received the basal diet (CON) while the remaining 5 cows received the basal diet plus 200 g/d of an encapsulated CLA supplement (CLA). On d 25 to 27 (REC), all cows were fed the basal diet. Milk samples were collected and fat was extracted from morning and evening milk and preserved for RNA extraction during d 17 to 25. Milk production, fat percent and fat yields were respectively 39.7 kg/d, 3.56%, and 1,357 g/d for CON and 42.8 kg/d, 3.13%, and 1,273 g/d for CLA during d 19 to 24. CLA supplementation increased milk production (P<0.01) and decreased milk fat percent (P<0.01) and yield (P<0.03) and treatment effects continued during the REC period. Relative expression of the genes ACACA, PPARG were reduced by 47% (P=0.08) and 32% (P=0.07) by CLA whereas there was no CLA effect on expression of FASN, GPAM, or SREBF1. We could not detect any day by treatment interactions for gene expression during CLA induction of moderate milk fat depression. The results of this experiment were consistent with the previous studies with moderate depression in milk fat that used mammary biopsy to study the mammary lipogenic gene expression. We concluded that RNA isolated from milk fat globules could be used to investigate lipogenic gene expression, with greater changes in milk fat to be able to measure temporal effects following diet change.

On-farm mortality and related risk factors in Estonian dairy cows

K. Reimus[1], T. Orro[1], U. Emanuelson[2], A. Viltrop[1] and K. Mõtus[1]
[1]Estonian University of Life Sciences, Institute of Veterinary Medicine and Animal Sciences, Kreutzwaldi 1, 51014 Tartu, Estonia, [2]Swedish University of Agricultural Sciences, Department of Clinical Sciences, P.O. Box 7054, 750 07 Uppsala, Sweden; kaari.reimus@emu.ee

Mortality reflects the health and welfare state of animals and causes financial loss for the farmer. The objective of this study was to identify risk factors associated with on-farm mortality in dairy cows. Data was retrieved from the Estonian Agricultural Registers and Information Board and Estonian Livestock Performance Recording Ltd for period between 1st of January 2013 and 31st of December 2015 of cows from herds with ≥20 cow-years. Analyses included data of 85,555 primiparous cows from 390 herds and 107,873 multiparous cows from 389 herds. The observation period for each cow started at the day of calving and ended at the date of death (mortality event as composed outcome including on-farm unassisted death and euthanasia) or censoring. Mortality rate was lower in primiparous compared to multiparous cows. In multiparous cows the mortality hazard was lowest at second parity and considerably higher from the fifth parity. Factors significantly associated with mortality in cows were early lactation period, Holstein breed, North-East region, larger herd size, decrease of herd size from 2013 to 2015 more than 15%, smaller herd average number of lactations, longer herd average interval from calving to insemination (for primiparous cows), low/high (for primiparous/multiparous, respectively) relative milk yield breeding value of a cow, older age at first calving, male sex of the calf (for multiparous cows), birth of twins/triplets, stillbirth and abortion, dystocia, low milk yield and high fat-protein ratio at first test-milking. Also, high milk somatic cell count at the last test-milking of the previous lactation in multiparous and the first test-milking of the ongoing lactation in primiparous cows were related with higher mortality hazard. Longer previous calving interval and higher herd average number of inseminations per conception were factors associated with increased mortality hazard in multiparous cows. It is important to pay more attention to herd management and good health of the cows and ensure easy calving in order to avoid mortality of cows.

Ruminal biohydrogenation of linolenic acid according to the lipid source: a meta-analysis approach
L.O. A. Prado[1], A. Ferlay[1], P. Nozière[1] and P. Schmidely[2]
[1]INRA, UMR1213 Herbivores, Theix, 63122, France; [2]AgroParisTech, INRA, UMR0791 Mosar, 16 rue Claude Bernard, Paris, 75231, France; lucas-de-ofeu.aguiar-prado@inra.fr

The improved prediction of milk fatty acid (FA) profile from feedstuffs requires the quantification of ruminal biohydrogenation (RBH) of polyunsaturated FA, and its variation according to dietary factors and rumen physico-chemical characteristics. As part of the renovation of the French feed unit systems (FUS) for ruminants, we propose new equations to predict more precisely the duodenal FA flows, and specifically linolenic acid (LNA) from a database composed of 119 publications (comprising 167 experiments with 452 *in vivo* treatments) obtained from a literature review. The equations derived from a variance-covariance model, in which experiment was considered as a fixed effect. Both LNA intake (int) and duodenal (duo) flow were expressed in g/kg of dry matter intake to avoid bias caused by the species effect. First, the prediction of duodenal LNA flows was depended on the variation of LNA intake and lipid source (oil/fat supplements including vegetable oils and animal fat, oilseeds and forage-based diets). Second, effects of the interfering factors on regression parameters were studied in order to identify the quantitative (diet and rumen fermentable characteristics) and qualitative (animal data, source and technological treatments of lipid supplement) factors. We obtained a linear relationship between duodenal LNA flow and LNA intake for oilseeds (LNA duo = 0.270 + 0.047 LNA int) while the relationship for oil/fat supplements was quadratic (LNA duo = 0.523-0.001 LNA int^2). The relationship for forage-based diets was not significant. However, the lipid source (oil/fat vs oilseeds) had a significant effect on slope (P=0.05). Concerning the LNA relationship with fat, we observed an effect of technological treatment (control vs free fat vs calcium salts vs formaldehyde) on residues. The formaldehyde treatment led to higher duodenal LNA flows (+0.53) than either the control diets (-0.30) or free oil/fat (-0.27), whereas calcium salts had no effect. Duodenal LNA flows were not previously predicted by FUS and the model obtained in this study will help predicting the RBH intermediates of LNA.

Effects of *Fusarium* toxin contaminated diets on dairy cow health and performance
G.E. Hop, G.H.M. Counotte, S. Carp-Van Dijken, K.W.H. Van Den Broek, K. Junker, F. Scolamacchia and A.G.J. Velthuis
GD Animal Health, Arnsbergstraat 7, 7418 EZ Deventer, the Netherlands; g.hop@gdanimalhealth.com

This study was conducted to determine the effects of an eight-week period mycotoxin intake on dairy cow health and performance through two different silage qualities. This study specifically looked at the effects of deoxynivalenol (DON) and zearalenone (ZEA) through naturally contaminated maize. After a four-week period of acclimatization, 32 high-yielding, pluriparous (parity 2 or 3) Holstein Friesian dairy cows in the first 70 days of lactation and free of antibodies against IBR, BVDV, neospora and paratuberculosis were divided into four groups. They received a total mixed ration based on good quality silage (good ration: GR) or moderate quality silage (moderate ration: MR) with mycotoxins (MT+) or without (MT-) for eight weeks. On average, MT+ rations contained 789 µg DON and 460 µg ZEA per kg ration. The following parameters were collected and analyzed: parameters related to the overall animal health, production characteristics, rumen function and digestion, energy and protein supply, kidney damage and function, liver damage and function, resilience and inflammation, fertility, and pathologic findings. Descriptive statistics and rank-sum tests were used to test for statistically significant differences among the four groups. Blood and milk research, clinical examinations of the animals and necropsy of the GRMT+ and MRMT+ cows did not reveal systematic effects of DON and ZEA on cow health and performance. Some parameters differed significantly between the MT+ and MT- groups, e.g. milk production, protein, fat and lactose content and various blood parameters. However, these parameters differed throughout the entire trial period and are, therefore, expected to be caused by the trial group composition. A few parameters were significantly different between the GR- and MR-groups, e.g. a lower feed intake and a less efficient utilization of feed. In this study the composition of the ration and the feed quality appeared to have more influence on cow health and performance compared to the intake of DON and ZEA. This study did not reveal effects of an eight-week period DON and ZEA intake on dairy cow health and performance through two different silage qualities.

A comparison of the *in vitro* ACE inhibitory capacity of dairy and plant proteins

C. Giromini[1], A.A. Fekete[2,3], A. Baldi[1], D.I. Givens[2] and J.A. Lovegrove[3]
[1]University of Milan, Department of Health, Animal Science and Food Safety, Via Trentacoste, 2, 20134 Milan, Italy, [2]University of Reading, Food Production and Quality Research Division, School of Agriculture, Policy and Development, Faculty of Life Sciences, RG6 6AP, Reading, United Kingdom, [3]University of Reading, Department of Food and Nutritional Sciences, School of Chemistry, Food and Pharmacy, Faculty of Life Sciences, RG6 6AP, Reading, United Kingdom; carlotta.giromini@unimi.it

Bovine milk proteins and their peptides have been shown to exert Angiotensin-Converting Enzyme inhibitory (ACEi) activity *in vivo* and *in vitro*, after gastro-intestinal digestion. To gain insight into dairy proteins digestion pattern and bioactive potential, a set of dairy proteins consisting of a range of casein and whey-based products and plant proteins including those derived from soya, peas and wheat was sequentially hydrolysed in simulated human physiological conditions. The total digesta was filtered using 3kDa membrane, in order to mimic the intestinal absorption. A permeate (absorbed fraction) and a retentate (intestinal fraction) were obtained. ACEi activity was measured as the ability of protein fractions (pre-digested, gastric, permeate and retentate) to decrease the hydrolysis of furanacroloyl-Phe-Glu-Glu (FAPGG) synthetic substrate for ACE enzyme. Results showed that permeate and retentate of dairy proteins exerted a significant ACEi activity ($27.05\pm2.99\%$; $P<0.05$) compared with the pre-digested dairy proteins ($16.26\pm1.53\%$). In particular, among dairy proteins tested, permeate of whey protein isolate and whey protein hydrolysed samples showed the highest ACEi capacity. Conversely, an opposite trend was observed for plant proteins. Plant protein permeate exhibited the lowest ACEi activity ($16.52\pm2.92\%$), compared with the all the other fractions. The ACEi results support the growing evidence that bioactive peptides within dairy and plant proteins are encrypted in an inactive form within parent protein sequences with hydrolysis needed for activity to be realised. The comparison of the *in vitro* effect of dairy and plant proteins performed in this study could provide valuable knowledge regarding their different bioactivities, promoting their use in the formulation of functional foods.

Effect of diet supplement in dairy cow with grape pomace on quality of milk and cheese

F. Castellani, A. Vitali, E. Marone, F. Palazzo, L. Grotta and G. Martino
Faculty of BioSciences and Technologies for Agriculture Food and Environment, University of Teramo, Balzarini 1, 64100, Teramo (TE), Italy; fcastellani@unite.it

The aim of the study was to evaluate the effect of dairy cow diet supplement with grape pomace (GP) on quality of milk and cheese derived. The GP, a biomass deriving from winery and distillery industries, is a source of polyphenols and unsaturated fatty acids. Twelve lactating Holstein Friesan cows were randomly divided in two homogenous groups for parity, milk yield and days in milk. Cows were fed during 60 days with conventional ration in the control group (CG) and with GP (10% on daily dry matter intake) as supplement of conventional ration in the experimental one (EG). Samples of both diet were collected for chemical analyses. Milk production was monitored during the trial and individual milk samples were collected at the end of the experiment for both groups. Milk samples were analyzed to evaluate quality parameters (protein, casein, lactose, lipids, fatty acids profile, urea and somatic cells). Milk produced the last day of the experimental period was collected for both groups and used to produce cheese. Cheese was sampled at 1 (T1) and 30 (T30) days over ripening and fatty acids profile and lipid oxidation (TBARs test) were assessed. Statistical analysis was carried out using GLM procedure of SAS. Chemical analyses performed on diets highlighted a higher content of linoleic acid in EG diet compared with the CG one (46.73 vs 40.11%, respectively). Milk yield did not differ between two groups during the experimental time (16.58 vs 15.29 kg/d, CG and EG group respectively). Vaccenic (0.95 vs 1.45%), rumenic (0.55 vs 0.91%) and linoleic (2.05 vs 2.60%) acids were significantly higher in EG milk compared to CG one. Fatty acids composition of cheese, in particular vaccenic, rumenic and linoleic acids, were higher in EG group. Malondialdehyde did not differ at T1 between groups while it resulted almost three times lower in EG cheese after 30 days of ripening (0.117 vs 0.043 µg MDA/g). Results indicate that GP diet supplement in dairy cows may enhance milk and cheese fatty acids composition. The lowest lipid oxidation observed in treated cheese may be related to antioxidant properties of the higher concentration of polyphenols in grape pomace.

A quarter level approach to monitor recovery after clinical mastitis

I. Adriaens[1], T. Huybrechts[1], S. Piepers[2], W. Saeys[1], B. De Ketelaere[1] and B. Aernouts[1]
[1]KU Leuven, Kasteelpark Arenberg 30, 3001 Leuven, Belgium, [2]UGent, Salisburylaan 133, 9820, Merelbeke, Belgium; ines.adriaens@kuleuven.be

Clinical mastitis (CM) is an important problem on modern dairy farms. The use of automated milking systems entails less frequent and thorough examination of the udder, causing shallower monitoring of udder health problems. To solve this, researchers focused mainly on milk parameters (MP) for detection systems. However, in this study we put forward that the MP also contain a lot of information during the recovery phase. For example, monitoring the MP on individual quarter level can distinguish between systemic illness and actual damage of the milk secreting tissue caused by the inflammation and the pathogen's toxins. In this study, first a reliable reference was established which can account for individual variability in the MP. Then the measured values are tested against this reference during the recovery phase of CM, quantifying damage and recovery in the four quarters. Modelling the MPs during this phase, we can predict recovery and bacteriological and physiological cure. A mixed model approach was proposed for the reference quarter milk yield as example for other MPs. Predictions for milk yield during a hypothetical recovery period of 21 days had an accuracy of about 10% with a mean RMSE of 0.38 kg, which is about the accuracy of the milk meters. The synergistic control technique was then used to distinguish between normal variability and variability caused by CM (detection phase). After detection, the out-of-control points can be integrated to quantify the effect of inflammation and tissue damage. MPs are linked with severity, treatments and the production during the remaining lactation period. The information extracted this way based on MPs allows to support farmers both in treatment and management (culling, dry-off) decisions without extra costs. Besides the economic reasons, monitoring recovery is also beneficial for optimization of antibiotic use and better welfare.

Microbiological quality of milk on farms, during transport, processing stages and in milk powder

L.F. Paludetti[1,2], A.L. Kelly[1] and D. Gleeson[2]
[1]University College Cork, School of Food and Nutritional Sciences, Cork, Cork, Ireland, [2]Teagasc, Animal & Grassland Research and Innovation Centre, Moorepark, Fermoy, Co Cork, Cork, Ireland; lizandra.paludetti@teagasc.ie

The microflora population in milk is an important determinant of milk processability and functional properties, as well as organoleptic qualities, shelf life and safety of dairy products. The objective of this study was to monitor the microbiological quality of milk from individual dairy farms, during transport, in factory silos and in milk powder subsequently manufactured. Milk was collected from 67 dairy farms by 11 tankers, which transported this blended milk to a factory, where milk was stored in a whole milk silo, pasteurised and skimmed, and used to produce milk powder. Milk samples were collected at each stage from farms to final product. Among the 12 microbiological analysis performed, samples were tested for: total bacterial count (TBC), psychrotrophic (PBC), thermoduric (LPC), thermophilic (THERM) bacterial counts and presumptive Bacillus cereus (BAC). Questionnaires were completed on farms in order to correlate hygiene and sanitation practices with the microbiological levels in milk. The average TBC and PBC for farm milks were 9,723 cfu/ml (range:450-80,000 cfu/ml) and 31,368 cfu/ml (range: 500-10^6 cfu/ml), respectively, and those levels increased from farm to whole milk silo. The average TBC in tankers and whole milk silo samples were $1.2\pm1.2\times10^4$ and $7.75\pm0.4\times10^5$ cfu/ml, while PBCs were $1.3\pm2.7\times10^4$ and 1.0×10^6 cfu/ml respectively. After pasteurisation, TBC and PBC decreased to 450 ± 238 cfu/ml and 50 ± 58 cfu/ml, respectively. Evaporation and the drying processing stage contributed to the lower bacterial levels observed in milk powder (TBC: 233 ± 49 cfu/g; PBC: 14 ± 9 cfu/g). After pasteurisation, no presumptive BAC colonies were identified; however the process was not efficient in reducing LPC and THERM, which did not vary greatly from farm to skim milk silo (LPC: 38 ± 10 cfu/ml; THERM: 69 ± 30 cfu/ml). Farms presented milks of a high microbiological quality and where inferior quality milk was observed this was related to inadequate cow and equipment cleaning protocols. This study can aid in monitoring sanitation practices and process controls to ensure manufacture of safe and high-quality dairy products.

Methane emission in Jersey cows during spring transition from in-door feeding to grazing

M. Szalanski[1], T. Kristensen[2], G. Difford[1], P. Løvendahl[1] and J. Lassen[1]

[1]Aarhus University, Center for Quantitative Genetics and Genomics, Dept. Molecular Biology and Genetics, AU-Foulum, 8830 Tjele, Denmark, [2]Aarhus University, Dept. Agroecology, AU-Foulum, 8830 Tjele, Denmark; mszalans@mbg.au.dk

In this study we evaluate the effects of gradual transitioning from indoor winter feeding to outdoor spring grazing, 8 to 11 h daily, on methane emission in an organic jersey herd in Denmark. A total of 151 mixed parity lactating jersey cows with free access to automated milking stations (AMS) were monitored from 30 d prior to barn release until 23 d post release. During each visit to the AMS, milk yield was measured and logged together with date and time. Methane (CH_4) monitoring equipment was installed in the feed troughs of AMS. Methane and carbon dioxide (CO_2) were measured continuously in ppm using a non-invasive 'sniffer' method, and the ratio of which was used as emission trait. This trait was averaged over milking. A small increase in the trait of interest was found with increasing grazing, along with increased yield. Herd average DMI (Dry Matter Intake) of TMR (Total Mixed Ration), AMS-feed and grazed grass were calculated for every day in the period and used as one of predictor variables in MME (Mixed Models Equations). Further analysis will quantify more effects of grazing on GHG emissions in these cows.

Proteasome activity and expression of E3 ubiquitin ligases in muscle of periparturient dairy cows

Y. Yang[1], H. Sauerwein[1], C. Prehn[2], J. Adamski[2], J. Rehage[3], S. Dänicke[4], B. Saremi[5] and H. Sadri[1]

[1]Institute of Animal Science, Physiology & Hygiene Unit, University of Bonn, 53115 Bonn, Germany, [2]Genome Analysis Center, Helmholtz Zentrum München, 85764 Neuherberg, Germany, [3]Clinic for Cattle, University of Veterinary Medicine Hanover, 30173 Hannover, Germany, [4]Institute of Animal Nutrition, Friedrich-Loeffler Institute, 38116 Braunschweig, Germany, [5]Animal Nutrition Research Group, Evonik Industries AG, 63457 Hanau, Germany; yyan@uni-bonn.de

Repartitioning of body protein in early lactation commonly occurs, as feed intake alone cannot meet the amino acids (AA) requirements. Skeletal muscle is the primary labile source of AA, but limited studies have investigated the regulation of protein metabolism in this tissue during late gestation and lactation. We hypothesized that the transition from late pregnancy to early lactation is associated with changes in muscle and plasma AA concentrations, which accompany with changes in expression and activity of the ubiquitin-proteasome system (the main proteolytic pathway in muscle) in dairy cows. Biopsies from M. semitendinosus and blood were collected from 11 pluriparous German Holstein cows on day (d) -21, 1, 21, and 70 relative to calving. AA and creatinine profiles in muscle and serum were quantified by targeted metabolomics using the Biocrates Absolute IDQ p180 Kit. A 20S proteasome assay kit was used for proteasome activity measurements. The mRNA abundance of 2 major muscle-specific E3 ubiquitin ligases, muscle RING-finger protein-1 (MuRF-1) and atrogin-1 was quantified by qPCR. The metabolizable protein (MP) balance of the studied animals was -429 and -71.9 g/d on d 1 and d 21, respectively. Cows returned to positive MP balance on d 70 (212 g/d). Muscle total AA remained unchanged, whereas those of serum were lower on d -21 and d 1 compared with d 21 ($P<0.05$). The muscle and serum creatinine were greater on d -21 and d 1 than other time-points ($P<0.001$). The proteasome activity was elevated on d -21 and 1 compared with the other time-points ($P<0.001$). The mRNA abundance of MuRF-1 and atrogin-1 increased from d -21 to d 1, remained unchanged until d 21, and then declined on d 70 ($P<0.05$). In conclusion, insufficient MP intake and thus AA deficiency resulted in muscle breakdown already ante partum.

Use of milk progesterone profile to predict reproduction status in dairy cows

I. Adriaens, T. Huybrechts, B. De Ketelaere, W. Saeys and B. Aernouts
KU Leuven, Kasteelpark Arenberg 30, 3001 Heverlee, Belgium; ben.aernouts@kuleuven.be

Optimal reproductive performance is a key factor contributing to the profitability of dairy farms. In order minimize reproduction-related losses, the accurate detection of heat, pregnancy and disorders is of primal importance. One possibility is the measurement of milk progesterone (P4), which provides a detailed description of the cow's reproduction status, allowing farmers to control their fertility management in detail and react quickly to aberrations. However, as the length, absolute level, and growth rates of P4 profiles vary both between and within cows, the currently used fixed threshold is suboptimal and the conversion of raw P4 data into useful information still needs improvement. Combining an online mathematical model with individual monitoring charts, allows for the identification of individual profile characteristics. This information can be translated into individualized actions, taking former and current knowledge of the milk P4 profile into account. A combination of sigmoidal functions was shown to properly describe the milk P4 cycle. Nevertheless, the corresponding profile characteristics still need to be translated into concrete actions for the farmers. A validation trail was conducted to verify the link between the profile characteristics and reproduction status. Milk P4 measurements were collected using the gold standard technique, ELISA. Additional information on milk fat and protein content was collected. The moment of ovulation was estimated through the analysis of the luteinizing hormone (LH) in the cow's serum and confirmed by ultrasonography. Serum samples were taken every two hours over a period of three days around estrus. Using this information, we were able to accurately verify the LH surge and, consequently, estimate the moment of ovulation. Next, the online mathematical model was fitted to the P4 data and the individual profile characteristics were related to the estimated moment of ovulation. Moreover, it was found that this individual approach performed slightly better relative to the use of a fixed threshold. Nevertheless, measures of the milk fat content, milk yield, and the milking interval were identified as not being useful for further refinement of the mathematical model.

Phenotypic and genetic correlations between workability and milk production traits in Polish HF cows

B. Szymik[1], P. Topolski[1], W. Jagusiak[2] and M. Jakiel[1]
[1]National Research Institute of Animal Production, Department of Genetics and Animal Breeding, ul. Krakowska 1, 32-083 Balice, Poland, [2]University of Agriculture, Department of Genetics and Animal Breeding, Al. Mickiewicza 24/28, 30-059 Krakow, Poland; bartosz.szymik@izoo.krakow.pl

The objective of this study was to estimate phenotypic and genetic coefficient correlations between workability and milk production traits in Polish Holstein-Friesian dairy cattle population: milking speed (MS) and temperament (MT) and five milk production traits: milk, fat and protein yield (kg), fat and protein content (%). Data were records of 11,901 cows milking from 2007 to 2014 and were collected in the database system SYMLEK. The cows were scored for workability at the second test-day of the first lactation. Phenotypic and genetic correlation coefficients between workability and milk production traits were estimated using REML method that has been implemented in the BLUPF90 package. The two trait linear model of observation was applied to estimate (co)variance components. The model included the fixed effects of herd-year-season (HYS), lactation stage, fixed regressions on percent of Holstein-Friesian genes and age of calving and random genetic effect. Phenotypic and genetic correlation coefficients between workability and milk production traits were estimated based on three traits models which always comprised both workability traits and one milk production trait. Estimated heritabilities were 0.12 for MT and 0.25 for MS. Coefficients of genetic correlations obtained between MS and milk production ranged from -0.34 for milk yield to 0.02 for fat content (%). Estimated coefficients of genetic correlations between MT and production were in range of -0.37 for protein content (%) to 0.71 for milk yield. Phenotypic correlation coefficients between MS and milk production traits ranged from -0.01 for fat content (%) to 0.11 for milk yield, and between MT and milk production were slightly lower from -0.01 for protein content (%) to 0.03 for milk yield. Obtained parameters were within the range of estimates published in the literature. Estimated coefficients of genetic correlations are large enough to be used as a tool of indirect selection for production.

Genetic parameters for calving traits in Polish Holstein-Friesian heifers

M. Jakiel, A. Żarnecki and B. Szymik
National Research Institute of Animal Production, Department of Animal Genetics and Breeding, ul. Krakowska 1, 32-083 Balice near Krakow, Poland; bartosz.szymik@izoo.krakow.pl

The objective of the study was to estimate heritabilities of direct and maternal genetic effects and the correlations between direct and maternal genetic effects for calving ease and perinatal mortality in Black-and-White Polish Holstein-Friesian heifers. Data consisted of 113,494 calving ease and perinatal mortality records from SYMLEK database and were restricted to single-calving heifers whose age at first calving was 18 to 48 months, gestation duration between 260 and 290 days, and records with information for both sire and maternal grandsire of calf. Heifers calved between 2006 and 2015. Calving ease was classified as unassisted, easy or difficult calving. Calves were classified live-born or dead (calves born dead or dying within 24 h of parturition). Genetic parameters for calving ease and perinatal mortality were estimated using linear and threshold sire- maternal grandsire models (I.Misztal-BLUPF90). Models included effects of herd-year of calving, year-season of calving, age of dam at calving, sex of calf, sire of calf, maternal grandsire of calf, birth year of sire and birth year of maternal grandsire. Effects of herd-year, sire of calf, maternal grandsire of calf and residual were random. A Gibbs sampling technique was used. Heritability estimates were larger in threshold than in linear models. When using the threshold model, the heritability for calving ease was 0.039 for direct effect and 0.027 for maternal effect. For perinatal mortality it was 0.054 and 0.135 for direct and maternal effects, respectively. When using the linear model, heritabilities for calving ease were 0.023 and 0.016 for direct and maternal effects, respectively, and for perinatal mortality they were 0.014 and 0.034. The genetic correlation between direct and maternal effects for calving ease was -0.27 for the threshold model and -0.35 for the linear model. The corresponding correlations for perinatal mortality were 0.24 and 0.17.

Rennet coagulation properties of bulk milk and its effect on Edam type cheese yield

I. Jõudu[1,2], M. Henno[2], M. Ots[1,2] and H. Mootse[2]
[1]Bio-Competence Centre of Healthy Dairy Products LLC, Kreutzwaldi 1, 51014 Tartu, Estonia, [2]Estonian University of Life Sciences, Institute of Veterinary Medicine and Animal Sciences, Kreutzwaldi 62, 51014 Tartu, Estonia; ivi.joudu@emu.ee

Research concerning cheese yield relationships with milk coagulation properties and protein content are contradictory and are usually carried out at the laboratory scale. The aim of this study was to investigate relationships between Edam type cheese yield and bulk milk protein content and rennet coagulation properties at the industrial scale. Milk samples (n=465) for the measurement of rennet coagulation and major compositional parameters were collected once a month throughout one year from 40 farms, which provided 80% of the raw material for the dairy plant. Cheese processing and vat milk data were obtained throughout a year from the dairy plant and the database contained information from 3,537 batches, each made from ~10 tons of milk. For statistical analysis, cheese batches were divided into four groups according to cheese fat content (30, 40, 45, and 50% in DM). Although the correlation between bulk milk protein content and curd firmness was moderately strong (r=0.54; P<0.05) this was not unambiguously found at the single farm level. There were several farms producing year-round milk with relatively high protein contents, but they also had poor coagulating ability. While on some farms that produced milk with moderate protein contents, milk curd firmness was good. Future study is needed to elucidate reasons for this. Cheese yield was significantly (P<0.05) affected by bulk milk coagulation properties. Milk consumption rate for cheese production was smaller when milk had better coagulation ability. This effect was more pronounced for cheeses having lower fat contents. Acknowledgments: EU Regional Development Fund financed the research in the framework of the Competence Centre Program of Estonia under Projects EU30002 and EU 48686 of the Bio-Competence Centre of Healthy Dairy Products LLC.

Genetics of protein fractions and free amino acids predicted by mid-infrared spectroscopy

G. Visentin[1,2], A. McDermott[1,2], D.P. Berry[1], M. De Marchi[2], M.A. Fenelon[3], M. Penasa[2] and S. McParland[1]
[1]*Teagasc, Animal and Grassland Research and Innovation Center, Moorepark, Fermoy, Co. Cork, Ireland,* [2]*University of Padova, Department of Agronomy, Food, Natural Resources, Animals and Environment (DAFNAE), Viale dell'Università 16, 35020 Legnaro (PD), Italy,* [3]*Teagasc, Teagasc Food Research Center, Moorepark, Fermoy, Co. Cork, Ireland; massimo.demarchi@unipd.it*

Milk protein fractions and free amino acids (FAA) are important traits for the dairy industry specialized in cheese and milk powder production. Despite this, they are not included in any breeding objective due to the lack of quick and cost-effective routine determination of these compounds. In the present study, mid-infrared spectroscopy prediction models previously developed on bovine milk samples in Ireland were applied to stored spectral data to predict both protein fractions and FAA concentrations. After edits, 134,546 test-day records from 16,166 lactations and 9,572 cows were available. Variance components were estimated using repeatability animal models including the fixed effects of contemporary group, milking time, cow breed proportion, general heterosis, recombination loss, and parity-by-stage of lactation. Random terms were additive genetic effect of the animal, and within- and across lactation permanent environmental effects. Means of α_{s1}-casein, α_{s2}-casein, β-casein, and κ-casein were 13.70 g/l, 3.66 g/l, 12.96 g/l, and 6.03 g/l, respectively. About 83% of total whey proteins (6.13 g/l) were represented by β-lactoglobulin. Concentrations of FAA exhibited average values that ranged from 8.09 µg/ml (glycine) to 1.52 µg/ml (valine), except for glutamic acid (30.93 µg/ml). Heritability estimates for protein fractions ranged from 0.36 (α_{s2}-casein and κ-casein) to 0.46 (β-lactoglobulin A), except for α-lactalbumin (0.19). The range of heritability estimates for FAA was smaller (0.15 for glycine to 0.36 for aspartic acid). Genetic correlations between casein fractions and whey proteins were all positive. Genetic correlations between FAA ranged from -0.44 (aspartic acid and lysine) to 0.97 (glutamic acid and total free amino acids). Breeding strategies for genetically improving protein fractions and FAA may represent a viable solution to meet the requirements of specialized dairy industry.

Imputation accuracy from a low density SNP panel in 5 dairy sheep breeds in France

H. Larroque[1], M. Chassier[1], R. Saintilan[2] and J.-M. Astruc[3]
[1]*INRA UMR1388 GenPhySE, Chemin de Borde Rouge, 31326 Castanet-Tolosan, France,* [2]*Allice, Domaine de Vilvert, 78350 Jouy-en-Josas, France,* [3]*Institut de l'Elevage, Chemin de Borde Rouge, 31326 Castanet-Tolosan, France; helene.larroque@inra.fr*

In France, genomic selection (GS) has been implemented in dairy sheep in Lacaune breed (LA) and more recently in Pyrenean breeds: Red-Faced Manech (RFM), Black-Faced Manech (BFM) and Basco-Béarnaise (BB). Genomic estimated breeding values (GEBV) are calculated on the basis of SNP50 Bead-Chip (50K) genotypes. Compared to dairy cattle, the relatively higher cost of genotyping limits the cost-effectiveness of GS in dairy sheep. However, a significant reduction of genotyping cost is forecast with the design of a low density (LD) chip by the International Sheep Genomics Consortium. Use of this chip requires harmonizing genotypes on LD panel and those existing on 50K panel by imputing missing SNPs. The aim of the study was to evaluate quality of imputation in 5 French dairy sheep breeds (with Corsica breed: CO) and its impact on GEBV. Genotypes of rams available in each breed (5,792 in LA, 1,741 in RFM, 470 in BFM, 556 in BB and 609 in CO) were split in two sets: a training set with oldest rams and a validation set with youngest rams (1,144 in LA, 430 in RFM, 80 in BFM, 140 in BB and 117 in CO). The 50K genotypes of validation set, after quality control, were 'pierced' in order to mimic LD genotypes. Out of 16,331 SNPs from LD panel, 9,822 were located on 50K panel. These genotypes were then imputed to the 50K panel, from the training population, using the FImpute software. Comparison of 50K imputed and true genotypes permitted to assess imputation accuracy. Concordance rates (CR: percentage of alleles properly imputed) per animal or per SNP were high, ranging from 96.6% to 99.1%, depending on breed. The squared correlations (r^2) between true and imputed genotypes ranged from 81.3% to 95%. On the 3 criteria, Lacaune breed had the higher scores followed by RFM, BB, BFM and finally CO breeds, in accordance with training population size. For validation rams, correlations between GEBV computed with true or imputed genotypes exceeded 0.99 for production and type traits whatever the breed. Consequently, the 4 breeds in GS have decided to use LD genotypes for pre-selection of young rams.

Correlation between genomic predictions based on female nucleus and traditional EBV in Sarda sheep

M.G. Usai, S. Casu, S.L. Salaris, T. Sechi, S. Sechi, S. Miari, P. Carta and A. Carta

AGRIS Sardegna, Settore Genetica e Biotecnologie, Loc. Bonassai, 07040, Olmedo, Italy; gmusai@agrisricerca.it

The use of female nucleus populations (NF) has been proposed as an alternative option to perform genomic selection for species or traits where large scale recordings are costly or difficult. Thus, Sarda sheep breeders set up a NF of approximately 1000 milked ewes with the aim of representing most of the genomic variability of the whole population by using, as sires, rams with a high genetic impact on the commercial flocks. A large range of traits are routinely measured on NF ewes and both ewes and rams are genotyped with the OvineSNP50 BeadChip. The aim of this study was to evaluate the correlation of genomic predictions for milk yield based on NF with the corresponding traditional EBV of a sample of Sarda rams. The NF consisted of 3,406 ewes with 11,585 lactation records. The validation sample consisted of 397 rams progeny tested in the Sarda herd book, 234 by artificial insemination (AI) and 163 by controlled natural mating (NM). One hundred and twelve AI rams directly contributed to NF as sires. The genomic predictions of rams were estimated by a repeatability animal model including records from NF ewes and the genomic relationship matrix among all animals (rams and NF ewes). Genomic predictions of rams were compared with the corresponding EBV from the official genetic evaluation 2016, estimated by a classical BLUP-animal model using a pedigree-based relationship matrix and lactation records just from the herd book population. Correlation between official EBVs and genomic predictions was 0.24 for the whole sample of rams. Different results were obtained for AI and NM rams with correlations of 0.44 and 0.17 respectively. Correlations within AI rams were 0.49 for the NF sires and 0.41 for the others. Correlations increased with the genomic relationship coefficients between NF and the different groups of rams. Results indicate that the use of a female nucleus is a promising strategy for genomic selection in species or traits where it is difficult to set up a traditional male training population. The main issue is to represent in the nucleus most of the genomic variability of the selected population. This study was funded by the project MIGLIOVIGENSAR, LR n.7/2007.

Genetic parameter estimations for feed intake in UK dairy goats

S. Desire[1], S. Mucha[1,2], R. Mrode[1,3], M. Coffey[1], J. Broadbent[4] and J. Conington[1]

[1]SRUC, W Mains Rd, EH9 3JG, United Kingdom, [2]Poznan University of Life Sciences, 33 Wolynska, Poznan, Poland, [3]ILRI, 30709 Naivasha Rd, Nairobi, Kenya, [4]Yorkshire Dairy Goats, St Helen's Farm, YO42 2NP, United Kingdom; suzanne.desire@sruc.ac.uk

Feed accounts for a large proportion of costs in dairy farming. Differences exist in how efficiently individual animals convert energy to milk, and genetic improvement of feed efficiency could play an important role in reducing feed costs per unit of output. This study aimed to estimate heritabilities for feed intake in dairy goats, and calculate genetic correlations between feed intake (FI), body weight (BW), and milk yield (MY). The data consisted of 9,970 test day records for FI (kg), MY (kg), and BW (kg) relating to 1,146 mixed-breed dairy goats. Animals were fed a digestible fibre-based blended feed *ad libitum* for the first 150 days of lactation, after which feed was restricted based on MY. Animals were the progeny of 64 sires and 909 dams, and the pedigree contained 6,644 animals. Breeding values were estimated using a univariate animal random regression model that included quadratic polynomials for random animal and permanent environment effects. Fixed effects were year-season of kidding, age at kidding, test day, feeding regime (*ad libitum* vs restricted feeding), and fixed lactation curves using third order Legendre polynomials. Each trait was included as a fixed effect for all other traits in the univariate analyses, but not in bivariate analyses used to calculate genetic correlations. An approximation of feed efficiency (FE) was obtained by including MY and BW as fixed effects. Heritability for FI, BW, and MY ranged from 0.04 (520 DIM) to 0.32 (124 DIM), 0.33 (37 DIM) to 0.68 (520 DIM), and 0.19 (513 DIM) to 0.36 (97 DIM), respectively. Genetic correlations at the start and end of lactation were as follows: FI and MY 0.52 (s.e. 0.25) to 0.91 (s.e. 0.05); FI and BW 0.08 (s.e. 0.21) to 0.47 (s.e. 0.21); BW and MY -0.24 (s.e. 0.27) to -0.25 (s.e. 0.22). The correlations indicate that selection for MY results in a rise in FI. The results suggest that there is sufficient genetic variation to select for FE in goats. We would suggest developing a multi-trait selection index due to the observed positive correlation between FI and BW.

Optimization of Swiss dairy goat breeding programs: modelling of different variations with ZPLAN

M.M. Hiltpold[1], B. Gredler-Grandl[2], A. Willam[3], B. Fürst-Waltl[3], J. Moll[2], S. Neuenschwander[1], E. Bangerter[4] and B. Bapst[2]

[1]*ETH Zürich, IAS, IAS, Animal Genetics, Tannenstr. 1, 8092 Zürich, Switzerland,* [2]*Qualitas AG, Chamerstr. 56, 6300 Zug, Switzerland,* [3]*University of Natural Resources and Life Sciences Vienna, Institut für Nutztierwissenschaften, Gregor-Mendel-Str. 33, 1180 Wien, Austria,* [4]*Schweizerischer Ziegenzuchtverband (SZZV), Schützenstr. 10, 3052 Zollikofen, Switzerland; maya@hiltpold.com*

Since 2010, breeding values (BV) are estimated for the Swiss dairy goat breeds Coloured Chamois, Saanen and Toggenburg. Traits considered are milk kg, fat % and protein % in 220 days of lactation. Nevertheless, selection is still based on a phenotypic minimum performance of buck dams for kg milk, % protein and conformation, which is rated in non-linear grades from 1 to 6. Because only natural service is used, living bucks rarely have enough daughters to obtain a published breeding value. For comparison, the current selection scheme and possible alternatives were modelled in ZPLAN. This deterministic approach allows a fast calculation of genetic gain and discounted profit by including costs and economic weights. Selection groups were milk performance recorded goats (MPG), natural service bucks (NSB), testing bucks (TB) and progeny proven bucks (PPB) in the breeding unit and unrecorded production goats (PG) in the production unit. PPB were selected in two-step selection out of best TB. Traits included in the selection index were the yield traits milk kg, fat % and protein %, the conformation traits breed character, format, fundament, udder, and teats and the fitness traits lactation somatic cell score (LSCS) and gastro-intestinal nematode resistance (GINR). However, these fitness traits are currently not phenotyped in Switzerland. Modelled alternatives were use of parent averages (PA), application of a progeny testing scheme (PTS) and consideration of fitness traits. Selection of NSB by PA almost doubled the accuracy of their BV to 0.40. Accuracy of PPB's BVs was 0.73 to 0.79 depending on traits phenotyped. The current breeding program has, based on the assumptions, very small profit. Simple measures like considering PA and including LSCS would increase the profit considerably. Phenotyping and breeding for GINR and PTS need more effort, but would additionally increase profit.

Genomic evaluation models including casein and/or SNP50K information in French dairy goat population

M. Teissier, H. Larroque and C. Robert-Granié

INRA, UMR1388 GenPhySE, 24 chemin de Borde Rouge, 31326 Castanet-Tolosan, France; marc.teissier@inra.fr

Genomic selection using single-step GBLUP (ssGBLUP) approach (based on female phenotypes) will be implemented at the end of 2017 in the two main French dairy goat breeds (Alpine and Saanen). Higher accuracy is expected by implementing approaches that include well-known major genes such as the αs1 casein gene. The αs1 casein is a multi-allelic gene located on chromosome 6 with significant effect on protein content in the two goat breeds. 6 alleles were identified in French goat populations and were each associated with a different level of αs1 casein synthesis. Genotypes of αs1 casein gene were available since 1982 for 4,203 bucks used for artificial insemination and 2,999 dams from Alpine and Saanen breeds. The aim of this study was to investigate methods of genomic evaluation including αs1 casein information to improve accuracies. Genomic evaluations were investigated both for the whole population (multi-breed pooling together Alpine and Saanen animals) and also for each breed separately. Several genomic models including 50k BeadChip genotypes and/or αs1 casein information were tested on protein content. A classical ssGBLUP approach using only SNP information was compared to (1) weighted ssGBLUP approaches allocating a weight to each SNP and used to construct the genomic relationship matrix (G), (2) a TABLUP approach based on some SNP selected to construct G and (3) a gene content approach using both αs1 casein and SNP genotypes in a multiple trait model. Cross validation analyses consisted into splitting the population of the 905 males genotyped with the goat 50K BeadChip into a training set of 554 males born before 2007 and at test set of 351 young males born between 2008 and 2012. Results on validation correlations were slightly better in multi-breed than within breed for all approaches. Selecting the most associated SNP to protein content with TABLUP approach did not improve validation correlations compared to ssGBLUP. Results of gene content approach and weighted ssGBLUP approaches were close and showed an improvement of the validation correlations (1% to 4.58% in multi-breed analyses, 0.4% to 3.10% in Alpine population and 2.90% to 5.15% in Saanen population) compared to ssGBLUP approach.

Genetic parameters of litter size and weight in Finnsheep ewes

M.-L. Sevón-Aimonen[1], K. Ahlskog[2] and A. Kause[1]
[1]Natural Resources Institute Finland (Luke), Myllytie 1, 31600 Jokioinen, Finland, [2]ProAgria Southern Finland, P.O. Box 97, 33101 Tampere, Finland, Finland; marja-liisa.sevon-aimonen@luke.fi

Finnsheep is an exceptionally prolific breed having an average litter size of 1.9 in the first parity and 2.5 in later parities. We estimated (co)variance components for litter size, litter weight and litter weight gain to be used in the breeding program for maintaining and improving the good productivity of ewes. A total of 108,304 lambing events were available between the years 1990 and 2016 from 43,327 ewes and 16,649 additional pedigree animals from the Association of ProAgria Centres. The fitted animal model contained birth season of ewes, size of birth litter of ewes, lambing age, time between parities, birth week of litter and herd-year as the fixed effects, and permanent and genetic effect of ewes as the random effects. The age at weighting day was used as a covariable for weight traits. The estimated heritability (and its SE) was 0.07 (0.01) for the total number of born, 0.04 (0.00) for the number of born alive, 0.03 (0.01) for the number of stillborn, 0.04 (0.01) for the number of alive at 3 d age, 0.03 (0.01) for the number of alive at 42 d, 0.05 (0.01) for the weight of litter at 3 d, 0.04 (0.01) for the weight of litter at 42 d, and 0.04 (0.01) for the weight gain from 3 d to 42 d. The common effect of ewes was low and varied from zero to 0.04. Genetic correlations r_g (SE) were highly positive between the total number of born and other litter size traits, varying from 0.67 (0.08) to 0.94 (0.02). There were unfavourable r_g 0.81 (0.04) between the total number of born and the number of stillborn. The r_g between litter size and weight traits were favourable varying from 0.32 (0.08) to 0.88 (0.05). The genetic correlations between number of stillborns and weight traits were low (no significant). The results indicate that genetic variation exists for litter productivity traits, although the heritability estimates were low. The favourable correlations between litter size and weight traits enhance the selection for better productivity of ewes. High unfavourable correlation between total number of born and number of stillborn demands special attention in breeding for larger litter size in order to prevent an increase in death rate of lambs. This research was funded by the EU iSAGE project.

Global experiences with genetic and genomic improvement in small ruminants

T.J. Byrne and P.R. Amer
AbacusBio Limited, 442 Moray Place, 9016 Dunedin, New Zealand; tbyrne@abacusbio.co.nz

Small ruminants support the prosperity of small and large populations of people, in developing and developed countries all over the world. As such, increasing the performance and profitability of these animals is hugely important. While numerous selection programmes exist for sheep and goats globally, they vary greatly in their efficacy. Unlike pig and poultry, and to a lesser extent dairy and beef cattle breeding, sheep and goat industries presents several practical challenges in the context of structured selection. The low-input nature of sheep and goat farming systems, which is in part a consequence of feeding systems based on pasture/low quality supplement and the relative lack of scale of individual farming enterprises, reduces the opportunity to exercise much control or integrate across industry sectors. Small ruminants have a low reproductive rate, which reduces selection intensity. The diversity of breeding goals (due to the range of products and farming environments) and therefore breeding objectives compromises adoption of genetic improvement technology. A lack of vertical integration (the connection of the entire industry from production to market) in terms of both ownership and price signals reduces the efficiency of application of genetic improvement technologies. All these practical challenges often result in technical and financial limitations to quantitative selection. Cultural and societal structures also play a significant role in defining the value of the animal. This paper highlights experiences with genetic and genomic improvement in small ruminants around the world and gives examples of innovation in the application of technology to overcome the challenges, and to support the success of these programmes.

From animal genetics to human medical research: day blind sheep as a model for vision gene therapy

E. Gootwine[1], E. Banin[2], E. Seroussi[1], H. Honig[1], A. Rosov[1], E. Averbukh[2], A. Obolensky[2], E. Yamin[2], R. Ezra-Elia[3], M. Ross[3], G.J. Ye[4], M.S. Shearman[4], W.W. Hauswirth[5] and R. Ofri[3]
[1]The Volcani Center, P.O. Box 15159, 7528809, Israel, [2]Hadassah-Hebrew University Medical Center, Jerusalem, 7691120, Israel, [3]Hebrew University of Jerusalem, Koret School of Veterinary Medicine, Rehovot, 7610001, Israel, [4]Applied Genetic Technologies Corporation, 14193 NW, 119th 13 Terrace, Suite 10, Alachua, FL 32615, USA, [5]University of Florida, Gainesville, FL 32610-0284, USA; gootwine@agri.gov.il

We reported two types of recessive mutations in the ovine CNGA3 gene causing day-blindness in Awassi sheep: a premature termination (p.Arg236*), and a deleterious substitution (p.Gly540Ser, Invest Ophthalmol Vis Sci, in press, 2017). Culling rams that carried the mutations eliminated the birth of affected lambs in commercial flocks. Alongside, we established an experimental flock comprising day-blind sheep for both mutations to serve as animal models for human achromatopsia (ACMH). To investigate gene augmentation therapy for ACMH, Adeno-Associated Viral vectors carrying the intact human or mouse CNGA3 gene under the control of a cone specific promoter were delivered unilaterally into the subretinal space of affected sheep. Animals were electrophysiologically and behaviorally assessed preoperatively and after treatment. Cone function was measured by electroretinography (ERG) following light adaptation (10 min., 30 cd/m^2). Responses to flash and flicker (10-80 Hz) stimuli were recorded at 4 intensities (1-10 cd×sec/m^2). Behavioral assessment included night time and day time maze testing, where passage times and number of fence collisions were recorded. Normal and untreated day-blind sheep were assessed as controls. While cone function was significantly depressed in affected sheep prior to surgery, there was a significant improvement following treatment. Untreated affected animals failed to navigate the maze during day time, yet, following treatment their ability improved, approaching that of normal controls. The ERG and behavioral improvement persisted for up to 5.1 year post-op without affecting animals' health. Based on these results, we plan to evaluate the ACMH gene augmentation therapy in a Phase 1/2 clinical trial.

Genetic evaluation of Awassi sheep for preweaning growth traits

K.I.Z. Jawasreh and F.H. Iye
Jordan University of Science and Technology, Animal Production, Irbid, Jordan, 3030 Al Ramtha, Jordan; kijawasreh@just.edu.jo

The main objectives of the present study were to calculate some of fixed effects (Sex, Type of Birth, Location and Year) and to estimate some genetic for pre-weaning growth traits birth (BWT) and weaning (WWT) preweaning daily (PWDG) gain and weaning age(WA)) of Awassi Lambs. A total number of 5,131 Awassi lambs, that were progeny of 181 sires and 927 dams were used in this study. Proc mixed and Restricted Maximum Likelihood (REML) procedures were used for conducting the analysis of the Fixed and Random effects that included in the Model. The overall means of birth, WWT, PWDG and WA were; 4.46±0.013 kg, 17.42±0.07 kg, 0.181±0.001 g and 64.89±0.282 d, respectively. The studied traits were significantly affected by Sex, Type of Birth, location and birth year. Heritability estimates were; 0.30 for BWT, 0.18 for WWT and PWDG and 0.19 for WA. Highly significant genetic correlation's were obtained between all studied traits, genetic correlation between traits were ranged between 0.62 and -0.42, positive genetic correlations were obtained between BWT and the other traits, while negative correlations were obtained between WWT and WA, PWDG and WAGE. Significant Phenotypic correlations between growth traits were obtained between the growth traits, ranged between 0.48 and -0.43. Positive phenotypic correlation between BWT and the other growth traits was obtained. Negative correlation between WWT and WA, between WA and average PWDG were obtained in this study. The highest additive genetic variance was obtained for WWT (34.581) while the lowest was estimated for the average daily gain (6.22×10^{-04}). Highest phenotypic variance obtained at WA (175.5), while the lowest value obtained at BWT (0.53). Maternal additive variance ranged between 0.12 and 0.034. The genetic trends were also estimated depending on the estimated breeding values that were negative while it was positive only for BWT. The opposite was obtained for WWT in term of genetic trend. It can be concluded that the absence of genetic selection programs that depends on the Breeding values estimates may explain this negative direction of genetic trend, and the selection should be conducted using the estimated breeding values for gaining more improvement in the two stations.

Identifying signature of selection in three Iranian sheep breeds

M. Pouraskari, T. Harkinezhad and M.B. Zandi
University of Zanjan, Animal science, Zanjan, Iran; mbzandi@znu.ac.ir

Identifying the genes and genomic regions associated with economic traits is one of the most important reasons to use of animal genetic dense markers. In this study, a genome-wide scan of three Iranian sheep breeds: Afshari, Moghani, and Ghezel were studied to identify genomic regions associated with fat storage by using 50K SNP genotyping results. Here unbiased F_{ST}, population differentiation statistics is used to identifying selection single on the genome. Five genomic regions on chromosomes 3, 13, 15 and 22 respectively in the 99.99 percentile of F_{ST} values were selected for further analysis. Linkage disequilibrium and Extended Haplotype Homozygosity(EHH) was performed to verifying the signature of the selection region. The results of unbiased F_{ST} combined with bifurcation diagrams strong verify the population differentiation and signature of selection for these three breeds. Finally, investigation QTL reported in the other species showed that these areas overlap with QTL for fat traits in cows. This result can provide a new insight about the fat storage in sheep to design a molecular breeding strategy.

Genomic structure of Iranian sheep breed by whole genome survey

P. Nabilo, M.B. Zandi and T. Harkinezhad
University of Zanjan, Animal science, Zanjan, Iran; mbzandi@znu.ac.ir

Molecular genetics studies allow a comparison of genetic diversity within and across breeds and make a new insight to reconstruction the breed history and history of ancestral populations. The aim of this study was to investigate the allele frequencies, genetic structure and genetic variation of indigenous and exotic breeds by using 50,000 SNP markers. Genotype data of indigenous breed(Afshari, Ghezel and Moghani) and exotic breeds(Dorper, Merinos and Romney) were obtained from the project OvinHapMap. Statistical analysis performed by population stratification, population multivariate statistics, and model-based approach. Differentiation and genetic diversity among Iranian indigenous and exotic breeds were recognized and all statistics methods showed a distinct structure for Iranian indigenous and exotic populations. The result showed that Iranian sheep breeds have more genetic similarity rather than exotic breeds so Iranian sheep breeds grouped as one category and Romney, Merinos and Dorper breeds were separated as distant groups. Although, when the indigenous breeds were studied separately, each breed had great of differentiation. Based on these results we can conclude that SNP array can be used for sheep population stratification analysis and Iranian indigenous sheep breed can be divided into three genetic groups separately, also we can conclude that the low contribution of genetic variation of Iranian indigenous sheep breeds to the high genetic diversity of exotic sheep breed in Globe.

The effect of divergent selection for a composite trait on genetic responses in components

S.W.P. Cloete[1,2], J.B. Van Wyk[3] and J.J. Olivier[1]
[1]Directorate Animal Sciences: Elsenburg, Private Bag X1, Elsenburg 7607, South Africa, [2]University of Stellenbosch, Department of Animal Sciences, Private Bag X1, Matieland 7602, South Africa, [3]University of the Free State, Department of Animal, Wildlife and Grassland Sciences, P.O. Box 339, Bloemfontein 9300, South Africa; vanwykjb@ufs.ac.za

It is generally accepted that reproduction traits are economically important in sheep. A composite trait such as lamb output per ewe joined, termed as net reproduction rate, was suggested to most closely resemble the breeding objective strived for. However, selection decisions may be complicated by the composite nature of this trait. Arguments were proposed in the literature that selection for such a composite trait will ensure that the contributing component traits remain in balance. This study reports genetic and phenotypic parameters and trends for composite as well as component reproduction traits in a Merino resource flock kept at the Elsenburg Research Farm near Stellenbosch. Two lines were divergently selected from the same base population from 1986 to the present, using maternal ranking values for number of lambs reared per joining. Data recorded from 1987 to 2007 were used in this study, as outside rams also contributed progeny in subsequent years. These outside sires were introduced to link this genetic resource with the broader South African Merino industry. Three component traits (ewes conceived per ewe joined, lambs born per ewe lambed and average lamb weaning weight) and three composite traits (number of lambs born and weaned as well as total weight of lamb weaned, all per ewe joined) were assessed. Most traits were variable and heritable and should respond to trait selection. Selection for number of lambs weaned resulted in genetic trends in the desired direction in those component traits considered here. In terms of composite traits, both lines responded as expected from the selection pressure applied. Expressed relative to the overall means, the responses in the composite traits were somewhat larger in magnitude compared to component traits. These results confirm a previous suggestion that composite trait selection for number of lambs weaned per ewe joined in ewes resulted in concomitant favourable responses in the component traits studied here.

Optimal contribution selection applied to the Norwegian Cheviot sheep population

M.H. Kjetså[1], J.H. Jakobsen[2], B.S. Dagnachew[1] and T.H.E. Meuwissen[1]
[1]Norwegian University of Life Sciences, Faculty of Biosciences, Aboretveien 6, 1432 Ås, Norway, [2]The Norwegian Association of sheep and Goat Breeders, Box 104, 1431 Ås, Norway; maria.kjetsa@nmbu.no

Managing inbreeding is essential for all breeding work, especially in small populations under selection. Optimal Genetic Contribution Selection (OCS) is a selection method that restricts inbreeding while maximizing genetic gain. The aim of this study was to apply OCS to the practical Norwegian Cheviot selection scheme, maximizing the genetic progress for total merit breeding value when rate of inbreeding is restricted to 1% per generation, and compare it to the realized selection. The ram selection process takes place in three stages within and across four ram circles. The ram circles selects test rams for progeny testing at 6 months of age and elite rams among previous year's progeny tested rams within each ram circle. Test- and elite rams are used for natural service only within their ram circle due to national restrictions on animal movement. The Norwegian Association of Sheep and Goat breeders selects rams for artificial insemination (AI) among the elite rams from the previous year across the ram circles. The data contained 1,751 pre-selected ewes and 156, 54 and 12 test-, elite- and AI ram selection candidates respectively. OCS was not able to handle both within and across ram circle selection simultaneously. Thus, four scenarios were explored: (1) within ram circle selection of all rams; (2) within ram circle selection of test and elite rams only; (3) across ram circle selection of AI rams only; or (4) across ram circle selection of all rams. All the scenarios kept the inbreeding restrictions. In scenarios 1,2 and 4 the average breeding values of the selected rams are significantly higher than the average breeding values of the realized selection(P<0.001). In scenario 3, there was no significant difference between OCS and realized selection. Scenario 3 would be the easiest scenario to implement in practice because the AI ram selection is already centrally controlled. However, scenario 2 would also be possible to implement and the results show that with scenario 2, OCS can increase the genetic gain in the population and at the same time keep the rate of inbreeding within preset limits.

Preliminary results for lifetime production of Pag sheep in Croatia
M. Špehar, M. Dražić, D. Mulc and Z. Barać
Croatian Agricultural Agency, Ilica 101, 10000 Zagreb, Croatia; mspehar@hpa.hr

Lifetime production (LP) is a trait of considerable importance in sheep husbandry which could be seen as a composite of production, health, and reproduction within specific environmental and farming conditions. Pag sheep is a Croatian autochthonous breed, primarily used for milk production. The objective of this study was to investigate the effects of breeding area corresponding to the side of the island and flock size on the LP of the Pag sheep in Croatia. Data included 15,131 records for 4,301 ewes recorded from 2005 to 2016. Analyses were performed using Proc Lifetest in SAS statistical package. Data were stratified by breeding area (1= northeast part, 2= southwest part) and flock size (small = 1 to 50 ewes, medium=51 to 100 ewes, and large = more than 100 ewes). The following null hypotheses were tested in Proc Lifetest: (1) there are no differences in the overall mean LP of ewes bred in the northeast part and southwest part of the island; and (2) there are no differences in the mean LP of ewes reared in the flocks having different size. Differences between curves for LP were tested using Kaplan-Meier estimator and the log-rank test. Significance was set at $P<0.05$. In general, approximately 50% of ewes (n=694 of 1,431) had LP of 10 years on the southwest part. However, 50% of survived ewes (n=3,176 of 5,895) on northeast part were age of nine years. Flock size influenced LP and approximate estimated 50% survival was 8, 9, and 10 years for large, small, and medium sized flocks, respectively. These preliminary results demonstrate the effect of environmental and farming conditions on LP.

Ewe reproduction affected by crossbreeding Corriedale and Dohne Merino
I. De Barbieri, F. Montossi and G. Ciappesoni
Instituto Nacional de Investigación Agropecuaria, Ruta 5. km 386, 45000, Tacuarembó, Uruguay; idebarbieri@tb.inia.org.uy

Given the changes observed in wool and sheep-meat world markets in the last two decades, Dohne Merino (DM) breed has grown in Uruguay, particularly used in crossbreeding schemes with the most popular breed (Corriedale; C). The focus of the study was to evaluate the effect of the crossbreeding between DM and C on ewe fertility, prolificacy and lambing percentage. We evaluated 530 records (223 hoggets and 307 ewes) from 384 animals of three genotypes: 100% C (100C), 50%DM×50%C (50DM), and 75%DM×25%C (75DM) during three years. Ewes from each genotype were assigned randomly to 16 DM sires, and managed together grazing native grasslands. Cervical insemination of ewes with fresh semen was performed in April. Ewe unfasted body-weight (BW) was assessed at the beginning of insemination. Ultrasound scanning was performed 45 days after the end of insemination and pregnancy status and number of foetuses per ewe were evaluated. During lambing, ewes were observed twice daily and number of lambs per ewe were recorded. Fertility (pregnant/inseminated ewes), prolificacy (number of lambs/pregnant ewes), and lambing percentage (number of lambs/inseminated ewes) were analysed using mixed models with repeated measures in a randomized block design with the GENMOD procedure. Year, genotype, and ewe age group were treated as fixed effects, while the animal as random effect, and BW was included as a covariate. Fertility was not affected (P>0.05) by the genotype with or without including the BW in the model, ranging from 0.83 to 0.87. Prolificacy was greater (P<0.05) in ewes from 50DM (1.26) group than in the other two groups, while no differences were observed between 100C (1.13) and 75DM (1.13). When including BW in the model differences in prolificacy between 50DM and 100C were not detected. Genotype affected (P<0.05) the lambing percentage, being greater in 50DM than 100C and 75DM, though this effect was not observed when considering BW as a covariate. We conclude that prolificacy and lambing percentage of C can be improved in the first cross by using DM. This effect is partially related with the BW differences between genotypes and will be not kept by increasing the percentage of DM within this crossbreed.

Selection signatures of fat tail in sheep

B. Moioli[1], F. Pilla[2], B. Portolano[3], S. Mastrangelo[3] and E. Ciani[4]
[1]Consiglio per la ricerca in agricoltura e l'analisi dell'economia agraria, via Salaria 31, 00015 Monterotondo, Italy, [2]Università degli Studi del Molise, Via De Sanctis, 86100 Campobasso, Italy, [3]Università degli Studi di Palermo, viale delle Scienze, 90128 Palermo, Italy, [4]Università degli Studi di Bari Aldo Moro, Via Amendola 165/a, 70126 Bari, Italy; bianca.moioli@crea.gov.it

The investigation of the genes with a role in lipid metabolism enjoy considerable scientific and commercial interest because of the strong correlations between fat deposition and the risk of cardiovascular disease. The fat tail characteristic of sheep is the adaptive response to harsh environment, and beyond representing a valuable energy reserve for facing future climate changes provides clues for elucidating the physiology of fat deposition. Studies on various sheep populations detected fat-tail signatures on chromosomes 2, 3, 5, 6, 7 and 13. Fat-tailed sheep represent about 25% of the world's sheep population, and the genes with a role in this phenotype are likely not the same for every breed, since the wild ancestor of sheep had a thin tail, and the fat tail was selected by humans in longstanding husbandry practices in different regions. In the present work, a genome-wide scan using ~50,000 Single Nucleotide Polymorphisms was performed to identify selection signatures for the fat tail in the Barbaresca sheep, an Italian breed originating from North Africa. Fst values of differentiation, and χ2 test of significance of allele frequency were calculated, for each marker, between the Barbaresca and each of 13 Italian thin-tailed breeds. Strong signals of selection were detected for all 13 breeds on chromosome 6, in a region encoding the *SLIT homolog* 2 gene, this gene acting as a molecular guidance cue in cellular migration. The signature on chromosome 7 was very strong only in some of the breeds used for comparison: the detected signal was located in proximity of the *Vertnin* gene, a candidate for variation in vertebral number, and was already revealed in Iranian and Mediterranean fat-tailed breeds, but not in the Chinese sheep, so confirming the complexity of the fat-tail phenotype, which is associated in some breeds to long and pendulous tail, while, in other breeds, to the short tail.

Phenotypic and genetic parameters of growth traits of the fat tail barbarine sheep breed

I. Ben Abdallah, A. Hamrouni and M. Djemali
National Institute of Agriculture of Tunisia, Animal Science, 43 Street Charles Nicole, Mahrajene, Tunisia, 1082 Tunisia, Tunisia; ichrak25121988@gmail.com

Information on pedigree is often lacking in small ruminants, particularly in developing countries. In Tunisia several authors have been interested in the estimation of genetic parameters of growth traits of the fat tail Barbarine sheep. Different results were reported according to methods and databases used. The objective of this work was to estimate the genetic parameters of live weights and gains of the Barbarine breed. A total of 193,371 lambs with growth records collected over 10 years from 2004 to 2014 was used in this study. This file was used to compute means, variation and phenotypic correlation of growth traits. The lack of connection between flocks was such that only two subgroups could be analyzed separately for the estimation of genetic parameters. The first contained 670 lambs with 70 sires and the second had 634 lambs from 46 sires. Growth traits were adjusted for sex of lamb, type of birth and age of dam. Genetic components were estimated using the REML method based on a sire model with fixed effects (farm-herd- year of birth and season). The main results showed that the average live weights were 5±0.4 kg, 6±1.04 kg, 10±2.02 kg, 16±3.3 kg, 18±3.6 kg, 182 g/d, 143±43 g/d, 135±40 g/d for birth weight, weight at 30 d, weight at 70 d, weight at 90 d, and weight gains ADGb-30d, ADG30-70d and ADG30-90d, respectively. Phenotypic correlations between the different live weights and live weight gains were all positive from 0.122 to 0.99. Heritability estimates of growth traits in the Barbarine breed ranged from 0.09±0.305 to 0.409±0.639.

Slovak dairy sheep: formation of composite breed and its characterization

M. Margetín[1,2], L. Mačuhová[2], M. Oravcová[2], M. Janíček[1] and J. Huba[2]
[1]*Slovak University of Agriculture Nitra, Tr. Andreja Hlinku 2, 94976 Nitra, Slovak Republic, [2]National Agricultural and Food Centre, Research Institute for Animal Production Nitra, Hlohovecká 2, 95141 Lužianky, Slovak Republic; macuhova@vuzv.sk*

In the beginning of 1990s, a programme for a new synthetic population of the Slovak Dairy sheep (SD) was launched. Some flocks under performance testing were considered to involve the establishment of breed was divided into two periods: first, crosses of indigenous breeds (Improved Valachian, Tsigai and Merino) with improving breeds (Lacaune, East Friesian) of various proportion were formed; second, close (inter se) breeding scheme within the crossbred population has been applied. Parents of the next generation are chosen on the basis of breeding values for milk yield and litter size. About 5,000 ewes are included in performance testing; about 30% yearling females enter the breeding scheme each year and about 50,000 ewes of SD are kept in commercial flocks at present. Milk and reproduction traits of SD were analyzed using mixed model methodology (SAS 9.2) that included fixed and random effects as follows: with milk traits analyses included 20,511 milk records belonging to 11,026 ewes kept in 52 flocks during the period between 1995 and 2015; with reproduction traits analyses included 30,034 litter size records belonging to 9,671 ewes kept in 26 flocks during the period between 1997 and 2015. The normalized milk yield (NMY) and average daily milk yield (ADMY) increased more than twice: 79.8±5.36 l (1995) vs 164.0±2.04 l (2015) and 495.1±33.5 ml (1995) vs 1,035.3±12.8 ml (2015), respectively. NMY in ten percent of the most productive ewes were equal to 257.6 l, ADMY equal to 1,604 ml and dry matter equal to 26.6 kg. Fat and protein contents decreased from 7.82±0.101% (1995) to 7.27±0.040% (2015) and from 5.83±0.043% (1995) to 5.69±0.017% (2015). Ewes of SD have good udder traits. Litter size in flocks of above average level of reproduction traits is more than 150%. Average daily gains from birth to weaning were 0.26 kg (female lambs) and 0.30 kg (male lambs). The population of SD is intended to be recognized distinctive breed in 2017. The study was supported by the projects APVV 0458-10, VEGA 1/0364/15 and KEGA 035 SPU-4/2015.

Purebreeding of Red Maasai and crossbreeding with Dorper sheep in different environments in Kenya

E. Zonabend König[1], E. Strandberg[1], J.M.K. Ojango[2], T. Mirkena[3], A.M. Okeyo[2] and J. Philipsson[1]
[1]*Swedish University of Agricultural Sciences, Department of Animal Breeding and Genetics, P.O. Box 7023, 75007 Uppsala, Sweden, [2]International Livestock Research Institute, P.O. Box 30709, Nairobi 00100, Kenya, [3]Hawassa University, School of Animal and Range Sciences, College of Agriculture, P.O. Box 5, Hawassa, Ethiopia; erling.strandberg@slu.se*

The aim of this paper was to study opportunities for improvement of the indigenous but threatened Red Maasai sheep (RM) in Kenya, by comparing purebreeding with crossbreeding with Dorper sheep (D) as a terminal breed, in two different environments (Env. A and a harsher Env. B), assuming genotype by environment interaction (G×E). Breeding goals differed between the environments and breeds. Four scenarios of nucleus breeding schemes were stochastically simulated. Overall, results showed an increase in carcass weight produced per ewe by more than 10% over 15 years. Genetic gain in carcass weight was 0.17 genetic SD/yr (0.2 kg/yr) across scenarios for RM in the best environment. For survival and milk yield the gain was lower (0.04-0.05 genetic SD/yr). With stronger G×E, the gain in the commercial tier for RM in the harsher Env. B became lower (nucleus was in Env. A). Selection of females also within the commercial tier gave slightly higher genetic gain. The scenario with purebreeding of RM and a subnucleus in Env. B gave the highest total income and quantity of meat. However, quantity of meat in Env. A increased slightly from having crossbreeding with D, whereas that in Env. B decreased. Crossbreeding of RM with D is not recommended for harsh environmental conditions due to differences in performance of the breeds in the different environments.

Genomic characterization of two Greek goat breeds using microarray chips

S. Michailidou[1,2], A. Argiriou[1], G. Tsangaris[3], G. Banos[2,4] and G. Arsenos[2]
[1]CERTH, Institute of Applied Biosciences, 6th km Harilaou-Thermis, 57001, Thermi, Greece, [2]Aristotle University of Thessaloniki, School of Veterinary Medicine, 54124, Thessaloniki, Greece, [3]Biomedical Research Foundation of the Academy of Athens, Proteomics Research Unit, Center of Basic Research II, 4 Soranou Ephessiou St, 11527, Athens, Greece, [4]SRUC, The Roslin Institute, Easter Bush, EH25 9RG, Scotland, United Kingdom; michailidouso@gmail.com

The Greek national herd of goats is the largest in the European Union comprising approximately 5 million animals. About 90% belong to the autochthonous breed Eghoria that is characterized by large variability in phenotypic characteristics. The second major autochthonous breed, Skopelos, is a highly homogenous population of about 11.000 animals. None of these breeds has been genetically characterized at genomic level. With this study we report, the genomic profiles of the two most important Greek goat breeds based on the analyses of Illumina's CaprineSNP60K genotyping microarrays. In total, 72 samples were genotyped and analyzed. Quality control on the 53,347 SNPs was performed by testing for Hardy-Weinberg equilibrium (a=1e-06), missing values on the genotyped SNPs (<0.1), SNP call rate (>0.95) and minor allele frequency (MAF<0.01). Thereafter, 50,130 remaining SNPs were further evaluated in R to assess genetic associations between breeds using the *SNPassoc* package. Investigation of the level of relatedness between breeds was also measured by calculating the pairwise F_{ST} value. One hundred and seventy six SNPs were statistically informative and could be applied to genetically separate the two breeds after Bonferroni correction of p values at level 1e-06. In addition, principal component analysis (PCA) was conducted and heat-map dendrograms were constructed in R, using the PCA3D and gplots packages, respectively. Results revealed that the Eghoria breed exhibits higher genetic diversity compared to the Skopelos breed. Moreover, an overlapping cluster was formed between two-breed crosses, suggesting that these animals should be excluded from the purebred pool for further breeding efforts. This study presents the first genome-based analysis of the genetic structure of the two Greek goat breeds and identifies markers that can be potentially used in breeding programmes.

Genomic characterization of three Greek sheep breeds using microarray chips

S. Michailidou[1,2], A. Argiriou[1], G. Tsangaris[3], G. Banos[2,4] and G. Arsenos[2]
[1]CERTH, Institute of Applied Biosciences, 6th km Harilaou-Thermis, 57001, Thermi, Greece, [2]Aristotle University of Thessaloniki, School of Veterinary Medicine, 54124, Thessaloniki, Greece, [3]Biomedical Research Foundation of the Academy of Athens, Proteomics Research Unit, Center of Basic Research II, 4 Soranou Ephessiou St, 11527, Athens, Greece, [4]SRUC, The Roslin Institute, Easter Bush, EH25 9RG, Scotland, United Kingdom; michailidouso@gmail.com

Dairy sheep is among the most significant livestock species in Greece with approximately 10 million animals. They are characterized by a large phenotypic variation and reared under various farming systems. The genetic structure of sheep breeds in Greece has never been studied with high-throughput whole-genome technologies. In this study, we used genome-wide genotyping to characterize the genetic diversity and population structure of three Greek dairy breeds of sheep; Boutsko, Karagkouniko and Chios. Ninety-six samples were genotyped with the Illumina OvineSNP50 microarray beadchip. Quality control on the 54,241 SNPs was performed for Hardy-Weinberg equilibrium (a=1e-06), missing values on the genotyped SNPs (<0.1), SNP call rate (>0.95) and minor allele frequency (MAF<0.01). The remaining 51,098 SNPs were further analyzed in R with four different genetic inheritance models (dominant, recessive, log-additive and codominant) in order to trace statistically significant SNPs that distinguish the breeds. In all cases, independently of the genetic model, Chios and Karagkouniko breeds shared the least number of discriminatory SNPs, indicating that they are more genetically related to each other compared to the Boutsko breed. Calculation of pairwise F_{ST} was used to explore the genetic relationship among breeds. Furthermore, principal component analysis (PCA) and heatmaps were constructed using R. All samples from the Boutsko breed clustered together and were genetically distant from the other two breeds. Concerning Chios and Karagkouniko, a distinct cluster was formed for each breed. Three animals of the Karagkouniko breed clustered with Chios samples revealing the two-breed cross and removed from the purebred pool. Our results provide data, for the first time, on the genetic background of the three local breeds to be further exploited in genetic improvement schemes.

Alternative breeding values for litter size in Romney sheep

J. Schmidova and M. Milerski

Institute of Animal Science, Přátelství 815, 104 00, Prague, Czech Republic; schmidova.jitka@vuzv.cz

Romney is the second most numerous sheep breed in the Czech Republic. The vast majority of animals of this breed are kept under extensive production systems with year-round stay in a pasture. The average litter size in this breed had increased over the last ten years from 1.47 to 1.64 in the Czech Republic. In connection with this the proportion of triplets and quadruplets had increased from 2.95 and 0.00% resp. in 2005 to 6.12 and 0.15% resp. in 2016. The occurrence of multiple-lamb litters is however unwanted by sheep owners, due to higher lamb mortality and especially greater burden for organism of the ewe under used extensive management. In this context the requirement for certain penalization of three and more numerous litters in genetic evaluation and attempts towards reduction of variability in litter size was raised. Two litter size evaluation variants, alternative to traditional one, were evaluated. Variant1 was based on the ratio between estimated economical values of litters of different size: 1 : 1.94 : 2.43 for 1, 2, 3 and more lambs in litter. In Variant2 three and more numerous litters were considered as twins. Coefficients of heritability of litter size estimated by REML were very low between all variants: 0.057 for traditional one; 0.055 for variant1; and 0.048 for variant2. Correlations between breeding values were r=0.992 for Variant1 vs traditional BV and r=0.965 for Variant2 vs traditional BV.

Involvement of the FecXGr mutation in the high prolificacy of the Flemish sheep breed in Belgium

R. Meyermans[1], L. Chantepie[2], S. Fabre[2], F. Woloszyn[2], J. Sarry[2], N. Buys[1], L. Bodin[2] and S. Janssens[1]

[1]KU Leuven, Livestock Genetics, Department of Biosystems, Kasteelpark Arenberg 30, 3001 Heverlee, Belgium, [2]Université de Toulouse, INRA, INPT, ENVT, GenPhySE, chemin de Borde-Rouge 24, 31320 Castanet Tolosan, France; steven.janssens@kuleuven.be

The Flemish sheep breed is of the dairy type and currently considered as a rare breed in Flanders (Belgium). The original breeding objective of the Flemish sheep was wool production (18th century) but it was also known for its high litter size and good maternal capacity. In order to support conservation efforts, pedigree data (n=5,095) and litter size information (2,385 litters on 1,224 ewes) were analysed. In the period 2010-2015, there were on average 38 sires and 238 ewes used for breeding annually. The average inbreeding coefficient was estimated at 0.055 and the effective population size was 124 (based on increase of coancestry) and 55 (based on increase of inbreeding). Average prolificacy (2002-2015) attained 2.02±0.87 (mean ±stand. deviation) with up to five lambs born. Using a linear animal model and a Gibbs sampling algorithm, a relatively high heritability coefficient of 0.25 was obtained for prolificacy This result warranted a screening of the breed for major fecundity (*Fec*) genes known to affect litter size. Fifty-eight individuals of the breed (20 rams and 38 ewes) were screened at the 4 *Fec* gene loci for all known mutations in ovine *BMP15*, *GDF9*, *BMPR1B* and *B4GALNT2*. Four individuals (3 ewes and 1 ram) were identified as carriers of the *FecX^{Gr}* mutation in the *BMP15* gene originally evidenced in the French Grivette breed and no other mutations were detected. This finding warrants: (1) a population wide screening of the breed for this mutation; and (2) an investigation into the effects of the mutation on prolificacy in this breed. This discovery also underlines the importance of conservation and investigation of endangered breeds because they may harbour valuable alleles for future use.

Identification of diacylglycerol acyltransfrase 1 (DGAT 1) gene polymorphism and association studies

N. Hedayat Evrigh and Z. Nourouzi
University of Mohaghegh Ardabili, Department of Animal Science, Faculty of Agriculture and Natural Resources, Daneshghah, 5619911367, Iran; nhedayat@uma.ac.ir

DGAT1 plays a fundamental role in triacylglycerol synthesis and existing SNPs in DGAT1 gene might provide important information in partially explaining variation of milk fat content. The main objective was to sequence the diacylglycerol transferase gene and undertake association analysis of SNP detected with economic traits in Iranian native khalkhali goats. DNA samples were taken from 100 goats and genotyped using PCR-SSCP. Different genotypes detected using pattern identification was sequenced by the Sanger sequencing method. The study of population genetic structure was investigated using BioEdit, MEGA, DnaSp and Network software. The association analysis between SNP and haplotype detected with important and economic traits was performed using SAS 9.2 software. Analysis of genetic diversity using DGAT1 genes led to identification of four mutations in two haplotypes in theKhalkhali goat population. The haplotype and nucleotide diversity were 0.667 and 0.0075 respectively. We observed four substitution including C21153T, C21154G, A21172C and A21194T (the position was defined based on accession code LT221856.1.). All of SNP identified in this study were homozygote genotype that causes two amino acid substitution. Those mutations create two haplotype that significantly affected on birth weight, litter size, milk yield and milk protein percent traits (P>0.05). but didn't influence milk fat, lactose or solid milk percent. Based on our funding in this study, it showed that we can use the haplotype observed in goat breeding strategy to improve Khalkhali goat populations.

Breeding Finnsheep in Finland

H. Mero and K. Ahlskog
ProAgria Association of ProAgria Centres, ProAgria Southern Finland, P.O. Box 97, 33101 Tampere, Finland; kaie.ahlskog@proagria.fi

ProAgria runs the National breeding programme for sheep, wich is arecognised breeding anconsulting organization for sheep farms. Finnsheep belong to the group of northern short tailed sheep breeds and have several different colours: white, brown, black and grey. In 2015, there were 10,040 Finnsheep ewes in the recording program and the number has increased since. The man live weight of adult ewes was 72 kg. Usually lambs were slaughtered at 20.4 kg carcass weight at an average age of 252 days with killing out % of 42. Carcasses were classified using the EUROP grid, and 69% belong to O class, 17% R and 8% P. At the moment, breeding values (EBVs) are calculated only for meat production traits. Estimates are based on phenotypic results in sheep recording: weighing, scanning and live EUROP-evaluation. Indexes are evaluad every other week. The Finnsheep herdbook was founded in 1922. Although only breeding values are calculated for meat traits, there are also many other important traits for breeding: growth, lamb production, fertility, adult weight, conformation, carcass quality and production and quality of wool. Breeding the Finnsheep for meat production is economically feasible. Litter sizes compensate for lower slaughter weight. The meat of Finnsheep has a special flavor. The breed is also very suitable for grazing in traditional rural areas and islands. Wool is fine and of a good lustre, ideally suited for handicraft the skin as well. ProAgria is participating in the work of developing new breeding indexes, improving recording systems and keeping an advisory organization for farmers.

Session 10

Welcome and introduction

J.-L. Peyraud
Animal Task Force, 149 rue de Bercy, 75012 Paris, France; info@animaltaskforce.eu

The demand, consumption patterns, consumers' engagement and perceptions of food are changing and diversifying. Besides a main stream, we see an increasing segmentation of the market. Beyond the nutritional and organoleptic qualities ('intrinsic value of food'), other criterion such as environmental footprint, animal welfare or the production of public goods (open landscape, image of naturalness…) are determining consumption choices. This is also called the 'extrinsic value of food'. Facing such new challenges and the necessity of attain added value from the export of animal products, food processing companies are now starting to develop husbandry guidelines for the supply chain. More than ever, a greater focus on animal derived food integrity (safety, authenticity and quality of products, but also their extrinsic value) is needed to secure Europe's role as a leading global provider for safe and healthy animal derived products and help European food systems earn consumer trust. Taking stock of the preparation of the EC-FOOD2030 strategy, the session would like to engage discussion with farmers, food processors and industries, retailers, nutritionists, scientists, but also with the society on the expected and possible contribution of the primary sector, in collaboration with the whole food chain, to support the animal derived food quality approaches and integrity. The most important findings of the session will be discussed with a panel. The outcomes of the session will be discussed in more details during the ATF seminar, in Brussels, on Oct. 26th 2017, where a large panel of European stakeholders will be invited. The Special Session aims to contribute to: (1) give insights on changes in consumption patterns and on the growing demand in terms of intrinsic and extrinsic qualities of animal products; (2) address how research and innovation can help the livestock sector contribute to food quality and integrity; (3) engage a dialogue between research, farmers, industry, decision makers, stakeholders from different backgrounds and citizens; (4) provide input for public policies if we want to secure Europe's role as a leading global provider for safe and healthy animal derived products.

Session 11

What is efficient livestock production and is it sustainable?

E. Röös[1], T. Garnett[2], D. Little[3], B. Bajželj[4], P. Smith[5] and M. Patel[1]
[1]Swedish University of Agricultural Sciences, Almas allé 8, 750 07 Uppsala, Sweden, [2]University of Oxford, Wellington Square, Oxford OX1 2JD, United Kingdom, [3]University of Stirling, Pathfoot Rd, Stirling FK9 4LA, United Kingdom, [4]University of Cambridge, Wilberforce Road, Cambridge, CB3 0WA, United Kingdom, [5]University of Aberdeen, 23 St Machar Drive, Aberdeen AB24 3UU, United Kingdom; elin.roos@slu.se

One of the most commonly voiced mitigation option in order to increase sustainability in livestock production is that of increased efficiency. However, what efficient livestock and efficient food production actually is differs among food system actors which also influences the solutions that different actors propose. This presentation describes different views on the concept of efficiency in animal production and consumption and the types of solutions for increased sustainability that logically follow. The first line of argument builds on the notion that animal production is inherently inefficient as humans could eat the plants used for animal feed directly, skipping a whole trophic level, and therefore greatly increasing the availability of food. Others see the use of pastures for food production using ruminants as the most efficient way of producing food – turning grass biomass into high value food products – requiring little or no fertilisers, irrigation or traction power. A third perspective highlights the high land requirements and greenhouse gas emissions of pasture-based ruminant production and see intensively reared chicken or fish as the solution as far less feed and land is required to produce food from these. Scenarios for future food production building on these perspectives show that they are all effective strategies for reducing land use and greenhouse gas emissions. An entirely plant based food system decrease greenhouse gas emission most effectively while a food system that uses pastures for meat and milk production and cropland for the production of foods for direct human consumption uses the least cropland. The preferred way forward differs among stakeholders and depends on their views on what is possible to change (in terms of both production and consumption patterns) and how wider effects of these strategies (e.g. for animal welfare, use of antibiotics) are factored in.

Official signs of quality in livestock production systems: multi-performance from farm to society

M. Benoit[1] and B. Méda[2]

[1]INRA, UMRH, 63122 Saint-Genès-Ch., France, [2]INRA, URA, 37380 Nouzilly, France; marc-p.benoit@inra.fr

Consumer demand is increasing for products with strong warranties regarding animal welfare, environmental respect or organoleptic quality. These warranties are generally given through official labels, transcripted into official specifications regarding farming practices. We studied one monogastric production (Label Rouge chicken) and one ruminant production (organic lamb) and evaluated the different ecosystemic services they provide. Similar elements of the specifications can be found in both productions, in particular a lower intensification (animal density, limited use of some inputs). These elements lead in particular to lower feeding efficiency and productivity which can increase some environmental impacts (e.g GHG emission) when expressed per kg of product. Yet, other properties of these systems can counterbalance this, such as the reduction of inputs (e.g. no chemical fertilisers in organic farming) and soil carbon sequestration in outdoor areas. Besides, results of the assessment of environmental impact (as GHG emissions) depend on the functional unit, and are, in these productions, lower when expressed by € of product. Products from both productions have indeed a higher added value compared to conventional ones: consumers are willing to pay for them as they return a positive image regarding animal welfare, organoleptic quality or the link to the 'terroir'. In order to maximise environmental and socio-economic services while improving 'provisioning' one, several solutions, at different scales, can be proposed. At animal scale, the main issues are the improvement of feed efficiency, adaptation to local conditions, and ewe productivity. The use of agroecology and territorial metabolism principles should help, at farm scale, to (re-)connect crop and livestock productions and, at regional scale, to imagine new synergies between farms. Collective organisation would indeed be a mean to improve the sustainibility of livestock systems by: (1) closing nutrient cycles by feed-manure exchanges; (2) organising selection of adapted genotypes (rustic breeds with low populations); (3) (re)building landscape with connected infrastructures and reinforced ecological regulations; (4) organising the sale of products, which can be strongly season-based for ruminants.

Efficiency of grassland based dairy farming systems in mountainous areas

M. Berton[1], M. Corazzin[2], G. Bittante[1], S. Bovolenta[2], A. Romanzin[2] and E. Sturaro[1]

[1]University of Padova, DAFNAE, Viale dell'Università 16, 35020 Legnaro, Italy, [2]University of Udine, DISA, Via Sondrio 2/A, 33100 Udine, Italy; marco.berton.4@gmail.com

The efficiency of mountain dairy systems (Italian eastern-Alps) was analyzed by using the following indicators: carbon footprint (CF, cradle-to-farm gate Life Cycle Assessment method) and gross energy and crude protein human-edible conversion ratios (computed as the ratio between the amount of gross energy and crude protein content in human-edible feedstuffs and the energy and protein content of milk, HeE_CR and HeCP_CR, respectively). Data were collected at farm level in 71 dairy farms (52±47 Livestock Unit – LU, 17.4±9.3 kg Fat Protein Corrected Milk – FPCM/cow/day on average). Two functional units were tested: 1 kg of FPCM (CF_{FPCM}, kg) and 1 m^2 of farm agricultural area (CF_{FAA}, m^2). Emissions due to herd and manure management, on-farm feedstuffs production, purchased feedstuffs, fuel and electricity were included within the system boundaries. Emission computation was based on IPCC Tier 1-2 procedure. Allocation between milk and meat was based on IDF method. CF_{FPCM}, CF_{FAA}, HeE_CR and HeCP_CR were analyzed with a GLM model with the fixed effects of the classes of farm size (LU) and milk productivity (kg FPCM/LU/day) – three classes computed on the basis of mean±0.5×SD (Low, Intermediate, High). Mean CF_{FPCM} was 1.1±0.3 kg CO_2-eq/kg and mean CF_{FAA} was 0.92±0.82 kg CO_2-eq/m^2, whereas mean HeE_CR and HeCP_CR were 1.27±0.82 MJ edible feedstuffs/MJ milk and 0.81±0.56 kg CP edible feedstuffs/kg CP milk, respectively. The main drivers of the impact were enteric fermentation (62%) and nitrous oxide volatilization from managed soils (14%). Farm size was significant in altering CF_{FAA}, HeE_CR and HeCP_CR (-43, -53 and -58% moving from class High to class Low, respectively), whereas productivity was significant for CF_{FPCM} (lowest values for class High). The results suggested that policies and strategies aiming at sustain the Alpine dairy system should focus on an efficient land management, taking into account both food security and maintenance of permanent grasslands.

Yield gap analysis of beef production systems: the case of grass-based beef production in France

A. Van Der Linden[1,2], S.J. Oosting[2], G.W.J. Van De Ven[1], P. Veysset[3], M.K. Van Ittersum[1] and I.J.M. De Boer[2]
[1]Wageningen University & Research, Plant Production Systems, Droevendaalsesteeg 1, 6700 AK Wageningen, the Netherlands, [2]Wageningen University & Research, Animal Production Systems, De Elst 1, 6708 WD Wageningen, the Netherlands, [3]INRA, UMR1213 Herbivores, 63122, Saint-Genès-Champanelle, France; aart.vanderlinden@wur.nl

Increasing livestock production per hectare agricultural land is a way to meet the increasing global demand for animal-source food in future decades. The bio-physical scope to increase livestock production is referred to as the yield gap. The aim of this research is to analyse yield gaps for the case of beef production systems in the Charolais region of France. Beef production per unit area should also account for the area used for all corresponding feed crops (feed-crop livestock system). The yield gap of feed-crop livestock systems is defined as the difference between the potential production (Y_P) or resource-limited production (Y_L) and the actual production (Y_A). Y_P is defined by climate and by the genotypes of beef cattle and of feed crops only. Y_L is defined by climate and genotypes too, but is also affected by feed quality, feed quantity, drinking water (cattle), irrigation water, and nutrients (feed crops). Mechanistic, dynamic production models for beef cattle and feed crops were combined to simulate Y_P and Y_L for twelve farm types in the Charolais region. Y_P was 2,377 kg live weight (LW)/ha/year, and Y_L was on average 664 kg LW/ha/year. The average Y_A was 354 kg LW/ha/year. Hence, relative yield gaps were 85% of Y_P, and on average 47% of Y_L. Farm types with limited concentrate use produced more human digestible protein (beef) than they required as feed input under resource-limited production. The magnitude of yield gaps was negatively correlated with the profit per hectare up to 2014. Improved grazing management and an earlier start of the grazing season may mitigate yield gaps and increase profitability under current and future conditions. We conclude that beef production systems in the Charolais region have a considerable scope to increase production from a bio-physical perspective, but yield gap mitigation was less profitable than farm expansion up to 2014.

Ingestive behavior components are they key-indicators of daily feed acquisition and LFS efficiency

M. Boval and D. Sauvant
INRA, PHASE, INRA AgroParisTech, MoSAR, 75005, France; maryline.boval@agroparistech.fr

Ingestive behavior (IB) of ruminants determines the dry matter intake (DMI) and reflects their adaptation to the feed resources, particularly in case of grazing livestock farming systems (LFS), with a strong influence of sward characteristics. Among the various components of IB, i.e. intake rate (IR), bite rate (BR), bite depth (BD), bite area (BA), the bite mass (BM) would be the pivot scale by which herbivore directly interact with the sward characteristics. A database was built from 62 papers (483 treatments and 142 experiments, nexp), in order to (1) synthesize quantitative knowledge about these behavioral components (expressed on a body weight (BW) basis) and (2) highlight their respective contributions to daily DMI, via a meta analytic approach. The BM follows a log-normal distribution (1.89±1.23 mg/kg BW, n=216), it is the most measured component, at very different time scales (from 10 min to 24 h), through a large panel of methods. Other parameters are less measured: BD (13.4±6.7 cm, n=116), BA (79.6±37.6 cm², n=105), BR (50.1±12.4 bites/min, n=244) or IR (70.2±12.5 mg DM/min/kg BW). The BM mainly results from BD and BA (BM = −0.28 + 0.066 BD + 0.019 BA, nexp=19, n=74, RMSE=0.66 mg/kg BW), and is positively influenced by sward height (SH=18.4±12.5 cm, n=333): BM=0.64 + 0.074 SH (nexp=26, n=80, RMSE=0.93). In contrast BM is negatively influenced by the forage bulk density (BD=1,711±1,251 g DM/m3, n=280), which is another major trait of the sward. By considering the intra-experiments effects, BM, IR and DMI are positively and closely related. BM is related to short-term IR (IR=218.7×(1-exp-0.27×BM; n=139, nexp=41, RMSE=8.5). On a longer term, BM is also linked to daily DM intake (DMI=8.46 + 43.7 [1-exp (-0.45 BM)], n=24, nexp=10, RMSE=1.4). Therefore, BM appears to be an appropriate indicator of feed acquisition at various time scales, especially since it is related to the size of the dental arcade, opening up possible pathways for genetic selection. The most practical and accurate method to estimate BM has to be highlighted. Analysis of the data base are underway to identify and model the roles of stocking rate and grazing time.

Environmental impacts of different innovative feeding strategies in pig and broiler farms

S. Espagnol[1], L. Dusart[2], B. Méda[3], D. Gaudré[1], A. Wilfart[4], P. Ponchant[5] and F. Garcia-Launay[5]
[1]IFIP, Pig Institute, 35651 Le Rheu cedex, France, [2]PEGASE, Agrocampus Ouest, INRA, 35590 Saint-Gilles, France, [3]URA, INRA, 37380 Nouzilly, France, [4]SAS, Agrocampus Ouest, INRA, 35000 Rennes, France, [5]ITAVI, Poultry institute, 37380 Nouzilly, France; sandrine.espagnol@ifip.asso.fr

Optimization of feeding strategies (S) is an effective option to reduce the environmental impacts of livestock. It can be performed, by using multi-objective feed formulation (MOF) to select feedstuffs according to their price and environmental impacts, and by adapting dietary composition to animal requirements during the rearing period. This study investigates the combination of those feeding strategies on the environmental impacts of pig (P) and broiler (B) productions. Different S for conventional farms were specified in B and for fattening period in P. In P, the S were: 2-phases feeding (S1); multiphase feeding with two (S2) or four (S3) pre-diets, mixed together in different proportions during time; and 2-phases feeding with a reduction of energy (S4) or amino acid (S5) content. In B, the S were: 3-phases feeding (S6) and 3-phases feeding with lysine and energy contents respectively increased and decreased (S7). For each S, feeds were formulated using least-cost optimization (LC) and MOF. The environmental impacts of the kg of BW at farm gate were assessed by Life Cycle Assessment and compared to the reference scenarios: LC-S1 and LC-S6 for P and B, respectively. S with LC formulation showed impacts between -19% and +13% compared to the reference impacts. MOF reduced impacts per kg of feed by up to 30%, and final impacts per kg of BW by 1 to 20%, except for Acidification and Climate Change (S4), and for Land Occupation (S7). Three criteria explain these contrasted results among S and LCA impacts at the kg of BW scale. The first one is the contribution of feed to impacts per kg of BW (34% to 98%). The second one is feed conversion ratio as it reflects the overall efficiency of the system, and thus impacts nutrient excretion and manure gas emission. The last one is nutritional balance of feed (protein, amino acids composition), explaining nitrogen excretion (S2, S3). This study underlines the need for optimizing S while considering simultaneously the consequences on economics and environmental performances at animal scale.

How to assess efficiency in animal production – different approaches from animal to farming systems

P. Faverdin
INRA, UMR PEGASE, Domaine de la Prise, 35590 Saint-Gilles, France; philippe.faverdin@inra.fr

Efficiency in livestock production is a measure of the extent to which resources are well used for producing animal products. Increasing the efficiency of livestock system is important to improve sustainability of animal production. It is an interesting way to decrease the competition between feed and food, to decrease environmental impacts and to increase competitiveness of animal products. However, if the concept is simple, it is difficult to use it in practice for animal production because several products are often dependent and several resources or impacts are concerned. Between the risks of wrong decision using too simple ratio and the impossibility to decide due to the complexity of multicriteria indicators, is there a possible way for efficiency indicators? This report presents methods and criteria are proposed to assess the efficiency considering the complexity of livestock farming system. The use of residuals of models or the methods of frontier analysis give new perspective to assess efficiency. With animal and farm scale examples, the advantages and limitations of the different approaches are discussed. At animal scale, the question of feed efficiency is addressed. The case of dairy cows is probably one of the most difficult due to the interactions between milk production, meat production, and diversity of diet. At farm scale, the difficulty to combine production and environmental impacts with the diversity of farming systems is also discussed. The Life cycle assessment is more and more recommended to assess environmental impacts of products, but it is often difficult to interpret its multiple indicators. The use of the whole farm as a functional unit, without any specific allocation of resources or impacts to each product, combined with data envelopment analysis is an interesting option to better consider the possible interactions within the farming system. The difficulty to find driving forces explaining the variations of the efficiency indicator through a large diversity of farms is also a challenge illustrated with an example with French dairy farms. If promising tools to assess efficiency are available, the questions of the definition of goals and of the access to the data required to use these tools is probably the next limiting step.

A path devising farm-specific adaptions to ensure economic viability and less environmental impact

J.O. Lehmann and T. Kristensen

Aarhus University – Foulum, Department of Agroecology, Blichers alle 20, 8830 Tjele, Denmark;
jespero.lehmann@agro.au.dk

A productive and efficient livestock production is key to securing the future supply of animal-derived foods, but environmental impacts and poor economic performance are challenging current production systems. It is pertinent that the individual farmer achieves better profitability to ensure future viability, but lesser environmental impact is crucial. Eco-efficiency, broadly defined as output per environmental impact, is an analytical concept developed to address the dynamics between technical, economic and environmental performance. Traditionally, eco-efficiency analyses have relied on surveys, official records, inventory data and simulation models, but there is an inherent compromise between comparability with other farmers and applicability for the individual farmer. Mostly, these analyses have been used to identify theoretically inefficient farms but with less emphasis on coupling with the specific production processes of a farm. We want to extend the eco-efficiency concept to couple farm specific performance with farm specific actions. In doing so, individual farmers can gain valuable information that can support a broad spectrum of improvements. Furthermore, the same analyses can be utilised by the political system to identify groups of farmers or farming strategies that will react similarly to new environmental restrictions. That should support a much more specific implementation of rules and technologies and thus an improved achievement of environmental targets. There are thus two perspectives with this approach. First, farmers and the individual farm, and, second, the political system and clusters of farms. Both perspectives have to consider that each farmer operates within local ecological boundaries and relies on natural annual patterns while managing available physical infrastructure as well as livestock and crop cycles. This requires a range of different types of data covering production processes, ecological and physical boundaries as well as environmental indicators and economic performance. Eco-efficiency analyses have typically relied on methods including data envelopment analysis and stochastic frontier analysis, and they may serve as a starting point for future analyses.

Downscaling whole-farm measures of livestock production efficiency to the individual field level

T. Takahashi, P. Harris, R. Orr, J. Hawkins, J. Rivero and M. Lee

Rothamsted Research, North Wyke, Okehampton, Devon, EX20 2SB, United Kingdom; taro.takahashi@rothamsted.ac.uk

With an increasing level of concern for global food security, the importance of improving livestock production efficiency has never been greater. In order to compare biophysical performance of multiple farming strategies, various forms of input-output ratios, such as total factor productivity, nutrient use efficiency and carbon footprint per unit of final product, have been developed and deployed to date. In the context of pasture-based livestock production systems, these metrics are typically designed for evaluation at the whole-farm level, each producing a single efficiency score representing the entire area of landholding. Recent research has shown, however, that grassland production has a significant level of spatial heterogeneity, both in terms of quality and quantity, even within a small geographical range. This finding suggests that the level of livestock production efficiency may also be different across multiple locations on farm, although the typical practice of rotational grazing poses practical challenges regarding quantification of such measures. Using the rich primary datasets from the North Wyke Farm Platform grazing trials in Devon, UK, this study proposes a novel method to evaluate livestock production efficiency for individual pastures. Whole-farm measures of animal production, such as herd-wide live weight gain, are allocated to each field according to its contribution through grazing, silage cuts and nutrient retention for the following season. The results of statistical analysis suggest that initial soil conditions and the timing of grazing are likely to be major determinants of this field-level production efficiency. We argue that the proposed method can facilitate an improvement in resource utilisation rate and demonstrate how this information could be incorporated into farm management planning.

Relationships between methane emissions and technico-economic data from commercial dairy herds

P. Delhez[1,2], B. Wyzen[3], A.-C. Dalcq[2], F.G. Colinet[2], E. Reding[3], A. Vanlierde[4], F. Dehareng[4], N. Gengler[2] and H. Soyeurt[2]

[1]*National Fund for Scientific Research, Egmont 5, 1000 Bruxelles, Belgium,* [2]*University of Liège, Gembloux Agro-Bio Tech, Passage des Déportés 2, 5030 Gembloux, Belgium,* [3]*Walloon Breeding Association, Champs Elysées 4, 5590 Ciney, Belgium,* [4]*Walloon Agricultural Research Center, Department of Valorisation of Agricultural Products, Chaussée de Namur 24, 5030 Gembloux, Belgium; pauline.delhez@doct.ulg.ac.be*

Considering economic and environmental issues is important for the sustainability of dairy farms. Regarding environment, direct methane (CH_4) emissions from cows are of increasing concern. Many studies examined CH_4 variation factors but often on a low number of experimental cows. Also, few studies linked CH_4 to economic aspects of dairy farms. The innovative aim of this study was to highlight technical factors associated with dairy cow CH_4 emissions and gain insight into the relationships between CH_4 and herd economic results by the use of large scale and on-farm data. A total of 525,697 individual CH_4 predictions from milk mid-infrared (MIR) spectra [MIR-CH_4 (g/day)] of milk samples collected on 206 farms during the Walloon milk recording were used to create a CH_4 proxy at the herd by year (herd×year) level. This proxy was merged with accounting data. This allowed a simultaneous study of CH_4 emissions and 56 technico-economic variables for 1,024 herd×year records from 2007 to 2014. Significant effects were detected from ANOVA analyses and correlations (r). MIR-CH_4 was weakly linked to technical variables considered individually (r<0.38), suggesting complex associations between variables. Lower MIR-CH_4 was associated with lower fat and protein corrected milk (FPCM) yield (r=0.18), lower milk fat and protein content (r=0.38 and 0.33, respectively), lower quantity of milk produced from forages (r=0.12) and suboptimal reproduction and health performances (e.g. higher calving interval (r=-0.21), higher culling rate (r=-0.15)). On an economic point of view, lower MIR-CH_4 was associated with lower gross margin per cow (r=0.19) and per litre FPCM (r=0.09). To conclude, this study suggested that low dairy cow CH_4 emissions tended to be associated with suboptimal and also less profitable herd management practices. Further research is needed to confirm and expand on these results.

How mastitis in dairy cows affects the carbon footprint of milk

P.F. Mostert, C.E. Van Middelaar, E.A.M. Bokkers and I.J.M. De Boer
Wageningen University, Animal Production Systems group, De Elst 1, 6708 WD Wageningen, the Netherlands; corina.vanmiddelaar@wur.nl

Efficient livestock production is of importance to fulfil the increasing demand for animal source food while reducing the environmental impact of production. Clinical mastitis (CM) in dairy cows results in inefficiency because of a decreased milk production and fertility, and an increased risk of culling. With an average incidence of 27%, CM is one of the main health and welfare problems in dairy cows in the Netherlands. The environmental impact of CM, however, is unknown. The objective of this study was to quantify the impact of CM on the carbon footprint (CFP) of milk at the animal and farm level, defined as the sum of greenhouse gas emissions per kg of milk produced, using the Dutch situation as a case study. In addition, milk losses are quantified as an indicator for the economic impact of CM. First, a dynamic stochastic simulation model was developed to study the dynamics and losses of CM at animal and farm level. Cows received a parity (1-5+), a potential milk production and a risk of CM. Based on the number of cases of CM and pathogen combination, cows had a reduced daily milk yield, discarded milk if treated with antibiotics, a prolonged calving interval, and an increased risk of culling. Second, a cradle-to-farm gate LCA was performed to quantify the impact of CM on emissions of carbon dioxide, methane and nitrous oxide related to feed production, enteric fermentation and manure management. Changes in emissions were summed into CO_2 equivalents and expressed per ton of fat-and-protein-corrected milk (CO_2-e/t FPCM). Preliminary results at animal level indicate that, per case of CM, the CFP of milk increased on average by 6.5% (57.9 kg CO_2/t FPCM), while milk losses (including discarded milk) summed up to 677 kg per cow per lactation. The CFP increased by more than 34% (>300 kg CO_2e/t FPCM) for cows that died on the farm because of CM. At farm level, emissions increased on average by 1.8% (15.6 kg CO_2e/t FPCM), and milk losses were estimated to be 18,000 kg per lactation, or 15,000 kg per year. In conclusion, reducing CM is an effective strategy to reduce the CFP of milk and to increase the efficiency of dairy production.

Scandinavian beef production models explored on efficiency on producing human edible protein

C. Swensson and A. Herlin
Swedish University of Agricultural Sciences, Biosystems and Technology, P.O. Box 103, 25355 Alnarp, Sweden;
anders.herlin@slu.se

Ruminants are superior at converting feed into the nutrient-rich foods such as milk and meat. But how effective are they? And are humans and ruminants competing for the same feed and or food. This study analysed the production of beef from two different systems, beef from beef breed cattle and six systems with beef from bull calves deriving from dairy production system. Three different systems from beef breed production were compared, a Danish extensive system based on high roughage intake from natural and permanent pastures and two more intensive systems from Denmark and Sweden. Two production models for steers were analysed. Four different production models using bull calves as the starting animal were compared with two veal calf models and two production scenarios for young bulls All models and feeding scenarios were obtained from Mogensen *et al*. Protein efficiency was calculated as the total protein efficiency and for the cattle's ability to turn feed protein to human edible animal protein using factors for different feeds derived from Wilkinson. Results show that more human edible protein is obtained than put into the production model from the extensive beef breed system (factor 2) and from the two steer production system (about factor 1.2) than from the other production models. These model are based on large pasture intake and large amounts of silage which results in high efficiency in converting feed to human edible protein. Mogensen *et al*. found that intensive beef production systems, especially veal and to some extent young bulls, had a much less carbon foot print than the extensive systems based on pasture and silage (steers and beef breed cattle). The management of the conflict of objectives between climate impact and protein efficiency is one of the key challenges for ruminant-based meat production.

Capturing livestock farms diversity in Uruguay: typology and classification of grazing management

I. Paparamborda and P. Soca
Facultad de Agronomía. Universidad de la República, DPAyP, Garzón 780, 12900, Uruguay; ipaparamborda@fagro.edu.uy

In Uruguay beef cattle and sheep production involves 13 mil farms are classified as family and most of them are oriented to beef cow-calf systems based on Campos grasslands. Our objective was to classify and characterize (Typology) livestock farming systems (LFS) by their structure (variables considered were: total area, area own/leased ratio, natural grassland area, stocking rate, sheep/cow ratio) and management practices (MP) (index that summarizes practices: differential feeding management according to BCS and physiological state, mating period, suckling control, month of weaning and pregnancy diagnosis). A survey was done to 244 LFS. Multivariate statistics including cluster analysis were performed. For 69 (from the 244) LFS a classification was made based on spatio-temporal grazing management (STGM), considering use of paddocks throughout the year by different animal categories, MP and stocking rate. Six groups were identified with different system structure and MP, but without differences in beef and sheep production/ha. Group 1 (24%) includes small area (147±120 ha) sheep systems, renters (0.04±0.1) with high stocking rate (1.1±0.43 AU), and low MP. Group 2 (11%) includes large area (416±183 ha) cattle systems, renters (0.22±0.24), with medium MP. Group 3 (14%) includes sheep systems, with intermediate area (229±122 ha), owners (0.9±0.16), with low MP. Group 4 (10%) includes cattle systems, with intermediate area (255±193 ha), owners (0.93±0.12). Group 5 (22%) includes cattle systems, renters (0.13±0.23), with high stocking rate (1.3±0.46 AU) and with medium MP. Group 6 (16%) includes small area (113±85 ha) cattle systems, with high stocking rate (1.2±0.36 AU) and low MP. However, we founded three groups by STGM: Managers (23%), Partial managers (25%), and Non-Managers (52%) with differences in production/ha between groups: Managers with 91 kg meat beef/ha/year vs Non-Managers with 70 kg/ha/year (P<0.05). We identify variables that allow us to capture the diversity of LFS and confirm an association between STGM and production/ha. This information is relevant to improve the resilience and sustainability of families related to livestock sector.

Permanent fences for temporary use in livestock protection against wolf predation

L. Pinto De Andrade[1], J. Várzea Rodrigues[1], J. Carvalho[1] and V. Salvatori[2]
[1]Instituto Politécnico de Castelo Branco, Escola Superior Agrária, Quinta da Sra de Mércules, 6001-909 Castelo Branco, Portugal, [2]Istituto Ecologia Applicata, Via Bartolomeo Eustachio, 10, 00161 Rome, Italy; luispa@ipcb.pt

The Iberian Wolf (*Canis lupus* signatus Cabrera, 1907) is found in Portugal in the border regions of Viana do Castelo and Braga districts, in Trás-os-Montes and part of the districts of Aveiro, Viseu and Guarda. In the study area (north of the Guarda District) there is a conflict caused by predation of the wolf on livestock and the resulting economic losses. Under Action C3 of Project LIFE11-NAT / IT / 069 MEDWOLF, 19 farmers have been involved and 34 permanent fences have been installed to prevent wolf attacks, distributed geographically by the Municipalities of: Almeida (27), Pinhel (4), Guarda (2) and Sabugal (1). These interventions involve attempting to protect a total herd of 2,951 animals (2,044 cattle, 881 sheep and 26 ostriches). The built fences have perimeters varying between 60 and 1,160 m. The justification to build the permanent fences for temporary use was the need to create conditions of protection of livestock, reducing the impact of predation by the wolf. Not being an objective is inherent to its existence the protection against other predators (eg dog) and the possibility of improving the management of herds. The permanent fences installed are intended to be rigid, solid, durable, easy to construct by farmers and with low maintenance. So far no problems associated with the use and efficiency of permanent fences have been detected. The solution is good in terms of reduction the attacks; Not great but satisfactory in economic terms and of work needs. The conservation and recovery of the Iberian wolf is possible as long as it is not equated and imposed in the detriment of the economic activities of the resident human population, especially the activities associated with animal production. The social and cultural aspects of conflicts cannot be neglected, beyond the easily understandable economic aspects. The survival of the wolf in the region, depends from the non-economic penalization of those living in the region and on the viability of existing productive activities, especially animal production.

Efficiency of permanent fences in the prevention of predation by wolves in livestock

J. Várzea Rodrigues, J. Carvalho, A. Galvão, P. Monteiro and L. Pinto De Andrade
Instituto Politécnico de Castelo Branco, Escola Superior Agrária, Quinta da Sra de Mércules, 6001-909 Castelo Branco, Portugal; luispa@ipcb.pt

The implementation of damage prevention measures (e.g. fences) has a high potential for producing a positive impact on the local population minimizing the conflict with wildlife predation due to economic losses. Their impact will be assessed both in terms of amount of damage and attacks suffered. In fact, such measures also contribute to the trust building process between local communities and local authorities, and to improve the livelihood of farmers. Under Action C3 of Project LIFE11-NAT / IT / 069 MEDWOLF, 19 farmers have been involved and 34 permanent fences have been installed in Center North of Portugal to prevent wolf attacks. The fences were built in farms with the highest incidence of attacks and only in farms with wolf attacks. The evaluation was done in 16 of the 19 farms involved (three farms present recent fence installations and were not considered). Steel mesh boards electrowelded with the following characteristics: 6×2.4 m with square 15×15 cm, thickness 6 mm for cattle and square 10×10 cm, thickness 5 mm for sheep and *I* - beams for supporting the steel mesh were used. These permanent fences are intended for temporary use (night protection or for animals in more sensitive physiological condition such as parturition and early suckling). Permanent fences for temporary use allow a significant reduction in the incidence of attacks on livestock herds (0.24 vs 0.04 attacks per farm and per month, before and after the fence functionality respectively). Also the average number of affected animals (including dead, wounded and missing) per farm and per month suffers a significant reduction (0.42 vs 0.05 before and after the fence functionality, respectively). Permanent temporary fences allowed a reduction of 83.3% in the average number of attacks per farm/month and a reduction of 88.1% in the average number of affected animals per farm/month. The results obtained are good and are consequence from the physical effect of the fence as a protection element, but also from improvements in animal handling and the possible deterrent effect resulting from a greater human presence.

Precision livestock farming and animal welfare: contradictions or synergies?

I. Veissier[1], M.M. Mialon[1], B. Meunier[1], M. Silberberg[1], P. Nielsen[2], H. Blokhuis[3] and I. Halachmi[4]
[1]INRA, UMR1213 Herbivores, 63122, France, [2]University of Copenhagen, Gronnegardsvej 8, 1870 Frederiksberg C, Denmark, [3]SLU, P.O. Box 7038, 750 07 Uppsala, Sweden, [4]Volcani Institute, Israel; isabelle.veissier@inra.fr

Precision Livestock Farming (PLF) techniques have been developed essentially to increase profitability and reduce workload by applying automatic processes to monitor animals and their environment. For instance: detecting oestrus allows timely insemination, while detecting lameness at an early stage or imbalance in the nutritional status or even abnormal ambiance parameters in the barn can help taking remedial actions quickly. PLF is often seen as an over-industrialisation of animal productions, leaving little room for the animal as a sentient being interacting with its environment including the farmer. However, the data generated by PLF sensors can support animal welfare. A system detecting health problems (e.g. mastitis, ketosis in dairy cows) can be part of welfare management. In addition and maybe more importantly, some PLF devices are based on animal behaviour detection directly, or indirectly through the position of animals: time spent feeding, ruminating, resting, walking, etc. Subtle changes in behaviour can indicate the mental state of an animal. Stressed animals may become hyper-reactive or on the contrary apathetic. Sick animals generally spend less time eating (and eat less) than healthy ones, they may also look for isolation. Changes in the daily rhythm of activity seem subtle signs of difficulties to adapt to the environment or of illness. Play behaviour and grooming may be affected by pain or fever. The proximity between animals and their social interactions can reflect the structure of the group, specially its cohesiveness. In this keynote speech, the authors will argue that PLF techniques offer a wide range of possibilities to use animal behavioural signs to address animal welfare in modern livestock farming, be the welfare related to health status, social relations, human-animal relationship or more general effects of a stressful environment. At the moment, these possibilities have been little explored and deserve more research.

Validation of an electronic herd control system for grazing dairy cows

G. Grodkowski[1,2], T. Sakowski[2], K. Puppel[3], K. Van Meurs[4] and T. Baars[1]
[1]Fundation of Stanisław Karłowski, Juchowo 54 A, 78-446 Silnowo, Poland, [2]Institute of Genetics and Animal Breeding of the Polish Academy of Sciences, Department of Animal Improvment, ul. Postępu 36A, Jastrzębiec, 05-552 Magdalenka, Poland, [3]Warsaw University of Life Sciences, Department of Animal Breeding and Production, Ciszewskiego 8, 02-786 Warszawa, Poland, [4]CowManager B.V, Gerverscop 9, 3481 LT Harmelen, Netherlands Antilles; g.grodkowski@ighz.pl

Currently dairy herds are increasing, whereas the amount of labour input per cow is decreasing. Therefor visual observation time of single animals is decreasing. Problems in large herds can be a proper heat detection, identification of lameness and metabolic diseases. CowManager Sensor® is a new device to support herd control. The system provides information on eating, rumination, physical activity and resting. The electronic ear tags were installed in 100 cows in two breeds (BS and HF) on an biodynamic farm in Juchowo (Poland). During the grazing season in 2016 validation of this system has been performed. Single animals were observed for one hour, and the main behaviour per minute was noted: grazing, ruminating, walking, laying down, standing, resting or being active. Observations were done in different times during the day in summer. Analysis of the first 40 h of observation delivered unacceptable results. A large number of flies resulted in increased head shaking and ear flapping of the cows, which gave poor correlations between observed behaviour and electronic outcome. After a system upgrade (V 7.0) the validation process proceeded, performing 56 h of observation. The correlation for grazing was $R^2=0.86$, for ruminating $R^2=0.97$ and for resting $R^2=0.94$. However, in cases, where Sensor® showed a (too) high physical activity, correlations were poor. After removal of all results with a high activity (>15%) detected, correlations for grazing, ruminating and resting further increased ($R^2=0.98$). The electronic device will be adopted further to improve the results. The authors acknowledge the financial support for this project provided by transnational funding bodies, being partners of the FP7 ERA-net project, CORE Organic Plus, and the cofound from the European Commission.

Associations between precision sensor data and subjectively scored cattle welfare indicators

M. Jaeger[1], K. Brügemann[1], B. Kuhlig[2], T. Baars[3], H. Brandt[1] and S. König[1]
[1] Animal Breeding and Genetics, University of Gießen, Ludwigstr. 21b, 35390, Germany, [2] Group agricultural engineering, Kassel University, Nordbahnhofstr. 1a, 37213 Witzenhausen, Germany, [3] Institute Genetics and Animal Breeding PAS, ul. Postepu 36 A, 05552 Magdalenka, Poland; maria.jaeger@agrar.uni-giessen.de

This study analyzed relationships among production, (grazing-) behavior and functional traits of native black and white dual-purpose cattle on phenotypic levels, including innovative sensor technology. Electronic sensor ear tags (CowManager B.V.) were implemented in dual-purpose cattle kept in participating research herds from eight EU countries (700 cows). Over a period of six months automatic sensor recordings measured individual animal activity traits (%/day): rumination, feeding, non-active and active. Subjectively scored health, welfare and (grazing) behavior indicators, according to official welfare recording guidelines, included body condition score (BCS), locomotion score (LS), hygiene score (HS), intra herd rank order (IHRO), aggressiveness (AGG) general temperament (GT) during milking, grazing speed (GRS), bite size (BS) and the preference (PREF) over herbs, legumes and grass. Grassland characteristics included sword height measured via rising platemeter, pasture composition and grass energy content. The sensor trait activity increased with days in milk (DIM), with highest activity (6%/day) beyond 200 DIM. A moderate positive correlation (0.41, $P<0.05$) was found between BCS and non-activity. Feeding was negatively correlated with BCS (-0.39, $P<0.05$), but positively correlated with test day milk yield (0.24). High yielding cows showed significantly higher feeding ($P<0.05$). Regarding GT and generalized linear mixed model applications, cattle temperament in the parlor was significantly related with feeding ($P<0.001$), indicating reduced feeding for nervous cows. Lowest IHRO were identified for non-active cows. Applying a threshold model, most important environmental descriptors influencing binary PREF were sword height and grass energy content (MJ NEL). Due to only moderate correlations with subjective welfare scores, automatically generated sensor data allows deeper insight into real cow behavior, with further applications for management and breeding.

Cow's behavioural activities and milk yield monitoring as early indicators of health events

A. Peña Fernández[1], K.H. Sloth[2], T. Norton[1], S. Klimpel[3] and D. Berckmans[1]
[1]KU Leuven, M3-BIORES, Animal and Human Health Engineering Division, Department of Biosystems, Kasteelpark Arenberg 30, 3001 Heverlee, Belgium, [2]GEA Farm Technologies GmbH, Nørskovvej 1B, 8660 Skanderborg, Denmark, [3]GEA Farm Technologies GmbH, Siemensstrasse 25-27, 59199 Bönen, Germany; tomas.norton@kuleuven.be

The behavioural activities and milk production time-evolution patterns of dairy cows provide an indication of their health status. Monitoring in real-time the dynamics of these behavioural activities and the milk yield allows detecting upcoming health event in individual dairy cows. Approximately 280 lactating cows from a Dutch commercial dairy farm, equipped with milking robots, have been monitored for this study. The Real-Time-Location-System CowView® (GEA®) provides the total distance travelled by each cow. Besides, through the location of the cow in the barn, it is possible to define the time the cow spent performing a specific behavioural activity. For a complete lactation period daily averages of the total distance walked by a cow, the time spent in the cubicles and at the feeding area were selected as behavioural activities, together with the daily milk yield. A monitoring algorithm was developed to study the dynamics present in the time-series evolution of the behavioural activities and the milk yield from each individual cow. The monitoring allows detecting sudden changes in the dynamic trends of the variables which are linked to the start of a health event. This technique worked successfully in 70% of the cows suffering from metabolic disorders and/or udder problems during the lactation period monitored. Moreover, these indications linked to an upcoming health event were detected, on average, 3 days sooner than the problem was logged by the farmer. Real-time monitoring of behavioural activities and milk yield of individual cows along a lactation period provides farmers a tool to check the health status of their herds.

Effects of body part determination or animal behavior on measuring traits from 3D data of dairy cows

J. Salau, J.H. Haas, W. Junge and G. Thaller

Christian-Albrechts-University Kiel, Institute of Animal Breeding & Husbandry, Hermann-Rodewald-Straße 6, 24118 Kiel, Germany; jsalau@tierzucht.uni-kiel.de

Precision livestock farming (PLF) has gained importance. Using cameras for monitoring farm animals has reduced labor and costs for farmers and handling stress for animals and has yielded various kinds of information. However, in many studies repeatability or accuracy were limited by unexplained variance due to inaccuracies in region of interest (ROI) determination or animal behavior. In this study, the recording unit was a framework with six Microsoft Kinect cameras through which the cows voluntarily walked without being led. The vertical distance from udder bottom to floor (UD) and rear leg angle (RL) were calculated from the recorded 3D data. UD and RL ranged from 42.4 to 61.2 cm (mean 49.8±5.3 cm) and 136.9 to 154.8° (mean 146.0±4.6°) with repeatabilities 76.9 and 47.4%, respectively. To address the effect of animal behavior on UD/RL measurement, it was observed how long it took the cows to pass the framework (run time) and how often they stopped. Averaged run times and number of stops differed between cows and ranged from 8.5 s to 51.6 s (mean 17.2±9.4 s), respectively, from 0 to 11 (mean 1.3±1.6). Measurement precision in UD and RL significantly depended on both behavioral traits. For unbiased measurements of body traits in PLF, behavioral effects (fear, curiosity) should be minimized e.g. by preferably unobtrusive recording units. Furthermore, PLF results are affected by the choice of ROI. For UD/RL calculation, a model based object recognition chose the data part which was best explained by a body part model. This induced variation due to shape or posture differences between animals or step phases. Therefore, a neural network (NN) was designed to use specific properties of each data point to decide about its affiliation. NN classified points in eight body regions: head, rump, back, udder, front and rear legs – both distinguished between facing and averted from the camera. Accuracies within regions ranged from 0.74 to 0.95 (mean 0.87±0.09). Exemplarily, points on the back were determined with 81% sensitivity and 95% specificity. The NN approach could improve accuracy in ROI determination, and thus the repeatability of trait measurements.

Development of an automatized method to improve heat detection in large dairy cattle herds

I. Jemmali[1], N. Lakhoua[1], I. Jabri[2], M. Djemali[3] and M. Annabi[2]

[1]Carthage National Engineering School of Tunisia, Research Unit of Signals and Mechatronic Systems, 45 Rue des Entrepreneurs, 2035, Tunisia, [2]National Superior School of Engineering, Ave Taha Hussein, Tunis, 1008, Tunisia, [3]National Institute of Agriculture of Tunisia, Animal Science, 43, Avenue Charles Nicolle Mahrajène 1082, Tunisia; mdejmali@webmails.com

The reproductive efficiency in dairy cattle depends on the accuracy of heat detection for cows to be inseminated on time. A variety of methods have been developed including pedometers, hormones concentration in milk and video surveillance. Beside their relatively high cost, they failed to automate heat detection. The objectives of this study were to develop an automated method based on cows observability able to: (1) recognize through image the most significant posture of a cow in heat; (2) track a cow's behavior when in heat and alert the dairyman by mail to inseminate the cow. A dairy herd of 200 Holsteins was observed by camera during two weeks from 6:00 am to 3:00 pm and a computer program based on camera images was developed. Focus was put on cows that were nervous and those that showed real signs of heat (immobilized when mounted). Cows were marked on their back with red markers. Recorded data on heat intervals, % heat intervals (18-24 days) and days between calving and first heat, calving and first AI and calving and fertilizing Insemination were used and analyzed. Main results showed that intervals between heats varied between 35 and 40 days, average % normal heat intervals was 16.5. Intervals between calving and first heat, calving and first AI and calving and fertilized AI were 77, 83 and 138 d, respectively. The accurate heat detected was 28%. This is why automated methods for heat detection are needed. The automated developed method is based entirely on image processing and Matlab programming. It allowed to track cows by using different types of filters, image subtract and bounding boxes and by calculating distances, barycenters and areas. It also allowed to save the images taken from the mounting steps in a dropbox file and when the cow is taken in an immobilized posture, emails were sent using SMTP.

The use of sensor technology to track and monitor animals in a group to reduce damaging behaviour

E.D. Ellen[1], E.N. De Haas[1], K. Peeters[2], B. Visser[2], L.E. Van Der Zande[3], H. Knijn[4], R. Borg[5] and T.B. Rodenburg[1]
[1]Wageningen University & Research, P.O. Box 338, 6700 AH Wageningen, the Netherlands, [2]Hendrix Genetics, P.O. Box 114, 5830 AC Boxmeer, the Netherlands, [3]Topigs Norsvin Research Center B.V., P.O. Box 43, 6640 AA Beuningen, the Netherlands, [4]CRV, P.O. Box 454, 6800 AL Arnhem, the Netherlands, [5]Cobb Vantress B.V., Koorstraat 2, 5831 GH Boxmeer, the Netherlands; esther.ellen@wur.nl

There is a tendency that livestock species are increasingly kept in large groups. To track and monitor animals in groups, novel technologies are needed. Examples of tracking technologies are ultra-wideband tracking, video imaging, or RFID. Tracking and monitoring of animals in groups is still difficult and it is unknown what the most efficient method is. Using tracking technologies will result in a large amount of spatial data. Analysing spatial data is a challenge and it is difficult to get an accurate and reliable phenotype. Furthermore, to reduce damaging behaviour in groups (like feather pecking (FP) in laying hens) it is important to monitor animal behaviour. Therefore, combining tracking with behavioural data is essential. The European COST Action GroupHouseNet aims to provide tools to prevent damaging behaviour in group-housed pigs and laying hens. One area of focus is how genetic and genomic tools can be used to breed animals that are less likely to develop damaging behaviour to their pen-mates. Breed4Food (B4F) is a consortium established by Wageningen University & Research and four animal breeding companies. One of the projects of B4F is to identify and develop tools to track individual animals in a group and to use this information to measure traits related to animal behaviour. The aim of this study is to investigate the most efficient method to track animals in a group and to investigate which method could be used to monitor behaviour, focussing on laying hens. Previous work shows relationships between FP, fearfulness and activity levels. Birds selected for high FP were found to be less fearful and highly active in a range of tests, measured automatically with ultra-wide band tracking. Further development of different measuring techniques will continue, focusing on automatic detection of high FP individuals in different group settings.

An app assessing animal welfare through animal-based measures

L. Michel[1], C. Guillon-Kroon[1], T. Doublet[1], L. Bignon[2], V. Courboulay[3], N. Bareille[4], R. Guatteo[4], M.-C. Salaun[5] and A. Legrand[6]
[1]Terrena Innovation, La Noelle BP 20199, 44150, France, [2]ITAVI, UMT-BIRD, Centre INRA Val de Loire, 37380 Nouzilly, France, [3]IFIP, BP 35104, 35651 LE Rheu Cedex, France, [4]BIOEPAR INRA Oniris, La Chantrerie, 44307 Nantes Cedex 03, France, [5]INRA, UMR1348 PEGASE, INRA Agro Campus, 35590 Saint-Gilles, France, [6]CIWF, 13 rue du paradis, 75010 Paris, France; lmichel@terrena.fr

TIBENA (Terrena animal-based welfare measures for a New Agriculture) is an innovative tool supporting farmers in their animal welfare improvement process. This diagnostic tool, based on the 5 Freedoms, helps identifying good practices and possible improvements, through 37 to 80 animal-based welfare indicators (depending on the species). The results are used to discuss every aspects of flock (or herd) management and welfare with the farmer: water and nutrition, comfort, health, stress level and animal behavior. Construction of the tool began with an exhaustive consultation process: bibliography (mainly Welfare Quality), interviews with scientists and professionals. Then, the trial app was assessed on the field including 20 to 30 farms per species, and with an expert committee, until the tool was considered reliable and robust enough to be used on large commercial scale. Reproducibility was tested, in order to select measures as objective as possible. Technical feasibility and scientific validity were also important selection criteria. By spending about one to three hours observing the animals, TIBENA gives an objective picture of the welfare state at the herd or flock level. Thanks to the clear and illustrated instructions, and the convenience of a smartphone app, anyone can use the tool and obtain a valid result. The app contains recommendations for welfare improvement. The app currently exist for pigs, broilers, rabbits and cattle, and is being developed for other farm species. During the period October 2015-December 2016, the TIBENA app was used on 54 Terrena farms to assess welfare and provide recommendations for improvement. The longer-term goal is to make the app available to the whole industry.

Research priorities on use of sensor technologies to improve dairy farming

J. Onyango
Innovation for Agriculture, Precision Livestock, Stoneleigh Park, Kenilworth, Warwickshire, CV8 2LZ, United Kingdom; josho@i4agri.org

Precision livestock farming (PLF) can be defined as the management of livestock farming by continuous automated real-time monitoring of the health and welfare of livestock and the associated impact on the environment. The benefits associated with PLF are far-reaching: improved animal welfare, improved profitability, improved product quality among others. A survey was developed to investigate the use of sensor technologies to improve productivity and profitability on dairy farms, in order to come up with research priorities for the Data Driven Dairy Decision For Farmers (4D4F) Horizon 2020 EU research project. The survey was sent to partners in different occupations (farmers, veterinarians, farm advisors and researchers) across eight countries in Europe. There were 103 replies. Analysis was carried out in R. The majority identified the following areas to be core; lameness, udder health, metabolic diseases, nutrition and reproduction. On experiences in the use of sensors, most respondents reported to have experience in more than one sensor while pressure sensor were least common on farms. The survey has highlighted the following areas as the top priority for sensor research; refining sensors to provide rapid information on lameness, mastitis, metabolic diseases and reproduction. The majority suggested that further research on wide spectrum and cost effective long term sensors, simplifying data so that it would be easy to understand in order to enable prompt action. Last but not least, having standard operation procedures for helping achieve efficiency, quality and uniformity, while reducing some of the failures were also research priority areas pointed out. In conclusion, the survey has highlighted several areas where research on sensor technologies should focus on in order to improve productivity and sustainability on dairy farms.

What do European farmers think about precision livestock farming?

J. Hartung[1], T. Banhazi[2], E. Vranken[3], M. Guarino[4] and D. Berckmans[5]
[1]University of Veterinary Medicine Hannover, Bünteweg 17p, 30559 Hannover, Germany, [2]University of Southern Queensland, West Street, 4350 Toowoomba QLD, Australia, [3]FANCOM B.V., P.O. Box 7131, 5980 AC Panningen, the Netherlands, [4] University of Milan, Via Festa del Perdono 7, 20122 Milano, Italy, [5]M3-BIORES, Kasteelpark Arenberg 30, 3001 Heverlee, Belgium; joerg.hartung@tiho-hannover.de

Precision livestock farming (PLF) is growing in modern animal production. However, little is known about the daily work and economic pressures the single farmer is exposed to and about his attitude to such increasingly complex production systems and monitoring technologies. In order to give the farmer a voice in the public debate about modern animal production methods or at least listen to his opinion a limited survey based on 21 farm visits (9 pig, 5 broiler, 7 dairy) in 10 EU countries (part of EU-PLF Project 2014/2016). The farmers were interviewed in personal free format face to face on their farms. Most pig and poultry farms were visited in 2014, after installation of a PLF technology and 2016 again. The dairy farms could be visited only once in 2016. All pig, broiler and cow farmers who get sufficient support from the providers developed a positive to very positive attitude to real-time monitoring of their animals. Broiler farmers were more open to PLF than pig farmers. Almost all farmers emphasized that the personal contact to the animals cannot be replaced by video cameras but PLF systems can be a great help in daily life. Farmers stated that problems can be significantly earlier recognized than with conventional methods. Some think PLF can improve animal welfare and may assist to bridge the gap between producers and consumers by transparency. Drawbacks are high prices for PLF equipment, sometimes poor maintenance service by the companies and the lack of wider experience in practice. One farmer responded after two years of experience with his PLF system that he understands his animals much better since he uses PLF monitoring. There is a demand for demonstration farms using PLF under practical conditions. It seems that PLF technologies can play an important role in the development of a future oriented, sustainable, animal-friendly, consumer accepted and efficient livestock production with healthy animals.

A field study to determine the incidence of subclinical and clinical ketosis in Finnish dairy farms

A.M. Anttila[1], J. Fält[1], T. Huhtamäki[2], S. Perasto[1] and S. Morri[1]
[1]ProAgria South Ostrobothnia, Huhtalantie 2, 60220 Seinäjoki, Finland, [2]ProAgria Centres, Urheilutie 6, 01301 Vantaa, Finland; anne.anttila@proagria.fi

The early lactation after calving is associated with a high incidence of production diseases and metabolic disorders. It is associated with the cow´s high increase of milk yield and negative energy balance (NEB) due to challenges in dry matter intake (DMI). MaMa ('From milk to milk') -project is conducted on 12-15 pilot dairy farms (50-150 cows/farm) in the western part of Finland (South Ostrobothnia). The focus of this project is to figure out the incidence of subclinical (SCK) and clinical ketosis (CK). The aim of this project is to study how management of feeding, daily work routines and conditions during transition period effect on cow´s health, the amount of ketone bodies in the milk or ketosis symptoms. The pilot farmers test twenty cows once a week until 50 days in milk (DIM) by taking the milk samples and using the Porta BHB ketosis test to analyze the ketone bodies in milk (µmol/l). The management and housing data is collected from these 20 cows as well. After the first test period the results will be summarized together and discussed which kind of actions on farm level are needed in management, housing or feeding to prevent ketosis. After a year there will be a second test period, and again 20 cows will be tested. The aim of the second test period is to study how the changes the farmers have done and the differences in feeds and conditions effect on the incidence of SCK and CK in the herd. In time for EAAP congress we will have the results of the first test period and some preliminary results for the second period. So far the level of ketosis has varied a lot between the pilot farms and especially different milking and feeding systems have explained the differences. All of the pilot farms have been very keen on working with this project to get results and also other farmers and veterinarians have been interested about this project. Even by now we have reached our goal to make both CK and especially SCK better known among the Finnish dairy farms. MaMa-project is conducted in co-operation with ProAgria Southern Ostrobothnia, Work efficiency institute and Association of ProAgria Centres.

The role of the health control questionnaires about diagnosis of metabolic diseases in dairy herds

L. Liepa and I. Sematovica
Latvia University of Agriculture, Faculty of Veterinary Medicine, Kr. Helmana street 8, 3004 Jelgava, Latvia; laima.liepa@llu.lv

In recent years, due to increasing of productivity in dairy cows it is very important to monitor the herd health regarding the feeding, hygiene and welfare as well as animal metabolic performance. In Latvia, SARA, ketosis and hypocalcaemia are metabolic problems that occur most often in dairy herds. The study was conducted as a part of the State Research Project (AgroBioRes) No. 2014.10-4/VPP-7/5; subproject AP4. The aim of this study was to prepare standardized questionnaires to assess the metabolic status of dairy cows in herd health audit. Studies were performed in four high-performance, freestall-housed dairy herds with 315-860 cows. Questionnaires consisted of two parts: general informative and special herd inspection questions. The general part included documented parameters about the herd and individual animals (parity, milk fat and protein level, milk fat/protein ratio, milk urea content). Questions in the special part addressed to assessment of the freestall buildings with regard to animal welfare, feeding, drinking management and quality. Individual animals were evaluated according to habitus and clinical signs (condition and cleanliness, lameness, chewing activity, rumen fill, stool consistence and digestibility using flushing method). Questionnaires of health audit included express analysis (rumen fluid analyses, blood β-hydroxybutyric acid, glucose or lactic acid concentration, ketone bodies in urine), or common serum biochemical analysis for 3-5 animals. All parameters had to be compared with physiological reference values and in connection to different periods of lactation. Information was compiled in the questionnaire forms to do a comprehensive analysis. A follow-up recurrent survey of herd health using these questionnaires reveals changes in the management performance, the improvement of cow health status, production, herd profitability. Excel program can be easily used for storing and comparison of the data in questionnaires. Conclusion. The results of the herd health visits obtained from questionnaires and stored in Excel improve and facilitate the work of the veterinarian in herd health surveillance.

Effect of breed, milkability and season on the temperature of the milk of dairy cows

M. Uhrinčat[1], J. Broušek[1], L. Mačuhová[1] and V. Tančin[1,2]
[1]NPPC – Research Institute for Animal Production Nitra, Hlohovecká 2, 951 41 Lužianky, Slovak Republic, [2]Slovak University of Agriculture, Tr. A. Hlinku 2, 949 76 Nitra, Slovak Republic; uhrincat@vuzv.sk

Evaluation of milk temperature, as a non-invasive treatment, could be an important data source of physiological responses of cows to ambient temperature. Therefore the aim of this study was to evaluate the milk temperature affected by different factors. Data from 1,413 cows of different breeds (Slovak Spotted Cattle, Holstein, Red Holstein crossbreeds with Pinzgau and Holstein) were collected from 19 farms in 4 seasons (spring, summer, autumn, winter). In our assessment, LactoCorder evaluating (milk flow, milk yield and milk temperature) was used during morning or evening milking. Significant differences between the groups were tested by a one-way ANOVA. For all possible pairwise comparisons of means the Scheffé's test was used (SAS, version 9.3). Breeds impact on the milk temperature was not confirmed. A similar effect with no significant differences was observed in type of flow. The temperature of milk (°C) of animals with bimodality was 37.86±0.03 and no bimodality was 37.83±0.02. We detected a significant influence of the season on the milk temperature. The lowest mean temperature was recorded during the winter season (37.41±0.04). During the spring (37.82±0.02) and autumn (37.72±0.02) period, there was only a slight difference between the mean temperatures, but significant never the less. During the summer season (38.29±0.03) we recorded the highest mean milk temperature. Mean of milk temperature for mornings milking for whole observation (37.71±0.02) was significantly lower than mean of evenings milking (37.96±0.02). With increasing milk yield (GR1 4-9 kg, GR2 9.01-12 kg, GR3 12.01-15 kg and GR4 15,01 kg and more) we found an increase in temperature of the milk (GR1 37.6±0.02, GR2 37.81±0.02, GR3 37.92±0.03 and GR4 38.04±0.03) with significant differences with each other. In conclusion, the temperature of the milk was significantly influenced by the season, i.e. ambient temperature. The existence of significant positive correlation between milk and body temperature is known and therefore the temperature of the milk could be used as an indicator of thermal comfort or discomfort of cows. This study was supported by APVV 15-0060.

Precision dairy monitoring opportunities and challenges

J.M. Bewley
University of Kentucky, Animal and Food Sciences, 407 WP Garrigus Building, Lexington, KY 40546-0215, USA; jbewley@uky.edu

Technologies are changing the shape of the dairy industry across the globe. In fact, many of the technologies applied to the dairy industry are variations of base technologies used in larger industries such as the automobile or personal electronic industries. Undoubtedly, these technologies will continue to change the way that dairy animals are managed. This technological shift provides reasons for optimism for improvements in both cow and farmer well-being moving forward. Many industry changes are setting the stage for the rapid introduction of new technologies in the dairy industry. Dairy operations today are characterized by narrower profit margins than in the past, largely because of reduced governmental involvement in regulating agricultural commodity prices. The resulting competition growth has intensified the drive for efficiency resulting in increased emphasis on business and financial management. Furthermore, the decision making landscape for a dairy manager has changed dramatically with increased emphasis on consumer protection, continuous quality assurance, natural foods, pathogen-free food, zoonotic disease transmission, reduction of the use of medical treatments, and increased concern for the care of animals. Lastly, powers of human observation limit dairy producers' ability to identify sick or lame cows or cows in heat. Precision dairy management may help remedy some of these problems. Precision dairy management is the use of automated, mechanized technologies toward refinement of dairy management processes, procedures, or information collection. Precision dairy management technologies provide tremendous opportunities for improvements in individual animal management on dairy farms. Although the technological 'gadgets' may drive innovation, social and economic factors dictate technology adoption success. These factors, along with practical experiences working with multiple technologies will be discussed in this presentation.

Computer-mediated communication applied to livestock farming

S. Rafaeli

The Center for Internet Research, Faculty of Management, University of Haifa, Israel; sheizaf@rafaeli.net

Digital technology is assuming a central position in all walks of life, and is pivotal in economics and the workplace too. The use of digital technology in agriculture is pervasive and rapidly growing. It includes smart sensing and mapping, climate monitoring and forecasting, livestock tracking and Geo-Fencing, monitoring of feeding and other procedures, monitoring equipment, predictives and machine learning. In this talk I will touch on the technologies available and those on the horizon, and ask about the implications for the future. Are our conceptions of Distance, Time and Center changing? How does the rapid replacement of Choice with Algorithms reshape the workplace and work experiences? Given the rapid evolution of research methods and the models of scientific reliability and validity, what are the implications for applied science? Are our basic conceptions and laws of managing and economics challenged in an era of bit replacing atoms?

Some thoughts about automation for livestock production from an automatic control perspective

K. von Ellenrieder

Florida Atlantic University, College of Engineering & Computer Science, Department of Ocean and Mechanical Engineering, 777 Glades Road, EW190, Boca Raton, FL 33431, USA; ellenrie@fau.edu

Session 13

Theatre 4

Future of precision livestock farming: the industry point of view

O. Inbar
SCR by Allflex,SCR Engineers Ltd, P.O. Box 13564 Hadarim, 42138 Netanya, Israel; oinbar@scrdairy.com

Precision farming is a big word and when it comes to livestock it carries many different interpretations. This invited keynote presentation, at the plenary session, will try to cover the state art in the area of livestock monitoring and possible future directions, as well as the main challenges in making precision livestock farming a reality that have a significant impact on food production.

Session 13

Theatre 5

How (some) technologies have influenced phenomics and animal breeding across species and across my life

L. Bunger
Sheep & Goat Genetics and Genomics, AVS, SRUC, Roslin Institute Building, Easter Bush, Midlothian EH25 9RG, Scotland, United Kingdom; lutz.bunger@sruc.ac.uk

Agriculture is an increasingly fundamental industry. The global challenge faced by the farming industry is to double food production and half the ecological footprint on the planet to feed 9 billion people by the year 2050. This requires farmers to use precision farming and breeders to use a very complex approach. They have to further enhance the genetic potential for yield but simultaneously broaden the breeding goals to include also product quality, health, welfare, environmental aspects to consider the consumer sector and maintenance of genetic diversity. However, it is estimated that worldwide, the productivity of farm animals is 30-40% below their genetic potential because of suboptimal conditions and health status. Therefore, management related innovations aiming to narrow the gap between genetic potential and performance at the farm level are needed (e.g. metabolic programming, microbiome manipulations) to meet national and global challenges. This paper will focus on genetic improvements and the role of phenomics, including associated technologies and their metrological aspects, although it is acknowledged that most described phenomics tools serve also the optimisation of on-farm-management. Eventually, carefully implemented phenomic, genetic and genomic technologies may provide new solutions to achieve ever more complex breeding objectives, where breeders are often faced with conflicting genetic desires for animal output and short-term profit vs sustainability. Improvement in the quality and quantity of animal performance data, including new and *difficult to measure* traits, as well as enhancements of genetic evaluations and the structure of selection indexes, are needed. New technologies are being used in animal production & -breeding and their use helps to further the development of these technologies. The implementation and effective use of Technologies depends on Teams and Transfer (TTT), where Teams comprise (international) networks and Transfer refers to using that knowledge. This paper will look at new (and current technologies for high-throughput, accurate and precise phenomics, the subsequent use of the data in breeding, and at the technology gap.

Experiences with trait by trait methodology for estimating breeding objectives in cattle

P.R. Amer and T.J. Byrne
AbacusBio Limited, P.O. Box 5585, Dunedin 9012, New Zealand; pamer@abacusbio.co.nz

Breeding objectives made up of economic weighting factors are widely used in cattle breeding around the world. Unlike more intensive farmed species such as pigs and poultry where breeding objectives tend to be controlled and operated at the breeding company level, breeding objectives for dairy and beef cattle are usually formulated more independently at a national, industry sector, or breed society level. Methodology and theory of calculating breeding objectives initially evolved from derivatives of relatively simple profit equations to complex bioeconomic modelling approaches. However, at least two waves of new sets of traits have arisen over recent years, that require updates to many national breeding objectives. Firstly, the advent of phenotypes related to functional performance that allow poorly heritable and poorly recorded traits such as longevity and lifespan to be better dissected into component traits related to specific diseases and health treatments. In these cases, where more appropriate weighting based on treatment and lost milk production can be incorporated into the breeding objective. Secondly, traits related to greenhouse gas emissions intensity and a revival of interest in selecting for feed efficiency are also initiating new requirements for selection index development. This paper describes a trait by trait approach to model the development for of economic weights, discusses its strengths and weaknesses, and provides some examples of how it can be used to address new challenges in the formulation of breeding objectives for cattle breeding programmes.

Economic selection indexes and realized genetic change in American Angus

S.P. Miller, K.J. Retallick and D.W. Moser
Angus Genetics Inc., Saint Joseph, MO 64506, USA; smiller@angus.org

The American Angus Association, established in 1873, is the largest beef cattle breed association in the USA with 334,607 new registrations in 2016, more than all other major US beef breeds combined. This dominant position means genetic change in Angus will have a significant impact on the commercial beef industry. Through the provision of genetic evaluations (Expected Progeny Differences EPD) and economic selection indexes, members select for improved profitability in the commercial sector. Genetic change is a result of the selection decisions of the combined 20,000 active breeders. Objectives were to analyze genetic trends of economic selection indexes and contributing growth and carcass traits (subset of all traits) to evaluate the realized genetic change accomplished and to compare trait changes before and after significant events such as the ultrasound evaluation of carcass traits in yearling cattle (2001) or adoption of genomic prediction (2010). Trends were evaluated by linear regression on average EPD by birth year divided by EPD SD (units expressed per year multiplied by 100). During the past 30 years, birth weight has decreased slightly (BW, -1.8) while increasing yearling weight (YW, 15), carcass weight (CW, 11), marbling (MB, 8.7), rib-eye area (RE, 8.2), and $B (15), where $B is the economic index for weaned calf performance including carcass revenue. Since 2001 with ultrasound, BW decreased faster (-4.0) while YW increased slightly (16) as did CW (12) with a most notable increase in RE (13) and intake (IN, 8) with $B decreasing slightly (14), possibly due to discovery of genetic conditions in families contributing to high $B, which were reduced in use. Since genomics (2010) most notable differences include a further increase in YW (18), CW (16) and RE (17), decreasing IN (6) and increase in $B (18). Trends post genomics also coincide with a focus among breeders on the economic indexes launched in 2004, which empasize carcass weight along with quality (MB). The downward trend in IN with increased weight is a suspected outcome of selection for the economic indexes. Realized genetic change in American Angus illustrates how breeders are utilizing the economic selection indexes to create more profitable cattle for the commercial sector.

Live weight, body condition score and reproductive performance of genetically diverse beef cows

S. McCabe[1,2], N. McHugh[3], N. O'Connell[1] and R. Prendiville[2]
[1]Queens University Belfast, School of Biological Sciences, Northern Ireland, United Kingdom, [2]Teagasc, Grange, Dunsany, Co. Meath, Ireland, [3]Teagasc, Moorepark, Fermoy, Co. Cork, Ireland; simone.mccabe@teagasc.ie

The Irish national beef replacement index was derived to identify animals suitable for breeding or selecting replacements on the basis of their genetic merit for maternal traits. The aim of this study was to compare live weight (LW), body condition score (BCS) at breeding and reproductive performance of high and low replacement index cows. Data were available over three consecutive breeding seasons; 96 and 83 high and low index cows, respectively. Cows were sourced from dairy (F_1) or suckler (S) herds and calved for the first time at 24 months of age. Breeding commenced in late April each year and lasted 13 weeks. Tail paint and vasectomised bulls were aids for heat detection. Pregnancy was confirmed using trans-rectal ultrasound imaging (Aloka 210D * II, 7.5 MH3). Effect of genetic merit (high or low), origin (F_1 or S) and the interaction between them on LW, BCS and reproductive performance was analysed using a linear mixed model in PROC HPMIXED; binary traits were investigated using logistic regression in PROC GENMOD (SAS Inst. Inc., Cary, NC). Sire of the cow was included as a random effect. Fixed effects included in all models were genetic merit, origin, year, parity and heterosis and recombination loss of the cow. Pregnancy rate to first service was similar for high and low animals; 52% and 56% (P>0.05), respectively. No difference in number of services was observed between high and low merit cows; 1.74 and 1.59 (P>0.05) respectively. Calving to conception interval was similar between genetic merit; 79 and 77 d (P>0.05) for high and low index. It was also similar (P>0.05) for cow origin; 77 and 78 for F_1 and S, respectively. The S cows were 63 kg heavier and 0.23 of a greater BCS than F_1 cows (P<0.001) at breeding. The F_1 high cows tended to have a greater submission rate (79%) in the first 24 days than F_1 low (62%; P=0.0579). The S high cows had a greater (P<0.05) pregnancy rate (94%) than S low (80%). Results for other reproductive variables investigated; calving to service interval, six week in-calf rate and calving interval were similar across cow origin and genetic merit.

A preference-based approach to deriving breeding objectives in cattle

M. Klopčič, A.D.M. Biermann and A. Ule
University of Ljubljana, Biotechnical Faculty, Dept. of Animal Science, Groblje 3, 1230 Domžale, Slovenia; marija.klopcic@bf.uni-lj.si

The aim of this research was to characterize the preferences of Slovenian dairy farmers for improvements in breeding goal traits. For this reason, a study was conducted on dairy farms using a questionnaire survey involving more than 200 dairy farms with Holstein, Brown or Simmental breed. We also collected a considerable number of herd characteristics on these farms. The most preferred traits improvement were longevity, udder health, fertility and milk production. For farmers with dual-purpose breeds (Simmental and Brown) also beef traits are important. The selection of traits in breeding goals was largely influenced by the farmer's herd breed and breeding objective (the 'profit cow' versus the 'management cow'). Farmers selected traits that are recognised as weaknesses of their corresponding herd, e.g. fertility, longevity, health (SCC, metabolic diseases, lameness, etc.). On the base of analysis we can conclude that farmers prefer to improve traits that are more problematic in their herd. Farmers are aware that they will need to pay in the future more attention to claw health, feed efficiency, fertility traits, reduce emissions and minimize the negative effects on the environment. The results of this study could be used for the future development of breeding goals for all three breeds in Slovenia and for the development of customized total merit indices based on farmer preferences and calculation of economic selection indices for various breeds.

Simulating consequences of choosing a breeding goal for organic dairy cattle production

M. Slagboom[1], A. Wallenbeck[2], L. Hjortø[1], A.C. Sørensen[1], J.R. Thomasen[1,3] and M. Kargo[1,4]
[1]*Aarhus University, Blichers Alle 20, 8830 Tjele, Denmark,* [2]*Swedish University of Agricultural Sciences, Ulls väg 26, 753 23 Uppsala, Sweden,* [3]*VikingGenetics, Ebeltoftvej 16, 8960 Randers SØ, Denmark,* [4]*SEGES Cattle, Agro Food Park 15, 8200 Aarhus N, Denmark; margotslagboom@mbg.au.dk*

Six different breeding goal (*BG*) scenarios were developed and simulated for organic and conventional dairy production, to study the effect on genetic change for specific traits. The first two BGs (*BG1* and *BG2*) were based on economic weights from SimHerd simulations for a conventional and an organic dairy system in Denmark. The third BG (*BG3*) was based on weights derived from a farmer preference survey for organic farmers. The fourth BG (*BG4*) was based on the four organic principles defined by the IFOAM, and on the results of a questionnaire that was sent out to farmers, researchers and experts in the area of organic dairy husbandry. The fifth and sixth BGs (*BG5* and *BG6*) were based on BG4, but altered to put more emphasis on roughage consumption and disease resistance, respectively. Each BG was simulated for 30 years using a stochastic simulation. A genomic breeding scheme using 100 young bulls yearly were simulated. Animals were selected based on a total merit index, including all 12 traits with different economic weights per scenario. The best animals were selected for breeding. Preliminary results show that some overall genetic gain in monetary units is lost when using BG3 (based on organic farmers' preferences) compared to using BG1 and BG2 (based on objectively derived economic weights). BG3 caused the least genetic decline in genetic standard deviation units in beef production, calving difficulty, cow and heifer fertility and cow mortality, but also the least genetic progress in genetic standard deviation units in milk production, roughage consumption and feed efficiency compared to BG1 and BG2.

Design of breeding strategies for feed efficiency and methane emissions in Holstein using ZPLAN+

K. Houlahan[1], S. Beard[1], F. Miglior[1,2], C. Richardson[1], C. Maltecca[3], B. Gredler[4], A. Fleming[1] and C. Baes[1]
[1]*Centre for Genetic Improvement of Livestock, University of Guelph, Department of Animal Biosciences, 50 Stone Road East, N1E 2W1, Canada,* [2]*Canadian Dairy Network, 660 Speedvale Ave West, Guelph, N1K 1E5, Canada,* [3]*North Carolina State University Campus Box 7621 Raleigh, Department of Animal Science and Genetics, Campus Box 7621, Raleigh, NC 27695, USA,* [4]*Qualitas AG, Chamerstrasse 56, 6300 Zug, Switzerland; cbaes@uoguelph.ca*

Feed efficiency (FE) and methane emissions (ME) are correlated traits currently undergoing intense research. Feed accounts for over 50% of the total costs on a dairy farm in North America, with feed prices expected to continue to rise in coming years. Methane has been identified as one of the most prevalent non-CO_2 greenhouse gases contributing to climate change. There is therefore a clear need to improve FE and reduce ME of dairy cattle. Studies have shown that there is variation in FE and ME between animals that have similar production. By exploiting this variation on a genetic level, there is an opportunity for greater and more permanent improvement in both FE and ME. Determining an optimal breeding strategy for improving FE and ME in dairy cattle would be helpful in reducing feed costs, while maintaining or increasing output in a sustainable way. Measuring FE and ME can be expensive and time consuming. There are different ways to express FE, including residual feed intake (RFI) and dry matter intake (DMI). Using genetic correlations among traits, along with phenotypic correlations and heritabilities, an optimal breeding strategy will be developed and analyzed using ZPLAN+. ZPLAN+ is a software program that aids in the modeling and calculation of complex breeding scenarios. This program will be used to assess the genetic gain, monetary gain, and costs associated with including FE and ME in a selection index for the Canadian dairy industry. Additionally, long-term effects of including FE and ME in the Canadian selection index on traits of economic interest such as milk yield, fertility and health will be analyzed. The results of this work will provide insight for the Canadian dairy industry as to the best method for including FE and ME into the existing selection index.

Do breath gas measurements hold the key to unlocking the genetics of feed efficiency in dairy cows?

G.F. Difford[1,2], Y. De Haas[1], M.H.P.W. Visker[1], J. Lassen[2], H. Bovenhuis[1], R.F. Veerkamp[1] and P. Løvendahl[2]
[1]Wageningen University & Research, Animal Breeding & Genomics, 6700 AH Wageningen, P.O. Box 338, the Netherlands,
[2]Center For Quantitative Genetics and Genomics, Department of Molecular Biology and Genetics, Aarhus University,
Blichers Alle, 8830, Denmark; gareth.difford@mbg.au.dk

Recording dry matter intake (DMI) in dairy cows is the precursor to determining feed efficiency, a highly profitable and desirable selection trait. However records on large numbers of animals are expensive and prohibitive under commercial conditions. Usually small research herds are recorded for DMI and then used to predict breeding values for DMI using other highly correlated traits like milk yield, body weight and chest width measured in commercial herds. Recent interest in greenhouse gases, such as methane (CH_4), has seen the development of tools for measuring gas concentrations in the breath of the cow during milking in automated milking stations or in concentrate feeders. This makes it possible to obtain records on large numbers of animals under commercial conditions. Since CH_4 production is a conditional by-product of DMI, it is not surprising that DMI is the single best predictor of CH_4 production. Traditionally, this relationship has been exploited to predict CH_4 production from DMI to aid in CH_4 related research. Here we turn this idea around and assess the effectivity of CH_4 and carbon dioxide breath measurements for the prediction of DMI and feed efficiency in Dutch and Danish Holstein cattle. Preliminary results on 1000 cows indicate a strong positive genetic correlation of 0.75 between CH_4 production and DMI and a strong negative genetic correlation of -0.68 between CH_4 concentration and RFI. Through the use of genomic tools we assess the added benefit of these strong correlations with large scale breath measurements for the prediction of DMI and RFI in dairy cows under commercial conditions without DMI records.

Response on claw health in breeding of Czech Holstein cattle

Z. Krupová, J. Přibyl, E. Krupa and L. Zavadilová
Institute of Animal Science, Přátelství 815, 10400 Prague Uhříněves, Czech Republic; krupova.zuzana@vuzv.cz

Claw disease is known as a serious animal issue with welfare, economic and socioethical aspects of animal production. Prevalence of this disease in the Czech Holstein farms resulted in direct economic loss (economic weight) of -100.08 € per incidence per cow and per year. Actually, the claw health is not used as breeding objective in the local cattle population, nevertheless some type traits of legs (rear leg rear view, foot angle, locomotion and legs) are defined there as criterion in the subindex 'legs'. Therefore, genetic and economic effect of three variants of the subindex based on (1) type traits, (2) on a claw health trait and (3) on both, type traits and claw health were calculated. The genetic parameters of the traits and general principles for construction of selection indices and calculation of selection response were considered. For claw disease and for all type traits reliability of input breeding values was set in two levels (0.20 and 0.50 vs 0.50 and 0.70). Based on the higher reliabilities of estimated breeding values, the subindex build only on type traits (index A) reduce the prevalence of claw disease by 0.26% (e.g. 0.26 €) per year. Genetic response for claw health as the only trait in the index (B) would gain profit 0.38 € per year. Claw health and type traits jointly included into index (C) would reduce the prevalence per claw disease up to 0.41%. The reliability of subindex of sires mentioned above were 23%, 50% and 56%, respectively. Genetic response and reliability of the subindex was reduced (e.g. 0.30% claw incidence and 32% reliability, both in index C) when lower reliability of breeding values was set in the study. Results showed that selection for legs traits reduced prevalence of claw diseases indirectly and common evaluation of health and type traits is of positive impact on the index reliability. Moreover, it is supposed that inclusion of claw health as a new breeding objective in the future would be beneficial for the local dairy population and genetic response for claw health along with reliability of selection index will increase with routine evaluation of disease incidence among daughters. Study was supported by project MZERO0714 and QJ1510139 of the Czech Republic.

Keep mixed models multitrait (MMM) to predict derived breeding values

T. Ådnøy, T.K. Belay and B.S. Dagnachew
NMBU – Norwegian University of Life Sciences, IHA, Arboretveien 6, 1433 Ås, Norway; tormod.adnoy@nmbu.no

Some phenotypes (e.g. fat%, …) are derived from multitrait information (milk mid-infrared spectra, …). Breeding values for 'fat%, …' are often calculated after first using regression for the multiple phenotypic traits in 'spectra' and then applying a mixed model to the derived 'fat%' phenotype. We call this Indirect Prediction (IP) of breeding values. An alternative Direct Prediction (DP) is to apply the mixed model to the multitrait information and predict the breeding values of the 'specter' components before calculating a breeding value for the derived 'fat%' based on the heritable part of the multitrait 'spectra'. In a milk spectra dataset for goats we showed that the prediction error variance of breeding values could be reduced (3-7%) using DP compared to IP. However, this result was found using the coefficient matrix of the Mixed Model Equations, since no independent milk content was available. In collaboration with the Polish Federation of Cattle Breeders and Dairy Farmers, we have had access to a dataset with 826 cows where blood BHB has been measured and there are milk spectra. Thinking that also the prediction of the phenotype would be improved by using a MMM by the DP method, we cross-validated, using 496 BHB measures to establish the model and predicting the 330 remaining BHB. However, no improvement was found using DP compared to IP for the BHB phenotype. The theoretical foundation for MMM to understand better why will be presented. BLUP-prediction of random effects is based on regression theory. When Y_1 and Y_2 are multinormally distributed the expected value is $E(Y_1|Y_2) = E(Y_1) + cov(Y_1,Y_2')var^{(-1)}(Y_2)(Y_2-E(Y_2))$, or as used in breeding value (U) prediction: $E(U|Y) = E(U) + cov(U,Y')var^{(-1)}(Y)(Y-E(Y))$, with $E(Y)$ replaced by its estimated value, and $E(U)=0$. This theory also applies to unobserved Y_1 (fat%) when Y_2 (spectra) are observed, and when Y is for many individuals or observations of many different traits. A challenge is to estimate the relevant trait (co)variances and account for the observation structure.

Application of combined decision models to investigate management strategies for local cattle breeds

J. Schäler[1], S. Addo[1], G. Thaller[1] and D. Hinrichs[2]
[1]Institute of Animal Breeding and Husbandry, Christian-Albrechts-University, Hermann-Rodewald-Straße 6, 24118 Kiel, Germany, [2]Albrecht Daniel Thaer-Institute of Agricultural and Horticultural Sciences, Humboldt-University, Invalidenstraße 42, 10099 Berlin, Germany; jschaeler@tierzucht.uni-kiel.de

The methodical combination of SWOT (Strengths, Weaknesses, Opportunities and Threats) and AHP (Analytic Hierarchy Process), so called SWOT-AHP analysis, is an appropriate tool to merge qualitative as well as quantitative decision making-approaches. SWOT-AHP analysis was applied to investigate effective strategic actions to counteract racial extinction and loss of genetic diversity in single local breeds. In this study, the methodical guideline of SWOT-AHP was elaborated and extended using two local cattle breeds in Northern Germany, the German Angler and Red and White dual purpose breed (DN). The internal and external analyses of the production system were extended in more detail. A multi-stakeholder approach to determine factors was used, whereas questionnaires of farmers (N_1=78, N_2=80) got a special focus in this hierarchical structure. An expert group of scientist and stakeholder ranked the SWOT strategies and inserted them into the decision making tool. A total of 96 individual SWOT strategies of action for both single cattle breeds were investigated. After pairwise strategy comparisons effective strategies of action were derived to achieve breed preservation and maintenance genetic diversity. For German Angler the Strength-Threat (ST) strategy was identified. Best combined SWOT strategy for the Red and White dual purpose breed (DN) was the Weakness-Threat (WT) strategy. Identified strategies of action are very promising for the future development within these single cattle breeds.

Corn yield and SPAD index in monoculture and integrated system in Estate of São Paulo, Brazil

F.F. Simili[1], P.M. Bonacim[1], G.G. Mendonça[2], J.G. Augusto[1] and C.C.P. Paz[1]
[1]*Instituto de Zootecnia, Rodovia Carlos Tonani, 14860000, Brazil,* [2]*FMVZ/USP Pirassununga, SP, Av. Duque de Caxias Norte, 225, 13635900, Brazil; flaviasimili@gmail.com*

Corn sowing was carried out in a no-tillage system in a intercropping with palisade grass (*Brachiaria brizantha* cv marandu) (integration systems), and in single sowing (monocrop). The study aimed to evaluate the effects of intercropping in grain yield and compare the level of Nitrogen based on SPAD index. The treatments were sown in December of 2015 at Instituto de Zootecnia, Sertãozinho, Brazil in an area of 15 ha. It was used a complete randomized block with three replications and five treatments: monocrop corn (C), corn and palisade grass sown simultaneously (CG), corn and palisade grass sown simultaneously plus herbicide (CGH), palisade grass sown at topdressing corn (CGT) and palisade grass sown at the line and inter-line of corn (CGL). The corn were mechanical harvested in May 2016. Nine samples were made per treatment, consisted of two lines of two linear meters each. All ears were harvested by hand and taken to the laboratory. Subsequently, the grain weight was determined and the data were transformed to grain yield per hectare (130 g/kg wet basis). The weight of 100 grains were evaluated from 30 plants per plot chosen at random in the usable area. To evaluate the level of leaf chlorophyll (SPAD index) the Minolta SPAD-502 equipment was used, fortnightly, in expanded leaves. The PROC MIXED by SAS was used to analyze the data. The intercropping sowing was not harmed by corn production (P=0.176), the final plant population (P=0.435) and weight of 100 grains (P=0.264). The results were: 12.02; 10.98; 12.06; 11.41 and 11.87 Ton grain/ha; 78; 75; 73; 73 and 77 (no. ×1000) plant/ha and 31.67; 32.33; 31.33; 31.63; 32.25 (W100) to C; CG; CGH; CGT e CGL, respectively. Significant interaction was observed among weeks of evaluation and treatments for SPAD index (P<0.001). The results were 55.68; 53.46; 54.87; 54.66 and 52.11 to C; CG; CGH; CGT e CGL, respectively. It was not observed in any treatment deficit of nitrogen. The presence of palisade grass in the integration systems did not affect the development of the corn crop. The integration systems can represent an alternative to vegetal and animal production.

Monoculture system production economic return as contrasted to the integrated crop-livestock system

G.G. Mendonça[1], J.G. Augusto[2], P.M. Bonacim[2], F.F. Simili[2] and A.H. Gameiro[1]
[1]*School of Veterinary and Animal Science, University of Sao Paulo, Avenida Duque Caxias Norte, 225, Jardim Elite., Pirassununga, São Paulo 13635-000, Brazil,* [2]*Institute of Animal Science, Rodovia Carlos Tonani, Km 94, Zona Industrial, Sertãozinho, São Paulo, Brazil; gabigeraldi@hotmail.com*

This study aimed to calculate the economic return of corn grain and beef cattle production in both monoculture and integrated crop-livestock systems. The experiment was developed at the Institute of Animal Science in Sertãozinho, Sao Paulo State, Brazil. Experimental treatments were: (1) corn monoculture; (2) Palisade grass (*Brachiaria brizantha* cv. Marandu) monoculture; (3) integrated corn and Palisade grass seeded simultaneously; (4) integrated corn and Palisade grass sown at corn top dressing fertilization period. All experimental plots were installed in December 2015 and were evaluated for a one-year period. Treatment area used encompassed 3 hectares. Variable costs included seeds, fertilizers, herbicides and insecticides; fixed costs were machinery and equipment depreciation, and labor. Revenues were calculated from corn productivity (average of 12 tons/ha for all treatments) and meat production (350 @/cycle for T2 and 190 @/cycle for T3 and T4 integrated treatments). The average exchange rate for November 2016 was R$ 3.3,420 for US$ 1.00. Total costs (R$/ha) for monoculture treatments T1 and T2 were, respectively, R$ 6,371.50 and R$ 11,321.80; for integrated systems T3 and T4 were, respectively, R$ 6,843.16 and R$ 6,889.90. In T2, earlier grass production allowed a higher stocking rate (animals/ha), which accounts for the higher cost of this system. Revenues (R$/ha) for T1 and T2 systems were respectively R$ 5,849.8 and R$ 15,820, whereas T3 and T4 revenues were R$ 15,026.80 and R$ 15,465.70, respectively. Economic returns (R$/ha) for treatments 1 to 4 were R$ -521.60; R$ 4,498.21; R$ 8,133.50 and R$ 8,575.80, respectively. Although input and final product price seasonality should be considered over a longer evaluation period, the results may suggest economic benefits of integrated systems. Integrated system proposed scenarios showed similar results; however, other aspects of this type of system should be evaluated from a productive and economic point of view.

Agronomic traits of corn in in monoculture and integrated system in Estate of São Paulo, Brazil
F.F. Simili[1], J.G. Augusto[1], G.G. Mendonça[2], P.M. Bonacim[1] and C.C.P. Paz[1]
[1]*Instituto de Zootecnia, Rodovia Carlos Tonani, 14860000, Brazil, [2]FMVZ/USP Pirassununga, SP, Av. Duque de Caxias Norte, 13635 900, Brazil; flaviasimili@gmail.com*

Corn sowing was carried out in a no-tillage system in a intercropping with palisade grass (*Brachiaria brizantha* cv marandu) (integration systems), and in single sowing (monocrop). The study aimed to evaluate the effects of intercropping on agronomic traits of corn: plant population (PP), plant height (PH) height of ear insertion (HEI), diameter of stem (DS), diameter of ear and (DE) length of ear (LE). The treatments were sown in December 2015 at Instituto de Zootecnia, Sertãozinho, Brazil in an experimental area of 15 ha. It was used a complete randomized block with three replications and five treatments: monocrop corn (C), corn and palisade grass sown simultaneously (CG), corn and palisade grass sown simultaneously plus herbicide (CGH), palisade grass sown at topdressing corn (CGT) and palisade grass sown at the line and inter-line of corn (CGL). The corn were mechanical harvested in May 2016. Nine samples were made per treatment consisted of two lines of two linear meters each. The plants were counted (PP) and all ears were harvested by hand and taken to the laboratory. The heights were measured with graduated ruler and the length of ear was made with tape measure and the diameters with Pachymeter. The PROC MIXED by SAS was used to analyze the data. Significant interaction was observed among weeks of evaluation and treatments for PH (P<0.0001). The HEI, DS and LE were different (P<0.0001) among treatments. There was no significant difference between PP (P=0.435) and DE (P=0.435). The intercropping sowing decreased the HP (2.26 m); HEI (1.07 m) and DS (18.3 mm) at CGL treatment, probably because of the greater amount of palisade grass, which may mean a greater competition between crops. However, the herbicide application at the CGH treatment controlled the growth of the palisade grass, providing greater HP (2.36 m); DS (20.2 mm) and LE (16.58 cm). The high PP obtained between the treatments, on average 75 (nox1000) plants/ha did not compromise the evaluated characteristics. The agronomics traits studied were not prejudiced at the monoculture system compared to the integrated systems.

Efficiency of selection indices for milk flow, production and conformation traits in Holsteins
T. Kaart[1,2], D. Pretto[2], A. Tänavots[1,2], H. Kiiman[1,2] and E. Pärna[1,2]
[1]*Bio-Competence Centre of Healthy Dairy Products, Kreutzwaldi 1, 51014 Tartu, Estonia, [2]Estonian University of Life Sciences, Institute of Veterinary Medicine and Animal Sciences, Kreutzwaldi 1, 51014 Tartu, Estonia; tanel.kaart@emu.ee*

The aim of this study was to estimate the annual genetic response for average milk flow (AMF), milk yield (MY), milk components, somatic cell score (SCS) and udder conformation traits in Estonian Holstein (EH) population under different selection indices. Current selection index (SPAV) includes fat (FY) and protein (PY) yields. Other selection indices are under study in order to improve SCS and conformation traits. Inexpensive electronically recordings of AMF from milking parlours and automatic milking systems are available in Estonia. Dataset from EH cows reared in 69 farms was used to estimate phenotypic and genetic parameters for all traits. The data of AMF (19,316 single records), milk production traits (302,629 repeated records) and udder conformation traits (11,143 single records) was collected by the Estonian Livestock Performance Recording Ltd from July 2010 to January 2016. Four generations pedigree with 66,370 animals was considered in animal models. Annual genetic response (GRy) for the abovementioned traits was estimated by using selection index theory and EH breeding program parameters. AMF had high heritability (0.50) and intermediate genetic correlations with MY (0.52), SCS (0.33) and udder conformation traits (the strongest -0.30 with teat length and 0.33 with rear udder height). Application of SPAV resulted with the increase of 0.14 genetic standard deviations per year (GSDy) for AMF, 0.19 GSDy for MY and 0.07 GSDy for SCS. Direct inclusion of AMF (25% of relative weight) in the index increased/decreased the GRy for yield traits and SCS by 0.01-0.03 GSDy, if there was selection for or against higher AMF, respectively. Direct inclusion (25% of relative weight) of SCS resulted with almost no increase in SCS (0.01 GSDy) and still relatively high increase in MY (0.18 GSDy). The inclusion of AMF has only slight effect on GRy of other traits, however the EBV of AMF can be used to exclude extreme animals from breeding. Estonian Ministry of Education and Research (grant IUT8-1) and Bio-Competence Centre of Healthy Dairy Products LLC (project EU48686).

Genetic associations of in-line recorded milkability traits and udder conformation with udder health

C. Carlström[1], E. Strandberg[1], K. Johansson[2], G. Pettersson[3], H. Stålhammar[4] and J. Philipsson[1]
[1]*Swedish University of Agricultural Sciences, Dept of Animal Breeding and Genetics, P O Box 7023, 75007 Uppsala, Sweden,* [2]*Växa Sverige, P O Box 7023, 75007 Uppsala, Sweden,* [3]*Swedish University of Agricultural Sciences, Department of Animal Nutrition and Management, P O Box 7024, 75007 Uppsala, Sweden,* [4]*Viking Genetics, Box 64, 532 21 Skara, Sweden; Erling.Strandberg@slu.se*

Milkability and udder conformation traits of Swedish Holstein (SH) and Swedish Red (SR) cows from 93 herds with automatic milking systems or conventional milking parlors, were used to study genetic relationships to lactation average somatic cell score (LSCS) and incidence of clinical mastitis (CM). Estimated genetic correlations between measures of milking speed, (average flow rate, milking time and box time) and LSCS ranged between 0.29 and 0.57 and showed that high milking speed is associated with increasing LSCS. Regressions indicated a curvilinear relationship. Genetic correlations between milking speed and CM showed similar values as for LSCS in SH cows, but were inconsistent in SR cows. Shallow udder and strong fore udder attachment were consistently correlated with good udder health. The unfavorable relationships between milking speed and udder health traits should be considered together with a few udder conformation traits when selecting for better milkability.

Weed management and accumulation of corn straw in integrated crop-livestock farming system

G.G. Mendonça[1], J.G. Augusto[2], P.M. Bonacim[2], A.P. Freitas[3] and F.F. Simili[2]
[1]*School of Veterinary and Animal Science, University of Sao Paulo, Av. Duque Caxias Norte, 225, Jardim Elite, Pirassununga,SP, 13635-000, Brazil,* [2]*Institute of Animal Science, Rodovia Carlos Tonani, Km 94, Zona Industrial, Sertãozinho, SP, Brazil,* [3]*Medical School, University of Sao Paulo, AV. Bandeirantes, 3900 Monte Alegre, Ribeirão Preto, SP 14049-900, Brazil; gabigeraldi@hotmail.com*

The objective of this study was to evaluate weed infestation, dry mass production of grass and straw accumulation after corn harvest in corn monocultures and corn plus palisade grass (*Brachiaria brizantha* cv. Marandu) integrated systems. The experiment was conducted at Instituto de Zootecnia, Sertãozinho. A randomized block design consisting of five treatments was used: corn monoculture (T1); corn plus Palisade grass (T2); corn plus Palisade grass plus herbicide (T3); corn plus Palisade grass sown during fertilization of the corn cover (T4); corn plus Palisade grass sown at the line and inter-line of corn plus herbicide (T5). All treatments were implemented in December 2015 and corn was harvested in May 2016. For the determination of grass dry mass, weed and straw production in the system, two squares of 1 m^2 each per plot were established and all material inside the plot was collected. The samples were sent to the laboratory for weighing of the collected material. Straw, weed and Palisade grass were then separated, weighed and dried in an air circulation oven for 72 h at 65 °C. MIXED procedure of SAS was used for statistical analysis. Dry mass production of Palisade grass was higher in T2 and T5 (1.13 and 1.18 t/ha, respectively) and differed significantly from T3 and T4 (0.68 and 0.33 t/ha, respectively). Weed infestation was lower in the treatments in which grass production was higher (0.32 and 0.28 t/ha in T2 and T5, respectively). It indicates that the presence of the grass competes with weed production, representing a natural weed control. The amount of straw did not differ among treatments, a finding that can be explained by the productivity of the corn crop which was similar for the monoculture and integrated systems. The possible control of weeds by the presence of palisade grass suggests economic and environmental advantages of integrated systems by minimizing the utilization of herbicides.

Possibilities to derive economic weights in local dual-purpose cattle for novel traits

S. König

Justus-Liebig-University Gießen, Institute of Animal Breeding and Genetics, Ludwigstr. 21b, 35390 Gießen, Germany;
sven.koenig@agrar.uni-giessen.de

A genetically and socio-economically balanced selection based on production (milk and beef) and functional traits in dual-purpose cattle requires correct economic weights. Derivation of economic weights is largely based on proper methodology, i.e. in terms of models including aspects of animal physiology, farm economics, social aspects, and appropriate assumptions of future production conditions. From the literature review, it is obvious that social aspects are generally not considered. Groen *et al.* defined the continuous integration of functional traits in dairy cattle breeding goals as a major challenge for animal breeders. Challenges also increase when extending those tasks to dual-purpose cattle kept in diverse pasture based production systems. This paper summarizes possibilities to derive economic weights for traits without direct costs and revenues in endangered breeds characterized by small population size, especially traits reflecting aspects of welfare, behaviour and also environmental impact (greenhouse gas emissions). Selection index calculations were applied to compare different breeding scenarios by altering traits in indexes and breeding goals, and to study the impact of derived economic weights on selection index evaluation criteria. In such a perspective, conventional and genomic dual-purpose cattle breeding programmes will be compared.

Phantom epistasis and why big data is not always better

G. De Los Campos[1], D. Sorensen[2] and M.A. Toro[3]

[1]Epidemiology & Biostatistics, and Statistics & Probability departments, Michigan State University, 909 Fee Road, East Lansing, MI 48824, USA, [2]Department of Molecular Biology and Genetics, Aarhus University, Blichers Allé 20, 8830 Tjele, Denmark, [3]Producción Animal, Universidad Politécnica de Madrid, Ciudad Universitaria, 28040 Madrid, Spain; gdeloscampos@epi.msu.edu

The biological processes underlying complex traits can involve multiple molecular interactions which in turn give rise to large epistatic networks. Discovering these networks has always been a central research goal. However, the data needed to achieve high power for detection of epistatic networks has been lacking. In the last two decades Genome Wide Association (GWA) studies have discovered large numbers of variants associated to important traits and diseases in plants, animals and humans. Most results published originate from additive models, predominantly single-locus analyses. Recently there has been an increased interest in using GWA to discover epistatic networks. Unfortunately, testing for interactions among markers with no causal effects can lead to wrong conclusions. For instance, we show that imperfect linkage disequilibrium (LD) between markers and QTL can create 'phantom epistasis': the emergence of epistasis at the observational level (i.e. non-null interactions between SNPs) in the absence of epistasis among causal variants. This can lead to highly elevated type-I error rates and wrong conclusions about the underlying trait architecture. The problem is exacerbated with Big Data, because the power to detect non-null interactions among markers with no causal effects increases with sample size. We use a simple model with one causal variant and two markers to characterize the problem analytically. We show that the existence of non-null interactions between the two markers depends on both 1st and 2nd order LD. Subsequently, we conducted simulations using data from the interim release of the UK-Biobank (N~100,000 unrelated Caucasian individuals) and report seriously elevated type-I error rates when testing interactions between loci that are within a distance of 2 Mb or less. Finally, we used our results to provide an alternative explanation of why non-additive models (e.g. kernel methods) may lead to higher predictive power even in the absence of epistatic interactions among causal loci.

Benefits of dominance over additive models for the estimation of average effects and breeding values

P. Duenk, M.P.L. Calus, Y.C.J. Wientjes and P. Bijma
Wageningen University and Research, Animal Breeding and Genomics, Droevendaalsesteeg 1, 6708 PB Wageningen, the Netherlands; pascal.duenk@wur.nl

Genomic prediction models that account for dominance tend to yield equal or slightly higher accuracies than additive models. This is surprising, since additive models should capture all heritable variation, including its dominance component. The mechanisms underlying the observed advantages of dominance models over additive models are still unknown. However, because estimated genomic breeding values are the sum of many estimated average effects multiplied by their marker genotypes, differences in accuracy between models must be related to differences in the accuracy of estimated average effects. The objective of this study, therefore, was to investigate bias and precision of estimated average effects (α-hat) in the presence of dominance, using either a single locus additive (A) model or an additive plus dominance (AD) model. We considered a finite sample from a large population in Hardy-Weinberg equilibrium (HWE), and calculated the bias and precision (by the root mean squared error) of α-hat for several sample sizes and allele frequencies with both models. Our results show that both the A-model and the AD-model yield biased estimates of average effects. With the A-model, the bias originated mainly from sampling deviations of genotype frequencies from HWE frequencies, and to a lesser extent from sampling deviations of allele frequencies. With the AD-model, the bias originated from sampling deviations of allele frequencies only, provided that all three genotype classes were sampled. We have proved that the sources and severity of bias are different between the two models by providing mathematical derivations. The root mean squared error of α-hat was always smaller with the AD-model compared with the A-model. In conclusion, the AD-model yields more accurate estimates of average effects, because it is more robust against sampling deviations from HWE frequencies than the A-model. This mechanism may explain the small observed advantage of dominance models over additive models in genomic prediction studies. Additionally, we show how these sampling errors can affect the accuracy of genomic estimated breeding values.

Genomic additive and dominance heritabilities for commercial traits in Nile tilapia

R. Joshi[1], J. Woolliams[1,2] and H.M. Gjøen[1]
[1]NMBU, IHA, Ås, 1433, Norway, [2]The University of Edinburgh, The Roslin Institute, Royal (Dick) School of Veterinary Studies, Easter Bush Campus, Midlothian, EH25 9RG, Scotland, United Kingdom; rajesh.joshi@nmbu.no

Relatively few studies have investigated non-additive genetic effects in aquaculture species compared to livestock animals, and these are limited to few species, especially Salmon and Trout, possibly due to the low power of commonly used designs to separate dominance, maternal and common environmental effects. Availability of SNP information has allowed us to separate additive and dominance effects easily, but it has not been properly utilized in aquaculture breeding programs yet; in contrast to what is seen in pig and poultry. The recent availability of a 58K SNP genotype array for tilapia has allowed us to separate these effects properly and for the first time report the genomic additive and dominance heritabilities in Nile tilapia. Genotypes (40,960 markers after cleaning for call rate <85 and MAF<0.05) and phenotypes of 1,119 offspring from 74 full-sib families (mean size 15.12), obtained from the factorial mating of the offspring of 20th generation GST® tilapia, were used to analyse the additive and dominance heritabilities for six commercial traits and to assess the impact of different additive and dominance relationship matrices on these parameters. GVCBLUP software was used to fit the GBLUP models, with both additive and dominance random effects, to estimate the variance components, SNP effects and the breeding values. Six different methods were used to construct the additive and dominance relationship matrices as given by Wang and Da. Our results showed the contribution of dominance on the phenotypic variation of body depth, body length and body weight at harvest, to be from 2% up to 5%. Similarly, the traits could be divided in two groups based on the magnitude of the additive heritability: body length, body thickness and fillet weight had heritabilities around 14%, whereas body depth, body weight at harvest and fillet yield had a little bit higher heritability, around 21%. Not much difference was found in the additive and dominance heritabilities using different definitions of relationship matrices.

Genomic model with correlation between additive and dominant genetic effects

T. Xiang[1,2], O.F. Christensen[2] and A. Legarra[1]
[1]INRA, UR1388 GenPhyse, CS-52627, 31326 Castanet-Tolosan, France, [2]Aarhus University, Centre for Quantitative Genetics and Genomics, Blichers Alle 20, 8830, Tjele, Denmark; olef.christensen@mbg.au.dk

Dominance genetic effect is rarely included in pedigree-based genetic evaluation, because large-scale datasets including a high proportion of full sibs are needed to estimate dominance effects accurately. With the availability of SNP markers and the development of genomic evaluation, estimates of dominance effects have become feasible. Usually, studies involving additive and dominance effects have ignored the relationships between them. However, mating in an inbred population will produce deviations from Hardy-Weinberg equilibrium, and these deviations generate correlations between breeding values and dominant deviations. Also, it has been often suggested that the magnitude of additive and dominance effects at the QTLs are related, but there is no existing applicable approach accounting for such correlation. Wellmann and Bennewitz showed that magnitudes of additive and dominant genetic effects in quantitative trait loci are related by the dominance coefficients $\delta=d/|a|$, and they put forwarded two ways of directional relationships between additive and dominance effects, such as $cor(|a|,\delta)=0$ (BayesD2) and $cor(|a|,\delta)>0$ (BayesD3). Nevertheless, these relationships cannot be fitted in individual scales in the animal model and they are not compatible with standard animal breeding software. In this study, we present a simple way of fitting the correlation between genotypic additive and dominant effects in individual scales by using a combined additive and dominance relationship matrix computed from marker genotypes. Then through a simulation study, we show that such correlations can easily be estimated by animal breeding software and accuracy and unbiasedness of prediction for genetic values would be significantly improved if such correlations are used in GBLUP.

Scanning the genomes of parents for imprinted loci acting in their ungenotyped progeny

I. Blunk[1], M. Mayer[2] and N. Reinsch[2]
[1]University of Rostock, Faculty of Agricultural and Environmental Science, Justus-von-Liebig-Weg 6b, 18059 Rostock, Germany, [2]Leibniz Institute for Farm Animal Biology (FBN), Institute for Genetics and Biometry, Wilhelm-Stahl-Allee 2, 18196 Dummerstorf, Germany; inga.blunk2@uni-rostock.de

Genomic imprinting effects, or more generally parent-of-origin effects (POE), occur when the alleles of imprinted quantitative trait loci (iQTL) are expressed differently depending on their parental origin. The mapping of iQTL within livestock genomes is of great interest as their relevance for the expression of agriculturally important traits has been demonstrated previously. IQTL mapping experiments are, however, challenging as heterozygote marker genotypes have to be distinguishable according to their parental origin (phasing) and animals possessing the phenotypes must deliver the genotypes or vice versa. In fact, phase of markers is often unknown and marker genotypes are usually unavailable for individuals with records, for e.g. slaughter animals. Thus, assuming that the regression of an animal's POE on its gene count is the imprinting effect, we propose to detect iQTL by using POE of parents to be regressed on their unphased genotypes, i.e. their gene counts. To validate this approach, we simulated two generations of genotyped parents without records. The markers were linked to iQTL and non-imprinted loci. Their expression in a third ungenotyped progeny generation delivered the phenotypic information. Using these data, a previously published imprinting model allowed the estimation of POE for all parents. De-regressed and weighted, the POE were then used as dependent variables to be regressed on the parental gene counts. As a result only markers linked to iQTL became apparent. Confirming the aforementioned assumption, iQTL and non-imprinted QTL could not be distinguished, when transmitting abilities were used instead of POE. In conclusion, the mapping of iQTL acting in ungenotyped progeny is possible merely based on the parental gene counts.

On the usefulness of the prediction of total genetic values in livestock breeding programs

J.W.R. Martini[1], T. Pook[1], V. Wimmer[2] and H. Simianer[1]
[1]*University of Goettingen, Albrecht Thaer Weg 3, 37083 Göttingen, Germany, [2]KWS SAAT SE, Grimsehlstraße 31, 37574 Einbeck, Germany; jmartin2@gwdg.de*

The discrepancy between the complex structures of biological mechanisms forming phenotypes and the assumption of additive marker effects which underlies the genomic best linear unbiased prediction (GBLUP) model, provides a motivation to investigate alternative methods. In this regard, several publications have illustrated that the prediction of phenotypes can be improved by 'epistasis' models that introduce some kind of non-additivity in the relatedness of individuals. In spite of the success that these models had in improving the prediction of phenotypes, their practical usefulness for livestock breeding programs is still being discussed. Animal breeders are often more interested in the additive genetic value of an individual, which reflects the increase in the mean phenotype of a reference population if the genes of the considered individual are enriched in the population, than in the phenotype or the total genetic value, that is the portion of the phenotype which is genetically determined. An illustrative argument against the use of the total genetic value is that favorable combinations of alleles of different loci may not be stable across generations due to segregation and recombination, and positive effects may therefore be reduced in the next generation. We simulated different breeding scenarios with traits of different genetic architecture to investigate under which circumstances an improved prediction of the total genetic value can be translated into extra selection gain. Our results indicate that also in scenarios with qualitative epistasis, that is that signs of effects can change with an altered genetic background, selecting for the additive genetic value is favorable for short-term improvement. However, an improved prediction of the total genetic value can be beneficial to reach a higher maximal value in the long-term response to selection. Both characteristics of the response to a selection for the additive genetic value -the strong early response and the reduced maximal selection gain- are caused by a relatively fast fixation of favorable alleles whose alternative is missing in the genetic pool when the genetic background has changed.

Bivariate genomic model for backfat thickness in Duroc and crossbred DLY pigs

O.F. Christensen[1], B. Nielsen[2] and G. Su[1]
[1]*Aarhus University, Center for quantitative genetics and genomcs, Department of molecular biology and genetics, Blichers alle 20, 8830 Tjele, Denmark, [2]Seges, pig breeding and genetics, Axeltorv 3, 1609 København V, Denmark; olef.christensen@mbg.au.dk*

The aim of this paper is to investigate purebred and crossbred performances in a genomic model including both additive and dominance genetic effects. Scanned backfat thickness on purebred Duroc and crossbred Duruc × (Landrace × Yorkshire), DLY, pigs at the same test station in 2014-2015 is analysed. DLY animals in the data set were genotyped with the 8.5K GGP-Porcine LD Illumina Bead chip, whereas Duroc animals were genotyped with either this chip or the Illumina PorcineSNP60 chip. The analysis was based on a bivariate model (considering the backfat thickness in the two populations as genetically correlated traits) with genomic inbreeding depression effects, additive genetic effects and dominance genetic effects. Marker-based additive and dominance relationship matrices were computed. The genetic variance-covariance matrices can be formulated using Kronecker products, making it possible to fit the model using standard genetic evaluation software. Breeding values of Duroc boars for DLY crossbred performance can be obtained from allele substitution effects for DLY crossbred performance. Results showed a statistically significant inbreeding depression effect in Duroc, but not in DLY. Genetic correlation between dominance effects in Duroc and dominance effects in DLY was 1, whereas the genetic correlation between additive genetic effects in Duroc and additive genetic effects in DLY was 0.95. Since the two populations are measured in the same environment, these results are an indication of the presence of epistatic genetic effects.

Single-step GBLUP using metafounders to predict crossbred performance of laying hens

J. Vandenplas[1], M.P.L. Calus[1], T. Brinker[1], E.D. Ellen[1], M.C.A.M. Bink[2] and J. Ten Napel[1]
[1]Wageningen University and Research, Animal Breeding and Genomics, P.O. Box 338, 6700 AH Wageningen, the Netherlands, [2]Hendrix Genetics, Research and Technology Centre, P.O. Box 114, 5831 CK Boxmeer, the Netherlands; jeremie.vandenplas@wur.nl

Several livestock production systems, including those for laying hens, are based on crossbreeding schemes. For these production systems, selection of purebred animals aims at optimizing crossbred performance, which is now feasible with the advent of genomic selection. The single-step genomic Best Linear Unbiased Prediction (ssGBLUP) is becoming the method of choice to predict genomic breeding values, because it enables simultaneous use of data from genotyped and non-genotyped animals by combining genomic and pedigree relationship matrices. While ssGBLUP is attractive, several issues exist. Firstly, genomic and pedigree relationships have to be made compatible. This is not always straightforward, especially for situations involving multiple lines and their crosses. Secondly, use of unknown parent groups (UPG) in ssGBLUP can lead to deteriorations of convergence rate, and estimates of UPG can be biased. Both issues can be tackled by the recently introduced concept of metafounders. A metafounder of a breed represents the ancestral population of this breed, and can be considered as a pseudo-individual included as the founder of the pedigree. The concept of metafounders can be viewed as a generalization of UPG. Therefore, the aim of this study is to assess the quality of the genomic prediction of crossbred performance of laying hens by ssGBLUP using metafounders. The pedigree includes 6,163 purebred White Leghorn individuals from the W1 sire line and the WA, WC, and WD dam lines, and 5,872 two-way crossbred (W1×WA, W1×WC and W1×WD) laying hens. Body weight at 402 days is available for all crossbred laying hens. Genotypes of 60k SNP are available for 2,616 purebreds and for 5,623 crossbreds. Results show that the use of metafounders allows to adjust pedigree relationships to better fit genomic relationships. Quality of genomic predictions will be reported through accuracies, bias and mean square errors.

Crossbreds in the genomic relationship matrix using linkage disequilibrium and linkage analysis

M.W. Iversen[1], Ø. Nordbø[1,2], E. Gjerlaug-Enger[1], E. Grindflek[1], M.S. Lopes[3] and T.H.E. Meuwissen[4]
[1]Topigs Norsvin, Storhamargata 44, 2317 Hamar, Norway, [2]GENO SA, Storhamargata 44, 2317 Hamar, Norway, [3]Topigs Norsvin Research Center, Beuningen, Beuningen 6641 SZ, the Netherlands, [4]Norwegian University of Life Sciences, Postboks 5003 NMBU, 1432 Ås, Norway; maja.iversen@norsvin.no

In pig breeding, the final product is a crossbred (CB) animal, and it is expected that incorporating CB data in genetic evaluations will result in higher genetic progress at the CB level. However, selection is performed at the purebred (PB) level using mainly PB data. This is because, in single-step genomic BLUP, which is the most commonly used method, genomic and pedigree relationships must refer to the same base, and this may not be the case when several breeds and CB are included. An alternative to overcome this issue may be to use a genomic relationship matrix (G matrix) that accounts for both linkage disequilibrium (LD) and linkage analysis (LA), called G_{LDLA}. This matrix has shown promise analysing PB data, but has not yet been applied to CB data. The objectives of this study were therefore to further develop the G_{LDLA} matrix approach to use PB and CB genotypes simultaneously, to investigate its performance, and the general added value of including CB genotypes in genomic evaluations. Data was available on Dutch Landrace, Large White, and the F1 cross between them. In addition to G_{LDLA}, and the pedigree-based relationship matrix, 6 other genomic relationship matrices were compared (PB alone, PB together, each PB with the CB, all 3 breeds together) for 3 maternal traits; total number born (TNB), live born (LB), and gestation length (GL). Bootstrapping was used to test for significant differences in prediction accuracies between matrices. Results show that G_{LDLA} gave the highest prediction accuracy of all the relationship matrices tested, and that including CB genotypes in general also increased prediction accuracy. However, in some cases, these increases in prediction accuracy were not significant (at P<0.05). To conclude, CB genotypes increases prediction accuracy for some traits, and some breeds, but not for all. The G_{LDLA} matrix had significantly higher prediction accuracy than the other G matrix that included genotypes from all 3 breeds in all but one case.

Optimum mating designs for exploiting dominance in genomic selection schemes for aquaculture species

J. Fernández[1], B. Villanueva[1] and M.A. Toro[2]
[1]INIA, Mejora Genética Animal, Ctra. Coruña Km 7,5, 28040 Madrid, Spain, [2]ETSIA, UPM, Producción Agraria, Av. Puerta de Hierro, 2, 28040 Madrid, Spain; jmj@inia.es

The potential benefits of exploiting dominance effects under genomic selection was investigated through computer simulations that mimicked an aquaculture breeding programme that included a nucleus (where selection is performed) and a commercial population (composed by individuals that are slaughtered) which is obtained from the nucleus each generation. The trait in the selection objective was determined by a large number of loci (1000) with additive and dominance effects. Every generation 50 full-sib families with 30 offspring each were created (1,500 evaluated animals). Phenotypes and genotypes for a variable number of SNP were available for all individuals. Genomic estimates of additive and dominance effects were obtained using GBLUP. Males and females (50 of each sex) with the highest estimated genomic breeding values were selected to create the next generation of the nucleus through a scheme of monogamous minimum coancestry matings (*MCM*). When generating the commercial population, selection and mating decisions were taken in a single step by using a 'mate selection' procedure. This was implemented to take advantage of the dominance effects by calculating for each potential couple of candidates the expected mean phenotype of their potential offspring. Then, the combination of couples that maximise the global value was chosen. The main result was that, for a moderate number of SNPs, equal or even higher mean phenotypic value for the selected trait are obtained by selecting on the GEBVs and then applying *MCM* than using the 'mate selection' strategy. This could be due to the fact that *MCM* leads to high levels of heterozygotes in all the genome, even for loci with effect on the trait that are not in linkage disequilibrium with any SNP. Contrarily, the 'mate selection' procedure promotes specifically heterozygosity in those SNPs with dominance effect. If this effect is poorly estimated it may lead to suboptimal results. In fact, when using the real QTL effects instead of the SNP effects the 'mate selection' strategy yielded better results than *MCM*.

Genomic prediction including external information on markers

M. Selle[1], E.F. Mouresan[2] and L. Rönnegård[2,3]
[1]Norwegian University of Science and Technology, Department of Mathematical Sciences, Department of Mathematical Sciences, NTNU, 7491 Trondheim, Norway, [2]Swedish University of Agricultural Sciences, Department of Animal Breeding and Genetics, Box 7023, 75007 Uppsala, Sweden, [3]Dalarna University, School of Technology and Business Studies, Dalarna University, School of Technology and Business Studies, 79188 Falun, Sweden; maria.selle@ntnu.no

Genomic selection aims to predict the genetic merit of an individual based on information on genomic markers. Most standard models in genomic prediction assume no prior difference between genetic markers. Models including prior knowledge have been proposed to improve predictions, and have been fitted using tailored software. However, user-friendly general methods for large-scale data have not yet been developed. We are developing a method for incorporating prior knowledge in genomic selection models by estimating different weights for SNP effects based on their biological function, for example SNP markers in genes, exons, pathways. We are extending current HGLM methods and thereby implementing a user-friendly algorithm useful for researchers and breeding organizations. Simulation studies show promising results, in that prediction of new populations improve when modelling SNP markers with weights based on prior assumptions compared to standard model assuming equal variance for all SNP effects. Future analysis of real data may also reveal which characteristics of the SNP markers are important and thereby increase our understanding of the biological mechanisms of quantitative traits.

Parental genetic scores to rank dairy crossbred cows in herds using ProCROSS® program

M. Berodier[1,2], M. Brochard[3], M. Kargo[4,5], P. Le Mezec[1], S. Minery[1], A. Fogh[4], H. Stålhammar[6], S. Borchersen[6] and D. Duclos[1]

[1]Institut de l'Elevage, 149 Rue de Bercy, 75595 Paris, France, [2]Coopex, 4, rue des Epicéas, 25640 Roulans, France, [3]Umotest-Coopex, Les Soudanières, 01250 Ceyzériat, France, [4]Knowledge Center for Agriculture (SEGES), Agro Food Park15, 8200 Aarhus N, Skejby, Denmark, [5]Aarhus University, Department of Molecular Biology and Genetics, Center for Quantitative Genetics and Genomics, Blichers Allé 20, P.O. Box 50, 8830 Tjele, Denmark, [6]Viking Genetics, Örnsro, 532 94 Skara, Sweden; marie.berodier@inra.fr

Until now, to the best of our knowledge, there is no official genetic tool dedicated to breeders to compute parental genetic scores, with the aim to rank dairy crossbred animals in a herd. In this study, we propose an original method to estimate parental genetic scores of female crossbreds from the official EBVs (Estimated Breeding Values) of their purebred ancestors for the three production traits: MILK YIELD, FAT YIELD and PROTEIN YIELD. This approach was tested using information from three dairy breeds: Holstein, Montbéliarde and Red Dairy Cattle, which are the breeds used in the ProCROSS® program. First of all, a parental genetic score requires EBVs expressed in a common reference base, which in this case was the HOL base of the crossbreds herd's country. Furthermore, crossbreds benefit from a non-additive genetic effect depending on their breed composition: the heterosis effect. Therefore, two parental genetic scores were calculated: (1) a pure additive genetic score, to rank the animals on their dam genetic potential and (2) a genetic score summing additive genetic value and heterosis effect, to rank the cows on their own total genetic potential. Finally, functional simulation tests on a simulated herd of 121 crossbreds were performed by deleting different percentage of pedigree information. This robustness tests allowed us to set a maximum of 5% of missing ancestors' information. Beyond this threshold, the crossbreds' ranking could be challenged because of the insufficient number of scores that could be calculated (between 15 and 20% of non-calculated scores when 5% of ancestors' information is missing).

Genomics to estimate additive and dominance genetic variances in purebred and crossbred pig traits

L. Tusell[1], H. Gilbert[1], Z.G. Vitezica[1], M.J. Mercat[2], A. Legarra[1] and C. Larzul[1]

[1]INRA, UMR1388, INPT ENSAT, INPT ENVT GenPhySE, 31326, Castanet-Tolosan, France, [2]IFIP/BIOPORC, La Motte au Vicomte, 35651 Le Rheu, France; helene.gilbert@inra.fr

This study aims at assessing the contribution of the additive and dominance genomic variances to the phenotypic expression of several purebred Piétrain and crossbred (Piétrain × Large White) pig performances. A total of 636 purebred and 720 crossbred male piglets were phenotyped for 22 traits that can be classified into the trait groups growth rate and feed efficiency, carcass composition, meat quality, behaviour, boar taint and puberty. Additive and dominance variances estimated in univariate genotypic models including additive and dominance genotypic effects and a genomic inbreeding covariate allowed us to retrieve the additive and dominance SNP variances for purebred and crossbred performances. These estimated variances were used, together with the allelic frequencies of the parental populations, to obtain additive and dominance variances in terms of genomic breeding values and dominance deviations. Estimates of additive genetic variances across traits were consistent with previous results without dominance indicating that additive and dominance genetic effects were non-confounded. Some traits showed a relevant amount of dominance genetic variance in both populations (i.e. growth rate 8%, feed conversion ratio 9-12%, backfat thickness 14-12%, lean meat 10-8%, carcass lesions 9%, in purebreds and crossbreds, respectively) or increased amounts in crossbreds (i.e. ham cut 8-13%, loin 7-16%, pH semimembranosus 13-18%, pH Longissimus dorsi 9-14%, dressing yield 5-15%, androstenone 5-13% and estradiol 6-11%). Results suggest that accounting for dominance in the models of these traits could lead to an increased GEBV accuracy and that using crossbred information can be beneficial to evaluate purebred candidates to selection for crossbred performance. Further research will compare additive and dominance marker effects between crossbred and purebred performances.

The participatory strategy

L. Kluver
Danish Board of Technology Foundation, Toldbodgade 12, 1253 København K, Denmark; LK@Tekno.dk

Biotechnology is not uncontroversial, and the biotech branch of animal breeding certainly not either. There are multiple reasons for this, based on the fact that society includes a diversity of views, including religious conceptions, taste and aesthetics, diversities of visions for the future, and a scientific discourse with its own controversies in and between fields of science. The core of the controversy is not a matter of illiteracy – quite the contrary – it is a consequence of the fact that there are many forms of interacting and competing literacies in our societies. Public participation projects already in the 1980's proved that deep and very well informed citizen engagement processes did not change the fundamental standpoints of the citizens, but they changed their ability to understand the position of other actors, and it made it much easier for them to reach compromises or even consensus on commonly accepted ways forward. By public participation processes, stubborn 'Yes/No' or 'Don't know' positions decreased while 'No, but …' and 'Yes, if …' positions increased. Since the early experiments with public participation more than 30 years ago methodology of this advanced form of communication has matured to a level where we can now reasonably say that public participation methods are available for all forms of situations and purposes. The societal discourse on animal breeding is not a discussion that can be 'won'. Any attempt to try to manipulate or convince whole-of-society is doomed. So are attempts to deem the position of others out by, for example, referring to low science literacy. And, so are attempts to go the power path by convincing politicians to take top-down decisions. The communication that fits a hyper-complex and par excellence controversial topic, such as new animal breeding technologies, is two-way, it is deep, it is honest and serious. In other words, it is participatory. The outcome of such an approach will probably not fully satisfy any of the involved actors when it comes to the content, because it will be a compromise. But the chance of that compromise being received well in society and among political decision-makers is high, because it enjoys legitimacy, it points out the least risky path forward and thus provides for new openings for action.

Responsible Innovation: from science in society to science for society, with society

P. Macnaghten
Wageningen University, Knowledge, Technology and Innovation Group, Hollandseweg 1, 6706 KN Wageningen, the Netherlands; philip.macnaghten@wur.nl

In this talk I provide a short review of the debate on responsible innovation and its mission to involve society in research and in innovation processes. First, I describe the ways in which responsibility in science is being reconfigured institutionally, from an internal focus on the provision of objective and reliable knowledge, to a more external view that embraces the ways in which it has an impact on society. Secondly, I introduce a framework for responsible innovation as a (partial) response to this shift, highlighting its constituent dimensions and the capacities and competencies that are needed to put it into practice. Thirdly, I set out how the framework of responsible innovation can be applied to animal science, using the technique of animal gene editing as an example. A prospective programme of research is set out in which animal scientists, social scientists and ethicists work together to anticipate possible impacts and implications, open up inclusive dialogue with stakeholders and wider publics, develop reflexive scientific and corporate cultures, and ensure that the science that develops is responsive to these processes.

Science communication to obtain social license for use of genome editing in animal breeding programs

A. Van Eenennaam

University of California, Department of Animal Science, One Shields Avenue, Davis, CA 95616, USA;
alvaneenennaam@ucdavis.edu

Genome editing enters the animal breeders' toolbox amidst a fierce global debate that has persistently focused on potential risks associated with genetic engineering in food producing species. Despite the fact that the types of genetic alterations enabled by genome editing are frequently quite distinct to the transgenic alterations typically associated with genetic engineering, there is already an attempt by some to conflate these two methods in the public discourse. Usually scientists have countered misinformation by throwing facts at people. The so-called 'information deficit' model of providing evidence-based data to audiences assuming this will convince them to change their minds, has shown to be ineffective. Social science research suggests a different approach to science communication is needed, one that focuses on shared values rather than data. Genome editing focussed on animal health and welfare could have tangible benefits to the animals themselves. Many in society would see this as a shared value, and preferable to treating sick animals with antibiotics, or continuing with necessary but painful animal husbandry procedures. Leading the discussion with WHY genome editing might be useful to address animal welfare and disease problems would be a different approach to explaining HOW the technology works. There are already some promising genome editing applications including pigs that are resistant to porcine reproductive and respiratory syndrome (PRRS), and dairy cattle that have been genetically polled using genome editing. These two examples have both obvious benefits to the animal, and a clear consequence associated with not using genetics to address these problems. Ideally framing the discussion in this way will lead to a more nuanced discussion of the various tradeoffs (i.e. risks and benefits) associated with all of the different approaches that can be used to address animal welfare and disease problems. Although this values-focused communication style does not come naturally to many scientists, and requires a change of mindset when communicating science to the public, it may be requisite if animal breeders wish to obtain social license for the use of genome editing in animal breeding programs.

Code EFABAR: code of good breeding and reproduction practice

J. Venneman

European Forum of Farm Animal Breeders (EFFAB), P.O. Box 76, 6700 AB Wageningen, the Netherlands;
jan.venneman@effab.info

The Code EFABAR is developed in 2005 to provide users with a practical guide to achieve sustainable farm animal breeding whilst delivering transparency for the society. The adoption of the Code EFABAR demonstrates that breeders carry out responsible breeding. The Code is applicable for breeders of cattle, pigs, poultry and fish, but the principles can also be adapted to other types of farm animals. The Code is based on European principles and legislation but can also be used in a more international context. The main objectives of the Code EFABAR are (1) to be the standard instrument for defining and maintaining sustainable and responsible farm animal breeding (2) to create transparency for the society. Sustainable breeding is defined as: the extent to which farm animal breeding, as managed by professional organisations, contribute to the production of sufficient, safe, nutritious and healthy food whilst taking care of genetic diversity, resource efficiency, environment, animal health and animal welfare to create 'a better world' for future generations'. Since 2005 the Code EFABAR is updated every three years. The latest update was carried out in 2016 which update resulted in the renewed Code 2017. This Code 2017 works with templates – instead of roadmaps – in which all societal concerns are addressed. When adopted, the Code EFABAR can be used by the companies to show their sustainability, responsibility and transparency to clients and to the society (policymakers, politicians, consumers, non-governmental organisations, etc.). In general, the Code EFABAR is used by EFFAB and its members in the 'political' society.

Precision livestock farming: genetic variants as biomarkers to evaluate feed efficiency in cattle

M. Cohen-Zinder
Agricultural Research Organization, Beef Cattle Section, Newe Yaar Research Center, 30095; Ramat Yishay, Israel;
mirico@volcani.agri.gov.il

Feed efficiency (FE) is a major component determining the profit of livestock production. Residual feed intake (RFI) is the most acceptable parameter to evaluate FE. Currently, methods for RFI measurements are complex and expensive, favoring the development of low-cost-non-invasive biosensors, to enable easily accessed phenotyping of FE. Precision livestock farming (PLF) approach integrates the development of animal sensing tools with individual performances. Such integration may lead to identification of biological markers, to be used in order to differentiate between most and least productive individuals in the herd. Recent advances in the field of animal genomics allows the development of genetic markers of different types (as SNPs, CNVs and indels), and their association with traits of interest, for detection of universal genetic biomarkers. In the current study, unique sequence-capture technology was performed on DNA samples of high or low RFI ranked Holstein calves to detect SNPs, associated with RFI phenotypic consistency. Of 48 significant polymorphisms identified, 11 were harbored by the fatty acid binding protein 4 (FABP4) gene. Some of the polymorphisms in FABP4 were consistent across different ages and diets, and thus, might be considered possible biomarkers for RFI-based selection for FE in the Holstein breed, following a larger scale validation. In the future, the integration of genetic markers (biomarkers) into PLF-based data, might enable a non-phenotypic prediction of FE.

A life cycle assessment of dairy farms in Northern Germany within a ten years course of time

J. Drews, I. Czycholl and J. Krieter
Institute of Animal Breeding and Husbandry, Christian-Albrechts-University of Kiel, Olshausenstr. 40, 24098 Kiel,
Germany; jaulrich@tierzucht.uni-kiel.de

Milk production is known to be one of the major contributors to greenhouse gas emissions in the agricultural sector. Life cycle assessment (LCA) is a globally accepted and standardized tool for the assessment of environmental impacts on livestock systems. Lately many studies on life cycle assessment are conducted globally, but mostly the number of farms is low or only data of a short time period is available. Hence the aim of the study was to assess the environmental impacts of individual farms in a region with similar conditions (e.g. soil, crop cultivation, precipitation) over a ten years course of time. Originating from a high variability in farm size, intensity and productivity of the farms hot spots are identified and mitigation strategies are developed. Furthermore the focus is set on the question if the farms themselves were able to reduce their environmental impacts during the investigated period. The study is based on the data of 5,543 operating branch settlements of dairy farms keeping German Holstein cows in Northern Germany from 2004 to 2014. The system boundaries were set from cradle to farm gate and co-products (surplus calves and culled cows) were included. Processes like the production of fertilizer, plant protection and feed were included as well as housing. 1 kg of energy-corrected milk (ECM) was defined as functional unit. As allocation method economic allocation was chosen. Three impact categories were considered: global warming potential (GWP), eutrophication potential (EP) and acidification potential (AP). Feed production (FP), feeding (F), housing (H) and milking (M) were identified as the most important processes associated with milk production. Preliminary results showed that FP (45%, 44% and 44%) and M (45%, 44% and 44%) are the processes mainly contributing to GWP, EP and AP respectively. F contributed least to GWP, EP and AP (<1% respectively). Different scenarios concerning feeding regime, productivity and efficiency will be used for further calculations. In order to quantify uncertainty of data, Monte Carlo analysis will be performed additionally.

Estimation of CH$_4$ emissions from milk MIR spectra using respiration chamber as reference technique

A. Vanlierde[1], N. Gengler[2], H. Soyeurt[2], F. Grandl[3], M. Kreuzer[4], B. Kuhla[5], P. Lund[6], D. Olijhoek[6], C. Ferris[7] and F. Dehareng[1]
[1]*Walloon Agricultural Research Centre, Gembloux, 5030, Belgium,* [2]*University of Liège, Gembloux Agro-Bio Tech, Gembloux, 5030, Belgium,* [3]*Qualitas AG, Zug, 6300, Switzerland,* [4]*ETH Zurich Institute of Agricultural Sciences, Zurich, 8092, Switzerland,* [5]*Leibniz Institute for Farm Animal Biology, Dummerstorf, 18196, Germany,* [6]*Aarhus University, Aarhus, 8000, Denmark,* [7]*Agri-Food and Biosciences Institute, Belfast, BT4 3SD, United Kingdom; a.vanlierde@cra.wallonie.be*

Reducing the methane (CH$_4$) from dairy cows is a challenging aspect of the cattle breeding. To permit large scale studies focusing on genetics and management, an equation has been developed to estimate individual CH$_4$ emission from milk MIR spectra. The existing equation is based on values obtained with the SF$_6$ technique. However, respiration chambers (RC) provide the gold standard for measuring CH$_4$ production. Hence, the purpose of this work was to develop a new equation based only on data collected from RC to compare its statistical performance with the existing SF$_6$ equation. Daily CH$_4$ production data linked with milk MIR spectra have been collected from Switzerland (60 data – 30 cows), Germany (115 data – 26 cows), Denmark (132 data – 19 cows) and Northern Ireland (24 data – 12 cows) yielding a total of 331 RC measurements from 87 cows. Cows were fed with different diets types and were at variable lactation stages. Measured CH$_4$ values ranged from 304 to 779 g/day (mean 504 g/day). A fivefold cross-validation (CV) was performed to evaluate the robustness of the equation. The statistics of the equation based on RC measurements showed an R^2cv of 0.62, a standard error of CV (SECV) of 60 g/day and a ratio performance deviation (RPD) of 1.6. In comparison, the equation based on 532 SF$_6$ measurements (different countries and cows) showed R^2cv of 0.70, SECV of 70 g/day and RPD of 1.8. Thus the SF$_6$ equation appears to be more robust. It might be explained by the greater number of measurements, cows (165 vs 87), lactation number and stage and diet types. Thus, greater variability was included in the SF$_6$ reference data set. Results obtained with the gold standard technique (RC) confirm the ability to estimate CH$_4$ emissions from milk MIR spectra.

Feasibility of FT-IR milk spectra to predict methane emissions in danish lactating dairy cows

N. Shetty[1], G. Difford[1,2], J. Lassen[3], P. Løvendahl[1] and A.J. Buitenhuis[1]
[1]*Aarhus University, Molecular Biology and Genetics, Center for Quantitative Genetics and Genomics, 8830 Tjele, Denmark,* [2]*Wageningen University & Research, Animal Breeding and Genomics, 6700 AH Wageningen, the Netherlands,* [3]*Viking Genetics, Ebeltoftvej 16, 8960 Randers SØ, Denmark; nisha.shetty@mbg.au.dk*

Enteric methane (CH$_4$) is naturally emitted by dairy cows during the microbial fermentation of feed components and it is a potent greenhouse gas that contributes substantially to global warming. In practice recording of CH$_4$ emission is expensive and difficult, therefore not applicable to commercial dairy farms. Milk Fourier transform mid-infrared (FT-IR) profile may be useful as indicator trait for CH$_4$ emission. The main advantage of using FT-IR spectroscopy is that it is rapid, easy to use and records phenotypes at population level and thereby has potential for practical application. However, the question of interest is how the CH$_4$ gas concentration from breath sample reflects in cow's milk composition and thereby in milk FT-IR profile. In general CH$_4$ emission is connected to milk fatty acid profile, so one can assume FT-IR spectral profiles could predict CH$_4$ emission. Feasibility and added value of FT-IR spectroscopy of milk to predict CH$_4$:CO$_2$ ratio and CH$_4$ production (l/d) in Danish lactating dairy cows was assessed. The data set consisted of ~4,000 records from ~200 Danish Holstein and ~100 Danish Jersey cows. Partial least squares regression was used to develop the prediction models and recursive partial least squares was used to identify relevant spectral regions. Models were developed using different combinations of predictor traits namely FT-IR spectra, milk yield (MY), breed, parity, herd, lactation stage and season. Use of FT-IR spectra alone showed low prediction accuracies when validated using external test sets. Integration with other factors like MY, breed, herd and lactation stage showed improvement in the prediction accuracy, however the added value of FT-IR was marginal along with these traits. Implying that CH$_4$:CO$_2$ ratio and CH$_4$ production models obtained using FT-IR spectra reflects mainly information related to MY, breed, herd and lactation stage rather than individual fatty acids relevant for CH$_4$ prediction. Therefore, it is not feasible to predict CH$_4$ emission based on FT-IR spectra alone.

Protocol for measuring CH$_4$ concentrations with the laser methane detector in the breath of cows

D. Sorg, S. Mühlbach, J. Kecman and H.H. Swalve
Martin Luther University Halle-Wittenberg, Institute of Agricultural and Nutritional Sciences, Theodor-Lieser-Str. 11, 06120 Halle, Germany; diana.sorg@landw.uni-halle.de

When taking profiles of the CH$_4$ concentration in the breath of dairy cows with the mobile Laser Methane Detector (LMD), a standardization of the protocol is essential in order to compare measurements. We tested different aspects of the protocol in regard to their influence on recorded CH$_4$ values. A profile length of 5 min, as suggested in the literature, seemed appropriate. Some CH$_4$ phenotypes, i.e. profiles reduced to a point measurement, changed when profiles were truncated to less than 2 min (n=10 profiles). Mean CH$_4$ values showed a trend to decrease with length of profile. When measuring in a range of between 0.4 to 2.5 m from the animal, mean CH$_4$ concentrations of the air, measured perpendicular to the animal, were not different (P>0.05) from the concentration at below 0.4 m. The plume of CH$_4$ seemed to dilute very fast and did not stretch far into the space around the cow (n=19). The pointing angle (front or side), the operator of the LMD, the LMD device (3 LMDs of the same model), and the time of day (morning or afternoon) each had an influence (P<0.05) on CH$_4$ phenotypes derived from the LMD profile. The phenotype 'number of peaks' of the CH$_4$ profile was not different (P>0.05) from the visually counted number of respiratory cycles (n=49) and was used as the respiratory rate (RR). However, for higher absolute values of the RR, the LMD tended to underestimate the RR, since some respiratory cycles were missed when the cow faced away from the operator. An estimate of daily CH$_4$ from the cow's tidal volume (estimated from weight), RR, mean respiratory CH$_4$ concentration, and a dilution factor from another method in the literature was 400 l (n=1,057) on average and similar compared to published values for dairy cows. However, the SD was high (173 l) and animals with implausible high and low values occurred (76-1,208 l). When measuring with the LMD, all conditions like operator, device number, pointing angle and time of day should be documented or standardized if possible. Distance seemed to have less influence than expected and 5 min was an acceptable profile length.

Heritabilities for methane concentrations measured by laser methane detector in the breath of cows

S. Mühlbach, D. Sorg, J. Kecman, F. Rosner and H.H. Swalve
Martin Luther University Halle-Wittenberg, Institute of Agricultural and Nutritional Sciences, Theodor-Lieser-Str. 11, 06120 Halle, Germany; sarah.muehlbach@landw.uni-halle.de

Breeding for highly efficient cows with low methane production is of interest for farmers and the public. The laser methane detector (LMD) is a hand-held device for on-farm measurements of methane concentrations in the breath of ruminants. For 344 dairy cows in two commercial herds 922 5-min profiles of the methane concentration in breath were measured with the LMD from a distance of 2 m. Each cow had one to three profiles recorded over three consecutive days. During the measurement cows were fixed in head gates and were standing idle without ruminating. The single methane values of each profile were divided into respiration and eructation values and peaks. From each profile several methane phenotypes were derived: the mean of all peaks (pmean), the mean of all eructation peaks (rpmean) and the mean of the maxima of the eructation events (remean), with an eructation event consisting of several consecutive eructation peaks. An estimate of daily methane emission (CH$_4$ l/d) was calculated from pmean, number of respiration cycles, bodyweight and a specific dilution factor. The LMD recorded methane concentrations from 21-327 ppm-m (mean=103±44 ppm-m), with ppm-m being the cumulative methane concentration along the 2 m laser path. For estimation of heritabilities of methane phenotypes a linear animal model with a random animal genetic and a permanent environmental effect of the animal was used. Fixed effects were the classes of days-in-milk, the combined effects herd visit × wind speed class, and LMD unit × operator, while body weight and energy corrected milk were included as covariates. Pmean, rpmean, remean and CH$_4$ (l/d) had estimates of heritability (± SE) of 0.12 (±0.10), 0.13 (±0.04), 0.18 (±0.04), 0.15 (±0.04). Our findings show that the LMD is a promising technique to collect methane phenotypes under commercial farm conditions.

Database construction for model comparisons of methane emissions by ruminants in relation to feed

X. Li[1], C. Martin[1], E. Kebreab[2], A.N. Hristov[2], Z. Yu[2], M. McGee[2], D.R. Yáñez-Ruiz[2], K.J. Shingfield[2], A.R. Bayat[2], C.K. Reynolds[2], L. Crompton[2], J. Dijkstra[2], A. Bannink[2], A. Schwarm[2], M. Kreuzer[2], P. Lund[2], A.L.F. Hellwing[2], P. Moate[2], N. Peiren[2] and M. Eugène[1]

[1]INRA, UMR1213, Route de Theix, 63122, France, [2]GLOBAL NETWORK, http://animalscience.psu.edu/fnn/news/2013/facce-jpi-proposal, 16802, USA; xinran.li@inra.fr

As part of the Global Network project, led by Alex Hristov (Penn State, USA), a dataset containing more than 6,000 individual animal data on methane (CH_4) and ammonia emissions, and other metadata (intake, diet composition, animal parameters) from ruminants was built. A sub-dataset (from 11 partners) was used to compare the performances of extant predictive models of CH_4 emission (built with different statistical approaches). In the present abstract, we report features of the sub-dataset. Data were discarded when CH_4 or dry matter intake (DMI) values were missing. Otherwise, missing values of metadata (with respect to dietary chemical composition) were estimated with the help of the respective national feed tables. The sub-dataset contained 3,134 observations that were collected from 110 studies using various experimental designs, animal species (dairy cattle, 66%; beef cattle, 20%; sheep, 12%; goat, 2%), 378 dietary treatments and three CH_4 measurement techniques (chamber, GreenFeed and SF6). There were 378 different dietary treatments within the sub-dataset, and these were classified into five CH_4 mitigating strategies (A to E) according to the purpose of the study. These were: (A) increased concentrate proportion (n=20 studies, 19% of data), (B) enhancing forage quality (n=48, 38%), (C) lipid supplementation (n=14, 9%), (D) plant extract supplementation (n=21, 10%), and (E) other additives (n=13, 7%). Control diet represented 18% of data. The mean (SD) for DMI was 14.7 (7.86) kg/d, CH_4 was 20.9 (5.90) g/kg DMI and milk yield was 28.8 (10.41) kg/d. For the chemical composition (g/kg DM) of the diets, crude protein was 160 (35.6), ether extract was 39.8 (17.64), NDF was 384 (119.2) and starch was 199 (102.6). The sub-dataset is heterogeneous and represents different contexts (feed, animal, production). Further analyses are in progress to evaluate the performances of extant predictive models of CH_4 emissions in ruminants.

Mathematical modelling of rumen pH in dairy cows

V. Ambriz-Vilchis[1,2], D. Shaw[3], M. Webster[1] and A. Macrae[3]

[1]BioSimetrics Ltd, Kings Buildings, West Mains Rd, EH93JG, Edinburgh, United Kingdom, [2]SRUC, Future Farming Systems, Kings Buildings, West Mains Road, EH93JG, Edinburgh, United Kingdom, [3]R(D)SVS and The Roslin Institute, The University of Edinburgh, EH25 9RG, Roslin, United Kingdom; virgilio.ambriz@sruc.ac.uk

Mathematical models have been used to predict many variables from nutrient and animal characteristics. Rumen pH affects digestion, microbial metabolism, and health. Rumen pH has been predicted from concentration of VFA, or animal and feed characteristics. However the ability of the models to predict rumen pH is low. Therefore the aim of the present study was to create mathematical models to predict rumen pH from data collected on-farm: animal and feed characteristics and continuously measured rumen pH (WellCow bolus). A database was constructed using data recorded from three on-farm trials with dairy cows conducted at the University of Edinburgh. To explore the relationship between animal, dietary factors and rumen pH the dataset was subjected to mixed effect model analysis, considering the random effect of each individual cow. The model simplification procedure proposed by Crawley was carried out and ANOVA with maximum likelihood method was used to determine the model using R and the nlme package. The obtained model was evaluated with an independent dataset on its ability to predict rumen pH (mean pH/d or time under pH 6.2 min/d) using the limits of agreement (LoA) methodology. Two models were obtained to predict daily mean pH (meanpH = 5.29-0.0035×DIM+0.0567×CP+0.044×uNDF) and to predict time rumen pH was below 6.2 (min/d pH<6.2=15740.54-28.32×DM-495.64CP-158.78×NDF). The mean rumen pH predicted by the model was 5.33±0.11 observed 5.96±0.14 predictions were on average 0.63 (95% CI -0.28 to 1.54) lower than those observed. The time rumen pH<6.2 was 603±63 min/d as predicted by the model and 943±194 recorded. Predictions were on average 363 minutes (95% CI -677 to 1403) shorter than observed. The results obtained in the present study were acceptable for mean rumen pH but the model to predicted time below the 6.2 threshold was less precise. Further work is required to be able to construct mathematical models capable to predict accurately such a dynamic system as rumen pH.

Trace elements in grasses and legumes in relation to cow requirements

A. Elgersma[1], K. Søegaard[2], S.K. Jensen[2] and J. Sehested[2]
[1]Independent scientist, P.O. Box 323, 6700 AH, the Netherlands, [2]Aarhus University, P.O. Box 50, Tjele, Denmark;
anjo.elgersma@hotmail.com

Trace elements are usually allocated as inorganic supplements in dairy cow feeding to avoid deficiency. To increase self-sufficiency of micronutrients on farms, knowledge and choice of plant species could help to improve and balance the mineral status of cow diets. Effects of grassland species on micronutrient profiles were studied during the season. Species were examined while growing in binary grass-legume mixtures. A range of herbage micronutrient contents is presented here, and discussed in relation to dairy cattle requirements and feeding.

Short term responses in feed intake and yield during concentrate regulation in dairy cows

J.C.S. Henriksen, L. Munskgaard and M.R. Weisbjerg
Aarhus University, Animal Science, Blichers Allé, 8830 Tjele, Denmark; juliec.henriksen@anis.au.dk

Modern milking systems automatically record individual cow data. The individual cow response might be useful to optimize production through the implementation of an individual concentrate strategy. The aim was to investigate the response in feed intake and yield of cows during regulation in concentrate offer in a milking robot from 3 kg to 6 kg, or opposite, with two concentrate types in a crossover, 2×2 factorial design. One concentrate was pelleted (S) and one was a mix of pelleted concentrate and steamrolled, acidified barley (O). The experiment was divided in two periods with half of the cows receiving type S and the other half receiving type O in the first period, all shifting type in the second period. Within period all cows were up- or down regulated twice over 6 days with 0.5 kg per day with a week of constant concentrate offer inbetween; half the cows were on 6 kg and half on 3 kg at the onset of the experiment. The 83 cows (42 Jersey, 41 Holstein) were balanced between treatments according to breed, parity and lactation stage. The mixed ration was fed *ad libitum*. The change in response during regulation was analyzed as a linear regression and reported as daily change (slope, β). The concentrate intake increased during the week of up regulation in daily concentrate offer, and decreased during down regulation (β=0.3 kg/day, β=-0.3 kg/day; $P<0.001$) with a higher intake when offered type O ($P=0.018$). The change in concentrate offer affected the mixed ration intake with a decrease during up regulation, and an increase during down regulation (β=-0.3 kg DM/day, β=0.06 kg DM/day; $P<0.001$). The daily eating time of mixed ration decrease during up regulation of concentrate and increase during down regulation (β=-1.1 min/day; $P=0.06$; β=1.3 min/day; $P<0.001$). The cows did not respond in milk yield during either up- or down regulation ($P=0.71$; $P=0.51$). When comparing results from the following week with constant concentrate offer, yield was highest at 6 kg vs 3 kg ($P=0.034$). In conclusion the cows responded both during up and down regulation in daily feed intake and eating time, but there was no change in milk yield during regulation. Still, yield was highest at 6 kg offer during the constant concentrate period.

Carbon footprint of meat from Holstein bull calves fed four different rations

L. Mogensen[1], A.L.F. Hellwing[1], P. Lund[1], N.I. Nielsen[2] and M. Vestergaard[1]

[1]Aarhus University, AU Foulum, Dept. of Agroecology and Dept. of Animal Science, Blichers Allé 20, 8830 Tjele, Denmark, [2]SEGES, Dept. of Livestock Innovation, Agro Food Park 15, 8200 Aarhus N, Denmark; lisbeth.mogensen@agro.au.dk

The aim of the present study was to quantify how much the composition of feed rations for bull calves affects the two major hotspots in carbon footprint (CF) of meat production from bull calves; greenhouse gas (GHG) emission related to feed production and enteric methane emission. CF per kg meat leaving the slaughterhouse was estimated by life cycle assessment (LCA) of dairy bull calves fed four different rations from 6 months until slaughter at 9 months. The four rations were all fed as TMR and were A: pelleted concentrate (rye, rapeseed meal, wheat, and dried distillers grain with solubles (DDGS)), and chopped barley straw; B: maize cob silage, wheat, dried sugar beet pulp, rapeseed meal, and soybean meal; C: rye, grass-clover silage, and DDGS, and D: grass-clover silage, barley and rapeseed meal. Choice of feed ration had a marked impact on emissions related to both feed production and enteric methane. GHG from feed production varied from 3.0 kg CO_2-eq./kg meat for ration D to 4.4 kg CO_2-eq./kg meat for ration C due to a high CF of DDGS (840 g CO_2/kg DM) and a low CF of grass-clover silage (250 g CO_2/kg DM) compared with a CF of e.g. barley of 520 g CO_2/kg DM. Enteric methane emission was measured by means of indirect calorimetri and was highest for ration D (2.7 kg CO_2-eq./kg meat) with 60% roughage compared with 1.8 kg CO_2-eq./kg meat for the ration based mainly on pelleted concentrate (A) and only 10% roughage. In total, CF per kg meat was lowest for bull calves fed ration A and D, 9.7 and 9.6 kg CO_2-eq./kg meat, respectively. For ration A, this was due to low GHG from both feed production and enteric methane emission. For the ration D, there was low GHG from feed production but a high enteric methane emission due to the high share of roughage. Bull calves on ration C had the highest CF of 10.5 kg CO_2-eq./kg meat. Thus, choice of feed and ration composition is an important factor in the production of 9 months bull calves as the overall CF per kg meat differs 9% from highest to lowest based solely on choice of feed ration fed during the last 3½ months before slaughter.

Enteric methane emissions from Holstein heifers and steers offered grass silage-based diets

H.P. Jiao and T. Yan

Agri-Food and Biosciences Institute, Hillsborough, Co. Down BT26 6DR, United Kingdom; tianhai.yan@afbini.gov.uk

Enteric methane (CH_4) emission from cattle is a considerable source of greenhouse gases responsible for global warming. The objectives were to investigate if the gender of growing Holstein cattle influenced CH_4 emissions and then use these data to develop prediction equations for CH_4 emissions. Ten Holstein male steers and 10 Holstein heifers at age of 12 months were used in a single period (28 days) study. Prior to the commencement of the study, animals were blocked into 10 pairs (steers vs heifers) according to age, live weight (268 to 355 kg) and body condition score. They were housed in a cubicle accommodation for the first 20 days, afterwards in metabolism units for the next 3 days, and finally transferred to indirect open-circuit respiration calorimeter chambers and remained there for 4 days, with CH_4 emissions measured in the final 3 days. All cattle were offered a mixed diet of grass silage and concentrates (ratio=0.77/0.23, dry matter (DM) basis). The grass silage was prepared from the first harvest of perennial ryegrass sward and ensiled without application of silage additives. All data were analysed using one-way ANOVA and some data were also analysed using linear regression techniques. The statistical programme used was Genstat 6.1. The gender had no significant effect on total DM intake or total CH_4 emissions (g/d). Although heifers had a higher ratio of CH_4 emissions per kg live weight (P=0.002) than steers, but the gender had no significant effect on CH_4 emissions per unit DM intake (24.3 vs 23.5 g/kg) or organic matter intake, or CH_4 energy output (CH_4-E) as a proportion of grass energy (GE) intake (0.068 vs 0.066 MJ/MJ) or metabolisable energy intake. Data from these two groups were also used to develop prediction for CH_4 emissions using DM intake or GE intake (Eq. 1 or 2). In conclusion, the gender of growing Holstein cattle has no significant effect on enteric CH_4 emissions. The present models can be used to estimate CH_4 emissions of young Holstein cattle where the measured data are not available. Eq. 1: CH_4 (g/d) = 16.74 DM intake (kg/d) + 47 (s.e.=6.87, r^2=0.77, P<0.001); Eq. 2: CH_4-E (MJ/d) = 0.048 GE intake (MJ/d) + 2.53 (s.e.=0.347, r^2=0.81, P<0.001).

Investigation of ruminal bacterial diversity in Hanwoo cattle fed different diets

M. Kim, J.Y. Jeong, H.J. Lee and Y.C. Baek
National Institute of Animal Science, Department of Animal Biotechnology and Environment, 1500, Kongjwipatjwiro, Iseomyeon, Wanju 55365, Korea, South; mkim2276@gmail.com

This study was conducted to evaluate the composition of ruminal bacterial communities as affected by different diets fed to Korean native Hanwoo steers. Ruminal samples were collected from 24 Hanwoo steers (average weight = 288 kg) fed one of two growing diets: (1) 12 steers fed the control diet composed of 35% Timothy, and 65% grains containing typical levels of CP and TDN; (and 2) 12 steers fed the treatment diet composed of 35% timothy, and 65% grains containing high levels of CP and TDN. Ruminal bacterial communities were investigated using 16S rRNA gene amplicon sequencing on the Illumina MiSeq platform. A total of 601,454 bacterial sequences were obtained from 24 steers with at least 22,000 sequences. *Bacteroidetes* and *Firmicutes* were dominant phyla in all ruminal samples irrespective of diets and accounted for more than 70% of the total sequences in collective data. The proportion of these two phyla did not differ between the two diet groups. At the genus level, *Prevotella* was the most dominant in all ruminal samples irrespective of diets and accounted for about 30% of the total sequences in collective data. Genera *Succiniclasticum*, *Paraprevotella*, *Selenomonas*, *Succinivibrio* and *Ruminobacter* accounted for more than 1% of the total sequences in collective data, where the proportion of *Paraprevotella* was greater in the control diet group than in the treatment diet group (P<0.05). Dominant unclassified groups that accounted for more than 1% of the total sequences were unclassified *Ruminococcaceae*, unclassified *Porphyromonadaceae*, unclassified *Succinivibrionaceae*, unclassified *Prevotellaceae*, unclassified *Lachnospiraceae*, unclassified *Planctomycetaceae*, and unclassified *Veillonellaceae*. The proportion of unclassified *Planctomycetaceae* tended to differ between the 2 diet groups (P<0.1), but the remaining 6 unclassified groups did not differ between the 2 diet groups. This study indicates that the composition of ruminal bacterial communities in Hanwoo steers was slightly affected by different levels of CP and TDN.

Thyme essential oil supplementation on performance and milk quality of lactating dairy cows

E.C. Silva Filho[1], L.C. Roma Junior[2], M.S.V. Salles[2], F.A. Salles[2], J.M.B. Ezequiel[1] and E.H.C.B. Van Cleef[1]
[1]São Paulo State University – UNESP, Acesso Prof. Paulo Donato Castelane, S/N, Jaboticabal, Sao Paulo State, 14884-900, Brazil, [2]São Paulo's Agency for Agribusiness Technology – APTA, Av Bandeirantes, 2419 Ribeirao Preto, Sao Paulo State, 14030-670, Brazil; lcroma@usp.br

The search for alternatives to antibiotics in dairy industry has stimulated recent researches focused on medicinal plants and essential oils supplemented for dairy cattle, trying to reduce somatic cell counts, improving milk quality. This study was carried out to evaluate the effects of *Thymus vulgaris* essential oil on the performance and milk quality of lactating dairy cows. Twenty Jersey cows (days in milk = 57.4±12.6, and milk yield = 21.4 kg/d) were fed for 28 days a corn silage-based total mixed ration. The experiment was conducted in a completely randomized design with two treatments: control (CON), and with the addition of 8 g/animal/d of essential oil (EOT). The essential oil was composed of vegetal extract from thyme (*Thymus vulgaris*). Animals were milked three times daily (06:00, 15:00, and 22:00 h). The roughage (corn silage) and concentrate ratio was 50:50. After third daily milking, the essential oil capsules were delivered using an esophageal probe. Dry matter (DM) of feed delivered and feed refused were recorded every day to calculate DM intake. Milk yield and milk quality data were obtained from the last five days of trial, in consecutive milkings. Milk samples were analyzed for fat, protein, non-fat solids, lactose, and urea nitrogen concentrations, as well as for somatic cell count. The EOT supplementation had no effect (P>0.05) on dry matter intake, milk yield, and milk composition. However, EOT decreased (P<0.01) electrical conductivity, and tended to decrease (P=0.10) somatic cell count (489,000 and 102,000 cells/ml, respectively for CON and EOT). In conclusion, the supplementation of lactating dairy cows with 8 g/d EOT does not affect milk yield and composition, but decreases somatic cell count. Further researches are needed to elucidate antimicrobial activity of Thymol and the antioxidant activity of melatonin constituents of Thymus vulgaris oil. (Financial support: FAPESP 2014/01212-4)

Novel water utilisation and drinking behaviour traits for turkey genetic selection programmes

I. Kyriazakis[1], J. Rusakovica[1], V. Kremer[2] and P. Rohlf[2]
[1]Newcastle University, School of Agriculture, Food and Rural Development, Tyne a, King's Road, Newcastle upon Tyne, NE1 7RU, United Kingdom, [2]Aviagen Turkeys Ltd, Tattenhall, Cheshire, CH3 9GA, United Kingdom; ilias.kyriazakis@newcastle.ac.uk

Defining, measuring and selecting for traits associated with water use and drinking behaviour is of interest from several perspectives. Analysis of such traits has received little attention due to the lack of access to appropriate technology to accurately measure drinking behaviour. We were able to collect such data on two genetic lines of turkeys using innovative water station equipment. Water intake records on individual visits to the water station were grouped in bouts, i.e. time intervals spent in drinking-related activity. This allowed us to identify biologically relevant drinking behaviour traits: number of visits per bout, water intake per bout, drinking time per bout, drinking rate, daily bout frequency, daily bout duration, daily drinking time and daily water intake. For these traits we estimated heritability and genetic correlations with performance traits and identified drinking behaviour strategies among individuals. Heritability estimates for most drinking behaviour traits were moderate to high, the highest ones being for drinking rate (0.49 and 0.50, for the two lines respectively) and daily drinking time (0.35 and 0.46). We estimated low genetic correlations between drinking behaviour and performance traits, except for the moderate correlations between daily water intake and weight gain (0.46 and 0.47). High breeding values for weight gain were found across the whole range of breeding values for daily water intake, daily drinking time and water intake per bout. We show for the first time that drinking behaviour traits are moderately to highly heritable. Low genetic and phenotypic correlations with performance traits suggest that breeding goals are independent from water use and drinking behaviour. Birds express a wide range of different drinking behaviour strategies which would be suitable to a wide range of environments and production systems.

Can we breed for improved maternal ability in sows?

M. Ocepek and I.L. Andersen
Norwegian University of Life Sciences, P.O. Box 5003, 1432 Ås, Norway; marko.ocepek@nmbu.no

The aim of this study was to find maternal behaviours important for piglet survival, to develop qualitative scores from continuous measures of those traits, and to study the relationship between maternal behavioural scores, piglet mortality and the number of weaned piglets. Results are presented from an experimental set-up with 38 sows of different breeds (L, LY, D), followed by a field survey with 900 LY-sows from 45 Norwegian loose-housed sow herds. Maternal behaviours scored were: nest building activities (NBA), sow communication (COS), and carefulness (CRS). The relationship between sow behaviours as continuous measures and as qualitative scores were explored by polyserial correlation, while the relationship between qualitative scores by polychoric correlation. In experimental set-up, piglet mortality was analysed using a GENMOD procedure, the number of weaned piglets with a GLM, whereas in the field data, analyses was conducted using a mixed model. There was a moderate positive correlation between the continuous measures and the qualitative score of NBA (r=0.47) and of COS (r=0.44), and the qualitative and quantitative behaviours similarly affected piglet survival. Since COS and CRS were highly correlated (r=0.88) and of similar effect on piglet mortality/survival, we tested the effect of those scores separately using two models (model 1: NBS, COS; model 2: NBS, CRS) and compared their relative predictive accuracies using AIC. Model 1 had better predictive accuracy than model 2 for all productive parameters. In model 1 for the experimental data, more piglets were weaned with increasing NBS and COS (P=0.043, P=0.004; respectively) and fewer piglets died (P=0.004, P<0.001, respectively) due to lower crushing (P<0.001, P<0.001; respectively). Our results demonstrated that all three maternal behaviour scores had a significant impact on piglet survival, and therefore we tested the same scores in a field survey. COS and CRS were still highly correlated (r=0.56). A model with COS and CRS had the best predictive accuracy for piglet mortality/survival. Thus, combining COS and CRS have the greatest potential to be tested in nucleus herds for calculation of genetic variation and heritability, and should be taken into account in a breeding program for more robust sows.

Can temperament be considered as the same trait in different milking systems?

K.B. Wethal[1] and B. Heringstad[1,2]
[1]Department of Animal and Aquacultural Sciences, Faculty of Biosciences, Norwegian University of Life Sciences, P.O. Box 5003, 1432 Ås, Norway, [2]Geno Breeding and A.I. Association, Storhamargata 44, 2317 Hamar, Norway; karoline.bakke@nmbu.no

More than one-third of the Norwegian milk is produced in herds with automatic milking systems (AMS), and the proportion are increasing. AMS presents new challenges when evaluating traits in the breeding program. Temperament, which has been included as a behavior trait in the breeding program for Norwegian Red (NR) since the 1970s, is an important trait in AMS as well. The aim of this study was to examine whether temperament (T), along with milking speed (MS) and leakage (L), are genetically the same traits in AMS as in conventional milking systems (milking parlor or pipeline systems). In this study, we used data from the Norwegian Dairy Herd Recording system on 316,341 first lactation NR cows that calved between 2001 and 2015. The data included one observation per cow: 53,042 records from AMS herds and 263,299 in milking parlor or pipeline herds. The traits T, MS and L were subjectively assessed by the farmer using a 3-point scale. We analyzed the three traits as different traits in the various systems using bivariate animal models. Heritability was lower for the traits in AMS compared to the parlor and pipeline systems: MS 0.23 vs 0.27, L 0.05 vs 0.15 and T 0.06 vs 0.10. Standard error (s.e.) of heritability estimates were ≤0.01 for all traits. The estimated genetic correlation (s.e.) between temperament in AMS and the other systems was 0.84 (±0.04), which indicates that this is not exactly the same trait genetically. There was a stronger genetic correlation between AMS and other milking systems for the traits MS and L, 0.98 (±0.008) and 0.90 (±0.03), respectively. These results suggest that the farmers evaluate milking speed and leakage similarly in different milking systems. We suspect that different aspects of a cow's temperament are more desirable in AMS compared to conventional milking systems.

Genetic assessment of intra-individual animal variation in milking order of Jersey cows

P. Løvendahl[1], U. Lauritsen[2], L.F. Hansen[2], J. Persson[2] and P.M. Sarup[1]
[1]Aarhus University, Molecular Biology and Genetics, AU-Foulum, 8830 Tjele, Denmark, [2]RYK, SEGES, Agro Food Park 15, 8200 Aarhus N, Denmark; peter.lovendahl@mbg.au.dk

It is a common observation that cows enter the milking parlor in a rather similar order and that each cow has a specific side preference. The milking order or side preferences may reflect behavioral strategies that may impact on the yield, but could also be of relevance to welfare aspects. When studying actual strategies of cows, this can be at two levels, first as the average of the chosen strategy (i.e. the average milking order), and next as the variation in the strategy (i.e. the changes in order from time to time). This second part is the 'intra-individual-variation', IIV. We studied milking order in 18 commercial herds of Jersey cows (n=5,101) with 58,076 milkings in total. Milking order was calculated from entry time stamps recorded by electronic milk meters (Tru-test) at test days occurring every 2 months, at morning and afternoon milkings. Milking order was standardized for herd sized to a scale from -1.00 (first cow) to +1.00 (last cow). Quantitative genetic analysis of milking order was based on ancestral relationships and used for estimation of variance components in mixed (animal) models, having cow-within-parity as permanent animal effect. Milking order was by itself repeatable (t=0.42-0.48), but IVV of milking order was much less repeatable (t=0.06-0.10). Repeatability estimates for daily milk yield was intermediate (t=0.53). Estimates of genetic parameters follow this pattern. The results indicate that IVV of milking order as a behavioral trait has some variation, but for a highly repeatable trait, the IVV is relatively smaller. The relationships with other behavior traits and with production traits will be included in further studies together with results from Holstein cows.

Models with indirect genetic effects depending on group sizes – a simulation study

M. Heidaritabar[1], P. Bijma[2], L. Janss[1], C. Bortoluzzi[2], H.M. Nielsen[3], B. Ask[3] and O.F. Christensen[1]
[1]Aarhus University, Aarhus, 8830 Tjele, Denmark, [2]Wageningen University, Wageningen, 6700 AH, the Netherlands, [3]Seges, Danish Pig Research Centre, Axeltorv 3, 1609 Copenhagen, Denmark; marzieh.heidaritabar@mbg.au.dk

With social interactions, the phenotype of an individual is influenced by the direct genetic effect (DGEs) of individual itself, as well as the indirect genetic effects (IGEs) of its group mates. With IGEs, the heritable variance and response to selection depend on the size of the interaction group (group size), which can be modelled via a 'dilution' parameter (d), which measures the magnitude of IGE as a function of group size. With d=0, IGEs are independent of group size, and with d=1 (full dilution), IGEs are inversely proportional to group size. Very little is known of the estimability of d and the precision of its estimate. The relevance of d estimation is due to its impact on the dynamics of response to selection and heritable variation. We simulated data with different group sizes and estimated d using social genetic models including d parameter. Scenarios were chosen based on average group size (4, 6 and 8) and variation in group size (CV ranging from 0.25 to 1.01). Scenarios consisted of either 2 or 3 group sizes. A design where individuals were randomly allocated to groups was used to estimate d. Three different values of d (0, 0.5 and 1) were simulated. Both genetic and environmental correlations between DGE and IGE were assumed to be zero. A moderate social heritability (h_s^2=0.3) was simulated. The data were analyzed using the DMU software. Results showed that it was possible to estimate d in data with varying group sizes. In general, standard errors on the estimates were smaller with larger CV of group size. Results suggest that d can be detected with random group compositions, a population size of 8,000 individuals and a combination of group sizes of 2 to 14. The most precise estimate of d could be detected when CV of group size was in range of 0.67 to 1.01, which corresponds to group sizes of 2,6,10 and 2,14, respectively. Estimation of the relationship between the degree of IGEs and group size would allow for proper interpretation of direct and social variance components that contributes to heritable variation.

Genome-wide association study for direct and indirect genetic effects on survival time in layers

E.D. Ellen, T. Brinker and P. Bijma
Wageningen University & Research Animal Breeding and Genomics, P.O. Box 338, 6700 AH Wageningen, the Netherlands; esther.ellen@wur.nl

Feather pecking is a worldwide economic and welfare problem in the commercial laying hen industry, which can lead to reduced survival time. Survival time of an individual depends on both the genotype of the individual itself (direct genetic effect; DGE) and on the genotypes of its group mates (indirect genetic effect; IGE). Previous studies showed that IGE contribute 33-76% of the total heritable variation in survival time. Knowledge of the genetic architecture of IGE is limited. So far most studies focussed on DGE only. The aim of this study was therefore to get more insight in the genetic architecture of IGE for survival time in crossbred laying hens. Therefore, we performed a single SNP genome-wide association study (GWAS) for DGE and IGE on survival time in crossbred laying hens. The dataset contained ~7,000 crossbred phenotypes and genotypes. Data were provided by Hendrix Genetics. Hens originated from one sire line and four dam lines. Hens has intact beaks and were kept in 5-bird family cages. We found a SNP for IGE associated with survival time. Interestingly, this SNP is located in the gene GABBR2 which is associated with abnormal behaviour in mice and aggressiveness in humans. Results varied across crosses which may reflect breed-of-origin effects, e.g. the effect may differ depending of the LD between SNP and QTL in the pure lines. Therefore, currently we are performing a single SNP GWAS based on breed-of-origin for direct and indirect SNPs for survival time. For approximately all crossbred hens both parents were known, therefore breed origin of alleles was easily determined after phasing of the genotypes. Results will give more insight in the effect of breed-of-origin on survival time in crossbred laying hens. This will help the breeding companies to improve survival time in commercial laying hens.

Genetic parameters of royal jelly production and behavioural traits of honey bees

F. Phocas[1], C. Le Bihan[2] and B. Basso[3]
[1]INRA, Animal Genetics, GABI, 78350 Jouy-en-Josas, France, [2]GPGR, 23 rue Jean Baldassini, 69364 Lyon, France, [3]ITSAP – Institut de l'Abeille, INRA – Domaine Saint-Paul, 84914 Avignon, France; florence.phocas@inra.fr

The French association of royal jelly producers implemented a breeding scheme at a national scale to improve a honey bee population for traits of economic interest for royal jelly production. This contribution aims to estimate genetic parameters for those traits because selective breeding requires knowledge of heritabilities and genetic correlations between the relevant traits. Data from 817 colonies were collected from 2011 to 2016 in 31 apiaries. Queens of those colonies were produced by 88 inseminated dam queens. Quantity of royal jelly of the first two harvests in the production season were collected for the 817 colonies and the average performance was the production trait studied (PROD). Three behavioural traits were recorded for 497 to 582 of those colonies at the beginning of the harvest season: gentleness (GENT), swarming (SWAR) and sanitary score (SANI). They were assessed by subjective scoring on a 4-mark scale with the mark 4 being the most favourable one. Genetic parameters were estimated with a multiple trait animal model considering the performance of the colony as a trait of the queen. The bee genetic (male haploïdy) and reproductive (polyandry) specificities were accounted for in the derivation of the relationship matrix. Heritabilities of PROD, GENT, SWAR and SANI were close to 30, 40, 40 and 10% respectively. Standard errors of heritability estimates ranged between 0.03 (SANI) to 0.13 (SWAR) while those of the estimates of genetic correlations were higher and ranged between 0.14 and 0.55 due to the small size of the dataset. Even if most of the genetic correlations were not significantly different from zero, a clear tendency toward unfavourable genetic correlations between SANI and all other traits was shown. The two correlations that were significantly different from zero were estimated to -0.71±0.14 between SANI and GENT and to -0.48±0.17 between SANI and PROD. These preliminary results have to be confirmed by a future analysis on a larger dataset. In conclusion, genetic improvement for royal jelly production is possible by selective breeding but attention should be paid not to deteriorate the hygienic behavior of the bees.

Visual assessment of plumage and integument damage in dual-purpose and conventional layers

M.F. Giersberg, B. Spindler and N. Kemper
University of Veterinary Medicine Hannover, Foundation, Institute for Animal Hygiene, Animal Welfare and Farm Animal Behaviour, Bischofsholer Damm 15, 30173 Hannover, Germany; nicole.kemper@tiho-hannover.de

The occurrence of severe damage and high mortality due to feather pecking and cannibalism is a major challenge in modern laying hen husbandry, especially when keeping hens with intact beaks in loose housing systems. Besides time-consuming direct behavioural observations and retrospective video analyses, the evaluation of plumage and integument condition provides useful information about the presence of behavioural disorders on flock level. The aim of the present study was to assess and compare the behaviour of dual-purpose breeds (Lohmann Dual, LD) and conventional layer hybrids (Lohmann Brown plus, LB+) by means of a visual scoring method (VSc). A total of 1,800 untrimmed hens per genetic strain was housed conventionally in four compartments of an aviary system. VSc was carried out on a weekly basis (20th-71st week of life) in five previously defined locations of the barn (n=200 hens/genetic strain and week). The hens' plumage and integument condition was scored on five body parts (head/neck, back, tail, wing, breast/belly) using a five and four point scale, respectively. Minor plumage losses on the back were detected in the LB+ flocks at the age of 25 weeks. Plumage damage increased to the 71st week, so that between 22.5% (wing) and 99.5% (back) of the LB+ hens showed feather losses to different extents. 38.0% of the LB+ hens with plumages losses on the back were assigned to the highest scoring category (>75% of the feathers of the body region missing). In contrast, only 2% of the LD hens showed minor feather losses (head/neck, breast/belly) which remained constant throughout the laying period. In 0.5% of the LB+ hens, minor integument damage (single injuries <0.5 cm length or diameter) started in week 46, reaching a peak in week 66 with 6% affected hens, of which 2% suffered from severe injuries (>1.0 cm length or diameter). However, integument damage decreased to 2.5% of injured hens at the end of the study. No injuries at all were found in LD hens. The results indicate that feather pecking and cannibalism only occurred in the LB+ flocks, though both genetic strains were kept under the same housing and management conditions.

Behavior and growth of pigs with divergent social breeding values

J.K. Hong[1], K.H. Kim[1], N.R. Song[1], H.S. Hwang[2], J.K. Lee[2], T.K. Eom[2] and S.J. Rhim[2]
[1]*National Institute of Animal Science, Rural Development Administration, Swine Science Division, 114 Shinbang-1-gil, Cheonan, 31000, Korea, South, [2]Chung-Ang University, School of Bioresource and Bioscience, 4726 Seodongdaero, Ansung, 17546, Korea, South; sjrhim@cau.ac.kr*

The present study investigated whether pigs chose high or low social breeding values (SBV) in relation to social behaviour and group growth. Positive (+) and negative (-) SBV groups of finishing pigs housed in 10 test pens (3.0×3.3 m, 7 pigs/pen) were observed with the aid of video technology for 9 consecutive hours on days 1, 15, and 30 after mixing. Moreover, pigs were weighed at approximately 90 kg body weight that was then calculated as the number of days to reach 90 kg. On day 1 after mixing, agonistic behaviour was significantly higher in the -SBV group than in the +SBV group (Mann-Whitney U test, Z=2.21, P=0.027). Feeding and mate feeding behaviours were significantly higher in the +SBV group on days 1 and 30 after mixing (Z=-5.11~-4.22, P<0.001). Moreover, group growth performance to reach 90 kg body weight was significantly faster in the +SBV group than in the -SBV group (t-test, t=3.20, P=0.002). The divergent SBV of pigs were affected by responses to social behaviour. Such social interactions among pen-mates might affect their growth rate and feed intake. Selection for SBV might be an indirect technique for improving the growth performance of pigs.

Effect of photoperiod and ambient temperature on reproductive behavior of Damascus goat bucks

E. Pavlou[1], G. Banos[2,3] and M. Avdi[1]
[1]*AUTH, Department of Animal Production, School of Agriculture, 54124, Thessaloniki, Greece, [2]AUTH, School of Veterinary Medicine, 54124, Thessaloniki, Greece, [3]The University of Edinburgh, The Roslin Institute, Easter Bush, EH25 9RG, Scotland, United Kingdom; elefpavlou@gmail.com*

Goat is widely known as an animal with marked seasonality of reproductive activity. The photoperiod is identified as the decisive factor of this phenomenon. The onset and the length of the natural breeding season are also affected by location, climatic conditions, breeding and feeding systems, genotype, and physiological stage. Research in indigenous goat bucks has revealed that there are periods during the spring and summer that bucks show no reproductive behavior. The aim of the current work was to study the effect of photoperiod and ambient temperature on sexual behavior of Damascus goat bucks in Greece. The experiment was conducted at the Aristotle's University of Thessaloniki (Greece) research farm, where all animals are raised under the same conditions and fed with regular diet. The reproductive behavior of 10 Damascus bucks was studied for a period of 28 months. The sexual behavior was assessed once a week during sperm collection sessions by recording latency to ejaculation and ejaculation rate. The influence of daylight and ambient temperature were analyzed using General Linear Models (PASW Statistics 18.0). As photoperiod was considered the day length and ambient temperature was the temperature of the surrounding environment where the study was conducted. Results revealed that the manifestation of the reproductive behavior of the Damascus goat bucks is significantly influenced by the photoperiod(P<0.05). Within the natural breeding season, 100% of bucks exhibited reproductive behavior. On the contrary, environmental temperature did not significantly affect the appearance of their sexual behavior. Although the percentage of bucks showing refusal to mate during semen collection increased outside the natural breeding season, there was a proportion of bucks that demonstrated reproductive behavior throughout the year. The presence of sexually active bucks can stimulate the ovarian activity of anoestrus goats and these bucks could be used in breeding schemes outside the natural breeding season.

The option space for feeding the world without deforestation: implications for livestock management

H. Haberl[1], K.-H. Erb[1], C. Lauk[1], T. Kastner[2], M.C. Theurl[1] and A. Mayer[1]
[1]Institute of Social Ecology, Schottenfeldgasse 29, 1070, Austria, [2]Senckenberg Biodiversität und Klima Forschungszentrum (BiK-F), Georg-Voigt-Straße 14-16, 60325 Frankfurt am Main, Germany; helmut.haberl@aau.at

Conserving the world's remaining forests is a high-priority goal. In this presentation, we discuss the biophysical option space for feeding the world in 2050 in a hypothetical zero-deforestation world. We systematically combine realistic assumptions on future yields, agricultural areas, livestock feed and human diets, and assess for each scenario (i.e. the unique combination of assumptions on each of these parameters) whether supply of crop products meets demand and whether grazing intensity stays within plausible limits. The presentation is focused on the role of different livestock products (ruminants vs monogastrics), feeding systems (grain vs roughage) and the level of animal products in diets. We find that many options exist to meet global food supply in 2050 without deforestation, even at low crop yield levels. Within the option space individual scenarios differ greatly in terms of biomass harvest, cropland demand and grazing intensity, mainly depending on quantitative and qualitative aspects of human diets. Grazing constraints strongly limit the option space in a world without deforestation. Without the option to encroach into natural or seminatural land, trade volumes will rise in scenarios with globally converging diets, decreasing the food self-sufficiency in many developing regions.

A systematic review of trade-offs between food-feed-bioenergy as a result of biomass competition

A. Muscat, R. Ripoll-Bosch and I.J.M. De Boer
Wageningen University and Research, Animal Production Systems, Zodiac De Elst 1 (Building 122), 6708 WD Wageningen, the Netherlands; abigail.muscat@wur.nl

The demand for food, in terms of both food crops but especially livestock products, is expected to further increase because of a growing and increasingly affluent population. Meanwhile, governments around the world are pushing bioenergy policies to move away from fossil fuels and help mitigate climate change. This is expected to result in an increased demand for biomass (for food, feed and bioenergy) and consequently, additional pressure on land use and natural resources. High levels of appropriation of biomass comes at the expense of ecosystems, whether within the country or through biomass imports through third countries. In this context, it becomes increasingly important to take into account the competition for biomass between feed for livestock, food for humans and bioenergy. This competition often results in difficult trade-offs between food, feed and fuel uses and in terms of social, economic and environmental impacts, but the full scale of these remains unknown. We conducted a systematic review to find out these trade-offs using search terms in Scopus. We inventoried the environmental impacts resulting from the competition for biomass acknowledged in literature and we ordered the impacts according to their relevance across spatio-temporal scales. We find that connections are not well established across scales. Therefore different frameworks/concepts might help to identify trade-offs and connect impacts. We argue that it may be useful to consider various frameworks such as socio-ecological metabolism. We argue that the metabolic approach has potential in (1) connecting information across scales (2) ability to show competition between different land uses and the problem of externalising environmental impacts in other countries (3) by connecting society with ecosystem can provide a better connection between production and consumption. Overall the socio-metabolic concept may be a useful concept to address research gaps.

Analyzing the spatial variation of crop and livestock balance on a nation-wide gradient

L. Puillet[1], T. Bonaudo[2], J.P. Domingues[2], M. Jouven[3], T. Poméon[4] and M. Tichit[2]
[1]UMR MoSAR, INRA, AgroParisTech, Université Paris-Saclay, Paris, France, [2]UMR SADAPT, INRA, AgroParisTech, Université Paris-Saclay, Paris, France, [3]UMR SELMET, INRA, Montpellier SupAgro, Montpellier, France, [4]US ODR, INRA, Toulouse, France; laurence.puillet@agroparistech.fr

Livestock production is increasingly affected by competition for land between food and feed and by the need to recycle by-products. A key area for innovation is to develop agro-ecological systems based on balanced interactions between crops and livestock. There is a need to better understand the diversity of interactions among crop, fodder and livestock populations. Objective of this study was to quantify the spatial variation of livestock production and its link to the local availability of feeding resources. The method relies on the use of different data source (statistical data from French government; data from industries and extension services) to calculate six indicators describing the balance between crop and livestock. Indicators were computed for 571 small agricultural regions (SAR). A principal component analysis (PCA) was performed on the 6 indicators followed by hierarchical ascendant clustering based on the coordinates of 571 SARs in PCA dimensions. Results showed four different types of crop/livestock situations. The first type corresponded to SARs (n=250) where livestock production was higher than national average and little feedstuff was locally produced. The second type corresponded to SARs (n=12) showing an even more extreme imbalance between livestock production levels and local feedstuff availability. The other two types (n=97; n=212) corresponded to SARs where a better balance between local feedstuff availability and livestock was achieved but at lower production level. The relationship between local feedstuff availability and livestock density revealed that above 1.5 LU/ha, the dependency to external feedstuff was inevitable. Beyond that threshold, livestock production was based on transforming imported feedstuff into the SAR rather than locally produced feedstuff. Analysing the spatial distribution of the four types of SARs enable to pinpoint that nearly 40% of SARs may require targeted adjustments of livestock densities in order to achieve sustainable balance between crops and livestock.

Brazilian livestock industry socioecological metabolism: N, P and K flow estimate at country level

A.H. Gameiro[1], T. Bonaudo[2] and M. Tichit[2]
[1]University of Sao Paulo, School of Veterinary and Animal Science, Av. Duque de Caxias Norte, 225, 13.635-900 Pirassununga SP, Brazil, [2]Université Paris-Saclay, UMR SADAPT, AgroParisTech, INRA, 16 Rue Claude Bernard, 75005 Paris, France; gameiro@usp.br

In the last two decades, Brazil became one of the largest players in livestock industry. This growth has probably been accompanied by an increase in the use of resources, such as Nitrogen (N), Phosphorus (P) and Potassium (K). In this paper, we estimate the evolution of their flows, as well as the evolution of N, P and K exchanges between Brazil and the rest of the world. N, P and K trends in the Brazilian livestock industry (BLI) were analyzed by a material flow accounting approach based on the socioecological metabolism conception. Accountability considered main animal feeds for bovine (meat and dairy), poultry (meat and egg) and pork. Annual accounts were estimated for a 22-year period (1992 to 2013). BLI's main flows were those linking natural resource extraction or imports, crop production and processing, animal feed, livestock products for final consumers or exports; material waste and environmental emissions. Results showed that the total N input mass in 1993 was 8.51 Tg, rising to 21.02 Tg in 2013, with a cumulative variation of 147%; the average annual compound growth rate (CAGR) was estimated at 4.50%. In the case of P, the total input was 1.57 Tg (1993) rising to 3.11 Tg (2013), a variation of 98% (CAGR 3.77%). For K, the total input was 4.43 Tg (1993), rising to 10.87 Tg (2013), a variation of 145% (CAGR 4.30%). As a reference, the Brazilian population in the same period had a variation of 33% and a CAGR of 1.36%. In terms of international exchange, the share of N imports over the total N inputs was between 6% and 7% in the early 1990s, rising to 16.6% in 2013; exports were between 12% and 13%, increasing to 22.2% (2013). In the case of P, the share of imports was around 25%, reaching 45.7%; exports were between 6% and 9% (1993) increasing to 17% (2013). For K, the share of imports was around 25%, rising to over 37% (2013); exports were around 6% of outputs and grew to more than 11% (2013). These results highlight the relevance of the socioecological metabolism conception to assess the dynamics of livestock industry in a country's scale.

Novel metrics to evaluate livestock farming systems based on dynamic flows of nutrients

T. Takahashi, L. Cardenas, P. Harris, A. Mead and M. Lee
Rothamsted Research, North Wyke, Okehampton, Devon, EX20 2SB, United Kingdom; taro.takahashi@rothamsted.ac.uk

Anthropogenic activities of agricultural production initiate complex physical and chemical interactions between the surrounding environments, including soil, water pathways and atmosphere. Because these interactions cannot easily be observed and their consequences on future environment cannot immediately be foreseen by either producers or policymakers, fair and just society needs quantified measures to evaluate the sustainability of various farming systems under which food and feed are produced. However, the vast majority of the metrics proposed to date are designed for single-season assessment and, as such, fail to acknowledge the long-term and dynamic nature of agricultural production; for instance, nutrients applied in the present season and stored in the soil until the next season are considered 'unutilised' under the current frameworks. Using the rich primary datasets from the North Wyke Farm Platform grazing trials in Devon, UK, this study proposes a novel method to evaluate the performance of various farming systems based on dynamic flows of nutrients beyond a single season. Under the proposed approach, a nutrient freshly introduced into the system, e.g. through fertiliser application, has three possible fates: instantaneously used for today's production (e.g. nitrogen content in grains), reserved within the system for tomorrow's production (e.g. organic nitrogen retained in soil), and lost without being used for production (e.g. nitrate leached). The system-wide sustainability is then measured by the dynamic time path of the ratio between these three categories. In an ideal society with no nutrient loss, the metric is simply be reduced to the ratio between the amounts of nutriens used and preserved; however, if the system incurs nutrient losses, there are multiple pathways to improve the production efficiency, with the two extreme cases being perfect intensification (achieved by increasing the plant uptake of the nutrient) and perfect extensification (achieved by trapping previously lost nutrient in soil). We argue that distinction between these two strategies is important for policy planning and demonstrate that their economic and environmental consequences are not identical.

The net contribution of livestock production to protein supply for humans

S. Laisse[1,2], R. Baumont[3], L. Dusart[4], D. Gaudré[2], B. Rouillé[1] and J.L. Peyraud[5]
[1]IDELE, Monvoisin, 35650 Le Rheu, France, [2]IFIP, La Motte au Vicomte, 35650 Le Rheu, France, [3]INRA, UMR Herbivores, 63122 Saint-Genès-Champanelle, France, [4]ITAVI, Domaine de l'Orfrasière, 37380 Nouzilly, France, [5]INRA, UMR PEGASE, 35590 Saint-Gilles, France; rene.baumont@inra.fr

Livestock may be perceived as less efficient and in competition with crops for the human food supply because they partly consume human-edible feedstuffs to produce milk, eggs or meat. The aim of this study is to estimate the net production of protein for human consumption of ruminant, pig and poultry livestock systems in France. The total protein Conversion Ratio (pCR = kg feed protein consumed/kg protein produced by milk, eggs or meat) does not differentiate human-edible from human non-edible feedstuffs (e.g. grass, byproduct of food and biofuels industries). Thus, we used also the human-edible protein Conversion Ratio (hepCR) that needs to estimate the human-edible protein fraction (hepF) of feedstuffs used in livestock diets. First, we built a table of hepF values for feedstuffs consumed by livestock, applied to the current market and technologies used in the food industries in France (e.g. hepF of wheat grain = 66%, hepF of rapeseed meal = 0%). Then, we estimated the pCR and hepCR values for contrasted feeding systems in ruminants, pigs and poultry. As expected, the pCR values are lower than 1, between 0.05 and 0.25 for ruminant systems, about 0.4 for pigs, 0.3 for laying hens and up to 0.5 for broiler chickens. In contrast, the values of hepCR show that all livestock systems can contribute positively to the human protein supply (hepCR~or>1). Grass-based dairy cow systems can produce up to twice the amount of edible protein they consume (hepCR~2). Pig or poultry systems can respectively produce up to 1.5 (for pig) and 1.3 (for broilers) time more human edible protein that the amount they consume, depending on the proportion of corn, rapeseed/sunflower/soybean meals and other byproducts in the diet. The utilization of grass in ruminant systems and of byproducts in all livestock systems are the key factors to improve the net contribution of livestock production to human food supply, in particular if we consider that advances in food industry technology and modifications of human eating habits can reinforce the feed/food competition in the future.

Dynamix, a serious game to design crop-livestock integration among farms: a case-study in France

J. Ryschawy[1], A. Charmeau[1], M. Moraine[2] and G. Martin[1]

[1]AGIR, Université de Toulouse, INPT, INRA, Chemin de Borderouge, 31320 Auzeville, France, [2]ISARA-Lyon, 23, rue Jean Baldassini, 69364 Lyon cedex 07, France; julie.ryschawy@inra.fr

Integrating crops and livestock allows maintaining production levels while limiting environmental impacts on soil and biodiversity. However, European crop-livestock farms declined due to globalized markets, agricultural policies and limited availability of workforce and skills. Exchanges of manure and crops among farms is an alternative to overcome these limiting factors. This study aimed to co-design with farmers agroecological scenarios of local integration between crop and livestock within two groups of farmers located in southwestern France. Both groups were seeking enhanced self-sufficiency for inputs through exchanges of grain, fodder, crop by-products and manure. The 30 crop farmers practiced conservation agriculture and were interested in manure as organic fertilizer and in crop diversification to better control pest outbreaks. The 39 livestock farmers sought cheap sources of animal feed to reduce purchase of animal-feed inputs. We applied a serious game called Dynamix (DYNAmics of MIXed systems) associating participatory sessions with the two groups of farmers to facilitate the design of crop-livestock integration scenarios and supply-demand balance model. We assessed scenarios through a multi-criteria framework including economic, environmental and social indicators at the individual farm and collective levels. Among the scenarios evaluated, the selected scenario considered both (1) the insertion of legumes as temporary grasslands (e.g. alfalfa) and within cereal-legume mixtures into crop rotations and (2) transfers of manure from livestock farmers to crop farmers. Overall gross margin increased while environmental indicators (share of soil covered, reduction of inputs) were improved. Still, work load and management complexity (logistics, knowledge acquisition) increased. The trade-offs between individual and collective performances appeared acceptable and resulted in greater self-sufficiency at the collective level. The serious game developed could be applied to other groups of farmers to design agroecological scenarios of integration between crops and livestock beyond the farm level.

Multiscale influence of feedstuff availability on environmental impacts of feed

S. Espagnol[1], F. Garcia-Launay[2], L. Dusart[3], S. Dauguet[4], A. Gac[5], D. Gaudré[1], B. Méda[6], A. Tailleur[7], A. Wilfart[8] and L. Morin[9]

[1]IFIP, institute, 35651 Le Rheu cedec, France, [2]PEGASE, Agrocampus Ouest, INRA, 35590 Saint-Gilles, France, [3]ITAVI, institute, 37380 Nouzilly, France, [4]TERRES INOVIA, 33600 Pessac, France, [5]IDELE, institute, 35652 Le Rheu cedex, France, [6]URA, INRA, 37380 Nouzilly, France, [7]ARVALIS, institute, 44370 La Chapelle Saint Sauveur, France, [8]SAS, Agrocampus Ouest, INRA, 35000 Rennes, France, [9]FEEDSIM, AVENIR, 35042 Rennes cedex, France; sandrine.espagnol@ifip.asso.fr

Multi-objective feed formulation (MOF) considers both economic and environmental criteria to decrease, through substitutions between feedstuffs (FS), the environmental impacts of feed with a moderate extra cost. Yet, the question of FS availability on the market is crucial. We studied the effect of MOF in different contexts for pig and broilers feeds, and concentrated feeds of cattle productions, in the North West of France (NW). First, feeds were independently formulated for each production with MOF by defining different lists of available FS and maximum incorporation levels, with a flexible availability of FS (NLIM) or a more constrained one (LIM). Then, all feeds for all productions were then simultaneously optimized (NWF) by considering current FS available volumes and total feed volumes produced in the NW territory. For each context (NLIM, LIM, NWF), Life Cycle Assessment impacts of tons of feed were assessed and compared to those of feeds optimized with least-cost formulation. In NLIM-MOF, Climate Change (CC) was reduced by up to 32, 22, 28 and 23% for pig, broiler, dairy and beef cattle productions respectively. In LIM-MOF, lower reductions were achieved (up to 14, 13, 12 and 11% respectively), due to reduced availability of low-impact FS such as sorghum or pea. In NWF, the global CC impact of feed production in the region is reduced by 7%, with a respective reduction of 5, 8 and 8% for pig, broiler and cattle feeds. Maize used in animal feed is partially replaced by wheat and soybean meal by rapeseed meal. The limited availability of FS leads to a huge competition for the choice of FS between feeds and shows the importance of considering the global livestock sector within a territory. Nevertheless, a significant reduction could be obtained with a limited extra cost (around +2% for NWF).

Session 19

Theatre 9

The key role of horses in the sustainability of territories

A.S. Santos[1,2,3] and L.F. Lopes[3,4]

[1]EUVG, Dep. Veterinary Medicine, Av. José R. Sousa Fernandes Campus Universitário, 3020-210 Coimbra, Portugal, [2]CITAB-UTAD, Quinta de Prados, 5001-801 Vila Real, Portugal, [3]Ruralidade Verde, Lda, Qta Engenheiros, Vila Marim, 5000-773 Vila Real, Portugal, [4]CETRAD-UTAD, Quinta de Prados, 5001-801 Vila Real, Portugal; assantos@utad.pt

The agro-ecological oriented bioeconomy stands for sustainable farming, forestry and related activities ensuring the supply of public goods and rural development. Closely associated with production systems based on a sustainable use of natural resources, horse breeding and other activities involving horses meet the present reforms and policies of the EU 'green' CAP, maintaining the rural landscape and ecological biodiversity of specific agricultural systems. Several studies conducted in Spain and Portugal showed that equines are able to include up to 30% of woody species in their diet and refer that equines can constitute a biological tool for controlling shrub encroachment and the re-growth of invasive plants, enhancing ecosystem biodiversity, and help keeping forest areas cleand reducing the ocurrence of forest fires that may cause serious environmental and economical losses. In several parts of the world, the terrain geography is so characteristic (and difficult), that its agricultural use is closely dependent on horse power. The total no. of working horses in the EU is estimated to be 1 million, and rising. The horse is being seen as an example of efficient, modern and sustainable technology. Recent Sweedish studies show that biodiversity is higher on small organic farms and that a self-suficient small farm of 11.5 ha could feed 65 people using a draugh horse. In forty years, equestrian tourism has gone from being a marginal activity to becoming one of the main touristic activities in some some areas. According to French statistics, the number of French territorial communities relying on horses passed from less than 20 in 2001 to more than 300 today. French statistics also refer that equestrian tourism has increased significantly in the last five years, representing one of the most important developments within the equine sector. Horses are partners in systems with local production, processing, transport and marketing, contributing to social cohesion, rural maintenance and preserving local products and cultural traditions.

Session 20

Theatre 1

COST action IPEMA: innovative approaches in pork production with entire males and immunocastrates

U. Weiler[1], C. Larzul[2], G. Bee[3], E. Von Borell[4], M. Font-I-Furnols[5], M. Škrlep[6], M. Aluwé[7] and M. Bonneau[8]

[1]Univ. Hohenheim, Garbenstr. 17, 70599 Stuttgart, Germany, [2]INRA, 24 ch. Borde Rouge, 31326 Castanet Tolosan, France, [3]Agroscope, Rte de la Tioleyre 4, 1725 Posieux, Switzerland, [4]Univ. of Halle, Theodor-Lieser-Str. 11, 06120 Halle, Germany, [5]IRTA, Finca Camps i Armet, 17121 Monells, Spain, [6]KIS, Hacquetova u. 17, 1000 Ljubljana, Slovenia, [7]ILVO, Scheldeweg 68, 9090 Melle, Belgium, [8]IFIP, La Motte au Vicomte, 35651 Le Rheu, France; michelbonneaupro@orange.fr

The overall objective of the COST action IPEMA is to find and disseminate general, region-specific or chain-specific solutions for the development of alternatives to surgical castration of male pigs: entire male production (EM) and immunocastration (IC). IPEMA is looking for integrated solutions taking into account meat quality (boar taint, water holding capacity, tenderness, fat quantity and quality), animal welfare, breeding, nutrition and economy. IPEMA aims to coordinate research between national programmes and fill knowledge gaps between different areas in Europe and between science and relevant stakeholders. The specific objectives are (1) to guide breeding programmes for pork production systems with EM (boar taint, other meat quality traits, behaviour), (2) to develop nutritional concepts for EM and IC (requirements, boar taint, fat quantity and quality), (3) to define EM specific housing, management, transport and slaughter conditions, (4) to evaluate innovations in grading and meat quality control systems (boar taint, other meat quality traits), (5) to assist innovations in the processing industry to adapt products to EM and IC characteristics and to valorise tainted meat, (6) to develop specific information strategies for European countries and export markets and harmonise sensory studies on boar taint. IPEMA started in October 2016 and will finish in September 2020. IPEMA membership currently includes institutions from 22 European countries. IPEMA wants to extend its membership by including more members from (1) the Eastern part of Europe, neighbouring countries and other international partners and (2) all levels of the pig chains: organisation representing the farmers as well as upstream and downstream industry. The IPEMA consortium acknowledges the financial support of the EU, COST action CA15215.

Producing and marketing entire male pigs
G.B.C. Backus
Connecting Agri and Food, Zilverpark 48, 5237 HN 's-Hertogenbosch, the Netherlands; g.backus@home.nl

This study presents the results of a five year research program that was aimed at evaluating: sensory consumer evaluation of meat from entire male pigs, measures to reduce boar taint prevalence, accuracy of boar taint detection, and the relationship between farm management and aggressive behaviour of boars. Using observational and experimental studies data were collected in various segments of the pork supply chain. The similarity of the rank order between consumer perception of odour and human nose scores, skatole and androstenone levels respectively was determined. Consumers evaluate meat that passed the boar taint detection test comparable to meat from gilts. Meat samples that did not pass the test were evaluated less favourable. Ranking AI boars on their genomic breeding values for low boar taint resulted in a reduction in boar taint prevalence of 40%. The skatole level is lower in boars fed via a long trough than in boars fed by a single space feeder. Few eating places, restricted feeding, insufficient water supply, a suboptimal climate and fear for humans were associated with higher levels aggressive behaviour and more skin lesions. A partly open pen wall, clean pens and pigs, wider gaps of the slats, and feeding by a long trough were associated with less sexual and aggressive behaviour and less skin lesions. Having more than 30 animals per pen was associated with a higher probability of high boar taint prevalence levels. Hygienic conditions were associated with lower boar taint prevalence levels. Assessing similarity of the rank order comparison between consumer perception and three boar taint detection parameters resulted in the highest Kendall's W values for the human nose scores. In conclusion, boar tainted meat was rated as less pleasant by consumers compared to meat of gilts and non-tainted boar meat. Breeding was an effective preventive measure to reduce boar taint. Farms with appropriate management, feeding and housing conditions have reduced levels of mounting and aggressive behaviour. Human nose scores were a better predictor of the rank order of consumer perception, compared to skatole levels and to androstenone levels.

CASTRUM project: a survey on male pig castration for traditional and conventional products in Europe
L. Fontanesi[1], A. Aluwé[2], M. Bonneau[3], L. Buttazzoni[4], M. Čandek-Potokar[5], V. Courboulay[3], S. Failla[4], M. Font-I-Furnols[6], B. Fredriksen[7], M. Škrlep[5] and E. Von Borell[8]
[1]University of Bologna, DISTAL, Viale Fanin 46, 40127 Bologna, Italy, [2]ILVO, Scheldeweg 68, 9090 Melle, Belgium, [3]IFIP, La Motte au Vicomte, 35651 Le Rheu, France, [4]CREA, Via Salaria 31, 00015 Roma, Italy, [5]KIS, Hacquetova ulica 17, 1000 Ljubljana, Slovenia, [6]IRTA, Finca Camps i Armet, 17121 Monells, Spain, [7]Norwegian Meat and Poultry Research Center, P.O. Box 396 Okern, 0513 Oslo, Norway, [8]MLU Halle-Wittenberg, Theodor-Lieser-Str. 11, 06120 Halle, Germany; luca.fontanesi@unibo.it

The two main objectives of the CASTRUM project were to: 1) identify and evaluate recognized methods for anaesthesia and/or prolonged analgesia at the time of male pig castration in Europe; 2) evaluate and review the alternatives to surgical castration for heavy pigs used in traditional pork products considering quality assurance systems, meat quality and animal welfare. The CASTRUM consortium included 11 partners from 10 countries and 7 associated national contact points (NCP). Information search was based on relevant literature, DOOR database on PDO, PGI and TSG products, on NCP's knowledge of regional situations and opinions from different actors and stakeholders of this field. Complete information was collected in 16 countries (AT, BE, BG, DE, DK, ES, FR, GB, HR, HU, IT, NO, PL, PT, SE, SI) and in part also for other 5 countries (CH, FI, LU, MT, NL). Large heterogeneity among countries was observed in the use of analgesia and/or anaesthesia for pain relief during surgical castration of male piglets. Regarding alternatives to surgical castration for heavy pigs used for traditional products, the surveys showed that besides situations based on EU protected products (PDO, PGI, TSG), there are many others that need special attention. Simplistic approach setting a threshold on weight/age at slaughter to prevent boar taint when producing entire males is not sufficient; assessment of risks for welfare, management and pork quality should be considered, including genotype, type of meat product and the way these products are consumed. The financial support of the EU DG SANTE is acknowledged. Special thanks are for all contacts who provided or helped collecting information used in the study and to all CASTRUM partners.

Evaluation of alternatives to surgical castration in heavy pigs for traditional products (CASTRUM)

M. Bonneau[1], M. Čandek-Potokar[2], M. Škrlep[2], M. Font-I-Furnols[3], M. Aluwé[4] and L. Fontanesi[5]
[1]IFIP, La Motte au Vicomte, 35651 Le Rheu, France, [2]KIS, Hacquetova ulica 17, 1000 Ljubljana, Slovenia, [3]IRTA, Finca Camps i Armet, 17121 Monells, Spain, [4]ILVO, Scheldeweg 68, 9090 Melle, Belgium, [5]DISTAL, Univ. of Bologna, Viale Fanin 46, 40127 Bologna, Italy; michelbonneaupro@orange.fr

A total of 272 situations in 17 European countries (AT, BE, BG, DE, DK, ES, FR, GB, HR, HU, IT, NL, NO, PL, PT, SE, SI) were analysed to assess whether they could use entire male pigs or other alternatives to surgical castration. The analysis was based on information taken from the DOOR database of EU protected products (PDO, PGI, TSG), from the bibliography and from a survey of relevant stakeholders. Entire males (EM) or immunocastrates (IC) were used in few situations and mostly to a small extent. More than 2/3 of stakeholders considered surgical castration as essential although being compulsory in only 1/3 of the situation. A small minority of stakeholders could consider using IC (17%) or EM (7%). The use of EM was deemed difficult or impossible in 92% of the situations for the following reasons: (1) castration is made compulsory by specifications (1/3), (2) sexual maturity at slaughter is likely to increase the incidence of boar taint and raise management issues (4/5), (3) product characteristics (high fat content, absence of masking ingredient, warm consumption) results in elevated perception of boar taint (1/3), (4) increased unsaturation of fat makes meat unfit for processing into dry products (3/5). A SWOT analysis was performed for the use of EM or IC in situations using heavy pigs for traditional products. On the whole, compared to standard production, there are less advantages and more disadvantages in using EM for heavy pigs aimed at high quality products. Immunocastration prevents most of (but not all) the disadvantages associated with EM, but stakeholder acceptability is still low in most countries. The CASTRUM consortium acknowledges the financial support of the EU, DG SANTE. Special thanks are extended to all the national contact persons who provided or helped collecting a large part of the information that was used in the present study.

Evaluation of information on methods for anaesthesia and analgesia for male pig castration (CASTRUM)

E. Von Borell[1], V. Courboulay[2], B. Fredriksen[3] and L. Fontanesi[4]
[1]Martin-Luther-University Halle-Wittenberg, Theodor-Lieser-Str. 11, 06120 Halle, Germany, [2]IFIP, La Motte au Vicomte, 35651 Le Rheu, France, [3]ANIMALIA Norwegian Meat and Poultry Research Center, P.O. Box 396 Okern, 0513 Oslo, Norway, [4]DISTAL University of Bologna, Viale Fanin 46, 40127 Bologna, Italy; eberhard.vonborell@landw.uni-halle.de

One of the main objectives of the CASTRUM project was to identify, specify and evaluate recognized methods for anaesthesia and/or prolonged analgesia at the time of male pig castration in Europe. The systematic use of anaesthesia with or without additional analgesia for pain relief during surgical castration of male piglets is currently limited to some countries (NL, NO, SE and CH). Our study evidenced a big heterogeneity among practices in different countries. The use of analgesia only (without concurrent use of anaesthesia) was reported in some other countries as part of their national assurance programmes. However, from scientific studies it is evident that the effectiveness of pain intervention during and after surgical castration is only given when anaesthesia is combined with preemptive analgesia. Pain interventions using CO_2/O_2 inhalation or ketamine/azaperone injection anaesthesia do not seem to meet the demand for a sustainable and welfare conform production system, considering the serious risks associated with these methods including aversiveness, limited safety margins, handling stress, practicability as well as economic feasibility. Some practices, such as local anaesthesia and inhalation anaesthesia with isoflurane/sevoflurane, both combined with analgesic preemptive treatment could be considered, as these alternative interventions seem to be superior to other methods considering effectiveness, drawbacks and risks. It should however be noted that the analgesics currently used have a limited half-life time of a few hours and that the application of analgesics and anaesthetics in general impose additional handling and stress on piglets. The survey also evidenced that in some countries the acceptance and likelihood of anaesthesia implementation will depend on authorisation of farmers to do the pain interventions after special training.

Performance and welfare assessment in entire males compared to anti GnRH vaccinated boars

S. Ebschke[1], M. Weber[2] and E. Von Borell[1]
[1]Inst. of Agr. and Nutr. Sci., MLU Halle-Wittenberg, T.-Lieser-Str. 11, 06129 Halle, Germany, [2]LLG ZTT Saxony-Anhalt, Lindenstr. 18, 39606 Iden, Germany; stephan.ebschke@landw.uni-halle.de

Immunization against GnRH is practiced as one of the alternatives to surgical castration and raising entire males (EM). The aim of the present study was to assess animal welfare and performance associated with this alternative as well as feeding options to affect organoleptic perception and consumer acceptance positively. A total of 185 male pigs were either fattened as EM with a standard protein/energy adapted diet for boars SBD (46) or were vaccinated (vac.) against GnRH (Improvac®) and fed with a boar diet SBD (45), an energy reduced boar diet ERD (47) and a diet suited for barrows BD (47). They were kept in groups of 12 in four replicates with identical randomly allocated pens. DFI was recorded individually. Animal based indicators like skin alterations, performance and carcass traits were evaluated on farm and abattoir. Furthermore, carcass quality, sensory testing at consumer level and animal behaviour (resting, locomotion, feeding, agonistic interaction) were estimated. DWG did not differ between groups until the 2nd vac. but subsequently increased for anti GnRH vac. boars ($P<0.05$). FCR was lower for both EM and vac. M fed the SBD in relation to ERD and BD ($P<0.05$). Lean meat percentage did not differ between groups but dressing percentage was at least 1.3% higher when fed with SBD. IMF was lowest in EM (1.15%) and differed from all other vac. groups ($P<0.05$). Juiciness was unaffected by treatment but tenderness was slightly reduced in EM vs vac. SBD ($P=0.046$). Flavor was consistently rated higher in all vac. groups ($P<0.05$) and did not differ within these groups. The overall consumer assessment of meat from EM was negatively perceived compared to vac. boars ($P<0.05$). Increased resting ($P<0.01$), less agonistic interaction ($P<0.001$) and riding events ($P<0.001$) were only observed in groups fed the BD. Skin damage alterations, especially on head and flank, were slightly higher in EM during the whole observation period. We conclude that fattening of EM may create animal welfare and consumer acceptance problems. The advantage of improved body and carcass performance in EM can be best utilized when providing them with a adapted boar diet and vac. against boar taint.

Influence of feeding level and dietary energy content on performance and behavior of entire male pig

N. Quiniou, A.S. Valable, N. Lebas and V. Courboulay
IFIP, BP 35104, 35651 Le Rheu cedex, France; nathalie.quiniou@ifip.asso.fr

Two trials were performed with 160 entire male pigs each to characterize the growth performance and boar taint risk (trials 1 and 2) and behavior (trial 2) when energy was provided *ad libitum* (A) or at a restricted level (R) using 2-phase diets either concentrated or diluted in net energy (NE, C=10.0 or D=9.4 MJ/kg, respectively). In each trial, 32 pens of five pigs each were allocated to one of the four treatments according to a factorial 2×2 design depending on the feeding level and the dietary energy content. In both trials, no regulation of spontaneous feed intake on the basis of dietary NE content was observed below 70 kg body weight. From the beginning of the fattening period onwards (20 and 28 kg body weight in trials 1 and 2, respectively), *ad libitum* fed pigs with C (called AC pigs) or D diets presented the same daily feed intake (DFI), that resulted in a higher NE intake for AC pigs compared to AD ones. Above 70 kg body weight, DFI of AC pigs was reduced and resulted in similar NE intake like AD in trial 1, but not in trial 2. At a given body weight, RC and RD pigs received the same amount of NE, that corresponded to 93% (trial 1) or 90% (trial 2) of *ad libitum* NE intake of AD pigs. Energy restriction with diet D lowered the daily body weight gain (ADG) without any effect on the feed conversion ratio (FCR) or the carcass leanness. With diet C, energy restriction was performed through a smaller feed allowance; feeding activities within the post-prandial hour were much more frequent in this treatment and may contribute to an increase in FCR observed in trail 1. Boar taint risks due to skatole or androstenone were very low in both trials that makes it difficult to draw any conclusion about the effect of the feeding strategy on this criterion. In conclusion, based on growth performance and behaviour, our results indicate that high NE diets can help to increase the NE intake of entire males with a low appetite at the beginning of the growing phase for example. They also demonstrate that feed restriction is not an interesting feeding strategy for entire male pigs, especially when a high energy diet is used. However, when feed restriction is implemented, a low NE diet must be preferred for entire male pigs.

Aggression, stress, leanness, feed, and their relation to boar taint

E. Heyrman[1,2], S. Millet[2], F. Tuyttens[2], B. Ampe[2], S. Janssens[1], N. Buys[1], J. Wauters[3], L. Vanhaecke[3] and M. Aluwé[2]
[1]*KU Leuven, Livestock Genetics, Department of Biosystems, Kasteelpark Arenberg 30, 3001 Heverlee, Belgium,* [2]*ILVO (Flanders research institute for agriculture, fisheries and food), Animal Sciences Unit, Scheldeweg 68, 9090 Melle, Belgium,* [3]*Ghent University, Faculty of Veterinary Medicine, Department of Veterinary Public Health and Food Safety, Laboratory of Chemical Analysis, Salisburylaan 133, B-9820 Merelbeke, Belgium; evert.heyrman@ilvo.vlaanderen.be*

The European pig sector has committed itself to ban surgical castration of male piglets by 2018. One alternative is raising uncastrated pigs (entire males) with as a major drawback the occurrence of boar taint. Boar taint is an off-odor that can occur in the meat or fat of entire males and affected carcasses should not reach the consumer. More insights are needed in factors affecting the prevalence of boar taint on farm. In this study 23 farms raising entire males participated and on average 5 slaughter batches per farm were included. Data on feed, management and genetics were collected per farm and per slaughter batch by means of questionnaires. Data on housing and pig behavior were collected through observations in the barns one week before slaughter. Carcass data were collected in the slaughterhouse. At slaughter backfat samples were collected (from 148 entire males per slaughter batch on average). These were scored on a 5-point scale (0-4) for boar taint by an olfactory panel of trained experts. Each sample was evaluated by 3 experts and the median score was taken as the final score. When this final score was 2 or higher, the sample was considered tainted. Factors were evaluated for their association with boar taint in univariable mixed binomial models The average boar taint prevalence per farm was 1.9% and the range was 3.3%. Skin lesion score in the barn and age at the start of the fattening period were positively correlated with the incidence of boar taint. Lean meat percentage, time spent in the lairage, single vs mixed sex rearing, lysine and fat content of the feed, and mortality were negatively correlated with the incidence of boar taint. The results suggest that reducing aggression and stress, breeding leaner pigs and adapting feed can lower the occurrence of boar taint.

Feeding adsorbents to reduce androstenone in plasma and backfat of boars

N. Canibe[1], J. Frederiksen[2], L.S. Brehmer[3] and B.B. Jensen[1]
[1]*Aarhus University, Animal Science, Blichers Alle 20, 8830 Tjele, Denmark,* [2]*PF&U Mineral Development ApS, Kullingade 31, 5700 Svendborg, Denmark,* [3]*IMERYS Industrial Minerals Denmark A/S, Koensborgvej 9, 7884 Fur, Denmark; nuria.canibe@anis.au.dk*

Boar taint is an off-odour and off-flavour meat trait released upon heating of meat from uncastrated sexually mature male pigs. The testicular pheromone androstenone and the tryptophan metabolite skatole are the two main compounds responsible for boar taint. Whereas the level of backfat skatole can be reduced by feeding, the impact of feeding on androstenone has been very scarcely investigated. Adsorbent materials, like activated coal and Tween, have been reported to reduce backfat- and plasma levels of androstenone in boars. The aim of these studies was to establish the impact of various adsorbents on the concentration of androstenone in backfat and plasma of boars. Initially, the impact of adding 5% activated coal to a standard diet was tested using 8 boars following a cross-over design with two periods of three weeks each. No impact of feeding activated coal on the concentration of androstenone in backfat or plasma was observed. Further, an *in vitro* study was conducted to test the impact of various adsorbent materials (three bentonites, two zeolites, hydrated sodium calcium aluminosilicate, two diatomaceous earth, and biochar) on the concentration of androstenone in digesta. Two bentonites, showing the highest ability to reduce androstenone concentration, were chosen to be tested *in vivo*. A study with 24 boars, 6 per treatment, was conducted. There were four groups: standard diet; standard + 5% Bentonite 1; standard+5% Bentonite 2; standard + 2.5% Bentonite 1 + 5% Bentonite 2. The animals were fed the experimental diets for four weeks. Preliminary results showed no effect of the adsorbing materials on backfat androstenone compared to the control treatment. In conclusion, the first results obtained in these *in vivo* studies showed no impact of the tested adsorbing materials on backfat androstenone concentration in crossbred boars under Danish conditions. Additional results of blood androstenone concentration in the experimental groups will be presented, as well as data suggesting an enterohepatic circulation of androstenone.

Expression quantitative trait loci reveals genes and pathways associated with boar taint in pigs

M. Drag[1], M.B. Hansen[1] and H.N. Kadarmideen[1,2]
[1]*University of Copenhagen, Department of Veterinary and Animal Sciences, Grønnegårdsvej 7, 1870 Frb C, Denmark,*
[2]*Technical University of Denmark, Department of Bio and Health Informatics, Kemitorvet, Building 208, 2800 Kgs.*
Lyngby, Denmark; markus.drag@sund.ku.dk

Boar taint (BT) is an offensive odour or taste of meat from a proportion of non-castrated male pigs due to skatole and androstenone accumulation in adipose tissue. Castration is an effective strategy to avoid BT but is currently under debate due to animal welfare concerns. This study aimed to functionally characterise and evaluate important expression quantitative trait loci (eQTL) to identify potential causal mechanisms of BT for use in optimised breeding for reduced BT without castration. Danish Landrace male pigs (n=48) with low, medium and high genetic merit of BT were slaughtered at 100 kg. Gene expression profiles were obtained by RNA-Seq and genotype data was obtained by Illumina 60K Porcine SNP-chip. Following quality control and filtering, 10,545 and 12,731 genes from liver and testis were included in eQTL analysis by the Matrix eQTL software together with 20,827 SNP variants. A total of 180 eQTLs containing 66 genes and 278 eQTLs containing 117 genes were mapped respectively in liver and testis with a significant (FDR<0.05) gene-SNP relationship and association (P<0.05) with BT phenotype. Functional enrichment test of eQTLs revealed biological processes within negative regulation of DNA replication, cell activation and DNA metabolic processes. By employing a multivariate Bayesian Hierarchical Model within Matrix eQTL, a total of 26 eQTLs containing 11 genes were mapped as multi-tissue eQTLs (FDR<0.05) and having significant association (P<0.05) to the BT phenotype. Enrichment test of multi-tissue eQTLs with known QTLs revealed 46 QTLs to have significant (P<0.05) overlap which were mainly traits categorised as meat and carcass QTLs, meat and carcass associations and production QTLs. The overlap of the QTL and eQTLs show potential causality for BT phenotypes. Finally, the unique 11 genes from the multi-tissue eQTLs were evaluated with respect to their role in BT and possible role as candidate biomarkers of BT in optimised breeding scenarios.

Stability of boar taint compounds during storage

E. Heyrman[1,2], L. Vanhaecke[3], S. Millet[2], F. Tuyttens[2], B. Ampe[2], S. Janssens[1], N. Buys[1], J. Wauters[3], K. Verplanken[3],
C. Van Poucke[4] and M. Aluwé[2]
[1]*KU Leuven, Livestock Genetics, Department of Biosystems, Kasteelpark Arenberg 30, 3001 Heverlee, Belgium,* [2]*ILVO*
(Flanders research institute for agriculture, fisheries and food), Animal Sciences Unit, Scheldeweg 68, 9090 Melle,
Belgium, [3]*Ghent University, Faculty of Veterinary Medicine, Department of Veterinary Public Health and Food Safety,*
Laboratory of Chemical Analysis, Salisburylaan 133, 9820 Merelbeke, Belgium, [4]*ILVO (Flanders research institute*
for agriculture, fisheries and food), Technology and Food Science Unit, Brusselsesteenweg 370, 9090 Melle, Belgium;
evert.heyrman@ilvo.vlaanderen.be

Boar taint is an unpleasant odor that can occur in meat and fat of uncastrated male pigs (entire males). The main compounds responsible for boar taint are skatole (SKA), androstenone (AND), and to a lesser extent indole (IND). Many studies on the subject of boar taint rely on sensory and chemical evaluation of boar taint. The latter consists of measuring the concentrations of SKA, AND, and IND in fat. In such studies fat samples are often kept frozen at -20 °C before they are analyzed. It is generally considered acceptable to do this as the boar taint compounds are assumed to be stable. This study involves two experiments to confirm this hypothesis. For each experiment, 20 non-tainted (NT) and 10 tainted (T) backfat samples were selected and divided in smaller samples (2×2 cm) which were kept at certain temperatures and analyzed repetitively at set times. Analysis consisted of sensory analysis by an olfactory expert panel as well as chemical analysis for boar taint compounds. In experiment I, storage temperatures were 4, -20 and -80 °C and timepoints were day 0, 1, 3, and week 1, 2, 4, 10, 20. In experiment II, storage temperature was -20 °C and the timepoints were day 0, 1, 2, 3, and week 1. In this second experiment the chemical analysis was done in 2 independent laboratories. In experiment I, a drop in SKA and AND concentration was seen in the first days for some samples, but this was not confirmed by experiment II. In experiment II there was considerable inter- and intra-laboratory variation. Storing fat samples at -20 °C had only limited effect on boar taint compound concentrations and sensory perception of boar taint.

Comparison of pig classification results between entire and castrated males

G. Daumas

IFIP-Institut du Porc, BP 35104, 35601 Le Rheu Cedex, France; gerard.daumas@ifip.asso.fr

Entire male is the alternative to pig castration chosen by some groups in Europe and in France by the leader cooperative. Although the advantage in carcass composition is well known precise estimates are not easily available and figures can vary a lot between samples. Conversely to most of European countries sex is registered online in France during pig classification. Statistics per sex are regularly published by the regional classification organisations. The aim of this work is to compare the national classification results of entire males with these of castrated males. Since 2013 the production of entire males in France has been growing up. In 2016 about 2.6 millions of entire males were classified, i.e. 11.7% of the pigs and 23% of the males. More than 95% of the entire males are classified with the classification method CSB Image-Meater® (IM) approved by the EU in 2013. This LM% (Lean Meat Percentage) prediction equation contains two fat depths (G3 and G4) and two muscle depths (M3 and M4). The analysis of the 2016 statistics published by the classification organisation Uniporc Ouest showed that entire males had less fat (-4.0 mm of G3 and -3.5 mm of G4) and less muscle (-2.0 mm of M3 and -2.6 mm of M4). Multiplying these differences by the respective depths coefficients in the LM% equation gave the contribution of each depth to the LM% difference between entire and castrated males. These contributions were +1.9, +0.5, -0.2, -0.1 respectively for G3, G4, M3 and M4. Summing fat depths on one hand and muscle depths on the other hand gave a fat contribution of +2.4 and a muscle contribution of -0.3. The balance, largely driven by fat, was thus of +2.1 LM%. The same type of calculation with the other classification method, called CGM, gave a balance of +1.8 LM%. The respective fat and muscle contributions were of +2.1 and -0.2. Nevertheless, as both IM and CGM equations were based on sample stratified with 50% of castrated males and 50% of females, and because of sex bias, the precise size of the LM% advantage for entire males is unsure. An update of the carcass classification equations would give better estimates. Moreover, sex biases could be removed by at least a different intercept in the prediction equations. This would contribute to a better efficiency in the pig chain.

Assessment of carcasses of pigs, slaughtered with different live weight

M. Povod[1], O. Kravchenko[2] and A. Getya[3]

[1]Sumy State National University, G. Kondratyeva str., 160, 40021 Sumy, Ukraine, [2]Poltava State Agrarian Academy, Skovorody str., 1/3, 36003 Poltava, Ukraine, [3]National University of Life and Environmental Sciences of Ukraine, Gen. Rodimtseva str., 19, 03041 Kyiv, Ukraine; getya@ukr.net

In modern pig production systems, the choice of optimal live weight of animal before slaughter belongs to the important issues. To clarify this question 80 heads of final hybrids Yorkshire×Landrace×Maxgro (gilts and castrates) with the age between 153 and 160 days were slaughtered. All animals, which were fattened under the condition of industrial pig farm, obtained the same ration wet feeding. Day before slaughter all animals were divided into different groups depending on their weight and sex: I group (live weight 85-95 kg), II group (live weight 95-110 kg), III group (live weight 110-120 kg). Carcass quality assessment was performed according to national Ukrainian and EUROP grading scheme using Fat-o-Meater S71 device. After assessment 8.75% of all carcasses were graded with E, 57.5% with U, 32.5% with R, 1.25% with O, while according to national norms all carcasses were ranked into the 2nd class. In the first group (light pigs) more than 80% carcasses of gilt and castrates were graded with E and U. The percentage of E and U carcasses in other groups was under 65%. Carcasses of gilts had slightly higher dressing percentage in all groups comparing with castrates. No relevant difference on lean meat content between castrates and gilts was found. The highest lean meat percentage was observed in group I: 52.27±0.71% in castrates and 52.61±0.71% in gilts. At the same time carcasses in group III had highest dressing percentage 75.23±0.46% and 76.24±0.55% in castrates and gilts respectively. Thus, increasing of live weight before slaughtering leads to increasing of dressing percentage of carcasses but causes the reduction of percentage of lean meat. This interaction need to be considered during introduction in Ukraine EUROP grading scheme instead of national norms, which don't allow classifying of all carcasses properly. Facing the forthcoming ban on chirurgical castration without anesthesia in EU the advanced study on fattening of entire mails and immunocastrated boars in Ukraine as well as on consumer behavior is required.

Use of genetic markers to reduce boar taint in Canadian pigs – validation in commercial trials

L. Maignel[1], F. Fortin[2], P. Gagnon[2], M. Jafarikia[1,3], J. Squires[3] and B. Sullivan[1]
[1]Canadian Centre for Swine Improvement Inc, Central Experimental Farm, Building 75, 960 Carling Avenue, Ottawa, ON, K1A0C6, Canada, [2]Centre de développement du porc du Québec Inc, Place de la Cité, tour Belle Cour, 2590, boul. Laurier, bureau 450, Quebec, QC, G1V 4M6, Canada, [3]University of Guelph, Department of Animal Biosciences, University of Guelph, 50 Stone Road East, Guelph, ON, N1G 2W1, Canada; laurence@ccsi.ca

Boar taint can occur in meat produced from entire male pigs. It is caused by concentrations of androstenne and/or skatole in fat tissues higher than specific levels. The objective of this study was to investigate the possibility of reducing the amount of androstenone and skatole in fat tissues of intact males using genetic markers. Fat samples were collected via biopsies or at the slaughter plant on a total of 3,474 boars. These animals were also genotyped for 97 SNP markers located in 40 candidate genes, from which 61, 80 and 83 genotyped SNPs were polymorphic in Duroc, Landrace and Yorkshire pigs, respectively. A two-step analysis was performed to examine the association of SNPs with measured androstenone and skatole levels. The results of this analysis were used to develop marker-assisted breeding values for androstenone and skatole levels. Active Duroc boars in three Canadian artificial insemination centres were genotyped with the same SNP set to identify boars with high and low genetic values for boar taint based on SNP marker genotypes. Commercial pigs born from these two groups were station-tested to validate the results previously found in purebred populations. Two commercial trials were carried out in Eastern Canada and one in Western Canada with two different feeding programs. Within each trial, females, intact males, castrates and males treated with Improvest were performance-tested from weaning to slaughter (at an average market weight of 130 kg) and tracked at the packing plants for carcass and meat quality evaluation. A subsample of each sire group and sex was used for sensory analyses by a trained panel.

Production is going to the far-east: potential contrasts in the application of PLF

T. Cohen
TH Milk Food JSC, www.thmilk.vn, nghe Son, Nghe Anh District, Nghe An, Vietnam; tal.cohen@thmilk.vn

The livestock production is gradually moving to emerging economies at the far east. Precision livestock farming (PLF) gains ground also there. However, 'tradition dies hard', the application of PLF in the far east requires experience and expertise. The keynote presentation will explore the following aspects: establishing mega-dairies in Vietnam 2010-2017, the Vietnam population, milk market, investor motivation, land and government support, environment protection. Design phase – constrains and opportunities that might affect the project for years – housing, groups size, milking parlors, herd management systems, level of automation, climate, habits. Local manpower availability and skills, relocating expert's recruitment. Feedstuff – import or local supply. Equipment to fit local conditions. First years – new cows, new workers, new young managers, unfinished construction – survival. Veterinary support and expertise, farm managers and workers recruitment and training. Herd management issues, cooling methods, overcrowding conditions. Feed supply from local resources, fields development, Maturity phase – adaptation to local conditions, local feeds and habits, develop company core and strengths. Protocols and SOP's adaptation, investment in Human resources start to bare fruits. Introducing Animal welfare codes in Vietnam. forage quality, milk quality, environment protection, manure to fields application o breeding policies and genetics progress. Technology and 'gadgets' – in perspective to large dairy farm in Vietnam. What works well and what was introduced too early or was missing. Technical workers training for 24/7 maintenance of the technology. Social aspect of large dairy project. Area development, business development, women encouragement, youth. Relationships with international expertise and with local academy.

A debate: how own the animal data ?

M. Coffey

sruc, sruc, sruc, United Kingdom; mike.coffey@sruc.ac.uk

Modern automation and sensor use on-farm is producing unprecedented volumes of data from many individual sources, for example from activity monitors, cameras, milking robots, pressure plates. The integration of those data produces insights and knowledge that the farmer wishes to use in day to day management. Problems are beginning to arise on ownership of raw data, access to unprocessed data from devices and ownership and access rights of information generated by integration of data. Who 'owns' the foreground IP and who can use it? How can companies whose main income is from devices that produce data facilitate the integration of data without losing competitive advantage? How can assimilated or generated data be used in national genetic evaluations? These questions and more will be considered in a challenge session involving 3, 5 minutes presentations followed by a debate on producing an environment under which solutions to these problems can be explored.

Vocalisations of weaner pigs as a potential tool to monitor ear biting behaviours

A. Diana[1,2,3], L. Carpentier[3], D. Piette[3], L. Boyle[1], T. Norton[3] and D. Berckmans[3]

[1]Teagasc, Moorepark, Co. Cork, Ireland, [2]UCD, School of Veterinary Medicine, Dublin, Ireland, [3]KU Leuven, M3-BIORES, Leuven, Belgium; tomas.norton@kuleuven.be

Ear biting (EB) is reported as a growing problem on commercial pig farms. However, such indicator of poor welfare is barely studied, hence the availability of PLF technologies may help to clarify this issue. During the EB event there are two subjects involved: the biter and the bitten pig. The latter can react through a type of vocalisation. The aim of the study was to identify the specific types of vocalisations in order to find reliable features to be used in conjunction with visual behaviours for the development of an algorithm. The study was conducted on an Irish commercial farm. Thirty-five growing pigs were video recorded 3 wks after weaning for a total of 2 h. Videos were scrutinized through a labelling software which allowed the labelling of both images and sounds. Three clear types of vocalisation were detected as the result of an EB event and classified as Grunt (GR, 30), Acute (A, 25) and Scream (S, 21). Feature values were extracted to identify acoustic differences between vocalisations. From the time domain the duration of GR (0.39 ± 0.06 s), A (0.27 ± 0.01 s) and S (0.81 ± 0.39 s) were extracted. Frequency (F) information of GR, A and S were obtained from statistical measurements on the aggregated spectrum of the labelled sounds. Specifically, vocalisations were found to have a mean F (GR: $3,375\pm134$, A: $4,660\pm142$ and S: $5,107\pm186$ Hz), peak F (GR: 353 ± 26, A: 488 ± 71 and S: 625 ± 67 Hz), median F (GR: $1,394\pm194$, A: $3,503\pm333$ and S: $4,451\pm424$ Hz), initial F (GR: 391 ± 24, A: 900 ± 105 and S: $1,281\pm184$ Hz), terminal F (GR: $5,578\pm387$, A: $7,935\pm197$ and S: $8,317\pm200$ Hz) and interpercentile F (GR: $5,187\pm375$, A: $7,035\pm173$ and S: $7,036\pm199$ Hz). In conclusion, bitten pigs seem to vocalise in 3 specific ways during an EB event. The acoustical features of the 3 vocalisations showed differences, hence they can potentially be used for an algorithm to distinguish between GR, A and S. Screaming is identified by literature as a pain-related call in piglets, hence it may be a reliable indicator to be used not only for the prediction of the occurrence of EB events and the further developing of automatic monitoring systems but also for the detection of the severity of such damaging behaviour.

Real time monitoring of small ruminants' drinking behavior and body weight; sensory system design

T. Glasser[1], N. Barchilon[2], V. Bloch[2], Y. Gal[2], Y. Godo[2], J. Grinshpun[2], Y. Lepar[2], H. Levit[2], L. Maor[2], E. Metuki[2], E. Ram[2], L. Rosenfeld[2], T. Schcolnik[2] and I. Halachmi[2]
[1]*Ramat Hanadiv Nature Park, P.O. Box 325, 30900 Zikhron Ya'akov, Israel,* [2]*Agricultural Research Organization (A.R.O.), PLF Lab., Institute of Agricultural Engineering, The Volcani Centre, Rishon LeZion, Israel; halachmi@volcani.agri.gov.il*

Drinking behavior and body-weight changes have a potential to indicate in the course of (1) nutrition alters, and (2) stress or illness events. The first requires animal group-level action; the second is animal-individual care. A new system comprises body-weight electronic weighing, combined with sensitive flow-meter sensors for animal drinking behavior and ultra high frequency RFID (radio frequency identification) was designated, built and validated for small-ruminants. The system was built at the Israeli precision livestock farming laboratory, ARO, the Volcani center, and installed at Ramat Hanadiv and Nordia commercial farms. One farm raises grazing dairy goat, the second farm intensively (zero grazing) grows meat-breed lambs. Daily average individual body weight (BW) in the new system was 64.8 kg, vs 65.2 in the manual gold reference. Individual values will be reported in the presentation. Average drinking per visit was 2,457 ml (std 1,195), differs, on average, between the new system and the reference was 89 ml (std 68). The presentation will report the individual values. In conclusion, a system for monitoring small ruminants' water intake and body weight was designed, built and validated. The system is ready for research and commercial applications. Applications will be presented at the EAAP 2017 conference.

Improving the efficiency of beef production: advances in technology

C.A. Duthie[1], G.A. Miller[1], J.J. Hyslop[1], D. Barclay[2], A. Edwards[3], W. Thomson[4] and D.W. Ross[5]
[1]*SRUC, Kings Buildings, West Mains Road, Edinburgh, UK, EH9 3JG, United Kingdom,* [2]*Innovent Technology Ltd, Markethill, Turriff, Aberdeenshire, UK, AB53 4PA, United Kingdom,* [3]*Ritchie Implements Ltd, Forfar, Angus, DD8 3BT, United Kingdom,* [4]*Harbro Ltd, Markethill, Turriff, Aberdeenshire, AB53 4PA, United Kingdom,* [5]*Agri-EPI Centre Ltd, Bush House, Midlothian, UK, EH26 0EB;, United Kingdom; Carol-anne.duthie@sruc.ac.uk*

The use of technology in the ruminant sector is gaining traction. A range of animal-mounted sensors, which can provide important behavioural classification (e.g. oestrus detection, onset of parturition, feeding behaviour, health monitoring) are now commercially available, and under continued development. Although the industry is beginning to embrace technology for making informed management decisions, there is still considerable inefficiency within the beef production sector. Large numbers of cattle are reaching abattoirs in a sub-optimal condition (both under- and over-finished) which has large economic and environmental implications. In order to overcome this inefficiency, automated monitoring of individual animal performance and carcass value is vital, whilst the animal is still live. By optimising cattle finishing times, farmers will be able to achieve maximum marketable yield and profit, combined with a reduction in production costs such as feeding and bedding, and improved efficiency of capital resources. Systems are currently under development, which use camera technology, originally derived from the gaming sector, to provide 3-dimensional topography of animals, and use this to estimate a number of parameters relating to performance and carcass yield. Underpinning these systems are auto-weighing platforms that integrate weighing at the water trough. These systems are now commercially available and offer instant growth information and predictions of carcass value at the individual animal level, to the farmer, via a computer or a phone application. Systems of this nature will also assist in identifying health and nutrition issues, which manifest through poor performance, allowing for quicker treatment or corrective management.

The effect of sample size on the accuracy of machine learning models in precision livestock farming

M. Pastell
Natural Resources Institute Finland (Luke), Koetilantie 5, 00790 Helsinki, Finland; matti.pastell@luke.fi

The sample size and and label noise both have an effect on the performance of machine learning models. Such models are frequently used in Precision Livestock Farming (PLF) applications to develop methods to e.g. predict the health status of an animal. The aim of this study was to evaluate the effect of sample size and label quality in the confidence of achieved model accuracy. Models were fitted to a simulated datasets with two classes. N samples (50%/class) were drawn from two multivariate gaussian distributions representing 2 features for 2 different classes, the means of the distributions were (1,1) an (2,2) with diag(0.5, 0.5) covariance matrix. Three classifiers (logistic regression (LR), linear support vector machine (SVM), SVM with radial basis kernel) were fitted to the training set. The accuracy of the classifiers was evaluated using independent validation set from same distribution. Simulation was repeated (n=1000) from 20 to 2,000 samples with label noise probabilities from 0 to 0.4 (code: https://github. com/mpastell/eaap2017). The classification accuracy (median±95% CI) obtained from simulations (linear SVM) with sample sizes 40, 100, 2,000 with no label noise were 82.5±12, 84±7 and 84.1±1.6% and with 20% label noise probability 70±12, 70±7 and 70±1.5%. Classifiers trained on very noisy data still achieved high classification accuracy when validated against a validation set of 20,000 samples with 100% correct labels e.g. a linear SVM achieved 83.7%±1.5% accuracy when trained on 2,000 sample data with 40% flipped labels. The results show that there can be a large difference between the results obtained from random sample as compared to the results on unseen samples from the same distribution with small and medium data and that the accuracy of labeling is crucial. It is often costly to collect large amounts of data to develop models for PLF applications. However, the effect of sample size should be taken into account in study designs and when considering the generalization ability of models developed on limited amount of input data. It is possible to fit accurate classifiers with noisy input data, but the performance needs to be validated against a good golden standard.

Integrating biomarkers into the electronic-sensors based PLF tool box

A. Shabtay
Agricultural Research Organization, Beef Cattle Section, Newe Yaar Research Center, 30095; Ramat Yishay, Israel; shabtay@volcani.agri.gov.il

Precision livestock production seeks to make use of large quantities of specific information about individual animals and locations to optimize performance. Precision livestock farming (PLF) utilizes continuous automated real-time monitoring to control production, reproduction, health and welfare of livestock. In addition, PLF can also be implemented to follow the effect of feeding regimes, such as integration of new feed stuffs or additives in the diet, on the individual's performance. Being non-invasive and of high throughput and in real-time capabilities, most of these monitoring systems exclusively involve the utilization of electronic sensing, with no additional input coming from the biochemical-physiological-genetic systems. Biological samples from various sources such as saliva, hair, feces and blood are easily attained and can be indicative of metabolic states and disorders. Moreover, they can predict the vulnerability to future diseases. Based on a series of studies in suckling Holstein calves, showing the association of bio-markers with production, environmental impacts and prediction of disease susceptibility, it is suggested that an integrated biological markers-electronic sensors approach could add valuable components to the PLF tool box, leading to successful management decision making on farm.

An android mobile application developed for record keeping in slaughter houses

Y. Bozkurt[1] and Ü.D. Uluşar[2]
[1]Suleyman Demirel University, Faculty of Agriculture, Department of Animal Science, Cünür, 32260, Isparta, Turkey,
[2]Akdeniz University, Faculty of Engineering, Department of Computer Science, Dumlupınar Blvd., 07058, Antalya,
Turkey; yalcinbozkurt@sdu.edu.tr

The goal of this study was to build an android mobile application to be used in smart phones for record keeping in slaughter houses for educational and research purposes, which can also be used to support and improve management ability of private and public sector in this field. The animal data such as ear tag number, the registered number of the farm, farm owner, breed, sex and age of weight of the slaughtered animal; including carcass weight can be stored and dressing percentage can be calculated. The mobile application was developed on the Java platform, using the Android Studio Software Development Kit (SDK) and Eclipse editors. Windows Communication Foundation (WCF) services were used to provide communication between the host software and the mobile application. It was divided into four components such as online data storage known as Google Drive, Android smartphone, RFID (Radio Frequency Identification) tags and reader. Each unique RFID tag can be linked to the database by either wireless local area networks (Wi-Fi) or WPAN (Wireless Personal Area Network, Bluetooth). If the RFID labeling system is available, the animal data is entered into the mobile device with the option 'Read RFID Tag' or 'Add Animal'. If 'RFID Read Label' is selected, the RFID reader on the weighing scale is activated, the information is read automatically from the RFID ear tagged of the live animal on the weighing scale. Android user interface (UI) act as an interaction platform which can transfer any information needed by a user from RFID system to the smartphone. Mobile phone will store the information gathered from RFID reader into the SQLite database, and further transfer the information into the Google Drive or host computer through the web providers. Therefore, this application will be very useful for improvement of managerial skills of the abattoirs and providing some contributions to the industry and will offer some help and guidance for further academic research.

Real time use of data for monitoring and evaluation and for adjusting settings

G.B.C. Backus
Connecting Agri and Food, Zilverpark 48, 5237 HN 's-Hertogenbosch, the Netherlands; g.backus@home.nl

Ever larger farms increase the need for tools providing overview and supporting control and transfer of tasks to employees. Web technology and data capture devices are now beginning to provide farm level solutions for production and management practices. Our objective of the study is to develop and evaluate an internet-based farm level management tool for real time use of data for monitoring and evaluation and for adjusting settings. We aim to enable pig farmers to relate real time sensor data to other standalone systems at the farm: water registration, feed computer, weighing system, and the farm management system. Sensor data capture inside and outside temperature, relative humidity, ammonia, and CO_2. Data are transferred wireless and battery operated every 30 minutes with the LoRa Alliance™ technology. A pig analytics dashboard is developed to be used on smart phones and tablets to provide a real time overview. The system gives an alert when data values are outside specified bandwidths. To improve the probability that farmers will make use of these new technologies, farm specific decision rules for the alert system are developed (and continuously being improved) to optimize between false positives and false negatives. The system is operating on several Dutch farms. Results indicate that farmers appreciate having real time overview, and do modify their management practices based on real time data.

Optimizing management of dairy goat farms through a platform based on individual animal data collect

A. Belanche[1], J. Fernandez-Álvarez[2], A.I. Martín-Garcia[1] and D.R. Yáñez-Ruiz[1]
[1]CSIC, Estación Experimental del Zaidín, Camino del Jueves s/n, Armilla, Granada, 18100, Spain, [2]CAPRIGRAN, Asociación Nacional de Criadores de Caprino de Raza Murciano-Granadina, Fuente Vaqueros, 18340, Spain; a.belanche@csic.es

The intensification process which is occurring in the dairy goat sector in Mediterranean countries aims to maximize productivity, efficiency and profitability. So far, this intensification process has mainly been based on the increase in the number of animals per farm and certain improvements in reproduction, health programs and milking practices. However, the implementation on precision farming practices could represent a step forward in the professionalization of this sector. Cabrandalucía Fedetation (Spain) is promoting a new concept of farming based on the use of the 'Eskardillo', a smartphone-based platform which relies on three main pillars: (1) systematic individual data collection from animals (milk yield and composition, genetic value, morphology, phylogeny, prolificacy), (2) big data processing and interpretation and (3) interactive feed back to the farmer to optimize decision making. In this study a number of farms belonging to the Murciano-Granadina goat breeding Association (Caprigran), which implemented the Eskardillo in 2014, were monitored in terms of farm performance. Changes on the genetic value, first kidding age, milk yield and dry period length were monitored over a 2 years period to determine the effectiveness of this platform. Preliminary results showed that Eskardillo allowed maximizing the genetic progress (4.5 times) by implementing a well-defined selection program. As a result, in 2016 farms using this platform increased milk production from 508 to 527 kg milk/lactation. Additionally, this platform implied a holistic and data-driven management, which helped to optimize the first kidding age, dry period length and culling strategy to minimize 'invisible loses' derived from unproductive periods. It could also provide additional benefits derived from the integration and interpretation of social, environmental and economic data able to asses farm sustainability. Thus, a more exhaustive study is required to fully evaluate the impact of Eskardillo platform on animal performance and farm profitability over the years to come.

Relationship between image analysis traits and consumer palatability score of Tokachi-wakaushi®

J. Hamanaka[1], T. Yoshikuni[2], R. Asa[1], K. Hagiya[1], T. Goto[1] and K. Kuchida[1]
[1]Obihiro University of Agriculture and Veterinary Medichine, 11, 2-sen, Inada-cho nishi, Obihiro, Hokkaido, 080-8555, Japan, [2]Tokachishimizu Foodservice Co.,Ltd, 419-79, Aza-shimizu, Shimizu, Kamikawa, Hokkaido, 089-0103, Japan; s25204@st.obihiro.ac.jp

Tokachi-wakaushi is young fatting cattle of Holstein in Shimizu, Hokkaido, Japan, and is produced from only 6 farms. They are branded cattle that has smooth, tender and lean meat, because they were slaughtered at young age (12 to 16 months). In this study, we investigated relationship between image analysis traits and consumer palatability of Tokachi-wakaushi. The carcass grading trait and image analysis traits were collected from 4,937 carcasses of Tokachi-wakaushi that slaughtered from April, 2015 to November, 2016. The samples for consumer sensory evaluation were 15 g of rib-eye steaks cut by 1 cm thickness at the processing plant of meat. Samples were evaluated by more than 3 panels. They scored taste, flavour, fat flavour, tenderness and juiciness at 3 points scale (1-3-5). We used averages of each sensory trait as individual scores. Also, image analysis traits including rib-eye area and marbling % of rib-eye were calculated by using Beef-Analyzer-G software. The iBCS, that can judge meat colour in detail, was calculated. Analysis of variance was performed using sensory traits as dependent variable, and production history, carcass grading and image analysis traits as fixed effect. Averages and standard deviations of carcass weight, marbling % and iBCS were 336.0±22.2 kg, 16.8±5.2% and, 3.4±0.7, respectively. Farm, slaughter month, aging period and marbling % had significant effects on all sensory traits ($P<0.05$). Scores of all sensory traits increased as increasing marbling % ($P<0.05$). Individuals slaughtered in September and October showed a tendency to increase scores of all sensory traits. Scores of fat flavour, tenderness and juiciness increased as advancing aging period. The iBCS had significance effects on fat flavour and juiciness ($P<0.01$), and their scores decreased if meat colour is comparatively dark. Interaction of farm and slaughter month had significance effects on flavour and fat flavour ($P<0.01$).

Rapid assessment of pork solid fat content using hyperspectral imaging

C. Kucha[1], M. Ngadi[1], C. Gariepy[2] and L. Maignel[3]

[1]McGill University, Bioresource Engineering, 21111 Lakeshore Road, Ste-Anne-de-Bellevue, QC, Canada, [2]Agriculture & Agri-Food Canada, 3600 Casavant Boulevard West, Saint-Hyacinthe, QC, Canada, [3]Canadian Centre for Swine Improvement, Central Experimental Farm, Building #75, 960 Carling avenue, Ottawa, ON, K1A 0C6, Canada; michael.ngadi@mcgill.ca

Solid fat content (SFC) is a measure of the solid-liquid ratio in fats under reference conditions of temperature. As a quality parameter, it impacts processing, textural, and sensory characteristics of pork. Conventional measurement techniques are destructive and time-consuming methods. The aim of this study was to developed a rapid and non-destructive technique based on Near Infrared (NIR) hyperspectral imaging for assessing SFC of pork fat. Samples of fats were excised from primal cuts namely ham, loin, belly and shoulder from 40 different pork carcasses to obtain a total of 160 samples (four samples from each carcass). NIR hyperspectral images (900-1,700 nm) were acquired from the samples. The images were processed and the mean spectra data were extracted and pre-processed using different techniques such as standard normal variate (SNV), multiplicative scatter correction (MSC) first derivative (FD), and second derivatives (SD). The reference SFC values (%) at 0, 5, 10, 15, 20, 25, 30, 35, and 40 °C were determined by using a DSC. The mean spectra data were correlated with the reference SFC values by Partial least square regression (PLSR), or least square-support vector machine (LS-SVM) algorithms for establishing calibration models. To build simpler models for developing a multispectral spectrometer, Successive projection algorithm (SPA), and regression coefficients (RC) were used to select important wavelengths with the high feature information. Independent prediction sample sets were used to validate the accuracy of the calibration models, and the coefficients of determination obtained in the prediction sets showed a good performance for predicting SFC in pork fat at the temperature range using both the full and selected wavelengths. The result indicates that NIR hyperspectral imaging technique could be a rapid and non-destructive alternative tool in the pork industry for real-time assessment of the SFC in pork fat.

Individual water recording on growing pigs

F. Fortin[1], L. Maignel[2], P. Gagnon[1], J.G. Turgeon[1] and B. Sullivan[2]

[1]Centre de développement du porc du Québec Inc, Place de la Cité, tour Belle Cour, 2590, boul. Laurier, bureau 450, Québec, QC, G1V 4M6, Canada, [2]Canadian Centre for Swine Improvement Inc, Central Experimental Farm, Building 75, 960 Carling Avenue, Ottawa, ON, K1A0C6, Canada; laurence@ccsi.ca

Water is an essential nutrient with many physiological functions, but little information exists about individual water intake in pigs. However, a better knowledge of individual water consumption can lead to many recommendations or applications to the swine sector. For example, water intake monitoring could lead to better performance through early detection of health problems, and could also be used as a predictor of behaviour. To obtain data on this new trait, the Centre de développement du porc du Québec Inc. (CDPQ) equipped the Deschambault swine testing station with a new system capable of measuring individual water intake. Now, every individual visit to the water dispenser and amount of water consumed is recorded for each pig throughout the entire grow-finishing period. The system developed by the CDPQ minimizes water waste and simultaneously considers the flow and quantity of water going through the nipple and out of the bowl. A total of 700 commercial pigs were monitored during two consecutive trials in 2014-2015, from approximately 25 to 130 kg live weight. From feeding phases 1 (25-50 kg) to 3 (75-100 kg), the average amount of water consumed per pig increased from 5 to 10 l/day, average time spent in the drinker increased from 11 to 14 min/day and average daily number of visits in the drinker decreased from 36 to 31. The variability of water intake at the pen level was mainly explained by sex, season and phase. However, variability of water intake at the individual level remained high even if these explanatory factors were accounted for. Using data available at the pen level and at the individual levels, daily water intake was analyzed as a potential predictor of feed intake, growth rate, feed conversion, carcass characteristics, meat quality and health problems.

SheepNet and iSAGE networking to improve efficiency in sheep and goat systems

J.M. Gautier[1] and G.I. Arsenos[2]
[1]*Institut de l'Elevage, BP 42118, 31321 Castanet Tolosan Cedex, France,* [2]*Aristotle University of Thessaloniki, Faculty of Veterinary Medicine, University campus, 54124 Thessaloniki, Greece; arsenosg@vet.auth.gr*

iSAGE and SheepNet are European projects aiming to make sheep and goat sectors in Europe competitive through innovation and communication. We want to know your experience, opinions and ideas on how to use innovation and communication better. Therefore, iSAGE and SheepNet have a challenge session at EAAP 2017 to briefly tell you how we aim to make sheep and goat sectors competitive and sustainable. We will also have a speaker from New Zealand describing how innovation and communication has been successfully used there. Then we will ask you for feedback: (1) what are your experiences of successful innovation or communication networks? (2) what are the barriers for innovation and communication (3) what you think the priorities are for Europe and how we can learn and share knowledge from other countries (4) How you could contribute to iSAGE and SheepNet innovation seeking and transfer? At the end of the challenge session we aim to have a plan of how we can improve innovation and communication and include you in the process. iSAGE: Innovation for Sustainable Sheep and Goat Production in Europe is finding solutions to socio-economic, demographic, ecological and market challenges. The iSAGE consortium has a strong industry representation with 16 research and 18 industry partners from diverse EU sheep and goat industries. This industry involvement ensures the solutions are practical to the diverse EU demands for productivity, sustainability and societal values. Therefore, we will accelerate the transfer of knowledge and adoption of research into practical applications to enhance the sustainability, competitiveness and resilience of the European Sheep and Goat sectors. SheepNet: a European wide network with partners from the six main EU sheep producing countries and Turkey. SheepNet includes stakeholders and sheep producers to develop practice-driven innovation to improve sheep productivity through efficient reproduction, efficient gestation and reduced lamb mortality. The SheepNet project will establish durable exchange of scientific and practical knowledge among researchers, farmers and advisors/consultants across Europe and overseas.

Increasing computational efficiency in ssGTBLUP by eigendecomposition

E.A. Mäntysaari[1], R. Evans[2], T.H.E. Meuwissen[3] and I. Strandén[1]
[1]*Natural Resources Institute Finland (Luke), Myllytie 1, 31600 Jokioinen, Finland,* [2]*Irish Cattle Breeding Federation, Highfield House, Newcestown Road, Bandon, Cork, Ireland,* [3]*Norwegian University of Life Sciences, Dept. Animal and Aquacultural Sciences, 1432 Ås, Norway; esa.mantysaari@luke.fi*

An equivalent computational approach for the original single-step GBLUP (ssGBLUP) called ssGTBLUP assumes that the genomic relationship matrix has form $G=ZZ'+C$. Here, Z is the (centered and scaled) marker matrix with size $n \times m$ (numbers of genotypes and markers), and C can be easily inverted. The inverse can be written as $G^{-1}=C^{-1} - TT'$ where T is an n by m matrix. In the preconditioned conjugate gradient (PCG) method, a matrix vector product $G^{-1}d$ needs to be computed. In ssGBLUP, this requires n^2 multiplications, but in ssGTBLUP, the product TT'd has 2 nm multiplications and $C^{-1}d$ has cn multiplications with c constant independent of n or m (in simplest case C is diagonal and c=1). With the aid of eigendecomposition of $Z'C^{-1}Z$ we can reduce the number of columns in the T matrix. We call this approximate approach ssGTBLUP(p) where p is the percentage of total variance explained by accepted eigenvalues. Although T for the approximate approach requires extra computations compared to ssGTBLUP, amount of these computations increase linearly in n. We investigated the performance of ssGBLUP, ssGTBLUP, ssGTBLUP(p) and the APY method using 50,000 (APY50) or 30,000 (APY30) animals in the core. We used the Irish beef and dairy cattle carcass evaluation which has a heterogeneous multibreed population. The pedigree had 13.3 million animals of which n=163,277 were genotyped. The external file read by the solver was 50 GB for ssGBLUP, 34 GB for ssGTBLUP, 23 GB for ssGTBLUP(99), 20 GB for ssGTBLUP(98), 14 GB for ssGTBLUP(94), 26 GB for APY50, and 17 GB for APY30. Correlations of breeding values for genotyped animals between ssGBLUP and ssGTBLUP(p) ranged from 0.999-1.000 for p=99, 0.998-1.000 for p=98, and 0.990-0.998 for p=94 but were 0.994-1.000 for APY50 and 0.969-0.997 for APY30. Computing times per iteration were 5.62 min for ssGBLUP, 4.31 min for ssGTBLUP, 2.94 min for ssGTBLUP(99), 2.74 min for ssGTBLUP(98), 1.87 min for ssGTBLUP(94), 2.27 min for APY50, and 1.29 min for APY30. Eigendecomposition in ssGTBLUP allowed a well-defined approach to approximate ssGBLUP and speed up computations.

Efficient single-step computations using equivalent ssGTBLUP approach

I. Strandén[1], R. Evans[2] and E.A. Mäntysaari[1]
[1]Natural Resources Institute Finland (Luke), Myllytie 1, 31600 Jokioinen, Finland, [2]Irish Cattle Breeding Federation, Highfield House, Newcestown Road, Bandon, Cork, Ireland; ismo.stranden@luke.fi

Single-step GBLUP genomic evaluation (ssGBLUP) offers a simple approach for including genomic information in genetic evaluations. The genomic information is accounted by the genomic relationship matrix G, and its inverse G^{-1} is included in the mixed model equations (MME). Iterative solving time of the MME increases quadratically in number of genotyped animals (n) when n is large. An equivalent alternative model ssGTBLUP assumes G to have form $G=ZZ'+C$ where Z is the (centered and scaled) marker matrix with size n × m (numbers of genotypes and markers), and C can be easily inverted. One choice for $C=wA_{22}$, where A_{22} is the relationship matrix of genotyped animals, and w is a suitable constant. Then Z includes scaling by square root of (1-w) as well. After applying matrix algebra, $G^{-1}= C^{-1} - TT'$ where T is an n by m matrix. In matrix $T'=L^{-1}Z'C^{-1}$, L is the Cholesky factor of $(Z'C^{-1}Z+I)$. With normal choices of C, solving time in ssGTBLUP increases linearly with increasing number of genotyped animals n. Moreover, the algorithm is very suitable for parallel computing. We considered $C=\varepsilon I$, i.e. $G=(ZZ'+\varepsilon I)$, where $\varepsilon=10-3$. We investigated performance of ssGBLUP and ssGTBLUP on a heterogeneous multibreed evaluation for the Irish beef and dairy cattle carcass trait evaluation. The pedigree had 13.3 million animals of which n=163,277 were genotyped. According to the results, the external file read by the solver was reduced from 50 GB for ssGBLUP to 34 GB for ssGTBLUP. Preprocessing time reduced from 12.3 to 7.7 h when 10 processors were used in both cases. The solver computing time per iteration reduced from 5.62 to 4.31 min. Number of iterations and final solutions were equal by both approaches. In conclusion, ssGTBLUP allows solving single-step MME efficiently when the number of genotyped animals is large.

Large-scale genomic prediction using singular value decomposition of the genotype matrix

J. Ødegård[1], U. Indahl[2] and T.H.E. Meuwissen[3]
[1]AquaGen AS, P.O. Box 1240, 7462 Trondheim, Norway, [2]Norwegian University of Life Sciences, Faculty of Science and Technology, P.O. Box 5003, NMBU, 1432 Aas, Norway, [3]Norwegian University of Life Sciences, Faculty of Biosciences, P.O. Box 5003, NMBU, 1432 Aas, Norway; jorgen.odegard@aquagen.no

Computational costs of genomic prediction increases with number of loci (marker effect models) and with number of genotyped individuals (genomic animal models). In the latter case, the inverse genomic relationship matrix (GRM) is typically needed, for which computing costs increases exponentially with number of individuals. There is therefore a great need for reduced-dimensionality models, capable of analyzing extremely dense genomic data (up to full sequence) from large numbers of individuals. In this study, we propose reduced-dimensionality genomic models utilizing singular value decomposition (SVD). Large populations of limited Ne have also a limited number of effective loci underlying the genetic variation, implying that relatively few principal components (PC), obtained through SVD of genomic data, can capture nearly all genetic variation. SVD can be performed by analyzing different chromosomes/genome segments in parallel and/or by restricting SVD to a smaller core sample, and use the chosen components for dimensionality reduction of the entire genomic data set. After SVD, fast genomic prediction can be performed by PC ridge regression (PCRR), resulting in a diagonal reduced-dimensionality equation system. Alternatively, a (single-step) genomic animal model with a PC-based GRM can be used, for which computation of inverse GRM is extremely undemanding. If needed, the inverse GRM can be obtained row-by-row for any number of animals, implying that the inverse GRM does not need to be stored explicitly, e.g. when using iteration on data. The SVD-based analyses were validated using simulated data. It was shown that PC-based prediction, utilizing chromosome-wise SVD of a limited core sample, is appropriate for genomic prediction in a larger population. The PCRR model and the animal model using a PC-based inverse GRM were equivalent and thus resulted in identical breeding values, which in turn were virtually identical to the predicted breeding values from the original full-dimension marker- or genomic animal models (r=1.000).

Assessment of phasing quality in highly related layer lines using simulated genomic data

N. Frioni[1], H. Simianer[1] and M. Erbe[2]
[1]University of Goettingen, Department of Animal Sciences, Albrecht-Thaer-Weg 3, 37075 Goettingen, Germany,
[2]Bavarian State Research Centre for Agriculture, Institute for Animal Breeding, Vöttinger Straße 38, 85354, Germany;
nicolas.frioni-garcia@uni-goettingen.de

Length and distribution of haplotype blocks and the number of independent chromosome segments are key parameters to assess genetic diversity and to evaluate the expected accuracy in genomic selection programs. With real single nucleotide polymorphism (SNP) data from genotyping arrays haplotype information is not available. Thus, haplotype phases must be reconstructed in-silico, but the quality of this process is not easy to measure as true haplotypes are unknown. The aim of this study was to assess haplotyping quality of two phasing programs using simulated genomic data reflecting a highly related layer line. The basis of this study was pedigree and real SNP data for chromosome 1 of 888 commercial brown layers with 65,177 SNPs after filtering. Derived haplotypes of these genotypes were used to build a library of haplotypes from which haplotypes for founders of the existing pedigree were drawn randomly. These haplotypes were dropped through the pedigree assuming no mutation and random crossing-over events. Finally, all individuals had simulated genotypes for which haplotype phases were known. The simulation process was repeated ten times. Simulated genotypes were processed with FImpute 2.2 and Beagle 3.3 for each replicate. The heterozygous loci (25,403±663) of the predicted phases were then compared with the true simulated haplotypes. The percentage phased correctly (%pc), the number of breakpoints between correct and incorrect phased parts of the chromosome (#bp) and the number of SNPs per incorrect phased part (ipp) were averaged over individuals within replicate and then over replicates. The median of %pc was 77.5 (75.4 to 80.1) for Beagle and 99.9 (99.9 to 99.9) for FImpute while #bp was 3.2±0.42 for Beagle and 3.1±0.99 for FImpute. The median of ipp was 134 (min=89, max=222) and 2,398 (min=2,138, max=2,631) for FImpute and Beagle, respectively. The results show that FImpute recovers largely the correct haplotype interspersed with few short inverted segments, while Beagle splits the chromosomes into several longer strands of consistently phased haplotypes.

Covariance between breeding values in the ancestral regression Gaussian model

R.J.C. Cantet[1,2], C.A. García-Baccino[2] and S. Munilla[2]
[1]CONICET, INPA, Av. San Martín 4453, 1417 Buenos Aires, Argentina, [2]Universidad de Buenos Aires, Facultad de Agronomía, Producción Animal, Av. San Martín 4453, 1417 Buenos Aires, Argentina; rcantet@agro.uba.ar

We have recently introduced the ancestral regression (AR) as a conditional model that generalizes the parental regression in the animal model. The breeding value (BV) of an animal in AR is regressed to the average of parental BV plus to a linear function of the BV of all four grandparents. This function is orthogonal to parental BV in the absence of inbreeding. Hence, AR reduces Mendelian residual variance and increases accuracy of prediction. The AR model includes two linear parameters per animal evaluated, which are path coefficients expressing the relative difference of contributions of grandsires over granddams (or viceversa) in the genome and BV of the individual. Segmental inheritance (non-independent genetic effects) is accounted for if probabilities of IBD at two loci are used to estimate the linear parameters. Whole genomic relationships between BV are then functions of these parameters and are also the elements of the covariance structure Σ. Asymptotic normality is employed to obtain the distribution of BV, and the resulting causal distribution is Gaussian and Markovian. This latter property allows inverting Σ as it is done with the additive relationship matrix because, conditionally on the BV of grandparents and parents, the BV of any animal is independent of anything else. Another plus of the AR is that covariances between BV are not limited to independent genetic effects but also to linked genes at the same gamete, and this allows including non-additive genetic effects that are transmitted as a part of the BV. Our goal here is to introduce the general expression for the covariance between relatives in the AR, and to present some important covariances such as parent-offspring, grandparent grand-offspring, full and half sibs. Whereas in the general case 36 elements of Σ are involved when calculating any covariance, the formula reduces to simple expressions in the absence of inbreeding because many of the elements are zero. When there is no genomic information available to identify the linear parameters, all of them are zero and covariances between relatives turn into the classical expressions.

Single-step genomic evaluation for uniformity of growth in Atlantic salmon (*Salmo salar*)

P. Sae-Lim[1], A. Kause[2], M. Lillehammer[1] and H.A. Mulder[3]
[1]Nofima, Breeding and Genetics, Osloveien 1, 1433 Ås, Norway, [2]Natural Resources Institute Finland, Biometrical Genetics, Jokioinen, 31600, Finland, [3]Wageningen University & Research, Animal Breeding and Genomics, P.O. Box 338, 6700 AH, Wageningen, the Netherlands; han.mulder@wur.nl

The heritability for uniformity of body weight is low, indicating that also accuracy of estimated breeding values (EBV) can be low. The use of genomic information could be one way to increase the accuracy and, hence, obtain higher response to selection. Genomic information can be merged with pedigree information to construct a combined relationship matrix (H matrix) for a single-step genomic evaluation (ssGBLUP), allowing realized relationships of the genotyped animals to be exploited, in addition to the numerator pedigree relationships for ungenotyped animals (A matrix). We compared the predictive ability of EBV for uniformity of body weight in Atlantic salmon, when implementing either A or H matrix in the genetic evaluation. We used double hierarchical generalized linear models based on an animal model (animal DHGLM) for both body weight and its uniformity. With the animal DHGLM, the use of H instead of A significantly increased the correlation between the predicted EBV and adjusted phenotypes, which is a measure of predictive ability, for both body weight and its uniformity (41.1 to 78.1%). When log-transformed body weights were used to account for a scale effect, the use of H instead of A produced a small and non-significant increase (1.3 to 13.9%) in predictive ability. The use of H significantly increased the predictive ability of EBV for uniformity when using the animal DHGLM for untransformed body weight. When using log-transformed body weights, the increase in predictive ability was only minor likely due to the lower heritability for uniformity of transformed body weight, a lower genetic correlation between transformed body weights and their uniformities. In conclusion, the use of ssGBLUP increases the accuracy of breeding values for uniformity of harvest weight and therefore is expected to increase response to selection in uniformity.

Starting breeding programs for mass spawning fish species using genomic selection

J.W.M. Bastiaansen, K. Janssen and H. Komen
Wageningen University & Research, Animal Breeding and Genomics, P.O. Box 338, 6700 AH Wageningen, the Netherlands; john.bastiaansen@wur.nl

Breeding programs for aquaculture species often rely on mass spawning in broodstock groups without the ability to control matings and with variable parental contributions to selection candidates. Selection within a broodstock can be based on 1) own phenotypes or, when candidates are genotyped, on 2) BLUP breeding values (EBV) with pedigree reconstruction. Genotyping also allows estimation of 3) genomic BLUP breeding values (GEBV) with the benefit to use information across broodstocks in the first generation. The three selection criteria were compared for response to selection (ΔG) and rate of inbreeding (ΔF) with mass spawning species. A breeding program was simulated with 6 broodstocks, each with 20 males and 20 females, randomly drawn from a base population with Ne of 200. Relative contributions of parents in a broodstock to a group of 400 selection candidates were based on a Gamma(0.75, 0.33) distribution, while limiting the number of matings to 40. Selection was for a trait controlled by 120 QTL with effects sampled from the exponential distribution, resulting in a normal distributed phenotype with a heritability of 0.35. Genotypes for GBLUP were used at a density of 0.5 N_e/Morgan. Average response to this initial round of selection was 0.96, 1.18, and 1.30 σ_G for phenotypic, EBV, or GEBV selection. Predicted ΔF (mean kinship of selected candidates of all broodstocks) was 0.87, 1.66, and 1.46, respectively. Variable parent contributions had limited effect on ΔG with a ≤5% reduction compared to equal contributions, but ΔF was doubled. GEBV estimated within 1 broodstock had accuracies of 0.81. When adding data from earlier broodstocks, accuracies increased only slightly to 0.82 after adding 1 broodstock and to 0.85 after adding all 5. Without phenotypes of selection candidates (selection of juveniles), accuracies ranged from 0.34 to 0.63 with reference populations of 400 to 2000 fish. GBLUP in a starting breeding program showed benefits in ΔG and ΔF compared to BLUP, but only in ΔG compared to phenotypic selection. Adding information from progeny of other broodstocks had little effect on accuracy. Controlling ΔF versus ΔG as well as their balance against genotyping cost needs to be optimized.

Massive sequencing reveals strong purifying selection in commercial chicken

S. Qanbari[1], C.J. Rubin[2], S. Weigend[3], A. Weigend[3], R. Fries[4], R. Preisinger[5], H. Simianer[1] and L. Andersson[2]
[1]Animal Breeding and Genetics Group, Department of Animal Sciences, University of Goettingen, 37075 Göttingen, Germany, [2]Department of Medical Biochemistry and Microbiology, Uppsala University, 75123 Uppsala, Sweden, [3]Friedrich Loeffler Institute, 31535, Neustadt, Germany, [4]Technical University of Munich, 85354, Freising-Weihenstephan, Germany, [5]Lohmann Tierzucht GmbH, 27472, Cuxhaven, Germany; sqanbar@gwdg.de

We carried out whole genome sequencing of 125 chicken including Red jungle fowl and multiple populations of broilers and layers developed in parallel to perform a systematic screening of adaptive changes in modern chicken (*Gallus gallus domesticus*). We uncovered >20 million high quality SNPs of which 31% are newly described variants. This panel comprises >78K amino-acid altering substitutions as well as 667 SNPs predicted to be stop-gain and -loss, several of which reaching higher frequencies. Signatures of selection were investigated through both analyses of fixation and differentiation to reveal selective sweeps of candidate mutations that have had a prominent role during domestication and breed development. We confirmed selection at the *BCO2* and *TSHR* loci and identified 52 putative sweeps co-localized with *KITLG, EPGR, IGF1, DLK1, GLI3, ZFRL1, MOXD1, HN1L, CRAMP1* and *ZSWIM6L* genes, among others. SNPs with marked allele frequency differences between wild and domestic chicken demonstrate a highly significant deficiency in the proportion of amino-acid altering mutations ($P<2.59^{-09}$). This implies that commercial birds have undergone rigorous purifying selection, and likely reflects human involvement in purging genetic variants contributing to phenotypic diversity during the establishment of commercial breeds. Enrichment analysis between groups of broilers vs layers further revealed highly differentiated missense variants localized in a panel of genes including *LEPR, MEGF10, CAPN2* and *FOXE1* suggestive of production-oriented selection in commercial chicken. These results are expected to contribute to the understanding major genetic switches that took place during the evolution of modern chickens and in poultry breeding.

Genome-wide uncorrelated trait analysis identify major pleiotropic markers for dairy cattle

R. Xiang[1,2,3], I.M. Macleod[1,2], S. Bolormaa[1,4] and M.E. Goddard[1,2,3]
[1]AgriBio Research Centre, Dept. Economic Development, Jobs, Transport & Resources, Bundoora, Victoria 3083, Australia, [2]AgriBio Research Centre, Dairy Futures Cooperative Research Centre, Bundoora, Victoria 3083, Australia, [3]University of Melbourne, Faculty of Veterinary & Agricultural Science, Melbourne, Victoria 3010, Australia, [4]Cooperative Research Centre for Sheep Industry Innovation, Armidale, NSW 2351, Australia; ruidong.xiang@unimelb.edu.au

Selection should favour alleles which increase profitability considering their effects across all important traits. Therefore, understanding pleiotropy is an important aim. Obviously if traits are genetically correlated they must share some causal variants but it is possible that even uncorrelated traits share some causal variants. Here we analyse 25 traits on Australian dairy cattle. The 25 raw traits (RTs), covering milk production, fertility, behaviour, mastitis and conformation, of 2,841 bulls were used to calculate uncorrelated principal components (PCs) and Cholesky transformation traits (CT). Multi-trait meta-analyses of single-trait genome-wide association studies (GWAS) for RT, PC and CT in these bulls were validated in 6,821 cows. We observed a positive relationship between heritability estimates and the number significant SNPs detected in RTs and CTs. However, there was no relationship between the phenotypic importance of PCs and the number of significant SNPs detected. The major dairy cattle locus DGAT1 not only affected dairy production traits, also had validated small effects on fertility, milk speed and temperament. Our results highlight the importance of using genetic information of all traits to maximise pleiotropy detection and prioritise multi-trait genetic markers for the dairy industry.

Validation of genomic and genetic evaluations: Nordic Red Dairy cattle 305 d production traits

M. Koivula[1], I. Strandén[1], G.P. Aamand[2] and E.A. Mäntysaari[1]
[1]Natural Resources Institute Finland (Luke), Myllytie 1, 31600, Finland, [2]NAV Nordic Cattle Genetic Evaluation, Agro Food Park 15, 8200 Aarhus N, Denmark; minna.koivula@luke.fi

As genomic selection has been used already for several years it has become evident that the validation of genomic evaluations relying on traditional animal models is becoming unsuitable. The GEBV validation test recommended by INTERBULL is cross-validation based on the forward prediction. It was designed at the time when the multi-step genomic evaluation was the standard method. The GEBV validation test is generally poorly suited for testing genomic animal models (single-step GBLUP). The aim of this study was to take a closer look on accuracy and stability of (G)EBVs. The validations for GEBVs were done using yield deviations (YD) or daughter yield deviations (DYD) calculated with single-step GBLUP instead of EBV model. Moreover, we studied the stability of (G)EBV estimations in consecutive evaluations. We used Nordic RDC 305d production data containing ca. 4 million cows with 8.8 million observations. The pedigree included 5.4 million animals. Genotypes were for 33,321 animals which had either records or offspring in the full 305d data. The test setup consisted of four data sets: the newest data, called data0, included calvings up to March 2016. Three reduced data sets were data-1, data-2, and data-3, from which one year of calvings was deleted at a time. Thus, e.g. data-3 included calvings up to March 2013. The data structure allowed studying the accuracy of predictions by production years, and also the stability of (G)EBV estimates across lactations. The bull validation was regression of DYD_{data0} on PAdata-3 or, for GEBV, regression of $ssDYD_{data0}$ on $GEBV_{data-3}$. The results indicated that after use of genomic selection the DYD from EBV model become biased and that GEBVs can be validated using DYDs from the ssGBLUP model. The validation reliability for protein GEBV (r^2) was 0.28 using DYD from EBV model and 0.36 using DYD from ssGBLUP. Similarly, when making cow validations, it would be better to use YDs calculated from ssGBLUP for validation of GEBVs. The r^2 in GEBV validations using YD from ssGBLUP were on average 14% units higher compared to validations using YDs from the EBV model.

Unraveling the genetic background of clinical mastitis in dairy cattle using whole genome sequence

J. Szyda[1], M. Mielczarek[1], M. Frąszczak[1], G. Minozzi[2], R. Giannico[3], J.L. Williams[4] and K. Wojdak-Maksymiec[5]
[1]Wroclaw University of Environmental and Life Sciences, Kozuchowska 7, 51-631 Wroclaw, Poland, [2]University of Milan, Celoria 10, 20133 Milan, Italy, [3]Fondazione Parco Tecnologico Padano, Einstein Albert, 26900 Lodi, Italy, [4]University of Adelaide, Adelaide, South Australia 5005, Australia, [5]West Pomeranian University of Technology, Piastów 17, 70-310 Szczecin, Poland; magda.a.mielczarek@gmail.com

Mastitis is an inflammatory disease of the mammary gland which has recently become one of the most important diseases of the dairy sector, mainly due to high economic importance and increased awareness of animal welfare. The purpose of this work was to characterize links between single base pair (SNP) and structural polymorphisms (CNVs) in whole genomes DNA sequence of thirty-two cows of the Holstein-Friesian breed and the incidence of clinical mastitis. The cows were selected from a dataset of 991 individuals based on a dedicated experimental design involving 16 paternal half-sibs with both sibs discordant for their susceptibility to clinical mastitis. Mastitis resistant cows had no incidence of clinical mastitis through their production life while mastitis prone cows undergone multiple clinical mastitis cases. Among 69 most significant SNPs ($P<1\times10^{-14}$) we identified five mutations located within exons, two of which resulted in an amino-acid change. One of those mutations was located within PPP1R3G gene coding for protein phosphatase 1 regulatory subunit 3G. The other SNP was located within HMX1 gene coding for H6 family homeobox 1 protein. In eight half-sib families the CNV of 2,500 bp length, located on BTA21 from 2,442,801 bp to 2,445,300 bp, was deleted in a mastitis prone sib, while it was present its healthy half-sister. RefSeq archive pointed out at a transcript homologous to the gene encoding for human ubiquitin protein ligase E3A (UBE3A). For six half-sib families a 1,300 bp CNV region on BTA18 was deleted in a sick half-sib and present in a healthy half-sib. It overlaps with GLG1 gene coding for Golgi glycoprotein 1. Both genes are part of immune response system as defined by Reactome and KEGG pathways.

The role of genotypes from animals without phenotypes in single-step evaluations

E.C.G. Pimentel, T. Shabalina, C. Edel, L. Plieschke, R. Emmerling and K.-U. Götz
Bavarian State Research Center for Agriculture, Institute of Animal Breeding, Prof.-Dürrwaechter-Platz 1, 85586 Poing-Grub, Germany; eduardo.pimentel@lfl.bayern.de

In a two-step genomic evaluation system, genotypes of animals without phenotypes do not take part in the training step and therefore do not influence genomic predictions of other animals. Once such animals are culled, their genotypes can be discarded in further evaluation runs. That might not be the case in single-step genomic evaluation systems. In this study we investigated the effects of including genotypes from culled bulls on the reliability of genomic predictions for other animals in single-step evaluations. Four scenarios with constant amount of phenotypic information and increasing numbers of genotypes from culled bulls were simulated and compared with respect to prediction reliability. With increasing numbers of genotyped culled bulls there was a corresponding increase in prediction reliability for other animals in the system. For instance, the reliability for selection candidates was twice as large when all culled bulls from the last four generations were included in the analysis. Single-step evaluations imply the imputation of all non-genotyped animals in the pedigree. We showed that this imputation was increasingly more accurate as an increasing amount of genotypic information from the culled bulls was taken into account. This resulted in higher prediction reliabilities. The extent of the benefit from including genotypes from culled bulls may vary from one genomic evaluation system to another, depending on the amount of genomic information already available. Improvements in prediction quality might be more relevant for small populations with low levels of reliabilities.

xbreed: an R package for genomic simulation of purebreds and crossbreds

H. Esfandyari and A.C. Sørensen
Center for Quantitative Genetics and Genomics, Department of Molecular Biology and Genetics, Aarhus university, Blichers alle 20, 8830 Tjele, Denmark; hadi.esfandyari@mbg.au.dk

Simulation has and will continue to play an important role in the study of genomic selection. xbreed is a simulation tool for purebred and crossbred genomic data as well as pedigree and phenotypes. It can be used for simulation of population with flexible genome structures and trait genetic architectures. xbreed can also be used to evaluate breeding schemes and generate genetic data to test statistical tools. Data simulated by xbreed is a good mimic to the real data in terms of genome structure and trait underlying genetic architecture. Furthermore, the package is handy, intuitive and has good performance properties in execution time and memory usage as main subroutines have been written in Fortran. In conclusion, xbreed would be a useful tool for the methodological and theoretical studies in the population and quantitative genetics and breeding.

Animal and single step genomic BLUP models in Romanian simmental cattle genetic evaluation

M.C. Rotar[1], D.E. Ilie[2], R.I. Neamt[2], F.C. Neciu[2], G. Saplacan[2], A.M. Gras[1] and H. Grosu[1]

[1]National Research Development Institute for Animal Biology and Nutrition, Calea Bucuresti No.1, Balotesti, 077015, Ilfov, Romania, [2]Research and Development Station for Bovine, Calea Bodrogului 32, 310059, Arad, Romania; danailie@animalsci-tm.ro

The aim of the current comparative study was to estimate the breeding values (EBV) of cattle using two mixed models: individual animal model (IAM) and single step genomic BLUP (ssGBLUP). Breeding value estimation can be done using phenotypical, genealogical and genomic information sources. Information quantity and the model used in estimation determine the precision and error effect across the model. The study was conducted on 131 Romanian Simmental cattle breed reared at Research and Development Station for Bovine from Arad. From the total number, 17 cattle (7 calves, 10 bulls) were without phenotypic information. For EBV we studied the contribution of one milk trait represented by the total protein content obtained during the normal lactation. For genomic information, altogether, 117 cattle were genotyped for the presence of A and B alleles of κ-casein (CSN3), β-lactoglobulin (LGB) and α-lactalbumin (LAA) by PCR-RFLP method. EBV display different minimums and maximums function by model. Therefore, ssGBLUP range between -6.61 kg and +13.98 kg and animal model recorded a minimum of -21.29 kg and a maximum of +20.74 kg. From the first twenty results (ranked after EBV) estimated using ssGBLUP, just 20% of them turn up when those were ranked using IAM estimation. Spearman's rank correlation calculus returned a low value (0.36), underlying the differences between models. The ssGBLUP is an early method for EBV, suitable for young genotyped animals with no performance. Also, ssGBLUP model generate more accurate breeding value estimations. Acknowledgements: This work was supported by a grant of the Romanian Ministry of Agriculture and Rural Development, throughout the projects ADER 5.1.5. and 5.2.4.

Relationship between estrus, efficiency and concentration of progesterone in milk of Polish HF cows

E.A. Bauer and J. Żychlińska-Buczek

University of Agriculture in Krakow, Department of Genetics and Animal Breeding, Al. A. Mickiewicza 24/28, 31-120 Kraków, Poland; e.bauer@ur.krakow.pl

Nowadays, the role of progesterone in maintaining pregnancy is well known in dairy cattle. The objective of this study was to estimate the influence of efficiency, lactation stage, intensity of external symptoms of estrus on the progesterone level of milk of Polish Holstein-Friesian (HF) cows. Production farm was located in southern Poland; 295 cows were kept in freestall barn system, divided in five technical groups (depended of stage of lactation). Cows were fed Total Mixed Ration (TMR) system, and the average yield was about 11,300 kg per 305-d lactation. The study was conducted using 29 selected dairy cows (n=29), according to age, lactation period and efficiency. The study was divided into three periods: control period 1 (estrous time), control period 2 (12 h after estrous), control period 3 (24 h after estrous). Control periods lasted three time a day, for 20 minutes each cow was observed, from September 2015 until January 2015. Milk samples were collected from all periods and data were analyzed with using FT-Multilyser Automatisierter®, to test the level of progesterone concentration. Factors like efficiency and progesterone concentration were analyzed using STATISTICA12® software. The progesterone level in milk was measured in nanograms per milliliter and for cows results were: 8.3 ng/ml after 24 h, ±7.9 ng/ml after12 h and ±5.2 ng/ml in estrus time. The milk simples for heifers showed ±7.8 ng/ml after 24 h, ±6.2 ng/ml after 12 h and ±3.7 ng/ml in estrus time. The milk simples from heifers had the lowest concentration of progesterone compared to cows samples, however, efficiency of dairy cattle did not have any influence on concentration of progesterone in milk. Results from the present study demonstrated that concentration of progesterone in milk was weakly correlated with syndromes of estrus (r=0.3). The knowledge of level of progesterone in milk could have practical use for farmers, to synchronize symptoms of estrus in dairy farms. We acknowledge The Cooperative Agricultural Production 'Diamond' in Otfinów.

What to expect for later-in-life breeding values when preselecting on early genomic breeding values?

M. Erbe[1], M. Schlather[2] and K.-U. Götz[1]
[1]Bavarian State Research Centre for Agriculture, Institute for Animal Breeding, Grub, Germany, [2]University of Mannheim, School of Business Informatics and Mathematics, Mannheim, Germany; malena.erbe@lfl.bayern.de

In genomic selection programs, genomic breeding values are used as selection criterion for young individuals. Only individuals chosen in this selection step can act as parents for the next generation and thus obtain a progeny based breeding value later (lateBV) that might be more reliable than an early genomic breeding value (earlyBV). In praxis, especially in cattle breeding, it is common to measure success or failure of genomic breeding value prediction by comparing the accordance of earlyBV and lateBV since comparison to true breeding value (TBV) is not possible. In this study we have tried to collect and derive statistical properties and measures for expected values of means and correlations of the different breeding values and for the probability that lateBV is greater/less than earlyBV in the group of individuals that were selected based on earlyBV. The general assumption in all derivations is a multi-variate truncated normal distribution of TBV, earlyBV and lateBV with variances depending on the true genetic variance and a covariance structure related to best linear unbiased prediction properties of the breeding values. Assumption of truncation is necessary because after selection the distribution is not normal anymore. Results show that in the group of individuals selected on earlyBV the expected mean of earlyBV equals that of TBV while the expected mean for lateBV can be considerably lower than that of earlyBV. The probability of lateBV being less than earlyBV is only 0.5 in cases in which lateBV is almost as reliable as TBV while it can be much higher especially if reliability of earlyBV is low, lateBV is only marginally more reliable than earlyBV or selection intensity is high. With an inauspicious combination of these factors, the probability might even exceed 0.9. The results show that the undesired effect of observing a drop in lateBV as compared to earlyBV more often than an increase in praxis can occur despite unbiased prediction models. Further analyses will investigate the influence of preselection on parent average or selective genotyping that might influence the distribution of earlyBV.

Genomic inference from relationship matrices in backcross experiments

A.A. Musa, J. Klosa, M. Mayer and N. Reinsch
Leibniz Institute for Farm Animal Biology, Institute of Genetics and Biometry, Wilhelm-Stahl-Allee 2, 18196 Dummerstorf, Germany; musa@fbn-dummerstorf.de

Conventional statistical methods in genomic evaluation of plants and animals are based on the assumption that molecular markers are independent, i.e. do not account for covariance between markers. This assumption may be violated in certain traits that are controlled by dependent multiple quantitative trait loci (QTL). In this study, we introduce a new method to obtain genetic values, covariance adjusted genomic best linear unbiased prediction (CAG-BLUP), and its equivalent linear marker model with single marker effects, which account for the covariance between markers. We also compare the ability of CAG-BLUP and a conventional method (genomic-BLUP) to provide realistic estimates of genetic variability and power to detect QTL carrying chromosomes by analysing a variety of simulated backcross experiments. The simulation consisted of scenarios with independent and dependent QTL. Each scenario had 100,000 individuals which were divided into 200 replicates with 500 individuals per replicate. Statistical analysis was carried out using ASREML-R and own written programs in R software. The results revealed that accounting for covariance between markers improved genomic prediction in dependent QTL scenarios. In these scenarios, CAG-BLUP improved the accuracy of estimated genetic values, QTL detection power and genetic variance estimates. Therefore, the choice of models for genetic analyses depends on the genetic architecture of the trait of interest.

The feed-a-gene project

J. Van Milgen

INRA, UMR1348 Pegase, 35590 Saint-Gilles, France; jaap.vanmilgen@inra.fr

The Feed-a-Gene project aims to better adapt different components of monogastric livestock production systems (i.e. pigs, poultry and rabbits) to improve the overall efficiency and to reduce the environmental impact. This involves the development of new and alternative feed resources and feed technologies, the identification and selection of robust animals that are better adapted to fluctuating conditions, and the development of feeding techniques that allow optimizing the potential of the feed and the animal. The project started on March 1st 2015 for a 5 year period and has been awarded a grant from the European Commission (Grant agreement no: 633531) under the EU Framework Programme for Research and Innovation Horizon 2020. The project will: (1) Develop new and alternative feeds and feed technologies to make better use of local feed resources, green biomass and by-products of the food and biofuel industry; (2) Develop methods for the real-time characterization of the nutritional value of feeds to better use and adapt diets to animal requirements; (3) Develop new traits of feed efficiency and robustness allowing identification of individual variability to select animals that are more adapted to changes in feed and environmental conditions; (4) Develop biological models of livestock functioning to better understand and predict nutrient and energy utilization of animals along their productive trajectory; (5) Develop new management systems for precision feeding and precision farming combining data and knowledge from the feed, the animal, and the environment using innovative monitoring systems, feeders, and decision support tools; (6) Evaluate the overall sustainability of new management systems developed by the project; (7) Demonstrate the innovative technologies developed by the project in collaboration with partners from the feed industry, breeding companies, equipment manufacturers, and farmers' organizations to promote the practical implementation of project results; (8) Disseminate new technologies that will increase animal production efficiency, whilst maintaining product quality and animal welfare and enhance EU food security to relevant stakeholders. For more Information about the project and registration at the stakeholder plateform: www.feed-a-gene.eu.

Effect of feeding regime and presence of antibiotics in diet on rabbit's microbial gut composition

M. Velasco, M. Piles, M. Viñas, O. Rafel, O. González, M. Guivernau and J.P. Sánchez

Institute for Food and Agriculture Research and Technology (IRTA), Torre Marimon, 08140, Spain; maria.velasco@irta.cat

Simultaneous *Bacteria* and *Archaea* detection by 16S rDNA-based MiSeq sequencing was performed to assess rabbit gut microbiota composition subjected to different factors: feeding regime (*ad libitum* (aL) vs restricted (R)), presence/absence of antibiotics in the diet, and the global environmental conditions defined by the farm. Caecum samples from 66-days-old rabbits (236 fed aL and 215 fed under R) were collected from two different farms (375 farm A and 94 farm B). 24 samples from farm B came from animals fed with the same standard feed but without antibiotics. Average daily gain (ADG) was recorded individually and its association with microbiome composition was studied. Globally, a total of 4,613 OTUs without singletons were clustered from 15,296,317 filtered contigs with a QIIME pipeline. Taxonomic assignment, based on Greengenes database gg_13_5_otus, revealed that intestinal microbiota was dominated by *Firmicutes* (79.2%), *Tenericutes* (6.6%), and *Bacteroidetes* (5.4%) phyla, *Archaea* domain was present at low percentage (1.5‰). PCoA based on weighted Unifrac distance matrix showed that farm exerted a more relevant effect than feeding regime or the presence/absence of antibiotics. Nevertheless, 3 OTUs belonging to *Clostridiales* order (*Firmicutes*) and *Methanobacteriaceae* family (*Euryarchaeota*) were overrepresented in samples from animals fed aL relative to those under R. Moreover, 57 OTUs were differentially represented in animals receiving diet with or without antibiotics. On the other hand, significant associations with ADG were found for 45 OTUs. This was negative for 13 families and 5 phyla, and positive for 3 families and 2 phyla. OTU richness was positively associated to ADG. Our results point to the fact that bacterial taxa abundance might explain an important percentage of ADG variability since the content in half of the detected phyla is associated to ADG. This study is part of the Feed-a-Gene project and received funding from the European Union's H2020 program under grant agreement n° 633531.

Feed efficiency and the faecal microbiome at slaughter weight in pigs

L.M.G. Verschuren[1,2], M.P.L. Calus[1], A.J.M. Jansman[1], R. Bergsma[2], E.F. Knol[2], H. Gilbert[3] and O. Zemb[3]
[1]Wageningen University & Research, Wageningen Livestock Research, Droevendaalsesteeg 1, 6700 AH Wageningen, the Netherlands, [2]Topigs Norsvin Research Center B.V., Schoenaker 6, 6640 AA Beuningen, the Netherlands, [3]INRA-INPT-ENSAT-Université de Toulouse, GenPhySE, Chemin de Borde Rouge 24, 31326 Castanet-Tolosan, France; lisanne.verschuren@topigsnorsvin.com

Feed efficiency (FE) is an important trait in the pig industry, as feed costs are responsible for the major part of production costs. Availability in the market and cost of feed ingredients dictate changes in feed composition. As a result, fibre level and composition can vary between pig diets. Microbiota in the gastrointestinal tract play an important role in fibre digestion, because they produce enzymes that break down fibre structures and deliver volatile fatty acids (VFA) to the pig. These VFA can be used as metabolic energy sources. As such, microbial fermentation could influence FE in pigs. The aim of this study was to investigate the association between FE and faecal microbiome in commercial grower-finisher pigs. Three-way crossbreed grower-finisher pigs (154) were either fed a diet based on corn/soybean meal (CS) or a diet based on wheat/barley/by-products (WB). Faecal samples were collected on the day before slaughter (mean body weight 122 kg), and sequenced for the V3-V4 16S ribosomal DNA regions. Sequences we clustered according to operational taxonomic units (OTU) for each individual. A partial least square regression was applied to the dataset, together with a discriminant analysis using principal components of FE extreme groups (10 high and 10 low FE animals for each diet by sex-combination). Pigs on different diets and males vs females had a very distinct microbiome, needing only two OTUs for diet (P=0.018) and 18 OTUs for sex (P=0.002) to separate the groups. Faecal microbiome was not related to FE groups fed the CS diet, but there were sex specific OTUs related to FE in male and female pigs in the groups fed the WB diet. In conclusion, our results show a diet and sex dependent relationship between the faecal microbial composition and FE in grower-finisher pigs at slaughter weight. This study is part of the Feed-a-Gene project and received funding from the European Union's H2020 program under grant agreement no. 633531.

Relationship between intestinal and blood metabolome and digestive efficiency in chicken

S. Beauclercq[1], C. Hennequet-Antier[1], L. Nadal-Desbarat[2], I. Gabriel[1], F. Calenge[3], E. Le Bihan-Duval[1] and S. Mignon-Grasteau[1]
[1]INRA, Animal Genetics, URA, 37380 Nouzilly, France, [2]Université François Rabelais, Département d'analyse chimique biologique et médicale, PPF Analyse des systèmes biologiques, 37000 Tours, France, [3]INRA, AgroParisTech, Université Paris-Saclay, Animal Genetics, GABI, 78350 Jouy-en-Josas, France; stephane.beauclercq@inra.fr

Digestive efficiency (DE) is an essential component of feed efficiency, especially in the context of increasing variety of feedstuffs with variable quality used in poultry diets. However, measuring fecal DE during balance trials is time-consuming and constraining as birds are placed in individual cages. Moreover, not all the mechanisms controlling DE are known. The aim of our study was thus to identify biomarkers of DE using intestinal and blood metabolomics. Our study used 60 chickens of an advanced intercross line (8th generation) between 2 broiler lines divergently selected for fecal DE, based on metabolisable energy corrected to zero nitrogen retention (AMEn). At 3 weeks, fecal AMEn and coefficients of digestive use of lipids, nitrogen, and starch were measured during a balance trial, ileal and caecal contents were sampled and blood collected. Metabolome was determined by proton high resolution NMR. Correlation models (canonical partial least squares) were fitted to assess the links between efficiency and metabolites of the 3 sample collection sites. Metabolites differences between animals with high or low levels of DE were mainly involved in amino acids metabolism (lysine, isoleucine, methionine) and energy metabolism (glutamate, glucose) in the 3 collection sites. High positive correlations were found between glucose in caecal content and AMEn and coefficient of DE of nitrogen, which is consistent with the large divergence found in the divergent lines on these criteria. This result suggests an effect of microbial fermentation on DE. These metabolic profiles provide information on the mechanisms implied in feed digestion in chickens. Further analyses will estimate if the blood metabolome can be used as an indirect criterion of selection of feed efficiency and DE. This study is part of the Feed-a-Gene project and received funding from the European Union's H2020 program under grant agreement no. 633531 and by the INRA program GISA-GALMIDE.

Molecular indicators of feed efficiency as proposed by a meta-analysis of transcriptomics data

F. Gondret, B. Koffi, J. Van Milgen and I. Louveau
PEGASE, Agrocampus-Ouest, INRA, 35590 Saint-Gilles, France; florence.gondret@inra.fr

Improving feed efficiency is an import challenge for pig production. This study aimed at proposing molecular traits able to predict feed conversion ratio (FCR) in growing pigs. A total of 71 pigs from two divergent lines selected for residual feed intake (RFI) and fed under different conditions (*ad libitum* or restricted) and different diets (low fat high starch or high fat high fiber) were considered, so that a broad range of FCR data was obtained. Transcriptomics data from the loin muscle and blood were obtained using porcine microarrays. The dataset (22,288 molecular probes per tissue and pig) was split into 70% for machine learning methods and 30% for cross-validation. Random forests were used to propose a reasonable set of 359 genes identified as very important predictors (VIP) of FCR. The FCR was well predicted (RMSE=0.16; R^2=0.63) by a model combining the expression levels of 50 genes in muscle (out of the 359 VIP). These genes were involved in various biological pathways, including the response to insulin, homeostatic processes, signal transduction, regulation of cell proliferation, apoptosis, protein metabolism, and inflammatory responses. About 82% of the muscle VIP were also expressed in the blood. The FCR was also predicted correctly (RMSE=0.21; R^2=0.52) by using the same model of genes expressed in blood. Technical validation is in progress to evaluate the predictive potential of the model when expression levels of these genes are measured by target methodology (qPCR) in blood of the same pigs. Further tests will be performed on blood samples taken at earlier growth stages to obtain early predictors and by using different pig populations to obtain generic predictors. In conclusion, identifying molecular traits related to feed efficiency could be helpful to identify important genomic regions and new biomarkers for genetic selection. This study is part of the Feed-a-Gene project and received funding from the European Union's H2020 program under grant agreement no. 633531.

Genetic and maternal effects on growth and feed efficiency in rabbits

H. Garreau[1], J. Ruesche[1], H. Gilbert[1], E. Balmisse[2], F. Benitez[2], F. Richard[2], I. David[1], L. Drouilhet[1] and O. Zemb[1]
[1]INRA, INPT, ENSAT, Université de Toulouse, GenPhySE, 31326 Castanet-Tolosan, France, [2]INRA, Pectoul, 31326 Castanet-Tolosan, France; herve.garreau@inra.fr

The aim of this study was to evaluate the significance of neonatal environment (ultimately including the microbiota composition) on feed efficiency. For that purpose, half of the rabbits of the G10 line, selected for 10 generations on residual feed intake (RFI), were fostered by does of a non-selected control line G0, and vice versa. In parallel, collaterals were adopted by mothers from their original line. Around 900 animals were produced in 3 successive batches and raised in individual or collective cages. Traits analyzed in this preliminary study were weights at weaning (32 days) and at the end of the test (63 days), average daily gain (ADG), feed intake between weaning and 63 days (FI), feed conversion ratio (FCR) and RFI. Line of the rabbit, type of housing (collective or individual cages: 2 levels) and batch (3 successive mating:3 levels) were significant effects for all traits. G10 does had a negative effect on FCR (+0.06, P=0.04), irrespective of the line of young rabbits. G10 animals were lighter than G0 at 32 days (-83 g) and at 63 days (-161 g). They also had a lower ADG (-2.36 g/day), FCR (-0.36), RFI (-548 g/day) and a lower FI (-839 g), confirming a better feed efficiency. Our results demonstrate that selection on feed efficiency was successful for direct effect but maternal effects were degraded by the selection. This study is part of the Feed-a-Gene Project, funded from the European Union's H2020 Programme under grant agreement no. 633531.

Social genetic effects on productive and feeding behavior traits in growing Duroc pigs

W. Herrera[1], M. Ragab[2] and J.P. Sánchez[1]
[1]*Institute for Food and Agriculture Research and Technology, Caldes de Montbui, Barcelona, 08140, Spain,* [2]*Poultry prod. Depart. Kafr El Sheikh University, Kafr El Sheikh, 33516, Egypt; william.herrera@irta.cat*

To explore the role of feeding behavior traits (FBt) and social genetic models for genetic evaluation of both FBt and performance traits (Pt), genetic parameters were estimated for daily gain (DG), daily feed consumption (DC), feed conversion ratio (FCR), backfat thickness (BF), feeding rate (FR), feeding frequency (FF), and occupation time (OT). Traits were recorded in 663 Duroc pigs. Two bivariate models were fitted: animal models (AM) and social interaction animal models (SAM). Estimations were done following Bayesian procedures. Heritability (h^2) estimates obtained with AM for all traits were medium-high, due to additional heritable variation captured by social genetic effects (SGE) higher estimates of the ratio of total genetic variance to phenotypic variance (T^2) were obtained with SAM. Only OT direct genetic effects (DGE) seem to be positively correlated with DGE of DG, DC and BF (0.34(0.14), 0.61(0.18) and 0.38(0.09), respectively), when AM was used the respective genetic correlations were not different from zero. With AM, unfavorable genetic correlation between BF and DG (0.64(0.15)) were estimated. With SAM either SGE or DGE correlations remained high and unfavorable, but the correlation between SGE of DG and DGE of BF was negative (-0.80(0.13)), being null that between SGE of BF and DGE of DG. Large estimation errors of within-trait direct-social genetic correlations prevented to properly define their sign, but they seem to be of low magnitude. The role of FBt to improve Pt genetic evaluations is limited, except for OT. Consideration of SAM allows disentangling the social origin of certain unfavorable AM correlations. SAM could be used to explore indexes combining SGE and DGE of different traits to take advantage of favorable genetic correlations that might exist between them. This study is part of Feed-a-Gene project and received funding from the European Union's H2020 program under grant agreement no. 633531.

Responses of pigs divergently selected for cortisol response or feed efficiency to an ACTH challenge

H. Gilbert[1], E. Terenina[1], I. Louveau[2], J. Ruesche[1], L. Gress[1], Y. Billon[3], P. Mormède[1] and C. Larzul[1]
[1]*INRA, INPT, ENSAT, Université de Toulouse, GenPhySE, 31326 Castanet-Tolosan, France,* [2]*INRA, AgroCampus Ouest, PEGASE, 35590 Saint-Gilles, France,* [3]*INRA, GenESI, 17700 Surgères, France; helene.gilbert@inra.fr*

Selection for feed efficiency can impair the animal's ability to respond to stress. A key driver of this response is the hypothalamo-pituitary-adrenal (HPA) axis, which releases cortisol in response to stressors. Injection of a normalized dose of adrenocorticotropic hormone (ACTH) to stimulate cortisol release by the adrenal cortex is a standardized method to evaluate the activity of the HPA axis independently of the animal's perception of stress. It has been used to select during three generations a highCortisol line and a lowCortisol line, that had divergent cortisol levels 1 h after ACTH injection (H1, peak of the response). A trial was set up to compare the responses to ACTH of these two lines and those of two other lines divergently selected for residual feed intake (RFI, measure of net feed efficiency) during nine generations. At 6 weeks of age, 48 pigs per line were tested. Blood samples were collected before and 1 (H1) and 4 h after injection (H4) when cortisol is expected to return to basal level. In the Cortisol lines, plasma cortisol was multiplied by 2 (P<0.0001) in highCortisol pigs compared to lowCortisol pigs at the 3 times. In both lines, cortisol was multiplied by 2.3 to 2.6 at H1 compared to H0. The highRFI (less efficient) line had a similar response to the ACTH injection as the Cortisol lines. The lowRFI line had higher cortisol levels at H0 (P=0.08) and H1 (P=0.0002) than the highRFI line. This difference was increased at H4 (P<0.0001), due to higher H4 than H0 cortisol levels in lowRFI pigs (P<0.0001). Blood counts, urea, glucose, IGF-I and free fatty acids (FFA) measurements were used to better understand the responses to ACTH, suggesting different underlying metabolisms. In contradiction with previous hypotheses, increased feed efficiency is not associated with a decreased HPA axis activity, but might be related to different dynamics of responses after stress. This study is part of the Feed-a-Gene Project, funded from the European Union's H2020 programme under grant agreement no. 633531.

Feed restriction effect on progeny of mice selected for birth weight environmental variability

N. Formoso-Rafferty[1], I. Cervantes[1], J.P. Gutiérrez[1] and L. Bodin[2]

[1]Universidad Complutense de Madrid. Facultad de Veterinaria, Avda. Puerta de Hierro s/n, 28040, Madrid, Spain, [2]INRA – GenPhySe, CS 52627, 31326, Castanet-Tolosan, France; n.formosorafferty@ucm.es

In line with aspects of the H2020 Feed-a-Gene project (grant agreement no. 633531) aiming to understand the genetic relationship between feed efficiency and robustness, we analysed the influence of mice feeding restriction on their offspring birth weights (BW) in two lines divergently selected for birth weight environmental variability. A total of 120 females (four full-sib females from 10 random different litters of the 12, 13 and 14 generations of selection) were chosen within high and low selected lines and split in four groups of feeding type combining restriction or not in two periods: from weaning at 21 to 77 days, and one week before mating to the 2nd parturition. Restriction consisted of feeding with 75, 90 and 85% of *ad libitum* consumed feed in the respective three studied generations. The data included 158 litters with 1,275 BW and 4,093 animals in the pedigree. A heteroscedastic model (using ASReml Release 4.1 software) was fitted to ascertain the genetic and environmental factors affecting the BW mean and its residual variance. The model included the diet type of the dam (restricted or not during the growing period and during the reproductive period), its line, generation, litter size where it was born and its parity, as well as the litter size of the progeny and its sex; including also all diet, line and generation interactions. The line (lower BW for the low variability line), and the generation effect for the progeny of dams restricted both during the growing and the reproductive period had significant effects on the progeny BW. Whereas the sex, the interaction diet-line (in dam pregnancy period) and the dam genetic effect have had effect on the BW variability. The interaction diet-line produced a decrease in environmental variability of 5 and 18% in low and high variability lines respectively. It seems that selection for BW variability has conferred a lower sensitivity of the dams to the environmental conditions, which could be interpreted as higher robustness.

Using metafounders to model purebred relationships in genomic prediction for crossbreeding

E.M. Van Grevenhof, J. Vandenplas and M.P.L. Calus

Wageningen University and Research, Animal Breeding and Genomics, P.O. Box 338, 6700 AH Wageningen, the Netherlands; mario.calus@wur.nl

Selection for feed efficiency in purebred pig and poultry breeding animals should ideally account for the performance of crossbred animals. As part of the Horizon2020 Feed-a-Gene project (grant agreement no. 633531), methodologies to account for crossbred data are being developed and optimized using simulated data. The advantage of using crossbred data instead of purebred data in genomic evaluations for feed efficiency will be evaluated to propose new selection schemes for feed efficiency in monogastric animals. Simulations were set up using a three-way crossbred breeding program, with 10 generations of purebreds (PB), and 5 generations of two-way and three-way crossbred animals. The last generation of PB are validation animals. True breeding values are simulated for each line/cross (5 traits). Genetic correlations between all PB lines were randomly sampled in the range 0.2-0.8. Heritabilities were randomly sampled in the range of 0.2-0.4. Each PB generation exists of 2,000 phenotyped individuals, and all genotyped animals were also phenotyped. Three scenarios will be tested with varying amounts of data available. In the first scenario 2,000 phenotypes and no genotypes will be available, in the second 2,000 phenotypes and 1000 genotypes, and in the third 2,000 phenotypes and 2,000 genotypes. The simulated genotypes follow the size and number of chromosomes of the pig genome (18 chromosomes). The number of SNP simulated is ~60k, with 4,500 QTL in total. Each simulation is replicated ten times. Breeding values of validation animals are predicted using pedigree-based BLUP, or single step GBLUP (SS-GBLUP) that either uses metafounders or not. Metafounders enable to estimate parental relationships between purebred lines using the genotypic information of different lines in the G-matrix (usually no relationship is assumed between PB lines). Results will contain estimated variance components and accuracies of breeding values of purebred selection candidates for crossbred performance. We hypothesize that the use of metafounders will lead to improved accuracies and more unbiased estimated variance components for SS-GBLUP.

A review of the purebred-crossbred correlation in pigs: theory, estimates, and reporting

M.P.L. Calus and Y.C.J. Wientjes
Wageningen University & Research Animal Breeding and Genomics, P.O. Box 338, 6700 AH Wageningen, the Netherlands;
mario.calus@wur.nl

Pig and poultry production relies on crossbreeding of purebred populations to produce production animals. Thus, the breeding goal is to improve crossbred performance, while selection typically takes place within the purebred populations. One of the objectives of the Horizon2020 project Feed-a-Gene (grant agreement no. 633531) is to develop selection strategies to select on crossbred instead of purebred performance. The genetic correlation between purebred and crossbred performance (r_{pc}) is known to be lower than unity for many traits. A low value of r_{pc} indicates that use of crossbred performance in selection is required to achieve sizable genetic progress. We aimed to 1) review estimates of r_{pc} in pigs, 2) review the different components of r_{pc} and their contribution to it, 3) give guidelines for future studies estimating rpc. In total, 195 r_{pc} estimates from 27 studies were used, published between 1964 and 2016. The r_{pc} estimates had an average value of 0.63, with 50% of the estimates between 0.43 and 0.88. Standard errors were on average 0.16, with 50% of the values between 0.07 and 0.2. Standard errors of r_{pc} reduced with increasing numbers of common sires between purebred and crossbred animals. For all different trait categories, e.g. growth, meat amount, meat quality, feed, and fertility, the average r_{pc} was below 0.8. The r_{pc} has three components: 1) genotype by environment interaction between nucleus and commercial herds, 2) differences in genetic background of purebred and crossbred animals, and 3) differences in trait definition or measurement used in purebred and crossbred animals. Genotype by environment interaction appeared to have a smaller contribution to r_{pc} than differences in genetic background. Across traits and studies, the r_{pc} did not show a relation with the heritability of the trait. Future studies estimating r_{pc} are advised to consider to keep both purebred and crossbred animals under nucleus and commercial conditions, report characteristics of the herd environments in detail, estimate separate r_{pc} for different pure lines, and genotype the animals under study.

Discussion with stakeholders

J. Van Milgen
INRA, UMR Pegase, 35590 Saint-Gilles, France; jaap.vanmilgen@inra.fr

Feed-a-Gene is a so-called multi-actor project, which means that the project '...needs to take into account that the objectives and planning are targeted to the needs, problems, and opportunities of end-users, and complementarity with existing research'. Partners from different sectors participate in the project, but we also need feedback from stakeholders about their expectations and on how project outcomes can be put into practice. What can and should we deliver to the livestock production sector and to society at large? The morning discussion deals with the following workpackages of the Feed-a-Gene project: (1) WP2: New animal traits for innovative feeding and breeding strategies; and (2) WP5: Use of traits in animal selection: genetic parameter estimations, genetic model developments, and evaluation of breeding schemes.

Cross links between feed efficiency parameters and gut microbiota in pigs
G. De La Fuente, A.R. Seradj, M. Tor and J. Balcells
Universitat de Lleida, Ciència Animal, Rovira Roure 191, 25198, Spain; gfuente@ca.udl.cat

The effects dietary CP content (normal or low) and pig type on productivity, nutrient digestibility, and GHG emissions, and gut microbiota were assessed in a 98 d, 3-phase feeding program. Thirty two castrated male Duroc (C-Du) and 32 entire male Pietrain (E-Pi) piglets were used. At the last day of each feeding phase and after 12 h of fasting, rectal spot feces and blood samples were taken. Feed and feces samples at each phase were analyzed for DM, CP, and NDF contents. Genomic DNA was extracted from stool samples with a Qiagen stool mini kit, and the libraries were prepared using V3-V4 amplicons from the 16s rRNA gene. Sequencing performed with Illumina Miseq, generating 902,131 paired-end reads. Sequence data was analyzed following the UPARSE protocol. OTUs were classified taxonomically using RDPII database and Multivariate analysis was conducted using package 'vegan' from R. Biodiversity indexes and Spearman correlations between biodiversity and performance traits were also calculated. *Firmicutes* was the most abundant phyla in both productive types (72.4±6.13% in C-Du; 73.5±4.81% in E-Pi), followed by *Bacteroides* and *Proteobacteria*. Among the main phyla, differences were observed between pig types in the relative titers of phyla *Bacteroidetes* (C-Du > E-Pi, $P<0.05$), *Proteobacteria* (E-Pi > C-Du, $P<0.01$), and *Actinobacteria* (C-Du>E-Pi, $P<0.1$). The effect of CP supply was phyla distribution was negligible. Thirteen genera presented changes in relative abundance between productive types and 6 among diets. Richness index of diversity showed significant differences between genotypes and diets, with the normal CP diet being more diverse than the low CP diet, and E-Pi more diverse than C-Du ($P<0.05$ in both cases). Microbial community structure showed to be highly affected by pig type ($P<0.001$), and by the pig type × diet interaction ($P=0.01$). The CP and starch digestibility, as well as ammonia and CO_2 levels were related with microbial community structure ($P<0.014$). Diversity correlated with ADG in C-Du and nutrients digestibility in E-Pi. This study is part of the Feed-a-Gene project and received funding from the European Union's H2020 program under grant agreement no. 633531.

Effect of pig type and dietrary protein level on the metabolomic pattern, feed effciency, and GHG
A.R. Seradj[1], D. Babot[1], M. Beckmann[2], J. Balcells[1] and G. De La Fuente[1]
[1]Universitat de Lleida, Ciència Animal, Rovira Roure 191, 25198, Spain, [2]Aberystwyth University, IBERS, Cledwyn Campus, SY23 3DA, United Kingdom; gfuente@ca.udl.cat

The effects of dietary CP content (normal or low) and pig type on productivity, digestibility of nutrients and GHG emissions were assessed in a 98 d, 3-phase feeding program. Thirty-two castrated male Duroc (C-Du) and 32 entire male Pietrain (E-Pi) piglets were used. A digestibility trial was conducted using chromic oxide as an external marker. At the last day of each feeding, rectal spot feces and blood samples were taken. The indoor environmental and outdoors fresh air samples were sampled using syringes, which were analyzed immediately for CO_2, N_2O, NH_3 and CH_4. Feed and feces samples at each experimental phase were analyzed for DM, CP, and NDF. Non targeted chromatographic analyses of both polar and global extracts from plasma at 18 wks of age were performed with a Hypersil GOLD C18 selectivity LC column. Metabolites were eluted using a gradient of formic acid and analyzed by an Orbitrap Fusion mass spectrometer. Data were acquired for 22 min in positive and negative modes; the full mass scan range was m/z 65-1,500 with a resolution of 240,000 and a maximum injection time of 1 ms. No differences were observed in the overall ADFI, but C-Pietrain presented a higher ADG ($P=0.01$). DM Digestibility was higher in the low CP diets ($P<0.001$), and CP digestibility increased in E-Pi pigs ($P=0.01$). Both types showed differences in the metabolic pattern in global extracts (PERMANOVA, $P<0.024$ in negative mode), although this was not considered significant in the polar extracts, suggesting that the non-polar fraction of metabolites could be responsible of the metabolomic differences between pig types. Diet also influenced global metabolomics pattern ($P<0.015$ in negative mode), but no effect was observed for the interaction between diet and pig type. Further work identifying key metabolites related with specific metabolomics patterns will help to deepen the understanding of the link between metabolism and productive types. This study is part of the Feed-a-Gene project and received funding from the European Union's H2020 program under grant agreement no. 633531.

Impact of a short term dietary challenge on growth performance and feeding behavior in finishing pig

D. Renaudeau[1], L. Brossard[1], B. Duteil[2] and E. Labussière[1]
[1]INRA, PEGASE, Domaine de la Prise, 35590 Saint-Gilles, France, [2]INRA, UEPR, 35590 Saint-Gilles, France; ludovic.brossard@inra.fr

Pigs are frequently facing environmental perturbations with subsequent short and long-term effects on their health and performance. The objective of this study was to evaluate the consequences of an acute dietary challenge on growth performance and feeding behavior in finishing pigs. A total of 160 pigs (average initial BW of 69.1 kg, 115 d of age) were used in two successive replicates of 80 animals with 4 treatments. In the control group (CC), pigs were fed a standard diet (7.6 g digestible Lys/kg and 9.6 MJ NE/kg) during the whole experimental period (55 days). The DC, CD, and DD groups were challenged with a 'diluted' diet for a 7-d period at 130 d, at 153 d, and at 130 and 153 d of age, respectively. Digestible Lys and NE contents were reduced by 20% in the diluted diet in comparison to the control diet, using wheat bran and sunflower meal as diluters. In both diets, essential amino acids were kept constant relative to Lys. Within each replicate, all pigs were equipped with a RFID ear-tag and housed in a same room. The room consisted of a resting area and 2 feeding areas separated by a weighing-sorting station, which allowed for continuous measurements of individual BW. The 2 feeding areas were equipped with 4 feeding stations each. These electronic feeder systems automatically distributed the chosen diet to each pig and recorded the visits to the feeder, with their time and the amount of feed consumed. The average daily BW gain and the BW at slaughter were significantly reduced in the CD and DD groups when compared to the CC group (1,052 g/d and 125.5 kg on average vs 1,119 g/d and 129.5 kg, respectively; P<0.01). Intermediate results were reported for DC group (1,073 g/d and 127.1 kg). Feed conversion ratio was higher in DC and DD groups than in CC and CD groups (3.13 and 3.15 vs 2.93 and 2.95 kg/kg, respectively; P<0.01). Over the total duration of the experiment, mean feeding behavior traits were not influenced (P>0.05) by treatments. Carcass traits were similar in the 4 experimental groups. This study demonstrates an age-of-exposure effect of a dietary challenge on growth performance in finishing pigs. This study is part of the Feed-a-Gene project and received funding from the European Union's H2020 program under grant agreement no. 633531.

Feed restriction on growth of mice divergently selected for birth weight environmental variability

N. Formoso-Rafferty Castilla[1], I. Cervantes[1], J.P. Sánchez[2], J.P. Gutiérrez[1] and L. Bodin[3]
[1]Universidad Complutense de Madrid. Facultad de Veterinaria, Avda. Puerta de Hierro s/n, 28040, Madrid, Spain, [2]IRTA – Torre Marimon, Caldes de Montbui, 08140, Barcelona, Spain, [3]INRA – GenPhySe, CS 52627, 31326, Castanet-Tolosan, France; n.formosorafferty@ucm.es

Nowadays, the selection for feed efficiency is one of the main aims in animal breeding to decrease the production costs. On the other hand, selection for less sensitivity with respect to environmental effects, as indicated by a low variation around the optimum trait value, may have benefits in terms of productivity and animal welfare. Therefore, the objective of this work was to analyze the influence of food restriction, understood as an environmental challenge, on weight at different ages in two lines divergently selected for birth weight variability in mouse lines with either a low variability (LV) or high variability (HV). A total of 40 females (four full-sib females from 10 random different litters from the 12, 13, and 14 generations of selection), were chosen within lines and fed either *ad libitum* or restricted from 21 to 77 days. Restriction consisted of feeding with 75%, 90%, or 85% of *ad libitum* feed consumption in the respective three studied generations. Weekly weights from 21 to 77 days were analyzed. The model was adjusted for the diet (restricted or *ad libitum*), mouse line, generation, and litter size, and included also the interaction between the line, generation, and the diet. The ASReml Release 4.1 program was used for the analysis. Animals fed *ad libitum* of the LV line had similar weights in all generations unlike those of the HV line, which had lower weights in successive generations. The feed restriction had a negative effect on the body weight of the animals but the interaction between line and diet was significantly different only after day 35, showing a differential response of the lines to the environmental challenge. Animals from the LV line were less sensitive to the feed restriction. This study is part of the Feed-a-Gene project and received funding from the European Union's H2020 program under grant agreement no. 633531.

Research prioritisation and coordination to face animal health challenges in modern farming

S. Messori[1], L. Dalton[2], E. Erlacher-Vindel[1] and A. Morrow[2]
[1]World Organisation for Animal Health (OIE), 12 Rue de Prony, 75017, Paris, France, [2]Department for Environment, Food & Rural Affairs (DEFRA), 17 Smith Square, Westminster, London SW1P 3JR, United Kingdom; s.messori@oie.int

Animal diseases can cause serious social, economic and environmental damage, impact on animal welfare and in some cases directly threaten human health. Intensive livestock production raises new challenges and needs adapted answers. The available control tools are often inadequate to effectively deal with animal populations in such systems. While research is fundamental to ensure the development of adequate disease prevention and control means, the level of funding in this area is currently low. However, more could be achieved, even with the current level of investment, through the coordination of research efforts and the sharing of results. To achieve this, an international forum of R&D programme owners/managers and international organisations was established to improve research collaboration and work towards common research agendas and coordinated research funding on major animal diseases, including zoonoses. The STAR-IDAZ International Research Consortium on animal health and its associated EU-funded secretariat (SIRCAH) aims to coordinate over $2.5 billion in research funding from the network's partners – which includes public and private funding bodies. The World Organisation for Animal Health (OIE) is member of the Consortium's Executive Committee and hosts its scientific secretariat. A prioritisation exercise to identify relevant diseases lacking adequate control tools was conducted in every continent. Interestingly, similar priorities emerged for Americas, Europe and Asia/Australasia, reflecting the global scale of challenges for intensive livestock production. Working groups are now being implemented on each of the priority diseases, in order to perform research gap analyses and draw research roadmaps, to serve as a basis for the common research agendas. An overview of the implemented methodology and of the first roadmaps will be presented.

A conceptual framework to promote integrated health management in monogastrics

L. Fortun-Lamothe, S. Combes, E. Balmisse, A. Collin, S. Ferchaud, K. Germain, M.H. Pinard-Van Der Laan, C. Schouler and N. Le Floc'h
INRA, GenPhySE, RIMEL an consortium including INRA GenPhySE, PEGASE, ISP, URA, GABI, EASM, GenESI, PECTOUL, CS52627, 31326 Castanet Tolosan, France; sylvie.combes@inra.fr

Reduce the use of antimicrobial compounds in livestock production is necessary to limit antimicrobial resistance development and to protect the health of animals, humans and ecosystems. This complex task may only be accomplished via the implementation of a systemic approach, which takes into account the management of animal health as a whole, avoiding simplistic strategies. To that purpose, 15 scientists working on different domestic monogastric species (pig, poultry, rabbit) formed the consortium 'RIMEL'. We propose an interdisciplinary conceptual framework for the evaluation and management of animal health. This conceptual framework listed 11 fundaments of animal health organized in two main dimensions. The first one, physical health, refers to the association of barriers (skin, mucosae and microbiota) and defenses (innate and acquired immunity, and nervous and endocrine systems). The second one considers the psycho-social health (mother-young interactions or amongst relatives, the expression of the innate behavioral repertoire, and the respect of circadian rhythm). We identified critical periods for health in each species related to the maturation dynamics of each fundament. Then, we associate these fundaments with various indicators of health status (zootechnical performance, behavior, metabolic or hormonal markers, etc.). Finally, the framework takes into account more than 30 practices applied in the long- or in the short-term to drive health in livestock systems (genetics, reproduction strategy and renewal management, feeding, hygiene and prophylaxis, living environment, etc.). We organized all the elements to describe the biological relays, to distinguish levers and risks, and to build a representation that took into account the lag between actions and results on a targeted population. This framework is a tool to (1) help and structure research on integrated management of animal health (2) characterize and analyze existing systems, (3) design and evaluate innovative practices and systems. Our ultimate goal is to support the transition of animal production towards systems with low antimicrobial inputs.

Gastrointestinal helminth infections in free-range laying hens under mountain conditions

K. Wuthijaree, C. Lambertz and M. Gauly
Faculty of Science and Technology, Free University of Bozen-Bolzano, Piazza Università 5, 39100, Bolzano, Italy;
kunlayaphat.wuthijaree@natec.unibz.it

Free-range egg production is a suitable farming activity for small-scale mountain farmers. However, it is questionable whether under these harsh environmental conditions infections with endoparasites are as severe as in lowlands. Therefore, the prevalence and burden of helminths in free-range laying hens kept under mountain conditions in South Tyrol, northern Italy was investigated. A total of 280 hens were collected from 10 conventional and 4 organic free-range farms at the end of laying period. After slaughter, gastrointestinal tracts were examined for the presence of helminths. Individual faecal samples were taken to estimate faecal egg counts (FEC) and oozyst counts (FOC). Almost all hens (99.3%) were infected with at least one helminth species. Average burden was 171±261 worms per hen. *Heterakis gallinarum* (95.7%), *Capillaria* spp. (66.8%) and *Ascaridia galli* (63.6%) were the most prevalent species. Tapeworms were found in 30.7% of the animals. The percentage of FEC-and FOC-positive samples was 55.5 and 14.4%, respectively. On average, 258±553 nematode eggs and 80±421 coccidia oocysts were excreted. Overall prevalence did not differ between farming systems (99.0 vs 98.8%; P≥0.05), but total worm burden was higher in organic than in conventional farms (319±396 vs 112±144; P≤0.001). Hens on organic farms had higher (P≤0.001) *H. gallinarum* and *A. galli* values than on conventional ones. Six out of the ten conventional farms used one or two anthelmintic treatments during the laying period. However, no differences were found regarding prevalence and infection intensity between farms. It can be concluded, that laying hens kept in free-range systems under mountain conditions are at high risk of nematode infection, especially in organic systems. The vast majority of hens were infected with multiple species, which suggests that the prevailing environmental conditions and free-range management systems are favorable, so that management practices that counteract parasitic infections should be considered, in particular from an animal welfare point of view. Strategies should include efficient measures in combination with anthelmintic treatments.

Health and welfare challenges in intensive dairy sheep farming in Greece

E. Sossidou, A. Kalogianni, A. Boura, E. Komi, C. Kamviliotis and A. Gelasakis
Hellenic Agricultural Organization-Demeter, Veterinary Research Institute, Hao Campus, 57001 Thessaloniki, Greece;
sossidou.arig@nagref.gr

Intensive production system has been recently adopted by several sheep farmers in Greece, in which the extensive farming is still by tradition the most popular one. The aim of this study was to define animal health and welfare risks in Greek dairy sheep farms. Three hundred and fifty animals raised intensively, were randomly selected from 7 farms sited in the North Greece. The AWIN protocol for Sheep was applied to individually assess various animal based indicators of health and welfare such as the lesions in the body, the Body Condition Score, the cleanliness of the wool, the quality of the wool, the renal discharge, the presence of cough, the presence of lesion/abscesses/ mastitis in the udder, etc. Data was summarized using the descriptive statistical methods and SPSS© v.21. More specifically, skin lesions, papilloma, asymmetry and abscess were found in 158 (45.1%), 25 (7.1%), 54 (15.4%), 63 (18%) animals, respectively. Hooves were overgrown and scored with 2, 3 and 4 in 76 (21.8%), 23 (6.6%) and 7 (2%), respectively. Fleece cleanliness was also assessed as indicator for good housing and found wet and dirty for 57 animals (16.2%). Mean Body Condition Score was 2.52 (±0.37). A pale color of mucosa, indicating the presence of anemia from parasitic infestation was observed in 7 (2%) of the animals. In conclusion, first results indicated that the dominant health and welfare problems in intensively sheep farming in Greece are related to the udder and leg pathology. This study is still in progress and is expected to conclude with relationships between animal health and welfare parameters and the parameters of milk quality.

Effects of maternal outwintering on spring born calf's growth, activity levels and learning ability

M. Fujiwara[1,2], M.J. Haskell[2], A.I. Macrae[1], K.M.D. Rutherford[2], D.J. Bell[1,2] and P.R. Hargreaves[2]
[1]University of Edinburgh, Royal (Dick) School of Veterinary Studies and the Roslin Institute, Easter Bush, Midlothian, EH25 9RG, United Kingdom, [2]SRUC, West Mains Road, EH9 3JG, United Kingdom; mayumi.fujiwara@sruc.ac.uk

Compared with traditional indoor winter housing, outwintering can reduce management costs whilst minimal effects on milk production or fertility have been reported. However, little is known on whether the outwintering of pregnant heifers affects prenatal development and subsequent postnatal progeny performance. To investigate effects of maternal outwintering on spring born calves, pregnant Holstein heifers (23.0±0.3 SD month old) went through the winter either housed (H), kept outdoors on deferred grass (G) or kept outdoors on kale (K) until four weeks before calving. Thirty-five calves (H: n=10, K: n=11, G: n=14) born to these heifers in straw bedded-calving sheds were kept in an individual hutch for the first week of life, and in a group pen with an automatic milk feeder for the second week. Calf body weight (BW) was measured at birth and at 14 days of age (d14). The average daily gain (g/day) from birth to d14 was calculated. Daily lying proportion (LP) in the hutch and in the first 24 h in the group pen was monitored using an accelerometer attached to a hind leg. After introduction to the group pen, calves were trained to drink milk from the automatic milk feeder. The number of trainings required for a calf to drink from the feeder on its own was recorded (training count). Data were analysed using General Linear Model (BW, daily gain) and Residual Maximum Likelihood (LP, training count) in GenStat. BW at birth and LP in the hutch and in the group were not different between H and G group (P≥0.1). However, H calves grew faster than G calves in the first 14 days (H: 290±65 g/day, G: -2.8±48 g/day, P≤0.01), and were heavier than G calves at d14 (H: 44.8±0.9 kg, G: 42.3±0.7 kg, P=0.03). H calves had a higher 'training count' (3.4±0.4) than G calves (2.1±0.4, P=0.04). There were no significant differences between K and H calves in all the parameters measured (P≥0.1). Results suggest that outwintering on deferred grass, but not on kale, can have a negative impact on the growth of offspring compared to indoor housing.

Genetic relation between antibody response and faecal shedding of MAP in dairy cattle

L.C.M. De Haer[1], M.F. Weber[2] and G. De Jong[1]
[1]CRV, AEU, P.O. Box 454, 6800 AL, Arnhem, the Netherlands, [2]GD Animal Health, P.O. Box 9, 7400 AA Deventer, the Netherlands; lydia.de.haer@crv4all.com

Paratuberculosis (Johne's disease) is an infectious disease of cattle caused by *Mycobacterium avium* spp. *paratuberculosis* (MAP). Breeding for disease resistance may be performed by selection on antibody response. However, the ultimate goal is reduction of faecal shedding of MAP. Selection on antibody response is only meaningful if it results in reduced shedding. Therefore, the aim of this study was to estimate the genetic relation between antibody response and faecal shedding. Two data sets consisted of results of laboratory tests performed by GD Animal Health on samples from Dutch dairy herds. The first data set (PA1) consisted of 517,672 individual milk samples of 109,213 cows from 5,938 herds tested by ELISA for antibodies against MAP between 2007-2010. The second data set (PA2) consisted of test results of 78,604 individual faecal samples of 52,348 cows from 435 herds. Faecal samples were tested between 1996-2015 by either modified Lowenstein-Jensen culture method, ESP-TREK culture system or qPCR assay. Heritabilities and genetic solutions for sires were estimated with a sire-maternal grandsire model with random permanent environment effect and with fixed effects herd×year, parity, birthyear and lactation period. In addition, for PA1 a covariable for milk production was included in the model, for PA2 a fixed effect for test method. Sire solutions were used to estimate MACE correlations (with correction for reliability) between PA1 and PA2. Sires with at least 15 daughters on 10 herds per trait were included in the evaluation, resulting in 446 sires for PA1 and 272 sires for PA2. Heritability for PA1 was 0.05 (0.003), for PA2 0.06 (0.008). Repeatabilities for PA1 and PA2 were respectively 0.42 (0.003) and 0.28 (0.006). The genetic correlation between PA1 and PA2 was 0.81. In conclusion, heritability and genetic correlation indicate that it is feasible to reduce faecal shedding of MAP by selection for low antibody responses against MAP.

The molecular basis of humpy-back disease in pigs

A. Clark[1], I. Kyriazakis[1], T. Giles[2], N. Foster[2] and G. Lietz[1]

[1]School of Agriculture, Food and Rural Development, Newcastle University, Newcastle Upon Tyne, NE1 7RU, United Kingdom, [2]School of Veterinary Medicine and Science, University of Nottingham, Loughborough, LE12 5RD, United Kingdom; a.clark@newcastle.ac.uk

The humpy-back disease, also referred to as kyphosis, has persisted within the European pork industry and the aetiology of the disease has remained inconsistent. A thorough understanding of the biological basis of the disease remains critical for the design of intervention strategies. The objectives of this study were to identify differential gene expression and how gene pathways are regulated in kyphosis, in order to understand the molecular basis of the disease. Samples of trabecular bone and intervertebral cartilage were obtained from 4 pre-weaning kyphotic pigs and 4 non-related age matched controls from the same commercial farms at 2 weeks of age. These samples were selected to generate pilot data; power calculations confirmed 51 samples per group were required to verify gene expression in follow-up qPCR tests. Bone was sampled from the last thoracic vertebrae, and cartilage sampled between the last thoracic and first lumbar vertebrae. RNA was extracted from tissues, labelled and hybridised on an Agilent 4*44k microarray. Genespring, with a moderated T-Test and $P \leq 0.05$, was applied to identify differentially expressed genes; Cytoscape was used to recognise gene pathways associated with kyphosis. False discovery rate in Cytoscape analysis was controlled using Benjamini-Hochberg method. A total of 1,196 and 348 genes were found to be differentially expressed in kyphotic bone and intervertebral cartilage ($P \leq 0.05$), respectively. In bone, processes such as fatty acid metabolism ($P \leq 0.05$) and bone turnover ($P \leq 0.01$) were upregulated, whereas organ morphogenesis ($P \leq 0.01$) and bone formation ($P \leq 0.01$) were downregulated. In cartilage, processes such as disc degeneration were upregulated ($P \leq 0.01$), whereas cartilage mineralisation was downregulated ($P \leq 0.01$). Our results indicate that critical molecular processes that govern tissue growth and development are abnormally regulated in kyphotic bone and cartilage at the pre-weaning stage, and that this abnormal regulation could be responsible for the defective growth and development of the spine resulting in kyphotic deformity.

Effect of in ovo feeding of vitamin C and E on hatchability and performance of broiler chicks

Ç. Yılmaz[1], U. Serbester[1], F. Yenilmez[2], Y. Uzun[1], M. Baykal[3], M. Baylan[1], M. Çelik[4], K. Tekelioğlu[5], H.R. Kutlu[1] and L. Çelik[1]

[1]Çukurova University Agricultural Faculty, Department of Animal Science, Balcalı, 01330 Adana, Turkey, [2]Çukurova University Vocational School of Tufanbeyli, Plant and Animal Production, Tufanbeyli, 01640 Adana, Turkey, [3]Çukurova University Vocational School of Yumurtalık, Yumurtalık, 01680 Adana, Turkey, [4]Çukurova University Fisheries Faculty, Fish Processing and Technology, Balcalı, 01330 Adana, Turkey, [5]Çukurova University Veterinary Faculty, Preclinical Science, Ceyhan, 01930 Adana, Turkey; ladine@cu.edu.tr

The aim of the present study was to evaluate the effect of in ovo injection with vitamin C or vitamin E on performance of posthatch chicks. Eggs were obtained from broiler breeder stock and divided into four similar groups (n=50) with equal weight frequency distribution (±5 g) and placed in the incubator. On E 16 day, 1st group received no treatment (negative control) while the 2nd was subjected to in ovo injection of 0.6 ml distilled water/egg as solvent (S) into the amnion fluid (positive control). The third one was given in ovo vitamin C at a concentration of 100 mg/egg and the fourth group was in ovo injected with 30 mg/egg of vitamin E. Hatching traits, body weight gain (BWG), feed intake, and feed conversion ratio (FCR) were determined during experiment. The results of the present study indicate that in ovo administration of vitamin C and vitamin E prehatch had no significant effects on hatching traits, body weight gain (BWG), feed intake, and feed conversion ratio, carcass parameters, in comparison with control groups during the entire period (1±6 weeks). It was suggested that vitamin C and vitamin E may not use as an in ovo feed additive for improving performance of the chicks.

Genetic relationship between clinical mastitis, somatic cell score, longevity and reproduction trait

M. Brzáková, L. Zavadilová, E. Kašná, M. Štípková and Z. Krupová
Institute of Animal Science, Přátelství 815, 104 00 Prague 10, Czech Republic; brzakova.michaela@vuzv.cz

The genetic relationship between health traits and functional traits were analyzed in Czech Holstein cattle. The cases of clinical mastitis (CM) from 30,882 lactations of 12,793 cows were recorded on 8 farms in the Czech Republic from 1996 to 2016. The genetic correlations were estimated between CM and somatic cell scores (SCS305), interval between calving and first insemination (INT), days open (DO), longevity (L) and functional longevity (FL). The CM was considered as an all-or-none trait (CM1_305) with values of 0 (no CM case) and 1 (at least 1 CM case), or as a number of CM cases in the whole lactation (CM2_305), or in periods 0-30 days in milk (CM2_30), 0-60 days in milk (CM2_60), 0-90 days in milk (CM2_90), and 0-150 days in milk (CM2_150). Factors included in the linear model were the parity, the effect of herd, the year of calving, the calving season, the fixed linear and quadratic regression of age at first calving, the permanent environmental effect of the cow, and the additive genetic effect of the cow (pedigree included 26,575 animals). The heritability of CM traits was in the range 0.03 (CM2_30) to 0.09 (CM2_305). The genetic correlations between CM1_305 and CM2_150 were 0.98±0.018, between CM1_305 or CM2_305 and CM2_30 were 0.84±0.05 or 0.73±0.06, respectively. CM traits showed negative genetic correlations to longevity in the range from -0.39±0.07 to -0.61±0.07. The genetic correlations between CM1_305 or CM2_305 and functional longevity were slightly larger than those with real longevity. For reproduction traits, we found the highest genetic correlations between CM2_30 and INS 0.18±0.11; between CM2_90 and DO 0.35±0.15. SCS305 showed the significant genetic correlation to all CM traits from 0.68±0.07 (CM2_30) to 0.83±0.04 (CM1_305). The breeding values and genetic trends for each CM trait were calculated. The significant rank correlations were found between CM2_150 and CM2_305 (0.90, P≤0.01). We assume that the first part of lactation (0-150 days in milk) can be used for genetic evaluation of CM traits instead of the whole lactation. The work was supported by the project QJ1510144 and the project MZERO0714 of the Ministry of Agriculture of the Czech Republic.

Efficiency of rearing native breed poultry in small organic farms

J. Walczak and W. Krawczyk
National Research Institute Of Animal Production, Department Technology, Ecology and Economics, Krakowska 1, 32-083 Balice, Poland; jacek.walczak@izoo.krakow.pl

A major limitation for small farms is the availability of own breeding stock, especially in the case of native breeds. The study material consisted of laying hens and native breed broiler chickens (600 Greenleg Partridge, Sussex and Rhode Island Red birds) obtained from them. Laying hens and broilers were kept in the free-range system and indoors with outdoor access. Chickens were mated naturally from time to time, and the collected eggs were hatched using a small combination setter/hatcher with a capacity of 100 eggs. Statistically significant differences (P≤0.05) in the level of egg production were found between the breeds, with Sussex achieving the best results at 28 weeks of egg production (28%) and Greenleg Partridge the worst (21%). This parameter did not differ within layer management systems. The statistically highest percent fertility (P≤0.05) was noted in RIR (84.4) and the lowest in Sussex (82.6%). Greenleg Partridge hens achieved significant highest hatching percentage (75.4, P≤0.05). Statistically highest weight gains were found in RIR and Sussex chickens (0.049 and 0.045 kg/day), but the latter consumed significantly more feed (0.181 vs 0.172 kg/day, P≤0.05). Due to breed-related statistically significant differences in conformation, Greenleg Partridges had lowest gains but consumed the least feed and water. This breed also had significant lowest level of mortality (0.9%, P≤0.05). Within each breed, the indoor system provided more favourable production conditions. However, vitamin E and A content was statistically significant (P≤0.05) higher in both eggs (51.68 vs 43.65 mg/g and 60.32 vs 35.77 mg/g for Sussex) and meat (0.79 vs 0.4 and 22.3 vs 13.8 mg/g for RIR, P≤0.05). Fatty acid profiles of meat were more beneficial for broilers from the free-range system (e.g. cholesterol 0.47 vs 0.52; PUFA 6/3 11.6 vs 13.1, P≤0.05). A closed production cycle with two management systems for egg and poultry meat production could be a viable alternative to small farms and the keeping of Sussex and RIR breeds. The Greenleg Partridge breed, which has been the least developed in terms of breeding traits, can be successfully reproduced on farms but only for free-range egg production.

Prevalence of helminths in laying hens under mobile housing conditions

K. Wuthijaree, C. Lambertz and M. Gauly
Faculty of Science and Technology, Free University of Bozen-Bolzano, Piazza Università 5, 39100 Bolzano, Italy;
kunlayaphat.wuthijaree@natec.unibz.it

The mobile housing systems are becoming more and more popular for small-scale poultry farmers mostly in mountain regions where parasitic infections might be altered by harsh environmental conditions. The aim of the present study was to investigate the prevalence of gastrointestinal parasites in free-range laying hens, raised under conventional conditions in a mobile house in northern Italy. The herd size was 225 hens and pasture areas were not used by laying hens prior to the study. Starting at the fourth month of the laying period, faecal samples were randomly collected from 20 to 40 individuals at monthly intervals until the end of the laying period (August 2015 until July 2016). Samples were analysed using a Modified McMaster method for the number of nematode eggs (differentiated for *Ascaridia galli/Heterakis gallinarum* and *Capillaria* spp.) and coccidian oocysts. Results are expressed as eggs (EPG) and oocysts (OPG) per gram of faeces ± standard deviation. Averaged over the whole laying period, more than 43 and 11% of the samples were EPG- and OPG-positive, respectively. The vast majority of the eggs quantified in the faecal samples were *A. galli* and *H. gallinarum*. The prevalence of *A. galli/H. gallinarum* rose from 5% (EPG=2±11) at the start to 67% (EPG=401±584) in November, decreased to 20% (EPG=32±71) during winter months and then increased again to almost 50% (155±426) at the end of the laying period. For *Capillaria* spp. a similar increase from 3% (EPG=1±8) at the start to 40% (EPG=65±94) in November was observed, before rates decreased to zero in winter and then increased to approximately 20% (EPG=21±58) at the end of the laying period. The prevalence of coccidia decreased from 36% (OPG=170±399) at the beginning to 9% (11±40) in November and then remained at 5% (OPG<20) throughout the laying period. In conclusion, endoparasitic infections might be altered by harsh environmental conditions in mountain regions, so that in combination with rotational pasture use infections can be substantiated at low levels throughout the entire laying period.

Impact of the cows' environment on their daily activity and milk production in an organic farm

P. Wójcik and J. Walczak
National Research Institute of Animal Production, Department Animal Genetics and Breeding, Krakowska 1, 32-083 Balice, Poland; piotr.wojcik@izoo.krakow.pl

The organic farming requirements are specific in that they demand the use of pasturing and give preference to native breeds. In conventional production, the effects of heat stress and cooling keep dairy cattle in barns. The purpose of the study was to determine the effect of thermal stress during cow pasture on behavior and production in organic farming. This study was performed in an organic farm with two groups of Holstein-Friesian (HF) and native Polish Black-and-White (ZB) cows, each having 30 animals. Animals had access to either outdoor area or pasture. Climatic conditions as well as cow activity using accelerometers were monitored. Cows that were pastured during the day were characterized by a higher distress indicator, lower total resting time, and higher milk yield per milking session. In the system with access to outdoor area, grazing time at an average temperature of 10-12 °C was characterized by higher resting frequency, higher total resting time, and lower milk yield of the animals. Highest activity in both groups occurred when wind speed was less than 10 km/h; beyond this value, resting frequency increas ed. ZB cattle were less active than HF cattle when the wind was light, but more active when the wind speed exceeded 30 km/h. ZB animals were more often willing to rest regardless of weather conditions. Unlike in ZB breed, no changes in milk production were observed in the HF breed as the wind strength increased. The cows were more active at both very low and high temperatures. Their activity became stable for the temperature range of 1-10 °C. The highest activity (156 steps) was noted when temperature exceeded 26 °C. At 10 °C, the activity varied between 124 and 139 steps. The increase in ambient temperature beyond 11 °C caused the milk yield to decrease. HF cows increased their activity when temperature rose from 11 to 25 °C, and ZB cows when it exceeded 26 °C. It was found that the national breed ZB definitely better than the HF tolerates high temperatures on the pasture, however, the wind restricting its milk production. Constant temperature monitoring of pasture is necessary to reduce losses in milk production.

National dairy cattle health recording in the Czech Republic

E. Kašná[1], L. Zavadilová[1], Z. Krupová[1], S. Šlosárková[2], P. Fleischer[2] and D. Lipovský[3]
[1]Institute of Animal Science, Genetics and breeding of farm animals, Přátelství 815, 104 00 Praha Uhříněves, Czech Republic, [2]Veterinary Research Institute, Hudcova 296/70, 621 00 Brno, Czech Republic, [3]Czech Moravian Breeding Corporation, Benešovská 123, 252 09 Hradištko, Czech Republic; kasna.eva@vuzv.cz

In January of 2017, a national monitoring system was initiated to provide information for health traits of dairy cattle in the Czech Republic. The Diary of diseases consists of an on-line health recording form for farmers, a simplified key of diagnoses based upon ICAR recommendations, and a database of registered medicaments. The system was designed to enable reliable data maintenance and storage as well as subsequent use for genetic evaluation of dairy cattle, analysis of antibiotic usage patterns, and herd health monitoring and management. In support of the new system, we carried out a pilot study from July 2015 through June 2016 that focused on the incidence of 20 common diseases and health disorders in Czech population of dairy cows. Data were gathered by means of an on-line survey, which contained information on the date, recipient, and nature of each diagnosis and the date, recipient, and medication of each utilization of an antibiotic. 46.4% of farmers responded to the survey. Collectively, they owned 78.3% of dairy cows in the Czech Republic at that time. Average number of diagnoses per farm was 238, with a maximum of 5,346 and a median of 143. There were 261,284 disease incidents reported from the 130,244 cows evaluated. 159,558 cows (55.1%) did not have any disease event recorded. Clinical mastitis was the most common diagnosis, accounting for 45.5% of all records. Reproductive disorders accounted for an additional 33.2%, with metritis (12.6%) and cystic ovary disease (8.5%) being the second and third most prevalent conditions in the set. Locomotory disorders made up 17.4% of all diagnoses, with foot and claw diseases (12.7%) and lameness (4.6%) being the most common. Metabolic disorders made up 4.0%, with ketosis as the most prevalent (2.7%). The study was supported by the project QJ1510217 and the project MZERO0714 of the Ministry of Agriculture of the Czech Republic.

Breeding values for clinical mastitis in dairy cattle, comparison of several models and traits

E. Kašná, L. Zavadilová and M. Štípková
Institute of Animal Science, Genetics and breeding of farm animals, Přátelství 815, 104 00 Praha 10, Uhříněves, Czech Republic; kasna.eva@vuzv.cz

The two linear animal models for the clinical mastitis genetic evaluation in Czech Holstein cattle were compared. The cases of clinical mastitis (CM) from 30,882 lactations of 12,793 cows were recorded on 8 farms in the Czech Republic from 1996 to 2016. The breeding values (BV) were estimated for CM traits considered as an all-or-none trait (CM1_305) with values of 0 (no CM case) and 1 (at least 1 CM case), or as a number of CM cases in the whole lactation (CM2_305), or the same traits in periods 0-150 days in milk (CM1_150, CM2_150). Single trait repeatability linear animal model (BVs) or two trait linear animal models including the first parity (BVt1) and the second and further parity (BVt2) were used for evaluation of breeding values. Factors included in the single model equations were the parity, the effect of herd, the year of calving, the calving season, the fixed linear and quadratic regression of age at first calving, the permanent environmental effect of the cow, and the additive genetic effect of the cow (pedigree included 26,575 animals). For two trait model, the equation for the first parity does not include the permanent environmental effect of the cow; the equation for the second parity does not include the regression of the age at first calving. The strong rank correlations were found between BVt1 and BVt2 (over 0.88 for CM305; over 0.78 for CM150). The rank correlations between BVs and BVt2 (over 0.97 for CM305; over 0.96 for CM150) were larger than those between BVs and BVt1. The rank correlations between BV for CM1_350 and BV for CM1_150 were 0.88 (BVs); 0.83 (BVt1) and 0.86 (BVt2). The rank correlations between BV for CM2_350 and BV for CM2_150 were 0.88 (BVs); 0.84 (BVt1) and 0.88 (BVt2). The genetic trends for evaluated BV and defined CM traits followed similar trends. Re-ranking among the top ten sires due to CM traits and due to employed models was substantial. According to our findings, the presented procedures for evaluation of breeding values are not interchangeable. The work was supported by the project QJ1510217 and the project MZERO0714 of the Ministry of Agriculture of the Czech Republic.

Effect of housing system and hygiene practices on neonatal diarrhea syndrome in dairy calves

P.D. Katsoulos and M.A. Karatzia
Clinic of Farm Animals, Faculty of Veterinary Medicine, Aristotle University of Thessaloniki, Aristotle University of Thessaloniki Campus, 54124, Thessaloniki, Greece; mkaratz@vet.auth.gr

Newborn calves are highly susceptible to neonatal diarrhea syndrome (NDS), which inhibits the absorption of essential nutrients, leading to long-term compromised performance and even death. The objective was to investigate whether housing system and hygiene management practices affect the occurrence of NDS. The study was carried out using 90 newborn calves from 3 commercial dairy farms located in Northern Greece. In each farm, 30 calves where observed from birth, until 15 days old. Newborn calves in Farm A were housed in group pens partially enclosed with concrete floor area. In Farm B, individual pens with concrete floor in a separate building were used. Long stem wheat straw bedding was laid, which was removed and replaced twice a week. All rails, gates, partitions, walls, feeders and floors were cleaned on a weekly basis and after each calf was removed from the building using a pressure water system and a broad spectrum disinfectant. Farm C housed newborn calves outdoors, in individual polyethylene hutches with an outside run and floor. Clean, dry straw bedding which was disposed of after each batch of calves was used in Farms A and C. Newborn calves were bottle-fed fresh colostrum from their respective mothers in all Farms. After day 4 all calves were fed a commercial milk replacer. Adequate passive immunity transfer was evaluated ($\geq$5.2 g/dl serum total protein concentration) on the age of 3 days and the incidence of diarrhea was recorded daily. Data were analyzed using SPSS$^\copyright$ v.21. The percentage of animals with adequate passive immunity was not significantly different among farms (90.3%, 93.3% and 93.3 for Farms 1, 2 and 3, respectively; P$\geq$0.05). The average serum total protein concentrations were also similar between groups (mean±SE: 6.027±0.184, 6.326±0.203 and 6.495±0.275 g/dl for Farms 1, 2 and 3, respectively; P$\geq$0.05). The incidence of diarrhea was significantly lower (P$\leq$0.05) in Farm B (71%) in comparison to Farms A and C (80 and 90% respectively). These results indicate that housing system and hygiene management practices may affect NDS occurrence.

Traditional farmers' practices in small ruminant management using wild relatives plants in Tunisia

I. Baazaoui[1] and S. Bedhiaf-Romdhani[2]
[1]Faculty of Sciences of Bizerte University of Carthage (Tunisia), Jarzouna, Zarzouna, Bizerte, 7021, Tunisia, [2]INRA-Tunisia, Labo-PAF, Rue Hédi Karray, El Menzeh Tunis, 1004, Tunisia; bedhiaf.sonia@gmail.com

International cooperation through the Consortium Research Program-Dry land system led by ICARDA and INRA-Tunisia launched the assessment of the antiparasite potential of some herbal medicines (medicinal and aromatic plants (MAPs)) on small ruminants' health. It was done through interviews of farmers in Zoghmar community (Center of Tunisia), secondary information and observations. Based on the community shepherds' practices of the MAPs uses for treating animals, the findings were clustered into groups of ailments: the first group was related to internal diseases: Shepherds treat diarrhea and belly ache with infusions of *Artemisia* spp. For intestinal obstructions are treated with seed decoctions of *Trigonella*. Kidney stones are treated by offering decoctions of rosemarie leaves. Shepherds mentioned feeding cereal grain (wheat, barley) at the mating period in order to increase reproductive performance. Twigs of pistacia and of olive cut and carried to lambs and young kids (goats) during the weaning period. *Pistacia lentiscus* foliage was associated with decreased nematode fecal egg count in lambs and young goat. The second group concerned the external parasites: Procedures against ecto-parasites: olive oil and salt; fig tree sap. The third group identified was for fevers and external wounds: lukewarm olive oil pasted and massaged on udder. Understanding how these plants influence sheep health and production can lead to better more sustainable uses of the rangelands and efficient conservation of these species.

Effect of conventional and organic production systems on hatching parameters of laying hens

E. Sosnówka-Czajka, I. Skomorucha and E. Herbut
National Research Institute of Animal Production, Department of Technology, Ecology and Economics of Animal Production, Krakowska Street 1, 32-083 Balice n. Kraków, Poland; ewa.sosnowka@izoo.krakow.pl

The aim of this study was to compare hatchability of native breed chickens raised conventionally or in organic system. Two native breeds of hens, Greenleg Partridge (Z-11) and Rhode Island Red (R-11), were kept in organic farming or in conventional barn system at a stocking density of 9 birds/m^2. At 34, 36, 38 and 40 weeks of age, eggs were collected to assess the quality of hatched chicks. The results were statistically analysed using analysis of variance and estimated by Fisher's test. For both the Z-11 and R-11 breeds, percent hatchability of set and fertile eggs was higher for the conventionally raised hens. In the Z-11 breed, hatchability of set eggs was 89.47% for conventionally raised hens and 79.12% for organically raised hens (P≤0.05); the respective values in the R-11 breed were 79.32% and 70.58% (P≤0.05). In the Z-11 breed, hatchability of fertile eggs was 90.87% for conventionally raised hens and 82.93% for organically raised hens (P≤0.05); the corresponding values in the R-11 breed were 80.04% and 72.99% (P≤0.05). Hatchability of Z-11 eggs was higher than that of R-11 eggs by 11% for set eggs (P≤0.05) and by 12% for fertile eggs (P≤0.05). At 1 and 7 days of age, R-11 chicks exhibited higher body weight (36.0 g and 62.3 g, respectively) compared to the chicks of the other native breed (38.9 g and 66.2 g, respectively) (P≤0.05). R-11 chicks hatched from organic eggs were characterized by lower body weight and distinctly better health compared to the chicks hatched from the eggs of conventionally raised hens (38.9 g and 39.8 g, respectively) (P≤0.05). Better yolk sac resorption was characteristic of chicks hatched from organic eggs. In 7-day-old Z-11 chicks, resorption of yolk sac was 98.99% in organically raised birds and 96.71% in conventionally raised chicks, and in R-11 chicks the respective values were 98.31% and 95.62% (P≤0.05). It was concluded that organically raised hens showed poorer hatchability compared to conventionally raised hens. Chicks hatched from organic eggs were characterized by better health compared to chicks hatched from conventional eggs.

Explaining differences in residual energy intake between lactating Holstein cows

A. Fischer[1,2], R. Delagarde[1] and P. Faverdin[1]
[1]INRA, UMR PEGASE La Prise, 35590 Saint-Gilles, France, [2]Institut de l'élevage, 149 rue de Bercy, 75595 Paris, France; philippe.faverdin@inra.fr

Residual feed intake, usually used to estimate individual variation of feed efficiency, requires frequent and accurate measurements of individual feed intake. Developing a breeding scheme based on residual feed intake in dairy cows is complicated, especially because feed intake is not measurable for a large population. Another solution could be to focus on biological determinants of feed efficiency which could potentially be broadband measurable on farm. Several phenotypes have been identified in literature as being associated with differences in feed efficiency. The present study aims to identify which biological mechanisms are associated with residual energy intake (REI) differences among dairy cows. Several candidate phenotypes were recorded frequently and simultaneously throughout the first 238 days in milk for 60 Holstein cows fed on a constant diet based on maize silage. The REI was defined as the residuals of a linear regression of the lactation average of net energy intake predicted from lactation averages for net energy in milk, metabolic body weight, lactation body condition score loss and gain. A partial least square regression (PLSr) was fitted over all recorded traits to explain REI variability. Linear multiple regression explained 93.6% of net energy intake phenotypic variation, with 65.5% associated with lactation requirement, 23.2% to maintenance, 4.9% to body reserves change. Overall measured traits contributed to 58.9% of REI phenotypic variability, with activity (26.5%) and feeding behavior (21.3%) being the major contributors. However, the level of contribution of each trait was dependent on the variables included in PLSr because REI variability was comparably explained by rumen temperature variation, body reserves change and digestibility when all traits associated with activity and behavior were excluded from PLSr. Drawing a conclusion on biological traits which explain feed efficiency differences among dairy cows was not possible due to this apparent confusion between traits. Further investigation is needed to characterize the causal relationship of feed efficiency with feeding behavior, digestibility, body reserves change, activity and rumen temperature.

Economic impact of the usage of AI bulls on Finnish dairy farms

E.P. Paakala[1,2], D. Martín-Collado[3], A. Mäki-Tanila[2] and J. Juga[2]
[1]Faba co-op, Urheilutie 6D, 01370 Vantaa, Finland, [2]University of Helsinki, Department of Agricultural Sciences, Koetilantie 5, 00014 Helsinki, Finland, [3]Centro de Investigación y Tecnología Agroalimentaria de Aragón, Avda. Montañana 930, 50059 Zaragoza, Spain; elina.paakala@helsinki.fi

AI bulls in the Nordic dairy breeding program are selected mainly on Nordic Total Merit (NTM) where the traits are weighted based on economic importance. Most AI bulls used in Finland are of Nordic origin but the amount of imported semen has increased. The aims of this study were to investigate (1) the variation and economic impacts of the usage of AI bulls among Finnish dairy herds, and (2) its association with herd characteristics. Data consisted of 104,670 insemination records of 1,279 Ayrshire (AY) herds and 544 Holstein (HOL) herds from 2015, estimated breeding values (EBV) of AI bulls (176 AY and 232 HOL) and farm characteristics. Herds' breeding profile was defined as the mean of the EBVs of the AI bulls used in a herd, weighted by the number of inseminations per bull. AI bulls were grouped according to their EBVs and herds to their breeding profile using cluster analysis, for both breeds. The economic implications of breeding choices were estimated, based on the traits 'genetic standard deviation' and 'economic value in Finland'. The result was compared to the expected response to the NTM based selection. Four bull and herd clusters in both breeds were identified. Herd clusters were named according to the main focus in the breeding choices; Pure AY herds: Yield and functionality (452 herds), Longevity, fertility and health (393), All-rounders (377), Conformation (57); Pure HOL herds: Longevity, health and fertility (280), Production (169), Yield, conformation and longevity (48), Conformation (47). The bull clusters followed mainly the same pattern. Conformation bulls were mostly from North America. In the first three herd clusters of both breeds, their average herd breeding profiles were close to the NTM based selection implicating that the vast majority of farmers followed NTM in their breeding decisions. In the Conformation clusters, the traits other than conformation were almost neglected, which expectedly lead to poorer economic result. The conformation focused herds were in a minority, while they were large herds with recent investments.

On-farm genomics in the Netherlands

W.M. Stoop, S.A.E. Eaglen and G. De Jong
CRV BV, Wassenaarweg 20, 6843 NW Arnhem, the Netherlands; marianne.stoop@crv4all.com

In the last decade, genomics has been introduced for Holstein in The Netherlands, first, by in-house genotyping within breeding companies to improve the breeding programs, followed by the introduction of genomic young bulls onto the Dutch-Flemish market. Nowadays, genomic young bulls dominate the national top lists, and their semen sales have an ever rising market share. Herdbooks have adopted genomic information for genetic defects and parentage identification. The CRV Holstein genomic system entails a training population of ~36,000 bulls for 75 traits, and aims at a turnaround time of 21 days from first genotyping requests to results. Reliabilities of genomic breeding values range between 60 and 75%. Farmers adopt the new technologies and incorporate these into their daily management. CRV has launched a new management tool to support their farmers in using genomic data to create an optimal farm-specific management. This includes advice on young stock selection, strategic mating advice for carrier matings (e.g. beta casein or polled), monitoring the genetic and phenotypic performance of the herd and advice how to optimize the two to reach full potential, and benchmarking of farm performance, both within-herd and against national data. Of course, the management tool also includes individual cow test results, genomic breeding values, tracking of individuals in the process of genotyping, a notification function for new test results, and the ranking on a farm-specific breeding goal, tied directly to the mating advice service for improved bull choices and semen sales based on genomic data. Continuous effort is on: (1) increasing genomic reliabilities by extending the training population with bulls and cows and by improved calculation methods; (2) improving chain control by a Track&Trace system that facilitates fast intervention when a sample drops out because of bad sample quality (new sample), failed DNA extraction (re-try), too low call-rate (re-run chip), or pedigree errors (automatic correction of database); (3) optimizing time needed between request and results, which is already down from 28 to 21 days, and is further enhanced by two lab-batches a week; and (4) improving the translation of genomic information into hands-on advice to increase herd profitability.

Sexed semen counteracts the increase in genetic lag in dairy herds managed for extended lactation

J.B. Clasen[1], J.O. Lehmann[2], J.R. Thomasen[3], S. Østergaard[4] and M. Kargo[1,5]
[1]*SEGES, Agro Food Park 15, 8200 Århus N, Denmark,* [2]*Århus University, Agroecology, Blichers Allé 20, 8830 Tjele, Denmark,* [3]*VikingGenetics, Ebeltoftvej 16, 8960 Randers SØ, Denmark,* [4]*Århus University, Animal Science – Epidemiology and management, Blichers Allé 20, 8830 Tjele, Denmark,* [5]*Århus University, Molecular Biology and Genetics, Blichers Allé 20, 8830 Tjele, Denmark; julieclasen@mbg.au.dk*

Extended lactation in dairy cows increases the generation interval in the dam population, leading to an increased genetic lag relative to the sire population. The objective of this study was to investigate the effect of breeding strategies using sexed semen to counteract the increase in genetic lag when managing cows for extended lactation. The SimHerd simulation tool (www.simherd.com) was used to simulate herd dynamics and the non-genetic consequences in 35 scenarios of interactions between strategies for extended lactation and the use of sexed semen. Seven scenarios described the strategy of extended lactation. A base scenario had calving intervals (CInt) set to 13 months, defining a system without extended lactation. Two scenarios had CInt adjusted to 15 and 17 months for all cows, two with CInt of 15 and 17 months for first parity cows only, and the last two scenarios had CInt of 15 and 17 months for older cows only. Within these seven scenarios, five scenarios were simulated with different proportions of sexed semen: no use of sexed semen (SS0), use of sexed semen on all heifers (SS100), 75% of the best heifers (SS75), 50% of the best heifers (SS50), and on 70% of the best heifers and additional 30% of the best first parity cows (SS7030). Output from SimHerd describing reproductive performance and numbers of selected heifers, first parity, second parity, and older cows were used as input in ADAM to simulate genetic changes in the 35 scenarios. The ADAM simulations are still in progress, but preliminary results show a tendency of decreased economic values of breeding when managing extended lactation without use of sexed semen. On the contrary, scenarios using sexed semen appear to counteract this tendency. Further analysis with ADAM will provide a deeper insight in to the genetic changes between different strategies for combining extended lactation and the use of sexed semen.

Breeding strategies to maximise genetic and economic benefits in the Irish dairy industry

D. Matthews[1,2], J.F. Kearney[1,2], F.S. Hely[3] and P.R. Amer[3]
[1]*GplusE Consortium, UCD, Dublin, Ireland,* [2]*ICBF, Bandon, Cork, Ireland,* [3]*AbacusBio, Dunedin, 9016, New Zealand; dmatthews@icbf.com*

The breeding decisions made in different subpopulations of dairy cattle all contribute to the rate of genetic gain for the wider industry. Options for selection within, and genetic contributions from, subpopulations within the Irish dairy industry were modelled to determine their impacts on the overall rate of industry genetic progress. Subpopulations were classified into groups including research herds, contracted breeding program herds, and commercial herds split according to whether they had historically provided a candidate bull calf for progeny testing. The model was initially parameterised with mean Economic Breeding Index (EBI) values for sires and dams, age distributions and sire usage statistics, and gene flows for animals that existed within the subpopulations in 2015. The model accounts for the delays and lags for genetic selection decisions at higher levels in the breeding structure to cascade sideways and downwards to other cows over time. Cumulative discounted benefits of a range of perturbations were calculated considering benefits over 20 years of selection. Increasing the EBI of stockbulls through genomic evaluation and culling brought about the largest genetic gain of all scenarios modelled resulting in €295m added value to the industry after 20 years. Alternatively, removing replacement heifers and stockbulls sired by stockbulls resulted in large economic benefits to the industry; €247m after 20 years. Removing contributions from Foreign sires, which tend to be unsuitable for seasonal and pastoral farming in Ireland, from all bull breeder subpopulations increased cumulative benefits by €223m. Increasing EBI reliability by 10% resulted in relatively modest benefits of €18m. However, increasing EBI reliability by 10% in conjunction with replacing all older, daughter-proven sires, with younger genomically-selected sires increases the industry's value by €99m after 20 years. Increasing sire contributions from sires bred in research herds or nucleus herds causes minus €19m and plus €21m contributions after 20 years. Accounting for subpopulation structure has been shown to be important when evaluating opportunities to accelerate genetic progress within the Irish dairy industry.

Recent advances of mid infrared spectroscopy applications to improve dairy industry profitability

M. De Marchi, A. Benedet, G. Visentin, M. Cassandro and M. Penasa
University of Padova, Department of Agronomy, Food, Natural resources, Animals and Environment, Viale dell'Università 16, Legnaro, Italy; massimo.demarchi@unipd.it

In the last ten years, many studies have proposed mid-infrared spectroscopy (MIRS) as a tool to predict innovative cow features and milk traits. For instance, MIRS has been used to predict pregnancy and body energy status, methane emission and β-hydroxybutyrate (BHB) as well as several milk processing traits. Despite many studies have demonstrated the utility of MIRS prediction models, only few of them are used by the dairy industry. This lack of field application is mainly related to: (1) scarce engagement of dairy industry, (2) difficulties to routinely implement these novel models in milk laboratory, and (3) restricted practical information on the potential of the novel prediction models. The routine application of novel prediction models is limited and can refer only to the prediction of milk fatty acids composition, coagulation and processing traits, acidity, and BHB. Milk coagulation and processing characteristics are related to the increase of milk added value and can be used as case study. In Italy, milk is mainly processed into cheese and in the last 5 years, several milk laboratories have implemented MIRS models to predict milk coagulation and acidity traits. One of the most recent examples refers to the biggest Italian dairy industry which collects milk from more than 600 dairy farms across 11 regions and processes milk in 15 dairy plants. More than 40,000 MIRS analyses of bulk milk samples per year are performed and results are used to (1) check processing ability of milk, (2) segregate milk according to its processing features (3), stimulate farmers to produce milk of better quality, and (4) implement new milk payment systems according to milk processing destination (fluid vs cheese). One of the main future challenge is represented by the on-field implementation of recent advances across dairy stakeholders. Several studies have been carried out on genetic aspects of novel phenotypes, but more effort should be placed into field applications (e.g. dairy industry engagement, milk laboratories guidelines, farmers training). Novel MIRS applications should lead to an enhancement of the dairy industry profitability and competitiveness.

Genetic and non-genetic effects on typical lactation curve parameters in Tunisian dairy cattle

N. Soumri and S. Bedhiaf-Romdhani
INRA-Tunisia, Laboratoire PAF, Rue Hédi Karray, El Menzeh, Tunis, 1004, Tunisia; bedhiaf.sonia@gmail.com

The aim of this work was to estimate the lactation curve parameters of Tunisian dairy cattle to determine the effects of the environmental factors on the lactation curve parameters and its characteristics, and estimate their genetic and phenotypic parameters. The individual lactation curves of 8,616 Tunisian Holstein-Friesian cows at their 1st, 2nd, 3rd, 4th, and 5th parity were analyzed by fitting the wood's incomplete gamma function to daily milk yields from monthly recording of 16,685 lactations collected between 2010 and 2014 from 510 dairy herds. Lactation curve traits included a scaling factor associated with yield at the beginning of the lactation (a), the inclining (b) and declining (c) slopes before and after the peak yield, the DIM at peak yield (T_{max}), the peak yield (Y_{max}), the persistency (S) and the total milk yield at 305 days (Y_{305}). The type of management sector, herd, calving season, parity, age at calving, days in milk at first test-day ($1^{st}DIM$) and duration of lactation (DL) were highly significant factors (P<0.001) influencing the most lactation curve traits. The origins of sires have a significant effect only on the inclining phase of the lactation curve. Winter is the most appropriate calving season to reach the highest 305-days milk yield. Cows with a high level of production at the beginning of the lactation have early reached their lactation peak. The first lactation cows have the lowest yields at the beginning of the lactation but are more persistent than older cows. Low values of heritability and repeatability have been found. Heritability estimates varied from 0.02 for a, peak time and persistency parameters to 0.06 for peak milk yield. Repeatability varied from 0.04 for peak time to 0.22 for 305-days milk yield. Genetic and phenotypic correlations among traits varied from -0.75 to 0.91. High positive genetic and phenotypic correlations have been found between the lactation peak and the 305-days milk yield. This suggests that selection for high peak yield will result in higher 305-day milk yield.

Effect of shortening the dry period length in two dairy breeds on plasma metabolites and fertility

E. Andrée O'Hara and K. Holtenius
Swedish University of agricultural sciences, Department of animal nutrition and management, Box 7024, 75007 Uppsala, Sweden; lisa.ohara@slu.se

Traditionally, farmers have aimed at about 8 week's dry period for their dairy cows. However, a shorter dry period has in some studies shown positive effects on fertility and energy balance without compromising over all milk yield. The hypothesis behind this study was that a dry period length of 4 weeks may ameliorate the metabolic load during the transition period and improve fertility. Cows (n=78) of both Swedish Red (SR) and Swedish Holstein (SH) breed were randomly subjected to a short dry period (4 weeks (4W)) compared to a control dry period of 8 weeks (8W). Blood was collected week -8, -4, -2, -1 *pre-partum* and week 1, 2, 3, 4, 6, 9, 12 *post-partum* and analyzed for NEFA, insulin, IGF-1, BHB and Glucose. Progesterone in milk was collected twice a week *post-partum* and the progesterone profiles were determined as normal or atypical (delayed, interrupted or prolonged luteal activity). The fertility as determined by the progesterone profiles did not differ significantly between treatments (P=0.13). Out of the 4W cows, 71% had normal progesterone profiles while 55% of the 8W cows had normal profiles. No overall treatment differences were found in any the plasma metabolites, but interactions between treatment and lactation week were found (P<0.05), indicating lower NEFA and higher insulin for 4W cows in early lactation, and also more stable IGF-1 levels during the weeks before and after calving. The two breeds responded equally to the short dry period, but regardless of dry period length, the SR cows had higher levels of insulin and IGF-1 than SH cows. The 4W cows produced virtually similar amounts of milk as the 8W cows when also including the milk produced during the 4 weeks lactation pre partum and 12 weeks postpartum, 3,724 vs 3,684 kg that the 8W cows produced during the same period. In conclusion, a shorter dry period may ameliorate the metabolic load during transition period and possibly also enhance fertility without losses in milk production during the first 12 weeks.

Relationship of calving interval and milk yield of Estonian dairy cows

A. Remmik[1] and U. Tuppits[2]
[1]Institute of Economics and Social Sciences, Kreutzwaldi 1a, 51014 Tartu, Estonia, [2]Institute of Veterinary Medicine and Animal Sciences, Kreutzwaldi 62, 51014 Tartu, Estonia; ulrika@veterinaarteenus.ee

Sales of milk and excess animals are the main income sources for a typical dairy farm. Reproductive performance of a herd affects both milk output and availability of heifers for sale, yet the relationship between reproduction and milk output is often unclear. Calving interval is a popular indicator for measuring reproductive performance of a dairy herd. It can be easily followed in Estonia through the national animal recording database VISSUKE on a monthly basis. We hypothesized that there is an inverse relationship between calving interval and milk production per day of calving interval. Data from ten Estonian dairy herds was used in the study and included all calving intervals between 01/01/2012 and 31/12/2016 for all milking cows that were present in the herds at the end of this period. The farms had 492 milking cows on average as of 31/12/2016 (herd size varied between 260 to 673 milking cows). In total, there were 6,008 calving intervals from 3,144 cows in the sample, including 2909 calving intervals from primiparous and 3099 calving intervals from multiparous cows. Relationship between calving interval and milk production was analyzed using linear regression model. We found statistically important (P<0.01) inverse relationship between calving interval and milk production per calving interval day. The other statistically significant factors were parity of the cows (first or later) and farm, as average milk yields at the farms varied to a sizable extent (87,15 to 11,035 kg per lactation). Calving interval is determined mostly by management decisions, e.g. voluntary waiting period, heat detection system, treatment and pregnancy diagnosing protocols and culling protocols. The results of our study confirm that reproduction management is an important part of effective dairy farming as optimal calving interval has a positive impact on milk yield and thus dairy farm income.

Working time analysis of mountainous dairy farms

I. Poulopoulou, C.M. Nock, S. Steinmayer, C. Lambertz and M. Gauly
Free University of Bolzano, Faculty of Science and Technology, Universitätsplatz 5, 39100, Italy;
ioanna.poulopoulou@unibz.it

Though studies are available on the situation of mountain dairy farms concerning productivity as well as production and management practices, data regarding man power expenditure (working time requirement) are missing. The aim of the present study was to: (1) estimate the labour input on dairy farms with loose and tie stall housing in South Tyrol, (2) propose strategies for the maintenance of holdings and improvement of farms' economic situation. The study was performed on 102 dairy farms, half with tie and half with loose housing systems. All daily and non-daily working activities, management practices and facilities used were surveyed with a standardized questionnaire. Additionally, 9 tie stall and 10 loose housing farms where selected for on-site measurements in order to determine the exact working time of each single activity and to validate questionnaire data. The average herd size of the tie stall (16.3 cows) and loose housing (23.2 cows) farms emphasized the representativeness for the region. Total working time requirement was 177 Manpower hours per cow and year (MPh/cow/year) for tie stall and 113 MPh/cow/year for loose housing. Milking estimated at 74 and 56 MPh/cow/year and feeding followed with 34 and 27 MPh/cow/year, for tie and loose housing, respectively. Labour costs per kg of milk were estimated at 32.6 for tie and 16.9 Euro cents for loose housing. Production per working hour was determined at 41.4 and 79.7 kg/MPh, respectively. In both housing systems, a close relation of working time and herd size was found. The comparison between questionnaire and on-site measurements showed that farmers are able to estimate the overall working time, but have difficulties to separate it into single activities. Moreover, results showed that the use of facilities, such as milking or feeding systems, which improve productivity and cost efficiency, is limited. Working time variation between farms was high and implicates that a more efficient organisation of working time is reasonable. Accordingly, investment in facilities (i.e. milking system) and adoption of new management practices might improve the economic situation of the farms.

Evaluation of the income over feed cost on North-West Portuguese dairy farms

I.M.L. Santos[1,2], A.C.M. Gomes[2], A.R.J. Cabrita[1] and A.J.M. Fonseca[1]
[1]REQUIMTE/LAQV, ICBAS, Universidade do Porto, Rua de Jorge Viterbo Ferreira 228, 4050-313 Porto, Portugal,
[2]Cooperativa Agrícola de Vila do Conde, CRL, Rua da Lapa 293, 4480-757 Vila do Conde, Portugal; isabel.santos@cavc.pt

The end of the EU milk quota regime on 31st March 2015, along with the Russia embargo on Western food imports, the reduction of milk consumption and the growth of the world dairy production, led to a strong falling milk price, with a particular negative impact on the Portuguese dairy sector (an EU peripheral country). In this scenario, it is crucial to monitor indicators that reflect farms profit, namely Income Over Feed Costs (IOFC), as feed is the greatest expense for milk production. The present study aimed to evaluate the effects of feed costs and inclusion of grass-silage (GS), on maize-silage (MS) based diets, on IOFC and milk production. Forty-two commercial dairy farms associated to the Vila do Conde Cooperative, representing different herd sizes and feeding strategies, were used to monthly determine, during 2015, the purchased real feed costs, home-raised feeds costs (market values), milk yield, IOFC and basal diet (MS vs mixtures of MS and GS). Herd sizes ranged from 22 to 159 milking cows, with a mean value of 65.8. Daily milk yield average 30.6 kg/cow (29.1 kg energy-corrected milk; ECM/cow). The IOFC ranged from 0.32 to 7.81 €/cow/day, with a mean value of 4.08±1.156 €. One-way ANOVA was conducted for the quartiles feed costs and the GS inclusion against to milk yield and IOFC. Forage costs did not affect milk yield, ECM nor IOFC. Conversely, these parameters were affected ($P<0.001$) by purchased and total feed costs; the average daily milk yield and IOFC being increased by purchased and total feed costs (28.7, 29.8, 31.2 and 32.8 kg/cow; and 3.81, 4.00, 4.23 and 4.30 €/cow/day, respectively, for the purchased feed costs quartiles: 1.72-3.14, 3.15-3.33, 3.34-3.47 and 3.48-4.51 €/cow/day). Although dietary inclusion of GS tended to increase milk yield ($P=0.054$), it did not affect ECM nor IOFC, suggesting a negative effect on milk composition and, consequently, on milk price. Overall, our results show that in order to dairy farmers to increase profit margins, they should opt dietary strategies that allow the cows to express their genetic merit, instead of reducing feed costs.

Grasscheck: grass growth monitoring and prediction to improve grassland management
N. Valbuena-Parralejo[1], S. Laidlaw[1], S. Gilkinson[2], A. Boyle[2], I. McCluggage[2], C. Ferris[1] and D. McConnell[1]
[1] Agri-Food and Bioscience Institute, Sustainable Dairy Systems, Large Park, Hillsborough, Co. Down, BT266DR, Virgin Islands (U.S.), [2] College of Agriculture, Food and Rural Enterprise, Greenmount Campus, CAFRE, Co. Antrim, UK, Greenmount Campus, CAFRE, Co. Antrim, UK, BT414PU, United Kingdom; nuria.valbuena@afbini.gov.uk

Grazed grass remains among the most cost-effective feedstuffs for lactating dairy cows. Northern Ireland (N.I) is one of the main European regions for forage production however the current performance of managed grasslands in N.I. remains sub-optimal. Grass growth rate forecasts are a useful tool to assist farmers in grassland management decisions throughout the season. The objective of this study was to evaluate the use of a grass growth prediction model to forecast seven and 14 day grass growth rates throughout the grazing season in N.I. The study was conducted between March and October 2016 at two permanent grassland sites in N.I. At each site two sets of nine plots (5.0×1.5 m) were established in perennial ryegrass pasture. Each plot was randomly allocated to one of three harvesting sequences beginning on 7 March, 14 March or 21 March. Following this plots were cut at three week intervals. All plots were assessed for soil fertility and corrected for phosphorus, potash, magnesium and sulphur, receiving 270 kg N/ha per annum. Weekly grass offtake, daily rainfall and daily air temperature were inputted to the model on a weekly basis. The forecasted grass growth was calculated using the GrazeGro model. Mean annual rainfall across both sites was 1,081 mm and mean monthly air temperature ranged between 4 and 15 °C. Annual herbage yield across the two sites was 13.6 tonnes dry matter per hectare. A significant linear regression (P<0.01, r=0.810) was established between the observed and forecasted grass growth rates, the model explained the 65% of the grass growth variability. Hence the GrazeGo model can be used as reliable predictor of grass growth rate, providing valuable information to livestock farmers however a wider spread of monitoring sites is required to overcome the high variability of grass growth within N.I.

The effect of the calving season on the lactation curve parameters in Tunisian dairy cattle
N. Soumri and S. Bedhiaf Romdhani
National Institute of Agronomic Research of Tunisia (INRAT), Lab-PAF, Rue Hédi Karray, El Menzeh Tunis, 1004, Tunisia; soumri@gmail.com

This study aimed to elucidate the effect of the calving season on the parameters of the lactation curve and its characteristics in Tunisian dairy cattle. The official dairy control database was used between 2010 and 2014. It contains 11,151 complete lactations of 8616 Holstein-Frisian cows belonging to 510 dairy farms. Five parity groups were considered: parity 1 with 3,976 lactations, parity 2 with 3,277 lactations, parity 3 with 2,199 lactations, parity 4 with 1,188 lactations and parity 5 with 511 lactations. Each parity group contains four groups according to the calving season: Spring (March-May), summer (June-August), autumn (September-November) and winter (December-February). The Wood's incomplete gamma function, described as $Y_t=at^be^{-ct}$ was used to fit the individual lactation curves. Y_t is the daily yield on day t and a, b and c are constants: a is the daily yield at the onset of lactation, b is the slope of the ascending phase of milk production and c is the slope of the descending phase. The lactation characteristics are: total milk yield of 305-days ($Y305=\sum_{(t=1)}^{305}Y_t$); milk yield at the lactation peak ($Y_{max}=a(b/c)^be^{-b}$); the days-in-milk at peak yield ($T_{max}=b/c$) and the persistency ($S=-(b+1)ln(c)$). Results showed that calving season had a highly significant effect (P<0.001) on a, b and c and the estimated lactation characteristics ($Y305$, Y_{max}, T_{max} and S) of the lactation, which suggests important intra-annual climate changes. The highest peak yield seemed to be reached in the 4th lactation. Total and peak yield were the lowest for cows calving in the summer season. First parity cows have the lowest total milk yield and the lowest milk yield at the lactation peak, but they have the highest days-in-milk at peak yield and the highest persistency. In general, cows that calve in spring, had the highest a parameter and reached sooner the lactation peak. While cows that calve in the fall season have the highest b parameter and the highest days-in-milk at peak yield. The adjustment of the calving season to parity seems to be a good way to improve dairy yield in Tunisia dairy farms.

Prediction of lactation curves after 305 days in milk by using curves within the first 305 days

T. Yamazaki[1], H. Takeda[2], A. Nishiura[2], K. Hagiya[3], S. Yamaguchi[4] and O. Sasaki[2]
[1]Hokkaido Agricultural Research Centre, NARO, Sapporo, 062-8555, Japan, [2]Institute of Livestock and Grassland Science, NARO, Tsukuba, 305-0901, Japan, [3]Obihiro University of Agriculture and Veterinary Medicine, Obihiro, 080-8555, Japan, [4]Hokkaido Dairy Milk Recording and Testing Association, Sapporo, 060-0004, Japan; yamazakt@affrc.go.jp

Lactation periods in dairy cows lengthen with increasing total milk production, so it is important to accurately predict individual productivities after 305 Days In Milk (DIM). Our objective here was to predict lactation curves from 306 to 450 DIM from those within the first 305 DIM at the first and later lactations of Holstein cows by using a random regression model. Test-day milk records through 450 DIM from 85,690 cows in their first lactations and 125,905 cows in their later (second to fifth) lactations were analyzed. Data within the first 305 DIM (M1) and from 306 to 450 DIM (M2) were analyzed separately by using different single-trait random-regression animal models. The covariates associated with DIM for random regression of animal effect and permanent environmental effect were second-order Legendre polynomials in M1 and first-order Legendre polynomials in M2. We then performed a multiple regression analysis of the regression coefficients of M2 on the regression coefficients of M1 for the random animal effect within the first lactation and later lactations. The coefficient of determination (R^2) for multiple regression of the zero-order regression coefficient of M2 on the regression coefficients of M1 was 0.778 in the first lactation and 0.706 in later lactations. R^2 for multiple regression of the linear regression coefficient of M2 on the regression coefficients of M1 was 0.593 in the first lactation and 0.513 in later lactations. The linear regression coefficients of M2, as predicted by using the multiple regression equation, were similar among cows with similar cumulative milk yields in M1. These results suggest that individual lactation curves after 305 DIM can be predicted by using curves for the first 305 days. Moreover, the slope of the curve after 305 DIM depends on the 305-day milk yield.

Genetic correlations between semen production traits and milk yields in Holsteins

K. Hagiya[1], T. Hanamure[2], H. Hayakawa[2], H. Abe[3], T. Baba[4], Y. Muranishi[1] and Y. Terawaki[5]
[1]Obihiro University of Agriculture and Veterinary Medicine, Obihiro, Hokkaido, 080-8555, Japan, [2]Genetics Hokkaido, Sapporo, Hokkaido, 060-0004, Japan, [3]Hokkaido Dairy Milk Recording and Testing Association, Sapporo, Hokkaido, 060-0004, Japan, [4]Holstein Cattle Association of Japan, Hokkaido Branch, Sapporo, Hokkaido, 001-8555, Japan, [5]Rakuno Gakuen University, Ebetsu, Hokkaido, 069-8501, Japan; hagiya@obihiro.ac.jp

We investigated the genetic correlations between yield or Days Open (DO) in cows and semen production traits in bulls by using a bivariate animal model. We used lactation records of milk, fat, and protein yields, and DO, from 386,809 first-lactation Holstein cows in Japan that calved between 2008 and 2014. Semen production records were collected between 2005 and 2014 and included semen production volume per ejaculate (VOL), sperm concentration (CON), Number of Sperm per ejaculate (NSP), progressive motility index of sperm (MOT), and MOT after freeze-thawing (F-MOT). NSP was log-transformed into an NSP score (NSPS). We obtained a total of 30,373 semen production trait records from 1,196 bulls. The pedigree file we used for our analyses contained records of 885,345 animals. Genetic correlations were estimated from the combined analysis of a bivariate animal model. A first trait was applied to the female records (yield or DO) and a second model was applied to the male traits. The GIBBS2F90 program was used for Gibbs sampling for the linear models. Posterior means of heritability were estimated for the semen production traits CON (0.12), MOT (0.08), and F-MOT (0.11). Moderate heritabilities were obtained in our posterior means for VOL (0.42) and NSPS (0.37). The genetic correlations between yield traits in cows and VOL, CON, or NSPS in bulls were weakly negative (ranging from -0.12 to -0.26 for VOL, from -0.20 to -0.29 for CON, and from -0.17 to -0.26 for NSPS). Our estimated genetic correlations were weak but negative; therefore, selection for increasing milk yield may gradually decrease semen production ability in bulls. Moderate negative genetic correlations between DO and MOT (−0.42) or F-MOT (−0.43) were estimated. Selection focused on MOT or F-MOT in bulls may therefore improve DO in cows.

The occurrence of hygienically important microorganisms in raw ewe's milk

M. Vršková[1], V. Tančin[1,2], M. Uhrinčať[1] and L. Mačuhová[1]
[1]NPPC – Research Institute for Animal Production Nitra, Department for Animal Breeding and Product Quality, Hlohovecká 2, 951 41 Lužianky, Slovak Republic, [2]Slovak University of Agriculture, Trieda Andreja Hlinku 2, 949 76 Nitra, Slovak Republic; vrskova@vuzv.sk

For the control of Raw Ewe's Milk (REM) quality under current legislation, the Total Plate Count (TPC) of mesophilic microorganisms is a major microbiological criterion (EC Regulation no. 1662/2006). The aim of our work was to determine the incidence of technologically and hygienically important species of microorganisms for REM quality in Slovak Republic. Samples of raw ewe's milk from bulk tank milk were taken in summer in 3 farms and autumn in 1 farm, which used unpasteurized milk for cheese production. Each farm represented different regions of Slovakia: Farm 1 was in Central Slovakia (with crossbred Lacaune and Tsigai sheep), Farm 2 was in Northern Slovakia (with crossbred Improved Valachian and Lacaune sheep) and Farm 3 was in West Slovakia (with Lacaune). TPC (norm STN EN ISO 4833), Psychrotrophic Microorganisms Count (PMC, STN ISO 6730) and Thermoresistant Bacteria Count (TBC) were cultivated on tryptic glucose yeast agar and the Coliform Bacteria Count (CBC, STN ISO 4832) were cultivated on violet red bile agar. The presence of Spore-forming Anaerobic bacteria (SPAN) were examined pouring liquid paraffin. Values of TPC in summer ranged from 187 to 964×10^3 cfu/ml and in autumn from 328 to 689×10^3 cfu/ml. About 67% of the total samples were above the legal limit of 500×10^3 cfu/ml for raw ewe's milk. In summer, we did not detect SPAN, but some samples were positive only in farm 2 in northern Slovakia in autumn. Values of TBC ranged from 18 to 560 cfu/ml in summer and only from 4 to 7 cfu/ml in autumn. For CBC (an indicator of udder hygiene and faecal contamination during milking), we found out values from 12 to 124×10^2 cfu/ml, indicating a poor level of hygiene observed farms in summer. We found a high incidence of PMC and CBC which indicate mostly insufficient culling and sanitation of bulk milk tank and machine device for milking. This experiment was conducted within the project APVV 15-0072 supported by the Slovak Research and Development Agency.

Approaches to evaluation of animal welfare in dairy production in Saxony-Anhalt (Germany)

P. Kühne[1], H. Scholz[1], T. Engelhard[2] and B. Taffe[3]
[1]Anhalt University of Aplied Sciences, LOEL, Strenzfelder Allee 28, 06406 Bernburg, Germany, [2]State Institute of Agriculture and Horticulture Saxony-Anhalt, Lindenstraße 18, 39606 Iden, Germany, [3]Saxony-Anhalt animal health service, Hegelstraße 39, 39104 Magedeburg, Germany; petra.kuehne@hs-anhalt.de

In recent years an increasing importance of animal welfare in agricultural livestock farming has been observed. According to the '§ 11 (8) Animal Protection Act', German farmers are obliged to carry out self-inspection by monitoring animals in animal welfare. However, proposals on the indicators to be used are not yet uniform. From this point of view, in 2015/16 in Saxony-Anhalt, a study on 31 dairy farms was conducted to determine an actual status of animal welfare in milk production. To identify suitable indicators of animal welfare, were recorded 26 register-based indicators (e.g. culling/mortality, udder health, performance figures), 11 animal-based indicators (e.g. cleanliness, lameness, body condition), 13 behaviour parameters (e.g. lying down behaviour, avoidance distance), 25 housing parameters (e.g. dimensions of cubicles, feed alleys) and 60 metabolism parameters (e.g. liver metabolism, protein metabolism). A total of 11,732 dairy cows has been analyzed. Herd size varied from 80 to 1,134 cows. The average productive lifetime was 3.2 lactations and 27,808 kg lifetime yield. The results of the investigation showed a great variation of the indicators between dairy farms. There was a significant correlation ($r=0.59$, $P \leq 0.001$) between whole register-based and whole animal-based indicators. For individual key figures of the two mentioned families, smaller and partly non-significant relationships could be determined. No significant correlations were found between register-based indicators to behaviour parameters ($r=0.31$, $P \leq 0.099$), housing parameters ($r=0.22$, $P \leq 0.227$) and metabolism parameters ($r=-0.37$, $P \leq 0.136$). Generally, register-based indicators provide a long term information about the development of a dairy herd and the animal-based indicator describe a 'right now' moment. This is a possibility for analysis of critical points in milk production and gives an opportunity to see changes in management and housing system. The analysis of the metabolism completes the analysis of critical points for the domains of feeding and management.

Using nonlinear quantile regression to describe the milk somatic cell count of Iranian Holstein cows

H. Naeemipour Younesi[1], M. Shariati[1], S. Zerehdaran[1], M. Jabbari Noghabi[2] and P. Lovendahl[3]
[1]*Ferdowsi University of Mashhad, Animal Science, Azadi Square, 9177948974, Mashhad, Iran, [2]Ferdowsi University of Mashhad, Statistics, Azadi Square, 9177948974, Mashhad, Iran, [3]Aarhus University, Molecular Biology and Genetics – Center for Quantitative Genetics and Genomics, Blichers Allé 20, 8830 Tjele Denmark, Denmark; peter.lovendahl@mbg.au.dk*

The main objective of this study was to compare the performance of different 'quantile regression' (QR) models evaluated at the τth quantile (0.25, 0.50, and 0.75) of somatic cell count (SCC) in Iranian Holstein dairy cows. Mathematical models used in fitting the lactation curve contribute towards better management, physiological and breeding decisions. QR is a flexible tool that allows a specific lactation curve at anyquantile of the trait and can be applied on data with non-normal distributions. Therefore, using QR can be more appropriate, for instance, where the pattern of lactation curve of the trait differs between high and low quantiles. This is more pronounced in trait SCC, where the distribution is not normal and cows with high levels of SCC are suspected to be mastitic cows. Data were collected by the Animal Breeding Center of Iran from 1991 to 2011, comprising 101,147 monthly milk yields of 13,977 cows in 183 herds. An exponential (Wilmink) and a polynomial (Ali & Schaeffer) functions were implemented in the quantile regression. The results showed that all parameters for SCC at the three quantiles Wilmink (a,b and c, parameters) and Ali & Schaeffer function (a, b, c, d and g, parameters) were significantly different from zero ($P<0.01$) and not significant, respectively. Parameters b (increasing slope parameter) and c (declining slope parameter) in Wilmink function had increased at the across quantiles and parameter a (SCS level at the beginning of lactation) was not similar. QR with Ali and Schaeffer function fitted the data better than the one with Wilmink function based upon Akaike information criterion and log-likelihood. Among quantiles, 0.25^{th} quantile showed best model fit with both functions. QR analysis of SCC, which is a non-normal trait with mixture distribution, provides more insight into the management decisions in dairy farms.

Connections between register-based and animal-based indicators for an assessment of animal welfare

P. Kühne[1], H. Scholz[1], T. Engelhard[2] and B. Taffe[3]
[1]*Anhalt University of Applied Sciences, LOEL, Strenzfelder Allee 28, 06406 Bernburg, Germany, [2]State Institute for Agriculture and Horticulture Saxony-Anhalt, Lindenstraße 18, 39606 Iden, Germany, [3]Saxony-Anhalt animal health service, Hegelstraße 39, 39104 Magdeburg, Germany; petra.kuehne@hs-anhalt.de*

Milk production is a multifactorial system, where a variety of factors have an influence on animal welfare. In results of Scholz *et al.* and Wallenberg, a significant correlation has been observed between register-based and animal-based indicators of animal welfare. So, both sources of information are useful to describe animal welfare. From this point of view, 26 register-based indicators and 11 animal-based indicators were recorded in 31 dairy farms with an herd size from 80 to 1,134 dairy cows. Then, the 37 indicators were reduced to 10 main indicators with the highest informative value and practical application for the stockman in terms of self-monitoring. Between the index of the six register-based (culling, mortality, culling of first-lactation dairy cows, culling till the first 30 days p.p., proportion of udder healthy cows, rate of mastitis of the first-lactation cows) and the four animal-based indicators (cleanliness of hindquarter and udder, swelling of tarsal joint, lameness, integument alterations), there was a significant correlation of 0.45. Especially, in the group of lactating cows, the indicator of lameness was significantly correlated with the index of register-based indicators (r=0.40). The group of non-lactating cows showed a significant correlation (r=0.41) between the indicator of swelling of the tarsal joint and the index of register-based indicators. Similar results were found by using a ranking system instead of an index calculation. The study showed that, in the group of 0-15% compared to the group of 16-20% of culling of first-lactation dairy cows, a 24% higher proportion of cows was without lameness. Also, there was a 27% higher proportion of cows without any swellings of tarsal joint in the group with less than 15% of culling of first-lactation dairy cows as in the group with more than 25%. Generally, the 10 register-based and animal-based indicators can be used by the stockman for a self-monitoring and further as an analysis of critical points of the milk production.

Evaluation of colostrum quality in Czech dairy herds

S. Staněk[1], S. Šlosárková[2], E. Nejedlá[1], M. Faldyna[2], R. Šárová[1], J. Krejčí[2] and P. Fleischer[2]
[1]Institute of Animal Science, Přátelství 815, 10400, Prague, Czech Republic, [2]Veterinary Research Institute, Hudcova 296/70, 621 00 Brno, Czech Republic; sarova.radka@vuzv.cz

Colostrum is essential for survival, health, and future productivity of calves. But only high-quality and uncontaminated colostrum can provide a sufficient nutrition and immunologic defense for calves. Therefore, the aim of our study was to test the quality of colostrum and to determine the measurement accuracy of refractometers. In 34 dairy herds (19 Czech Fleckvieh ; C; 12 Holstein; H; 3 C and H), colostrum samples were taken from August 2015 to December 2016. Farm staff took a sample from the whole first colostrum obtained after the calving into sterile 30 ml vessels. The samples were frozen and transported into the laboratory where they were slowly thawed. The colostrum IgG levels (g/l) were estimated using radial immunodiffusion (RID) in 1,265 samples. Subsequently, the samples were evaluated using an optical and two digital refractometers (Optical OPT – 0 to 32% Brix; Simple digital DIGI – 0 to 85% Brix; Digital MISCO – 0 to 85% Brix). The data were analysed using Statistica CZ 10.0. The mean (±SD) of IgG in colostrum estimated using RID was 83.2±43.1 g/l (median 76.9; min 5.9 and max 205.4 g/l of IgG) and 77.3% of colostrum samples contained IgG≥50 g/l (min requested value). The mean (±SD) of %Brix in OPT refractometer was 22.9±4.7% (median 22.9%), in DIGI 18.6±5.3% Brix (median 18.9%), in MISCO 22.8±4.8% Brix (median 22.9%). Spearman's correlation between RID (IgG g/l) and Brix (%) for the three types of refractometers were: RID × OPT r_s = 0.64; RID × DIGI r_s = 0.58; RID × MISCO r_s = 0.64 (P<0.01). Spearman's correlation between the three refractometers were: OPT × DIGI r_s = 0.90; OPT × MISCO r_s = 0.99 and MISCO × DIGI r_s = 0.90 (P<0.01). The differences in estimating colostrum quality with digital refractometers (MISCO × DIGI) may be due to the different numbers of detector elements and PPI resolution (pixels per inch) or also due to the fact, that more precise types of refractometers (e.g. MISCO) are equipped with a protective evaporation cover to reduce the measurement error (distribution of light and refraction in the liquid). This study was supported by projects QJ1510219 and MZERO0717.

Factors affecting lactation performance of Hungarian buffalo cows

B. Barna and G. Holló
Kaposvár University, Guba S. street 40., 7400 Kaposvár, Hungary; hollo.gabriella@sic.ke.hu

The aim of this study was to analyse some factors (morning/evening milking, lactation stage, age) affecting lactation performances of organic buffalo herds. The milk production of herds was recorded in test-day analysis (TD1-TD16); the lactation milk yield based on test day recorded milk yield (morning+evening) was calculated. The effect of age, lactation stages or rather the effect of morning and evening milking were evaluated by using GLM method (SPSS 20.0). Time of milking (am or pm) had a significant influence on milking time and milk yield, evening milking was not recorded in case of some buffaloes. The milkability of animals of the young age category was lower than for animals from age categories from 6 to10 years old. The average milk production per lactation of buffaloes (n=19) was 1,115 kg. The highest milk production was detected in the 6th lactation, whilst the lowest in the 2nd lactation. This period is shorter than the standard lactation period proposed by ICAR. The daily milk yield was the same in all lactations (4.8 kg). The mean fat kg was the highest in the 6th lactation (107 kg), and the lowest in the 2nd, 8th and 10th lactations (94 kg). The overall means of non-fat solids and proteins were on average 113 kg and 49 kg, resp. These values increase with the lactation number until the 6th lactation. The milk quantity values did not significantly differ between lactations. The mozzarella yield index increased until the 6th lactation. The highest milk yield and milk quantity (fat, protein, non-fat solids) of buffaloes from 6 to 10 years of age were measured. The lactation length is longer in older animals. However, concerning milk quality traits, the protein and non-fat solids percentages were the highest in young animals (<6 years) and, at the same time, the oldest age group (>10 years) showed the highest fat percentage. The fat percentage increases, whilst protein percentage decreases, with increasing age. The intermediate age group (6-10 years) showed the highest mozzarella yield index (299 kg). The highest milk yield and milk quantity (fat, protein, non-fat solids) of buffaloes from 6 to 10 years of age were measured. The lactation curve was different in the various age classes; the maximum daily milk yield was from 60 to 75 days of lactation in younger animals, but at 100 days of lactation in older animals. The persistency was the most favourable for buffaloes between 6 and 10 years of age.

Session 27

Theatre 1

Improved conditions in the animal gut

V. Bontempo

Università degli Studi di Milano, Via Trentacoste 2, 20134 Milano (MI)Link verso un sito esterno, 20134 Milano, Italy; valentino.bontempo@unimi.it

Gastrointestinal diseases can cause large economic losses in animal production. Gut health has been extensively controlled by the use of drugs; however, the risk of antimicrobial resistance and public opinion discourage the use of antibiotics and suggest a reassessment of measures to improve the health conditions of the gut. Gut health is a very complex system where a number of interactions, positive and negative, occur immediately after weaning or at any change of diet. Dietary interventions as protein and carbohydrates modulation are fundamental strategies to improve nutrient absorption and reduce production of toxin metabolites. Use of raw materials of high quality and digestibility also reduce intestinal microbial challenge. A strategic supplementation of feed additives, alone or in combination, may help to enhance gastrointestinal structure and function. The review will also focus on current knowledge pertaining feed additives supplementation in order to improve nutrient utilization (enzymes, emulsifiers), to modulate positively GI microbiota (probiotics, prebiotics, essential oils) promoting intestinal health.

Session 27

Theatre 2

Effect of selenium sources in weaned piglets subjected to an immune challenge

M. Briens[1], X. Guan[2], E. Eckhardt[1], Y. Mercier[1] and F. Molist[2]
[1]Adisseo France S.A.S., Antony, 92160, France, [2]Schothorst Feed Research B.V., Lelystad, 8200 AM, the Netherlands; mickael.briens@adisseo.com

Selenium (Se) is biologically active through selenocysteine in selenoproteins playing redox and immune functions. This study aimed to compare the effect of dietary mineral (sodium selenite, SS) or organic (OH-selenomethionine, SO) Se sources on immune and redox responses of weaned piglets subjected to systemic challenges with *Escherichia coli* lipopolysaccharide (LPS). Forty-eight 26 day old piglets were randomized in a 2×3 factorial design comprising a challenge (LPS or vehicle: phosphate-buffered saline, PBS) and three diets (negative control; not supplemented with Se (NC); NC + SS or NC + SO; both supplemented at 0.3 mg Se/kg feed). Piglet were fed the three diets *ad libitum* for 21 days and were subjected on days 7 and 21 to LPS or PBS injections. Weekly growth performances were recorded. Blood samples, rectal temperature and tissues were taken at various times. Loin Se content, immune biomarkers (IL-6, TNF-α, IL-10, acute phase protein, natural antibody) and oxidative stress (total glutathione, glutathione peroxidase (GPx) activity, protein carbonyls, malondihaldehyde (MDA)) were measured. Data were analyzed by two-way ANOVA. Rectal temperature and pro-inflammatory cytokines reflected robust stimulation of the immune system by LPS but no dietary effect was observed. On day 21, piglets on the SO diet had higher serum levels of the anti-inflammatory cytokine IL-10 3 h after LPS challenge compared to piglets on SS and NC diets (P<0.05). The SO diet tended to raise the level of the natural antibody IgM. Growth performances were not affected by challenge or diet. However, LPS challenge tended to increase liver MDA over PBS (P=0.066), whereas SO diet tended to reduce MDA over NC and SS diets (P=0.059). SO diet significantly increased loin Se concentrations on day 22 compared to NC and SS diets (P<0.05). Serum GPx activity of LPS challenged piglets was significantly improved with SO diets compared to SS diets at day 7 and 3 h post-challenge and day 14 (P<0.05). These results indicate that OH-selenomethionine, compared to selenite and control, may positively affect immune responses by promoting anti-inflammatory cytokine production and alleviate oxidative stress damages during immune stimulation.

Effects of *Scutellaria baïcalensis* on sow mammary epithelial cells

M.H. Perruchot[1], F. Gondret[1], H. Quesnel[1], F. Robert[2], L. Roger[2], E. Dupuis[2] and F. Dessauge[1]
[1]Pegase, Agrocampus-ouest, INRA, Saint-Gilles, 35590, France, [2]CCPA Group, ZA du bois de Teillay, Janzé, 35150, France; marie-helene.perruchot@inra.fr

In the context of sow hyper-prolificacy, the amount of milk available per piglet may be limiting for piglet survival, health and growth performance, especially for the lightest piglets. Findings from an *in vivo* study suggested that inflammation occurring the days after farrowing had a negative impact on sow milk production in hyperprolific sows. Therefore, any feed ingredients with anti-inflammatory properties may be helpful to increase milk production and piglet performance. Recently, *Scutellaria baïcalensis*, a widely spread labiate in Russia, China and Japan, has been shown to increase piglet growth rate from birth to weaning. Baicalin, the major compound found in *S. baïcalensis* extract, is a flavone glycoside with strong antioxidant and anti-inflammatory effects. The aim of the present study was to characterize the effects of baicalin on porcine mammary epithelial cells (MEC). MEC were purified from mammary glands collected collected in sows at peak lactation and cultured in the presence of increasing concentrations of baicalin (from 0 up to 200 µg/ml). We demonstrated that baicalin stimulated MEC proliferation when added to culture media at low concentrations (+60% at 10 µg/ml, $P<0.001$). The effect of baicalin on oxidative stress was investigated through reactive oxygen species (ROS) production by MEC after 1.5 h or 24 h of incubation. A marked anti-oxidative effect was observed on MEC at low as well as at high concentrations of baicalin during a 1.5 h or 24 h incubation (+90% ROS on average with baicalin at 200 µg/ml, $P<0.001$). When hydrogen peroxide, a pro-oxidative reagent, was added in culture media, the anti-oxidative effects of baicalin were also observed. In conclusion, these results suggest that effect of *S. baïcalensis* on piglet growth rate could be related to the anti-oxidative effects of baicalin on mammary epithelial cell features, which may have favored milk production.

Effect of amino acid supplementation and milk allowance on liveweight and body composition in calves

S.A. McCoard, F.W. Knol, C.M. McKenzie and M.A. Khan
AgResearch Grasslands, Palmerston North, 4442, New Zealand; sue.mccoard@agresearch.co.nz

Prior studies indicate that arginine (Arg) or glutamine (Gln) supplementation or increasing milk allowance (MA) may enhance calf performance. However, the impact of Arg or Gln supplementation on growth and organ development in non-disease challenge models is limited and their interaction with milk allowance is poorly understood and was therefore the focus of this study. Forty calves (4±1 d) were randomly allocated to 4 groups (n=10/group), and individually fed whole milk powder at 125 g/l at low (10% d 0 BW/d; LA) or high (20% d 0 BW/d; HA) allowance with Arg or Gln added at 1% of milk DM or no AA (Ctrl) in combination with LA or HA using automatic feeders. Meal and water were offered *ad libitum*. Live weight, eviscerated carcass weight (ECW) and mammary, liver and peri-renal fat weights recorded at slaughter (35±1 d). Data were analysed with a linear mixed model (using REML) with diet (Arg, Gln and MA) as the fixed effect and calf parameters (farm source, sex, age, and d 0 BW) as random effects, and ECW as a covariate for evaluation of organ parameters. Live weight and ECW was greater in HA than LA calves irrespective of AA supplementation ($P<0.001$). An interaction between AA and MA was observed for the mammary gland, liver and peri-renal fat weights ($P<0.01$). Mammary size was increased in response to higher MA with additional increases in response to Arg and Gln in HA calves only. Liver weight increased in response to higher MA with additional increases in response to Arg in both HA and LA groups, and to Gln in the LA group only relative to HA and LA controls. Peri-renal fat mass increased in response to higher MA with additional benefits observed with Gln in the HA group relative to HA-Ctrl calves. These results indicate that while MA has a positive effect on mammary gland, liver and peri-renal fat mass, additional increases in the mass of these organs can be achieved in response to Arg or Gln depending on the level of MA. Studies to evaluate the impact of these changes in organ growth on calf performance and underpinning mechanisms are underway to explore the potential utility of these feeding strategies to optimise nutritional management of artificially-reared calves.

Physicochemical characterization of feed grade zinc oxide sources

D. Cardoso[1,2], Y. Chevalier[1] and A. Romeo[2]
[1]*University of Lyon, LAGEP, 43 bd 11 Novembre, 69622 Villeurbanne, France, [2]Animine, 332 chemin du Noyer, 74330 Sillingy, France; dcardoso@animine.eu*

The bioavailability of a trace mineral source is related to its *in vivo* solubility, which in turn is determined by its physicochemical properties. It is still not clear which characteristics are more relevant in affecting solubility and bioavailability of feed compounds. Zinc Oxide (ZnO) is a common feed additive used to supplement zinc in the diet of monogastric animals. However, different sources have shown different responses in animal bioavailability. This project aims to characterize different feed grade ZnO to better clarify their fate in the digestive tract and significant differences in bioavailability. Over 25 samples of ZnO have been collected from the feed industry from Americas, Europe and Asia. 4 representative samples were analyzed and compared for particle size, morphology and specific area using low-angle light scattering, electron microscopy (SEM and TEM) and BET nitrogen adsorption isotherms. SEM suggested that ZnO-1, ZnO-2 and ZnO-3 were grinded dense materials, while ZnO-4 showed aggregates having a sponge internal structure. TEM demonstrated ZnO-1 and ZnO-2 were made of platelets, not forming aggregates. ZnO-3 presented rod-like crystals forming aggregates. ZnO-4 showed platelets forming aggregates. Light scattering provided the size distributions of aggregates and agglomerates. ZnO-4 presented large agglomerates and small aggregates, while the other samples did not differ from aggregates and agglomerates. ZnO-4 showed a porous high specific area, while ZnO-1, ZnO-2 and ZnO-3 have a very low specific area and not porous. The different physicochemical characteristics of zinc oxides sources can affect their solubility resulting in variable bioavailability. A systematic study is being developed to understand this relationship.

Effects of starch type and starch to fibre ratio on rumen *in vitro* starch and NDF degradability

S. Malan, E. Raffrenato and C.W. Cruywagen
Stellenbosch University, Animal Sciences, Private bag X1, 7602 Matieland, South Africa; emiliano@sun.ac.za

Proportions of starch type (amylose/amylopectin) impact fermentability of grains in the rumen. The objective of this study was to isolate and determine the specific effects of amylose and amylopectin on rate and extent of starch and fibre digestion, when combined with two forages and at different starch to fibre ratios. Two different starch sources were used with different amounts of amylose, Hylon VII (74% amylose starch) and Amioca (98% amylopectin starch). Both starch types were combined with either lucerne or oat hay in order to create combinations of either high (2:1) or low (1:1) starch to NDF ratios. The samples were analysed for *in vitro* starch and NDF digestibility. All the following statements of significance correspond to a $P<0.05$. Amioca had the greatest starch digestibility and the addition of forages increased starch digestion. Rate of starch digestion was 0.275 and 0.181 h^{-1} for Amioca and Hylon, respectively. Rates were influenced by starch and forage, but not by starch level. Neutral detergent fibre digestion (NDFd) was influenced by forage and starch type. Lucerne had the greatest NDFd and the addition of starch reduced NDFd of forages. Forages in combination with Amioca had the lowest NDFd. The NDF rates of digestion were 0.073 h^{-1} for lucerne and 0.039 h^{-1} for oat hay. The rate of NDF digestion for oat hay was not influenced by starch amount or type and resulted in 0.031 h^{-1} for Amioca and 0.033 h^{-1} for Hylon. Rate of NDF digestion for lucerne was significantly reduced by the addition of starch, resulting in 0.070 h^{-1} for Amioca and 0.058 h^{-1} for Hylon. By knowing the exact proportions of each starch type in grains we can better characterise the starch fraction digestion characteristics. When other factors affecting starch digestion are known, a well-defined amylose-amylopectin ratio can be obtained to better quantify, and adjust if needed, the speed of starch digestion in the rumen, which can easily affect rumen health in dairy cows.

Improvement of feed and nutrient efficiency in pig production, value of precision feeding

L. Brossard and J.Y. Dourmad
INRA, Domaine de la Prise, 35590 Saint-Gilles, France; ludovic.brossard@rennes.inra.fr

Recent studies indicate that feed efficiency is a key factor for both the economic and the environmental sustainability of pig production. The efficiency of energy utilization by sows and fattening pigs depends mainly on their reproductive and growth potential, and on the rate of deposition of body lipids and protein. These factors also affect the efficiency of use of protein, amino acids and minerals, and consequently their excretion. In fact, both the oversupply and the undersupply of nutrient may induce a decrease in the efficiency of nutrient use, which makes it difficult to handle in practice because of the changes in nutrient requirements over time and their variability among animals. In this context, the availability of new technologies for high throughput phenotyping of pigs and their environment, and of innovative feeders that collect information and allow the distribution of a combination of different diets, offers opportunities for a renewed implementation of prediction models of nutrient requirements, in the perspective of improving feed efficiency and reducing feeding costs and environmental impacts. The approach is to better adapt the amount and composition of the feed distributed to a group of animals or to each individual animal, according to the expected performance over the forthcoming period. The ultimate goal is to feed each individual pig, each day with an optimized tailored ration, corresponding to its expected performance. This requires (1) the real-time collection of information on animal performance and housing conditions, (2) the development of a decision support system for the determination of expected performance and nutrient requirements and (3) the mixing and the distribution by an automated feeder of the optimal ration to be fed to a given pig or a given pen. This represents a real change in the paradigm of animal nutrition for the coming years. Recent experimental studies in the literature confirm the interest of precision feeding, as well for fattening pigs and gestating sows, with potential reductions of feeding cost and nitrogen and phosphorus excretion of up to 10 to 20%. However further technological developments may still be needed before practical implementation.

Amino acid concentration in different seaweed species and availability for ruminants

M.R. Weisbjerg[1], H.S. Bhatti[1], M. Novoa-Garrido[2], M.Y. Roleda[3] and V. Lind[3]
[1]Aarhus University, AU Foulum, 8830 Tjele, Denmark, [2]Nord University, Bodø, 8049 Bodø, Norway, [3]NIBIO, Bodø and Tjøtta, 1431 Ås, Norway; martin.weisbjerg@anis.au.dk

Seaweeds have great potentials as feed component for ruminants. Earlier we have shown that crude protein in some raw seaweeds may be protected against rumen degradation and still have a high intestinal digestibility of rumen escape protein. This study aims to examine the amino acid (AA) composition and concentration, and rumen degradability and total tract digestibility of AA. Three red seaweed species (Mastocarpus stellatus, Palmaria palmata, Porphyra sp.), four brown (Alaria esculenta, Laminaria digitata, Pelvetia canaliculata, Saccharina latissima) and two green (Acrosiphonia sp., Ulva sp.) were used in this study. All seaweeds were collected in four harvests in Bodø, Northern Norway, during spring and autumn in 2014 and 2015, except for Ulva, which was only sampled in autumn both years, and Saccharina which was not sampled in spring 2014. All samples were studied for AA concentration, but only six species were selected for the *in situ* study. The content of total AA in crude protein in the seaweed species ranged from 67.1 in Laminaria to 90.2 gAA/16 g N for Ulva. No difference was observed between seasons for the proportion of AA N in total N (AAN/N) (P=0.80). However, the effect of species on AAN/N was profound (P<0.01) and ranged from 0.57 to 0.77. An effect of species was seen on the *in situ* AA degradation profile (P<0.05). However, there was no effect of season on the ruminal AA degradability except for alanine and proline (P<0.03) and the interactions between species and season were found not to be significant with the exception of proline (P=0.03). The total tract degradability showed that some of the AA were rumen protected and the protection seemed to be acid labile giving high total tract digestibility. Total AA digestibility in the total tract varied from 419 g/kg for Laminaria spring samples to 917 g/kg for Porphyra autumn samples. The study showed that AAN proportion of total N is comparable with terrestrial forages, and for some seaweeds AA protein is protected against rumen degradation, without invalidating total tract digestibility, making them interesting sources for bypass protein supply.

Excess methionine supply during a short period improves technological and sensory pork quality

B. Lebret[1], D.I. Batonon-Alavo[2], M.H. Perruchot[1], Y. Mercier[2] and F. Gondret[1]
[1]INRA, Pegase, Agrocampus-Ouest, 35590 Saint-Gilles, France, [2]ADISSEO France SAS, 2 rue Marcel Lingot, 03600 Commentry, France; benedicte.lebret@inra.fr

Both meat industries and consumers claim for better pork quality. Production factors including nutritional strategies affect muscle properties before pig slaughtering, which might improve technological and sensory meat quality traits. However, many of those factors deteriorate growth performance, limiting their application in pig production. Methionine (Met) is the second limiting amino acid in pigs, and is an essential component for glutathione (GSH) synthesis, the main non-enzymatic cellular antioxidant. Reduced protein and lipid oxidations during meat storage may limit drip loss and improve the stability of meat color. This study aimed at evaluating whether dietary Met provided in excess might improve pork quality. During the last 14 days before slaughter, crossbred [(LWxLD)xPietrain] pigs were individually fed one (n=15 pigs/diet) of three pelleted finishing diets (13.6% protein, 5.8% fat, 2,488 kcal/ kg Net energy, 0.73% Lys) supplemented with 0 (CONT), 5 (Met+) or 10 (Met++) g/kg of OH-Met (HMTBA). Dietary Met content was 0.22% (growth requirement), 0.66% (3X) and 1.1% (5X) in CONT, Met+ and Met++ diets, respectively. After 14 days of feeding, pigs were slaughtered. Growth rate, feed intake and feed conversion ratio were not significantly different between groups. Carcass weight, carcass composition and lean meat content were similar in the 3 groups. In the Longissimus muscle, GSH content was greater (P<0.001) in Met++ pigs than in the two other groups. Ultimate pH in the loin and ham muscles was also higher in Met++ pigs (+0.10 to 0.15 units, P<0.01). Drip losses, meat lightness and hue angle (yellow color) of loin after 7 d storage at 4 °C were the lowest (P<0.10) in Met++ pigs. Technological meat quality index of ham was improved by Met supplementation in a dose-dependent manner (P<0.01). Thus, dietary Met in excess (5X requirement) improves the technological quality and color stability of pork during storage, without any adverse effects on pig performance or carcass traits.

The effect of the supplementation of Bacillus subtilis RX7 and B2A strains in laying hens

W.L. Zhang[1], H.M. Yun[1], J.H. Park[1], I.H. Kim[1] and J.S. Yoo[2]
[1]Swine Nutrition and Feed Technology 421-1 Department of Animal Resources Science Dankook University, Department of Animal Resource & Science, Dankook University, #29 Anseodong, Cheonan, Choognam, 330, #119, Dandae-ro, Dongnam-gu, Cheonan-si, Chungnam, 330-714, Korea, Korea, South, [2]Daehan Feed Co., Ltd, 13, Daehan Feed Co., Ltd 13 Bukseongpo-gil, Jung-gu, Incheon 400-201, Bukseongpo-gil, Jung-gu, Incheon 400-201, Korea, South; yunhm822@naver.com

This study was conducted to investigate the effect of the supplementation of *Bacillus subtilis* RX7 and B2A strains on egg production, egg quality, blood profiles and excreta *Salmonella* in laying hens. A total of 288 ISA-brown laying hens (40-week-old) were selected for a 5-wk feeding trial. Hens were randomly allocated into 6 treatments with 4 replications per treatment, and 12 hens per replication, according to a completely randomized design. Treatments consisted of, basal diet (No antibiotic, No *B. subtilis*), PC, NC + 0.05% antibiotic (virginiamycin) diet; A, NC + 0.05% *B. subtilis* RX7 1.0×10^9 cfu/g; B, NC + 0.05% *B. subtilis* B2A 1.0×10^9 cfu/g. All data were arranged to evaluate by analysis of variance following the GLM procedure in a completely randomized design using the SAS software program. The treatment effect was observed significant with the P<0.05. Egg production was higher (P<0.05) in treatment B (96%) than in NC (90%) and PC (91%) treatments at 3 wk. Eggshell thickness in treatments B (41.17 mm^{-2}) had higher (P<0.0.5) than NC (40.48 mm^{-2}), PC (40.51 mm^{-2}) and A (40.64 mm^{-2}) treatments at 3 wk. Excreta *Salmonella* counts was higher (P<0.05) in treatment NC (2.59) than in PC (2.23), A (2.28) and B (2.28) treatments. In conclusion, the supplementation of *B. subtilis* RX7 and B2A strains improved egg production, eggshell thickness and lower excreta salmonella in laying hens.

Relative bioavailability of selenium sources for beef cattle using selenium serum level
J.S. Silva, J.C.C. Balieiro, J.A. Cunha, T. Oliveira, B.F. Bium and M.A. Zanetti
University of Sao Paulo, Animal science, Duque de Caxias Norte, 225, Pirassununga, SP, 13630-500, Brazil;
mzanetti@usp.br

Studies with beef cattle conducted in Brazil have shown that high levels of dietary selenium may reduce cholesterol in meat. In this experiment it was compared the bioavailability of high levels of organic and inorganic selenium using serum selenium concentration. It was used 63 Nellore cattle with approximately 24 months of age and 390 kg live weight, in a feedlot study during 84 days, in individual pens. The animals (nine per treatment) were submitted to one of the seven diets: control diet without additional supplementation of selenium; Control + 0.3 mg Se/kg DM in the form of sodium selenite; Control + 0.3 mg Se/kg DM in the form of organic selenium; Control + 0.9 mg Se/kg DM in the form of sodium selenite; Control + 0.9 mg Se/kg DM in the form of organic selenium; Control + 2.7 mg Se/kg DM in the form of sodium selenite and Control + 2.7 mg Se/kg DM in the form of organic selenium. The organic selenium used was yeast selenium. Diets were formulated according to NRC and roughage: concentrate ratio was 30:70. As roughage, it was used corn silage and the concentrate was a mixture of corn grain and soybean meal. The control diet had 0.065 mg of Se/kg of DM. Animals were weighed at the beginning of the experiment and every 28 days. Food intake was monitored daily and offered in amount to occur 10% of orts. At day zero, 28, 56 and 84 blood was sampled for selenium analysis according Whetter and Ullrey. The bioavailability was calculated by the technique of slope ratio assay. Linear regression was performed with the general linear models procedure (Proc GLM) of SAS to characterize serum Se concentrations. The slopes estimation with standard errors were: 0.031 ± 0.002 for the sodium selenite and 0.041 ± 0.002 for the yeast source. The difference between the slopes was significant ($P<0.0001$). The relative bioavailability estimated by serum selenium concentration for the yeast selenium in relation to the sodium selenite (100) was 1.32 or 136%, when using diets high in concentrate and with high selenium levels. Acknowledgments: FAPESP.

Different sources of resistant starch *in vitro* show contrasting fermentation and SCFA profiles
J. Leblois, J. Bindelle, A. Genreith and N. Everaert
Gembloux Agro-Bio Tech (ULg), Precision Livestock and Nutrition Unit, Passage des Déportés, 2, 5030 Gembloux,
Belgium; julie.leblois@ulg.ac.be

Resistant starch (RS) is well known to be fermented in the caecum and the colon of animals, increasing the production of short-chain fatty acids (SCFA), especially butyrate. The latter is health-promoting, exhibiting anti-inflammatory effects on the gut and hence it has been postulated that dietary strategies should aim for an increased intestinal butyrate production in pig production. In this study, five different purified sources of RS (high-amylose maize, potato and pea starches) have been tested *in vitro* for their fermentation kinetics and SCFA profiles as a preliminary step to include one of these substrates in the diet of sows to modulate intestinal microbiota and fermentation patterns. Briefly, after an *in vitro* hydrolysis with porcine pepsin and pancreatin, undigested residues recovered by dialysis (1000 kD) were fermented *in vitro* for 72 h in a gas test using sows faeces as microbial inoculum. Six vials per RS source were fermented, 3 were used for determination of SCFA profiles at five consecutive time-points and 3 for the fermentation kinetics based on the monitoring of gas volume at regular intervals. SCFA production and profile was measured by high performance liquid chromatography, while fermentation kinetics was mathematically modelled to allow proper comparison between RS sources. All statistical analyses were performed with the MIXED procedure of SAS and repeated measurements for SCFA. All investigated parameters were influenced ($P<0.05$) by the RS source. Pea starch showed the highest butyrate level at each time point (>15% of total SCFA from 12 h after the beginning of the fermentation onwards) and exhibited an extensive and rapid fermentation (highest final gas volume (A) and gas production rate and lowest time to reach A/2) while high amylose maize produced the lowest butyrate proportion (<6% of the total SCFA) during the slowest and lowest fermentation. Therefore, pea starch appears to be the most promising to be used in pig nutrition to modulate intestinal fermentation.

In vitro evaluation of exogenous proteolytic enzymes in monogastric feed

L. Coppens and Y. Beckers
Gembloux Agro-Bio Tech (ULg), Precision livestock and nutrition, Passage des déportés, 2, 5030 Gembloux, Belgium;
lcoppens@ulg.ac.be

Poultry and pig meat consumption will continue to increase in the future. However, these animals are competitive with humans, both for energy and protein sources. It is therefore important to develop feed alternatives to reduce their dependence on eligible products for human needs. In this context, proteolytic enzyme supplementation is justified on nutritional, economic and environmental aspects. Nowadays, a limited number of proteolytic enzymes are allowed on the European feed market. The aim of this research is to develop fast and accurate *in vitro* methods for the evaluation of proteolytic enzymes eligible in monogastric diets in order to improve protein digestibility. Fifteen proteolytic enzymes were characterized and examined for their effectiveness to produce peptides and amino acids from wheat gluten. The tested enzymes showed different optimal temperature and optimal pH, range from 40 °C to 90 °C and from pH 4,5 to pH 10,5, respectively. An *in vitro* two-step method was performed simulating digestion in the stomach and small intestine of wheat gluten by each tested enzyme. Supernatants were recovered and alpha-amino nitrogen and total nitrogen contents were determined using ninhydrin assay and Kjeldhal method, respectively. The proteolytic hydrolyses performed on wheat gluten showed a correlation between alpha-amino nitrogen released and total nitrogen according a ratio of one to ten. PH-sensitivity was also evaluated. Some proteolytic enzymes present a decrease in their efficiency after incubation at pH 3, while others acted only under acidic conditions. All tests will be judged on their ability to discriminate proteolytic enzymes. *In vivo* tests will be done to confirm or infirm results obtained from *in vitro* methods. This will be done by a growing *in vivo* test, *in vivo* protein digestibility and N-balance trials with poultry.

Selenium and vitamin E in the cow's diet improve milk's components before and after pasteurization

M.S.V. Salles[1], A. Saran Netto[2], L.C. Roma Junior[1], M.A. Zanetti[2], F.A. Salles[1] and K. Pfrimer[3]
[1]APTA, DDD, A. Bandeirante 2419, 14030670, Brazil, [2]USP, FZEA, Duque de Caxias norte,225, 13635-900, Brazil, [3]USP, FMRP, Av. Bandeirantes, 3900, 14049-900, Brazil; marciasalles@apta.sp.gov.br

Healthy nutrition is a concern of most of the world's population, thus the importance of animal science studies to improve milk nutrient composition. The aim was to study the vitamin E and selenium (antioxidants) with sunflower oil (SFO) added to the diet of lactating cows to improve the nutrient profile of milk before and after pasteurization. Twenty-eight cows were allocated in four treatments, as follows: C (control diet); O (4% of SFO in dry matter (DM) diet); A (3.5 mg/kg DM of organic selenium + 2,000 IU of vitamin E/cow per day); OA (4% of SFO in DM diet + 3.5 mg/kg DM of organic selenium + 2,000 IU of vitamin E/cow per day). Cows were fed with 0.50 of concentrate, 0.42 of corn silage and 0.08 of coast-cross hay (DM). Milk samples from each cow were taken in the last week of trial, than milk samples per treatment (pool from cows) were taken after pasteurization and analyzed for selenium and alpha-tocopherol. Data were analyzed as a RCBD with a factorial treatment structure (GLM/SAS). The addition of antioxidants in the cow's diet increased selenium and alpha-tocopherol on raw milk (C=0.011, O=0.027, A=0.235 and OA=0.358 mg/l, P<0.0001, 0.033 SEM of the selenium and C=2.27, O=1.56; A=3.08 and OA=2.89 mg/l, P=0.0088, 0.36 SEM of the alpha-tocopherol). An interaction (P=0.0036) was observed between antioxidants and SFO treatments for selenium on pasteurized milk (C=0.029, O=0.016, A=0.286 and OA=0.349 mg/l, 0.009 SEM of selenium).The addition of antioxidant in the diet increased alpha-tocopherol on pasteurized milk (C=1.71, O=1.79, A=2.08 and OA=2.15 mg/l, P=0.0014, 0.09 SEM of alpha-tocopherol). The antioxidants added to the diet of lactating cows improved the nutritional profile of raw milk, but only selenium remained in the pasteurization process. Vitamin E concentrations decreased in the processing, but the treatments with supplementation still remained higher than the control. Biofortificated milk with these nutrients is important for health and human nutrition. Financial support: 2012/12667-7.

Effect of a novel association of yeast fractions on broilers sensitivity to an acute heat stress
F. Barbe, A. Sacy, B. Bertaud, E. Chevaux and M. Castex
Lallemand SAS, 19, rue des Briquetiers, 31702 Blagnac, France; fbarbe@lallemand.com

Periods of high temperature represent a major stress in poultry, inducing downgraded performance and increased mortality. 360 day-old chicks (ROSS PM3) were randomly allotted in 2 groups, housed in a total of 18 pens of 20 animals each and fed during 35 days, either with a control diet (C, n=9) or a diet enriched with a novel association of specific yeast fractions (Y, n=9) during the starter phase (0-10 d, 800 g/T) and the grower phase (10-25 d, 400 g/T). Both groups received a control non supplemented diet in the finisher phase (25-35 d). The period of heat stress consisted in a sudden room temperature rise (from 24 to 30 °C during 5 h, in the night between 19th and 20th day of age). The initial weight at D0 was not significantly different between both groups (C: 40.4 g vs Y: 40.6 g, P=0.666). During the starter phase, zootechnical performance were significantly improved in Y group (bodyweight at D10 = C: 346 g vs Y: 359 g; P=0.011), which appeared also more homogeneous than C group (CV = C: 8.9% vs Y: 7.6%). The heat stress induced higher mortality in C group (27%), compared to Y group (12%, P=0.001). The autopsy of dead animals following the heat stress confirmed the difficulty for the animals to evacuate the extra-heat (as observed by the affluence of blood in exchange organs: epidermis and lungs) and highlighted an increased sensitivity of the most feathered and the lightest animals. At D25, it appears therefore that Y group presented more animals which were lighter in average than C group (C: 1,437 g vs Y: 1,374 g; P=0.245), with an increased number of small broilers, inducing a numerical increase of heterogeneity in Y group (CV = C: 9.9% vs Y: 10.8%). However, profitability with this novel association of specific yeast fractions was increased by 8% at the end of grower phase (C: 19.7 kg/m² vs Y: 21.3 kg/m²) and by 14% at slaughtering (C: 33.7 kg/m² vs Y: 38.3 kg/m²). These results demonstrate that this novel association of specific yeast fractions allows improving the resistance of broilers in case of acute heat stress.

Effects of alternative feedstuffs on growth performance and carcass characteristics of growing lambs
M.S. Awawdeh[1], H.K. Dager[2] and B.S. Obeidat[2]
[1]Jordan University of Science and Technology, Department of Veterinary Pathology and Public Health, Faculty of Veterinary Medicine, P.O Box 3030, Irbid 22110, Jordan, [2]Jordan University of Science and Technology, Department of Animal Production, Faculty of Agriculture, P.O Box 3030, Irbid 22110, Jordan; dr.awawdeh@gmail.com

The objective of this study was to investigate the effects of dietary inclusion of alternative feedstuffs (AF) on growth performance, hematological and biochemical parameters, and carcass characteristics of growing Awassi lambs. Lambs (n=27, 19.7±0.5 kg body weigh; BW) were randomly assigned to one of three dietary treatments (9 lambs/treatment). Diets contained, respectively, 0 (CON), 25 (25AF), or 50% (50AF) of the selected AF (dry bread, carob pods, olive cake, and sesame meal). Dietary treatments had no effects (P≥0.54) on dry matter (DM) digestibility and feed conversion ratio. Dietary treatments decreased neutral detergent fiber (NDF) and acid detergent fiber (ADF) and increased ether extract digestibilities. Lambs fed the 50AF had the least (P<0.05) nutrient (DM, NDF, crude protein, and metabolizable energy) intake with no differences between the CON and 25AF diets. Lambs fed the 50AF diet tended to have lower final BW, hot and cold carcass weights, retained nitrogen, and average daily gain than the CON diet with no differences between the CON and 25AF diets. No substantial differences among dietary treatments were observed in carcass and non-carcass cut weights, eye muscle width and depth, rib-eye area, leg composition (muscle, fat, and bone %), and meat quality parameters. Meat form lambs fed the 25AF diet had higher carcass, rib, and leg fat depths than the CON diet, with no differences between the CON and 50AF diets. Hematological and biochemical parameters were not significantly (P≥0.07) affected by dietary treatments, except for blood urea nitrogen which was lowest in lambs fed the 50AF diet. Blood cortisol was lowest in lambs fed the 50AF diet. Cost of gain decreased in the 25AF and 50AF diet. Dietary inclusion of AF at 25 or 50% decreased production cost with similar feed conversion ratio. However, at high level (50%) AF could negatively affect nutrients intake, digestibility, and growth rate.

Inclusion of protected feed additives on performances and FMD antibody titer of growing pigs

H.M. Yun[1], K.Y. Lee[2], B. Balasubramanian[1,2] and I.H. Kim[1]
[1]Dankook University, Department of Animal Resource & Science, Cheonan, 31116, Korea, South, [2]Morningbio Co., Ltd, Floor 4 Semi B/D, 9-1 Dujeongyeokdong4-gil, Seobuk-gu, Cheonan, 31111, Korea, South; yunhm822@naver.com

A total of 150 growing pigs [(Landrace×Yorkshire)×Duroc] with an initial body weight (BW) of 25.61±0.01 kg were used in a 4-wk trial to evaluate the effects of protected feed additives on growth performance, blood metabolites, fecal microflora counts and immune responses. In this study, commercially available mixed protected feed additive (MARK) was used that contains protected organic acids (fumaric acid, citric acids, malic acid), nucleotide, omega-3 and amino acids (threonine, lysine, methionine). Pigs were allotted to one of five diets with basal diet supplemented with 0, 0.35. 0.75, 1.5 and 3% of MARK (6 replications with 5 pigs/pen). Diets were formulated to comply with NRC recommendations of nutrient requirements for swine. The results of growth performance indicated that inclusion of MARK had a linear trend on BW (P=0.085) at week 4 and an overall average daily gain (ADG, P=0.064), however there was no significant effects on average daily feed intake and gain:feed ratio during the entire experiment period. When more MARK was added into the diets, reducing linear effects were observed on cortisol (P=0.032) and LDL/C (P=0.008) at week 4 with no difference (P>0.05) on HDL/C and total cholesterol among dietary treatments. There was a linear trend on fecal score at week 3 and 4 (P=0.068, 0.023 respectively), increased number of *Lactobacillus* (P=0.099) and reduced *Escherichia coli* (P=0.067) counts at week 1 and 4 and increased *Lactobacillus* (P=0.014) and *E. coli* (P=0.043) counts at week 4 and 1, respectively. Furthermore, the antibody titer (percentage inhibition [PI] titer) of 1.5% MARK supplemented diet of growing pigs after FMD vaccination was maintained at a higher level (80%) among the other dietary treatments. This result indicated that an improved ADG, reduced cortisol, LDL/C, beneficial fecal microflora counts, decreased diarrhea and rapid increase in levels of antibodies at 4 weeks after vaccination by supplementation of MARK in growing pigs diet.

N-3 fatty acids transfer from hempseed-rich diet to sow milk and changes in plasma lipid profile

M. Habeanu, E. Ghita, A. Gheorghe, M. Ropota, N. Lefter, T. Mihalcea and G. Ciurescu
National Research Development Institute for Animal Biology and Nutrition, Animal Nutrition, 1 Calea Bucuresti, 077015 Balotesti, Ilfov, Romania; elena.ghita@ibna.ro

This study aimed to investigate how a diet rich in n-3 fatty acid (FA) by hempseed addition changed n-3 FA composition in sow's milk and certain plasma parameters (cholesterol and triglycerides). The correlation between these parameters was determined, too. The trial was conducted on 8 sows assigned for 21 d to two groups, namely a control (C), and an experimental (E) with the additional inclusion of hempseed (50 g/kg). The diets were isonitrogenous, isoenergetic and had a similar content in fat (4.5%). Sow milk samples were collected by hand 1st d after farrowing (AF), at 7 and 21 days, respectively. Blood samples were taken in the first, 7th and 21st days of lactation, by puncture in the jugular vein. The FAs composition was determined by gas chromatography. The fat content of the milk was 7.71% at 7 d and 6.8% at 21 d AF. The E group had 1.6 times higher level of alpha-linolenic FA (ALA) in the diet than C group. This was positively reflected in milk ALA composition whatever the day of sampling, although the higher level was noticed at 21 d (1.47 times higher in E diet compared to C diet). The dietary addition of hempseed had more pronounced effects on ALA (P=0.001). This increase of ALA concentration led to significantly higher levels of n-3 FA and to lower n-6: n-3 ratio. DHA level was higher in E group compared to C group (1.87 times at 7 d, 2.33 times respectively at 21 d). At 7 d AF, the plasma triglycerides level increased, then decreased until 21 d, while the cholesterol level increased linearly during lactation whatever the diet. A negative correlation was noticed between milk ALA and plasma triglycerides and a positive correlation between cholesterol and triglycerides. We concluded that 5% hempseed incorporated in the diet markedly changed the n-3 FA in milk. A positive correlation with cholesterol and triglycerides was found to be not significant.

Effects of different n-6 to n-3 polyunsaturated fatty acids ratio on gestation-lactating sows

J. Yin[1], H.M. Yun[1], I.H. Kim[1], K.Y. Lee[2] and J.H. Jung[2]
[1]Swine Nutrition and Feed Technology 421-1 Department of Animal Resources Science Dankook University, Department of Animal Resource & Science, Dankook University, #29 Anseodong, Cheonan, Choognam, 330, Dankook University #119, Dandae-ro, Dongnam-gu, Cheonan-si, Chungnam, 330-714, Korea, Korea, South, [2]Morningbio Co., Ltd, Floor 4 Semi B/D, 9-1 Dujeongyeokdong4-gil, Seobuk-gu, Cheonan, Chungnam, 31111, Korea, 9-1 Dujeongyeokdong4-gil, Seobuk-gu, Cheonan, Chungnam, 31111, Korea, Korea, South; yunhm822@naver.com

The polyunsaturated fatty acids (PUFA) can benefit pregnant and lactating sows under catabolic conditions with outcomes of improving fetal growth, neonatal health, and lactation performance. This study was conducted to evaluate the effects of dietary ratios of n-6:n-3 PUFA on reproductive performance, fecal microbiota and nutrient digestibility of gestation-lactating sows and suckling piglets. Fifteen primiparous sows (Landrace × Yorkshire) were randomly allotted into three treatments. Treatments consisted of different ratios of omega 6: omega 3 (20:1, 15:1, and 10:1, respectively) supplementation in the diet. Differences among all treatments were separated by Duncan's multiple range test. No differences were detected among the treatments for average daily feed intake (ADFI) of sows and the back fat levels during lactation (P>0.05). Body weight (BW) loss of sows after farrowing to weanling were greater (P<0.05) in 10:1 treatment compared with 15:1 and 20:1 (4.5 vs 5.7 kg, 5.8 kg). In piglets, a linear increase in BW was observed at 4 weeks (7.248, 7.426 and 7.845 kg, respectively, P<0.01). Furthermore, average daily gain (ADG) of piglet in 10:1 treatment was higher than 15:1 and 20:1 (229 g/d vs 216 g/d, 210 g/d, respectively, P<0.05). No difference was observed among treatments in nutrient digestibility of sows (P>0.05). Sows fed 10:1 and 15:1 treatment had decreased fecal *Escherichia coli* counts than those fed 20:1 (6.11 log10 cfu/g, 6.13 log10 cfu/g vs 6.31 log10 cfu/g, respectively, P<0.05). Moreover, 10:1 and 15:1 treatment had increased (P<0.05) fecal *Lactobacillus* counts compared to those fed 20:1 (7.45 log10 cfu/g, 7.38 log10 cfu/g vs 7.14 log10 cfu/g, respectively, P<0.05). In conclusion, altering the ratio of n-6:n-3 PUFA in gestation-lactating sow diet can partially improve reproductive performance.

The effect of supra-optimal inorganic, organic or nano selenium on blood selenium status of ewes

M.R.F. Lee[1,2], H.R. Fleming[2], C. Hodgson[2] and D.R. Davies[3]
[1]University of Bristol, School of Veterinary Science, Langford, BS40 5DU, United Kingdom, [2]Rothamsted Research, Sustainable Agriculture Science, North Wyke, EX20 2SB, United Kingdom, [3]Silage Solutions Ltd, Pontrhydygroes, SY25 6DP, United Kingdom; michael.lee@rothamsted.ac.uk

Selenium (Se) is a trace element essential for cellular function. In Europe Se is often deficient in livestock diets due to the low Se status of soil. Supplementation of diets with organic selenomethionine is common due to its greater bioavailability over inorganic Se. Recent studies have shown that the addition of L. plantarum and sodium selenite to grass during the silage making process results in silage containing Nano-selenium (NSe) and small traces of selenocysteine which may act as an alternative delivery mechanism. The aim of this experiment is to examine the effect of feeding Se in three different forms on the Se status of ewes. Suffolk × Mule ewes (24; mean weight 76±4.5 kg, mean age 5±1.7 years) were enrolled onto the 6-week study. Animals were split into four blocks of six according to live weight and allocated: Grass silage baled with the addition of *Lactobacillus plantarum* (control; Con); or with sodium selenite added prior to feeding (ISe); or with selenomethionine (Sel-plex®) added prior to feeding (OSe); or Grass silage baled with the addition of *L. plantarum* and sodium selenite (NSe). Total Se and speciation (organic and inorganic species) were assessed using HPLC-UV-HG-AFS. Se intake for Con, NSe, ISe, OSe were 0.28, 2.17, 3.81 and 4.99 mg/kg DM. Blood selenomethione levels after 6 weeks were: 4.2, 7.9, 10.3 and 15.3 mg/l with blood selenocysteine 4.5, 6.9, 7.9 and 6.3 mg/l for Con, NSe, ISe and OSe, respectively. The active site of the main Selenoenzymes (GPX and Iodinase) contains SeCys, whereas storage Selenoproteins consist of SeMet. The control showed no difference from week 0 to 6. However, for NSe, ISe and OSe there is a significant increase in SeCys with no difference between treatments, despite the higher levels of Se intake in ISe and OSe. However, SeMet is highest in OSe, followed by ISe and NSe, reflecting the differences in intake and increase in storage Selenoprotein formation.

In vitro fermentation and acidification potential of several carbohydrates sources with two inocula
Z. Amanzougarene, S. Yuste, A. De Vega and M. Fondevila
Universidad de Zaragoza – CITA, Departamento de Producción Animal y Ciencia de los Alimentos, Instituto Agroalimentario de Aragón, Miguel Servet 177, 50013 Zaragoza, Spain; mfonde@unizar.es

The high rate and extent of rumen microbial fermentation in high cereal diets for young ruminants frequently leads to acidosis. This in vitro work studies the differences in fermentation of different carbohydrate sources and their capacity to modulate environmental conditions. Six carbohydrate feeds (barley, B; maize, M; brown sorghum, S; sugarbeet pulp, BP; citrus pulp, CP; and wheat bran, WB) were tested in an in vitro semi-continuous culture system under a poorly buffered medium from 0 to 6 h, and allowed pH to rise to around 6.5 from 8 to 24 h. The rumen fluid was obtained from six lambs, three fed with concentrate and barley straw (concentrated inoculum, CI), and three fed with alfalfa hay (forage inoculum, FI). Gas production and incubation pH were recorded at 2, 4, 6, 8, 10, 12, 16, 18 and 24 h, and dry matter disappearance (DMd) was determined at 24 h. With CI, incubation pH and in vitro gas production were affected by the substrate ($P<0.01$). Lowest incubation pH was recorded after 6 h, and averaged 5.96 ± 0.2. After this time, pH increased to an average of 6.64 ± 0.02 at the end of incubation. Among substrates, from 2 to 18 h, CP always recorded the minimum incubation pH ($P<0.05$). At every time of measurement, the volume of gas recorded with CP was highest ($P<0.05$), followed by B and WB, whereas the lowest volume was recorded by S. Similarly, with FI, minimum incubation pH was reached after 6 h (6.25 ± 0.2), and increased to 6.63 ± 0.04 at 24 h. The lowest pH was recorded with CP, and the highest with S ($P<0.05$). From 6 to 24 h, the volume of gas produced was highest from CP and lowest for S ($P<0.05$). Gas production results were supported by DMd for both inocula. Magnitude and extent of in vitro microbial fermentation were higher with CI than FI. The capacity of rumen and substrate acidification are parameters which play an important role on in vitro microbial fermentation.

Glutathione peroxidase (GPx) activity in beef cattle supplemented with sodium selenite or Se yeast
M.A. Zanetti, J.S. Silva, T. Oliveira, B.F. Bium, S.M.P. Pugine and M.P. Melo
University of Sao Paulo, Animal Science, Duque de Caxias Norte, 225, 13635-600, Brazil; mzanetti@usp.br

Recent research conducted at FZEA-USP Pirassununga, Brazil, demonstrated that it is possible to reduce the cholesterol in bovine meat using high levels of organic selenium. This study aimed to compare the glutathione peroxidase activity in beef cattle supplemented with high levels of sodium selenite or selenium yeast analyzing muscle concentration. It was used 63 Nellore cattle with approximately 24 months of age and 392 kg live weight, in a feedlot study during 84 days, in individual pens. The animals (nine per treatment) were submitted to one of the seven diets: control diet without additional supplementation of selenium; Control + 0.3 mg Se/kg DM in the form of sodium selenite; Control + 0.3 mg of Se/kg DM in the form of selenium yeast; Control + 0.9 mg Se/kg DM in the form of sodium selenite; Control + 0.9 mg Se/kg DM in the form of selenium yeast; Control + 2.7 mg Se/kg DM in the form of sodium selenite and Control + 2.7 mg Se/kg DM in the form of selenium yeast. Diets were formulated according to NRC and roughage: concentrate ratio was 30:70. As roughage, it was used corn silage and the concentrate was a mixture of corn grain and soybean meal. The control diet had 0.065 mg of Se/kg of DM. Animals were weighed at the beginning of the experiment and every 28 days. Food intake was monitored daily and offered in amount to occur 10% of orts. At the end of the experiment (84 days) the animals were slaughtered and muscle samples were taken for glutathione peroxidase activity, according Paglia & Valentine (1967). Statistical analysis was performed with the general linear models procedure (Proc GLM) of SAS (2004), and means compared by Tukey test ($P<0.05$). There was treatments difference ($P<0.001$). All supplemented levels had GPx higher than the control ($P<0.05$). There was not difference between sources ($P>0.05$), but the highest GPx level was for the sodium selenite supplementation at the level of 2.7 mg/kg. Acknowledgments: FAPESP.

Effect of chitooligosaccharides on egg production, egg quality and blood profiles in laying hens

Y. Jiao, Y.M. Kim, H.Y. Sun, J.E. Bae and I.H. Kim
Swine Nutrition and Feed Technology 421-1 Department of Animal Resources Science Dankook University, Department of Animal Resource & Science, Dankook University, #29 Anseodong, Cheonan, Choognam, 330-7, Dankook University #119, Dandae-ro, Dongnam-gu, Cheonan-si, Chungnam, 330-714, Korea, Korea, South; 958425592@qq.com

This study was conducted to evaluate the effects of dietary supplementation of chitooligosaccharide (a natural product obtained through fermentation with different bacterial and yeast strain) on egg production, egg quality and blood profiles in laying hens. A total of 192 Hyline Brown laying hens (46 weeks) were allocated to 4 treatments for 8 weeks trial. Dietary treatments consisted of CON, basal diet; PK1, CON+0.1% chitooligosaccharide; PK2, CON+0.2% chitooligosaccharide; PK3, CON+0.3% chitooligosaccharide. Laying hens were randomly assigned to 4 treatments with 8 replications per treatment and 6 cages as replication. Orthogonal comparisons were conducted using polynomial regression to measure the linear and quadratic effects of increase in the dietary supplementation of chitooligosaccharide. After the dietary supplementation of chitooligosaccharide, linear improvement in HDL levels (P=0.0023) was observed at subsequent weeks (47.9, 55.8, 55.1, 58.2 mg/dl). In week 1, linear (P=0.0010) improvement was observed in shell color (11, 12, 12, 12) and linear increase (P=0.0446) in haugh unit (89.4, 90.2, 90.8, 91.7) in week 7 were observed with the increase in the dose of chitooligosaccharide. In addition, linear effects (P<0.05) on shell thickness at week 2 (44.3, 44.8, 45.2, 46.3 mm^{-2}), week 4 (44.5, 44.2, 45.1, 46.9 mm^{-2}), week 5 (44.6, 45.0, 45.1, 46.3 mm^{-2}), week 6 (44.0, 44.7, 45.5, 46.1 mm^{-2}) and week 7 (43.9, 44.3, 45.3, 46.3 mm^{-2}) were detected as the chitooligosaccharide level increased. In conclusion, the supplementation of chitooligosaccharide in laying hen diets with different doses linearly improved egg shell thickness. It also had positive effect in HDL/C concentration. Thus, chitooligosaccharide has the potential to be used as a substitute to antibiotics.

Evaluating the impact of 25-hydroxyvitamin D3 on sow performance and their litters

J.Y. Zhang, K.T. Kim, J.E. Bae, Y.J. Jeong and I.H. Kim
Swine Nutrition and Feed Technology 421-1 Department of Animal Resources Science Dankook University, Department of Animal Resource & Science, Dankook University, #29 Anseodong, Cheonan, Choognam, 330, Dankook University #119, Dandae-ro, Dongnam-gu, Cheonan-si, Chungnam, 330-714, Korea, Korea, South; inhokim@dankook.ac.kr

The study was to evaluate the impact of dietary supplementation of 25-hydroxyvitamin D3 on sow performance and reproduction. Sixteen multiparous sows [(Landrace Yorkshire) Duroc, average parity=3.79±0.32] were randomly allotted to 2 treatments to give 8 replicates per treatment. Treatments as follow: 1) none-active 25-hydroxyvitamin (0.36 mg cholecalciferol/g) D3 (ND); 2) active 25-hydroxyvitamin D3 (AD) diet (0.36 mg cholecalciferol/g) during pregnancy. The results of this experiment were observed at the gestation of d58-75, d76-95, d96-110, and d111-115. Corn-soybean meal-based diet was formulated to meet or exceed the nutrient requirements recommended by NRC. Data were statistically analyzed by ANOVA, using general linear model (GLM) procedure of the SAS program (SAS Inst. Inc., Cary, NC, USA), for a randomized complete block design. Mean values and standard error of means (SEM) are reported. Statements of statistical significance are based on P<0.05. Results indicated that the sows' farrowing duration was shortened (5.38, 4.71 h) and the number of mummies was significantly decreased (0.5, 0.1) in AD treatment compared with ND treatment (P<0.05), meanwhile, the birth weight (1.18, 1.44 kg) was significantly improved (P<0.05). There were no significant effects on healthy performance at the gestation phase, and the total birth, stillbirth, totally alive, and survival rate of the litter at farrowing duration (P>0.05). This study suggests that sows reproduction performance was improved when active 25-hydroxyvitamin D3 (0.36 mg/g) was fed during pregnancy phase.

Evaluation of feeding periods of dietary conjugated linoleic acid supplement in finishing pigs

N.H. Tran, H.M. Yun, S.I. Lee, J.K. Kim and I.H. Kim
Swine Nutrition and Feed Technology 421-1 Department of Animal Resources Science Dankook University, Department of Animal Resource & Science, Dankook University, #29 Anseodong, Cheonan, Choognam, 330, Dankook University #119, Dandae-ro, Dongnam-gu, Cheonan-si, Chungnam, 330-714, Korea, Korea, South; yunhm822@naver.com

One approach for improving pork quality is the supplementation of naturally occurring feed additives, such as conjugated linoleic acid (CLA), in the growing-finishing diet. The study was conducted to evaluate the diet with 1% level of CLA at different periods prior to finish. A total of 90 finishing pigs ([Yorkshire×Landrace] × Duroc) with an average body weight (BW) of 83.13±1.91 kg were used in this 4-wk trial. Pigs were allotted to 3 dietary treatments based on 6 pens/treatment, 5 pigs/pen. Each dietary treatment consisted of supplementing the basal diet with 1.0% CLA during 2, 3 and 4 weeks prior to finish. All data were statistically analyzed using the GLM procedure of the SAS program. Differences among the treatment means were determined by using the Tukey's test with P≤0.05 indicating significance. Pigs fed CLA supplemented diet 2 weeks before finish had a higher average daily gain than those fed during 4 weeks before finish in overall (P<0.05) (936 vs 892 g). In the overall, pigs fed CLA before 2 weeks of finish had a higher Low-density Lipoprotein (LDL) than those fed CLA during 3 and 4 weeks before finish (72 mg/dl vs 52, 65 mg/dl). However pigs receiving CLA diet 4 weeks before finish had greater triglyceride levels (P<0.05) than those fed CLA during 2 and 3 weeks before finish (71 mg/dl vs 58, 56 mg/dl). The color and firmness of meat based on sensory evaluation was reduced (P<0.05) in pigs fed CLA before 2 weeks of finish compared with those fed CLA diet 3 and 4 weeks before finish(3.19 vs 3.40, 3.48) and (2.90 vs 3.25, 3.29) respectively. Our results indicated that supplementation 1% of CLA diet 2 weeks prior to finish improved ADG overall, reduced triglycerides level but increased LDL in blood profiles. However, supplementing the diets with 1% of CLA 4 wk before finish improved colour and firmness of meat quality in finishing pigs indicating supplementation of CLA 4 weeks prior to finish is better in terms of improving meat quality.

Efficacy of dietary supplementation of conjugated linoleic acid in finishing pigs

S.D. Upadhaya, H.M. Yun, S.Q. Huang, I.H. Kim and S.O. Jung
Swine Nutrition and Feed Technology 421-1 Department of Animal Resources Science Dankook University, Department of Animal Resource & Science, Dankook University, #29 Anseodong, Cheonan, Choognam, 330, Dankook University #119, Dandae-ro, Dongnam-gu, Cheonan-si, Chungnam, 330-714, Korea, Korea, South; yunhm822@naver.com

In pork industry, one of the strategies to improve pork quality is the inclusion of conjugated linoleic acid (CLA). Various aspects of meat quality are directly influenced by the type of fat and fatty acids in the adipose tissue or muscle of the animal. The present study was conducted to evaluate the efficacy of dietary supplementation of 0.5 and 1% of CLA on growth performance, apparent nutrient digestibility, serum lipid profile, meat quality and fatty acid profiles in muscle and adipose tissue in finishing pigs. A total of 90 finishing pigs ([Yorkshire ×Landrace]× Duroc) were used in 5-wk trial. All data were subjected to the statistical analysis as a randomized complete block design using the GLM procedures of SAS and the pen was used as the experimental unit. Differences among the treatment means were determined by using the Tukey's test with P≤0.05 indicating significance and P<0.10 indicating trends. The lean percentage was greater (P≤0.05) in pigs fed CLA2 than CON and CLA1 diet (51.7% vs 50.8% and 51.3% respectively) whereas the 2-Thiobarbituric acid reactive substances values were reduced (P<0.05) during the storage days 2, 5, 8 and 14 in CLA supplemented groups. The CLA supplementation improved (P<0.05) C14:0, C 16:0, C18:0, saturated fatty acid, CLA and omega 3 levels and reduced (P<0.05) C18:1, C18:2, unsaturated fatty acid and omega 6 levels in muscle and subcutaneous adipose tissue. In conclusion, CLA supplementation tended to reduce total serum cholesterol and improved lean percentage as well as meat value through enrichment of CLA and omega-3 fatty acid in the muscle and adipose tissue.

Commercial cuts and loin characteristics of lambs fed diets with increasing levels of whey permeate
L.M. Alves Filho, D. Sandri, T. Rodrigues, R.D. Kliemann, H. Maggioni, W.G. Nascimento, S.R. Fernandes and A.F. Garcez Neto
Federal University of Paraná, Animal Science, Rua Pioneiro, 2153, Jd. Dallas, Palotina, Paraná, 85950000 Palotina, Brazil; williangoncalves@ufpr.br

The evaluation of carcass from the longissimus lumborum muscle allows identifying animals with muscular development and fat deposition suitable for consumer markets. Maximum width (LEW), maximum depth (LED) and area of loin eye (LEA), and the fat thickness on the loin eye (LFT) are measurements usually affected by nutrition. The use energetic feedstuffs as dried whey permeate (WP), which contains 90% of lactose on DM basis, may be an alternative to increase that development and deposition. The study was carried out to evaluate the commercial cuts and the loin characteristics of feedlot lambs fed diets with increasing levels of WP. Twenty four crossbred ½ Dorper× ½ Santa Inês non-castrated male lambs with four months of age and 24.1±3.2 kg BW were used. It was used a completely randomized design with four treatments (substitution of ground corn by WP in the diet at the levels of 0.0, 5.0, 12.5 and 25.0% DM) and six replicates. Lambs were fed *ad libitum* with isoproteic (15.8% CP) and isoenergetic (67.0% TDN) diets composed by 64% of *Cynodon* spp. hay and 36% of concentrate. After cooling, the carcasses were longitudinally divided and the left half was sectioned into six commercial cuts: neck, breast, ribs, shoulder, loin and leg. Weight and yield of each cut were registered. In the longissimus lumborum muscle, LEA, LEW, LED and LFT were measured. The weight and yield of cuts were not affected (P>0.05) by the WP levels and showed mean values of 0.65, 0.88, 1.12, 1.44, 0.84 and 2.34 kg for weight; and 9.0, 12.0, 15.4, 19.9, 11.4 and 32.2% for yield of neck, breast, ribs, shoulder, loin and leg. LED, LEA and LFT were not influenced (P>0.05) by the WP levels, with mean values of 3.10 cm, 12.28 cm^2 and 2.23 mm. LEW increased linearly (P<0.05) with the WP (5.65 and 6.25 cm in diets without and with 25.0% DM of WP). Since WP is a carbohydrate of fast fermentation, it was expected better conversion of diets into muscle and fat with the levels of WP. However, the substitution of ground corn by WP up to 25.0% DM of diet does not improve the cuts and loin traits in feedlot lambs.

Dietary organic acids blend supplementation on performance and gut health in sows and their litters
T.S. Li[1], Y.M. Kim[1], S. Serpunja[1], J.H. Cho[2] and I.H. Kim[1]
[1]Swine Nutrition and Feed Technology 421-1 Department of Animal Resources Science Dankook University, Department of Animal Resource & Science, Dankook University, #29 Anseodong, Cheonan, Choognam, 330, Dankook University #119, Dandae-ro, Dongnam-gu, Cheonan-si, Chungnam, 330-714, Korea, Korea, South, [2]Department of Animal Science, Chungbuk National University, Cheongju, Chungbuk, South Korea 28644, Chungbuk National University, Cheongju, Chungbuk, South Korea 28644, Korea, South; inhokim@dankook.ac.kr

This study was conducted to evaluate the effects of protected organic acids blend (OB) by lipid base matrix coating on growth performance, nutrient digestibility, fecal microbiota and fecal score in sows and their litters. A total of 36 sows and their litters were used in a 35-d experiment. On day 108 of gestation, sows were assigned to 3 dietary treatments to give 12 replicates per treatment based on parity. Dietary treatments were: CON, basal diet; MC1, CON + 0.1% OB; MC2, CON + 0.2% OB. All data were analyzed with SAS 2003 using the mixed GLM procedure. At week 2 and 3, there was a linear increase (P<0.05) in body weight (4.10 vs 4.36 vs 4.40, 5.79 vs 6.09 vs 6.18, respectively) and average daily gain (211 vs 221 vs 225, 236 vs 247 vs 255, respectively) of sucking and weaning pigs associated with the supplementation of OB in sows diet. On parturition day, there was a linear increase (P<0.05) in fecal Lactobacillus counts associated with the inclusion of OB (7.30 vs 7.33 vs 7.39). On day 21 of lactating, there was a linear increase (P<0.05) in fecal Lactobacillus counts (7.35 vs 7.45 vs 7.51) and a linear decrease (P<0.05) in fecal *Escherichia coli* counts (6.43 vs 6.35 vs 6.30) of sows associated with the inclusion of OB. The weaning pigs also had a linear increase (P<0.05) in fecal Lactobacillus counts (7.27 vs 7.32 vs 7.39) and a linear decrease (P<0.05) in fecal *E. coli* counts (6.50 vs 6.45 vs 6.41) associated with the inclusion of OB in sows diet. There was a linear decrease (P<0.05) associated with the inclusion of OB in sows diet on week 3 (3.21 vs 3.13 vs 3.05). In conclusion, dietary supplementation of OB in the lactating sows improved the ADG and lower fecal score in sucking and weaning pigs, as well as shifted microbiota by increasing Lactobacillus and decreased *E. coli* counts in sows and their litters.

Effects of drinking water within Ca and Mg on production performance and egg quality of laying hens

X.X. Liang[1], H.M. Yun[1], J.S. Yoo[2], J.Y. Cheong[2] and I.H. Kim[1]
[1]Swine Nutrition and Feed Technology 421-1 Department of Animal Resources Science Dankook University, Department of Animal Resource & Science, Dankook University, #29 Anseodong, Cheonan, Choognam, 330, Dankook University #119, Dandae-ro, Dongnam-gu, Cheonan-si, Chungnam, 330-714, Korea, Korea, South, [2]Daehan Feed Co., Ltd, 13, Bukseongpo-gil, Jung-gu, Incheon 400-201, Bukseongpo-gil, Jung-gu, Incheon 400-201, Korea, South; yunhm822@naver.com

Calcium (Ca) and Magnesium (Mg) are critical in energy-requiring metabolic processes, as well as protein synthesis. This study was executed to investigate the effects of drinking water within Ca and Mg on production performance and egg quality in laying hens. A total of 384, 62-week-old layers (Hy-Line Brown) were used in this six-week trial. Birds were randomly allotted to four dietary treatments each with four replicates and 24 hens per replicate (6 hens/cage, 114×50×40 cm wire cage). The dietary treatments consisted of: (1) CON (basal diet), (2) TRT1 water hardness (WH) is 300 (CON + Ca 0.0002% + Mg 0.0025%), (3) TRT2 WH is 600 (CON + Ca 0.0004% + Mg 0.0051%), and (4) TRT4 WH is 900 (CON + Ca 0.0005% + Mg 0.0076%). All Data were statistically analyzed by analysis of variance, using Duncan's multiple range test of the SAS program (SAS, 2013). No significant differences ($P>0.05$) were observed in feed intake, average daily feed intake, egg production and egg broken rate in laying. In the 2nd wk, eggshell strength of TRT1 (3.869), TRT2 (3.878), and TRT3 (3.884) were significantly higher ($P<0.05$) than CON (3.79). Egg shell strength of TRT3 (3.858) was significantly higher ($P<0.05$) than CON (3.769) in the 5th wk. In the 6th wk, eggshell strength (3.869) and egg shell thickness (47.3) of TRT3 were significantly higher ($P<0.05$) than those of CON (3.782, 45.4). There was no significant difference ($P>0.05$) on egg quality in storage period. In conclusion, Ca and Mg had a positive impact on eggshell quality.

Effects of dietary energy levels on growth performance in lactating sows and piglets

S.Q. Huang, J.K. Kim, S.I. Lee and I.H. Kim
Swine Nutrition and Feed Technology 421-1 Department of Animal Resources Science Dankook University, Department of Animal Resource & Science, Dankook University, #29 Anseodong, Cheonan, Choognam, 330, Dankook University #119, Dandae-ro, Dongnam-gu, Cheonan-si, Chungnam, 330-714, Korea, Korea, South; inhokim@dankook.ac.kr

Twenty five sows and 265 piglets (Landrace×Yorkshire) were used to evaluate the effects of dietary energy level on pre-weaning, post-weaning performance of piglets and first parity sows. Sows with an average initial BW of 217.54±25.47 were randomly assigned with 2 treatments. Treatments consisted of a TRT1 diet containing 3,100 kcal and TRT2 diet containing 3,400 kcal of metabolizable energy (ME)/kg respectively. Data were analyzed using Duncan statements to test the effect of dietary energy levels on growth performance in lactating sows and piglets. In this study, sows fed lower energy diet had higher number of weaned piglets (10.1 vs 9.4) per litter ($P<0.05$). Piglets born from sows fed the TRT2 diet had higher ($P<0.05$) body weight than those fed the TRT1 at weaning (7.772 vs 7.461 kg), meanwhile they had higher total average daily gain ($P<0.05$) compared to TRT1 (246 vs 233 g). Sows fed TRT2 diet had higher average feed intake than those fed the TRT1 during both gestation (2.49 vs 2.43 kg) and lactation (6.70 vs 6.11 kg). There were no significant differences on the litter size or litter birth weight. No differences ($P>0.05$) were noted in piglets survival as well as in backfat thickness and body weight loss in sows. In conclusion, these results indicated that high-energy diets had no effect on the body weight and backfat thickness of sows during gestation and lactation, but influenced the weanling pig's body weight and average daily gain during lactation period.

The effects of protein sources on growth performance, apparent ileal digestibility in weaning pigs

Y. Yang, Y.M. Kim, H.M. Yun and I.H. Kim

Swine Nutrition and Feed Technology 421-1 Department of Animal Resources Science Dankook University, Department of Animal Resource & Science, Dankook University, #29 Anseodong, Cheonan, Choognam, 330-7, Dankook University #119, Dandae-ro, Dongnam-gu, Cheonan-si, Chungnam, 330-714, Korea, Korea, South; yunhm822@naver.com

Two experiments were conducted to evaluate pre-digested chicken byproduct meal as a new animal protein source on growth performance, blood profiles, fecal score, ileal digestibility of amino acids in weanling pigs. In Exp. 1: a total of 112 [(Yorkshire × Landrace) × Duroc] pigs were randomly allocated to 4 treatments (28 pigs per treatment) for a 35 days metabolic trial. The dietary treatment of Exp. 1 was SBM (soybean meal), FM (fish meal), PD-CBM (Pre-digested chicken byproduct meal) and PD-SH (pre-digested swine hair). Four barrows [(Yorkshire × Landrace) × Duroc] with average initial BW of 15.82±0.64 kg were allotted to a 4×4 Latin square with four diets and four periods per square. In Exp. 2, 4 cannulated barrows were fed a cornstarch-based diet containing (1) SBM (soybean meal), (2) FM, (3) PD-CBM and (4) PD-SH. ADG and G/F were significantly (P<0.05) higher in PD-CBM treatment than SBM treatment. BUN was higher (P<0.05) in PD-CBM treatment than that in SBM treatment. The apparent ileal digestibility (AID)of DM and N in SBM treatment were the lowest among the treatment (P<0.05). Pigs fed FM had higher (P<0.05) AID of leucine and serine than that in other treatments. The AID of isoleucine and alanine were lower (P<0.05) in SBM than that in FM and PD-CBM. The AID of methionine and tyrosine in SBM were lower (P<0.05) compared with that in FM. Pig fed PD-SH had highest (P<0.05) AID of phenylalanine. The AID of histidine was lower (P<0.05) in SBM than that in PD-SH. The AID of lysine was lower (P<0.05) in SBM than that in PD-CBM. The results provide information on the amino acid composition and digestibility of a new protein source in weanling pigs.

Effects of dietary 1,3-diacylglycerol supplementation for growing pigs fed low energy diets

W.C. Liu[1], J.K. Kim[1], J.H. Cho[2], I.H. Kim[1] and S.O. Jung[1]

[1]*Swine Nutrition and Feed Technology 421-1 Department of Animal Resources Science Dankook University, Department of Animal Resource & Science, Dankook University, #29 Anseodong, Cheonan, Choognam, 330-7, Dankook University #119, Dandae-ro, Dongnam-gu, Cheonan-si, Chungnam, 330-714, Korea, Korea, South,* [2]*Department of Animal Science, Chungbuk National University, Cheongju, Chungbuk, South Korea 28644, Chungbuk National University, Cheongju, Chungbuk, South Korea 28644, Korea, South; inhokim@dankook.ac.kr*

The objective of the present study was to investigate the effects of 1,3-diacylglycerol (DAG) supplementation on growth performance and crude fat digestibility in growing pigs fed low energy diets. A total of 80 crossbred pigs [(Landrace × Yorkshire) × Duroc] with an average initial BW of 27.50±1.42 kg were used in this 6 wk feeding trial. Pigs were allotted to 4 dietary treatments based on their initial BW and sex (5 replications; 4 pigs per pen with 2 barrows and 2 boars). Dietary treatments consisted of: 1) LE, low energy diets (3,400 kcal/kg metabolizable energy); 2) LE+0.075% 1, 3-DAG; 3) LE+0.10% 1,3-DAG; 4) LE+0.15% 1,3-DAG. Pigs were weighed on initial and at the end of wk 6 while feed consumption was recorded to calculate ADG, ADFI and G:F. One week before the end of experiment, Cr_2O_3 was used as an indigestible marker and supplemented to the diets at a level of 2 g/kg. At the end of trial, fresh fecal samples from each pen were collected for apparent total tract digestibility (ATTD) of crude fat. The method for determining the ATTD of crude fat was according to AOAC. All data were analyzed by using GLM procedure of SAS. orthogonal polynomial contrasts were used to test the linear and quadratic effects of the increasing levels of 1,3-DAG. Statements of statistical significance were based on P<0.05. Dietary 1,3-DAG supplementation in LE diet linearly increased (P<0.05) the ADG (677, 679, 698 and 709 g/d) and G:F (0.398, 0.404, 0.409 and 0.415). At the end of the experiment, dietary 1,3-DAG supplementation in LE diet linearly improved (P<0.05) the ATTD of crude fat (58.75, 61.55, 61.42 and 61.29%). In conclusion, dietary 1,3-DAG supplementation had positive effects on improving growth performance and fat digestibility in growing pigs fed low energy diet.

Effect of controlling feeding time and additional water on finishing pigs under heat stress

J. Hu, J.Y. Zhang, J.K. Kim and I.H. Kim
Swine Nutrition and Feed Technology 421-1 Department of Animal Resources Science Dankook University, Department of Animal Resource & Science, Dankook University, #29 Anseodong, Cheonan, Choognam, 330-7, Dankook University #119, Dandae-ro, Dongnam-gu, Cheonan-si, Chungnam, 330-714, Korea, Korea, South; inhokim@dankook.ac.kr

A 12-week experiment was conducted to investigate the effect of controlling feeding time and additional water supply on growth performance, nutrient digestibility, blood stress hormone profile and meat quality on finishing pigs under heat stress (34 °C). A total of 120 crossed finishing pigs [(Landrace × Yorkshire) × Duroc] were assigned to 1 of 4 treatment in a 2 × 2 factorial arrangement according to initial body weight (BW) (50.10±1.38 kg) with controlling feeding time [*ad libitum* feeding vs restricted feeding (from 19:00 to 07:00)] and additional water supply [*ad libitum* vs additional water supply (*ad libitum* + additional water cup from 12:00 to 18:00)]. All data were analyzed by ANOVA using a 2 × 2 factorial arrangement of treatments. A P<0.05 was considered as significance. Overall, pigs given an additional water supply had significantly heavier BW (119.15, 116.55 kg) and higher average daily feed intake (2,254, 2,212 g) compared with *ad libitum* water supply (P=0.026, P=0.045, respectively). The concentrations of cortisol (5.37, 6.47 µg/dl), epinephrine (53.6, 59.8 pg/dl) and norepinephrine (72.5, 77.4 pg/dl) was significantly reduced in pigs receiving an additional water supply compared with *ad libitum* water supply (P=0.034, P=0.003 and P=0.003, respectively). Additionally, there was significantly decreased in pH (5.27, 5.06) in pigs receiving an additional water supply compared with *ad libitum* water supply (P<0.001). However, there were no differences in white blood cell, lymphocyte, blood urine nitrogen, glucose, and nutrient digestibility. In conclusion, these results showed that adopting the *ad libitum* feeding method and simultaneously give an additional water supply from 12:00 to 18:00 could effectively improve the growth performance and PH, decrease the blood stress hormone concentration.

Effects of feeding olive meal and supplemental dietary enzyme on carcass characteristics of broiler

A.R. Seidavi[1], S. Sateri[1], M. Bouyeh[1], P. Neumann[2], M. Kutzler[2], V. Laudadio[3] and V. Tufarelli[3]
[1]Rasht Branch, Islamic Azad University, Pole-Taleshan, 41335-3516, Rasht, Iran, [2]State University, Corvallis, OR, 97331, USA, [3]Section of Veterinary Science and Animal Production, University of Bari 'Aldo Moro', Valenzano, Department of Emergency and Organ Transplantation, Italy; alirezaseidavi@iaurasht.ac.ir

Food by-products are commonly used in the broiler industry as a financially and environmentally favorable way to provide nutrients to birds. Olive meal is a by-product of the olive mill extraction process and is high in lipids and fiber. This study investigated the carcass characteristics of broiler chicks fed 0, 2, 4, 6 or 8% olive meal in their diet. In addition, this study investigated the use of enzymatic feed supplements (such as β-glucanase, phytase, and hemicellulose) to enhance digestibility of this high fiber diet. At the age of 42 days after 4 h of fasting for complete evacuation of the gut, one bird from each replicate was euthanized. Care was taken to choose the most representative male birds with respect to body weight compared to the group mean body weight. These animals were used for measuring carcass yield and distribution of meat and gastrointestinal tract characteristics. Birds were fully pecked by dry pecking method. Feet were separated from the carcass in the tibio-tarsal joint. Neck, wingtips, gut and liver were removed and the empty or edible carcass was weighed and intestinal segments dimensions were recorded. Birds fed 4% olive meal had significantly higher wing, proventriculus, crop, and testes weight. The addition of the supplemental enzymes to the diet had minimal impact the measured values. The data collected in this study supports literature suggesting that olive meal is a suitable feedstuff for use in the broiler production industry. The addition of olive meal by-products into poultry diets has the potential to provide an economic and environmentally friendly alternative to traditional commercial broiler diets. In addition, this study identified that the addition of dietary enzyme supplements are not necessary to achieve optimal carcass quality of birds fed a diet containing less than 9% olive meal.

Efficacy of biocholine in finishing pigs

Y. Jiao[1], S.Q. Huang[1], H.M. Yun[1], J.H. Cho[2] and I.H. Kim[1]

[1]Swine Nutrition and Feed Technology 421-1 Department of Animal Resources Science Dankook University, Department of Animal Resource & Science, Dankook University, #29 Anseodong, Cheonan, Choognam, 330, Dankook University #119, Dandae-ro, Dongnam-gu, Cheonan-si, Chungnam, 330-714, Korea, Korea, South; [2]Department of Animal Science, Chungbuk National University, Chungbuk National University, Cheongju, Chungbuk, South Korea 28644, Chungbuk National University, Cheongju, Chungbuk, South Korea 28644, Korea, South; 958425592@qq.com

Choline acts as an important methyl donor, a component of lecithin and acetylcholine in animals, which is usually added to animal diets in the choline chloride form. This study was conducted to evaluate the effect of choline chloride alone or in combination with propylene glycol or β-1,4-mannanase on the growth performance, nutrient digestibility, blood profiles, meat quality and backfat in finishing pigs. A total of 120 crossbred ((Landrace×Yorkshire)×Duroc) pigs with an average initial BW of 75.15±1.94 kg were randomly assigned by body weight and sex. This 8-week experiment consisted of 4 treatments with 6 replications per treatment and 5 pigs per pen. The experimental treatments included: CON, basal diet; TRT1, CON+0.05% choline chloride; TRT2, CON+0.05% Biocholine P (chloride chloride and propylene glycol); and TRT3, CON+0.05% Biocholine N (choline chloride and β-1,4-mannanase). All data were subjected to the statistical analysis as a randomized complete block design using the GLM procedure SAS. Tukey's tests using the GLM procedure. The level of significance was set at $P<0.05$. The BW was higher ($P<0.05$) overall in pigs fed Biocholine P diet compared with the CON diet (120.13 vs 118.00 kg). The total tract digestibility of nitrogen increased ($P<0.05$) during week 4 (73.37 vs 68.03%) and backfat thickness was higher ($P<0.05$) in week 6 (19.3 vs 18.4 mm) and 8 (21.8 vs 21.1 mm) in pigs fed Biocholine P diet than CON diet. Overall, pigs fed TRT1 diet had higher ($P<0.05$) average daily feed intake than CON diet (2,373 vs 2,326 g), meanwhile pigs fed TRT1 diet had higher marbling than CON and TRT3 diets (2.38 vs 2.22, 2.25). In conclusion, Biocholine P had some positive influences than other treatment diets on growth performance and nitrogen digestibility whereas TRT1 diet had positive effect on marbling score.

Effect of activated yeast on *in vitro* fermentation of barley in diets for intensive beef production

Z. Amanzougarene, S. Yuste, A. De Vega and M. Fondevila

Universidad de Zaragoza -CITA, Departamento de Producción Animal y Ciencia de los Alimentos, Instituto Agroalimentario de Aragón, Miguel Servet 177, 50013 Zaragoza, Spain; mfonde@unizar.es

Responses of adding live yeast in diets for ruminants are often inconclusive, maybe because dry product is directly mixed with feed without being activated, and growth conditions (temperature, presence of oxygen) of yeasts differ from rumen environment. Thus, three 24 h *in vitro* incubation series were carried out, using low-buffered medium (pH 6.2). Bottles included 500 mg of substrate, either alone (CTL) or mixed with three levels (0.7, 1.4 and 2.1 mg/g; L, M and H) of activated *Saccharomyces cerevisiae* yeast (AY), were incubated using inoculum from beef steers receiving a concentrate diet. Before incubation, yeast was activated by incubation with growth substrate for 24 h at 30 °C under aerobic conditions. Gas production (GP) was recorded at 2, 4, 6, 8, 10, 12, 18 and 24 h. Incubation pH was determined at 8 and 24 h, and dry matter disappearance (DMd) and microbial mass in the liquid fraction were estimated at 24 h. Throughout all the incubation period, inclusion of AY enhanced barley microbial fermentation measured as the volume of gas produced, which increased quadratically ($P<0.05$) from 2 to 18 h, and tending ($P=0.054$) to be significant at 24 h. Such pattern of fermentation is explained by the lack of differences between M and H levels. Comparing treatments with the control, gas production results were numerically supported by a 4-5 percentage points increase in DMd, although differences were not significant. The pH at 8 and 24 h diminished linearly ($P<0.001$) with the addition of AY, although at 24 h (it must be considered that buffer was exhausted after 10-12 h) differences after with unsupplemented barley (pH 5.91) were below 0.16 pH units. No differences were recorded among treatments on total volatile fatty acids (VFA) concentration at 8 h incubation, nor on molar proportions of their major VFA. The inclusion of AY increased the extent of barley fermentation *in vitro*, but without a major impact on pH as an index of rumen environmental conditions.

A life-cycle approach to identifying local sources of emissions with global implications
E. Kebreab
University of California, Davis, 2111 Meyer Hall, 95616, USA; ekebreab@ucdavis.edu

Livestock production contributes to climate change through greenhouse gas emissions and nutrient loading in the environment. The demand for livestock products is expected to grow substantially, creating even more environmental pressure. Measurement or estimates of emissions tell only part of the story because the total environmental impact of food production should consider all factors that contribute to production of specific animal source food. For example, the production of non-ruminant meat such as pork or poultry will require estimates of emissions from preparation of feed (i.e. feed production, transport, and milling), animal husbandry (e.g. energy for hatching, lighting, cooling, ventilation, and feed distribution) and manure management. The basic system boundary is also affected by feed conversion ratio (FCR) and may involve more emissions due to land use change (e.g. soybean use in Europe from Latin America). Local production systems will vary greatly and affect emission estimates. For example, in areas where specialty feed ingredients are used, feed conversion ratio is improved, which is reflected in lower emission intensity (i.e. emissions per unit of product). Similarly, in ruminant production systems (milk and meat), the system boundary should include enteric methane emissions on top to those described above that are variable based on the type and amount of diet the animals consume. For example, dairy cattle in North America, on average emit 5.6% of their gross energy intake while in Europe the estimates are greater at about 6.3%. The feed conversion ratio for milk production in some regions of the world such as California is greater than pasture based systems, which brings the emission intensity down but may involve greater cost in feed production and transport. Therefore, to assess sustainability of livestock systems, local sources of emissions need to be properly estimated within a life-cycle assessment approach. This will allow for a fuller understanding of livestock's impact on the environment globally.

Ideal future dairy farm: a Walloon breeders' point of view
A.-C. Dalcq[1], H. Soyeurt[1], T. Dogot[1], Y. Brostaux[1], P. Delhez[1], F. Vanwindekens[1,2], E. Froidmont[2], P. Rondia[2], B. Wyzen[3], A. Masure[4], C. Bauraind[5] and Y. Beckers[1]
[1]ULg-GXABT, Agrobiochem, Passage des Déportés, 2, 5030 Gembloux, Belgium, [2]CRA-W, Rue du Liroux, 8, 5030 Gembloux, Belgium, [3]AWE, Rue des Champs Elysées, 4, 5590 Ciney, Belgium, [4]FWA, Chaussée de Namur 47, 5030 Gembloux, Belgium, [5]Collège des Producteurs, Avenue Comte de Smet de Nayer 14, 5000 Namur, Belgium; anne-catherine.dalcq@ulg.ac.be

This research aims to characterize the dairy breeders regarding their idea of the ideal future farm ensuring them an income, in order to highlight their present situation and the ways to advise them towards their wished dairy model. The 245 answers to a survey of breeders, conducted between November 2014 and February 2015 provided information about, amongst others, their wishes concerning the intensification, the specialization, the technological innovation, the kind of workforce, structure, market and milk production (standard vs differentiated quality milk). Based on this information, a Multiple Correspondence Analysis allowed to create 4 groups of breeders with a similar view of their ideal farm: global-based intensive (GBIb), local-based extensive (LBEb), intermediate and no-opinion breeders. The relationships between these groups and the other recorded qualitative variables as formation needed, obstacles and advantages of breeders organization, of diversification and so forth were studied using Chi Square tests and Correspondence Canonical Analysis. A moderate link was observed between the ideal future farm and the current situation of the respondent. This suggested that not all the breeders were in the production system that they considered as most profitable. As a brake to the transformation and diversification, GBIb tended to be more numerous to speak about the uncertainty of the customer loyalty (P=0.07) and LBEb pointed out the size of the investments (P=0.05). LBEb asked more for administrative (P=0.04) and transformation and diversification formations (P=0.03) while GBIb looked more for finance and management formation (P=0.02). In conclusion, there were different ideal dairy farm models considered by the breeders. Their needs were not similar and indicated which tools must be developed and which domain must be studied to support them.

Balancing local and global challenges in farm-level sustainability assessment tools

E.M. De Olde, E.A.M. Bokkers and I.J.M. De Boer
Wageningen University & Research, Animal Production Systems Group, P.O. Box 338, 6700 AH Wageningen, the Netherlands; eveliendeolde@gmail.com

Global environmental, social and economic challenges urge agriculture to develop towards more sustainable modes of production. To support decision making towards sustainable development, a large number of sustainability assessment tools have been developed. These tools operationalize the concept of sustainable development at farm-level by assessing the performance of farms on a wide range of indicators. The assessment results could be used by farmers to identify local solutions to global sustainability challenges. Although the number of tools is rapidly increasing, concerns are raised whether current sustainability assessments actually contribute to sustainable development in practice. This study compared tools in practice to gain insight into their practical requirements, procedures and relevance perceived by farmers. Of 48 indicator-based sustainability assessment tools, only four tools (RISE, SAFA, PG and IDEA) complied to the selection criteria and were applied to assess the sustainability performance of five Danish farms. The comparison revealed differences between tools in the assessment time, data requirements, and scoring and aggregation methods. Critical factors in the farmers' perception of tool relevance were context specificity, user-friendliness, complexity and language use. Moreover, a match between value judgements of tool developers and farmers on what can be considered as sustainable agriculture, is critical for the acceptance and implementation of conclusions derived from sustainability assessments. Farmers in this study emphasized the importance of a context-specific approach to farm-level sustainability assessments, in other words, a tool that is sensitive to regional sustainability challenges and norms. Although context-specific assessments are expected to provide outcomes that match the context in which the farmer is operating, thereby, stimulating farmers in taking action to improve the sustainability performance of their farm, such an approach risks neglecting global sustainability issues. Further research is needed to identify approaches to balance global and local sustainability issues while maintaining farmer's interest and motivation towards sustainable development.

Goals and behaviours of farmers in mountain dairy cattle farms

G. Faccioni[1], A. Bernués[2], M. Ramanzin[1] and E. Sturaro[1]
[1]Università di Padova, DAFNAE, Viale dell'Università 16, 35020 Legnaro (PD), Italy, [2]Instituto Agroalimentario de Aragon, Centro de Investigación y Tecnologia Agroalimentaria de Aragon, Avda. Montañana 930, 50059 Zaragoza, Spain; georgia.faccioni@phd.unipd.it

The study analysed the goals and their relationships with behaviours of dairy cattle farmers in a mountain area (north-eastern Italian Alps). We performed a qualitative approach giving at 46 farmers face to face questionnaires. They scored a list of statements regarding their goals for their farming career using a 5-point Likert scale. Next, they answered questions on actions that they had performed in the past 5 years. Data on their farm structure and management were also analysed. We performed a principal component analyses (PCA) and a cluster analysis on the goals answers. The relationships between clusters and behaviours were tested with a Kruskall-Wallis test. Three factors resulted from the PCA and they were named 'life quality', 'environmental values' and 'economic values'. Using these factors, we identified three clusters of farmers: entrepreneurial farmers (cluster 1, 7 farmers), traditionalist farmers (cluster 2, 14 farmers) and planner farmers (cluster 3, 25 farmers). The results showed that cluster 1 grouped farmers interested in improving the quality of life through the diversification of their activity, whereas farmers assigned to cluster 2 gave a high importance to environmental problems and to the self-sufficiency of their farm. Finally, farmers of Cluster 3 have broader point of view on the management of the farm, holding in high esteem all the aspects reported. Nonetheless, few behaviours resulted significant among clusters: taking holidays, improving of facilities and machineries, and modification of the amount of concentrates per cow. From the analysis of the farm management, significant differences among clusters were among variables related with the territory (stocking rate and ha of meadows/livestock Unit). The willingness to achieve a set of goals can be affected and delayed by many issues that reduce the differences among farmers actual behaviours. The identification of the heterogeneity of farmers' behaviour is a relevant starting point to achieve the sustainable development of the mountain farming system and for the application of participatory approaches.

Effect of paddock design on sow excretory behaviour in a pasture-based system with poplar trees

H.M.-L. Andersen, A.G. Kongsted and M. Jakobsen
Aarhus University, Department of Agroecology, Blichers Allé 20, 8830 Tjele, Denmark; heidimai-lis.andersen@agro.au.dk

Free-range pasture systems for organic pigs comply well with the organic principles and the organic consumers' expectations to organic livestock production. However, the systems as practiced in the Northern countries are associated with high risks of nutrient losses. This is caused by inputs of concentrate feed combined with poor vegetation cover due to the pigs' rooting behaviour. Well-established trees like e.g. poplar are more robust to the pigs rooting behaviour and have deep root systems with nutrient uptakes across a long growing season. Thus, implementing trees in the paddocks is expected to reduce risk of nutrient leaching. In order to achieve an effect, it is crucial that the pigs are motivated to excrete nearby the trees. The objective of the study was to investigate, the effect of spatial arrangements of trees, hut and feed on sow excretory behaviour in a pasture based system. The experiment was carried out during spring and autumn 2016 on an organic farm in Denmark. It included twenty-four sows housed in individual farrowing paddocks (36×13 m). Each paddock was divided into five zones. Zone 1-3 consisted of grass clover and zone 4-5 each included two rows of four poplar trees. A wallow was located in zone 2. The sows were randomly allocated to the six treatments: Feed trough located in zone 1 (farthest away from the poplar trees) or in zone 3 (closest to the poplar trees), and the hut located in zone 1, 2 or 3. The sows' location and activity (e.g. lying, standing and rooting) was observed by scan sampling, and elimination behaviour by all occurrences observations. Observations were carried out from sunrise to sunset, three days in the third- and in the sixth week of the lactation. In general, the main part of the elimination occurred in zone 1 (26%) and 5 (26%) and least in zone 2 (12%) and 3 (13%). Preliminary results indicate a tendency to interaction between hut location and zone (P=0.07), with less elimination in the hut area, whereas no effect of feed location was observed. A more detailed data analysis will be performed, focusing on sub-areas within the five main zones and the frequency of activity to further investigate the effect of paddock design on excretory behaviour.

'Clean, green and ethical' management of small ruminants: frontiers in the 'green'

G.B. Martin and P.E. Vercoe
The University of Western Australia, UWA Institute of Agriculture, 35 Stirling Highway, Crawley 6009, Australia;
graeme.martin@uwa.edu.au

We have long used the term 'clean green and ethical' to describe our teaching and research in livestock industries. 'Clean' drives us to reduce our reliance on chemicals, drugs and hormones, and 'ethical' drives us to maximise animal welfare. 'Green', the focus of the present paper, drives us to reduce our environmental footprint. Nowadays, the societal (and therefore market) credentials of a ruminant industry include output of greenhouse gas (methane) and impact on biodiversity. We present two solutions that encompass these issues and also provide other benefits. (1) Alternative forage shrubs, such as *Eremophila glabra*, that evolved in our landscape but were then largely ignored by European colonists, a fundamentally flawed strategy. Including these plants in a forage system offers six advantages: (a) It reduces methane emissions; (b) As a native Australian plant, it contributes to the maintenance of biodiversity; (c) As a perennial, it provides nutritious green feed during the dry season when traditional annual fodder species are dead; (d) It is deep-rooted so helps lower the water table and prevent salinity; (e) It inhibits gastrointestinal helminths; (f) It improves neonatal survival by providing shelter in inclement weather. In other countries, there must be comparable forage plants that can improve livestock production systems. We only need to look. (2) Genetics: the heritability of methane production in sheep and cattle seems to be about 0.3, so we can now develop breeding strategies to mitigate methane emissions. However, we need to remember that ruminants effectively carry two genomes, the host and the rumen microbiome, both of which contribute to production efficiency. Consequently, selection decisions need to be based on interactions between the animal and its external environment, between the microbiome and the rumen environment, and between the two genomes. Importantly, it seems highly likely that reducing methane production in the rumen will increase production efficiency. Therefore, with clever forage systems and genetics, we can improve both the environmental credentials and the profitability of ruminant industries.

What Corsican bee-keeping can teach us about how beekeeping can face a diversity of challenges?

A. Adam[1], A. Lauvie[2], J.M. Sorba[3], L. Amzil[4] and G. Michon[1]
[1]IRD UMR GRED, Agropolis, 34394 Montpellier cedex 1, France, [2]INRA UMR SELMET, pl. Viala, 34060 Montpellier cedex 01, France, [3]INRA LRDE, Grossetti, 20250 Corte, France, [4]Univ. Mohammed-V, Agdal, Rabat, Morocco; anne.lauvie@inra.fr

Beekeeping is facing more and more global challenges worldwide, from climate change to varroa expansion (and other sanitary problems), exposure to pesticides, or loss in bee diversity. Generic approaches are developed to address those challenges, but they do not allow understanding how beekeepers integrate those global challenges with more specific challenges they are facing in their practices and lands. To address this issue, based on a study conducted on Corsican beekeeping, we elaborate on the importance of diversity (of practices and patterns) existing in local beekeeping systems. Our hypothesis is that diversity represents an asset to tackle both global local and challenges (such as the uncertain dynamics of bees colonies or of plant flowering from a year to another, or market issue in island conditions) that beekeeping is facing. We first develop an approach that, through a 'farming style' frame, allows to underline and comprehend this diversity and, based on interviews and observations, describe seven different beekeeping 'styles'. In a second step, we discuss how these 'styles' provide elements for understanding how beekeepers, as individuals and as a coherent group, develop multiple strategies to cope with various challenges. We consider the place given in each 'style' to such strategies as (1) the mode of renewal of bee colonies (colony splitting or queen rearing), (2) the extent and sites of hives transhumance, (3) the range of honey types produced, (4) the adhesion to the Corsican PGO guidelines (5) the diversification of activities. We also analyse how far practices related to these different strategies are intertwined. we conclude on the interest to consider beekeeping as a true (though specific) farming activity, and therefore to analyse it through a systemic approach, as commonly practiced in researches concerning other farm animal species. This approach, combined with a territorial perspective, allows highlighting diversity in beekeeping and elaborating on the importance of this diversity in a context of increasing environmental and economic difficulties.

The importance of livestock in crop-livestock regions: competition with cash crops or opportunities

J. Ryschawy[1], M. Benoit[2], C. Delfosse[3], N. Hostiou[4] and B. Dumont[2]
[1]AGIR, INPT ENSAT, Université de Toulouse, INRA, 31324 Castanet-Tolosan, France, [2]INRA, VetAgro Sup, UMR1213 Herbivores, Theix, 63122 Saint-Genès-Champanelle, France, [3]Université de Lyon, Laboratoire d'études rurales EA3738, 69363 Lyon, France, [4]INRA, Université Clermont Auvergne, AgroParisTech, Irstea, VetAgro Sup, UMR1273 Territoires, 6300 Clermont-Ferrand, France; julie.ryschawy@inra.fr

Integrated crop-and-livestock systems (ICLS) making use of the complementarities between arable crops and animals are to supposed to allow for the highest provision of economic and environmental performance (by improving soil quality and biogeochemical cycling (nitrogen, carbon), and enhancing habitat diversity). Still, livestock production has suffered from competition with crop agriculture, which has been more strongly supported by markets backed by significant public subsidies. This study aimed at highlighting the diversity of services bundles provided in areas where crop and livestock production still coexist and suggest levers toward agroecological ICLS. Our transversal analysis is based on a literature review and expertise on three French case-studies. This study is part of a national research collective expertise on livestock role and services in France and Europe. We highlighted first that the animal stocking rate should be adapted as regards to the potential of the agricultural area. Combining crops and livestock favours the autonomy in inputs through two main technical levers: diversification of crop rotations and organic fertilisation of crops and grasslands through animal waste. More diversified rotations, including cash crops, cover crops or trees can increase the feed self-sufficiency of livestock farms, particularly through the inclusion of legume species. Where technical and organizational barriers inhibit the reintroduction/maintaining of animals on farms, local exchanges between specialized livestock farms and specialized crop farms could be considered. Such coordinations seek to favour environmental service provision, without introducing new labor demands. Specific organisational and political levers should be explored to favour crop-livestock integration at the farm and local level.

Local sheep rearing system characterization in the Brazilian northeastern region

J.K.G. Arandas[1], M.N. Ribeiro[1], A.G.C. Alves[1] and O. Facó[2]
[1]*Universidade Federal Rural de Pernambuco, Animal. Science, Av. D. Manoel de Medeiros, 52171900, Brazil, [2]Empresa Brasileira de Pesquisa Agropecuária, Goat and Sheep, Sobral, Ceará, 62030270, Brazil; ribeiromn1@hotmail.com*

The paper aim to characterize the *Morada Nova* sheep breed rearing system and the rearing system prevalent in crossbred herds in the state of Ceará, Brazil. The study was carried out in the center of origin of the *Morada Nova* sheep breed, the municipality of Morada Nova. Purebred *Morada Nova* breeders (n=13) and crossbreed *Morada Nova* breeders (n=48) were interviewed in order to characterize the rearing system profile of the two groups. It was applied a semi-structured interview and a data file was constructed based on the answers of the interviewed, for which descriptive statistics and analysis of variance were done, followed by a tukey test. Multivariate techniques such as factorial analysis based on principal components, clustering and discriminant analysis were applied. The extensive rearing system was the most usual among the evaluated groups. The animal feeding base consisted of Caatinga resources and supplementation with maize and sorghum silage in times of low food availability. The diversified management system based on livestock plus tillage was the most common in the two studied groups. Based on the factorial analysis, the use of supplementation, time of supplementation, breeding age, age of the breeders, herd size and the main activity were the most important traits to characterize the local rearing system for the two studied groups. Both pure-breed and crossbred breeders adopted very similar management measures. This information may be useful in defining sustainable development programs for local sheep breeds, for instance the *Morada Nova*.

Ethnozootechny and the conservation of domestic and wild animal diversity

A.G.C. Alves[1], M.N. Ribeiro[1], R. Bozzi[2], P. Bruschi[2] and R.G. Costa[3]
[1]*Universidade Federal Rural de Pernambuco, Av. D. Manoel de Medeiros, 52171900, Brazil, [2]Università Degli Studi di Firenze, Scuola di Agraria, P.zza S.Marco, 4, Firenze, 50121, Italy, [3]Universidade Federal da Paraíba, Animal Science, Bananeiras, Paraíba, 58220, Brazil; ribeiromn1@hotmail.com*

Efforts to conserve biodiversity in agroecosystems also include domesticated plants and animals. A survey was carried out in Tuscany, Italy, with shepherds of the Zerasca and Pomarancina breeds. Both are endangered autochthonous breeds, belonging to the apenine trunk and meat producers. The aim was to describe and analyze the local knowledge regarding the care of the life and health of sheep, with emphasis on the use of plants in the treatment of diseases of these animals. The concept of 'farming styles' was used as a basis, which allows to relate the cultural, technical and environmental dimensions of agricultural production. Initially, a survey was carried out in the field to identify relevant issues and problems in local sheep farming. There was a tendency to disconnect from the ancestral practices of veterinary use of plants, characterizing, at least apparently, an 'extinction of experience' situation. Still in reference to the protection of life and animal health, there was a strong conflict between the shepherds and the wolves (*Canis lupus italicus* Altobello, 1921), the main sheep predators in the region. About half of the farms were run by women; Labor was almost exclusively family; About 75% reported wolf attacks on their herds. This has caused changes in management, with an increase in production costs, according to the sheherds´ perception. Rumors about the intentional reintroduction of wolves can be seen as folk social science, that is, explanations of the world that represent a cultural resistance against the dominant narratives and the structures of power that sustain them. Ethnobiological approaches, possibly associated with the scientific study of 'rumors' may contribute to the mediation of these conflicts. (Funding bodies: Capes / PGCI, Science without Borders, CNPq).

Carcass characteristics of Istrian bulls in beef production
A. Ivanković[1], G. Šubara[2], J. Ramljak[1], E. Šuran[2] and M. Konjačić[1]
[1]*University of Zagreb, Faculty of Agriculture, Department of Animal Science and Technology, Svetošimunska cesta 25, 10000 Zagreb, Croatia, [2]Agency for Rural Development of Istria, Ulica prof. Tugomila Ujčića 1, 52000 Pazin, Croatia; aivankovic@agr.hr*

Istrian cattle are one of three endangered autochthonous cattle breeds in Croatia. Regional government has supported programs of economic reaffirmation of the breed through beef production and marketing. The aim of this research was to investigate production indicators of the Istrian cattle for beef production. The research included 60 pure bred Istrian bulls divided into two numerically equal groups, young bulls (20 to 24 month) and older bulls (25 to 28 month). Animals were kept on small households, in traditional housing and feeding conditions. They were slaughtered according to the standard procedure in local slaughterhouse and after 24[h] post-mortem, the carcass EUROP conformation score and subcutaneous fat thickness were determined. Slaughter live weight in the group of young bulls was 588.1 kg and in the group of older bulls 645.8 kg. Young bulls achieved higher average daily gain (839 g) than older bulls (789 g). Cold carcass weight in group of young bulls was lower (324.4 kg) than in the group of older bulls (344.7 kg). Young bulls achieved significantly higher average net daily gain (471 g/day) than older bulls (429 g/day). Chilling loss in group of young and older bulls was similar (1.71%, 1.73%). Dressing percentages of young bulls were higher than in the group of older bulls (56.1%; 54.9%). Slaughtered animals in group of older compared to younger bulls have had some favorable EUROP carcasses score class. Based on this research we conclude that Istrian cattle have a favorable potential for beef production, especially in the technology of low to moderate intensity level.

Effects of restricted feeding on carcass and meat characteristics of Lithuanian pig breeds
V. Razmaitė
Lithuanian University of Health Sciences, Institute of Animal Science, R. Žebenkos 12, 82317 Baisogala, Radviliškis district, Lithuania; violeta.razmaite@lsmuni.lt

The objective of the study was to examine the effects of restricted feeding on the carcass characteristics and meat quality of conserved fatty Lithuanian pig breeds. The study was carried out on 24 Lithuanian White and 18 Lithuanian indigenous wattle pigs, 28 females and 14 castrated males, divided into homogeneous groups, denominated 'Standard-*ad libitum*' and 'Restricted' (80% *ad libitum*) after 60 kg of live weight. All pigs received complete concentrate feed. The data were subjected to the analysis of variance in general linear (GLM) procedure in SPSS 17. The model included the fixed effect of feeding level, breed and gender. LSD significance test was used to determine the significance of differences of means between the groups. Model calculations revealed that the live weight of the pigs was affected (P<0.001) by the feeding regime, carcass weight – by the feeding regime and breed, whereas carcass fatness and flare fat were affected by feeding regime, breed and gender. Although no differences were found in the chemical composition of meat, pH, EZ drip and cooking losses, and colour between the groups in relation to either to restricted and *ad libitum* feeding regimes or the gender, thawing loss in Longissimus dorsi muscle was significantly higher (P<0.05) with *ad libitum* feeding regime than with restricted feeding. The highest differences in meat quality characteristics were found between the breeds. Longissimus dorsi muscle of Lithuanian White pigs showed higher redness (P<0.05) and higher content of protein (P<0.010), and the same muscle of Lithuanian Indigenous wattle pigs showed higher thawing loss (P<0.05). Semimembranosus muscle of Lithuanian Indigenous wattle pigs showed higher pH1 (P<0.001) measured in hot carcass. This research was funded by the H2020 project Treasure (grant no 634476).

Regenerative practices: a sustainable alternative for land-based livestock

N. Mandaluniz, A. Pascual, J. Arranz, E. Ugarte and R. Ruiz
NEIKER, Animal Production, P.O. Box 46, 01080, Spain; rruiz@neiker.eus

Dairy sheep production in the Basque Country has been traditionally based on a pasture-based farming system with a local dairy breed. Land use and grazing management practices have changed during the past few decades as a result of the intensification of traditional pasture-based systems. Some of the consequences of these changes are directly related to environmental impacts and ecosystem services. The regenerative agriculture is an agricultural vision based on scientific and/or empirical knowledge of natural processes, the biology of soil, the physiology of plants and their relationship with animals. By maximizing these relationships, it is possible to increase the sustainability of the system. The objective of LIFE REGEN FARMING project was to determine the effect of some regenerative practices on the sustainability of land-based livestock. Different regenerative practices were tested in 3 different agroclimatic areas with diverse livestock production systems: NEIKER (conventional flock of dairy sheep), INTIA (organic flock of dairy sheep) and Urdunederra Rural Development Agency (four farmers with organic beef-cattle herds). The project implemented during 3 consecutive years direct sowing of pasture with perennial plant species, organic fertilizations and 'grazing plans' that consist of planning grazing with high livestock unit loads and enough resting periods for the soil and vegetation to recovery. The regenerative practices tested demonstrated that they enhance both environmental and socio-economic aspects. The most relevant results showed that the practices implemented maintained livestock production and increased carbon fixation in pasture (+10%), improved soil fertility (+7% particulate organic matter) and, finally, increased the botanical diversity of grasslands (+3%) and grass production (+10-15%). The project has been carried out by following a work dynamic of participatory research action (a space in which farmers, researchers and rural development structures work together). Throughout the project, a major effort was made to combine scientific knowledge, with the fine tuning of simple, rapid and inexpensive evaluation methods that may work in the field for farmers or technicians, combined with other laboratory techniques.

Session 29

Theatre 1

An overview on alternatives to antibiotics and zinc oxide for prevention of diarrhea post weaning

N. Canibe, O. Højberg and C. Lauridsen
Aarhus University, Animal Science, Blichers alle 20, 8830, Denmark; nuria.canibe@anis.au.dk

Post-weaning diarrhea is a significant enteric disease causing considerable economic losses for the pig industry. Among several etiological risk factors, enterotoxigenic *Escherichia coli* (ETEC) is considered to be a major cause. After having been used routinely for several decades, the use of antibiotics at sub-therapeutic concentrations was banned in the European Union in January 2006 due to the increasing prevalence of bacterial resistance to antibiotics. The removal of in-feed antibiotics from diets had negative economic consequences as it resulted in increased morbidity and mortality due to ETEC as well as increased use of antibiotics for therapeutic purposes. Addition of pharmacological levels of zinc oxide to the diet is a strategy used in some countries to reduce post-weaning diarrhea in piglets, but high zinc excretion can have a negative impact on the environment. Thus, strategies to control ETEC infections in piglets post-weaning and thereby reduce the use of antibiotics and high levels of heavy metals are searched for. A number of nutritional, genetic, and management strategies have been proposed as alternative means of preventing ETEC infections. In this presentation, we will pay special attention to the possible mode of action by which feed components and manipulation of the gut microbiota and immune system can prevent and/or reduce ETEC-related post-weaning diarrhea in pigs.

Genetic parameters and breeding strategies for novel litter traits in weaner production systems

S. Klein, H.R. Brandt and S. König

Institute of Animal Breeding and Genetics Justus-Liebig-University Giessen, Ludwigstrasse 21 b, 35390 Giessen, Germany; sebastian.klein@agrar.uni-giessen.de

Sustainable piglet production implies an optimization of litter size, while increasing piglet birth weight and reducing piglet losses. The aim of this study was the implementation of a new record recording system for litter quality traits, which was used (1) for the estimation of genetic (co)variance components, and (2) the ongoing evaluation of breeding strategies. Litter quality traits included three traits 'Average Piglet Birth Weight' (APBW), 'Litter Evenness' (LE), and 'Piglet Vitality' (PV), all scored on a linear scale from 1 to 4 in increments of 1. The litter quantity trait was 'Number of Piglets Born Alive' (NPBA). The dataset included 2,602 litters from 1,102 crossbred sows (German Landrace × German Large White), kept in six herds in North-West Germany. For the estimation of genetic parameters, a linear multiple trait animal model was applied, including the fixed effects of the herd and litter number. Heritabilities were 0.09 for LE, 0.14 for PV and 0.21 for APBW. Favourable genetic correlations among litter quality traits were identified, but their genetic relationships with NPBA were of antagonistic nature. Selection index methodology was applied in order to define an index and a breeding goal including APBW, LE, PV and NPBA. The optimal index variant in terms of selection response and prediction accuracy included two litters of the potential replacement gilt, three litters from their dam and ten litters of their half-sibs. For an economic weight of 7.5 € for NPBA, optimal economic weights were 20 € for LE, APBW and PV, allowing accumulation of genetic gain for litter quality traits while keeping NPBA on the current level.

Looking to biological variables predicting growth during the first three weeks after weaning in pig

A. Buchet[1,2,3], E. Merlot[3] and C. Belloc[1]

[1]ONIRIS, INRA, UMR 1300 BIOEPAR, 44307 Nantes, France, [2]Cooperl Arc Atlantique, BP 60238, 22403 Lamballe, France, [3]INRA, UMR 1348 PEGASE, 35590, Saint Gilles, France; arnaud.buchet@inra.fr

Pig production needs to conciliate the reduction of the use of antibiotics and animal welfare while maintaining profitability. The early detection of animals with high risks of diseases could allow the implementation of individual care and/or treatments. This study aims to look for the association of health of piglets after weaning with biological variables measured before or immediately after weaning. Growth was used here as a rough indicator of health. Piglets (n=270) coming from 15 commercial farms, weaned at 28 days of age, were blood sampled at 26 and 33 days of age and weighed at 26, 33, 47 days of age. Biological parameters (n=19) concerning immunity, stress, oxidative stress or nutrients metabolism were measured. Considering the strong correlation between average daily gain (ADG) and live weight, the relative 26/47 ADG, calculated as the ratio of ADG between 26 and 47 days of age and live weight at 26 days of age, was used. A linear model was built for each biological variable, testing its explanatory role on relative 26/47 ADG. Then a Bonferroni correction was implemented and all variables with significance below 0.1 were included together in a final model. This model included farm effect, plasma day 26 hemoglobin and day 33 creatinine, non-esterified fatty acids (NEFA) and vitamin A plasma concentrations. The model explained 43% of the variance of relative 26/47 ADG with 31% accounting for farm effect and 6% for creatinine concentration. Apart from farm effect, piglets having a low relative 26/47 ADG displayed low hemoglobin concentration before weaning, and high protein (high creatinine) and energy (high NEFA) mobilization and low antioxidant concentration of vitamin A in blood after weaning. This study gives promising insights of the early detection of piglets with low relative growth after weaning.

Can piglets that are small for their size at birth or weaning catch up?

A.M.S. Huting[1], I. Wellock[2] and I. Kyriazakis[1]

[1]Newcastle University, Agriculture Building, NE1 7RU Newcastle Upon Tyne, United Kingdom, [2]Primary Diets, Melmerby, HG4 5HP Ripon, United Kingdom; a.m.s.huting1@ncl.ac.uk

Small (S) piglets are a major concern for pork production systems as they contribute to batch inefficiency. Although most piglets born or weaned S remain S throughout life, some are able to catch up growth. This study investigated whether different S piglets either being born S or having suffered from a growth check pre-weaning, differ in their ability to catch up. Piglets were weighed at birth (BiW), weaning (WW, week 4), grower (week 9) and finisher (FIN, week 15), and grouped and managed according to normal farm practises, which included cross-fostering for litter uniformity pre-weaning. Based on BiW, WW and FIN weight, pigs were categorised into 8 live weight (LW) classes (C) using percentiles, resulting in 0.960, 1.18, 1.33, 1.43, 1.53, 1.63, 1.77, and 1.97 kg for respectively BiW C1-8 and 4.85, 5.92, 6.45, 6.92, 7.37, 7.92, 8.57, and 9.67 kg for respectively WW C1-8. Pre-weaning mortality was significantly affected by BiW C ($P<0.01$) with piglets of C1 (26%) and C2 (16%) having a considerable higher mortality rate than others (~4.0% per C). A weak positive correlation ($P<0.01$, $r=0.49$) was found between LW at BiW and WW. Piglets of BiW C1-4 were weaned significantly lighter ($P<0.05$) than piglets of C$\geq$5, especially those in C1 and 2 (5.97 and 6.47 kg) who remained S throughout the productive period ($P<0.05$) weighing 2.7 (C2) to 4.3 (C1) kg below FIN average. Similar results were seen for piglets of WW C1-2 weighing 3.4 (C2) to 6.2 (C1) kg below FIN average. The probability to change C between BiW and WW was significantly ($P<0.01$) different within BiW C. The probability of piglets born S to change C decreased with increasing WW C, suggesting that under the experimental conditions piglet ability to compensate was limited. The probability of piglets born heavy (C8) to change C increased with increasing WW C, with the highest probability for WW C7-8. On the other hand, the probability to change C between WW and FIN was not significantly different within WW C. Pre-weaning ADG affected piglet probability to change C between WW and FIN, especially those in WW C1. It is suggested that S pigs at weaning are managed in a manner that allows compensation to be achieved.

Health parameters of group-housed sows and piglets during lactation

C.G.E. Grimberg-Henrici[1], I. Czycholl[1], O. Burfeind[2] and J. Krieter[1]

[1]Institute of Animal Breeding and Husbandry, Christian-Albrechts-University, Olshausenstr. 40, 24098 Kiel, Germany, [2]Futterkamp, Chamber of Agriculture of Schleswig-Holstein, Gutshof 1, 24327 Blekendorf, Germany; cgrimberg@tierzucht.uni-kiel.de

Since permanent fixation of pregnant sows is banned in Europe it has become a matter of discussion whether permanent fixation of lactating sows is longer acceptable. The aim of this study was to compare sows (n=59) from a group housing system (GH) with sows (n=47) from a conventional single housing system (SH) with regard to health and reproductive traits. Data were collected in 3 batches out of 9 planed batches with 20 GH sows and 16 SH sows per batch. All sows moved 1 week before farrowing to each housing system, respectively. GH sows were housed in a group of 10 sows and were separated in their pens 3 days a.p. until 6 days p.p. Each GH sow had an individual pen. A running area was shared among all sows and piglets. GH sows had significantly higher total piglet losses (GH: 5.4±0.44 vs SH: 2.8±1.88; $P<0.05$) and more crushed piglets compared to SH sows (GH: 4.9±0.41 vs SH: 1.2±0.19; $P<0.05$). The general high number of piglets losses in both housing systems can be explained by the high number of piglets born alive (GH: 16.6±0.38; SH: 17.2±0.58). With regard to health parameters GH sows had a better body condition after weaning ($P<0.05$), which can be explained by *ad libitum* feeding. Furthermore, less udder lesions were documented for GH sows after weaning (GH: 34.5% vs SH: 70.2%; $P<0.05$). Regarding piglet condition GH piglets were in tendency slightly lighter at weaning (GH: 7.1±0.06 vs SH: 7.5±0.07; $P<0.10$). The health parameters showed that GH piglets had less facial lesions 1 week p.p. but more body lesions due to interaction with other litters 4 weeks p.p. (GH: 29.9% vs SH: 14.1%; $P<0.05$). To conclude, allowing sows to move freely reduced the safety of piglets which resulted in higher piglet losses. However, multi-suckling of GH sows did not result in impaired sow condition and reduced the incidence of udder lesions. In addition, no alarming health problems were found for GH sows and piglets. Additional batches are planned and a reconstruction of the individual GH pens to improve pre-weaning mortality.

Effects of a free farrowing system on reproductive performance and welfare indicators

R.Y. Ladewig[1], I. Czycholl[1], J. Krieter[1] and O. Burfeind[2]
[1]Institute of Animal Breeding and Husbandry, Christian-Albrechts-University, Olshausenstr. 40, 24098 Kiel, Germany, [2]LVZ Futterkamp, Chamber of Agriculture, Schleswig-Holstein, Gutshof 1, 24327 Blekendorf, Germany; rladewig@tierzucht.uni-kiel.de

The conventional farrowing crates restrict the freedom of movement and natural behaviour of the sow during farrowing and lactation. The aim of the present study was to compare a conventional farrowing crate (2.0×2.6 m) and two types of loose farrowing pens (P1: 2.5×2.8 m; P2: 2.7×2.75 m) with regard to reproductive performance and welfare of sow and piglets. Eight batches will be included in this study from which three batches were finished so far (92 single housed sows and 1,109 weaned piglets). Recorded traits per litter included the number of piglets born alive and weaned, piglet losses, and individual birth and weaning weight. In addition, one week ante partum and four weeks post-partum the body condition of the sows was recorded and their gait, shoulders and teats were scored. Furthermore, in the first and fourth week of lactation facial lesions, skin lesions of the carpus and ear and tail injury of the piglets were scored. There were no significant differences in comparison of pens and crate between the number of piglets born alive (P1: 16.8±3.90 piglets; P2: 16.5±3.84 piglets; Crate: 17.3±3.95 piglets) as well as the birth and the weaning weight of piglets. Litters from loose pens had a significant higher pre-weaning mortality compared to litters from crates (P1: 24.4%; P2: 23.3%; Crate: 16.9%). In loose pens, crushing caused 64.4% of the piglet losses compared to 42.8% in crates. There were no significant difference between the body condition of sows in loose pens and sows in crates (P1: 2.95±0.25; P2: 2.89±0.32; Crate: 2.96±0.42). The number of sows with shoulder lesions did not differ between the housing systems. However sows in crates had more teat lesions compared with sows in loose pens. In pens 95.2% of the sows showed a normal gait four weeks post-partum and in crates 76.1%. In crates piglets had more facial lesions and skin lesions of the carpus compared with piglets in loose pens. The number of piglets with ear and tail injury did not differ between the farrowing systems. Fife additional batches will follow and the pen design will be changed due to the high piglet losses.

Ten weeks of lactation in organic pig production – a case study

A.G. Kongsted[1], M. Studnitz[2], K. Knage-Drangsfeldt[2] and H.M.-L. Andersen[1]
[1]Aarhus University, Agroecology, Blichers allé 20, 8830 Tjele, Denmark, [2]SEGES, Organic Farming, Agro Food Park 15, 8200 Aarhus N, Denmark; Anneg.kongsted@agro.au.dk

In Denmark the majority of organic piglets are weaned at seven weeks of age. Despite the markedly longer suckling period compared to conventional pig production, health problems at weaning is also a challenge in some organic pig herds. In the current conventional and organic practice legal pharmacological doses of zinc are used in diets to treat weaning diarrhea. This may increase the risks of zinc accumulation in the soil and antimicrobial resistance. Extending the lactation beyond the seven weeks is practiced in a few organic pig herds in Denmark to improve piglet robustness at weaning. This is expected to reduce health problems at weaning. On the other hand it decreases the potential number of weaned litters per sow per year unless it is possible to induce and utilise lactational oestrus. A case-study was initiated on a private organic pig farm with approximately 350 sows to explore sow and piglet performance as well as overall herd efficiency. The sows were kept in pasture systems all year round during gestation and lactation. The piglets had access to supplementary piglet feed from approximately 14 days of age. At ten weeks of age they were weaned and moved to an indoor-housing system with a concrete outdoor area. Based on the farmers daily recording, herd data was collected every 14th day during one year. The piglets were weighted batch-wise at weaning. Preliminary results indicate that the sows (n=505) in average weaned 12.1 (SD=1.4) piglets per litter after 70.4 (SD=6.1) days of lactation with a mean weaning weight of 26 kg ranging from 24-28 kg between the 14 batches. Given 1.8 weaned litters per year this corresponds to 566 kg weaned piglets per sow per year. There were no indications of negative side-effects on sow body condition at weaning. Further analyses of sow reproduction, piglet growth and health after weaning and total feed use in lactation and in the entire fattening period are needed before it is possible to evaluate the overall herd efficiency when employing ten weeks of lactation.

Bioactive compounds as alternative to antibiotics to prevent coliform diarrhea in weaned piglets

M. Girard, N. Pradervand, P. Silacci and G. Bee

Agroscope, Tioleyre 4, 1725 Posieux, Switzerland; marion.girard@agroscope.admin.ch

Post-weaning diarrhea (PWD) has a multifactorial etiology but is often related to infection by enteropathogens, such as enterotoxigenic *Escherichia coli* (ETEC). In 2016 in Switzerland, ETEC was detected in 42.5% of weaned pigs suffering from diarrhea. However, the massive use and misuse of antibiotics, especially for prophylactic treatments, has led to the development of antimicrobial resistances. This increased occurrence of resistances strongly incites the development of alternative solutions. Besides genetic selection or vaccination, nutritional approaches could represent a good alternative as well. Bioactive compounds, such as polyphenols, naturally present in plants can be good candidate. Some recent studies showed that polyphenols could inactivate enterotoxins produced by ETEC *in vitro* and *in vivo* and ultimately reduce PWD in piglets. Among the wide diversity of polyphenols, in this review tannins will be taken as an example of promising phytonutrients displaying antimicrobial activities. In the literature, tannins are often mentioned to have anti-nutritional as well as antimicrobial properties. The primary role of tannins in plants is to defend them against predation by herbivores and insects and to prevent infections by pathogens. Tannins can be classified into two groups according to their chemical structure. Condensed tannins are polymers of flavan-3-ols units while hydrolysable tannins are polyesters formed by a sugar surrounded by one or several phenolic compounds such as gallic (gallotannins) or ellagic acid (ellagitannins). Owing to their hydroxyl and phenolic groups, tannins can form complexes with protein, lipids, polysaccharides and metal ions. Some *in vitro* studies showed that tannins could inhibit the growth of bacteria including *E. coli.* However, the mode of action of tannins remains unclear. In order to show that tannins can have beneficial effects on PWD and ETEC infection, the results of several experiments conducted on artificially infected piglets and supplemented with hydrolysable tannins from a chestnut wood extract (Silvafeed Nutri P/ENC for Swine, Silvateam, Italy) will be presented.

Influence of a dynamic meal feeding approach on weaning performance and health of piglets

E. Royer[1], N. Lebas[2], M. Lapoujade[2] and R. Granier[2]

[1]Ifip-institut du porc, 34 bd de la gare, Toulouse, France, [2]Ifip-institut du porc, Rte de la Mathébie, Villefranche-de-Rouergue, France; eric.royer@ifip.asso.fr

A total of 936 weaned piglets were used in three experiments to study the effects of a dynamic meal feeding at weaning. Piglets were affected by post-weaning colibacillosis and edema disease (ED) of *Escherichia coli* O141:K85 in Exp. 1 and 2, whereas this infection was controlled in Exp.3. In Exp.1 and Exp.2, a phase 1 diet was offered *ad libitum* with free access to hoppers (control), or given in long troughs seven times daily from 08:00 am to 7:30 pm in quantities adjusted each day to appetite (7M), or restricted (R-7M). In Exp.3, the treatments were (1) control, (2) eleven meals in troughs from 07:00 am to 10:30 pm with quantities adjusted after each meal to appetite (11M), and (3) 11M with a mixing strategy at weaning limiting the number of litters per pen (11M-FAM). The dynamic phase duration was 13, 9 and 10 d in Exp. 1, 2 and 3, respectively, after which, using only hoppers, all piglets received *ad libitum* the phase 1 diet to d 14. In Exp.1, a too restrictive adjustment of feed supply from d0 to 14 resulted in lower ADFI for 7M or R-7M pigs than for control pigs (305, 294 and 349 g/d, respectively; P<0.001). However, ED outbreaks did not differ between treatments (mean=3% of piglets). In Exp.2, ADFI was reduced for 7M pigs (-4%; *P*>0.05) and for R-7M pigs (-8%; P=0.02) compared to control pigs (279, 257 and 291 g/d, respectively), but was unaffected from d 9 to 14. ED outbreaks from d 12 to 15, resulted in an equal number of dead or sick pigs for 7M and R-7M treatments (7%), which was higher than for control (3%). In Exp.3, ADFI from d0 to 10 was similar for piglets offered 11M, 11M-FAM and control treatments (289, 270 and 283 g/d, respectively). Video monitoring showed that the number of animals with feeding intake by meal period was increased at d 1 for 11M and 11M-FAM piglets compared to control piglets (75, 81 and 67%, respectively; P<0.01). By reproducing the behavior of suckling piglets, the distribution of meals in long troughs may briefly enhance the feed intake during the first two days after weaning. The study did not confirm that decreasing or distributing the feed intake may control the edema disease.

Effect of prenatal flavours exposure on creep-feed acceptance and performance of suckling pigs

J. Figueroa[1,2], S. Guzmán-Pino[1] and I. Marchant[1]

[1]*Universidad de Chile, Fomento de la Produccion Animal, Av. Santa Rosa 11735, La Pintana, Santiago, Chile, 8820000, Chile,* [2]*Pontificia Universidad Católica de Chile, Ciencias Animales, Av. Vicuña Mackena 4860, Santiago, Chile, 6904411, Chile; figueroa.jaime@uc.cl*

Prenatal learning of flavours preferences has been described in pigs. However, when flavours are included in creep-feed, pigs don't present preferences over unflavoured diets, even if their mothers received those flavours prenatally. This study evaluate pig's feed acceptance and performance when their mothers received the same or a different flavour during late gestation. A total of 24 sows were fed a flavoured diet; garlic (n12) or aniseed (n12) from d91 until the end of gestation (d114). Litters received creep-feed diets with the same flavour as their mothers did (CS+, n8), a different flavour from the one their mothers experienced (CS-, n8), or a diet with no artificial flavours added (Control, n8) from d10 until the end of lactation (d21). Flavours were added at 0.075% and counterbalanced in the experimental design. Pigs were identified and weighted at d4 and d20 after birth to calculate their average daily gain (ADG). Creep-feed intake was measured between d14 and d18 and expressed by the average daily feed intake (ADFI) of a pig. Data was analysed with ANOVA considering creep-feed (CS+, CS- or control) as the main factor, by using the statistical software SAS. Creep-feed diets didn't affect pigs ADFI (P=0.587), body weight (P=0.372) or ADG (P=0.768). Nevertheless, pigs that received CS+ presented a numerically higher ADFI than did pigs that received CS- (8.23 vs 7.71 g, P=0.881) or control diets (8.23 vs 7.01 g, P=0.557). Moreover, pigs that received CS+ were numerically heavier than pigs that received CS- (6,392 vs 6,005 g, P=0.564) or control diets (6,392 vs 5,850 g, P=0.375), and they gained numerically more weight than did pigs that received CS- (252 vs 247 g, P=0.905) or control diets (252 vs 240 g, P=0.748). Creep-feed acceptance was numerically higher when pigs received the same flavour than sows, probably explained by an associative learning between the flavour and the hedonic effect of amniotic fluid during gestation. Nevertheless, the huge variability of intake among litters did not allow to establish statistical differences in this study..

Effect of early feeding strategies on feed intake of suckling piglet's

J.F. Greydanus and J. Zonderland

De Heus Animal Nutrition, R&D and Concept Development Pigs, Rubensstraat 175, 6717 VE Ede, the Netherlands; jgreydanus@deheus.com

An adequate level of feed intake immediately after weaning is one of the determinant factors for good piglet performance and production efficiency after weaning and in later life. However, in practice, we commonly observe a low level of feed intake directly after weaning. Furthermore, about 10% of the piglets do not consume feed during the first 48 h post-weaning. Pre-weaning provision of creep feed next to sow milk facilitate the transition from liquid to solid feed and has a direct impact on how fast piglets start eating after weaning. This experiment was conducted to explore the effect of three different feeding strategies on early feed intake. A total of 90 litters, equally split over the treatments, were used to evaluate the effect of feeding piglets 1) a milk-based pelleted diet during the first 7 days of life followed by a complex diet until weaning (MP7), 2) a milk-based pelleted diet during the first 14 days of life followed by a complex diet until weaning (MP14) and 3) a complex diet from day one until weaning (CD). Creep feed was offered 2-3 times a day, in two different feeders, depending on the feed intake of the individual pen. Feed intake was registered per pen at day 7, 14, 21 and at weaning. The ANOVA procedure was used to estimate the effect of feed intake. Piglet breed, sow parity number class and treatment were included in the model. For the analysis of feed intake from day 21 until weaning, the number of days from day 21 till weaning was used as a covariate. Although no feed intake differences were found for the total period (377, 389 and 336 g/piglet for MP7, MP14 and CD respectively; P=0.22), at day 14 litters of the MP14 feed strategy showed a higher feed intake (58 g/piglet; P=0.01) compared to litters of the MP7 and CD feed strategy (44 and 39 g/piglet, respectively). This study suggests that early feeding strategies have an impact on pre-weaning feed intake of piglets.

Digestive capacity and gut health of Norwegian Landrace pigs fed a high-fiber rapeseed diet

M. Pérez De Nanclares, L.T. Mydland, N.P. Kjos and M. Øverland
Norwegian University of Life Sciences, Animal and Aquacultural sciences, P.O. Box 5003, 1432 Aas, Norway;
marta.perez.de.nanclares@nmbu.no

The European pig industry is heavily dependent on imported feed ingredients, especially soybean meal (SBM) as a protein source. Increased use of local protein sources, such as rapeseed meal (RSM), could alleviate the dependency on imports. Improvements in feed efficiency (FE) when using RSM could enhance production profitability. Digestive capacity and gut health are key factors influencing FE. Alternative protein feedstuffs such as RSM could affect these factors, thus the FE. Our aim was to study the digestive capacity and gut health of 40 Norwegian Landrace weanling pigs fed a high fiber rapeseed (RS) co-product diet (RSF) or a SBM control diet. Nutrient digestibility, enzyme activities, intestinal morphology, gut microbiota, and metabolomics profile of digesta, liver, and plasma were measured. Feeding RSF resulted in reduced apparent ileal digestibility (AID) and apparent total tract digestibility (ATTD) of DM, OM, CP, NDF, ADF, P, and energy, and ATTD of starch. The reduced AID of CP coincided with a decrease in trypsin activity in the jejunum. Large variation in nutrient digestibility was observed among pigs within both dietary treatments. RSF exposure markers including sinapine, sinapic acid, and gluconapin, and SBM exposure markers such as daidzein were identified in the digesta by an unsupervised principal component analysis model and metabolite characterization. Feeding RSF increased the levels of multiple oxidized metabolites and aldehydes in liver and serum, but decreased the level of ascorbic acid in the liver, thereby indicating a disruption of the redox balance status of the pigs. Feeding RSF did not affect gut morphology or cause intestinal dysbiosis as showed by bacterial culture, metagenomics analysis, and unaffected levels of microbial metabolites such as short chain fatty acids in the digesta. In conclusion, there was large variation in the digestive capacity of pigs between and within dietary treatments, indicating differences in the ability to digest high fiber rapeseed diets. This suggests that there is potential for developing more robust pigs that can better utilize alternative protein sources.

Effects of reducing dietary crude protein and metabolic energy in weaned piglets

J.H. Lee[1], W. Yun[1], S.D. Liu[1], W.G. Kwak[1], C.H. Lee[1], H.J. Oh[1], H.Y. Park[2], S.H. Choi[1] and J.H. Cho[1]
[1]Chungbuk National University, animal science, 1, Chungdae-ro, Seowon-gu, Cheongju-si, Chungcheongbuk-do, Republic of Korea, 28644, Korea, South, [2]Foundation of Agri. Tech. Commercialization & Transfer, 126, suin-ro, Gwounseon-gu, Suwon-si, Gyeonggi-do, 16429, Korea, South; junenet123@naver.com

The objective of this experiment was to determine the effects of a pure reduction in the dietary crude protein (CP) and metabolic energy (ME) content on growth performance, nutrient digestibility, blood profile, fecal microflora and odor gas emission (ammonia, hydrogen sulfide and volatile fatty acid) in weaned piglets. A total of 80 weaned piglets [(Landrace × Yorkshire) × Duroc] with a mean initial BW of 6.8 ± 0.5 kg were randomly allotted to 4 treatments with 2 replicate pens of 10 piglets per pen (based on average BW) for 45 days. The dietary treatments were: (1) CON: basal diet, (2) LME: -10% ME then CON, (3) LCP: -10% CP then CON, (4) MECP: -10% CP and ME then CON. During overall period of experiment, there were no significant differences ($P>0.05$) between treatments in growth performance such as average daily gain (ADG), average daily feed intake (ADFI) and feed efficiency (G:F). Apparent total track digestibility (ATTD) of CP and gross energy (GE) were not affected ($P>0.05$) in LME and LCP compared to CON piglets. The concentration of BUN decreased ($P<0.05$) in LME, LCP and MECP compared with CON piglets. Fecal NH_3, H_2S and VFA emission was lower ($P<0.05$) in LME, LCP and MECP compared with CON piglets. In conclusion, these results indicate that reduction in dietary CP and ME content didn't decrease growth performance and nutrient digestibility but had rather a positive effect by decreasing BUN and gas emission.

Effect of 1-day old experimental supplementation on piglet mortality and performance
N. Van Staaveren[1], M. Bertram[2], D. McKilligan[3], J. Marchewka[4] and J.A. Calderón Díaz[4,5]
[1]University of Guelph, Department of Animal Biosciences, ON, Canada, [2]First Choice Livestock, Des Moines, IA, USA,
[3]Techmix International, Stewart, MN, USA, [4]Institute of Genetics and Animal Breeding, Polish Academy of Sciences,
Magdalenka, Poland, [5]Teagasc – Pig Development Department, Moorepark, Fermoy, Ireland; j.marchewka@ighz.pl

This study aimed to assess the effect of a supplement, a proprietary blend of glucose, medium chain triglycerides and plant extracts designed to provide energy to encourage colostrum intake, in piglets on their performance and mortality. The oral supplement was randomly given to 1,208 piglets at 1-day old (SUPP, n=588) or not (CON, n=620) balanced according to sow parity, birth weight and piglets born alive. Weaning weight (n=1,061) and weight at the end of nursery (n=407) were collected and average daily gain from birth to end of nursery (ADG) was calculated. Mortality records including no. of and reasons for mortality were recorded. Piglet performance was analysed using mixed model equations while mortality data were analysed using logistic regression models. SUPP piglets were heavier than CON piglets at weaning (4.6±0.17 vs 4.4±0.05 kg, P<0.05) and the end of nursery (30.3±3.24 vs 22.1±0.47 kg, P<0.01). ADG tended to be 0.06±0.04 kg higher in SUPP than CON piglets (P=0.09). A total of 145 piglets died during lactation. SUPP piglets were 2.2 times more likely to die during lactation (P<0.001), but were 0.3 times less likely to die due to crushing (P<0.01) and 0.2 times less likely to die due to Streptococcus infection (P<0.01) than CON piglets. Nevertheless, all piglets with no recorded death reason (n=48) were SUPP piglets. Piglets born in larger litters and piglets with lighter birth weight were at greater risk of mortality (P<0.001). Results indicate that providing an experimental proprietary supplement to 1-day old piglets improves performance throughout lactation which carries over to the nursery period. Although a higher number of SUPP piglets died, results suggest that the supplement reduces the risk of death due to crushing and disease. Further research is needed to determine whether supplementation could cause side effects which could account for the piglets that died due to unknown reasons.

Effect of emulsifier (sodium stearoyl lactylate) in weaning pigs
J.Y. Zhang, J. Hu, J.S. Lee, K.T. Kim and I.H. Kim
Swine Nutrition and Feed Technology 421-1 Department of Animal Resources Science Dankook University, Department of Animal Resource & Science, Dankook University, #29 Anseodong, Cheonan, Choognam, 330, Dankook University #119, Dandae-ro, Dongnam-gu, Cheonan-si, Chungnam, 330-714, Korea, Korea, South; inhokim@dankook.ac.kr

This study aimed to evaluate the effect of emulsifier on growth performance, blood profile, faecal performance and nutrient digestibility in weaning pigs. A total of 120 weaning pigs [(Landrace × Yorkshire) × Duroc] with an average initial BW of 7.57±0.94 kg were used in the 6 weeks-trial. Pigs were randomly allotted to four experimental treatments according to initial BW. There were 6 replicates per treatment with 5 pigs per replicate. The treatments included 0 (TRT1), 0.05% (TRT2), 0.075% (TRT3), 0.1% (TRT4) emulsifier in the basal diet. All data were analyzed with the GLM procedure of SAS and using orthogonal and polynomial contrasts (P<0.05) to assess linear and quadratic effects of the increasing levels of emulsifier. Results indicated that from 0-3, 3-6, and 0-6 wk, ADG (395, 441, 462, 465 g; 504, 552, 581 588 g; 450, 496, 521, 526 g), ADFI (509, 514, 538, 564 g; 835, 847, 860, 868 g; 672, 681, 699, 716 g) and FCR (0.777, 0.857, 0.858, 0.826; 0.604, 0.652, 0.676, 0.677; 0.669, 0.729, 0.745, 0.735) linearly increased (P<0.05) with increasing emulsifier level compared with TRT1. At wk 6, serum concentration of total cholesterol (93.8, 72.8 mg/dl), triglyceride (49.8, 30.8 mg/dl) and LDL (60.3, 44.0 mg/dl) was reduced (P<0.05) in TRT4 compared with TRT1. Digestibility of DM (80.57, 82.04%) and fat (80.05, 83.52%) was greater (P<0.01) in TRT4 than TRT1. In conclusion, added emulsifier improved growth performance, reduced serum cholesterol, triglyceride and LDL levels and improved digestibility of DM and fat in weaning pigs.

Session 30

Theatre 1

Diversity of European horse breeds – a review
J. Kantanen
Natural Resources Institute Finland, Green Technology, Myllytie 1, Alimentum, 31600 Jokioinen, Finland;
juha.kantanen@luke.fi

The present review focuses on the current status of diversity of horse breeds in Europe and molecular genomic applications to conserve horse genetic resources. According to the FAO's Global Databank, over 450 horse breeds or different varieties within breeds exist in Europe and the Caucasian region, of which 370 breeds have data on population size. A total of 67 horse breeds are known to have become extinct. Particularly several pony and work-type breeds have been lost. In addition, 121 horse breeds have a critical population status (≤100 breeding females and/or ≤5 breeding males) and 89 breeds are classified as endangered, i.e. there are 100-1000 breeding females and/or 5-20 breeding males within a breed. A total of 97 European breeds (~26%) having population size data are classified not at risk to become extinct. Among the domesticated mammalian species, horse has relatively high number of breeds currently at risk. The changes in the European horse culture have led to extinction of several horse breeds. Currently in Europe, the species is mainly used for sports and as hobby animals and has a minor importance as providers of draught power in agriculture, forestry and other primary industries. Records in the FAO's Global Databank indicate that only around 15% of the endangered horse breeds have a conservation program, cryobanking of genetic materials (semen and embryos) and preservation of living animals (*in vivo* conservation). There is an urgent need to strengthen conservation activities and sustainable utilization of rare horse breeds in order to preserve their genetic resources and cultural heritage. Genetic diversity of horse breeds has been assessed by maternally inherited mitochondrial DNA, paternally inherited Y-chromosomal markers, autosomal microsatellites and whole-genome SNP-marker data. In addition, there exist whole-genome sequence and transcriptome data sets to investigate structural and functional variations associated with adaptation and phenotypic characters. Genomic data can provide new insights into horse domestication, development of the breeds and demographic factors that have shaped the gene pools. These studies may guide conservation decisions aiming to preserve horse genetic resources in Europe.

Session 30

Theatre 2

Shaped for the survival – the early history of Finnhorse
A. Bläuer
Natural Resources Institute Finland, Green Technology, Itäinen Pitkäkatu 3, 20520 Turku, Finland; auli.blauer@luke.fi

The native domestic animal breeds are a valuable resource of genetic diversity. Their unique genetic features are the result of a long local adaptation and selection process. To understand the value of these animals for the future agricultural and societal needs, it is important to know their past and understand the factors – both environmental and anthropogenic – that have shaped their genetic makeup. This presentation summarizes the early history of horses in Finland from the Bronze Age to the Post-Medieval period (ca. 1700 BC-1900 AD). Horses were important part of Finnish society during prehistoric and historical periods. Their role, use and status have changed several times during the past few thousands of years they have been bred in Finland. Moreover, during this period horses had to adapt to the local environment and to the animal husbandry system. The Finnish climate – long, dark and cold winters and warm, short and productive summers – presented a challenge to the early animal husbandry in Finland. Domestic animals had to adapt to tolerate inferior quality and quantity feeding during the c. 7 months long winter period and to be able to exploit the short but abundant summer grazing season. This study is multidisciplinary and utilizes archaeological, zooarchaeological, historical and ethnographical sources. During the prehistorical period horses were utilized for their power and meat. They were also important ritual animals. The Christianization in the beginning of the medieval period ended the utilization of horse meat in Finland. Although during the historical period horses were regarded as valuable animals, dead horses were considered impure. During this period growing human and animal populations and increasing colonization made the winter feeding of animals more and more challenging. The importance of forests as source of animal fodder also increased as indicated by the isotopic studies.

Admixture and recent migration in three Czech draught horse breeds as revealed by microsatellites

L. Vostry[1], H. Vostra-Vydrova[2], N. Moravcikova[3], B. Hofmanova[1], Z. Vesela[2], J. Schmidova[2], I. Majzlik[1] and R. Kasarda[3]
[1]Czech University of Life Sciences Prague, Kamycka 129, 16521 Prague, Czech Republic, [2]Institut of Animal Science, Pratelstvi 815, 10400 Prague, Czech Republic, [3]Slovak University of Agriculture in Nitra, Tr. A. Hlinku 2, 94976 Nitra, Slovak Republic; schmidova.jitka@vuzv.cz

This study presents a valuable insight into the genetic structure and diversity of three Czech draught horse breeds: Silesian Noriker (genetic resource), Noriker and Czech-Moravian Belgian (genetic resource), using 13 microsatellite markers. All microsatellite markers were highly polymorphic in all breeds. A total of 104 alleles in 13 microsatellite loci were detected in 777 horses. Allele number, observed and expected heterozygosity, polymorphic information content were estimated. The average number of alleles per locus was the highest in Czech-Moravian Belgian horse (7.31) and the lowest in Silesian Noriker (6.77), whereas the observed and expected heterozygosities per breed ranged from 0.673 (Czech-Moravian Belgian) to 0.709 (Noriker) and 0.677 (Silesian Noriker) to 0.710 (Noriker), respectively. The estimates of Wright's F_{ST} between each pair of breeds indicated low intra-breed divergence (FST values ranged from 0.01 to 0.07). Although the Czech draught horse breeds are diverse, our data suggest a low level of differentiation as well as a high gene flow between them as indicated by the tests of genetic differentiation and assignment of individuals to populations. The low F_{ST} values especially between SN and N breeds can be explained by crossing between these breeds. The migration rate indicates a continuous gene flow between the Silesian Noriker and Noriker breeds (24%) (crossing). Our results show high genetic variability, low inbreeding and low genetic differentiation, especially between Silesian Noriker and Noriker, which is caused by the high level of admixture. The analysis revealed the genetically wrong assignment of many individuals to their breed. Individuals with high proportion (80%) of Silesian Noriker genes have been included in Noriker breed and vice versa. The results of this study will be applied in the conservation program of genetic resources of draught horses in the Czech Republic. Supported by the project QJ1510141.

Multifactorial analyses of body measurements in Polish Konik horses from Poland

E. Pasicka
Wrocław University of Environmental and Life Sciences, Biostructure and Animal Physiology, Kożuchowska 1/3, 51-631 Wrocław, Poland; edyta.pasicka@yahoo.com

The Polish Konik is one of the younger horse breeds, originating directly from tarpans. The Polish population of Koniks consists of approximately 150 mares and 40 stallions in national preservation breeding centres, and around 550 mares and 130 stallions in the wild. According to FAO regulations, included in the Program for the Management of Farm Animal Genetic Resources based on the Convention on Biological Diversity, number of females of Polish Koniks classifies this breed as 'endangered with extinction'. On a global scale, the Polish Koniks can be considered as a natural relic, and so are included in the aforementioned program, as well as in the conservation breeding. Therefore, in order to complete and update the existing state of knowledge on the morphology of our native horse breed, the objective of this study is to demonstrate morphometric diversity of 172 Koniks from the largest Polish breeding centres. Analyzed animals of both genders aged from 3 to 24 years and each horse was characterized with 40 metric features. Applied multifactorial analyses showed exterior differentiation and development of morphometric types within the studied population. Due to a growing interest in Polish Konik horses for riding, more and more emphasis is put on the evaluation of movement of these horses. This is reflected in performance tests as well as exhibitions of stable-bred Koniks. This study has established – based on discriminant function analysis – that the metric traits with high discriminant power are i.a. sections constituting segments of limbs. Elongation of segments, particularly pelvic limb which serves driving function, can be associated with deliberate actions of breeders who aim at improving motor abilities of these horses. Koniks are a conservative race, what creates an obligation of possibly comprehensive assessment of their current suitability for various forms of saddle and draft use. Nonetheless one must not forget, that the primary objective of preservation breeding of Polish Koniks is to keep the breed's primeval type.

Factors affecting foaling rates in three native Norwegian horse breeds

S. Furre[1] and O. Vangen[2]
[1]Norwegian Horse Centre, Starumsvegen 71, 2850 Lena, Norway, [2]Norwegian University of Life Sciences, P.O. Box 5003, 1430 Ås, Norway; siri.furre@nhest.no

Factors affecting fertility in three of Norway's native horse breeds was investigated to shed light on factors affecting foaling rates in the breeds. The Dole is a heavy draft type, the Fjord is a slightly lighter type while the Nordland/Lyngen is a sturdy pony type. Population sizes are 3,800, 5,800, and 2,300 respectively, and inbreeding levels are between 6 and 12%. The Norwegian Equestrian Centre has recorded all coverings and resulting foals since 1989, and the current foaling rate is between 58 and 70%. Declaration of matings is reported by the stallion holder while declarations of foals are reported by the mare owner, and data contained information on 8,332 coverings for the Dole, 13,171 for the Fjord and 4,898 for the Nordland/Lyngen horse. As artificial insemination is not in wide spread used in the breeds, mares is covered either by hand or by entering a herd with one stallion and from one to 40 mares. The herds graze on cultured pastures close to the farm, or on mountain pastures. The model included the year of covering, herd size, age of mares and stallions, if it was the stallions first season or not, and the mares' status (if it was her first season, if she previously had carried a foal to birth, was infertile the previous season, or had had a foal earlier in life). Preliminary results showed that the year of covering, the stallions experience and the age and status of the mare had an effect on the fertility. The relationship between the sire and dam and individual inbreeding level will be included, as well as the type of pasture and other management information.

Garrano: the native mountain pony of northern Portugal

A.S. Santos[1,2,3] and L.M.M. Ferreira[3]
[1]Ruralidade Verde, Lda, Qta dos Engenheiros, Vila Marim, 5000-773 Vila Real, Portugal, [2]EUVG, Av. José R. Sousa Fernandes Campus Universitário, 3020-210 Coimbra, Portugal, [3]CTAB-UTAD, Quinta de Prados, 5001-801 Vila Real, Portugal; assantos@utad.pt

The Garrano is a Portuguese native feral pony. These horses are considered native in the North of the Iberia Peninsula since the middle palaeolithic. This horse descends from the glacial paleolithic fauna and from equids introduced by the Celts in the North of the Iberia Peninsula (*Equus caballus celticus*). In present days, these animals are represented in the Minho region (Portugal) by the Garrano breed, and in Spain by the breeds Galego do Monte (Galiciano), Asturcón, Jaca-Navarra, among others. Garrano is characterized by a strait or concave head profile, main coat is bay with a dense main and tail, and their average withers height is 1.30 m with an average weight of 300 kg. Although these horses had an important role in the past in farming work and transport in Minho rural areas, their use has decreased dramatically due to agricultural abandonment. Currently, the breed's population is around 1,500 registered animals, 1,300 of which are mares. In present days the Garrano mainly exists in a free ranging system, some of the herds are completely wild, representing a very important link in the mountain ecosystem. Herds walk freely in the National Peneda-Gerês Park (PNPG), the only Portuguese National Park, located in the northwest of Portugal. PNPG hosts a distinctive and numerous fauna and flora, including the Iberian wolf (*Canis lupus signatus*), that interacts closely with the Garrano by hunting the foals (it is estimated that about 80-90% of foal mortality is due to wolf predation). Oak forest, riparian vegetation, humid bushes and several kinds of grass fields form the continuously changing habitat of PNPG where the Garrano survives. The Garrano horse is recognized as Portuguese socio-cultural heritage, and its present uses are mainly related to leisure (equestrian tourism, riding centers). However, given their unique breeding system, and adaptation capacities, the Garrano under controlled management, can also play an important role in the maintenance of PNPG biodiversity and, simultaneously, producing environmental and socio-economic benefits to rural populations in these areas.

Fat and somatic cell content of milk in indigenous Slovenian draft horse

B. Luštrek and K. Potočnik
University of Ljubljana, Biotechnical Faculty, Department of Animal Science, Jamnikarjeva 101, 1000 Ljubljana, Slovenia; klemen.potocnik@bf.uni-lj.si

Mare's milk has been long known but recently rediscovered as functional food, especially concerning the western parts of the world. The aim of the study was to assess the possibility of providing high quality mare's milk that complies with food safety requirements for human consumption. In the present study, milk of Slovenian draft horse (SDH) was analysed for the first time, and its fat content (FC) and somatic cell count (SCC) were studied. Altogether 137 milk samples were taken by hand (type A; first jets of milk; n=69) or by machine milking (type B; representative sample form 1 milking; n=68) in 8 test days, at 3 consecutive daily milkings, from 3 routinely milked mares, within the stage of lactation of 49-241 days. Noteworthy, significant correlation was obtained between the FC and the SCC (A: r=0.64; P<0.0001; B: r=0.41, P=0.0006) in both sample types, as well for the FC (r=0.68; P<0.0001), and for the SCC (r=0.64; P<0.0001) between the sample types. Significant effects (P≤0.05) of sample type, consecutive milking, stage of lactation and mare on the FC and the SCC were estimated by the analysis of variance. Type B samples contained significantly (P<0.0001) higher FC (1.04%) and lower SCC ($3.01 \times 10/ml^3$) than type A (0.32% and $4.17 \times 10/ml^3$, respectively). The milk of SDH mares was shown to have low FC and SCC. In accordance with these results and the statements on dietetic quality of mare's milk found in literature, mare's milk can be perceived as safe food. Furthermore, SDH mares could be used for commercial milk production, which would present an additional motivation to preserve the indigenous breed.

The effect of the horse assisted activities on the children

R. Gumilar[1], M. Tušak[2] and K. Potočnik[1]
[1]University of Ljubljana, Biotechnical Faculty, Department of Animal Science, Groblje 3, 1230 Domžale, Slovenia,
[2]University of Ljubljana, Faculty of Sport, Gortanova 22, 1000 Ljubaljana, Slovenia; klemen.potocnik@bf.uni-lj.si

The positive effects of the equine assisted activities and therapies on the human well-being are generally known. The aim of the study was to determine how horse assisted activities effects on the children's sleeping quality (SQ), anxiety (AN), general health (GH) and on the feeling about the contentment with the life (CF). The horse has a special position among animals, and it is widely believed that it has the strongest positive effect on the emotional status of the people, especially children. The design of the trial was made in the collaboration of four partners, two faculties, one private company and one private farm. The practical part was conducted in the Educational and Research Centre for Horse Breeding Krumperk, Biotechnical Faculty, University of Ljubljana. Each child participating in the trial took part in 12 sessions (one per week). In the session, a collaborate horse or pony, two children and a moderator were included. The children were from three to nine years old, nine boys and eleven girls. The session had three blocks of 15 minutes. First block were when children prepared the horse for riding. In the second block, children were alternately riding or walking close to the moderator. In the last block, children prepared horse for resting. Children's parents were interviewed three times: before the first therapy session, after the first session and after the last one. Summed scores of 6 questions for SQ, 20 for AN, 12 for GH and 5 questions for CF were analysed using the SAS/STAT software. The child's gender did not have a significant effect on the studied traits. The prior contact with the horse had significant effect on the SQ and AN. The age significantly affected the SQ and the GH. The SQ significantly increased during 12 weeks of the therapy by 4.65 points (on the scale from 6 to 30 points), AN significantly decreased by 6.3 points (on the scale from 20 to 80 points) and the CF significantly increased by 3.75 points (on the scale from 5 to 35 points). GH showed a decreasing trend, but it was not statistically significant.

Causal relationships between fertility and type traits and selection response in IHDH breed

R. Mantovani[1], G. Pigozzi[2] and C. Sartori[1]
[1]Dept. of Agronomy Food Natural Resources Animals and Environment, Viale dell'Universita' 16, 35020 Legnaro (PD), Italy, [2]Italian Heavy Draught Horse Breeders Association, Via Verona 90, 37068 Vigasio (VR), Italy; roberto.mantovani@unipd.it

Relationships of causal dependence may occur among traits, meaning that the expression of a phenotype may affect another one. Lifetime fertility (LF) in Italian Heavy Draught Horse (IHDH) is a novel indicator of mare reproduction efficiency. It is expressed as foaling rate (i.e. occurred/expected foals) after 6 reproductive seasons (RS). This study has aimed at disentangling possible causal effects of type traits (TT, target of selection for heavy draught and meat) on LF. Furthermore, indirect response to selection (R) on LF has been valued considering a matrix of genetic (co) variance obtained using either the traditional multi-trait model (MTM) or a structural equation model (SEM) for causal effects. The TT were linearly scored on adult mares using a 9 point scale (from 1 to 5 with half points). Dataset included 7,633 mares, 2,394 of which had both LF and TT data and 11,672 animals in pedigree. Mares with <6 registered RS (n=1,413) had LF predicted through a set of equations accounting for different age at first foaling, no. of previous foals and no. of registered RS. The occurrence of causal effects (λ) of a trait on another was investigated running Gibbs sampling algorithm for both MTM and SEM. A low causal effect on LF was found for head size and expression (HS; λ=0.10), frame size (λ=0.11) and fore diameters (λ=0.07), suggesting that the overall appearance of animals might have played a slight role on the foaling rate. Including the causal effect, the genetic correlation (rg) between HS and LF decreased from 0.18 to 0.01. However, the positive rg between LF and fleshiness, fore diameters and rear diameters (all included in selection index) were maintained also using SEM (rg=0.31, 0.25, 0.39, resp.; i.e. 0.08 mean points lower than using MTM). The indirect selection response for LF was thus still positive also considering causal relationships (R=0.19 vs R=0.33 using MTM (co)variances). The study indicates that the current selection for draught and meat in IHDH allows also a positive R for LF.

Conservation or improvement of indigenous horses breeds – where are the golden means?

G.M. Polak
National Research Institute of Animal Production, National Focal Point, Wspolna Str. 30, 00-930 Warsaw, Poland; grazyna.polak@izoo.krakow.pl

Preservation of horse genetic resources is based on an assumption that several horse breeds which are under conservation possess valuable traits, which may be important in the future. However, given that the breeds are subject to constant, inevitable changes due to the environment influence, changing of human needs or genetic drift, we cannot talk about their immutability. Changes are also necessary in order to maintain the usefulness of the breed in a changing world. In Poland, currently under protection there are 7 breeds of horses including 2 breeds of cold blooded horses. The number of cold blooded mares in both breeds is about 2,200, which are used mainly for the breeding purposes and production of slaughter foals. In 2016, the survey carried out in a group of 500 farmers, showed that only 20% of the farmers answered used the horses as working animals. This number continues to decline compared to previous researches (30% in 2011, 25% in 2014). Also, in the assessment of performance the breeders association is not putting so much emphaise on the original use of cold blooded horses. In addition, preferences are given to weight and muscles of body parties, which improves slaughter performance. Increasingly less importance is attached to the horses character and temperament, movement skills, strength, ability to collaborate with people. The question arises: what degree of changes that would adapt the breeds to the current socio-economic situation can we accept in native breeds?

Phenotypic and morphological traits and conformation indices of Barb and Arab-Barb horses in Tunisia

B. Gharbi[1], A. Hamrouni[1], M. Haddad[2], A. Najjar Maatoug[1] and M. Djemali[1]
[1]*National Institute of Agriculture of Tunisia, Animal Science, 43, Avenue Charles Nicolle, Tunis, Mahrajène, 1082, Tunisia, [2]National Fonndation for the imporovement of the Horses Breeds, Sidi Thabet, 2020, Tunisia; abirturki@yahoo.fr*

The Barb horse has a Northern African origin and the breed is occassionally mistakenly considered as Arabian horse. The Arab-Barb horse is a crossbred between the Arabian and Barb breeds. The objectives of this study were to: (1) describe the conformation traits of these two breeds; (2) estimate phenotypic correlations between their body measurements; and (3) standardize their morphological traits. A total of 20 conformation measurements of 33 horses with pedigree records from the two breeds were collected in June 2016. Averages, variation, phenotypic correlations and body indices were computed. The main results showed that average body measurement fits within the two breeds standard norms. The Barb horse weighed 489.69 ± 126.64 kg while the Arab-Barb horse weighed 458.23 ± 55.36 kg. Computed body indices showed that the Tunisian Barb and Arab-Barb were medium sized horses with a body index equal to 1 (a 'square' horse). For the corpulence index, the mean of the studied samples showed 1.24 and 1.19 for the Barb and the Arab-Barb, respectively. The average of the compactness index ranged from 2.89 to 3.11 for Arab-Barb and Barb, respectively. Phenotypic correlations ranged from -0.63 to 0.92. The highest phenotypic correlation was 0.92 between body length and thoracic perimeter for the Arab-Barb. For the Barb, this correlation was 0.62.

Conservation and reaffirmation of Gidran horse breed in the future

J. Ramljak[1], A. Ivanković[1] and J. Trogrlić[2]
[1]*University of Zagreb Faculty of Agriculture, Department of Animal Science and Technology, Svetošimunska cesta 25, 10000 Zagreb, Croatia, [2]Croatian Association of Breeders of Gidran horses, Franje Vidovića bb, 43240 Čazma, Croatia; jramljak@agr.hr*

Croatia has a long tradition in horse breeding. In long-term, the preservation of native breeds is directed through process of reaffirmation like sporting activities, tourism and others programs. In Croatia one part of horse breeds arrived from neighbouring countries during migration or during the war, after which they are accepted as part of the tradition. Gidran is one of such breeds with breeding activities in two centuries and is related to the political turmoil, firstly, of Habsburg Empire, then Austro-Hungary and other neighbouring countries. At the end of the First and Second World War, number of Gidran horses has been drastically reduced and survival of the population in Croatia has come into question. At the beginning of the 20th century reaffirmation of the breed began, primarily by formation of historical unit 'Bjelovar Border Troops – Hussars 1756' which uses only Gidran horse in the cavalry. The aim of this study was to assess the exterior of Gidran population in Croatia and compare it with the neighbouring population of Gidran horses (Hungary). The second aim was to provide guidance for the utilisation of Gidran breed in the future. 10 body measurements were taken from 31 breeding animals older than three years. Results of the performances at the National Championships in show jumping, dressage and endurance were obtained from the archives of Croatian Equestrian Federation. Statistical analysis was carried out using the SAS software package. The Gidran population in Croatia shows population homogeneity and uniformity as a result of a succesful breeding program. Compared to Hungarian Gidran population, Gidrans from Croatia have had lower values for body measurements. In sport competitions Gidran horses are mostly represented in endurance. The Gidran population in Croatia plays an important role in activities of social significance like traditional performances, events, and folklore manifestations. The population of Gidran horses also represents important genetic heritage thus collaboration at the international level is necessary. In that sense Gidran could be considered as transboundary breed.

The assessment of risk status of endangered horses breeds in Poland

G.M. Polak

National Research Institute of Animal Production, National Focal Point, Wspolna Str. 30, 00-930 Warsaw, Poland; grazyna.polak@izoo.krakow.pl

Assessment of the risk status of endangered local breeds is one of the main indicators for their effective preservation. For many years, investigations have been carried out in order to develop the best early warning systems. The genetic resources conservation programs conducted in Poland for the conservation of two horse breeds: Sztumski and Sokolski cold blooded horses led to a significant increase of their population sizes: from about 600 mares in 2008 to approx. 2,400 in 2016. The study took into account the 1,194 Sokólskie and 1,202 Sztumski mares enrolled in programs in 2016. The analysis based on the method of E. Verrier *et al.*, which consideres six factors affecting the status of the risk of a livestock population: basic factor – the number of broodmares (Nf) as well as five additional factors: population growth in the last 5 years (T5), the percentage of cross-breeding (C), the effective population size (Ne), breeders organizations and technical assistance (OTS), socio-economic aspect (SEC). The factors were evaluated on a scale from 0 to 5 points, assuming 0; no risk, 5; the maximum risk. It was found that based on 3 factors (T5, C, Ne), there was no risk of extinction. However, factors such as number of broodmares (Nf), assistance by breeding organizations and socio-economic aspect indicated negative impact on the future of the breeds. The final result – the average of 6 factors in the five-point scale was X=1.5, but the main factors – the number of broodmares = 4, determined the status of both breeds as 'at risk'.

The biology of adipose tissue in non-ruminants: renewed interests in oxido-reduction pathways

F. Gondret[1], M.-H. Perruchot[1], K. Sierzant[2] and I. Louveau[1]

[1]PEGASE, Agrocampus-Ouest, INRA, 35590 Saint-Gilles, France, [2]Wroclaw University of Environmental and Life Sciences, Faculty of Biology and Animal Science, 51630 Wroclaw, Poland; florence.gondret@inra.fr

White adipose tissue (WAT) comprises large fat depots interspersed in the body and plays a major role in energy homeostasis. This primary function related to lipid metabolism and mainly ensured by cytosolic enzymes acting in lipid synthesis or triacylglycerol hydrolysis is well documented in the literature. WAT is also characterized by a low number of mitochondria compared with brown adipose tissue, a fat depot specialized in heat production. Therefore, the relevance of mitochondria in WAT metabolism and how their metabolism could impact whole-body energy homeostasis have received only limited interests, especially in species such as pigs lacking UCP1, the hallmark of brown adipocytes. Recently, several high-throughput studies have revealed that mitochondrial genes playing roles in fatty acid beta-oxidation, energy oxido-reduction or response to oxidative stress were regulated in WAT following genetically- and environmentally-induced variations in animal's growth efficiency and body fat mass. Activities of some of the encoded enzymes were also modulated, suggesting actual modifications in adipocyte physiology. Moreover, the abundance of proteins acting in oxidation or antioxidant processes exhibited significant differences along WAT development in pig fetuses. Finally, experiments on porcine preadipose cells harvested from WAT showed that adipogenesis and reactive oxygen species production in adipocytes can be modulated by exogenous antioxidant substances. Thus, a renewed interest in mitochondrial oxido-reduction metabolism in WAT may be useful for a better control of adipose tissue features in animal production.

Proteomics profiling of cow adipose tissue in relation to energy balance around calving

C. Piras[1], A. Soggiu[1], I. Alloggio[1], V. Greco[2], L. Bonizzi[1], A. Urbani[3] and P. Roncada[4]
[1]*University of Milan, Veterinary Medicine, via Celoria 10, 20133, Italy,* [2]*Fondazione Santa Lucia – IRCCS, Rome, Italy, Via del fosso di Fiorano 64, Roma, 00143, Italy,* [3]*Università Cattolica, Roma, Italy, Istituto di Biochimica e Biochimica Clinica, Largo Agostino Gemelli, Roma, 00136, Italy,* [4]*Istituto Sperimentale Italiano L. Spallanzani, TechnologieS srl., via Celoria 10, 20133, Italy; cristian.piras@unimi.it*

The genetic selection of cows for higher rates of milk production is negatively related to body condition score (BCS) and to reproductive performance. The major problem seems to be related to the increasing gap between milk yield and feed intake which turns in the occurrence of negative energy balance status (NEB). The worst period for cow NEB is experienced after parturition. In this period, proteins and fats are mobilized in order to compensate the energy requirements due to high amounts of milk production. The aim of this study was to analyze the differential proteomics profiling of adipose tissue from cows with different NEB scores. The analysis was performed in a total of 18 samples (3 per group), the first group was composed by cows with low NEB scores and the second group was composed by cows with high NEB scores (hNEB). Both groups were analyzed at three different time points, one month before calving, one week after calving and 16 weeks after calving. Four key proteins involved in adipocyte metabolism were differentially expressed between groups in all time points and, 22 other proteins were differentially expressed when considering each single comparison between the two groups in each time point. These differences highlighted the differentially regulated pathways among hNEB cows. The differential expression of proteins and pathways was consistent one month before calving suggesting that the individual variability in adipose tissue metabolism/mobilization/energy availability could be linked to the different levels of EB among dairy cows.

Regulation of 17bHSD12 mRNA expression in adipose tissue of dairy cows with different body condition

K. Schuh[1], S. Häussler[1], C. Urh[1], C. Koch[2], D. Frieten[3], G. Dusel[3], H. Sadri[1] and H. Sauerwein[1]
[1]*University of Bonn, Animal Science, Katzenburgweg 7, 53115 Bonn, Germany,* [2]*Educational & Research Centre, Hofgut Neumühle, 67728 Münchweiler/Alsenz, Germany,* [3]*University of Applied Sciences Bingen, Animal Nutrition, Berlinstr. 109, 55411 Bingen, Germany; sauerwein@uni-bonn.de*

Amongst other tissues, adipose tissue (AT) expresses 17β-hydroxysteroid dehydrogenase type 12 (17bHSD12) which converts estrone into estradiol, i.e. potentiating estrogenic effects, and is also involved in the fatty acyl elongation cascade, including the synthesis of eicosanoids. Estrogens but also eicosanoids (PGE2) are known to stimulate the release of leptin (Lpt). Hence, we were interested in testing the relationship between the mRNA expressions of 17bHSD12 in subcutaneous (sc) AT and serum Lpt in dairy cows with high or normal body condition score (BCS). Holstein cows (n=38) were preselected 15 wk before calving and allocated to either a high (HBCS; BCS>3.75) or normal (NBCS; BCS<3.5) BCS-group. Biopsies from scAT (tail head) and blood were sampled on d -49, 3, 21, and 84 relative to calving. 17bHSD12 mRNA abundance in scAT was assessed by qPCR. Fatty acids (FA) and β-hydroxybutyrate (BHB) in serum were quantified photometrically. Serum Lpt was assayed by ELISA. Data were analyzed using the mixed-model procedure followed by Bonferroni correction. Relations were tested by Pearson correlation (P<0.05; SPSS 21). Early grouping of the cows resulted in BCS differences at all time points (P<0.001). Serum concentrations of Lpt were 2.3 fold higher in HBCS cows on d -49 (P<0.001) and d 3 (P<0.05), and FA were increased by 24% at both d 3 and d 21 (P<0.05), when compared to NBCS cows. The BHB concentrations were also 41% higher for HBCS cows at d 21 (P<0.05). In both groups, mRNA abundance of 17bHSD12 peaked at d -49, while the expression of 17bHSD12 was 32% higher in HBCS compared to NBCS cows (P<0.01). On d -49, positive relations were observed between 17bHSD12 mRNA and serum Lpt (P<0.05; r=0.37) and BCS (P<0.05; r=0.43). The effects of body condition on 17bHSD12 were limited to d -49 and were also reflected by serum Lpt; the changes of 17bHSD12 expression with time indicate a decreased activity in estrogen conversion and FA elongation around parturition.

Identification and characterization of adipokines and myokines involved in fat deposition in cattle

E. Albrecht, L. Schering, Y. Liu, K. Komolka and S. Maak
Leibniz Institute for Farm Animal Biology, Institute of Muscle Biology and Growth, Wilhelm-Stahl-Allee 2, 18196 Dummerstorf, Germany; elke.albrecht@fbn-dummerstorf.de

Adipose tissues in cattle and other species are not homogenous in their cellular and molecular characteristics. Visceral adipose tissue is metabolically more active than subcutaneous or muscle associated adipose tissues (intermuscular adipose tissue; IMAT and intramuscular fat; IMF). Large differences between breeds can be observed in the amount of fat in different depots. However, transcriptome analyses of adipose tissues revealed larger differences between tissues within a breed than between breeds within a tissue. The cellular environment of adipocytes, including extracellular matrix and capillary network, is crucial for their development. It is obvious that IMAT and IMF are directly influenced by skeletal muscle via bioactive molecules like myokines which are secreted by muscle cells. Myostatin is the best known myokine affecting muscle and fat development in cattle. New myokines can be discovered with different approaches. We provided recently a list of putative myokines using a cross-species comparison of transcriptome and secretome data. Those candidates need further investigations. Adipocytes secrete also bioactive molecules termed adipokines, which may act in an auto-, para- or endocrine manner. Leptin and adiponectin are well known examples of secreted molecules, but many more exist and need to be identified and characterized. Cross-talk between adipocytes and muscle cells may influence nutrient partitioning and accretion in different depots. Therefore, it is important to determine how adipokines and myokines are involved in this process. Expression analysis of the encoding gene in purified cell collections, localization of the protein in tissue sections with immunohistochemistry, detection of the protein in cell culture supernatant, and database search were used to prove or exclude a role as adipokine, myokine, or adipo-myokine. Results for agouti signaling protein, myostatin, and thrombospondin 4 are presented exemplarily for each group. Extending our knowledge about the complexity of communication between adipocytes and skeletal muscle cells can help to develop strategies for modulation of body composition and sustainable production of high quality meat.

Adiposity and expression of lipid metabolism genes in skeletal muscle in two beef cattle breeds

O. Urrutia, A. Arana, B. Soret, L. Alfonso, A. Purroy and J.A. Mendizabal
Public University of Navarre., Campus Arrosadía, 31006, Pamplona, Spain; olaia.urrutia@unavarra.es

There is growing interest in developing alternative technologies to alter lipid deposition to selectively enhance marbling fat in meat animals and it is also interesting to gain knowledge regarding fat development. This work aimed to study the adiposity and expression of some markers involved in lipid metabolism in *longissimus thoracis* (*LT*) and *masseter* (*MS*) muscles from Friesian and Pirenaica bulls. Eight Friesian young bulls slaughtered at an average age of 297 ± 3.5 days and a carcass weight of 230 ± 5.6 kg and 8 Pirenaica young bulls slaughtered at an age of 389 ± 4.4 days and a carcass weight of 348 ± 4.3 kg were used. The adipocyte size distributions in *LT* muscle in both breeds were unimodal (Bimodality Coefficient, BC<0.555), which showed a largest proportion of adipocytes with a mean size of 19.27 ± 0.31 and 20.27 ± 0.26 μm in Friesian and Pirenaica breeds, respectively. This suggest that the development of this depot may be mainly owing to hyperplasia, which corresponds to early phases of fat accretion, and concurs with the fact that intramuscular fat deposition is late-developing. In the *MS* muscle of both breeds, the adipocyte size distributions were bimodal (BC>0.555). A population of small adipocytes was observed, suggesting a predominant hyperplasic process, and a second population of 45-55 μm, which may indicate an initiation of a hypertrophy process. Expression of LPL in both breeds was upregulated in *MS* muscle compared to *LT* (P<0.01). This is in agreement with higher hypertrophy of *MS* adipocytes, which was not observed in *LT* muscle. An overexpression of PPARG was only observed in *MS* of Pirenaica bulls (P<0.05), which probably indicate a regulation of LPL by PPARG enabling fatty acid release from lipoproteins. FABP4, ACACA and SCD exhibited no differences between muscles in both Friesian and Pirenaica young bulls (P>0.05). In conclusion, the results of the present study indicate that fat development differed between the studied muscles. The higher of adipocyte hypertrophy observed in *masseter* muscle may be indicative of early fat development of this muscle.

Immune stimulation in adipose tissue and its consequences for health and performance in the pig

K. Ajuwon

Purdue University, Dept. of Animal Sciences, 915 W State Street, West Lafayette, IN 47907, USA; kajuwon@purdue.edu

Recent discoveries on the existence of an intact innate immune system in pig adipose tissue suggests that adipose tissue can have a functional role in the regulation of health and growth performance in the pig. Although the pig has been a domesticated animal for about 13,000 years, we still do not fully understand the whole range of factors that regulate its immune system and the impact of those factors on animal performance. However, it is well known that chronic immune activation and inflammation in pigs is a major mitigating factor against optimal health and growth performance. Therefore, strategies for improving health status of pigs must involve minimization of unnecessary and wasteful immune activation. Therefore, it is exciting to explore mechanisms of immune regulation in porcine adipose tissue as novel and practically useful information for optimizing animal health and performance may be obtained. The adipose tissue expresses innate pattern recognition receptors by which it is able to recognize conserved pathogen associated molecular structures. Evidence also indicates that preadipocytes possess phagocytic properties that may contribute to the clearance of pathogens from circulation. Acute phase proteins such haptoglobin and inflammatory cytokines, e.g. IL-6, IL-15 and TNF alpha are highly expressed in adipose tissue as a result of immune activation. The role of factors that regulate immune activation in adipose tissue will be discussed using examples from human, rodent and porcine studies.

Adipose-immune cross-talks in two pig lines divergently selected for feed efficiency – part 1

M.H. Perruchot, E. Merlot, I. Louveau and N. Le Floc'h

PEGASE, Agrocampus Ouest INRA, Saint-Gilles, 35590, France; marie-helene.perruchot@inra.fr

Adipose tissue is now recognized as an endocrine organ with a role in organ cross-talk. It produces a number of factors including inflammatory cytokines such as Tumor Necrosis Factor α (TNF-α). This factor regulates also adipocyte differentiation and adipokine secretion. The aim of this study was to determine if different levels of immune stimulation influenced the features of adipose tissue cells in two lines of pigs, divergently selected for residual feed intake (RFI), a measure of feed efficiency. Growing pigs (n=24) of low (RFI-) or high (RFI+) RFI were reared in clean or dirty housing conditions for 6 weeks, to obtain healthy and chronically inflamed animals, respectively. After subcutaneous adipose tissue collection, adipocytes and cells of the stromal vascular fraction (SVF) were isolated. Gene expression levels were determined in freshly isolated adipocytes and SVF cells. TNF-α, Toll-Like Receptor (TLR) 2 and 4 mRNA were expressed in SVF cells but were not detected in adipocytes. Levels of TLR2 were higher (P<0.05) in RFI+ than in RFI- pigs in both housing conditions. The IL6 gene was expressed in both adipocytes and SVF cells with no difference between the experimental groups. SVF cells were cultured in differentiation medium in the absence or presence of TNF-α (1 and 10 ng/ml). TNF-α inhibited the accumulation of lipid droplets in SVF cells of the different groups, indicating an impaired differentiation (P<0.05). Analysis of gene expression in differentiated cells demonstrated that TNF-α impaired Malic Enzyme (ME) (P<0.05). In conclusion, the expression of the TNF-α gene in SVF cells and the response of these cells in culture to TNF-α did not differ between healthy and chronic inflamed animals. The finding of a higher TLR2 expression in RFI+ suggests difference in the immune response between the two pig lines that needs to be further investigated. Research has received funding from the EU FP7 program (PROHEALTH, grant agreement no. 613574).

Effect of two different omega-3 fatty acid sources in the prepartum diet on performance of calves

A. Ataozu[1], F.E. Gursel[2], K. Oztabak[2], L. Celik[1] and U. Serbester[1]
[1]Cukurova University Agricultural Faculty, Department of Animal Science, Adana, 01330, Turkey, [2]Istanbul University Veterinary Faculty, Department of Biochemistry, Istanbul, 34320, Turkey; ugurserbester@gmail.com

Flaxseed oil (FOS) and fish oil (FO) contain omega-3 fatty acids. Flaxseed oil is characterized by a high concentration of alpha linolenic acids (ALA) while fish oil is rich in eicosapentaenoic acid (EPA) and docosahexaenoic acid (DHA). ALA can be converted through a series of desaturation and elongation reactions into EPA and DHA. However, these reactions have a poor efficiency in mammals. Therefore, the hypothesis was that maternal omega-3 fatty acids would influence differently gestation length, birth weight, weaning weight and average daily gain, and blood parameters of calves. Twenty five multiparous Holstein cows were used in an incomplete randomized block design with repeated measurements during summer 2016. Cows were blocked according to parity and last lactation milk yield and were randomly assigned to 1 of 2 prepartum total mixed ration supplemented with 5.4% FO or FOS during the last 21 d of gestation. Calves were fed colostrum within 2 h of birth from their own dam or from a dam fed the same treatment. After the colostral period, they were fed 4 L of whole milk and *ad libitum* alfalfa hay until weaning. Body weights were measured weekly, blood samples were taken at biweekly intervals. There were no difference in gestation length (274.5 vs 272.0 d for FO and FOS, respectively) and body weight at partum (594.6 vs 593.5 kg for FO and FOS, respectively) among the groups. Omega-3 fatty acid sources did not affect birth weights of calves (36.6 vs 34.1 kg for FO and FOS, respectively). Also, the source of omega-3 fatty acids in the diets did not influence the mean serum concentrations of alanine aminotransferase, aspartate aminotransferase, gamma-glutamyltransferase, alkaline phosphatase, glucose, total cholesterol and triglycerides. The results indicate that inclusion of FO or FOS to prepartum diet shows similar effect on the calf performance including metabolic profile.

Cholesterol metabolism and performance in Holstein cholesterol deficiency carrier dairy cows

A.-C. Schwinn[1], J.J. Gross[1], C. Drögemüller[2], R.M. Bruckmaier[1], F. Schmitz-Hsu[3], A. Barenco[4] and T. Neuenschwander[5]
[1]Veterinary Physiology, Vetsuisse Faculty, University of Bern, Bremgartenstrasse 109a, 3012 Bern, Switzerland, [2]Institute of Genetics, Vetsuisse Faculty, University of Bern, Bremgartenstrasse 109a, 3012 Bern, Switzerland, [3]Swissgenetics, Meielenfeldweg 12, 3052 Zollikofen, Switzerland, [4]Swissherdbook cooperative, Schützenstrasse 10, 3052 Zollikofen, Switzerland, [5]Holstein Association of Switzerland, Route de Grangeneuve 27, 1725 Posieux, Switzerland; ann-catherine.schwinn@vetsuisse.unibe.ch

The recently detected cholesterol deficiency (CD) in Holstein cattle is associated with reduced cholesterol in calves and bulls and caused by a loss of function mutation of the *APOB* gene. We set out to investigate if this mutation influences also the metabolism and performance of dairy cows. Twenty pairs of full sibling cows from same farms were investigated. Each pair contained a heterozygous carrier (CDC) and a non-carrier (CDF) of the disease associated *APOB* mutation. Blood samples were taken at 3-4 weeks and 4-6 months in lactation to measure cholesterol metabolites. A paired t-test was used to compare both groups. Total and free cholesterol (TC), high density lipoproteins, low density lipoproteins, phospholipids and cholesterol esters, but not triacylglycerides and very low density lipoproteins were lower in CDC cows compared with matched CDF siblings at both time points (P<0.05). Concentrations of glucose, free fatty acids, beta-hydroxybutyrate, and aspartat-aminotransferase and gamma-glutamyl-transferase activities did not differ between both groups (P>0.05). Milk yield, milk protein and milk fat in the previous and current lactation as well as days open did not differ between groups (P>0.05). Although cholesterol and associated fractions were reduced in heterozygous carriers of the CD causing *APOB* mutation, performance within matched sibling pairs did not differ in both early and mid-lactation.

Association of gene expression of ASIP and its receptors with fat deposition in cattle

Y. Liu, E. Albrecht, L. Schering and S. Maak
Leibniz Institute for Farm Animal Biology, Institute of Muscle Biology and Growth, Wilhelm-Stahl-Allee 2, 18196
Dummerstorf, Germany; liu.yinuo@fbn-dummerstorf.de

Transcriptome analyses of bovine muscle tissue differing in intramuscular fat (IMF) content identified agouti signalling protein (ASIP) as candidate gene with increased expression in the high IMF group. The protein is secreted from adipocytes and may serve as a signalling molecule influencing IMF and body fat deposition. Known receptors for ASIP are the melanocortine receptors (e.g. MC4R) and attractin (ATRN). The present study investigated the expression of ASIP and its receptors in different bovine tissues to determine associations with fat deposition. Samples of adipose tissues, liver and muscle (M. longissimus dorsi) were collected from 247 F_2 crossbred bulls (Charolais × Holstein) and RNA was extracted with standard methods. The expression levels of target genes were measured with RT-qPCR. Analysis of subcutaneous fat (SCF) revealed increased ASIP levels in 17 bulls. A long interspersed element (LINE, L1-BT, Exon2C) inserted in the ASIP gene caused this overexpression. Correlation analysis of normalized mRNA values of SCF with carcass traits showed significant associations with fat related traits in bulls without Exon2C. Retrospectively, three groups were assigned (Exon2C, n=17; high carcass fat (HCF), n=20; low carcass fat (LCF), n=20) and analysed. Expression of ASIP could be detected in liver of Exon2C bulls in contrast to HCF and LCF bulls, indicating ectopic expression of ASIP if Exon2C is abundant. Increased amounts of ASIP mRNA were also observed in other investigated tissues of the Exon2C group. Highest ASIP mRNA levels were measured in SCF. Higher values (1.6-fold, P<0.05) were observed for HCF than for LCF bulls in SCF, but not in other tissues. The mRNA of MC4R and ATRN could be detected in all investigated tissues. The level of MC4R was higher and of ATRN was lower in several tissues of LCF compared to HCF bulls. Bulls of the Exon2C group had lower ATRN mRNA values than HCF in perirenal and intestinal fat, whereas it was 1.25-fold higher (P<0.01) in muscle. Both receptors were expressed in liver, where ASIP mRNA could not be detected. This may imply that liver is one of the target organs for secreted ASIP influencing lipid metabolism.

Effect of monensin on milk fatty acid profile and blood metabolites in transition dairy cows

P. Karis, M. Henno, T. Ariko, H. Jaakson, K. Ling and M. Ots
Estonian University of Life Sciences, Department of Animal Nutrition, Kreutzwaldi 46, Tartu, 51006, Estonia;
priit.karis@emu.ee

Excessive mobilisation of body reserves in early lactation often leads to ketosis, one of the most frequent metabolic diseases of high producing dairy cows. To prevent ketosis, monensin, a pharmaceutical product that alters the ruminal fermentation pattern towards an increased production of propionate is proposed for use in multiparous dairy cows. As a non-invasive method, milk fatty acid (FA) profile is suggested for the identification of mobilisation of body fat and of rumen fermentation. The objective of this study was to evaluate the influence of monensin on milk FA profile and on metabolic status in multiparous high production dairy cows. Starting from 25.4±5.4 days before parturition, 18 experimental (M) and 16 control (C) multiparous dairy cows were fed close-up TMR (Total Mixed Ration) followed by post-partum TMR up to four weeks after calving, and subsequently fed peak-lactation TMR. The metabolisable energy and crude protein contents per kg dry matter of the TMRs were 10.8 MJ and 141 g, 11.7 MJ and 170 g, 11.8 MJ and 171 g, respectively. Monensin controlled-release capsules (Kexxtone; Elanco®, Bad Homburg, Germany) were administered to 18 M cows 24.7±6.2 days before expected calving. Blood samples were taken at the beginning of the trial, on the first or second day after parturition and hence forward weekly, up to the sixth week post partum, and were analysed for glucose, AST, NEFA and BHB. Milk samples were taken weekly and analysed for fatty acid composition. Higher NEFA (week 3 and 5) and BHB (at parturition, week 1 and 4) concentrations in the C cows evidenced more intensive fat mobilisation compared to the M cows. Higher proportions of total trans FA (including C18:2 cis-9, trans-11; CLA), total branched-chain FA and total de novo synthesized odd-chain FA were found in the M cows' milk compared to the C cows' milk. The proportions of de novo synthesised FAs and C18:1 cis-9 were not significantly different between the groups. The results suggest that differences in milk FA profiles resulted probably from the shift of the ruminal fermentation towards production of propionic acid for M cows.

Relationship between major muscle constituents in longissimus muscle from Hungarian Simmental bulls

G. Holló[1], B. Húth[1], I. Anton[2] and I. Holló[1]
[1]Kaposvár University, Guba S. street 40., 7400 Kaposvár, Hungary, [2]NARIC Research Institute for Animal Breeding, Nutrition and Meat Science, Gesztenyés street 1., 2053 Herceghalom, Hungary; hollo.gabriella@sic.ke.hu

The three major muscle constituents (protein, fat, connective tissue) influence both nutritional and sensorial meat quality components. The aim of this study was to analyze the association between intramuscular fat level and protein content as well as hydroxyproline/connective tissue proportion in longissimus muscle of Hungarian Simmental bulls. The tissue composition was analyzed using X-ray computed tomography (CT) and laboratory examination. We collected randomly, altogether 39 muscle samples in a commercial Hungarian abattoir. The animals were slaughtered at 492±228 days of age and 504.6±117.1 kg of weight. After 24 h chilling, rib samples were taken from the right half carcass longissimus thoracis (LT) cut at the 12th rib. The CT-examination of rib cuts was performed using a 16-slice CT system (Siemens Somatom Sensation Cardiac, slice thickness: 5 mm). Volumetric intramuscular fat (-200-19 CT value) and muscle tissue (20-150 CT value) percentage were determined with the use of MANGO (3.8, 2016) software. After CT examination, a chemical analysis was performed (Weendei analysis). Hydroxyproline measurement was used to determine connective tissue (collagen) content in meat. For statistical analysis, the SPSS 20.0 program package was used. Close correlation was detected between CT measured fat content and laboratory fat (r=0.93, P<0.001). The CT measured fat percentage showed moderate negative relationship with hydroxyproline/connective tissue proportion (r=-0.38, P<0.05), but a weaker association was detected between chemical fat and hydroxyproline/connective tissue content (r=-0.20, P<0.05). The protein content correlated negatively with fat content (r=-0.55, P<0.01) and with hydroxyproline/connective tissue proportion (r=-0.27; P<0.05). Further research should investigate the modulation of these muscle properties that determine the major components of meat quality in cattle. We gratefully acknowledge support from the Hungarian Scientific Research Fund (Project 111645).

Influence of carnitine and fat on performance, carcass characteristics and serum lipids of broilers

M. Bouyeh and H. Akhondzade
Department of Animal Science, Agriculture Faculty, Rasht Branch, Islamic Azad University, 4193963115 Rasht, Iran; mbouyeh@gmail.com

This study was carried out to determine the effect of adding three levels of L-carnitine (0, 200 and 400 ppm/kg) and three levels of sunflower oil (0, 2.5 and 5%) to basal diet containing 1.3% crude fat on performance, carcass characteristics and composition of blood lipids in broiler chickens. A completely randomized design in a factorial 3×3 arrangement with 9 treatments, 3 replicates and 10 chicks in each box using 270 Ross 308 broiler one-day-old chicks was done with a trial period of 42 days. Performance, some carcass and blood traits and body composition were measured in this experiment. Feed and water were provided *ad libitum* and lighting system was provided according to Ross 308 recommendations. The data were collected weekly or at the end of experiment. Using SAS software, analysis of variance and Duncan test were used to compare the means on a value of P<0.05. The results showed the significant effect of L-carnitine on serum triglyceride concentration (TG) and abdominal fat pad (AFP) so that the highest level decreased both parameters (P<0.05). Fat levels did not significantly affect the studied parameters (P>0.05) except feed conversion ratio (FCR), AFP, carcass and intestine weights and European performance efficiency factor (EPEF). So that treatment groups with 5% fat were the best in comparison to 0 and 2.5% (P<0.05). The significant effect of L-carnitine × fat on carcass weight, EPEF, AFP and liver percentage and also TG was observed so that the highest performance in carcass and liver weights and TG belonged to treatment group 400 ppm carnitine with 5% fat and in EPEF and AFP to treatment group 200 ppm L-carnitine with 5% fat. The lowest performance in the mentioned parameters belonged to the control group (0 ppm L-carnitine with 0% fat). Results reported here support the hypothesis that it is possible to produce more healthy and economic poultry meat by adding L-Carnitine and suitable level of fat to broiler diets.

Expression of perilipin 2 and leptin genes in muscle and backfat tissues of pigs

K. Tempfli[1], K. Szalai[1], E. Lencses-Varga[1], Z. Simon[2] and A. Bali Papp[1]
[1]*Szechenyi Istvan University, Department of Animal Sciences, Mosonmagyarovar, 9200, Var 4., Hungary,* [2]*Olmos & Toth Ltd, Debrecen, 4025, Hatvan st 6, Hungary; tempfli.karoly@sze.hu*

Perilipin 2 (PLIN2) and leptin (LEP) are potentially involved in the regulation of several traits of economic importance, such as average daily gain, feed conversion ratio, and ham weight. PLIN2 plays roles in the formation of lipid droplets and in their protection from cytosolic lipases, whereas LEP is involved in the regulation of feed intake and energy homeostasis in pigs. Our aim was to identify possible differences in the expression of PLIN2 and LEP genes between phenotypically different pig breeds and crosses, since differentially expressed genes can explain phenotype variations. PLIN2 and LEP expression was analysed in muscle and backfat tissues of fat-type Mangalica (M; n=12), Mangalica × Duroc (MD; n=12), and lean-type Hungarian Large White (HLW; n=10), and Pietrain × Duroc (PD; n=4) pigs with a similar body weight (126.1±9.2 kg), using quantitative reverse transcription PCR and the 2-ΔΔCt method. Beta-actin normalized PLIN2 expression was higher ($P<0.05$) in muscle samples of M and MD compared with HLW and PD pigs. Backfat PLIN2 expression of the four groups was not different ($P>0.05$). Elevated PLIN2 expression in M and MD pigs can contribute to increased intramuscular fat (IMF) accumulation, which results in the production of marbled pork. IMF content is an important factor affecting meat quality traits and sensory characteristics. In each breed group, expression of LEP was higher in adipose tissue samples compared to muscle. Highest LEP mRNA level was observed in the backfat tissues of M pigs ($P<0.05$), whereas muscle LEP was also higher in M and MD compared with HLW and PD pigs. The increased LEP expression in M does not suppress appetite due to probable LEP resistance, which is a possible mechanism leading to the development of the obese phenotype. Based on the detected variations in their expression, both analysed genes are considered candidate factors contributing to differences in fat metabolism-related traits in pigs.

Adipose-immune cross-talks in two pig lines divergently selected for feed efficiency – part 2

E. Merlot, M.-H. Perruchot, I. Louveau and N. Le Floc'h
INRA, PEGASE, Agrocampus-Ouest, 35590 Saint-Gilles, France; marie-helene.perruchot@inra.fr

Chronic immune stimulation deteriorates adipose tissue accretion through the release of pro- and anti-inflammatory cytokines. In turn, adipokines modulate immune cell activity. It has been shown that pig genetic selection for residual feed intake (RFI), a measure of feed efficiency, influences their energy and protein metabolism, both in healthy and inflammatory states. This study aimed to determine if this selection also interfered with immune-adipose cross-talks. Male and female growing pigs of low (RFI-, n=28) or high (RFI+, n=25) RFI were reared in clean or dirty housing conditions for 6 weeks (n=26 and 27), to obtain healthy and chronically inflamed animals, respectively. Gene expression levels were determined in adipocytes and in cells of the stromal vascular fraction (SVF) of subcutaneous adipose tissue, and in whole blood cells (WB) cultured for 20 h in the absence or presence of lipopolysaccharide (LPS). Levels of adiponectin were greater ($P<0.05$) in adipocytes of RFI+ than RFI- pigs but did not differ with the line in SVF cells. In WB, there was a greater expression in RFI+ compared with RFI- pigs of adiponectin receptor 1 (AdipoR1, $P<0.05$) and Transforming Growth Factor-beta (TGF-β, $P=0.05$ and 0.09) in the 2 culture conditions, and of interleukin (IL)-1 ($P<0.05$), IL-10 ($P=0.06$) and Tumor Necrosis Factor-alpha (TNF-α, $P<0.01$) in LPS-stimulated cultures only ($P<0.05$). The line effect in WB was mainly due to 5 of the 12 RFI+ females who expressed dramatically greater levels of nearly all tested genes. When excluding them, the AdipoR1 gene was still more expressed in RFI+ pigs ($P<0.05$) whereas the expressions of AdipoR2, leptin receptor and IL-6 were not affected by the line. Unstimulated WB from dirty pigs expressed more AdipoR1 and TGF-β ($P<0.05$), and tended to express more Toll-like receptors 2 and 4 ($P<0.1$). To conclude, RFI- pigs might have a lower sensitivity of immune cells to adiponectin, which exerts regulatory immune functions, and a decreased expression of cytokines suppressing adipogenesis such as IL-1, TNF-α and TGF-β. Research has received funding from the EU FP7 program (PROHEALTH, grant agreement no. 613574).

Development of a mature adipocyte culture system

P.Y. Aad and S. Bou Karroum
Notre Dame University Louaize, Sciences, S224, Zouk Mosbeh, Lebanon; paad@ndu.edu.lb

Adipose tissue is a major endocrine organ, composed of white (WAT) or brown (BAT) adipocytes. Mature adipocytes are linked to many endocrine disorders. The Awassi sheep stores WAT mainly in its tail, an easily accessible WAT source. Characterization of this fat and development of a cell culture system from mature adipocytes (WMA) is needed to determine the responsiveness of this tissue to various stimuli. Fat samples from the tail of 9 male Awassi animals (2-3 years) were collected in the slaughterhouse, were paraffin embedded (hematoxylin eosin staining) or used for optimization of a culture system for pre-(WPA) and mature (WMA) adipocytes. Variations in the fixative (Bouin's, zinc formalin or both), dehydration time, clearing agent (xylene or isopropanol) and trimming thickness were tested to optimize paraffin embedding of fat tail (6 experiments (exp) of 4 replicates each) for later immunohistochemistry (IHC). For digestion of fresh fat tail, collagenase (2 mg/ml) or trypsin (2 mg/ml or 20 mg/ml) were used for up to 1 h. Preadipocytes (WPA) from the pellet and supernatant WMA were suspended in DMEM:trypsin; WPA were seeded in plates with DMEM:F12 containing 10% Fetal bovine serum (DMEM:F12:FBS), while WMA were cultured in DMEM:F12:FBS for 4 d in inverted flasks (ceiling culture) to allow cell attachment, then regularly afterwards. Each cell culture exp was repeated 3 times with 3 replicate plates/bottles each time in 11 exp. For embedding,1:1 Bouin's/ Zinc formalin was the best fixative with isopropanol cut at 10 μm. Paraffin embedding showed a predominance of WAT to BAT in fat tail of Awassi sheep. With collagenase, there was a very low recovery of adipocytes. The higher trypsin concentration at WPA cell isolation showed fewer days (13.1 d) to reach 100% confluence, with no difference (P>0.05) between exp. WMA reached static confluence and shape after 4 weeks in culture. Oil Red O stain confirmed the presence of an apparent lipid droplet in or around each WMA. In this study, for the first time, we developed successful paraffin embedding of fat tail tissue, cultured pre-adipocytes, and subcultured them 3 times. However, WMA culture further requires a specific media for dedifferentiation and growth.

From estimation of inbreeding toward complex modelling of negative consequences of inbreeding

I. Curik[1], M. Ferenčaković[1] and J. Sölkner[2]
[1]University of Zagreb, Faculty of Agriculture, Department of Animal Science, Svetošimunska cesta 25, 10000 Zagreb, Croatia, [2]University of Natural Resources and Life Sciences Vienna, Department of Sustainable Agricultural Systems, Division of Livestock Sciences, Gregor Mendel Strasse 33, 1180 Vienna, Austria; icurik@agr.hr

While negative consequences of inbreeding in livestock populations were noticed from the beginning of domestication, systematic approach to the quantification of those effects has started in the 20th Century. Thus, for example, regression of phenotypes on the pedigree inbreeding coefficients (F_{PED}) has been the major method used in quantifying negative effects of inbreeding in livestock populations for over 70 years. Here, we first present restrictions and limits of the method that result from definition of F_{PED} per se. We further describe properties of the recently introduced runs of homozygosity inbreeding coefficient (F_{ROH}), a measure of realized autozygosity that can be further partitioned to chromosomal segments or even single SNPs, as an efficient tool in quantifying negative effects of inbreeding. Particular stress is given to two approaches that are related to estimation of inbreeding effects at regional scale, up to the single SNP partition. The first approach, ROH-based mapping of loci contributing to inbreeding depression, is frequently used in human research as well as in several studies related to livestock populations. The second approach has recently been proposed by the authors of this review and describes SNP-wise dissection of inbreeding depression in terms of the quantitative genetics concept. At the end, perspectives of new technologies, such as are prediction of detrimental load from the whole-genome sequences and gene editing, in reducing detrimental load in livestock populations are discussed.

Intra-chromosomal estimates of inbreeding depression on fertility in the Finnish Ayrshire population

K. Martikainen and P. Uimari
University of Helsinki, Department of Agricultural Sciences, Latokartanonkaari 5, 00014 University of Helsinki, Finland;
katja.martikainen@helsinki.fi

The fertility of dairy cows has been decreasing during the past decades. Impaired fertility reduces profitability, because the lifetime milk production of the cow decreases and the costs related to inseminations and veterinary treatments increase. It is well known that inbreeding has a negative effect on fertility. As the genetic diversity varies substantially across the genome, detecting the regions with lower genetic diversity could indicate inbreeding depression. The objective of this study was to estimate the chromosomal and intra-chromosomal inbreeding coefficients based on runs of homozygosity (ROH) and to estimate inbreeding depression on female fertility traits in the Finnish Ayrshire population. Genomic information from 13,712 Finnish Ayrshire females with fertility phenotypes was analysed. Genotypes were obtained with the Illumina BovineLD v.2 BeadChip low-density panel and imputed to 50K density. ROH-based chromosomal inbreeding coefficients (F_{ROH}) were used as a covariate in the mixed model equation. The dependent variables consisted of cow records pre-adjusted for the most important fixed effects. Based on the analyses, a 10% increase in F_{ROH} on chromosome 2 or 18 was associated with 1.9 or 1.0 days longer interval from first to last insemination (IFL) for heifers, respectively (P<0.01). Similarly, F_{ROH} estimated on chromosome 15 was associated with 2.9 days longer IFL for the second parity cows (P-value<0.01). To locate more precise region on chromosomes that show inbreeding depression, a sliding window approach was applied where intra-chromosomal F_{ROH} was calculated as a number of SNPs in ROHs belonging to the window divided by the total number of SNPs in the window. Using a window size of half of the chromosome length, the first window on chromosome 2 showed significant inbreeding depression: 10% increase in F_{ROH} increased IFL by 1.6 days for heifers ($P=6.6\times10^{-06}$). The next step is to run the same analyses for chromosomes 15 and 18 and then further narrow down the window size to pinpoint the regions associated with IFL related inbreeding depression.

Estimation and characterization of individual genomic inbreeding in Belgian Blue beef cattle

M. Solé[1], A.-S. Gori[1,2], P. Faux[1], M. Gautier[3] and T. Druet[1]
[1]Unit of Animal Genomics, GIGA-R, University of Liège, 4000 Liège, Belgium, [2]Association Wallonne de l'Élevage, 4, rue des Champs Elysées, 5590 Ciney, Belgium, [3]UMR CBGP, INRA, 34988 Montferrier-sur-Lez, France; msole@ulg.ac.be

With the developments of genomic tools, inbreeding estimators obtained from the genomic relationship matrix (GRM) or from runs of homozygosity (ROH) are widely used in livestock species. Recently, we proposed a hidden Markov model (HMM) that estimates both global and local inbreeding and makes use of marker allele frequencies, marker positions and genotyping error rates. In addition, the method partitions inbreeding in different identity-by-descent (IBD) classes with different ages (measured as the size in generations of the IBD loop). Our goal was to apply our HMM to Belgian Blue Beef (BBB) cattle and to compare estimates obtained with different marker densities and with other approaches. To that end, pedigree information of 634 BBB bulls and genotypes for 7,713 (Low Density; LD), 34,008 (50K) and 613,005 (High Density; HD) SNPs were considered after data cleaning. Results obtained with our HMM and the HD genotypes indicate that a large fraction of the total inbreeding (F_G) is associated with ancient IBD classes but that the most recent classes account for most of the individual variation. With lower marker densities, only the most recent inbreeding is captured but correlations with HD estimates are high (0.87/0.98 with the LD/50K). Estimates of our model with the HD panel are highly correlated with those obtained with excess homozygosity (r=0.95) or with ROH (r=0.93); these correlations are lower at lower marker densities. Comparisons with ROH indicate that with the HD panel our model captures roughly ROH larger than 100 kb (and larger than 500 kb to 1 Mb with LD or 50K panels). Correlations with GRM based estimators were lower (0.02 or 0.26 according to the method). With pedigree-based measures, our model presented overall moderate correlations (0.46 for individuals born after 2000) but these were higher (up to 0.56) if we focused only on the most recent IBD-classes. Finally, we show that our model proved efficient to estimate global and local inbreeding with low-density genotyping arrays in livestock populations.

Genomic selection and inbreeding in North American Holstein cattle

M. Forutan[1,2], S.A. Mahyari[2], C. Baes[1], N. Melzer[3], F. Schenkel[1] and M. Sargolzaei[1,4]
[1]*University of Guelph, 50 Stone Road East Guelph, Ontario, N1G 2W1, Canada,* [2]*Isfahan University of Technology, Isfahan, 84156-83111, Iran,* [3]*Leibniz Institute for Farm Animal Biology, Wilhelm-Stahl-Allee 2, 18196 Dummerstorf, Germany,* [4]*Semex Alliance, 5653 Highway 6 North Guelph, ON, N1H 6J2, Canada; mforutan@uoguelph.ca*

Genomic selection (GS) is expected to reduce the rate of inbreeding per generation by capturing Mendelian sampling variation more accurately than traditional BLUP selection. However, in dairy cattle, GS may lead to higher rate of inbreeding per year due to the reduced generation interval. There is limited information on the effect of GS on the distribution of autozygosity across the genome. Ongoing stretches of homozygous genotypes (ROH) provide a good estimate of autozygosity. A gene dropping simulation was performed to obtain true inbreeding in a population undergoing different selection strategies (random, pedigree-based, and genomic selection). Inbreeding coefficients based on ROH, pedigree, and the genomic relationship matrix were estimated and compared. ROH-based inbreeding was calculated using PLINK and SNP1101 (with window sizes of 20, 35 and 50 SNPs), and BCFtools. Furthermore, inbreeding coefficients were estimated in a sample of genotyped North American Holstein animals born from 1990 to 2016 using 50K chip data. Results confirmed that pedigree inbreeding could result in biased estimates of autozygosity in selected populations, whereas ROH detected with SNP1101 with a window size of 20-50 SNPs (approximately 1-2.5 Mb) and BCFtools provided inbreeding values closest to true inbreeding. The number of ROH increased as a consequence of selection. GS resulted in the most number of ROH segments. In real data, noticeable differences in ROH frequency and distribution were observed before and after implementation of GS in 2010. The increase in the rate of inbreeding per year was faster after implementation of GS. Therefore, managing inbreeding has become more important. Results confirmed that ROH is a better estimator of inbreeding and the type of selection plays an important role in determining and shaping distribution of ROH.

Investigating runs of homozygosity in dairy cattle using imputed genomic information

A. Fleming[1], M. Forutan[1,2], B. Makanjuola[1], N. Melzer[3], F.S. Schenkel[1], M. Sargolzaei[1,4] and C. Baes[1]
[1]*CGIL, University of Guelph, 50 Stone Rd E, Guelph, ON, N1G 2W1, Canada,* [2]*Isfahan University of Technology, Department of Animal Science, Isfahan, Iran,* [3]*Leibniz Institute for Farm Animal Biology, Wilhelm-Stahl-Allee 2, 18196 Dummerstorf, Germany,* [4]*Semex Alliance, 5653 Hwy 6 N, Guelph, ON, N1H 6J2, Canada; cbaes@uoguelph.ca*

Dairy cattle genotypes are routinely imputed to a higher density for use in genomic evaluations. Due to the increased selection intensity in dairy populations, pedigree inbreeding levels have steadily increased over the past decades. To enable more accurate estimation of inbreeding levels, genomic information is now commonly used instead of pedigree information. Inbreeding levels can be determined by finding stretches of homozygous loci (runs of homozygosity, ROH) using single nucleotide polymorphism (SNP) arrays or next-generation sequencing (NGS) data. The analysis of ROH in genome-wide SNP data allows detection of autozygosity (i.e. stretches of homologous chromosomes within an individual likely derived from the same ancestor). It is unclear how well ROH can be detected using imputed data, especially short ROH. The objective of this study is to determine the accuracy of ROH identification using imputed high-density genotype data. The HD genotypes of around 1000 animals out of 3,000 will be masked to mimic those of 50K data. These genotypes will then be imputed back up to HD. As well, the NGS genotypes of 100 animals out of 400 will be masked to mimic HD/50k data, and then imputed back up to NGS. The correlation of ROH in both real and imputed HD and NGS datasets will be calculated, and overlaps in ROH for short, medium, and long segments will be compared and analyzed. Other aspects of imputation accuracy will also be discussed (reference size, relationships, SNP density in target animals, etc). The use of genomic data to detect and characterize ROH can provide insights into autozygosity and effects of intense selection for traits of economic importance in livestock. Before using imputed data for these analyses, however, it is important to understand how well ROH detected in imputed data correlate to those detected in real data. Resulting information will add confidence to future work on ROH analyses using imputed data.

Extent of runs of homozygosity in a wide set of phylogenetically diverse chicken populations

S. Weigend[1], A. Weigend[1], J. Sölkner[2], D. Malomane[3] and H. Simianer[3]
[1]Friedrich-Loeffler-Institut, Institute of Farm Animal Genetics, Department of Breeding and Genetic Resources, Höltystrasse 10, 31535 Neustadt, Germany, [2]University of Natural Resources and Life Science Vienna, Department of Sustainable Agricultural Systems, Gregor-Mendel-Straße 33, 1180 Vienna, Austria, [3]Georg-August-Universität Göttingen, Department of Animal Sciences, Albrecht-Thaer-Weg 3, 37075 Göttingen, Germany; steffen.weigend@fli.de

Runs of homozygosity (ROH) are long stretches in an individual's DNA sequence which show continuous homozygosity. ROH provide insight into the degree of inbreeding, population history and the relationship between genetic diversity and effects of intense selection or genetic drift at the molecular level. Long ROH may emphasize recent inbreeding, but may also be affected by recent strong directional selection, while crossbreeding contributes to an increase in diversity by disrupting long ROH in the genome. The extent and distribution of ROH has been studied in most livestock species, such as cattle, swine, sheep, goat and horse, but to a much lesser extent in chickens. Within the framework of the SYNBREED project a wide range of chicken breeds were sampled across four continents. The 'Synbreed Chicken Diversity Panel (SCDP)', which encompasses more than 3,200 individuals of 175 populations including two wild chicken populations and commercial purebred lines, was genotyped with the 580K SNP Affymetrix Axiom® Genome-Wide Chicken Genotyping array. Independent of some variation in the criteria that were used to define ROH, chickens from Africa and Asia show a lower proportion of genomic regions in ROH, while chicken populations sampled in Europe display a wide variation ranging from 4 to about 70% of the genome being included in ROH. Regarding the commercial lines, the genome of white layer lines was least polymorphic, while broiler lines clustered at the polymorphic end of the SCDP spectrum, and brown layers showed a medium degree of variability. Assuming that ROH is a genomic measure of individual autozygosity, white layers displayed a much higher degree of inbreeding than brown layers, and even more so than broilers. Detailed analyses are underway to evaluate the length of ROH stretches in the genome as well as their distribution across chromosomes.

A generalized approach to calculate expectation and variance of kinships in complex breeding schemes

T. Pook[1], S. Weigend[2] and H. Simianer[1]
[1]University of Goettingen, Animal Sciences, Albrecht-Thaer-Weg 3, 37075, Germany, [2]Friedrich-Loeffler-Institut, Institute of Farm Animal Genetics, Hoeltystraße 10, 31535 Neustadt, Germany; tpook@gwdg.de

An accurate prediction of inbreeding in a population is a topic of interest breeding programs. To model this, the concept of kinship for individuals was introduced by Malécot in 1948, which is easy to use and allows an iterative calculation of pairwise kinships for individuals with known relationship. Oftentimes one is not only interested in the characteristics of individuals but in a cohort of animals with similar age, gender and genetic origin. The goal of this study is to extent the definition of kinship to cohorts of animals by combining the gene-flow method with the concept of kinship. Other than in the approach of Sitzenstock *et al.* we differentiate between chromosomes of two animals (identity-by-descent) and the two chromosomes of one animal (homozygosity-by-descent). By doing this we can properly model the presence or absence of selfing and therefore obtain unbiased expected kinships for any group size. As possible transition options between cohorts we consider aging, reproduction and migration and discuss corresponding transition probabilities. Additionally to the basic calculation of expected kinship we also discuss how this changes under selection, mutation and in dynamic breeding programs. Furthermore we discuss how to estimate the variance of the expected kinship. We apply our methods to a conservation breeding scheme based on cock rotation in chicken. We compare our results with the ones of Sitzenstock *et al.*, an exact simulation of the breeding program with averaged individual kinships (R-package: RekomBre, unpublished work so far) and genotype data of a real conservation program.

The Rambouillet sheep breed: a unique chance to investigate inbreeding at genome level

C. Danchin[1], A.C. Doublet[2] and G. Restoux[2]
[1]Institut de l'Elevage, Génétique et Phénotypes, 149 rue de Bercy, 75012 Paris, France, [2]AgroParisTech/INRA UMR 1313, 16 rue Claude Bernard, 75005 Paris, France; coralie.danchin@idele.fr

The negative effects of inbreeding (higher occurrence of genetic defects, inbreeding depression…) are well known by breeds' managers. However we still have a poor insight of how the genome is impacted by inbreeding. For over a 200 years period, a French sheep breed called the Rambouillet has been kept in a single flock without any introgression. Pedigree information was fully recorded since the 1870s. For the last fifty years, minimizing the breed's inbreeding level has been the only goal of its management plan. It reached 52% for nowadays animals according to simulation data based on pedigrees. Despite this level, the breed's weight, prolificacy and fertility have not changed over the last forty years. Altogether this breed gives a unique opportunity to explore the consequences of inbreeding at the genome level. To do so, we used a high density chip (600 K SNP) on a panel of 46 individuals. Among these animals, 38 were born between 2006 and 2012, and eight are older males born in the 1980s whose samples come from the French National Cryobank. First results showed that the genome of the breed was highly structured with a very low level of diversity. However the genomic inbreeding level based on run of homozygosity (ROH) was about half the estimation based on pedigree information, and a very weak inbreeding depression was detected. Therefore, we investigated more precisely at the chromosome level for 'cold spots' of diversity based on ROH. When compared with two related breeds, the Mourerous and the Arles, the Rambouillet showed much more ROH segments, but the total length of ROH is actually lower than expected. Linkage disequilibrium (LD) was also calculated at chromosome level and the genome structure appeared to differ between each chromosome. This information suggests a mutational load purge that occurred throughout the years. Ongoing research seems to confirm this hypothesis since hot spots of diversity are found in regions with genes involved in major development patterns (such as cells division). Finally, the Rambouillet sheep proves to be a valuable source of investigation in the understanding the genome aspects of inbreeding.

Alternative genomic relationship matrices in genomic optimum optimun contribution selection

G.T. Gebregiwergis[1], T.H.E. Meuwissen[1], A.C. Sørensen[2] and M. Henryon[3,4]
[1]Norwegian University of life Sciences, Animal and Aquaculture Sciences, Arboretveien 6, Postboks 5003, 1433 Ås, Norway, [2]Aarhus University, Molecular Biology and Genetics, Blichers Allé 20, 8830, Tjele, Denmark, [3]The University of Western Australia, School of Animal Biology, Crawley WA 6009, 35 Stirling Highway, Australia, [4]SEGES Pig Research Centre, Axeltorv 3, 1609 København V, Denmark; gebreyohans.tesfaye.gebregiwergis@nmbu.no

A stochastic simulation study was carried out to investigate genetic gain and true genomic inbreeding in breeding schemes that applied genomic optimum contribution selection(GOCS) with different genomic relation matrices. Three genomic-relationship matrices were used to predict the genomic breeding values based three different information sources: markers (GM), QTL (GQ), and markers and QTL (GQM). Two genomic-relationship matrices were used to constrain rates of inbreeding: GM and GQM. Two genetic architectures were simulated: with 7,704 and 1000 QTL and 54,218 markers. Selection was for a single trait with additive genetic variance 1.0 and heritabilities 0.05, 0.2, 0.5, and 0.99. All selection candidates were phenotyped and genotyped before selection. With 7,704 QTL, there were no significant differences in rate of genetic gain at the same rate of true inbreeding using different genomic relationship matrices in GOCS across all heritibilities. However, as the number of QTLs was reduced to1000, prediction of genomic breeding values (GEBVs) using a genomic relationship matrix constructed based on GQ and constraining inbreeding using GM gave 28% higher genetic gain than using GM to both prediction and constraining inbreeding. These findings indicate that with large numbers of QTL, it is not critical what information is used to construct genomic-relationship matrices. However, it becomes critical with small numbers of QTL. This highlights the importance of using appropriate genomic-relationship matrices in GOCS, where the relationships used to predict GEBVs may differ from that used to constrain the inbreeding.

Inbreeding management in honeybees using DNA-based methods

G.E.L. Petersen[1,2], P.F. Fennessy[1] and P.K. Dearden[2]
[1]*AbacusBio Ltd, P.O. Box 5585, Dunedin 9016, New Zealand,* [2]*University of Otago, Department of Biochemistry, 710 Cumberland Street, Dunedin 9016, New Zealand; gpetersen@abacusbio.co.nz*

Lack of genetic diversity and inbreeding can present a problem in a number of livestock species, and is usually managed by the means of relationship constraint based on a known pedigree of both parental animals. In the Western Honeybee, *Apis mellifera*, inbreeding affects the population on two different levels, due to the insects' peculiar system of sex determination based on a single gene called *complimentary sex determiner* (*csd*). In the development of honeybee embryos, the number of present *csd* alleles determines the sex of an individual, leading unfertilised eggs to develop into haploid male drones and fertilised eggs with two different *csd* alleles into diploid female workers and queens. If the *csd* alleles on both chromosomes of a diploid individual are identical, this gives rise to non-viable diploid males. As a result, besides an autosomal inbreeding effect on honeybee colonies, a low number of *csd* alleles present in a population has direct and devastating effects on worker brood viability. Honeybee pedigree records are complicated by the fact that queens engage in extreme polyandry, mating on the wing with 6 to 25 drones during their nuptial flight and that limiting contribution to the drone cloud that queens are exposed to requires considerable management efforts. In addition, low numbers of effective paternities in the work force of a honeybee colony limit colony productivity as well as the ability to react to disease occurrences. Due to these limitations, a way to manage inbreeding in honeybee populations taking both present *csd* alleles and autosomal (whole-genome) relationships into account is needed. Methods to characterise *csd* have been described previously, as has Genotyping-by-Sequencing for honeybees. Combining *csd* records with whole-genome Genotyping-by-Sequencing results presents a powerful tool to enable efficient inbreeding management for commercial honeybee operations.

Genetic rescue of inbred dog populations

J.J. Windig
Wageningen University and Research, Animal Breeding and Genomics, P.O. Box 338, 6700 AH Wageningen, the Netherlands; jack.windig@wur.nl

Dog population frequently suffer from excessive inbreeding rates and associated genetic defects and inbreeding depression. Several measures can be taken to enlarge the effective population size. We evaluated the effect of such measures in a number of breeds in the Netherlands (the Golden Retriever, Saarloos Wolfhond, Markiesje, Stabij, French Bulldog, Dutch Smous hond, Scottish Collie and Irish Setter). The most effective measure is to restrict breeding of dogs with a high mean kinship to all other dogs in the population. In some breeds kinship levels are so high that no relatively unrelated dogs can be found, and genetic defects may be so common that mating carriers of genetic defects is unavoidable within the breed. Outcrossing, e.g. crossing with another breed, is seen as a last resort to overcome high inbreeding levels. We investigated the effect of outcrossing in the Saarloos Wolfhond with computer simulation. Mating individuals with unrelated dogs from outside the population produced non-inbred litters. Backcrossing, however, reduces the effect, and if no other measures are taken inbreeding rates quickly return to the high levels of before the outcross. A better strategy is to cross now and then a small part (<5%) of the population with unrelated dogs, such as is done with the Dutch Smoushond.

Genomic comparison of Western Finncattle and Nordic Red Cattle in Finland

T. Iso-Touru, D. Fischer, I. Stranden, J. Vilkki and M. Tapio
Natural Resources Institute Finland, Green technology, Myllytie 1, 31600 Jokioinen, Finland; miika.tapio@luke.fi

Western Finncattle is the most productive Finnish native cattle breed. It has been claimed to be among the most productive landraces worldwide. Maintaining breed competitiveness is challenged by the small population size. Breed may have accumulated inbreeding and deleterious homozygotic variants. Small population size limits the available reference population for genomic selection. It is desirable to use production data from Nordic Red Cattle in Finland to increase the reference population. To compare the genetic basis of the two breeds, pre-existing whole genome data of 30 Finnish Nordic Red Cattle bulls were compared to generated data for 17 Western Finncattle bulls. Mapping Illumina NGS reads were done using BWA. Variant detection was done following GATK best practices protocol. One Finncattle bull deviated from the other Finncattle. Comparison of shared and unique variants demonstrated a degree of unique variants including differences in homozygotic variants. The produced data can be used to impute full genome variation for gene mapping and genomic selection purposes.

Analysis of the inbreeding level in the active population of pigs of different breeds in Poland

G. Żak, M. Szyndler-Nędza and R. Eckert
National Research Institute of Animal Production, Krakow, Department of Animal Genetics and Breeding, ul. Sarego 2, 31-047 Krakow, Poland; grzegorz.zak@izoo.krakow.pl

Pig breeders in Poland have made extensive use of the BLUP-AM method since 1999. The application of this method for evaluating the breeding value of pigs improves their breeding value in nucleus herds and in the general population. On the other hand, the use of many sources of information (including pedigree information) in the calculations considerably increases the risk of inbreeding. The objective of the study was to estimate the level of inbreeding in sows and boars from the domestic pig population, which have been selected for more than ten years based on the BLUP-AM estimated breeding value. The study involved Polish Large White, Polish Landrace, Hampshire, Duroc, Pietrain and Puławska sows and boars born between 2007 and 2011. Data on a total of 16,842 sows and 977 boars representing the 6 breeds raised in Poland were collected for the pedigree analyses. During the analyzed period, the coefficient of inbreeding in sows ranged from F=0 to F=0.25 (14 animals). Most of the animals were non-inbred (70% in the PLW and Puławska, over 80% in the PL, Duroc and Pietrain and 100% in the Hampshire). For the boars, the coefficient of inbreeding varied from 0 to 0.1875 (1 animal). The proportion of non-inbred boars was 77% in the PLW, 91% in the PL, 63% in the Puławska, and about 80% in the Duroc and Pietrain breeds. The results point to the appropriate structure of the pig population in terms of inbreeding, which shows that the selection of animals for mating has been correct and there have been no negative consequences of using the BLUP method.

Estimating the genetic variability of the Letelle sheep breed with the use of microsatellite markers

T. Magwaba[1], L. Van Der Westhuizen[2,3], J.P. Grobler[1], H. Bindeman[1], C. Du Plessis[1], E. Van Marle-Kösterd[4] and F.W.C. Neser[3]

[1]University of the Free State, Genetics, P.O. Box 339, 9300 Bloemfontein, South Africa, [2]ARC, P Bag X2, 0062 Iren, South Africa, [3]University of the Free State, Animal Wildlife and Grassland Sciences, P.O. Box 339, 9300 Bloemfontein, South Africa, [4]University of Pretoria, Animal and Wildlife Sciences, P Bag x20, 0028 Pretoria, South Africa; neserfw@ufs.ac.za

The purpose of the study was to obtain insight into the genetic variability of the Letelle sheep breed that will reveal vital information regarding the genetic resources and future breeding policies. The Letelle is a South African developed dual purpose sheep breed which falls under a Merino type breed but with a Spanish Merino background. The breed exhibits exceptional wool characteristics and high quality mutton. For the past 25 years no new genetic material was allowed in the breed and a breeding policy of line-, family- and inbreeding was applied. Multiple sire mating was practiced to prevent a single sire from having a too large influence on the national flock. Samples were collected from 10 commercial- and 11 seed stock flocks. A total of 210 animals from these flocks have been genotyped and a total of 17 microsatellite markers were used during the study. Unbiased heterozygosity varied between 0.579±0.058 and 0.676±0.064, observed heterozygosity varied between 0.524±0.038 and 0.647±0.037 and the mean number of alleles varied between 3.765±1.480 and 5.235±2.195. The Analysis of Molecular Variance, STRUCTURE and unrooted Neighbour-joining tree revealed that no genetic structure exists within the breed. Consequently, seed stock- and commercial flocks can be grouped together. The average within flock- (F_{IS}) and within breed (F_{IT}) inbreeding coefficients were 10.1 and 14.5%, respectively. Although high, this level of inbreeding along with the moderately high genetic diversity estimates and a carefully draft breeding policy, will allow the Letelle to adapt and successfully respond to future climate fluctuations.

The assessment of genetic diversity in sturgeons aiming at aquaculture stocks improvement

A. Dudu, A. Burcea, I.E. Florescu, G.O. Popa, M. Costache and S.E. Georgescu
University of Bucharest, Department of Biochemistry and Molecular Biology, 91-95, Splaiul Independentei, 050095 Bucharest, Romania; tn_andreea@yahoo.com

Sturgeons are the caviar producers and extremely valuable from commercial point of view. Various anthropic factors have severely affected the populations all over the world. Romania is one of the few countries from Europe with wild populations that are still reproducing in the Lower Danube. Different native sturgeon species or exotic ones are also raised in the Romanian fish farms. The aim of our study was to assess the genetic diversity of aquaculture stocks in stellate sturgeon (Acipenser stellatus) and beluga sturgeon (Huso huso) from four Romanian fish farms in order to develop selective breeding programs. The genetic variability analysis is important in order to avoid the inbreeding depression and genetic drift that might have serious negative consequences. Six aquaculture strains of stellate sturgeons (n=120) and two strains of beluga sturgeon (n=40) were analyzed by using eleven microsatellites (LS19, LS34, LS39, LS54, Aox27, AoxD161, AoxD234, AnacC11, AnacE4, AciG198 and Spl106). Following multiplex PCR and capillary electrophoresis, the genotype data were statistically analyzed with GENETIX 4.05 and FSTAT 1.2 software. The allelic richness (Ar), the inbreeding coefficient (Fis), the observed (Ho) and expected heterozygosity (He) were estimated within strains and stocks, while the genetic distance (D) and the gene flow (Fst) were calculated between strains and stocks of the same species. In stellate sturgeon, the overall genetic diversity of the stocks and strains was moderate. One strain only showed low genetic diversity and inbreeding. The genetic differentiation and the gene flow between the stocks and strains are also moderate in this species. In beluga sturgeon both strains showed a small degree of inbreeding and the genetic diversity within stocks was lower than in stellate sturgeon. The genetic differentiation and the gene flow between the stocks are moderate in beluga sturgeon, but higher than in stellate sturgeon. The assessment of genetic diversity along with the recording of economically important traits (growth, caviar yield) will support the selection of breeders included in selective breeding programs.

Differences in morphology and genetic structure associated to the coat colour in PRE horses

S. Negro[1], M.J. Sánchez[1], S. Demyda[2] and M. Valera[1]
[1]University of Seville, ETSIA, Seville, 41013, Spain, [2]UNLP – CONICET, La Plata, 1900, Argentina; z12neras@uco.es

Mutations in the genes influencing melanocytes are associated not only with the horse coat colour variability, but also with their behaviour. However, their influence in morphological traits was not yet described. In this study, we assessed 23 morphological measurements (MM) on a population of 34,402 PRE horses and its association with the four basic coat colours of the breed (grey, brown, chestnut and black). Additionally, a large-scale molecular data set (17 STR markers from 174,036 individuals) was analysed in order to also determine their genetic diversity and structure. The historical population was composed by 304,289 PRE: 169,987 grey, 28,801 black, 98,507 brown and 6,994 chestnut horses. Phenotypic analysis showed significant differences between the subpopulations and morphological traits ($P<0.01$). 22/23 traits were highly different ($P<0.01$) in mares and 12/23 traits were statistically significant in stallions, being chestnut horses significantly smaller than the rest. Differences were also detected using molecular data. The observed and expected heterozygosity values were low in grey horses (0.64 and 0.67 respectively) compared with the other three subpopulations (0.68-0.69 and 0.70 respectively). Nei distances demonstrated a clear division among groups, being grey horses the most separated (0.04 with brown and chestnut horses, and 0.02 with black horses). On the contrary, brown, chestnut and black horses showed a decreased genetic distance (0.003) among them. Same results were obtained estimating the gene flow (number of migrants per generation, Nm), which showed the lower values in grey horses (5.61 in grey-brown pair and 5.97 in grey-chestnut pair) compared with brown-chestnut pair (78.86). These findings agree with the breed origins in which chestnut horses were mainly influenced by Arab breeds (normally smaller) during the Muslim occupation and black and brown horses which are descendants of the German horses using during the XVI century (Carlos V and Felipe II reigns). These results demonstrated a clear differentiation between grey and black, brown and chestnut PRE-horses, in both morphological traits and genetic structure.

Nutritional evaluation of dehulling and thermal treatments for production of expeller soybean meal

P. Carré[1], E. Royer[2], P. Wikselaar[3], A. Quinsac[4] and P. Bikker[3]
[1]OLEAD, 11 rue Monge, Pessac, France, [2]Ifip-institut du porc, 34 bd de la gare, Toulouse, France, [3]Wageningen University & Research, De Elst 1, Wageningen, the Netherlands, [4]Terres Inovia, 11 rue Monge, Pessac, France; eric.royer@ifip.asso.fr

The objective of this study was to determine the influence of different processes on the nutritional quality of expeller soybean meal (SBM). Extrusion or cooking processes in combination with dehulling and pressing were used to produce 4 partly defatted SBM with low residual trypsin inhibitor (TI) activity. European soybeans were ground using a roller mill, then dehulled or not with a cleaner separator. The 2 products were either extruded using single-screw extrusion at 140 °C at around 100 kg/h (EP), or cooked at 150 °C during 60 minutes (FCP) after flaking. All beans were then pressed to extract the oil. Extrusion allowed a higher oil extraction than cooking (residual oil: 4.9 vs 8.6 g/100 g DM in whole EP and FCP meals, and 5.2 vs 6.4 g/100 g DM in dehulled EP and FCP meals, respectively). The dehulling step resulted in an increase of almost 3 g/100 g protein (58.8 and 58.4 g/100 g for dehulled EP and FCP meals, respectively, and 56.0 and 55.8 g/100 g for whole EP and FCP meals, respectively, on a fat free DM basis). The KOH protein solubility was increased by dehulling for EP (70 vs 76%) and FCP processes (82 vs 89%, for whole and dehulled meals, respectively). The TI values were 2.6, 3.5, 3.6, and 7.6 TIU/mg for whole and dehulled EP and FCP meals, respectively. The lower dryer outlet temperature measured for the dehulled beans compared to whole beans (90 vs 97 °C) may explain the lower TI inactivation for the dehulled FCP meal. The amino acid and reactive lysine content and the *in vitro* rate of degradation based on hydrolysis of protein bonds (pH-stat) are determined and related to the chemical composition to evaluate the nutritional value of the meals. All 4 processes may produce good quality SBM with 46-52 g/100 g crude protein, 4-8 g/100 g residual oil, and a TI content below 8 TIU/mg. Extruded-expelled SBM produced in medium-sized crushing plants from local and GMO-free crops have a nutritional and economic potential in Europe. This study is part of the Feed-a-Gene project and received funding from the European Union's H2020 program under grant agreement no. 633531.

Development of a decision support system for precision feeding application in pigs and poultry

L. Brossard[1], N. Quiniou[2], M. Marcon[2], B. Meda[3], L. Dusart[4], V. Lopez[5], J.-Y. Dourmad[1] and J. Pomar[5]
[1]PEGASE, INRA-Agrocampus Ouest, 35590 Saint-Gilles, France, [2]IFIP, Institut du Porc, 35651 Le Rheu, France, [3]URA, INRA, 37380 Nouzilly, France, [4]ITAVI, Domaine de l'Orfrasière, 37380 Nouzilly, France, [5]Department of Agricultural Engineering, Universitat de Lleida, Av. Alcalde Rovira Roure 191, 25198 Lleida, Spain; ludovic.brossard@inra.fr

Precision feeding is a promising way to improve feed efficiency and thus economic and environmental sustainability of livestock production. A decision support system (DSS) was built to determine in real-time the nutritional requirements of animals and feed characteristics (i.e. composition, amount) for an application of precision feeding in pig and poultry commercial farms. This tool, dedicated to animals managed individually or in groups, is designed with a modular structure for adaptation to different feeder devices, species and production stages. The modules are built to perform specialized tasks in a cooperative way. It includes a data management module with a proper characterization of data by meta-data definition for precision feeding. It ensures standard encoding to allow data interoperability from any platform. Other modules are dedicated to data verificatrion and correction for inclusion in a database, prediction of most probable body weight (BW) gain and feed intake (ad libitum or restricting feeding), and the calculation of nutritional requirements. The BW and feed intake prediction is based on dynamic data analyses. For that, specific methods have been evaluated and selected depending on the number of available data, data type (BW or feed intake), and recording frequency. The calculation of nutritional requirements is performed using nutritional models specific for a species or production stage. These two modules are currently designed for healthy animals and will be refined to extend prediction to a larger range of field situations (e.g. health problems, climatic conditions) with nutritional models in development/refinement in other workpackages of the project. The general specifications of this DSS and dynamic data analyses will be illustrated for growing pigs. This study is part of the Feed-a-Gene project and received funding from the European Union's H2020 program under grant agreement no. 633531.

Development of a decision support tool for precision feeding of pregnant sows

J.Y. Dourmad[1], L. Brossard[1], C. Pomar[2], J. Pomar[3], P. Gagnon[4] and L. Cloutier[4]
[1]INRA, UMR Pegase, 35590 Saint-Gilles, France, [2]Agriculture and Agri-Food Canada, 2000 rue Collège, Sherbrooke J1M 0C8, Canada, [3]Universitat de Lleida, Dep. Agricultural Engineering, 25198 Lleida, Spain, [4]CDPQ, 2590 bd Laurier, Québec G1V 4M6, Canada; jean-yves.dourmad@inra.fr

Nutritional studies indicate that nutrient requirements for pregnancy differ largely among sows and according to the stage of pregnancy, whereas in practice the same diet is generally fed to all sows in a given herd. In this context, the availability of new technologies for high throughput phenotyping of sows and their environment, and of innovative feeders that allow the distribution of different diets, offers opportunities for a renewed and practical implementation of prediction models of nutrient requirements, in the perspective of improving feed efficiency and reducing feeding costs and environmental impacts. The objective of this study was thus to design a decision support tool that could be incorporated in automated feeding equipment. The decision support tool was developed on the basis of InraPorc® model. The optimal supply for a given sow is determined each day according to a factorial approach considering all the information available on the sow: genotype, parity, expected prolificacy, gestation stage, body condition (i.e. weight and backfat thickness), activity, and housing (i.e. type of floor and ambient temperature). The approach was tested using data from 2,500 pregnancies on 540 sows. Energy supply was calculated for each sow to achieve, at farrowing, a target body weight established based on parity, age at mating and backfat thickness (18 mm). Precision feeding (PF) with the mixing of two diets was then simulated in comparison with conventional (CF) feeding with a single diet. Compared to CF, PF reduced protein and amino acid intake, N excretion and feeding costs. At the same time, with PF, amino acid requirements were met for a higher proportion of sows, especially in younger sows, and a lower proportion of sows, especially older sows, received excessive supplies. This study is part of the Feed-a-Gene project and received funding from the European Union's H2020 program under grant agreement no. 633531. The data used for the simulations were issued from a project conducted within the AgriInnovation Program from Agriculture and Agri-food Canada.

Preliminary model to predict P-requirement of growing pigs

V. Halas[1], G. Dukhta[1], G. Nagy[1] and G. Kövér[2]
[1]*Kaposvár University, Department of Animal Nutrition, Guba S. 40., Kaposvár, 7400, Hungary, [2]Kaposvár University, Department of Mathematics and Informatics, Guba S. 40, Kaposvár, 7400, Hungary; halas.veronika@ke.hu*

Phosphorus is an essential nutrient and, as phosphate, is involved in metabolic activities as well as in bone formation. There is evidence that a long term P deficiency reduces the growth rate of animals. However, an oversupply results in high rate of excretion that is critical from an environmental point of view. Modelling P metabolism allows to improve our understanding on the main factors affecting the P requirement. The aim of the work was to develop a mathematical model predicting the dynamics of P partitioning and retention in growing and fattening pigs over time. The model is a comprehensive description of the underlying mechanisms of digestible P and Ca utilization. Input parameters are related to the diet and include dry matter, the Ca and P content of the feed, as well as Ca and P digestibiliies, and to the animal such as daily feed intake, protein and fat deposition rate. The model presents the distribution of true digestible P and Ca within the body. Absorbed P and Ca are used for maintenance purposes, soft tissues (i.e. muscle and backfat), and bone tissue development. Surplus P and Ca is excreted via the urine. Retention of P in the body is the sum of P retention in soft tissues and the skeleton. It is presumed that soft tissues have a priority in utilizing the absorbed P. Thus, an insufficient P supply results in a reduction in or even negative bone ash retention to ensure P for the development if soft tissues. However, there is a limit for rate of demineralization and, under the threshold level, the absorbed P has priority to be retained in bone over soft tissues. The bone formation depends on P bioavailability and Ca supply, and limited by potential bone P retention. The model is able to predict P retention, urinary P excretion and digestible P requirement of swine at different body weights and with a different body composition. The results show that the P requirement depends on growth rate and particularly on protein deposition of the pigs. This study is part of the Feed-a-Gene project and received funding from the European Union's H2020 program under grant agreement no. 633531.

Re-parametrization of a swine model to predict growth performance of broilers

G. Dukhta[1], J. Van Milgen[2], G. Kövér[3] and V. Halas[1]
[1]*Kaposvár University, Department of Animal Nutrition, Guba S. 40, 7400, Kaposvár, Hungary, [2]PEGASE, Agrocampus Ouest, INRA, 35590, Saint-Gilles, France, [3]Kaposvár University, Department of Mathematics and Informatics, Guba S. 40, 7400, Kaposvár, Hungary; galyna.dukhta@ke.hu*

The aim of the study was to investigate whether a pig growth model is suitable to be modified and adapted for broilers. As monogastric animals, pigs and poultry share many similarities in their digestion and metabolism, many structures (body protein and lipid stores) and the nutrient flows of the underlying metabolic pathways are similar among species. For that purpose, the InraPorc model was used as a basis to predict growth performance and body composition at slaughter in broilers. Firstly, the backbone of InraPorc was translated to Excel and examined whether the equations were suitable for growing birds by evaluating the parameters and model behavior. After determining of differences the model was modified for predicting growth in broilers in relation to the nutrient supply. The model core is very generic in terms of representing the most important nutrient flows and the transformation from absorbed nutrients to protein and fat accretion during growth. The idea of nutrient utilization is mainly based on concepts used in net energy and ideal protein systems. The model is driven by feed intake, the partitioning of energy between protein and lipid deposition, and availability of dietary protein and amino acids. Parameters of the Gompertz function were adjusted using literature data to express maximum feather-free body protein deposition. A separate Gompertz equation was used to estimate feather growth and protein content of feather was assumed to be a nonlinear function of age. The amino acid loss with feathers was considered as a part of the maintenance requirement and the fasting heat production was used as the NE requirement for maintenance ($460 \text{ kJ}/(\text{kg BW})^{0.70}$/day). Feed intake for broilers was expressed on a NE bases and estimated by a Gamma-function (which expresses the NE intake as multiples of maintenance), as in InraPorc. Parameters of the Gamma function were adjusted to experimental data from different broiler studies. This study is part of the Feed-a-Gene project and received funding from the European Union's H2020 program under grant agreement no. 633531.

Characterizing animal response to environmental challenges: new traits of more efficient animals
M. Taghipoor, L. Brossard and J. Van Milgen
INRA, AgroCampus Ouest, 35590 Saint-Gilles, France; masoomeh.taghipoor@inra.fr

Farm animals are constantly facing perturbations due to changing environmental and farm conditions. Characterization of the animal's response when it is facing perturbations that influence performance and health is of main concern to ensure sustainable livestock production. Indeed, a better understanding of the adaptation mechanisms used by the animal to cope with challenges (through resistance and resilience) is a prerequisite to propose adequate farm management strategies. Several experimental studies have been conducted to investigate the influence of the environment on animal performance. Mathematical models can be used consider and to quantify the systemic aspects of the animal's response to a perturbation. Existing models of farm animal performance have accounted to a limited extent for environmental perturbations. With novel monitoring technologies, it is now possible to evaluate the impact of these perturbations on the animal in real time and with a high frequency. We propose a mechanistic model to describe the influence of a generic perturbation of unknown origin on feed intake of growing pigs. The model is based on two sub-models: InraPorc, a model to describe the performance of the growing pig in a standard environment, and the well-known spring-and-damper system used in physics to describe the behavior of a system in presence of an external force. The InraPorc model was used to describe the phenotypic performance of the animal in the absence of acute perturbing factors. The spring-and-damper system included two parameters to characterize the adaptive response of animals when facing a perturbation. The main interest of this characterization is to define new standards to rank animals based on feed efficiency together with their adaptive capacity, and to identify potential correlation between performance and robustness traits. We can then propose model-derived traits for genetic selection of more efficient and robust animals. Currently, the model is able to simulate the performance of animals facing with a single perturbation. Future development of the model will include successive perturbations of known or unknown origin. This study is part of the Feed-a-Gene project and received funding from the European Union's H2020 program under grant agreement no. 633531.

Modelling individual uncertainty and population variation in phenotypical traits of livestock
J.A.N. Filipe and I. Kyriazakis
Newcastle University, School of Agriculture Food and Rural Development, Newcastle upon Tyne, NE1 7RU, United Kingdom; joao.filipe@newcastle.ac.uk

Characterising between-animal variation and its population-level consequences is key to effective livestock management and selective breeding. The aim may be to predict trait development (e.g. performance) from early growth or estimate unobserved traits (e.g. maximum growth or maturity parameters). A usual sequence of steps is: (1) to develop a mathematical model of chosen animal-level traits; (2) to estimate individual parameters across a sample of animals; (3) to estimate a population distribution of parameters; (4) to generate a population distribution of traits by simulating the model across distribution 3. The challenge is in the parameter estimation given typical limitations. We use a Bayesian inference methodology to estimate the population distribution of predicted traits. The approach: (i) accounts for individual-level uncertainty in parameters (2) due to their correlation and data limitations, e.g. short growth span or infrequent records, and (ii) does not invoke distributional assumptions and estimation of variance-covariance parameters (3). We present results derived from individual data with usual limitations. Results include distribution of growth parameters within breeds and across species (pigs, chicken, rabbits); they extend the literature by showing the extent of uncertainty and variation in parameters and by comparing variation not only across species but against that within breeds. We show distributions of protein and lipid growth parameters and metabolic heat production (HP) estimated across animals and species and predicted distributions of dynamic body composition. Literature body composition estimates usually condition on input of average HP data; by estimating both jointly, their individual-level correlation is included and no metabolic data is needed. HP estimates distributed about 0.7 MJ/kg/d in pigs in line with literature, and body fat content variation was much larger than that of body protein. We suggest this approach has general application in model parameterisation and prediction of trait development in populations using limited individual data. This study is part of the Feed-a-Gene project and received funding from the European Union's H2020 program under grant agreement no. 633531.

Discussion with stakeholders

J. Van Milgen
INRA, UMR1348 Pegase, 35590 Saint-Gilles, France; jaap.vanmilgen@inra.fr

Feed-a-Gene is a so-called multi-actor project, which means that the project '...needs to take into account that the objectives and planning are targeted to the needs, problems, and opportunities of end-users, and complementarity with existing research'. Partners from different sectors participate in the project, but we also need feedback from stakeholders about their expectations and on how project outcomes can be put into practice. What can and should we deliver to the livestock production sector and to society at large? The afternoon discussion deals with the following workpackages of the Feed-a-Gene project: (1) WP1: Alternative feed ingredients and real-time characterisation; (2) WP3: Modelling biological functions with emphasis on feed use mechanisms; (3)WP4: Management systems for precision feeding to increase resilience to fluctuating environments and improve feed efficiency.

Rules and guidelines for successful funding and research: opportunities for young scientists

C. Lambertz[1,2], A. Smetko[2,3] and P.Y. Aad[2,4]
[1]Faculty of Science and Technology, Free University of Bozen-Bolzano, Bolzano, Italy, [2]youngEAAP, Via G. Tomassetti 3/A1, 00161 Rome, Italy, [3]Croatian Agricultural Agency, Ilica 101, Zagreb, Croatia, [4]Faculty of Natural and Applied Sciences, Notre Dame University-Louaize, Zouk Mosbeh, Lebanon; paad@ndu.edu.lb

Competitiveness in research depends on writing successful grant research proposals. In Europe, the European Commission's Horizon 2020 research programme is one of the biggest EU Research and Innovation programmes with a fund of nearly €80 billion available since 2014. This programme is competitive, however, with a success rate for funding of just one in nine. Additionally, application procedures are becoming more complicated. Therefore, it is difficult for a young scientist with little experience to be competitive in European research programs. We have prepared a hands-on session with experts and successful grantees to prepare young scientists for writing European research grants. During this session we will answer questions including where and how to apply for grants and how to cater to stakeholders and funding agencies. youngEAAP asked Hans Spoolder, Gus Rose and Claire Wathes to share their experiences. Hans Spoolder, a senior scientist and EU account manager at Wageningen Livestock Research in Wageningen, The Netherlands and coordinator of international EU projects. He will present the Horizon 2020 program, the preparation of proposals and structures of consortia, and the latest developments of Horizon2020 and current calls. Gus Rose, scientist, project manager and WP4 leader of the iSAGE project, will share his views and experiences on EU projects write-up and management. Claire Wathes, Farm Animal Health and Production Group of the Royal Veterinary College, London UK and project co-coordinator of the WP5 of the GplusE project, will present her FP7 experiences and provide an overview on Marie Sklodowska-Curie Funding for individual fellowships. The audience will have the opportunity to interact with the panel on the intricacies of the application for research grants process as well as the dos and dont's for acquiring and executing research projects successfully.

Characterisation and development of the bovine respiratory microbiome

A. Thomas[1,2], M.R. Lee[1,2], M. Bailey[2], A. Finn[3] and M.C. Eisler[2]
[1]Rothamsted Research, North Wyke, Devon, EX20 2SB, United Kingdom, [2]University of Bristol, School of Veterinary Sciences, Langford, BS40 5DU, United Kingdom, [3]University of Bristol, Bristol Childrens Vaccine Centre, BS2 8AE, United Kingdom; amyc.thomas@bristol.ac.uk

We aimed to characterise development of colonisation with commensal, occasionally opportunistically pathogenic bacteria in healthy beef suckler calves using qPCR. We collected short nasopharyngeal swabs from 30 housed calves within their first week of life and at approximately monthly intervals thereafter until 10 months of age. Using published PCR gene targets and 16s rRNA sequences, we developed three qPCR assays for detecting and quantifying *Pasteurella multocida* (Pm), *Mannheimia haemolytica* (Mh) and *Histophilus somni* (Hs). Confidence intervals for proportions and differences between proportions were calculated using the Wilson score and Newcombe-Wilson hybrid score methods respectively. Preliminary findings reveal marked differences in colonisation patterns of commensal Pasteurellaceae. Within the first week of life, 8/30 calves were colonised with Hs, and carriage remained at similar rates up to 3 months of age. The proportion of calves positive for Hs was similar at 1 week (8/30; 95% CI: 0.139-0.462), 2 months (5/29; 95% CI: 0.0653-0.365) and 3 months (8/29; 95% CI: 0.134-0.478); differences not significant (P>0.5). Carriage was transient, and no calves positive at 1 week sustained carriage to 2 months. However, 2 calves positive at 1 week were re-colonised to detectable levels at 3 months. In contrast, colonisation with Pm was not detected until 3 months with one calf (1/29; 95% CI: 0.00180-0.196) co-carrying Hs. Prevalence of Hs and Pm increased with age (56% and 94% at 10 months respectively). The distribution of Hs density at 10 months showed 94% of positive calves had Ct values between 29 to 35, indicating low-density carriage (similar trend seen in neonates). Results suggest that bacterial colonisation in the upper respiratory tract is dynamic, with different organisms predominating as cattle age. While it is not yet clear how these colonisation dynamics affect or are affected by host mucosal immune responses, early life colonisation could be an important factor in determining later respiratory health.

Performance of calves fed a fixed amount of milk replacer with or without restrictions on meal size

A. Jensen, C. Juhl, M.B. Jensen and M. Vestergaard
Aarhus University, Department of Animal Science, Foulum, 8830 Tjele, Denmark; mogens.vestergaard@anis.au.dk

In rosé veal calf production, the transition period from a liquid-based milk replacer diet to a solid feed-based diet poses a potential risk of lag in growth. Calves are typically purchased at 2-4 weeks of age and fed limited (5-6 Liters/day) amounts of milk replacer (MR). In this type of production, it is important to obtain a high growth rate after weaning and to avoid diseases to be able to market these calves before 10 months of age. The hypothesis was that a reduction in milk allowance and meal size would encourage calves to eat more concentrate during and after gradual weaning. The objective was to investigate the pre- and post-weaning intake of solid feed and performance through either a conventional flat-rate (CON) or a step-down (STEP) milk feeding protocol combined with either a restricted (RES) or an unrestricted (URES) meal size offered from the automated milk feeder. A total of 32 calves (in blocks of 16) were purchased from three dairy herds at the age of 12.4 (±1.2) days and 47.7 (±1.4) kg LW. Each block comprised 2 pens, allocated to either CON or STEP. Within each pen of 8 calves, 4 calves were allocated to RES (min. 4 meals/d) and 4 to URES (min. 1 meal/d). Until weaning at 8 weeks of age, all calves were offered a total of 224 L MR (21% CP, 20% fat). Calves had free access to a pelleted concentrate (19% CP) and to artificially-dried, chopped, grass-hay (14% CP, 13% sugar). MR and concentrate intake were recorded individually via automated feeders. Starter intake did not differ between treatments in the pre- or post-weaning period, increasing from 1.33 in week 8 to 2.11 kg/d in week 10. Hay intake was 0.11 and 0.13 kg/d in calves fed CON and STEP, respectively, in week 3-8 and 0.14 and 0.17 kg/d in week 3-10. ADG was numerically higher in calves fed RES (0.73 kg/d) compared to URES (0.65 kg/d) (P<0.12) in week 2-8. ADG was numerically higher in calves fed STEP (1.34 kg/d) compared to CON (1.22 kg/d) in week 9-10 (P=0.20). Whether feeding MR through a step-down strategy stimulates the consumption of solid feed, and whether smaller volumes of MR distributed in more meals per day compared with no restriction on milk meal sizes have a positive effect on calf performance require further studies.

Comparison of methods to assess the diet ingested by grazing cattle, consistency with weight gain

A. Agastin[1], M. Naves[1] and M. Boval[2]

[1]INRA, GA, URZ, INRA Antilles, Guadeloupe, 97170, Guadeloupe, [2]INRA, PHASE, UMR MoSAR, INRA Agroparistech, Université Paris-Saclay, 75005, France; maryline.boval@agroparistech.fr

Reliable and flexible methods to assess the nutrition of grazing animals are needed for better management of livestock and natural grasslands. We aimed at comparing two methods of predicting basic diet characteristics of Creole heifers: organic matter digestibility (OM), OM intake (OMI) and digestible OMI (DOMI), by using two methods based on the faecal CP content, the analysis of fecal samples, by a near-infrared spectrometer (NIRS). The comparison of methods was made with data coming from 2 independent and simultaneous trials in 2 feeding systems with Digitaria sp. (offered in stalls, or directly grazed) with two groups of 4 heifers per feeding system, being supplemented or not, during 4 experimental periods (n=64). In a first step, the estimates obtained from the two methods were compared to the *in vivo* measurements in stall-fed conditions, of OMD, OMI and DOMI. The best prediction of OMD was obtained from the NIRS method (RSD=0.03, R^2=0.61, n=32). The best predictions of OMI and DOMI were obtained considering the OMD directly predicted by the NIRS method and the fecal output (FO). Consideration of the concentrate had no significant effect on any of the regression parameters. In a second step, the estimated values of DOMI by the 2 methods were compared to live weight gain (LWG) measured both in stalls and at pasture. In stalls, the LWG predicted with DOMI estimated via the fecal CP content method has provided the lower RSD of prediction to assess the LWG in stalls (RSD=0.74 g/kg LW, R^2=0.68, n=32). At pasture, the best prediction of LWG has been predicted by the DOMI directly estimated from the NIRS method (RSD=0.90 g/kg LW, R^2=0.53, n=32). Thus, fecal NIRS provided the most reliable estimates of OMD. This OMD combined with FO, provided the most accurate values of OMI and DOMI, compared to *in vivo* values measured in stalls. However, due to the difficulty to recover the feces at pasture and measure FO, direct predictions by using NIRS give accurate estimates for OMD and satisfactory for OMI and DOMI, consistent with the LWG achieved in stalls as at pasture. This method may therefore contribute to better management of grazing animals.

GHG emission calculation methodologies for livestock sector in South Korea and Indonesia

E. Nugrahaeningtyas, J.Y. Park, J.I. Yun and K.H. Park

Kangwon National University, Chuncheon-si, Gangwon-do, 24341, Korea, South; eskanugrahaeningtyas@gmail.com

The aim of this study was to compare the calculation methodologies for greenhouse gas (GHG) emissions from South Korea and Indonesia and to cast suggestion regarding to Indonesia's inventory method. Main livestock in Indonesia is chicken. Others are beef cattle, dairy cattle, sheep, goat, and duck with varied farm scales. Changes toward increased consumption of animal product due to the increase of human population and improvement to the living standard result in the increase of livestock population in Indonesia. Whereas, South Korea's livestock industry has grown in the past 50 years causing more livestock and fewer farms. According to Paris Agreement in 2015, all countries are encouraged to prepare, communicate, and maintain Nationally Determined Contribution (NDC) every five years. South Korea and Indonesia submitted its first NDC in 2015 and 2016. Currently, South Korea is using mainly Tier 1 and partially Tier 2 methods for livestock sector in Intergovernmental Panel on Climate Change (IPCC) 1996 Guideline (GL), and by 2023, South Korea would use 2006 GL. Indonesia is using 2006 GL Tier 1 method and would use Tier 2. The calculation methods and activity data from both countries were compared. South Korea has all activity data needed to calculate the emission from enteric fermentation and manure management, such as the number of population and utilization rate of manure management process, but Indonesia does not have the manure management statistic. However, Indonesia has begun to develop country-specific emission factor for methane (CH_4) from enteric fermentation in beef cattle and volatile solid (VS) to estimate the CH_4 emission from manure management in swine. Indonesia needs to improve its statistics, especially for the utilization rate of animal waste management. Indonesia can adopt methods from South Korea that makes proper documentation for the data needed for report. Indeed, South Korea has discussed those matters among any governmental bodies to get the proper statistical data and do survey. Moreover, the nitrogen excretion and VS data available from grey literature can be used to calculate with Tier 2 method. Therefore, those data can increase the certainty of GHG inventory and reflect the actual condition of Indonesia.

Economic effects of participation in animal welfare programmes: is it worthwhile for farmers?

H. Heise, S. Schwarze and L. Theuvsen

Georg-August University of Goettingen, Agricultural Economics and Rural Development, Platz der Goettinger Sieben 5, 37073 Goettingen, Germany; hheise@gwdg.de

Public demand for higher animal welfare standards are on the rise. In response, some programmes that market products of animal origin produced under higher animal welfare standards have been established in recent years. However, the market segments for products from so-called animal welfare programmes (AWPs) remain small. Farmers are considered an important stakeholder group for higher market shares of more animal-friendly products. Farmers' decisions to adapt their production to the requirements of AWPs are multidimensional, but always linked to financial incentives. Since little is known about financial attractiveness of higher animal welfare standards in livestock farming, this study investigates the perceived economic success of 579 conventional farmers who keep livestock. The survey data was analysed using propensity score matching to assess the *average treatment effect of the treated* for participation in AWPs on a farm's perceived profitability, liquidity and stability from farmer's point of view. We found that participation in AWPs had no significant effect on farmers' perceived economic success. The implications of this result are twofold. On the one hand, it suggests that it is particularly important to create further financial incentives to encourage even more farmers to take part in these programmes. On the other hand, it shows that farmers' concerns that the costly and highly specific investments required by participation will not pay off are unfounded since farmers who participate in AWPs rate their own financial situation just as highly as their collegues who do not participate in AWPs.

Generation and distribution of productivity gains in beef cattle farming over the last 36 years

P. Veysset[1], M. Lherm[1], J.P. Boussemart[2] and P. Natier[2]

[1]INRA, UMR1213 Herbivores, INRA, VetAgro Sup, 63122 Saint-Genès-Champanelle, France, [2]Univ. Lille, CNRS, IESEG School of Management, UMR 9221-LEM, 59000 Lille, France; patrick.veysset@inra.fr

As a major source of growth, productivity is a key variable in economics. Total Factor Productivity (TFP) is an indicator of technical and managerial efficiency. TFP measures the growth from quantity changes, while Productivity Surplus (PS) measures shares from price variation simultaneously. The surplus accounting method makes it possible to evaluate the productivity gains (output and input volume variations) and the combined effects of output and input price variations for a specific economic sector or for a panel of enterprises, if we have their individual data, over a long period. We applied this method on a database of 164 suckler-cattle farms in the Charolais area from 1980 to 2015, i.e. 3,127 farms-years (or observations). Over the 36-year period, the TFP increased at a rate of +0.17%/year (the cumulative volume of outputs increased little more than the cumulative volume of inputs used). This small increase in the TFP was linked to the constant increase in labor productivity, while the productivity of the other factors (intermediate consumptions, land, and equipment) decreased. Despite the huge increase in labor productivity (+2.03%/year), farmers' incomes remained stable. We observed a small drop in intermediate consumption prices, land rent and financial costs, and a very high increase in public aids and subsidies. Over the 36 years, with a share of nearly 64% of the total economic surplus, the downstream of the beef sector (from the farm gate to the consumer) appeared as the main beneficiary of these productivity gains through a decrease in animal product prices; 22% of this total economic surplus being absorbed by the downstream of the other sectors (mainly cereals). Because of the direct payments introduced to compensate for European intervention price reductions, French and European taxpayers were the main funders of this fall in agricultural product prices. The stability of farmers' income indicated that the farmers were the losers in the surplus distribution in spite of the improvement of their labor productivity, their structural adaptation and their investment in new equipment.

How does peri-implantational subnutrition affect red blood cell parameters in two beef breeds?

A. Noya[1], B. Serrano-Pérez[2], D. Villalba[2], I. Casasús[1], E. Molina[2], I. López-Helguera[2], J. Ferrer[1] and A. Sanz[1]
[1]Centro de Investigación y Tecnología Agroalimentaria (CITA) de Aragón, Avda. Montañana 930, 50059 Zaragoza, Spain, [2]Universitat de Lleida (UdL), Avda. Alcalde Rovira Roure 191, 25198 Lleida, Spain; anoya@cita-aragon.es

Undernutrition is common in extensive beef cattle farming systems at some stages of the production cycle. A poor nutrient diet during the peri-implantation period can interfere with the correct foetal development. The aim of this study was to analyse the effects of peri-implantational undernutrition on red blood cell parameters in dams and calves of two beef breeds. Seventy-four lactating Parda de Montaña (PA) and 40 Pirenaica (PI) multiparous cows were artificially inseminated and randomly allocated to a control (CONTROL, n=52) or nutrient-restricted (SUBNUT, n=62) group, which were fed at 100 or 65% of their estimated energy requirements during the first 82 days of pregnancy, and thereafter received a control 100% diet until parturition. Red blood cell count (RBC), haemoglobin content (HGB) and haematocrit (HCT) were determined on day 19 post artificial insemination and one month before parturition for dams, and once on the first days of life (between 1 and 11) for calves. At the beginning of pregnancy, PI dams showed higher values than PA dams for RBC (6.8 vs 6.1×10^6 counts/mm^3, for PI and PA respectively, P<0.001), HGB (12.6 vs 10.8 g/dl, for PI and PA, P<0.001) and HTC (37.2 vs 32.1%, for PI and PA, P<0.001). These differences were maintained one month before parturition for RBC (6.41 vs 5.7×10^6 counts/mm^3, for PI and PA, P<0.01), HGB (11.5 vs 10.3 g/dl, for PI and PA, P<0.01) and HTC (33.2 vs 30.2%, for PI and PA, P<0.05). No differences in haematological profiles were found due to undernutrition (P>0.05). In calves, neither breed nor feeding treatment influenced the red blood series profiles (P>0.05). A negative correlation between calf age and haematological parameters was observed only in CONTROL calves (R^2=-0.37 for RBC, P=0.069; R^2=-0.47 for HGB, P<0.05; R^2=-0.52 for HTC, P<0.01), suggesting an earlier maturation of the haematopoietic system in these calves. More studies during gestation and other phases of calf development are needed to assess the effects of undernutrition during the peri-implantation period.

Plasma and muscle responses to pre-slaughter mixing of suckler bulls

A.P. Moloney[1], E.G. O'Riordan[1], N. Ferguson[1], M. McGee[1], J.B. Keenan[2] and M.H. Mooney[2]
[1]Teagasc, Animal & Grassland Research and Innovation Centre, Grange, Dunsany, County Meath, Ireland, [2]Institute for Global Food Security, Queen's University Belfast, 18-30 Malone Road, Belfast, Northern Ireland, BT9 5BN, United Kingdom; aidan.moloney@teagasc.ie

Meat colour is an important influence on the purchase decision of the consumer; 'bright red' is preferred. Dark, firm, dry beef with ultimate pH (pHu)>5.9 is typically ascribed to pre-slaughter stress but the relationship between animal interactions pre-slaughter and dark beef is uncertain. The objective was to determine the impact of mixing unfamiliar bulls, the day before slaughter, on stress-related plasma variables and beef pH and colour. Prior to mixing, Charolais-sired suckler bulls (mean (SD) live weight 671 (71.6) kg and age 17.4 (2.01) months) were housed indoors in 7 slatted floor pens (4 to 5 bulls/pen, 2.5-3.0 m^2/bull) and offered a barley-based ration and grass silage *ad libitum*. Two pens of 5 bulls were chosen as controls. From the other pens, 18 bulls were selected and moved to a single new pen (outdoors, bedded with wood chip, 6 m^2/bull) 18 h before slaughter. Bulls were then transported (45 min) without further mixing to an abattoir and slaughtered on arrival. Blood was collected for plasma preparation and muscle pH was recorded periodically post-mortem. At 48 h post-mortem, the left half of the carcass was cut at the 5/6th rib interface and muscle pH and lightness (L*) measured. Muscle was collected for measurement of glycolytic potential (GP) and drip loss (DL). Data were analysed by one-way ANOVA. Mixing increased plasma creatine kinase activity (14,556 vs 145 U/l, P<0.01) but cortisol, lactate and creatinine concentrations were not affected. Muscle GP was decreased (84 vs 179 µmol lactate equivalents/g, P<0.001) and pH at 1.5, 3, 4.5 and 6 h post-mortem increased (P<0.05) by mixing. Muscle from mixed bulls had higher pH (5.98 vs 5.55, P<0.05), was darker (lower L*; 25.5 vs 39.9, P<0.05) and had lower (P<0.05) DL (g/kg). While the maximum pH in muscle from control bulls was 5.60, six mixed bulls had muscle pH≤5.60. It is concluded that mixing of bulls increased some stress-related plasma indicators. That not all mixed bulls had high muscle pHu illustrates the complexity of the relationship between pre-slaughter stress and muscle biochemistry.

Improving beef cattle management through better breeding

S.P. Miller, K.J. Retallick and D.W. Moser

Angus Genetics Inc., Saint Joseph, MO 64506, USA; smiller@angus.org

Improvements to beef cattle management to improve production efficiencies, such as reduced labour and improved animal health and welfare, often focus on animal husbandry practices such as feeding and housing and inputs such as nutrition and pharmaceuticals. Genetic improvement can also play an important role in improving the same targeted outcomes. Objectives were to outline how genetic selection in American Angus cattle has enabled a more efficient and profitable animal while reducing labour associated with calving difficulty, reducing culling for fertility and improving animal handling. Analyses of genetic trends in American Angus cattle reveal significant productivity gains with no increase in birth weight while reducing dystocia. Comparing mean expected progeny differences (1/2 estimated breeding value) from 1986 to 2016, the difference in weight at 365 days increased 34 kg while birth weight decreased 0.14 kg directly contributing to a 7% increase in unassisted births in heifers. The genetic increase in growth rate also increased carcass weight (20 kg) and mature cow size (8.2 kg). Besides calving difficulty, heifer pregnancy, also the target of improved management, was improved genetically with 6% more heifers becoming pregnant during their first breeding season. Docility was improved with 6% more animals having desired temperament. Considering the combined economic impact of genetic change during this 30 year period, profitability (USD) increased $69 per cow per weaned calf and $128 per calf post-weaning including carcass revenue. These improvements in management traits, while increasing productivity was accomplished through dedicated producer recording coupled with a multiple trait genetic evaluation. Multiple trait evaluation is needed to overcome many of the antagonistic genetic relationships present such as the inherent positive genetic relationship between birth weight and growth. Reducing calving difficulty through genetic selection is very important in beef cattle production under extensive grazing conditions where cow management at calving is not practical or very difficult. These results in American Angus demonstrate how genetic selection can develop cattle to match different production environments while improving productivity and profitability.

The threshold approach does not improve the eating quality of beef

S.P.F. Bonny[1,2], D.W. Pethick[2], G.E. Gardner[2], I. Legrand[3], J. Wierzbicki[4], P. Allen[5], R.J. Polkinghorne[6], J.F. Hocquette[1] and L. Farmer[7]

[1]INRA VetAgro Sup, UMR1213, 63122 Theix, France, [2]Murdoch University, School of Veterinary and Life Sciences, WA 6150 Murdoch, Australia, [3]Institut de l'Elevage, Service Qualité des Viandes, MRAL, 87060 Limoges, France, [4]Polish Beef Association, Ul. Kruczkowskiego 3, 00-380 Warszawa, Poland, [5]Teagasac Food Research Centre, Ashtown, Dublin 15, Ireland, [6]Birkenwood Pty Ltd, 431 Timor Road, Murrurundi, NSW 2338, Australia, [7]Agri-Food and Biosciences Institute, Newforge Lane, Belfast BT9 5PX, United Kingdom; s.bonny@murdoch.edu.au

Variable beef eating quality is a major driver of declining beef consumption. To combat this, a 'Threshold' type quality assurance system could be used to exclude animals or carcasses on the basis of pre-slaughter factors or abattoir measurements. This study demonstrates the impact of applying a system of this type and the associated cost of non-compliance. A total of 18 different cuts from 774 carcasses which represented a cross section of the beef production industries from France, Poland, Ireland, Northern Ireland and Australia, underwent standard MSA sensory testing using four different cooking methods by over 15,000 untrained consumers. These consumers rated the beef samples as one of four options: unsatisfactory (fail), good everyday (good), better than everyday (better) or premium (best). For the striploin samples, 19% received a fail grade by consumers. A set of threshold criteria was then applied to the data. 42% of the striploin samples from the database did not meet the threshold. Of the remaining 58% that did meet the threshold criteria, the proportion of samples rated as unsatisfactory was unchanged remaining at 19%. Similar results were seen in other muscles in the database. The inability of the threshold criteria to increase the average quality of beef demonstrates the need for more accurate quality prediction systems within the European beef industry to guarantee eating quality for the consumer.

Selection of grasses and use of nitrogen to optimise grass yield and quality

E. Genever[1], D. McConnell[1], J. Matthews[2] and S. Kerr[2]
[1]Agriculture and Horticulture Development Board, Stoneleigh Park, Kenilworth, CV8 2TL, United Kingdom, [2]NIAB-TAG, Huntingdon Road, Cambridge, CB3 0LE, United Kingdom; liz.genever@ahdb.org.uk

The Recommended Grass and Clover Lists of England and Wales (RGCL) is an annual independent system that provides essential information on grass and clover varieties. This enables the selection of grass and clover varieties that are suited to British production systems, maximising the productivity and efficiency of these systems. A weakness is the management of RGCL trials can be different from that which occurs at farm level. Within the RGCL testing protocol, the grasses are tested under 400 kg of nitrogen per hectare per year compare to less than 200 kg of N/ha per year being used on beef and dairy farms. Plots of three diploid perennial ryegrass (PRG) (AberGreen, Premium and Rodrigo) and three tetraploid PRG varieties (Aubisque, Montova and Seagoe) were sown at three locations in 2011 and two locations in 2012. Three nitrogen treatments were applied: 100, 200 and 400 kg of nitrogen per hectare (kg N/ha) per year over three years. During the first and third years, the plots were cut every six weeks to represent conservation management and in the second year the plots were cut every three weeks to represent grazing management. Yields (tonnes of dry matter per hectare (t DM per ha)) were collected at every cut and grass samples were collected for quality measurements. Data were analysed by analysis of variance and other appropriate statistical tests. Nitrogen fertilisation regimes significantly ($P<0.001$) influences grass yield under conservation and grazing management. The greatest yield was expressed under the 400 kg N/ha input system. The first year conservation harvests averaged 8.6, 11.1 and 16.7 t DM per ha, 11 for 100, 200 and 400 kg N/ha respectively. Late season digestibility value quality analysis demonstrated significant ($P<0.01$) quality improvement with higher N levels. This is potentially attributable to physiological changes within the plants at differing N levels. There was no significant N × variety interaction with annual or individual cuts yields. This suggests that the performance of varieties in terms of yield and quality was not significantly affected by N rate.

Optimum test duration for residual feed intake and related traits in young Nellore bulls

J.N.S.G. Cyrillo, M.Z. Marzocchi, M.E.Z. Mercadante, S.F.M. Bonilha, B. Pires and E. Magnani
Instituto de Zootecnia, Centro APTA Bovinos de Corte, Rodovia Carlos Tonani – km 94, 14174000, Brazil; mercadante@iz.sp.gov.br

This study was performed to determine an appropriate period of time required for accurate evaluation of Residual Feed Intake (RFI) and its components, Dry Matter Intake (DMI), Metabolic Body Weight (MBW) and Average Daily Gain (ADG). Data were collected from 242 Nellore bulls, tested at Instituto de Zootecnia, São Paulo, Brazil, for 84 days in 2012 (n=126), 2013 (n=58) and 2014 (n=58). Animals had on average 283±24 days of age and 263±2.2 kg of BW at the beginning of the test and were weighted every 14 days during the test. Diet consisted on corn silage, hay, ground corn, soybean meal and mineral salt. Daily feed intake was determined by electronics feed intake equipments GrowSafe® (Growsafe Systems Ltd, Airdrie, Alberta, Canada) and Intergado® (Intergado Ltd, Contagem, Minas Gerais, Brazil). Records intervals (15; 29; 43; 57; 71 and 84 days) were analyzed using PROC MIXED of SAS, with a repeated-measures model, including the fixed effects of contemporary group (year – equipment), random effects of animal and residual error. Pearson and Spearman correlations were estimated among records intervals. Residual variances and relative changes in residual variance (percentage difference between the variance from the previous period and the current period divided by the variance from the first period), showed no clear tendency for DMI, and MBW over the test, with values ranging from -5.17, -2.0, 0.58, -1.32 and 1.15% and -2.79, -3.59, -3.01, -2.21 and -2.27%, respectively, for periods of 29; 43; 57; 71 and 84 days. For ADG and RFI, the residual variance decreased with increasing test period, with changes from 41.1; 16.64; 5.88; 5.88 and 0% and 2.66; 2.66; -4.00; -2.66 and 0%, for the same periods. Pearson and Spearman correlations above 0.95 were observed between 84 and 57 days for DMI, 84 and 15 days for MBW, 84 and 71 days for ADG, and 84 and 57 days for RFI. Although for RFI the correlation coefficient between 84 and 57 days was 0.98, at least 71 days of testing are required for a sufficient accuracy of all variables involved in the calculation.

Effect of peri-implantational undernutrition on the dam and offspring performance in two beef breeds

A. Noya, I. Casasús, J.A. Rodríguez-Sánchez, J. Ferrer and A. Sanz
Centro de Investigación y Tecnología Agroalimentaria (CITA) de Aragón, Avda. Montañana 930, 50059 Zaragoza, Spain; anoya@cita-aragon.es

Undernutrition in early pregnancy, which is common in extensive beef cattle systems, can affect foetal development and postnatal performance. This study analyzed the effect of dam undernutrition in the first third of pregnancy on dam and calf performance in two beef breeds. Forty-nine Parda de Montaña (PA) and 32 Pirenaica (PI) multiparous cows were inseminated and allocated to a control (CONTROL, n=35) or nutrient-restricted (SUBNUT, n=46) diet, which met 100 or 65% of their estimated energy requirements during the first 82 d of pregnancy, and thereafter received a control 100% diet until parturition. Animals were weighed fortnightly. On d 23 post-calving, dams were machine-milked to determine milk yield and composition. On d 25 and 120 (weaning), calf milk intake was determined by the weigh-suckle-weigh technique. Dam weight at calving was not influenced by breed or nutritional treatment (599 kg, P>0.05), but body condition score was higher in PI cows (2.66 vs 2.87 in PA and PI, P<0.001). Peri-implantational undernutrition did not affect calf birth weight, but PA calves were heavier (45.8 vs 40.9 kg, P<0.001). At d 23 milk yield tended to be higher in PA than PI cows (9.7 vs 8.6 kg/d, P=0.06), and PI had greater fat, fat-free dry matter and lactose content (P<0.001) than PA. Surprisingly, milk fat content was lower in CONTROL than SUBNUT cows (4.4 vs 4.9%, P=0.02). Calf milk intake was higher in PA than PI calves (P<0.001) both at d 25 (9.3 vs 7.9 kg/d) and at weaning (7.3 vs 5.7 kg/d), but it was not affected by dam nutrition. The correlation between the yield obtained by machine milking and calf milk intake in the first month was 0.46 (P<0.001), and the latter was correlated with calf gains in the whole lactation (0.65, P<0.001), calf milk intake being a good predictor of weaning weight. Calf gains during lactation did not differ between nutritional treatments in PA calves, but gains of PI-CONTROL were higher than those of PI-SUBNUT (0.88 vs 0.70 kg/d). In conclusion, peri-implantational undernutrition increased milk fat content in dams, and reduced offspring growth in PI breed, probably due to a higher sensitivity of this breed to maternal undernutrition.

Performance of weaned dairy bull calves fed concentrate or free choice of concentrate and TMR

A. Jensen, C. Juhl, N. Drake, M.B. Jensen and M. Vestergaard
Aarhus University, Department of Animal Science, Foulum, 8830 Tjele, Denmark; mogens.vestergaard@anis.au.dk

In rosé veal calf production, transition from a liquid-based milk replacer diet to a dry feed-based diet poses a potential risk; calves not only change diet but are often shipped, mixed, or re-grouped before weaning. Some rosé veal farms feed calves a total mixed ration (TMR) from weaning, while others feed pelleted concentrate or a combination of both. The objective of this study was to test how weaned dairy bull calves performed from 10 to 18 wk of age when offered either a pelleted concentrate (PELL) or free choice between pelleted concentrate and TMR (FREE). A total of 32 calves were purchased from 3 dairy herds at 12±1 days of age and 48±1 kg LW (Mean±SE). Until weaning at 8 wk of age, calves were offered a total of 224 L milk replacer (21% CP, 20% fat). Until 10 wk of age, calves had free access to a pelleted concentrate (19% CP) and artificially-dried chopped hay (14% CP). Milk and concentrate intakes were recorded individually via automated feeders. At wk 10 (95±2 kg LW), calves were assigned to either PELL or FREE. The pelleted concentrate and the TMR had (kg DM basis) similar NE (7.9 MJ), CP (218 g), and starch content (335 g), but varied in DM (87 vs 70%). A total of 31 calves completed the experiment. During wk 10-18, hay comprised 8.8% of DMI for PELL and 2.6% for FREE, while TMR intake comprised 43% of DMI from pellets+TMR for FREE calves. Concentrate intake was 4.25 and 2.76 kg DM/d (P<0.001) for PELL and FREE, respectively. TMR intake was 1.98 kg DM/d for FREE calves. LW at 18 wk of age was 181 and 175 kg (P<0.32) and ADG from wk 10 to 18 was 1.54 and 1.41 kg/d (P<0.31) for PELL and FREE, respectively. The lower concentrate intake of FREE may pose a risk of a lower growth rate compared with PELL but this could not be confirmed in the present study. Any long-term consequences for growth, feed utilization, and carcass value also need to be studied.

Beef characteristics predicted by NIRS

M. Corazzin[1], E. Saccà[1], R. Bozzi[2], G. Ferrari[3], R. Negrini[4,5] and E. Piasentier[1]
[1]University of Udine, Department of Agricultural, Food, Environmental and Animal Sciences, Via Sondrio 2A, 33100 Udine, Italy, [2]University of Florence, Department of Agrifood Production and Environmental Sciences, Piazzale delle Cascine 18, 50144 Firenze, Italy, [3]BÜCHI Italia S.r.l., Via G Galilei 34, 20010 Cornaredo, Italy, [4]Associazione Italiana Allevatori, Via G Tomassetti 9, 00168 Roma, Italy, [5]Università Cattolica del Sacro Cuore, Istituto di Zootecnica, Via Emilia Parmense 84, 29100 Piacenza, Italy; edi.piasentier@uniud.it

The importance of objective, on-line evaluation of the nutritional and sensory characteristics of fresh meat is increasing for both selective and commercial purposes. NIRS (Near-Infra-Red Spectroscopy) may allow to reach these objectives. The aim of this study was to evaluate the possibility to predict by NIRS the chemical composition and physical properties of beef. The spectra were collected by using NIRFLex N-500 (Büchi, Switzerland) on samples of *Longissimus dorsi* m. dissected at slaughter from 280 young bulls. The spectral acquisition was obtained in a range from 4,000 to 10,000 cm^{-1} on entire fresh meat, kept at 4 °C for 48 h. The proximate and fatty acids (FA) analysis, color, cooking loss, Warner Bratzler shear force (WBSF) were carried out on the same samples. The chemometric analyses were performed using Buchi NIRCal 5.5. The PLS models predicted humidity with a coefficient of determination of calibration (R^2_c) of 0.682 (standard error of calibration, SEC, 1.09%, n=173) and a coefficient of determination of validation (R^2_p) of 0.479 (standard error of validation, SEP, 1.11%, n=85), ether extract (R^2_c=0.594; SEC=0.59%, n=110; R^2_p=0.550; SEP=0.54%, n=47), ash (R^2_c=0.760; SEC=0.06%, n=205; R^2_p=0.787; SEP=0.05%, n=73), lightness (R^2_c=0.555; SEC=1.90, n=122; R^2_p=0.557; SEP=1.89, n=58), WBSF (R^2_c=0.781; SEC=5.09N, n=134; R^2_p=0.740; SEP=5.15N, n=57), cooking loss (R^2_c=0.693; SEC=2.84%, n=96; R^2_p=0.714; SEP=2.80%, n=45), monounsaturated FA (R^2_c=0.756; SEC=1.83%, n=119; R^2_p=0.738; SEP=1.82%, n=49), polyunsaturated FA (R^2_c=0.854; SEC=2.25%, n=115; R^2_p=0.869; SEP=2.21%, n=56). In conclusion, NIRS was able to reliably predict proximate and FA composition, color, cooking loss, and WBSF of entire fresh meat. Consequently, it may be a useful technology for an on-line application. This project was funded by CRITA of FVG Autonomous Region.

Pasture-based finishing of early- and late-maturing breed suckler bulls at 19 months of age

M. Regan[1,2], C. Lenehan[1,2], A.P. Moloney[1], E.G. O'Riordan[1], A. Kelly[2] and M. McGee[1]
[1]Teagasc, AGRIC, Grange, Dunsany, Co. Meath, Ireland, [2]School of Agriculture, University College Dublin, Belfield, Dublin 4, Ireland; maeve.regan@teagasc.ie

Pasture-based finishing of cattle is economically attractive however achieving a commercially adequate level of carcass fatness (6, scale 1-15) with grazing suckler bulls is difficult. Performance of early- (EM) and late-maturing (LM) sired spring-born suckler bulls finished at pasture, with or without concentrate supplementation, at 19 months of age was evaluated. Sixty yearling bulls (live weight 399 (SD 42.5) kg and 446 (SD 41.2) kg, for EM and LM, respectively) previously offered grass silage *ad libitum* and supplementary concentrates were blocked within breed by weight, and randomly assigned to a 2 (breed type: EM and LM) × 2 (finishing strategies: grass only (G0) or grass + 3.2 kg dry matter (DM) barley-based concentrate daily (GC)) factorial arrangement. They were turned out to pasture on 7 April and slaughter occurred 192 days later. Concentrates were introduced 97 days post-turnout for GC. Bulls rotationally grazed *Lolium perenne*-dominant swards. Data were statistically analysed using ANOVA with terms for breed, finishing strategy and their interactions, and block in the model. There were no (P>0.05) breed × finishing strategy interactions. Mean estimated herbage dry matter intake (kg/day) during the supplementation period for EM G0 was 8.7 and for GC, 7.2. Corresponding values for LM were 10.0 and 8.2. LM had higher slaughter weight (697 v. 632 kg, P<0.001), carcass weight (400 v. 355 kg P<0.001), kill-out proportion (575 v. 561 g/kg, P<0.01), carcass conformation (9.7 v. 8.3, scale 1-15, P<0.001), and lower carcass fat (6.8 v. 5.6, P<0.01) scores than EM. Supplementation increased average daily gain (P<0.001) during the grazing season (1.42 v. 1.21) and supplementation phase (1.44 vs 1.05), slaughter weight (683 v. 646 kg, P<0.01), carcass weight (391 v. 364 kg, P<0.001), kill-out proportion (573 v. 564 g/kg, P=0.07), carcass fat score (6.8 v. 5.7, P<0.01), subcutaneous fat 'L', (P=0.06), 'a', 'b', and muscle 'b' values (P<0.05) and, reduced muscle 'L' (P<0.01). In conclusion, EM carcasses were lighter but adequately finished, with or without concentrates, whereas the heavier LM carcasses were only adequately finished when supplemented.

Correlation among feed efficiency, fertility and fat thickness traits in beef cattle
M.E.Z. Mercadante, R.J. Ferreira Júnior, S.F.M. Bonilha, A.L. Guimarães and J.N.S.G. Cyrillo
Instituto de Zootecnia, CP 63, Sertãozinho, SP, Brazil; mercadante@iz.sp.gov.br

Genetic correlations were determined between residual feed intake (n=1,156, RFI mean: $0.61 \times 10^{-11} \pm 0.605$ kg MS/ day) of Nellore purebred males and females at 328±35 days of age, and scrotal circumference (n=3,062, SC mean: 23.0±2.73 cm), backfat (n=2,725, BFT mean: 1.76±1.37 mm) and rump fat (n=2,261, RFT mean: 4.96±2.38 mm) thickness at yearling, and repeated days to calving (n=13,467 of 2,848 females, DC mean: 343±36.1 days). To this end, a multi-trait analysis was performed using AIREMLF90. Data were from a selection experiment for body weight (2 selection lines and a control line). RFI was recorded during the feedlot performance tests of 86±16 days. SC, BFT and RFT were recorded at the end of the performance test. Days to calving were recorded from natural mating in breeding seasons of 90 days. Cows which failed to calve were included as such if they had calved 21 days after the highest DC in the contemporary group. The models included several fixed effects of contemporary group class (year of birth, test facility and sex for RFI; line and year of birth for SC, BFT and RFT; line and year of breeding season for DC), month of birth class (for RFI, SC, BFT and RFT), previous reproductive status class (for DC), and dam age and animal age (for RFI, SC, BFT and RFT), plus random effects: permanent animal environment effects (for DC) and animal additive genetic effects. The number of animals in the matrix was 6,473. The heritabilities were 0.16, 0.50, 0.29, 0.34 and 0.10 for RFI, SC, BFT, RFT, RFI and DC, respectively. The genetic correlations between RFI and all of the traits studied were of antagonistic sense, close to zero with SC (0.07), and of moderate magnitude with BFT, RFT and DC (0.53, 0.36 and -0.47, respectively). SC (0.17 and 0.15) and DC (-0.09 and -0.21) were weakly and favorably genetically correlated with fat thickness (BFT and RFT, respectively). All phenotypic correlations were below 0.10, except between BFT and RFT (0.35). There is evidence of moderate genetic antagonism between RFI and the female fertility trait, without evidence of genetic relationship between RFI and the male fertility trait. There is also evidence of low genetic synergism between fertility and fat thickness traits.

Effect of age on intramuscular and intermuscular fat ratio in Japanese beef
M. Ohtani, K. Sakoda, R. Asa, T. Goto and K. Kuchida
Obihiro University of A &VM, Inada-cho, Obihiro-shi,Hokkaido, 080-8555, Japan; s24074@st.obihiro.ac.jp

Japanese Shorthorn (JS) is one of Japanese beef cattle breeds (so-called Wagyu) and is characterized by lean meat. JS cattle are well known in Japan as suitable for grazing because of their ability to perform on a high roughage diet. Although JS cattle can finish maturation and fattening earlier, they have a predisposition to laying down larger quantities of intermuscular fat and subcutaneous fat. The aim of this study was to compare the intramuscular fat ratio and intermuscular fat ratio of 6th-7th cross section in JS cattle at different slaughter ages (early and normal) by using image analysis. The carcass data was collected from 32 JS steers slaughtered in 4 prefectures from 2014 to 2016. In this study, animals were divided into two groups, which was the early slaughter age group (n=11, slaughter age less than 26 months) and normal slaughter age group (n=21, slaughter age greater than 26 months). Rib-eye images of the 6th-7th rib cross section were collected, and image analysis determined rib-eye area, intermuscular fat ratio, muscle area and intramuscular fat ratio of three individual muscles (*M. longissimus thoracis* (LT), *M. trapezius* (TR), and *M. spinalis dorsi* (SD). To investigate effects of slaughter age, analysis of variance was carried out for each individual muscle and also fattening farm location. Individual muscle size averages in the early and normal slaughter age groups were: LT (43.8 cm^2 and 46.0 cm^2), TR (32.7 cm^2 and 35.0 cm^2), SD (34.4 cm^2 and 37.5 cm^2), and rib-eye area (258.1 cm^2 and 275.4 cm^2), respectively. Averages of intramuscular fat ratio in early and normal slaughter age groups were: LT (24.0% and 21.4%), TR (20.8% and 21.2%), SD (29.2% and 29.3%), and intermuscular fat (34.0% and 35.8%), respectively. There was no significant difference between slaughter age groups. In one prefectural group, LT area of early slaughter group was exceptionally larger than normal group. In other areas except for LT, intramuscular fat ratios were similar to intermuscular fat ratios in early and normal groups. These results suggested that muscle area tends to increase as advancing age, but both intramuscular and intermuscular fat ratios are not altered by slaughter age.

Comparison of short-term and traditional fattening period using image analysis in Japanese Black
K. Kitamura, R. Asa, T. Goto and K. Kuchida
Obihiro University of Agriculture and Veterinary Medcine, Inada-cho, Obihiro, Hokkaido, Japan, 080-8555, Japan;
s25105@st.obihiro.ac.jp

Japanese Black cattle are a major beef breed in Japan. Although shortening the fattening period of beef cattle will be financially beneficial in reducing rearing costs, meat quality of short-term fattening steers may be lower than that of the traditional fattening period. In this study, we compared meat quality between short-term fattening and traditionally fattened using image analysis in Japanese Black steers. The carcass grading traits and image analysis traits were collected from 392 Japanese Black steers slaughtered in four prefectures (A24, B24, C28 and D28) in 2016. There were two fattening strategies. Farmers in A and B and in C and D are aiming for producing short-term fattening steers (slaughtered at 24 months) and traditionally fattened steers (slaughtered at 28 months), respectively. In addition, we added the steers in D24 (slaughtered at 24 months) under the traditional fattening regime, in order to compare two fattening strategies at the same age. The images from A24, B24 and C28 were obtained from cross section of rib cuts, whereas the images from D28 (D24) were obtained from the 6th to 7th rib cross sections of the carcasses. Analysis of variance was performed using carcass grading traits and image analysis traits as dependent variable, and fattening prefectures and slaughtered age (month) as fixed effects. Averages and standard deviations of carcass weight in A24, B24, C28, D24 and D28 were 490.5±5.3, 428.2±28.6, 460.8±42.1, 434.4±57.7 and 491.1±52.1 kg, respectively. Those of marbling % were 53.2±6.9, 53.6±7.0, 53.9±11.8, 45.7±15.4 and 54.4±7.0%, respectively. In carcass weight, the steers from B24 were significantly lighter than the steers slaughtered at D28 (P<0.05). However, there were no significant differences between the steers from A24 and the others. In image analysis traits e.g. marbling level, there were no significant differences among fattening areas and slaughtered age. These results suggested that marbling level can be maintained even if the cattle are fattened for shorter periods. Furthermore, it was suggested that the carcass weights also can be stabilised by using the other factors, such as fattening areas and pedigree.

Body condition score and fat thickness measured *in vivo* and post slaughter in feedlot lambs
D. Sandri, R.D. Kliemann, E.M. Nascimento, C.I.S. Bach, H. Maggioni, W.G. Nascimento, S.R. Fernandes and A.F. Garcez Neto
Federal University of Paraná, Animal Science, Rua Pioneiro, 2153, Jd. Dallas, Palotina, Paraná, 85950000 Palotina, Brazil; williangoncalves@ufpr.br

The *in vivo* evaluation of lamb carcass traits is an approach which may allow the selection of animals with carcasses of better post slaughter yield. Currently, the evaluation of body condition score (BCS) is the one of most used techniques to estimate the carcass fat deposition. However, the BCS is a subjective measurement and depends on the farmer expertise and on the animal breed. In this context, the ultrasonography has been a more accurate and reliable method to estimate the fat deposition on the carcass than BCS. The aim of this study was to evaluate the relationship among BCS, subcutaneous fat thickness measured *in vivo* by ultrasonography (SFT_{US}) and post slaughter in the loin (LFT) in lambs finished in feedlot. Twenty four crossbred ½ Dorper × ½ Santa Inês non-castrated male lambs with four months of age and 24.1±3.2 kg BW were used. Lambs were fed *ad libitum* with diets composed of 64% of Vaquero hay (*Cynodon* spp.) and 36% of concentrate, and were isoproteic (15.8% CP) and isoenergetic (67.0% TDN). The BW, BCS and SFT_{US} were measured at slaughter, and the LFT was measured in the loin 24 h after slaughter. Data were submitted to correlation analysis among carcass traits (P<0.05). The mean values for BW, BCS, SFT_{US} and LFT characteristics were 35.7 kg, 2.9 points, 1.55 mm and 2.23 mm, respectively. BCS was not correlated to SFT_{US}, which may be explained by the ultrasound device characteristics and operator technique affecting the latter measurement. Also, the low fat deposition in the carcass lead to a decrease in the accuracy of SFT_{US} compared to the LFT due to the difficulty to analyze the fat thickness in the ultrasonographic images. This was confirmed by the moderate correlation between SFT_{US} and LFT (r=0.48, P<0.05). High correlation was detected between BCS and LFT (r=0.76, P<0.05) which may be explained by the better visualization of fat covering in the carcass than that in the ultrasonographic images. Thus, the BCS is a more suitable technique than SFT_{US} to evaluate the fat deposition before slaughter in lambs with lean carcasses.

Effects of feeding extruded linseed on milk production and composition in dairy cow: a meta-analysis

T. Meignan[1,2], C. Lechartier[3], G. Chesneau[1] and N. Bareille[2]
[1]*Valorex, La Messayais, 35210 Combourtillé, France,* [2]*BIOEPAR, INRA, Oniris, La Chantrerie, 44307 Nantes, France,* [3]*URSE, Univ Bretagne Loire, Ecole Supérieure d'Agricultures (ESA), 55 rue Rabelais, 49007 Angers, France; thomas.meignan@oniris-nantes.fr*

The objectives of this study were to quantify the effects on production performance and milk fatty acid (FA) profile of feeding dairy cows extruded linseed (EL), a feed rich in α-linolenic acid, and to assess the variability of the responses related to the dose of EL and the basal diet composition. This meta-analysis was carried out using only data from trials including a control diet without fat supplementation. The dependent variables were defined by the mean differences between EL supplemented groups and control groups. The data were processed by multivariable regression testing the effect of each potential interfering factor associated with the dose effect, and then stepwise regression with backward elimination procedure with all potential interfering factors retained in previous steps. This entire strategy was also applied to a restricted dataset including only trials conducted within a practical range of fat feeding (only supplemented diets with less than 60 g fat/kg DM and supplemented with less than 600 g fat from EL). The whole dataset consisted of 17 publications representing 21 control diets and 29 EL supplemented diets. The daily intake of fat from EL supplementation ranged from 87 to 1,194 g/cow/d. No effect of dose or diet was identified on dry matter intake, milk yield or milk protein content. Milk fat content decreased when EL was supplemented to diets with a high proportion of corn silage in the forage (-2.8 g/kg between low and high corn silage-based diets in the restricted dataset) but did not decrease when the diet contained alfalfa hay. Milk trans-10 18:1 proportion increased with EL supplementation in interaction with corn silage proportion in forage. A shift in ruminal biohydrogenation pathways, from trans-11 18:1 to trans-10 18:1, probably occurred when supplementing high dose of EL with high corn silage-based diets, related to a change in the activity and/or composition of the microbial equilibrium in the rumen. Finally, EL supplementation increased linearly proportions of potentially human health-beneficial FA in milk.

Effect of solid intake stimulation of suckling rabbit on survival and growth performance

S. Combes[1], S. Ikken[2], T. Gidenne[1], E. Balmisse[3], P. Aymard[3] and A. Travel[2]
[1]*INRA, GenPhySE, INRA Toulouse, 24 chemin de Borde Rouge, CS 52627, 31326 Castanet Tolosan, France,* [2]*ITAVI, URA, l'Orfrasière, 37380 Nouzilly, France,* [3]*INRA, UE PECTOUL, INRA Toulouse, 24 chemin de Borde Rouge, CS 52627, 31326 Castanet Tolosan, France; sylvie.combes@inra.fr*

The objective of our study was to stimulate the solid intake of suckling rabbit by modulating the age of access to pelleted solid feed (8 days vs 18 days) or protein-starch ratio of the pellet. Three groups of 17 litters were used. In group 1, suckling rabbits had access to P + A- pellet (13.5% digestible protein, 6.8% starch) as early as 8 days in the nest. In group 2, the suckling rabbits had access to the same pelleted feed but only from 18 days in feeders. In group 3, the suckling rabbits were fed from 18 days with a P-A + pelleted feed in feeders (digestible protein: 9.9%, starch 8.9%). At weaning (35 days), the young rabbits of all 3 groups were fed at 80% of the voluntary intake. The health status was monitored on a daily basis, the milk production of the does was measured twice a week. Live weight and feed consumption of the young rabbits, were measured in the nest between 8 and 18 days, and after nest removal until slaughter weight (18-70 days). The mortality rate was not different between the groups (2.8% between 8 and 35 days, and 0.4% between 35 and 70 days). In group 1, the ingestion of pelleted feed in the nest between 8 and 18 days was 0.75 pellet/rabbit/day. This early consumption of pelleted feed did not affect the milk production of the does or the growth until weaning of the young rabbits. During the period 35-70 days, group 3 had the highest growth (39.5±0.4 g/d), the lowest was observed for group 2 (37.6±0.4 g/days) while group 1 has an intermediate value (38.8±0.5 g/d). These results confirm the ability of young rabbits to ingest a solid feed with no adverse effect on health. In our favorable sanitary conditions, the lowest protein-starch ratio feed shows an interest in post-weaning growth of young rabbits.

New approach for polyphenol oxidase-mediated rumen bypass polyunsaturated fatty acids

N. De Neve[1], F. Gadeyne[1], B. Vlaeminck[1], E. Claeys[1], P. Van De Meeren[2] and V. Fievez[1]
[1]Ghent University, Laboratory of Animal Nutrition and Animal Product Quality, Coupure Links 653 BW13, 9000 Ghent, Belgium, [2]Ghent University, Particle and Interfacial Technology Group, Coupure Links 653 BW08, 9000 Ghent, Belgium; nympha.deneve@ugent.be

Previously, *in vitro* ruminal protection (>80%) of PUFA was obtained by (1) emulsifying a PUFA-rich oil with protein extract from potato tuber peel, containing polyphenol oxidase (PPO), and (2) inducing enzymatic cross-linking of the interfacial protein by addition of the PPO substrate 4-methylcatechol (4MC). In the current study, it was hypothesized better protein emulsifiers can adopt the emulsifying function, and PPO extract can be added after emulsification together with its substrate. Linseed oil (10 wt%) emulsions were prepared using a protein emulsifier (1 wt%) (soy glycinin, gelatin, whey protein isolate, bovine serum albumin or sodium caseinate) and were diluted five times with PPO extract that was obtained from potato tuber peel by acetone precipitation. A linseed oil (2 wt%) emulsion with PPO extract only was used as the positive control. Protein adsorption behavior was analyzed by non-reducing SDS-PAGE and zymography. Emulsions were incubated for 24 h with 0 (negative control) or 40 mM 4MC. Droplet size distributions were characterized by laser diffraction and light microscopy. Rumen metabolism was simulated by 24 h *in vitro* incubation with rumen fluid. All emulsions were prepared and characterized in triplicate. Results were analyzed by the MIXED procedure of SAS. Several treatments, including the control, showed an increase in droplet size after cross-linking in presence of 40 mM 4MC, due to droplet aggregation (P<0.001). In all negative control emulsions (0 mM 4MC) PUFA were extensively biohydrogenated (94-96%). In presence of 40 mM 4MC, compared to the positive control (81%) protection efficiencies were equal or higher for all treatments (P=0.017, 86-93%). Partitioning of PPO isoforms was similar for all treatments, showing low molecular weight PPO isoforms in the adsorbed protein fraction. Protein emulsifier can successfully adopt the emulsion function of PPO for ruminal protection, but even when PPO was added after emulsification, some PPO was present at the droplet interface.

Impacts of specialty ingredients on improving sustainability: an overview

J.L. Peyraud
INRA, Domaine de la Prise, 35590 Saint-Gilles, France; jean-louis.peyraud@rennes.inra.fr

Using specialized supplements to maximise forage utilisation and performance in ruminants

A.S. Chaudhry
Newcastle University, Agriculture, Food & Rural Development, Agriculture Building, NE1 7RU, United Kingdom;
abdul.chaudhry@ncl.ac.uk

Grass, fresh or conserved, is the most economical source of nutrients for ruminant animals. However its quantity and nutritional value varies with variety and maturity in different seasons worldwide. This creates challenges for animal health and welfare that are vital for sustainable animal production. Correct supplements are needed to fill the deficits in grass to improve nutrient use efficiency in forage consuming ruminants. A series of *in vitro* and *in vivo* studies tested supplements having either low (120 g= LCP) or high (250 g= HCP) crude protein but the same energy (13 MJ ME) per kg DM for their use with grass. The *in vitro* factorial study tested the effect of 0 to 180 g/ kg forage (Amount) of either LCP or HCP on the disappearance of grass nuts (Grass) and barley straw (Straw) as contrasting forages (Forage) over 0 to 72 h of incubations with buffered rumen fluid under anaerobic conditions. The Grass and Straw residues were washed and dried to estimate dry matter disappearance (DMD). The statistical analysis of DMD data revealed that the main effects of Forage and Amount were significant (P<0.01) at most hours but the Supplement effect was only significant at 72 h where HCP showed greater DMD than LCP (P<0.05). All 2 and 3 way interactions were non-significant. Grass had significantly higher DMD than Straw at all incubation times (P<0.001). DMD increased significantly (P<0.05) at most times with increased Amount having more DMD at 0 to 90 g/kg than at 90 to 180 g/kg. The LCP supplement was then tested as feed blocks for ewes either grazing grass or consuming silage based diets during their full breeding cycle from their joining with rams to scanning and lambing. It appeared that the sheep consuming supplement utilised grass better than the control sheep as reflected in their condition, service, lambing % and overall performance. The studies imply that the nutrient use efficiency and animal performance can be enhanced by using speciality ingredients but their suitability as supplements needs to complement the grass quality and quantity in various situations. The success of similar speciality ingredients may help us promote sustainable animal production for the supply of safe food for human consumption.

Molecular characterization of the rumen microbiome in lambs fed hydrolysable and condensed tannins

S.A. Salami[1,2], B. Valenti[1], L. Campidonico[1], G. Luciano[3], M.N. O'Grady[2], J.P. Kerry[2], E. Jones[4], A. Priolo[1] and C.J. Newbold[4]
[1]University of Catania, Di3A, Animal Production Science, 95123 Catania, Italy, [2]University College Cork, School of Food and Nutritional Sciences, Cork, Ireland, [3]University of Perugia, Department of Animal Science, 06121 Perugia, Italy, [4]IBERS, Aberystwyth University, SY23 3DA Aberystwyth, United Kingdom; s.salami@umail.ucc.ie

The application of meta-omics techniques to examine the impact of different tannin sources on the rumen microbiome could enhance the selection of tannins achieving a target response on rumen function. This study investigated the performance, ruminal fermentation and microbiome in lambs fed vegetal sources of hydrolysable tannins (HT) and condensed tannins (CT). Forty-four lambs (19.56±2.06 kg) were randomly assigned to either a commercial concentrate diet (CON, n=8) or CON supplemented with 4% of two HT [chestnut (HT-c) and tara (HT-t)] and CT [mimosa (CT-m) and gambier (CT-g)] commercial extract sources (n=9) for 75 days pre-slaughter. Rumen contents were collected immediately after slaughter. Lambs fed HT-c had lower growth performance characteristics compared to the CON group. Supplementation of tannins did not influence ruminal fermentation traits. Quantitative PCR demonstrated that tannins did not affect the absolute abundance of ruminal bacteria and fungi. However, CT-m (-12.8%) and CT-g (-11.5%) reduced methanogens while HT-t (-20.7%) and CT-g (-20.8%) reduced protozoa abundance. Metagenomic analyses further revealed that tannins caused changes in the phylogenetic structure of the bacterial community and had a minimal effect on the methanogen community. Notably, both HT and CT sources exhibited comparable inhibitory effects on the abundance of the fibrolytic bacterium, *Fibrobacter*. Canonical correspondence analysis illustrated that the bacteriome in lambs fed HT-c appeared to promote greater butyrate concentration, acetate/propionate ratio and bacterial diversity. In conclusion, HT-t, CT-m and CT-g extracts displayed antimicrobial activity against methanogens and protozoa without compromising lamb productivity in a long-term feeding trial. Overall, results demonstrate that both HT and CT sources could impact the ruminal microbiome when supplemented at moderate levels.

Use of feed-grade amino acids in low protein diets: towards a more sustainable broiler production?

B. Méda[1], P. Belloir[1,2], W. Lambert[2], E. Corrent[2], H. Juin[3], M. Lessire[1] and S. Tesseraud[1]
[1]INRA, URA, 37380 Nouzilly, France, [2]Ajinomoto Eurolysine, SAS, 75017 Paris, France, [3]INRA, EASM, 17700 Saint-Pierre-d'Amilly, France; bertrand.meda@inra.fr

The aim of this study was to investigate the effect of decreasing dietary crude protein (CP) in finishing broiler on animal performance, meat quality and environmental impacts. PM3 Ross male broilers were reared together between 1 and 20 d of age. At d21, they were randomly distributed in 24 floor pens (8 pens per treatment; 38 birds per pen) and fed until d35 (slaughter) with diets formulated with an amino acid (AA) profile based on the ideal protein concept. More specifically, the minimum AA:Lys ratios proposed by Mack et al. were used with modifications for Thr and Arg (Thr:Lys ratio increased from 63 to 68% and Arg:Lys ratio decreased from 112 to 108%, respectively). CP contents of experimental diets were 19, 17.5 and 16%, respectively. With CP reduction, the proportion of soybean meal in diets decreased (28, 24 and 18%, respectively), while those of corn and feed-grade AA increased (for AA: 0.23, 0.48 and 1.19%, respectively). CP reduction did not affect body weight (BW) gain, feed efficiency or breast meat yield but abdominal fat increased. Meat quality criteria responded to dietary CP reduction with higher ultimate pH and lower lightness and drip loss, but these variations were considered acceptable for meat preservation or processing. Concerning environment, nitrogen excretion decreased with CP (-12% per CP point) and so did volatilization. Based on these results, a life cycle analysis was carried out to compare the environmental impacts of one kg of BW at farm gate, with 19 or 16% CP in finishing diet respectively. CP reduction decreased climate change, eutrophication and energy use impacts by 8, 7 and 1%, respectively. For acidification, the impact per ton of finishing diet increased by 4%, but the final impact decreased by 5%, due to decrease in nitrogen excretion and volatilization. In conclusion, CP reduction in finishing broilers is possible using an adapted AA profile and feed-grade AA. Such a feeding strategy could improve the sustainability of broiler production with similar animal performance, no detrimental effect on meat quality, lower environmental impacts and a reduced use of GMO soybean meal.

Effects of *Enterococcus faecium* and the blend of organic acids in finishing pigs

X.J. Lei[1], J.S. Lee[1], I.H. Kim[1] and K.Y. Lee[2]
[1]Swine Nutrition and Feed Technology 421-1 Department of Animal Resources Science Dankook University, Department of Animal Resource & Science, Dankook University, #29 Anseodong, Cheonan, Choognam, 330, Dankook University #119, Dandae-ro, Dongnam-gu, Cheonan-si, Chungnam, 330-714, Korea, Korea, South, [2]Morningbio Co., Ltd, Floor 4 Semi B/D, 9-1 Dujeongyeokdong4-gil, Seobuk-gu, Cheonan, Chungnam, 31111, Korea, Seobuk-gu, Cheonan, Chungnam, 31111, Korea, Korea, South; xjlei1988@hotmail.com

The study was conducted to evaluate the effects of *Enterococcus faecium* and the blend of organic acids (OAs) on performance, nutrient digestibility, and meat quality in finishing pigs. A total of 120 pigs (initial body weight: 51.04±1.82 kg) were randomly allocated to 5 treatments according to initial BW and sex with 6 replicates per treatment and 2 barrows and 2 gilts per pen. The diets included: (1) control, basal diet (CON); (2) CON + 0.05% blend of OAs (OA1); (3) CON + 0.1% blend of OAs (OA2); (4) CON + 0.05% blend of OAs + 0.002% *E. faecium* (OAEF1); (5) CON + 0.1% blend of OAs + 0.002% *E. faecium* (OAEF2). The blend of OAs contained 17% fumaric acid, 13% citric acid, 10% malic acid, 0.6% capric acid, and 0.6% caprylic acid). *E. faecium* contained 1.0×10^{10} cfu/g of *E. faecium* DSM 7134. All data were subjected to the General Linear Models procedures of SAS. Differences among treatments were separated by Tukey's range test. During 0 to 5 weeks, the average daily gain (ADG) and gain to feed ratio (G:F) in pigs fed OA2 (840 g and 0.379), OAEF1 (860 g and 0.380), and OAEF2 (853 g and 0.375) diets were higher (P<0.05) compared with those fed CON (798 g and 0.357) diet. During 5 to 10 weeks, the ADG of pigs fed OAEF1 (909 g) and OAEF2 (885 g) diets was greater (P<0.05) than that of pigs fed CON (824 g) diet. During 0 to 10 weeks, pigs fed OA2 (855 g), OAEF1 (885 g), and OAEF2 (869 g) diets had increased ADG compared with those fed CON (811 g) diet. The OAEF1 treatment had higher (P<0.05) apparent total tract digestibility (ATTD) of nitrogen (71.57%) than CON (68.60%) treatment. Pigs fed OAEF1 diet had higher longissimus muscle (LM) area (65.70 cm^2) compared with those fed CON (59.42 cm^2) diet. In conclusion, dietary supplementation of the combination of *E. faecium* DSM 7134 and the blend of OAs improved growth performance, ATTD of nitrogen, and LM area.

Seaweeds in ruminant diets: a long-term evaluation of *in vitro* rumen fermentation and methanogenesis

M.R.G. Maia[1,2], A.J.M. Fonseca[2] and A.R.J. Cabrita[2]
[1]REQUIMTE, LAQV, DGAOT, Faculdade de Ciências, Universidade do Porto, Porto, Portugal, [2]REQUIMTE, LAQV, ICBAS, Instituto de Ciências Biomédicas de Abel Salazar, Universidade do Porto, Porto, Portugal; mrmaia@icbas.up.pt

Increasing interest on seaweeds as feed ingredients has emerged over the last years due to their content in organic minerals, complex carbohydrates, proteins, vitamins, volatile compounds, pigments, and bioactive substances. Additionally, seaweeds have been shown to successfully mitigate methane emissions by ruminants. Recently, our group revealed the potential of *Ulva rigida, Saccharina latissima,* and *Gracilaria vermiculophylla* on ruminal methane mitigation when incubated *in vitro* with two substrates (meadow hay and corn silage) for 24 h. The present study aimed to evaluate the long-term effects of seaweeds inclusion on gas and methane production and on fermentation parameters in a rumen simulation technique (RUSITEC) system. Green (*U. rigida*), red (*G. vermiculophylla*) or brown (*S. latissima*) seaweeds produced in an integrated multi-trophic aquaculture system were included at 25% (dry matter basis) in a total mixed ration diet and incubated in a RUSITEC system along with a control (no seaweed). Each experimental treatment were allocated to two vessels with different rumen inocula (from maize silage or hay silage fed donor cows), and incubated for 15 days, in two independent trials. Seaweeds inclusion at 25% did not affect fermentation pH, daily total volatile fatty acid production nor methane production while tended to reduce the daily gas production (P=0.092) compared to control. Conversely, seaweed inclusion decreased redox potential of fermentation media and the proportions of butyric and valeric acids (P<0.001), tended to decrease propionic acid (P=0.079), and increased acetic acid (P<0.001) and the acetic acid to propionic acid ratio (P=0.015). Overall, the effects of seaweed on long-term *in vitro* rumen fermentation parameters were moderate and independent of the diet of the donor cows. Financial support from the EU (FEDER funds through COMPETE) and National funds (FCT) through projects EXPL/CVT-NUT/0286/2013-FCOMP-01-0124-FEDER-041111 and UID/QUI/50006/2013-POCI/01/0145/FEDER/007265 and individual grant to MRGM (SFRH/BPD/70176/2010) are acknowledge.

Effects of biochar inclusion on *in vitro* rumen fermentation parameters

A.R.F. Rodrigues[1], M.R.G. Maia[1], A.R.J. Cabrita[1], H.M. Oliveira[2], M. Bernardo[3], I. Fonseca[3], H. Trindade[4], J.L. Pereira[4,5] and A.J.M. Fonseca[1]
[1]REQUIMTE, LAQV, ICBAS, Instituto de Ciências Biomédicas Abel Salazar, Universidade do Porto, Porto, Portugal, [2]INL, International Iberian Nanotechnology Laboratory, Braga, Portugal, [3]REQUIMTE, LAQV, Faculdade de Ciências e Tecnologia, Universidade Nova de Lisboa, Caparica, Portugal, [4]CITAB, Universidade de Trás-os-Montes e Alto Douro, Vila Real, Portugal, [5]ESAV, Instituto Politécnico de Viseu, Viseu, Portugal; anaferodrigues@gmail.com

New strategies to reduce enteric methane emissions from ruminants remain a hot topic due to its negative impact on the environment and energy loss to the animal. Biochar had been used to amend soils, but few studies evaluated its effects on rumen fermentation. Therefore, this study evaluated the effects of two sources of biochar (agro-forestry residues (AF) and potato peel residues (PP)) at three inclusion levels (0; control, 5 and 10%, dry matter, DM, basis) on rumen fermentation parameters in 24-h *in vitro* incubations. Hay-silage and corn silage were used as substrates in four independent runs, using rumen inocula from fistulated cows fed different diets. Biochar decreased total gas production (ml/g DM) when included at 10% (P<0.001), having no effect (P>0.05) on fermentation pH, propionic and butyric acid proportions and acetic:propionic ratio. Total volatile fatty acids (VFA; mmol/g DM) and acetic acid (mmol/g DM) decreased (P<0.001) with biochar inclusion, 10% AF promoting the lowest values. Ammonia-N (mg/g DM) increased with biochar, 10% PP presenting the highest value (P=0.042). Methane production (ml/g DM) was affected by additive × substrate interaction (P=0.001), biochar inclusion leading to higher productions when incubated with corn silage than with hay-silage. Overall, biochar affected rumen fermentation mostly at 10% inclusion level, differences between AF and PP being only observed for total VFA and acetic acid. Effects of different sources of biochar and dietary inclusion levels will be addressed in the future. Authors acknowledge the financial support from FCT and FEDER under Program PT2020 (project 007265 – UID/QUI/50006/2013), and the individual grants to ARFR (PDE/BDE/114434/2016), MRGM (REQUIMTE2016-34) and MB (SFRH/BPD/93407/2013).

Commercial oil of *Nigella sativa* as growth promoters in sheep rations

S. El-Naggar[1], G.A. Abou-Ward[1], A.M. Ali[2] and A.Y. El-Badawi[1]
[1]National Research Center, Animal Production, Doki, Giza 12311, Egypt, [2]Faculty of Agriculture, Cairo University, Animal Nutrition, Giza 12613, Egypt; soadelnaggar75@gmail.com

This work aimed to investigate the effect of supplementing different levels of *Nigella sativa* oil in growing lambs ration on nutrients digestibility and growth performance. Eighteen growing Ossimi lambs (29.04 kg average live body weight and 6-7 months old) were randomly divided according to their live body weight into three feeding groups (6 each) in growth and metabolism trials for 104 days. Animals in all groups were fed total mixed ration consisted of 60% concentrate: 40% roughage was offered by 3% of live body weight during the growth trial for 90 days, whereas the first group was control, R1, while *N. sativa* oil was supplemented by 0.1 and 0.2% from DM in R2 and R3, respectively. Results were statistically analyzed via general liner model for one way analysis and indicated that there was a significant (P<0.05) increase in the digestibility of DM, OM, CP and NFE by 12.8, 9.67, 2.18 and 12.1% for R2 and by 18.3, 11.6, 5.9 and 12.7% for R3, respectively compared with R1. While there was insignificant (P<0.05) differences in the digestibility of EE and CF among groups. This improvement in nutrient digestibility reflected on improve in the nutritive value as TDN and DCP, meanwhile the average daily gain was higher in R2 and R3 (217 and 226 g/h/d., respectively) compared with R1 (167 g/h/d.). Meanwhile, it can be concluded that supplementation of *N. sativa* oil in growing lambs ration enhance nutrients digestibility and growth performance.

A life cycle assessment of high sugar grasses on pasture-based dairy farms

A. Foskolos[1], J. Gibbons[2], A. Gonzalez-Mejia[2], J. Moorby[1] and D. Styles[2]
[1]Aberystwyth University, Institute of Biological, Environmental and Rural Sciences, Campus Gogerddan, SY23 3EE, Aberystwyth, United Kingdom, [2]Bangor University, School of Environment, Natural Resources & Geography, Deiniol Road, Bangor, LL57 2UW, Wales, United Kingdom; anf20@aber.ac.uk

Pasture-based dairy systems are associated with polluting emissions, such as methane (CH_4), ammonia (NH_3) and nitrous oxide (N_2O), while the relatively low use of concentrate feeds constrains scope for dietary intervention. In this context, high sugar grasses (HSG) have been proposed to enhance nitrogen (N) utilization efficiency in the rumen and consequently reduce N excretion. The use of HSG could thus reduce the carbon and N footprints of milk production. We use life cycle assessment (LCA) integrated with a detailed dietary model to investigate the potential for using HSG to reduce polluting emissions and improve the environmental footprint of milk. A previously-constructed LCA model describing dairy farm production on pasture was employed. Two feeding scenarios were considered, using either conventional perennial ryegrass or HSG. A meta-analysis provided data to formulate a diet with a daily dry matter intake of 2.6 kg of concentrate feed/cow, and provided N excretion estimates. The Cornell Net Carbohydrate and Protein System was then used to simulate the feeding scenarios and to calculate data on CH_4 emissions for lactating cows fed on pasture during the grazing season (6 months per year). The HSG scenario resulted in a 19% reduction in N excretion (g/kg of milk), leading to similar reductions in eutrophication and acidification burdens. Carbon dioxide equivalent greenhouse gas emissions from the HSG scenario were 3.1% higher per hectare owing to a 1.5 kg/cow higher milk yield (not statistically significant), but emissions intensities (g CH_4/kg milk) were 3.4% lower. In conclusion, our results suggest that the use of HSG may reduce nitrogen excretion of cows, together with associated eutrophication and acidification footprints of milk production, with more minor effects on the carbon footprint. Acknowledgements: Support was provided through the Sêr Cymru NRN-LCEE project Cleaner Cows.

Milk production of dairy cows fed with high protein sunflower meal

B. Rouillé[1], B. Beaumont[2], J. Legarto[1] and M. Gele[3]
[1]*Institut de l'Elevage, 149 rue de Bercy, 75595 Paris, France,* [2]*Chambre d'agriculture des Landes, 55 Avenue Cronstadt, 40000 Mont-de-Marsan, France,* [3]*Institut de l'Elevage, 9 rue André Brouard – CS 70510, 49105 Angers cedex 02, France; marine.gele@idele.fr*

Sunflower meals are known to be 'poor' meals with few energy and protein contents for both classic and semi decorticated meals. Due to this statement, they are not used as protein concentrates in dairy cows diets. However, they are usually used in processed feeds at a 10 to 12% rate in energy and protein equilibrated feeds. Sunflower meal from highly decorticated grains (HDSM) contains around 20% of crude fiber and less than 35% of total fiber (NDF). Its crude protein content reaches 38% of dry matter. 1 kg of HDSM is an interesting substitute to 0.65 kg of soya meal. A trial on 34 dairy cows verified that it is possible to substitute more than half of the soya meal by 2.4 kg DM of HDSM without decreasing milk production and modifying milk fat and protein content. Both diets were based on maize silage (67% of total dry matter) and protein concentrates. Milk production per cow has been the same: 27.1 versus 27.2 kg/cow/day for control and HDSM diet respectively. Fat and protein contents are also statistically the same; around 40 g/l for fat and 34.5 g/l for protein. Based on these results, HDSM is economically interesting as soon as its price is lower than 80% of the soya meal one. Moreover, environmental indicators such as nitrogen excretion and enteric methane emissions are not increased by using HDSM. Our conclusion is clear: it is possible to consider HDSM as a serious technical and economical alternative to soya meal.

Effect of dietary slow release calcium supplementation on egg quality in Hy-line brown layers

H. Shi[1], J.H. Park[1], K.Y. Lee[2], I.H. Kim[1] and J.H. Jung[2]
[1]*Swine Nutrition and Feed Technology 421-1 Department of Animal Resources Science Dankook University, Department of Animal Resource & Science, Dankook University, #29 Anseodong, Cheonan, Choognam, 330, Dankook University #119, Dandae-ro, Dongnam-gu, Cheonan-si, Chungnam, 330-714, Korea, Korea, South,* [2]*Morningbio Co., Ltd, Floor 4 Semi B/D, 9-1 Dujeongyeokdong4-gil, Seobuk-gu, Cheonan, Chungnam, 31111, Korea, Seobuk-gu, Cheonan, Chungnam, 31111, Korea, Korea, South; inhokim@dankook.ac.kr*

In the current study, slow release calcium which consists of 45% calcium carbonate, 50% carrier, and medium chain fatty acid was used to evaluate the effects of slow release calcium supplementation on egg production, egg quality and blood profiles in laying hens. A total of 144 Hy-line brown layers (25 weeks old) were randomly divided into 3 treatments (CON, EP1 and EP2) in 8 replicate pens with 6 hens per pen in this 10-week study. Treatments were: (1) CON, Basal diet; (2) EP1, CON + 0.5% slow release calcium; (3) EP2, CON + 1% slow release calcium. All data were statistically analyzed by ANOVA using the GLM procedure of SAS. Orthogonal comparisons were conducted using polynomial regression to measure the linear and quadratic effects of dietary concentrations of supplemental slow release calcium. The results showed that egg broken rates were reduced linearly ($P<0.05$) with the increase in the level of slow release calcium at week 10 (2.60%, 1.25% and 0.60%). A significant linear increase ($P<0.05$) on Haugh Unit were shown at week 5 (84.29, 85.42 and 86.88); and the eggshell strength was also enhanced significantly (linear effect, $P<0.05$) by increase in supplementation level of calcium at week 10 (3.99, 4.11 and 4.12 kg/cm^2). The blood phosphorus concentration was higher ($P<0.05$) in the morning (6.45, 6.73 and 7.08 mg/dl) and afternoon (6.53, 6.90 and 7.00 mg/dl) at week 5 and calcium content of blood was increased (linear effect, $P<0.05$) in the morning at week 10 (18.30, 19.08 and 19.20); and phosphorus density also increased linearly (6.45, 6.90 and 7.05 mg/dl) at week 10 ($P<0.05$) with the increase in inclusion level of calcium in the diet. In conclusion, the addition of 1% slow release calcium (EP2) can promote the absorption of calcium and phosphorous which can decrease egg breakage and enhance the eggshell characteristics.

Inclusion of xylanase on nutrient digestibility and ileal digestibility of amino acids in pigs

H.M. Yun[1], K.Y. Lee[2], B. Balasubramanian[2] and I.H. Kim[1]
[1]Dankook University, Department of Animal Resource & Science, Cheonan, 31116, Korea, South, [2]Morningbio Co., Ltd, Floor 4 Semi B/D, 9-1 Dujeongyeokdong4-gil, Seobuk-gu, Cheonan, 31111, Korea, South; yunhm822@naver.com

This study aims to determine the effects of xylanase supplementation on apparent total tract digestibility (ATTD) of dry matter (DM), nitrogen (N), energy (E), apparent ileal digestibility (AID) of amino acids and fecal microflora counts in twelve cannulated growing pigs [(Landrace×Yorkshire)×Duroc)] fed with corn-soybean meal based diet. Pigs having weights 24.32±1.77 kg were surgically equipped with T-cannulas of approximately 15 cm prior to the ileo-cecal junction and randomly allotted to one of four dietary treatments of basal diet supplemented with 0, 0.005, 0.01, and 0.02% xylanase with 3 replications per treatment for 21-day trial period. The pigs were fasted for 16-20 h prior to surgery. Anesthesia was induced using Stresnil® and Virbac Zoletil® 50 injections. After surgery, pigs were transferred to individual pens (1.2×0.6 m) in environment controlled room. Each pig used as experimental unit with orthogonal comparison was conducted using polynomial regression to measure the linear and quadratic effects of increasing the dietary supplementation of xylanase. The enzyme preparation used in this study was Nutrase-Xyla™ (Nutrex Nv, Lille, Belgium), a specific endo-1, 4-β-xylanase (IUB: EC 3.2.1.8; 9,000 U/g) with an optimal activity at neutral pH derived from fermentation by *Bacillus subtilis*. Inclusion of xylanase in the diet showed linearly improved ATTD of DM, N and E (P=0.001, 0.004, 0.001, respectively) and AID of histidine, lysine, methionine, threonine and asparatic acid (P=0.016, 0.017, 0.020, 0.035, 0.053, respectively). In addition, dietary inclusion of xylanase increased fecal and ileal *Lactobacillus* (P=0.007, 0.01, respectively) and reduced *Escherichia coli* counts (P=0.002, 0.05, respectively). In conclusion, dietary supplementation of xylanase showed better digestibility of amino acid, improved nutrient digestibility and ileal digestibility of amino acids, shifted microbiota by increasing fecal *Lactobacillus* counts and reduced *E. coli* in growing pigs.

Effects of different levels of poultry by-product meal on commercial performance in growing chickens

R. Hosseini[1] and A. Karimi[2]
[1]Orum Gohar Daneh Co., Product Management, Miandoab, 59716-56711, Iran, [2]University of Tabriz, Ahar Faculty of Agriculture and Natural Resources, Tabriz, 5166616471, Iran; akkg2000@yahoo.com

About 60-70% of poultry production costs involve in feeding and protein and its cost are one of the most important parts of broiler feeding. The animal sources of protein are notable due to their high protein content and competitive cost. This experiment was conducted to investigate the effects of different levels of inclusion of poultry by-product meal (PBP) on broiler chicken commercial performance. A total of 1,600 unsexed day-old Ross 308 boiler chicks were reared until day 20 of age with similar diet. Then, they were used for feed evaluation of poultry by-product meal in four levels of 0 (control), 20 (PBP2%), 40 (PBP4%) and 80 (PBP8%) gr/kg for the growing period (21-42 day of age) following a completely randomized design with 4 treatments and 10 replicates. Grower diets were iso-energetic and iso-nitrogenous based on Ross 308 guide. During the experimental period chicks fed control and PBP2% diets had significantly higher feed intake than PBP8% diet (P<0.05); there was no notable difference between PBP4% and PBP8% or PBP4% and other diets in feed intake (P>0.05). The results showed higher final body weight (gr) and body weight gain (gr/bird/day) in control group versus other experimental groups (P<0.05); these parameters were statistically similar in other groups (P>0.05). The best FCR (feed conversion ratio) was obtained in control group that was higher than PBP2% group (P<0.05); birds fed on PBP4% and PBP8% showed no significant difference on FCRs with control group (P>0.05). The best PEF (productivity Efficiency Factor) was observed in control group versus others (P<0.05). Comparison of total rearing cost ($/kgbw) declared consumption of PBP8% (80 gr PBP/kg diet) lead to lower cost than PBP2% (P<0.05); although PBP8% wasn't significantly different with control or PBP4% groups (P>0.05). It seems although the PEF of control group was the best but the lowest cost ($/kg BW) was observed in group fed 80 gr poultry by-product/kg diet (PBP8%).

Fungal treatment effect on the chemical composition of matured forages for their use inruminant diet

O.J. Bolaji[1], A.S. Chaudhry[1] and J. Dolfing[2]
[1]Newcastle University, School of Agriculture, Food and Rural Development, Newcastle University, NE1 7RU, United Kingdom, [2]Newcastle University, School of Civil Engineering and Geosciences, Newcastle University, NE1 7RU, United Kingdom; abdul.chaudhry@newcastle.ac.uk

This experiment determined the ability of 2 white rot fungi in improving the chemical composition of selected matured forages for their use in ruminant diet. *Pleurotus ostreatus* and Ceriporiopsis rivulosus were cultivated on *Brachiaria decubems*, Andropogon gayanus, *Lolium perenne* and Triticum aestivum straw using two forage:liquid ratios (1:3 & 1:5). The treatment of forages with fungi was done alongside the controls (with or without autoclaving) for 28 days using a solid state fermentation technique. The fungus reduced the dry matter (DM), organic matter (OM), neutral detergent fibre (NDF), total phenols (TP) and total tannin (TT) contents and increased the ash and crude protein (CP) contents of the forages with *P. ostreatus* having a higher effect on the forages. Forages with liquid ratio of 1:3 recorded a significantly (P<0.05) higher DM (972.182 g/kg), Ash (63.10 g/kg DM) and CP (102.45 g/kg DM) contents but lower NDF (746 g/kg DM) content than those forages with the ratio of 1:5. However the TP and TT values of the forages were not affected by the 2 ratios used in combination with liquid. Amongst forages, *L. perenne* recorded the highest CP (153.93 g/kg DM), TP (35.74 g/kg DM) and Ash (80.23 g/kg DM) contents; *T. aestivum* straw recorded the highest DM (973.22 g/kg) and OM (950.13 g/kg DM) and *A. gayanus* recorded the highest TT (17.16 g/kg DM) content. It is concluded that the fungal treatment increased the soluble fraction and decreased the fibre content of the mature forages. However higher DM and OM losses were in the forages which indicated that fungi were able to convert some soluble contents into high quality protein in these forages. The substrate: liquid ratio of 1:3 appeared to have worked better for the fungal growth and so it will be used in the subsequent studies.

Effect of dietary fermented and unfermented grape skin on broiler chickens

M. Nardoia[1], S. Chamorro[2], A. Viveros[3], I. Arija[3], C. Ruiz-Capillas[4] and A. Brenes[2]
[1]University of Molise, Agricultural, Environmental and Food Sciences, via de Sanctis, snc, 86100 Campobasso, Italy, [2]Instituto de Ciencia y Tecnología de Alimentos y Nutrición (ICTAN-CSIC), C/José Antonio Novais, 10, 28040 Madrid, Spain, [3]Universidad Complutense de Madrid, Departamento de Producción Animal, Facultad de Veterinaria, Ciudad Universitaria, 28040 Madrid, Spain, [4]Instituto de Ciencia y Tecnología de Alimentos y Nutrición (ICTAN-CSIC), Department of Products, (ICTAN-CSIC), C/José Antonio Novais, 10, 28040 Madrid, Spain; m.nardoia@izs.it

Grape skin is a source of polyphenols with antioxidant and antimicrobial properties. Little information is available regarding its application in animal feeds. An experiment was conducted to investigate the effect of inclusion of fermented (FS) and unfermented (UFS) grape skin at different doses (30 g/kg, FS30 and UFS30, and 60 g/kg, FS60 and UFS60) and of α-tocopheryl acetate (200 mg/kg) in broilers fed a corn-soybean diet. Growth performance, ileal protein digestibility, ileal and excreta content, digestibility of total polyphenols and tannins, thigh meat lipid oxidation (at 1 and 7 days of storage) and intestinal microflora in one hundred and fifty 21-day-old broiler chickens were determined. Growth performances were negatively affected by dietary FS60 and UFS60. Protein digestibility decreased in birds fed UFS60. The UFS supplementation increased ileal and excreta total polyphenols content; whereas ileal and excreta tannins content increased in all experimental groups. Birds fed UFS30 showed a higher value of polyphenols digestibility. Excreta digestibility of total polyphenols increased in birds fed FS and UFS compared to the control group, with higher value in birds fed UFS in comparison to those fed FS. Oxidative stability of thigh meat increased with the dietary addition α-tocopherol, while grape skin did not exhibit any protective effect on meat lipid oxidation. Intestinal microflora was not affected by dietary treatment. In conclusion, grape skin (at 60 g/kg) impaired chicken growth performance and protein digestibility. Total polyphenols content were more bioavailable in the large intestine than in the small intestine. The antioxidant potential of FS and UFS grape skin was not effective as vitamin E.

Session 37

Theatre 1

Future cattle housing with smart ventilation design and control

G. Zhang
Aarhus University, Inge Lehmanns Gade 10, 8000, Denmark; guoqiang.zhang@eng.au.dk

Global climate warming and environmental impacts are two major challenges for cattle and dairy production housing. Facing to the challenges, smart design and control of ventilation system is essential. The most interested zones of indoor climate and air quality control is within the occupant space since the indoor air space of farm animal housing is generally not perfect mixed and the thermal conditions around animal directly affect animals' wellbeing. To achieve optimal control of indoor environment, especially in animal occupant zone (AOZ) requires smart and complex ventilation, including precision zone ventilation concept. Objectives of the paper are to provide an overview of the technologies and methods used in precision zone climate control including inlet air jet momentum control, precision air supply and precision air exhaust; integrated Earth-Air Heat exchange; inlet design and setups; complex system design & control, local cooling & heating and smart floor for varied climate requirements. Potential of future technology developments and challenges are addressed. The paper provides also a few of examples on floor air speed and AOZ thermal comfort control in automatically controlled natural ventilation of cattle building to increase air speed in AOZ in summer and ensure low air speed in AOZ in winter, as well as pollutant source zone air exhaust to ensure optimal indoor air quality and transport the most pollutant air for cleaning treatment.

Session 37

Theatre 2

Assessments of earth-air heat exchanger (EAHE) for precision zone cooling in dairy housing

X. Wang and G. Zhang
Aarhus University, Inge Lehmanns Gade 10, 8000, Denmark; guoqiang.zhang@eng.au.dk

The application of earth to air heat exchanger (EAHE) as a supplementary partial air in-taking in natural ventilated cattle barn was investigated. A steady numerical model was developed and validated with published experimental data. A parametric analysis was carried out to find out the effects of air velocity, inlet air temperature, diameter of tube and length of tube on the thermal performance of EAHE. A model equation on the investigated four parameters was derived using response surface methodology after a series of numerical simulations. Given a certain requirement of temperature difference, an optimal design was proposed with the diameter as 0.3 m, the velocity as 2.5 m/s, and the length as 35 m. Under such situation, each tube could provide amount of airflow rate of 640 m^3/h that equivalent to the capacity required for 7 cows with a partial air in-taking of 20% designed total ventilation capacity in a Danish dairy housing.

Animal response (rumen-tempratue) based cattle cooling method ease heat stress

S. Goldshtein[1,2], H. Levit[2], S. Pinto[3], I. Halachmi[2] and Y. Parmet[1]
[1]Ben-Gurion University of the Negev., Industrial Engineering and Management., P.O. Box 653, Beer-Sheva 84105, Israel, [2]Institute of Agricultural Engineering, Agricultural Research Organization (A.R.O.), The Volcani Center, Bet Dagan, Bet Dagan, Israel, [3] Leibniz Institute for Agricultural Engineering an, Department of Engineering for Livestock Management, Leibniz Institute for Agricultural Engineering a, ATB, Potsdam 14469, Germany; halachmi@volcani.agri.gov.il

Heat stress in dairy cows can be a major issue resulting: lower food intake, reduced milk production, and lessen fertility causing framers losing income. One way of indication of heat stress can be the cow's vaginal temperature, if reaches 39.3 Celsius for over than 30 minutes. In order to predict those temperatures an experiment was conducted at the ARO Volcani research barn, Israel (August 2016). 13 cows of Holstein breed with an average of 3rd lactation and 141 days in milk in average were picked for the experiment. The cows sensrors were: vaginal temperature (as the dependent variable with a rate of 10 minutes), rumen temperature, ruminant time, steps, lying time, standing time, food intake, milk yield, milk fat and milk protein. Also climate data was collected: area temperate, relative humidity, wind speed and global radiation. All other variables were used as the predictors for the dependent variable (Vaginal temperature). Looking into each density plot of the vaginal temperature (for each cow) clued that in order to predict the letter an individual model is needed. We used three types of models: Multiple Linear Regression, LASSO Regression and Linear Mixed Models, the first two would produce 13 models for each specific subject and the last one will contain one model including all cows, this model also allows us learning new subjects as opposed to the first two models. a K-fold method was used in order to evaluate the best one, all models were measured with the RMSE (root mean square error) metric. Multiple Linear Regression RMSE (averaging all 13 metrics): 0.161 (°C), LASSO Regression RMSE (averaging all 13 metrics): 0.162 (°C) and Linear Mixed Models RMSE: 0.196 (°C). The first two singular models performed quite well predicting the vaginal temperature as opposed to the general model. But, they lack the function of learning a new subject with LMM can do that task.

Modeling the effects of heat stress of dairy cattle at farm scale

E. Galán[1], G. Pardo[1], E. Sanchis[2], F. Estellés[2], S. Calvet[2] and A. Del Prado[1]
[1]Basque Centre for Climate Change (BC3), Edificio Sede no.1, Planta 1ª, Parque científico de UPV/EHU, Barrio Sarriena s/n, 48940 Leioa, Spain, [2]Institute of Animal Science and Technology, Universitat Politècnica de València (UPV), Camino de Vera, s/n, 46022 Valencia, Spain; elena.galan@bc3research.org

Climate change will exacerbate the effects of warm climates on welfare, health, yield and survival of farm animals. There are few farm models that integrate impacts of climate change, GHG mitigation and adaptation. In the framework of the ERANET + OptiBarn project, our objective is to develop a module of heat stress effects in dairy cows permanently housed in the SIMSDAIRY model to use local climate data projections. The SIMSDAIRY model will be used to make scenarios of different economic situations (price changes of milk or concentrates) and different possibilities of adaptation (conditioning of stables or changes in the seasonality of births) and their relation with GHG and NH_3 emissions and welfare. To develop the module, we have reviewed the heat stress effects described in literature. At the productive level, we have discussed studies and models that relate the temperature-humidity index (THI) with yield decreases, fat and protein contents, as well as in fertility. At the level of welfare, the revision has been directed to modify the current welfare index of SIMSDAIRY according to THI ranges. In addition, we have modified the SIMSDAIRY sub-model in relation to the management of manure to make it more sensitive to temperature and ventilation according to the new SIMSWASTE model. We used current climate data from Valencia to test the performance of the equations selected from literature. The result is a farm-level production curve that declines in warm months and fertility decline that lengthens calving intervals. Also, warm temperatures affect animal excretion and NH_3 and CH_4 emissions from animal and manure management. In a scenario of seasonal calving we have decreased part of this effect. This method is already used in Bétera (Valencia), which is one of the study cases of the OptiBarn project, to avoid peaks of lactation curves in summer. The SIMSDAIRY model allows the door to be left open in the future to adapt the heat stress module to systems that include access to grass, in addition to the effects of acclimatization.

Influence of environmental climate conditions on animal welfare criteria of lactating dairy cows

T. Siemens[1], T. Amon[1], S. Pinto[1], J. Heinicke[1], S. Hempel[1], G. Hoffmann[1], C. Ammon[1] and I. Halachmi[2]
[1]Leibniz Institute for Agricultural Engineering and Bioeconomy (ATB), Department of Engineering for Livestock Management, Max-Eyth-Allee 100, 14469, Germany, [2]Agricultural Research Organization (A.R.O.) – The Volcani Centre, Institute of Agricultural Engineering, HaMaccabim Road 68, Rishon LeZion, Israel; tamon@atb-potsdam.de

Naturally ventilated barns (NVB) are particularly affected by climate-related changes. Considering the animals as 'biological sensors' the animal comfortable zone in a barn can be assessed as a condition for developing smart barns. This study was carried out at a NVB in Germany with on average 45 high lactating Holstein Friesian cows (first to eighth lactation, with 38.2 kg average milk yield). Cows were fed at 07:00 h and 11:00 h. Relative humidity and ambient temperature were measured at 10 places in at least 10-min intervals and used to calculate an average Temperature Humidity Index (THI) according to NRC. Respiration Rate (visually estimated in 30 seconds, hourly or twice a day between 07:00 h to 15:00 h), Rumination Activity (with a microphone recording rumination time in periods of 2 h) and Activity Behavior (IceTag3D™-Pedometer at the hind leg recording body posture every second) were used to characterize heat stress and its effects on lactating dairy cows. Cow-related factors and the THI influenced the respiration rate (RR). The mean RR differed between THI intervals. The RR in lying cows at normal and alert THI were greater than in standing cows at the same THI categories. An increase of RR was observed with increased yield. The rumination activity followed the feeding times, but was also affected by environmental climate conditions. Rumination time decreased strongest with increasing THI in the late afternoon. The duration of lying per day decreased by increasing THI and the duration of standing rose, while the duration of moving remained constant. The strongest changes in behavior were observed in the afternoon.

Microclimate monitoring as basis for smart cattle barns

S. Hempel, M. König, D. Janke, S. Pinto, T. Siemens, J. Heinicke, C. Ammon, B. Amon and T. Amon
Leibniz Institute for Agricultural Engineering and Bioeconomy (ATB), Department of Engineering for Livestock Management, Max-Eyth-Allee 100, 14469, Germany; tamon@atb-potsdam.de

Naturally ventilated barns, as often used for dairy cattle housing, are particularly vulnerable to climate change. Increased climate variability will result in a sub-optimal thermal environment impairing production and welfare of animals. In order to permit a targeted and efficient control of the microclimatic conditions, we use case study measurements and modelling to identify relevant variables and uncertainties in the assessment of microclimatic conditions. Measurements were conducted in case studies under on-farm conditions in Germany, Israel and Spain since 2015. Temperature and relative humidity were logged at least every 10 minutes simultaneously at different positions approximately 3 m above the barns' floor. The microclimatic conditions were evaluated using the temperature humidity index (THI) as introduced by National Academy of Science. Taking the device uncertainty estimated in lab experiments and the spatial variability into account a Monte Carlo approach was used to estimate a probability distribution for the measured THI values. Our study indicated that the uncertainty in the monitoring of microclimatic conditions inside naturally ventilated dairy barns is determined by the accuracy in the humidity measurements (±15% deviation from the spatial average were observed at individual sensor positions, while temperature varies only by approximately ±2 °C). In addition, we observed that devices used over several months under on-farm conditions are biased towards higher relative humidity values. Considering the estimated THI distributions it has to be noted that THI measurement conducted at individual positions inside naturally ventilated barns must be attributed with an uncertainty up to ±2 THI-scores. This value varies over time depending on the outdoor conditions (e.g. wind).

Compost bedded pack barns as a lactating cow housing system for the Southeast United States

J.M. Bewley, R.A. Black, F.A. Damasceno, E.A. Eckelkamp, G.B. Day and J.L. Taraba
University of Kentucky, Department of Animal and Food Sciences, 407 W.P. Garrigus Building, Lexington, KY 40546-0215, USA; jeffrey.bewley@uky.edu

A compost bedded pack barn is a lactating dairy cow housing system consisting of a large, open resting area, usually bedded with sawdust or dry, fine wood shavings. Bedding material is composted in place, along with manure, when mechanically stirred on a regular basis. Recently, the popularity of compost bedded pack barns has unquestionably increased in the Southeast (at least 80 compost bedded pack barns have been constructed in Kentucky. Because of warm climates, the compost bedded pack barn fits the Southeast particularly well. Galama suggested that compost bedded pack barns fit within goals of sustainable agriculture because of benefits to the cow (space, rest, exercise, and social interaction), the farmer (low investment, labor-extensive, reduced manure storage costs), and the environment (reduced ammonia and greenhouse gas emissions, odor and dust emissions, reduced energy consumption). Producers report reduced incidence of lameness and improved hoof health resulting from greater lying times and a softer, drier surface for standing. Cows may be more likely to exhibit signs of estrus because of improved footing on a softer surface, leading to improved heat detection rates. Compost bedded pack barns reduce the need for liquid based manure storage systems and provide producers with the option to economically transport nutrients in a dry, concentrated form to areas where there is an off-farm demand for nutrients. The initial investment costs of a compost bedded pack barn are lower than for traditional freestall or tie-stall barns, because less concrete and fewer internal structures (stall loops, mattresses) are needed, This system represents a viable entry option for smaller, start-up dairies. Proper composting increases the bedding temperature and decreases the bedding moisture by increasing the drying rate. Keeping the top layer of bedding material dry is the most important part of managing a compost bedded pack barn. The pack should be stirred at least two times per day. Stirring is typically accomplished while the cows are being milked, using various types of cultivators or roto-tillers. Poor management may lead to undesirable compost bed conditions, dirty cows, SCC, and increased clinical mastitis incidence. Proper management of compost bedded pack barns includes facility design, ventilation, timely addition of fresh, dry bedding, frequent and deep stirring, and avoidance of overcrowding.

Innovative housing for dairy cows

P.J. Galama, H.J. Van Dooren, H. De Boer and W. Ouweltjes
Wageningen University and Research, Livestock Research, De Elst 1, 6708 WD, the Netherlands; paul.galama@wur.nl

Different designs of new housing systems for dairy have been created with groups of farmers and experts on animal welfare and environment. Some are tested and monitored on practical farms or on research station Dairy Campus at Leeuwarden in the Netherlands. To introduce new housing systems it is important to reduce the emission of ammonia and green house gasses in the stable. Therefore new facilities are built on Dairy Campus in 2016. In six units of each 15 cows case control research is being carried out on different floor types. Emissions are being measured on a bedding with woodchips, but also a very new floor in a free walk housing system, an artificial floor. The first one is built in 2014 on a practical farm. It separates the faeces from the urine to reduce the ammonia emission. The faeces are picked up by a robot. Also floor types in cubicle stables and treatment of slurry are being measured on Dairy Campus. Results of low emission floors will be shown. A summary will be given of the research on 10 farms with a compost bedded pack barn about different aspects of sustainability: animal welfare, longevity, ammonia emission, N-losses, manure quality, milk quality and economics. They differ in farm design, type and management of the bedding and aerating system. Around 50% use wooden chips as bedding material. They compost this in the stable together with the faeces and urine of the cows. The heat of the composting process will stimulate the evaporation of the moisture. They control the composting process with an aerating system by blowing or sucking air through the bedding. If they manage the composting process well this is a sustainable housing system. In general the yearly costs for bedded pack barns with 15 m^2 bedding area per cow are higher than a cubicle stable, but depends a lot on the prices of bedding material. However with 10% lower cow replacement the yearly costs are lower. Future images of innovative housing systems will be displayed, as a result of different design processes with different stakeholders, with Free walk as base.

Free walk housing systems in development

A. Kuipers[1], P. Galama[2] and M. Klopčič[3]
[1]Expertise Centre for Farm Management and Knowledge Transfer WUR, P.O. Box 35, 6700 AA Wageningen, the Netherlands, [2]Wageningen Livestock Research, Postbus 338, 6700 AH Wageningen, the Netherlands, [3]University of Ljubljana, Animal Sciences Department, Biotechnical faculty, Groblje, Domzale, Slovenia; abele.kuipers@wur.nl

A 3-year EU-project is under way to research and further develop free walk cattle farming systems, which improve animal welfare and soil structure, utilize waste products and have public support. As innovative housing system, the compost bedded pack barn and a synthetic floor system (so called cow garden or high welfare floor) is applied with a completely free walking and lying area and is compared with cubicle barns for reference. The bedded pack barn uses organic waste materials as bedding, cultivated daily to improve the composting process. The synthetic floor separates urine from the manure and is cleaned by a robot. These systems have the potential to elevate the welfare and longevity of animals and improve soil quality. Integration with grazing provides a year round free walk farming system including outside exposure. Enthusiastic farmers' groups and NGO's are promoting the system, however, it has not yet been scientifically evaluated and questions are raised by local authorities and processors about effect on environment and product quality. Our approach will deliver an integrated assessment of case farms in 8 European countries, using experimental and modelling methods to evaluate systems performance. Housing as part of intensive and extensive farming will be examined in a holistic context, encompassing the whole farm: bedding, animal welfare, health, manure, soil, C P and N-balances, and product quality. Greater insight in the composting process plays a crucial role in the success of the system. Societal appreciation of housing, farming system and products will be assessed. Higher capital efficiency is studied by keeping other species and horticulture on the compost bedding during the grazing season. To integrate the results of the various research activities, a systems analysis and economic evaluation will be performed. The outlay of this project will be presented and field examples of and preliminary experiences with innovative free walk housing and other systems will be shown.

Climate change related heat stress impact on milk yield of dairy cattle in the United Kingdom

A. Foskolos[1], C.F.E. Topp[2], J.M. Moorby[1], C.H. Foyer[3] and N. Fodor[3]
[1]Aberystwyth University, Institute of Biological, Environmental and Rural Sciences, Gogerddan, SY23 3EE, Aberystwyth, United Kingdom, [2]Scotland's Rural College, Crop and Soil Systems Group, West Mains Rd. Edinburgh, EH9 3JG, Edinburgh., United Kingdom, [3]University of Leeds, Faculty of Biological Sciences, University of Leeds, LS2 9JT, Leeds, United Kingdom; anf20@aber.ac.uk

Climate change is expected to cause increased temperatures (T) affecting animal's welfare and productivity. Pasture-based dairy systems, such as those in the UK, are vulnerable to environmental changes. Our objective was to quantify the impact of climate change induced heat stress on milk production of lactating cattle on pasture. Further, we assessed the adequacy of modelling methodology to quantify cattle responses. Climate data was obtained from the UKCP09 project. The 11-member projection ensemble is based on the SRES A1B medium emission scenario, and contains daily climatic data in a 25 km spatial resolution. The daily milk loss was calculated for each grid cell as follows: (1) A T humidity index (THI) was calculated using either T and dew point T or T and relative humidity, (2) Milk losses were quantified at either a daily or sub-daily time scale, and (3) THI thresholds were set at 68, 70 and 72. The annual milk loss (AML) showed a large spatial variance across the UK. Currently, the average AML is around 1 and 40 kg/cow in the north and south, respectively. However, AML will increase by the end of the century, exceeding 150 kg/cow in SE England. Both daily and sub-daily step methods showed an increase of AML through time, but the sub-daily step methods projected a much more significant raise of AMLs, which in present situations was unrealistically high. In conclusion, small AML were projected at the end of the century suggesting that moderate adaptation measures may be needed. However, some regions, and in relation with increased occurrence of extreme weather conditions, may experience severe heat stress, and more sophisticated adaptation measures should be considered. Further, the daily step method seemed to be more adequate for British conditions. Acknowledgements: Support was provided through the Sêr Cymru NRN-LCEE project Cleaner Cows.

Effect of monochromatic light on the performance and egg quality of laying hens

F. Yenilmez[1], N. Saber[2], U. Serbester[2] and L. Celik[2]
[1]*Cukurova University, Vocational School of Tufanbeyli, 01640 Tufanbeyli, Adana, Turkey, [2]Cukurova University, Agriculture Faculty, Department of Animal Science, 01330 Balcalı Sarıçam, Adana, Turkey; fyenilmez@hotmail.com*

The present study was conducted to investigate whether monochromatic light affect feed intake, egg production and egg quality of laying hens. In this study, it was used white and green fluorescent lamps to illuminate Brown-Nick hens for 8 wk. Light sources were equalized to a light intensity of 15 lx. 47 weeks old layers were divided into 2 groups of similar mean weight comprising 16 birds each and housed in individual cages. A 16:8 hours light:dark photoperiod was employed. The birds were housed in an air-conditioned room at a conventional ambient temperature (15-25 °C) with a relative humidity of 50-60% for 24 h per day. Laying performance was assessed by recording feed intake, egg weight, egg production daily and egg quality weekly of the experiment. The results indicated that green light application did not have significant (P>0.05) effects on feed intake, feed conversion efficiency, egg production, total number of eggs, total egg weight, average egg weight and egg quality criteria on laying hens. However, the green light application affected egg yolk colour a (redness) (P<0.01) and b (yellowness) values (P<0.05). The results suggest that green light application may improve egg yolk colour score of laying hens.

The effect of housing system and gender on performance and carcass traits of growing rabbits

L. Zita[1], Z. Volek[2], M. Jeníková[1], L. Volková[2] and O. Krunt[1]
[1]*Czech University of Life Sciences Prague; Faculty of Agrobiology, Food and Natural Resources, Department of Animal Husbandry, Kamýcká 129, 165 00, Prague – Suchdol, Czech Republic, [2]Institute of Animal Science, Department of physiology of nutrition and quality of animal products, Přátelství 815, 104 00, Prague – Uhříněves, Czech Republic; volek.zdenek@vuzv.cz*

The aim of this work was to study the effect of different housing systems (wire-net cage vs pen; 0.15 m² per rabbit) and gender (male vs female) on the growth performance and carcass traits of fattening rabbits. A total of 110 Hyplus rabbits (PS 19 × PS 39) of both sexes (1:1) were weaned at 36 days of age (on average 908±11 g). The rabbits were provided with a standard fattening diet. They were randomly divided into 4 experimental groups (by housing system and by gender). The final rabbit's live weight (at 80 d of age), as well as daily weight gain were affected by the housing system (P≤0.001) and gender (P≤0.05). Feed intake and feed conversion ratio were not affected by housing systems and gender. The females (gender effect) and cage-housed rabbits (housing system effect) showed the higher slaughter weight (P≤0.05) compared to other rabbits. There was a higher hind part to reference carcass ratio in females (by +1.0 percentage points; P≤0.05) than in males. There was a higher hind part to reference carcass ratio in pen-housed rabbits (+1.6 percentage points; P≤0.001) than in cage-housed rabbits. Similarly, the pen-housed rabbits demonstrated the both significantly higher thigh to reference carcass ratio (+0.7 percentage points) and thigh muscle to reference carcass ratio (+0.5 percentage points). It can be concluded that regardless of gender, the productive performance was better in cage-housed rabbits. The housing of rabbits in the pen affected the carcass characteristics in a beneficial way. This research was supported by the Ministry of Education, Youth and Sports of the Czech Republic ('S' Project) and Ministry of Agriculture of the Czech Republic (Project No. MZER00714).

Reduction of emissions from storage of natural fertilizers in dairy farming

J. Walczak and P. Wójcik
National Research Institute of Animal Production, Technology, Ecology and Economics, Krakowska 1, 32-083 Balice,
Poland; jacek.walczak@izoo.krakow.pl

According to the Nitrate Directive and NEC Directive guidelines, it is necessary to determine efficient methods for reducing the accumulation of biogenic compounds as well as gas emissions from cattle raising. Fertilizer composition (N, P) and gas emissions in dairy cattle farms were compared. A total of 960 t of fertilizers were used in the study. Three sites of storage were investigated between February and April. Manure was stored on a manure pad (GK), a sheet-covered pad (GD1) and in a heap aerated by turning (GD2). Slurry was stored in an open tank (GK1), a tank tightly covered with sheet, (GD3), a tank with a height/diameter ratio of 3/1 (GD4), and a tank with a natural surface crust (GD5). All the groups were kept under natural conditions. Continuous measurements based on FID meters were made using the wind tunnel method. The higher concentration of N-NO$_3$ in manure GD1 (4.81 vs 3.3 kg/t) resulted from the lower emission mainly of ammonia (0.072 vs 0.24 kg/t GK). Statistically significant differences were found for N$_2$O emission (0.012 kg/t). Anaerobic conditions caused methane emissions to increase to 7.64 kg/t. In GD1, lower release of phosphorus from the manure heap was observed. Systematic turning of the heaps (GD2) increased NH$_3$ and N$_2$O emissions, and the differences were significant. The emissions resulted in a lower concentration of nitrogen compounds in the manure (2.93 kg/t). The study showed the highest reduction effect for NH$_3$ when the tank was tightly covered (GD 3). A reduction in the emission of this gas from 0.45 kg/t (GK) to 0.095 kg/t translates into an emission reduction of as much as 80%. This method was equally effective for N$_2$O (0.047 vs 0.062 kg/t). The expected effect of emission reduction was the higher nitrogen concentration in the slurry (4.3 vs 3.5 kg/t). However, CH$_4$ emission increased from 18.45 to 16.2 kg/t. The effect of GD4 was confirmed for all the three gases analyzed. Development of a natural crust layer on the tank surface was less effective than covering, but more efficient than increasing the height of the tank. It is possible to reduce GHG and NH$_3$ emissions simply and at a low cost, but this increases the load of N and P applied to the soil.

Relationship between herd life and housing type in Holsteins in Japan

Y. Nakahori[1], S. Yamaguchi[2], H. Abe[2], S. Nakagawa[2], T. Yamazaki[3], T. Baba[4] and K. Hagiya[1]
[1]Obihiro University of Agriculture and Veterinary Medicine, Obihiro, Hokkaido, 080-8555, Japan, [2]Hokkaido Dairy
Milk Recording and Testing Association, Sapporo, Hokkaido, 060-0004, Japan, [3]Hokkaido Agricultural Research
Center, NARO, Sapporo, Hokkaido, 062-8555, Japan, [4]Holstein Cattle Association of Japan, Hokkaido Branch, Sapporo,
Hokkaido, 001-8555, Japan; logi-universe@myyn.sakura.ne.jp

Extending the productive life of cows makes dairy farming more effective. In Japan, Herd Life (HL) has been used as indicator of longevity in the genetic evaluation of dairy bulls since 2006. In previous studies, longevity traits typically were incorporated as single traits regardless of management practices. However, many styles of cow management are used in Japan, and many traits, including Milk Yield (MY), type traits, and longevity traits, can vary according to housing type. In this study, we managed HL as different traits that reflected three housing types (tie stall [TS], free stall [FS] and grazing [GZ]) and estimated the genetic correlations among HL, MY, somatic cell score (SCS) and 6 type traits (feet and legs, chest width, angularity, udder support, udder depth, and front teat placement) of 46,241 Holstein cows that experienced first calving during 1993 through 2008 in Hokkaido, Japan. The HL was defined as the total number of months from birth until culling at 84 months of age. We used a multiple-trait animal model and the GIBBS2F90 program for the analysis. The model included herd year, age at first calving, month of calving, lactation stage, and herd × classifier × classified-day as fixed effects and the animal's breeding value and residual error as random effects. Genetic correlation but not residual covariance was considered among HLs. The mean HL was 67.6 months in TS, 64.4 months in FS and 69.6 months in GZ. The genetic correlation among HLs was 0.65 for TS × FS, 0.33 for TS × GZ and 0.22 for FS × GZ. In addition, genetic correlations between HLs and the traits of MY, chest width and udder support differed significantly among housing types. Our results indicate that HL in various housing types should be treated as different traits in genetic evaluations.

The role of advanced genetics in the equine sector

H.P. Meier

Verband Schweizer. Pferdezuchtorganisationen, Baumgaertliweg 17, 3322 Urtenen-Schoenbuehl, Switzerland;
mail@hanspetermeier.ch

Thanks to the ecologically beneficial employment of the horse in agriculture, forestry, transportation and the army in earlier times, horse breeding was of major importance. However, today's geneticists, progressive breeders and veterinarians consider the state of horse breeding as outdated. We therefore have to reconcile the possibilities of modern genetics with the valuable traditional experience. First steps in this direction already showed flaws of conventional methods, and the breeders are conscious of the fact that existing selection procedures could have hardly been in relation to gene expression. Requirements for the utilization of modern genetic methods are given in so far, as the necessary data relating to identity and performance of the animals are available with many breeding organizations (though in different qualities). The heritabilities of many traits have also been estimated already, so we have some basis for assessing where to employ advanced genetic and genomic methods sensibly. In that case, it is to be expected that they will play a central role. However, forensic and ethic aspects also have to be considered from the beginning. In consideration of the efficiency of today's possibilities, these requirements and the principles of breeding must be taken care of more strictly than ever. Breeding goals must be clearly defined and binding in crucial parts. We may assume that mainly the ability for physical performance will be required, both of sporthorses and working equids. The most important preconditions are soundness, longevity and durability, and it is highly advisable to impart the necessary knowledge to the breeders right now and to aim at such targets in the future. Additionally, optimal environmental conditions have to be provided to make best use of their great influence on the development of the progeny. Contrary to the situation in the field of livestock breeding, the fertility in horse breeding is still modest and significant progress cannot be expected for the present. However, advanced genetics may facilitate improvements here, too. In any respect, a scientifically based cooperation between breeders, veterinarians and geneticists will be of greatest importance, and for the promotion of progress, open-mindedness is essential in the whole equine industry.

Genetic analysis of Selle Français conformation and gaits data

M. Sabbagh[1], A. Ricard[1,2], S. Danvy[1], G. Blanc[1] and B. Chaigne[3]

[1]Institut Français du Cheval et de l'Equitation (IFCE), La jumenterie du Pin, 61310 Exmes, France, [2]INRA, UMR1388 – Genetics, Physiology and Livestock Systems, 24, Chemin de Borde Rouge BP 52627, 31326 Castanet Tolosan Cedex, France, [3]Stud Book Selle Français, 56 Avenue Henri Ginoux, 92120 Montrouge, France; margot.sabbagh@ifce.fr

Between 2005 and 2014, the Selle Français studbook collected 34,604 scores for conformation and gaits on 21,114 young horses. These horses descended from 12,058 mares and 1,407 stallions. The horses could be judged at 2 and/ or 3 years of age and several times the same year at local, regional and national shows. The aim of this study was to estimate genetic parameters for judgements during shows. The horses were rated on 10 conformation, 3 free gaits, and 4 free-jumping criteria, and for 3-year-old horses additionally on 4 criteria for presentation under saddle. Several fixed effects were included in the model: age in days (4 classes), sex (3 classes), year of judgment (10 classes), category of the test (6 classes), scale of competition (3 classes) and the competition, i.e. combination of: place, date and judges (1,082 classes). The heritability estimates ranged from 0.08 (behavior under saddle) to 0.43 (chic). Among the 21 heritability estimates obtained, only 2, for neckline and chic, were high: 0.42 and 0.43, 13 values were between 0.20 and 0.40, and the other 6 were weak, below 0.20. The criteria associated with presentation under saddle were quite heritable: balance 0.14, style 0.17 and availability 0.08, as were the criteria describing the stands 0.11 and joints (front 0.12 and back 0.19). This first analysis of Selle Français data will allow to publish individuals' indices on morphological traits and gaits.

Genetics of linear traits for specifying and enhancing breeding programs for sport horses

K.F. Stock[1], I. Workel[2] and W. Schulze-Schleppinghoff[2]
[1]*IT Solutions for Animal Production (vit), Heinrich-Schroeder-Weg 1, 27283 Verden (Aller), Germany,* [2]*Oldenburger Pferdezuchtverband e.V., Grafenhorststrasse 5, 49377 Vechta, Germany; friederike.katharina.stock@vit.de*

With the wider implementation of linear profiling of conformation and performance in sport horses, refined phenotypic information have become routinely available for the studbooks. Genetic proofs for linear traits can valuably support targeted selection decisions and can retrospectively used for monitoring realized breeding strategies. For this study, results from the prototype genetic evaluation for linear traits of the Oldenburg studbooks were available and allowed characterizing distinct groups of sires defined by discipline (dressage, show jumping) and popularity (numbers of progeny). Breeding values (BV) were estimated in multiple trait linear repeatability animal models using linear data collected from foals, mares and stallions in 2012-2016. Depth of the 12,931 linear profiles varied depending on the assessment conditions, with maxima in performance tested mares and stallions presented for approval with testing under saddle. Total numbers of traits in genetic analyses were 46 for conformation, 22 for gaits, 13 for jumping, and 5 for special remarks and behavior. Of the 1,812 sires with on average 7 linearly described progeny, 931 sires were represented with foals and 1,268 sires with mares and/or stallions. Analyses of BV distributions of the stallions revealed clear indications of more extensive use of stallions with genetic potential for certain conformational characteristics like long legs and arched neck shape. Patterns of BV for performance related linear traits showed differences between disciplines, with obvious selection effects in for example freedom of shoulders and hind limb activity in trot and in free jumping characteristics. The results of this study indicated that different breeding goal related aspects of conformation and performance varied considerably in their genetic progress and impact on the breeding use of stallions. With the availability of genetic and future genomic proofs for linear traits, more targeted breeding progress can be achieved in the breeding programs for sport horses.

Managing the non-placed horses records in endurance performance genetic evaluation

S. García-Ballesteros[1], J.P. Gutiérrez[1], A. Molina[2], M. Valera[3] and I. Cervantes[1]
[1]*Departamento de Producción Animal, Universidad Complutense de Madrid, Avda. Puerta de Hierro s/n, 28040, Madrid, Spain,* [2]*Departamento de Genética, Universidad de Córdoba, Ctra. Madrid-Córdoba km 396a, 14071, Córdoba, Spain,* [3]*Departamento de Ciencias Agroforestales, Universidad de Sevilla, Ctra. de Utrera km 1, 41013, Sevilla, Spain; icervantes@vet.ucm.es*

The racing time and rank at finish traits are commonly used for endurance horses breeding programs as measure of their performances. Endurance competitions are carried out cross country where the adapted physical and metabolic conditions of the horses are essential to reach the goal. Often, only horses which finish the race are included in the genetic evaluation models. However, the exclusion of non-placed horses from the dataset could influence the prediction of individual breeding values. The objective of this work was to develop a multiple trait model including time and a binary trait (placed, 2 or not, 1) to improve the genetic evaluation in endurance competition in Spain. The database contained 3,286 records from 933 horses. The percentage of horses that completed the ride was 69%. The pedigree contained 8,733 animals (47% Arab Horses). All models included gender, age and race as systematic effects and interaction horse-rider, animal and residual as random effects. Kilometres per race was including as covariate. TM program was used to run univariate, bivariate and trivariate models. The placed or not trait had a heritability of 0.09, the time 0.10 and the rank 0.27. The correlation between time and the binary trait was positive with a low-medium magnitude (0.35). The combination of these three traits to predict the breeding values of endurance horses is proposed. This could allow increasing the accuracy in the breeding values. The binary trait will be used as the probability of the horse finishing the race or its breeding value could valuably supplement time and rank in the current genetic index.

Imputation of high density genotypes from medium density genotypes in various French equine breeds

M. Chassier[1], E. Barrey[1], C. Robert[1], S. Danvy[2], A. Duluard[3] and A. Ricard[1]

[1]GABI, INRA, AgroParisTech, Université Paris Saclay, 78352 Jouy-en-Josas, France, [2]IFCE, Direction Développement et Recherche, La Jumenterie du Pin, 61310 Exmes, France, [3]Le Trot, 7 rue d'Astorg, 75008 Paris, France; marjorie.chassier@inra.fr

The objective of the study was to estimate the efficiency of imputation of genotypes of the Affymetrix Axiom Equine genotyping array, including 670,806 SNP, from genotypes of the Illumina Equine SNP50 BeadChip including 54,602 SNP and the Illumina Equine SNP-74k-chip including 65,157 SNP. Genotypes were avalaible from 5 breeds: Arabs (AR, 1,207 horses), Selle Français (SF, 1,978 horses), Trotteur Français (TF, 979 horses), Anglo-Arabs (AA, 229 horses), and various foreign sport horses (FH, 209 horses). In AR, 15% of horses were genotyped with the high density chip, in SF 57%, in TF 30%, in AA 10% and FH 15%. A validation set equal to the third of the sample was drawn in horses genotyped with high density chip and their genotypes blinded. Sampling was made to mimic the parental structure of the target population (the one genotyped with medium density), i.e. at random in SF and AR and with a genotyped sire for TF. Two strategies were compared with a reference population of all breeds and with only the same breed as the validation set. The software FImpute was used. First results in TF showed better concordance with a validation set of the same breed: mean of 98.0% of correctly imputed genotypes (s.d. 2.3%) versus 96.0% (s.d. 5.6%) with a validation set of all breeds. Theses results were compatible with the use of imputed data for future GWAS and genomic evaluation for these data.

Genomic structure of the horse MHC class II region resolved using long-read sequencing technology

A. Viļuma[1], S. Mikko[1], D. Hahn[1], L. Skow[2], G. Andersson[1] and T.F. Bergström[1]

[1]Swedish University of Agricultural Sciences, Department of Animal Breeding and Genetics, Ulls väg 26, 75007 Uppsala, Sweden, [2]College of Veterinary Medicine, Texas A&M University, Department of Veterinary Integrative Biosciences, College Station, TX 77843, USA; agnese.viluma@slu.se

The Major Histocompatibility Complex (MHC) region harbors multi gene families that encode class I and class II antigen presenting molecules. The classical MHC class II molecules (DR, DQ and DP) bind and present exogenous peptides to circulating CD4[+] T-helper cells to initiate immune response against pathogens. Specific alleles of the human MHC class II genes have been associated to susceptibility and resistance to infectious, inflammatory and autoimmune disorders. Susceptibility to Equine Sarcoids, Summer Eczema and Equine Recurrent Uveitis has been associated to genetic markers located in MHC class II region. The structural complexity of the MHC has made it difficult to construct a reliable reference sequence of the horse MHC region. In this study, we used long-read single molecule real-time sequencing technology from Pacific Biosciences to sequence eight Bacterial Artificial Chromosome (BAC) clones spanning the horse MHC class II region. Obtained BAC clone assemblies were annotated using the gene prediction software AUGUSTUS with incorporating publicly available cDNAs, ESTs, RNA-seq data and homologous protein sequences. The final assembly resulted in a 1,165,328 bp continuous gap free sequence with 35 manually curated genomic loci of which 23 were considered to be functional and 12 to be pseudogenes. In comparison to the MHC class II region in other mammals, the corresponding region in horse showed extraordinary copy number variation and different relative location and directionality of the *Eqca-DRB*, *-DQA*, *-DQB* and *-DOB* loci. High synteny of the non-classical MHC class II genes was observed. This is the first long-read sequence assembly of the horse MHC class II region with rigorous manual gene annotation, and it will serve as an important resource for association studies of immune-mediated equine diseases and for evolutionary analysis of genetic diversity in this region.

Patterns of gene flow in the Slovak Lipizzan population

N. Moravčíková, R. Kasarda and O. Kadlečík
Slovak University of Agriculture in Nitra, Department of Animal Genetics and Breeding Biology, Tr. A Hlinku 2, 94976
Nitra, Slovak Republic; nina.moravcikova1@gmail.com

The aim of this study was to evaluate the current state of genetic diversity, population subdivision, migration events and effective population size of the Slovak Lipizzan population with respect to the populations (studs) of six other countries representing the gene pool of European Lipizzan horses. The genotyping data based on 13 microsatellites have been collected from a total of 951 individuals. The collected sample of 103 Slovak animals represents the nucleus of Lipizzan horses kept in Slovakia. Across all microsatellite loci the average number of alleles was 5.74 and effective allele number was 3.14. Population-level statistics indicated sufficient levels of genetic diversity with expected heterozygosity ranging from 0.63 (Italy) to 0.69 (Slovakia). The average value of Shannon's information index (I=1.28) confirmed the high degree of within population genetic diversity. The number of private alleles, providing the evidence for unique diversity in individual populations, ranged from 0.62 (Slovakia) to 0.08 (Croatia). Evidence of potential inbreeding based on the Wrights F_{IS} index was not obtained. The LDNe method was able to estimate effective population size (N_e) with finite C.I. for each of evaluated populations. Obtained low estimates of N_e (ranging from 32.4 to 131.5) suggested small founder populations or extensive losses of genetic variation through random genetic drift. However, the analysis of bottleneck did not show significant putative bottleneck events. The genetic differentiation at the inter-population level tested by pairwise F_{ST} coefficients and Nei's genetic distances indicated the formation of three main clusters in respect to population's genetic background. The Hungarian population showed highest level of genetic similarity to the Slovak horses resulting also from the obtained values of recent migration rates. Generally, the strongest evidence of recent migration was identified between Austrian and Italian populations. Understanding connectivity among fragmented populations is of paramount importance in breed conservation because it determines their long-term viability and helps to identify and prioritize populations to conserve.

Mapping genes that regulate trotting performance using a unique Nordic horse model

B. Velie[1], K. Fegraeus[1], J. Meadows[2] and G. Lindgren[1]
[1]Swedish University of Agricultural Sciences, Department of Animal Breeding & Genetics, Ulls vag 26, 75007, Sweden,
[2]Uppsala University, Department of Medical Biochemistry & Microbiology, 75236 Uppsala, Sweden; brandon.velie@slu.se

The origin of the Coldblooded trotter (CBT) provides a unique opportunity to identify genes influencing trotting performance. The CBT originates from the North Swedish draught horse (NSD) and these two breeds retain high levels of genetic similarity (Fst=0.08). However, prior to the introduction of paternity testing in 1969, crossbreeding with Standardbreds (SB) was used to improve CBT performance. We hypothesize that the gains in CBT performance over the last 50 years may in part be explained by the maintenance of favorable genetic variants originating from the SB. As such, the aim of the current study was to compare the genetic makeup of these three breeds and to identify genetic footprints of athletic performance. A preliminary study using a sliding window Delta Fst analysis was performed across all breeds using data generated from the equine SNP50K array (CBT, n=11; NSD, n=19; SB, n=12). Analyses revealed five key regions where the CBT and SB were genetically similar, but both differed from the NSD. These regions included the tsukushi (TSKU) and F-box and leucine rich repeat protein 4 (FBXL4) genes. Based on the relationship between the CBT and the NSD as well as the intensive selection for harness racing performance in the CBT, we believe that these genes are important for trotting performance. This proposed association is currently being followed up in 475 randomly selected CBTs, and the results for these genes and any other potential candidate genes will be presented at the meeting. The results of this study are likely to provide important information regarding the genetic regulation of trotting performance and will undoubtedly assist in the selection of CBTs better suited for harness racing.

More from less: reaping benefits from biology and management at all farming system levels
L. Puillet
UMR MoSAR, INRA, AgroParisTech, Université Paris-Saclay, Paris, France; laurence.puillet@agroparistech.fr

Producing more from less is a key issue for animal production. It is even more challenging for small ruminant production: these farming systems are generally located in harsh areas where environmental conditions are constraining and variable. Thus, in addition to being efficient, small ruminant systems also have to be resilient in order to cope with perturbations. Efficiency and resilience are complex and dynamic properties, emerging from a great diversity of mechanisms, operating at various scales, from a single cell until network of farms within a territory. This report illustrates how we can reap the benefits from this diversity with examples concerning modelling approaches at animal, herd and farm level. These examples highlight three key aspects of efficiency and resilience. The first aspect is the need to clearly define the level of organization at which efficiency and resilience are considered. Being efficient or resilient at a low level, for instance the animal, doesn't imply that the property will be conserved when upscaling at upper level, for instance the farm. The second aspect is the need to consider both biological and management levers to favour regulating mechanisms and compensate environmental perturbations (adaptive capacities at animal level, individual variability at herd level, complementarity between crop and livestock components at farm level). The third aspect is the need to consider the time dimension: options for enhancing efficiency or resilience in the short term can be detrimental in the long term. Finally, examples presented in this report highlight that achieving 'producing more from less in a changing environment' is not an issue of producing knowledge per se but rather an issue of combining knowledge across disciplines.

Dairy sheep farmers cope with production and adaptation through breeding management at flock level
L. Perucho[1], A. Lauvie[2], C. Ligda[3], J.C. Paoli[1] and C.H. Moulin[4]
[1]INRA, LRDE, Quartier Grossetti, 20250 Corte, France, [2]INRA, SELMET, 2 Place Pierre Viala, 34060 Montpellier, France, [3]Hellenic Agricultural Organization, VRI, 57001, Thermi, Greece, [4]Montpellier SupAgro, SELMET, 2 Place Pierre Viala, 34060 Montpellier, France; lola.perucho@inra.fr

Local breeds or specific adaptation traits of breeding animals are commonly considered as levers to handle constraints and opportunities of the farming activity. However, the breeding management at the flock level, either intra-breed or through crossbreeding, is rarely studied. The objective of this work is to understand how breeding management could be a lever used by dairy sheep farmers to cope with the issues perceived as major ones in their activity. The Mediterranean regions of Corsica (France) and Thessaly (Greece) were chosen for the data collection. We performed semi-structured interviews with farmers (n=40 in Corsica and n=46 in Thessaly) to characterize farming conditions and flock management, deepened through follow-up on 7 case-studies in Corsica and 15 surveys on changes in practices through time in Thessaly. Data treatment consisted in building typologies of breeding management, that enable joint analysis of the farmers' practices and the issues tackled through these practices. Labour management, balance between the different sources of income and feeding inputs minimization were the main issues that farmers try to cope with through their breeding management practices. In Corsica, trade-off between rigor of replacement and culling processes and selection on udder traits were levers for labour management in large-sized flocks or hand-milked flocks. In Thessaly, the switch from local breeds towards crossbred flock including high productive breeds was a lever to cope with changing opportunities of income brought by meat, milk and crop products. In both field, minimization of stored feed was pursued by exploiting the abilities of local breeds: ability to reproduce after transhumance, low feed intake, etc. The orientation of the breed, studied through on-farm practices at the intra-breed and inter-breed level, questions how farmers balance production and adaptation to the main issues of their farming activity and gives elements of reflection for the selection goals considered in selection schemes.
hier

Updating requirements and multiple responses to diets of dairy goats to better tackle resilience and

D. Sauvant and S. Giger-Reverdin
INRA-AgroParisTech, 16 Rue C. Bernard, 75005 Paris, France; sauvant@agroparistech.fr

Inra nutrients requirements and allowances previously proposed for feeding goats have been updated noticeably to better predict the animal multiple responses to nutrients, a key aspect to model animal resilience to dietary variations. This update was first based on a complete digestive model which was applied to the database 'Caprinut' of experiments carried out on goats (158 experiments, 462 treatments). All the new values were obtained through meta-analyses of datasets. The energy maintenance requirements were updated with data from calorimetric studies. Simultaneously, efficiency of Metabolizable Energy (ME) conversion to only Milk Net Energy (NEL) (kls) was computed in a common approach pooling goat and cow results. A similar approach was used for the coefficient ktg, the common efficiency of ME into Body Reserves (BR) and that of BR into NEL. Moreover, an entirely new function was derived to account for the evolution of BR driven by homeorhesis along the cycle gestation-lactation. For metabolizable protein (PDI), non-productive requirements were entirely updated. The efficiency of PDI into milk protein is no longer assumed constant but varies with a common value applied to all the processes of protein synthesis occurring in the body. Taking into account all these proposals, it was possible to predict faecal and urinary losses of N within the process of diet formulation. For energy and protein requirements of gestation and growth, entirely new approaches were also proposed to take into account recent data including slaughter trials. Beyond the covering of requirements, a set of equations was proposed to predict the response of intake, milk yield and milk composition to variations in concentrate, or energy, or PDI supplies centred around the requirements in energy or protein. For grazing and range animals, available information were pooled to provide a consistent framework of proposals to manage more precisely goat feeding at pasture. In conclusion, most of the aspects of goat nutrition were renewed to be included in diet formulation or in models of improved goat management.

Are mixed cattle-sheep farms more sustainable than specialized farms under economic risks?

C. Mosnier, P. Veysset and C. Verdier
INRA, Theix, 63122 Saint Genès Champanelle, France; patrick.veysset@inra.fr

France has the largest herd of suckler cows in Europe and the third flock of suckler ewes. The number of suckler ewes is declining since the 1980s while the number of suckler cows has raised until 2000 and is now rather stable. Since 1990, sheep prices have declined and have been variables in loewer extend comprd to beef prices. Concomitantly, the Common Agricultural Policy has implemented different policies to reduce market risks and to support livestock production, favoring sometimes sheep production, sometimes beef production. Diversification is generally seen as a way to stabilize income and use more efficiently farm resources. Beef and sheep were frequently managed in the same farm, namely in the north of the French Massif Central, however, farms are specializing more and more. The objective of this study is to analyze the impacts of the degree of sheep and beef mix on farm sustainability over the period 1995-2015. Simulations have been performed with the bio-economic model Orfee. It represents an annual farm production at equilibrium but with economic results calculated annually, for each market and policy situation that occurred between 1990 and 2015. Beef, sheep and crop productions, buildings and machines are optimized under economic risks to maximize average net profit balanced with its standard deviation or its average value below a target level. The Scenarios tested are: (S1) only beef can be raised, (S2) the livestock units must be half beef and half sheep; (S3) only sheep can be raised; (S4) sheep and beef productions are freely optimized. The production context refers to the north of the French Massif Central. Farm land endowment is set at 125 ha of which 95 ha are permanent grasslands. The results show that before the 2010 CAP reform, beef production brings higher average profit per worker unit than sheep production. Nonetheless since 2010, it is the opposite situation. The mixed livestock system (S2) provides economic results more stable than specialized ones. The optimal herd composition over the studied period is around 20% of sheep livestock unit to value grasslands in winter and to balance average profit and risk.

Supplementation of natural or synthetic vitamin E on α-tocopherol concentration in lamb tissues

L.N. Leal[1,2], S.K. Jensen[3], J.M. Bello[4], L.A. Den Hartog[1,2], W.H. Hendriks[1] and J. Martín-Tereso[2]
[1]Wageningen University, Department of Animal Science, De Elst 1, 6708 WD Wageningen, the Netherlands, [2]Trouw Nutrition Research and Development, Veerstraat 38, 5831 JN Boxmeer, the Netherlands, [3]Aarhus University, Department of Animal Science, Blichers Alle 20, 8830 Tjele, Denmark, [4]Nanta S.A., Ronda de Poniente 9, 28460 Tres Cantos, Spain; leonel.leal@trouwnutrition.com

Based on the international unit system a difference in efficacy of 1.34 is used to discriminate between RRR- and all-rac-α-tocopheryl acetate. The current study aimed to determine the effect of supranutritional doses of RRR-α-tocopheryl acetate or all-rac-α-tocopheryl acetate on α-tocopherol deposition in lamb tissues. One hundred and eight Rasa Aragonesa breed lambs were fed diets supplemented with four levels of all-rac-α-tocopheryl acetate (250, 500, 1000 and 2,000 mg/kg compound feed), four levels of RRR-α-tocopheryl acetate (125, 250, 500 and 1000 mg/kg compound feed) or the basal diet without vitamin E supplementation. The diets were fed the last 14 days prior to slaughtering at 25.8±1.67 kg BW. Within 20 min after slaughter samples of muscle, heart, liver, brain and spleen were placed on a foil bag and frozen at -20 °C until α-tocopherol analysis. A non-linear regression procedure (exponential response curve) was carried out to quantify the effect of vitamin E supplementation on α-tocopherol deposition in tissues. Increased supplementation of vitamin E (for both sources) led to a significant increase (P<0.001) in α-tocopherol concentration in the selected tissues and organs fitting curves of diminishing response towards a horizontal asymptote. Moreover, when comparing the maximum slopes of the response (at dose=0), the relative response to RRR-tocopheryl acetate as compared to all-rac-α-tocopheryl acetate was 1.43 in muscle (P<0.001), 0.91 in liver (P<0.001), 1.16 in spleen (P<0.001), 0.73 in brain (P=0.12) and 1.41 in heart (P<0.01). This study illustrates that the generally accepted equivalence value between these sources of 1.34 is inadequate to predict differences in α-tocopherol deposition in tissues of lambs fed supranutritional doses of these vitamin E forms.

Effect of melatonin on reproductive performance in small ruminants

B. Alifa, A. Hamrouni, A. Najjar and M. Djemali
National Institute of Agriculture of Tunisia, Animal Science, 43 Avenue Charles Nicolle, Tunis, 1082, Tunisia; abirturki@yahoo.fr

Melatonin is a molecule naturally synthesized by the brain. It is produced only during the night period. The placement of melatonin implants on sheep allows, at certain times of the year to, artificially simulate short days, trigger heat in females and stimulate the sexual ardor of rams. The aim of this work was to study the effect of melatonin implants on the reproductive performance of three sheep breeds in Tunisia. A total of 400 ewes of the Barbarine breed, 400 ewes of the Black Thibar breed and 200 ewes of the Queue Fine breed were used in this study. Two experimental batches: each from each breed (The Barbarine breed: 200 treated and 200 control, the Black Thibar, 400: 200 treated and 200 control and the Queue Fine, 200 ewes (100 treated and 100 controls). Out of a total of 995 ewes recorded, 821 gave birth (82.51% fertily rate) and 174 were empty (17.49%). The Black Thibar breed had the highest fertility rate (93.45%) followed by the Barbarine breed (78.89%) and the Queue Fine breed (68%) (khi2,P<0.001). Fertility rate in the treated groups showed 86.6%, 72% and 68% for Black Thibar, Barbarine and Queue Fine, respectively. Untreated groups showed 100%, 86% and 71% fertility rate for the Black Thibar, Barbarine, and Queue Fine, respectively. The overall prolificacy rate was 129.60. By breed (treated and control); prolificacy rates were 106.37, 155.53 and 112.5, respectively for Barbarine, Black Thibar and Queue Fine. For (treated groups vs control), prolificacy rates were (108.7 vs 105.02),(157.80 vs 153.5) and (121.54 vs 104.23), respectively for the Barbarine, Black Thibar and the Queue Fine breed. The period of lambing was reduced for treated groups. A reduction of 15 d, 24 d and 44 d, for the Barbarine, Black Thibar and Queue Fine, respectively.

Effect of omega-3 fatty acids during late gestation on performance of dairy does and their kids

B. Habitouche[1], U. Serbester[1], I. Erez[1], M. Kose[2], N. Koluman[1], K. Oztabak[3] and M. Gorgulu[1]
[1]*Cukurova University Agriculture Faculty, Department of Animal Science, Adana, 01330, Turkey,* [2]*Dicle University Veterinary Faculty, Department of Obstetrics and Gynecology, Diyarbakir, 21280, Turkey,* [3]*Istanbul University Veterinary Faculty, Department of Biochemistry, Istanbul, 34320, Turkey; ugurserbester@gmail.com*

Essential fatty acids, especially omega-3 fatty acids, have anti-inflammatory properties, and also are needed during peri-natal growth and development of mammals. This study was conducted to determine if inclusion of fish oil as a omega-3 fatty acid source in late gestation diet improved performance of dairy does and their kids. Thirty German Fawn × Hair cross breed does were blocked according to parity and body weight. All groups were fed with same total mixed ration containing 2.8% protected fat between mating and 76th of gestation. Afterwards one of the group continued to receive the protected fat including TMR while other one received TMR including 2.8% fish oil throughout kidding. Total mixed ration contained 60% roughage (75% alfalfa hay and 25% wheat straw). After kidding, does and kids were fed common lactation diet and starter diet, respectively, until 60 days of lactation. Morning and afternoon milk yields were determined weekly by separating kids from dams in alternate days. Milk samples were collected for determination of total solids, fat, protein, casein, lactose, and urea. Feed intakes were determined weekly. Body weight and condition of does were taken biweekly. The fish oil supplementation during late gestation did not affect live weight, milk yield, and milk composition of the does. However, feed intake increased during pregnancy, while it caused a decline in the lactation period for the does fed TMR containing fish oil. The kids fed fish oil maternally consumed less feed compared to those receiving protected fat maternally. There was no effect of maternal fish oil on daily gain and feed conversion. In summary, inclusion of fish oil to maternal diet during late gestation affected performance of neither does nor kids.

Effects of rearing system, sex and age on the muscles fatty acid profile of Damascus goat kids

E. Tsiplakou[1], G. Papadomichelakis[1], D. Sparaggis[2], K. Sotirakoglou[3], M. Georgiadou[1] and G. Zervas[1]
[1]*Agricultural University of Athens, Nutritional Physiology and Feeding, Iera Odos 75, Athens, 11855, Greece,* [2]*Agricultural Research Institute, Nicosia, P.O. Box 22016, 1516, Cyprus,* [3]*Agricultural University of Athens, Plant Breeding and Biometry, Iera Odos 75, Athens, 11855, Greece; gzervas@aua.gr*

The effects of source of suckled milk, sex and age on the intramuscular fatty acid (FA) profile was studied using samples of the *Semimembranosus proprius* (SP), *Longissimus dorsii* (LD) and *Triceps brachii* (TB) muscles from 40 goat kids of Damascus breed. The goat kids were assigned into two groups balanced for body weight and sex. The first group (n=20), (10 male and 10 female goat kids) underwent natural rearing and received only maternal milk until weaning. The second group (n=20) (10 male and 10 female goat kids), was subject to artificial suckling with a commercial milk replacer. At weaning (49 days of age), 10 animals from each group were weighed and slaughtered. The remaining goat kids of both groups, after weaning were fed daily 100 g barley hay per animal and a commercial concentrate diet *ad libitum* up to 98 days of age, where they were also weighted and slaughtered. Analysis of variance was performed by a General Linear Model (GLM) procedure for the analysis of the fatty acids composition. The model included fixed effects ascribed to sex, rearing system, age and their interactions. The muscles FA composition was related with that of suckled milk (natural or artificial). Significant differences were found among the three muscles (SP, LD and TB) of goat kids for several individual FAs. The naturally reared goat kids had better muscles FA profile at younger age before a shift from milk to a fodder diet whereas the opposite happened in the case of artificially reared goat kids. Finally the sex did not affect the FA profile of goat kids.

Effect of photoperiod and ambient temperature on plasma testosterone concentration of Damascus bucks

E. Pavlou[1], G. Banos[2,3] and M. Avdi[1]

[1]AUTH, Department of Animal Production, School of Agriculture, 54124, Thessaloniki, Greece, [2]AUTH, School of Veterinary Medicine, 54124, Thessaloniki, Greece, [3]The University of Edinburgh, The Roslin Institute, Easter Bush, EH25 9RG, Scotland, United Kingdom; elefpavlou@gmail.com

Goat bucks in the Mediterranean basin exhibit a seasonal pattern of reproductive performance and differences are observed in the seasonality between distinct breeds and locations. The onset and length of the natural breeding season in goats depends on a number of factors including location, climate, breed, physiological stage, breeding system but mostly the photoperiod. The present study aimed at assessing the impact of photoperiod and ambient temperature on testosterone levels of bucks. The experiment was conducted at the AUTH's research farm, where all animals are raised under the same conditions and fed with regular diet. Blood samples were collected once a week for 12 months from 10 mature Damascus bucks. The concentration of plasma testosterone was measured with enzyme-linked immunosorbent assay (ELISA). The photoperiod was considered as the day length and the ambient temperature was the temperature of the surrounding environment in the farm. The effect of photoperiod and ambient temperature on testosterone was assessed using General Linear Models (PASW Statistics 18.0). Results showed that plasma testosterone concentration was significantly affected by photoperiod ($P<0.05$), while it was not significantly affected by temperature ($P=-0.004$). It was found that mean plasma testosterone concentration demonstrates a seasonal variation with highest value in August (7.32 ± 3.53 ng/ml) and the lowest values in April and May (1.09 ± 1.12 and 0.82 ± 0.73 ng/ml, respectively). Even though the effect decreased after August, testosterone levels remained high in September and October (6.06 ± 4.4 and 4.24 ± 3.53 ng/ml, respectively). Results from this study suggest that Damascus bucks plasma testosterone exhibit a significant seasonal variation, with the highest values recorded during the natural breeding season. These observations could be applied in selective breeding programs, in order to improve annual reproductive performance and make bucks sexually active throughout the year that can provoke reproductive activity in goats in their anoestrus season.

Effects of diet energy source on fertility, body weight, milk yield and composition in postpartum ewes

S. Aldabeeb[1], M. Shehab-El-Deen[1], M. Al-Saiady[2] and H. Kamel[3]

[1]Qassim University, Animal Production and Breeding, P O Box 6622, Buraidah 51452, Saudi Arabia, [2]Arabian Agricultural Services Company (ARASCO), P.O. 53845, Riyadh-11593, Saudi Arabia, [3]Alexandria University, Animal & Fish Production, P.O. 21545, Alexandria, Egypt; saldabeeb@yahoo.com

In the tropics and subtropics summer heat stress negatively affects sheep fertility, milk yield and composition. Limited energy intake due to depressed appetite is one of the mechanism through which heat stress exerts its negative effects. This study was aimed to examine the effects of dietary energy source during heat stress on postpartum ewe's fertility, milk yield and composition. Twenty healthy multiparous Naimi ewes in postpartum period during summer heat stress were divided into two groups. Each group was given either glucogenic or lipogenic diet. Body weight change, milk yield and composition were determined weekly from lambing until week 5 postpartum. Pregnancy rate was calculated in 4-month period. For statistical analyses, general linear model with animal as random factor, and diet and time postpartum as fixed effects was used. A two-tailed t-test was used to test the differences between the two diets at each time point during the postpartum period. There were no significant effects of diet energy source on body weight change or pregnancy rate. However, diet energy source affected ($P<0.05$) the milk yield (451.0 ± 62.7 and 814.4 ± 89.7 g) and also milk fat percentage (4.12 ± 0.33 and $3.18\pm0.28\%$) in glucogenic and lipogenic diets, respectively, during first week postpartum and vice versa for lactose. Milk production was enhanced in heat-stressed ewes fed Lipogenic diet. It was concluded that both diet can be used during summer heat stress with no effects on body weight change or pregnancy rate, however, ewes produced more milk when fed lipogenic diet.

Effect of mixed rearing system on milk yield and composition of Damascus goat

C. Constantinou, D. Miltiadou and O. Tzamaloukas

Cyprus University of Technology, Department of Agricultural Sciences, Biotechnology and Food Science, P.O. Box 50329, 3603 Lemesos, Cyprus; ouranios.tzamaloukas@cut.ac.cy

Thirty Damascus goats, in their second lactation, were allocated, as they kidded, to the following two homogenous groups of 15 animals: (1) no-kid group, NK, kids were weaned immediately after birth and their dams were machine milked twice a day, and (2) mixed-rearing group, MIX, goats were machine milked twice daily and suckled their kids (1.4 kids/goat) for a 7-week period. All goats after weaning (at 49±3 days) were machined milked twice and milk yield was determined until the 31st week of lactation. In both groups, the feeding regime for does included forage and concentrate supply at-libitum. Body-weight measurements were taken every two weeks, while milk samples were collected for composition analysis every week for the first fourteen weeks of lactation and at monthly intervals thereafter. Commercial milk yield of NK goats (94.8 l) during the pre-weaning period was higher (P<0.001) than MIX goats (55.3 l). However, the commercial milk produced post weaning till the 31st week of lactation was higher (P<0.05) for MIX compared to NK goats (271.2 and 200.9 l, respectively). As a result, no differences were found in total commercial milk produced from the two groups till 31st week (mean values of 326.5 and 295.8 l for MIX and NK goats, respectively). Interestingly, fat composition of commercial milk collected from suckling goats during the pre-weaning period was higher (P<0.05) for the MIX group compared to milk collected by NK group in every weekly measurement (mean values for fat percentage were 4.2 and 3.6, respectively). This difference in fat content of commercial milk was not observed after weaning of the suckling group. Other milk constituents (protein, lactose or solids non-fat) was not affected by the rearing method throughout lactation. In dual purpose breeds, as the Damascus goat, the galactopoietic effect of additional suckling under mixed rearing system may provide some benefits to the farmers.

Feed intake, nutrient content and digestibility of diets selected by goats grazing alfalfa and oat

L.R. Gaytán, J. Mendoza, G. Arellano, M.A. De Santiago, V. Contreras and M. Mellado

Universidad Autónoma Agraria Antonio Narro, Departamento de Medicina Veterinaria, Periférico Raul López Sanchez sin numero, 27054, Mexico; zukygay_7@hotmail.com

The aim of this study was to determine the chemical composition, dry matter intake and in vitro digestibility of diets selected by goats grazing oat aftermath and alfalfa aftermath unrooted by plowing. Goats kept on rangeland were taken to a recently harvested oat an alfalfa field. Samples were obtained from the diet of 10 multiparous goats that were separated and led by a rope tied to the neck with a non-slip knot, the forage was taken directly from the oral cavity before being swallowed by manually opening the jaws, this process was repeated approximately every 10 minutes for 4 h. Variables studied were feed intake, dry matter degradability and nutrient content. Feed intake was evaluated by collection of fecal samples (24 h) in canvas bags and harnesses. In vitro dry matter digestibility was determined using the Ankom® Daisy II digestion method (ANKOM-DAISY Procedure). The estimates of dry matter intake (DMI) were calculated from the total fecal yield in 24 h and dry matter digestibility estimates were calculated using the formula: DMI = kg fecal yield/1-the coefficient of dry matter disappearance. The data analysis was performed using analysis of variance (ANOVA) with the PROC GLM procedure of SAS. The goat was considered the experimental unit. Dry matter intake (P<0.05) was higher in goats grazing the oat residue field (2.86±0.09% body weight) that that in goats grazing on the alfalfa residues field (2.77±0.08); dry matter degradability of alfalfa residues selected by goats (44.0±3.4%) was higher (P<0.01) as compared to oat residues (37.3±2.7%), also crude protein content of oat residues (9.4%) was higher (P<0.01) as compared to alfalfa residues (7.9±1.2%). Oat residues contained less (P<0.01) neutral detergent fiber (53.6±3.9%) than alfalfa residues (57.0±1.9%); alfalfa residues contained more (P<0.01) Ca (1.18±0.18%) as compared to oat residues (0.42±0.2%). It was concluded that both oat and alfalfa residues represent a moderate quality temporal feeding resources for goats kept on rangeland in years of low forage availability.

Innovative marketing of high quality dairy products from transhumant sheep and goat farms in Greece

A. Ragkos[1], A. Theodoridis[1], K. Zaralis[2], G. Rose[1], V. Lagka[3] and G. Arsenos[1]
[1]*Aristotle University of Thessaloniki, School of Veterinary Medicine, University Campus, 410, 54124, Thessaloniki, Greece,* [2]*Organic Research Centre, Elm Farm, Hamstead Marshall, RG20 0HR, Newbury, Berkshire, United Kingdom,* [3]*Alexander Technological Educational Institute of Thessaloniki, Department of Agricultural Technology, Sindos, 100, Thessaloniki, 57400, Greece; gus@gusrose.com*

Transhumant sheep and goat systems are still common in Greece accounting for about 7.5% of the national flock. Recent scientific evidence demonstrated that the milk produced by such flocks is of premium quality which is directly related to the biodiversity of mountainous grasslands where the flocks graze from April/May to September/October. Nonetheless, the milk produced during this period is not rewarded in markets up to its true potential. The absence of local infrastructure for milk processing is a major obstacle that forces farmers to sell their milk to dairies based in lowlands, which mixed the milk of transhumant farms with that produced in other farms (i.e. intensive). Dairy products made exclusively from milk of transhumant flocks are very limited and are produced either on-farm or in small local dairies. This paper presents initial results of an on-going survey of the economic, management and marketing implications of three different approaches in summer milk marketing in transhumant farms. The first approach involves farmers selling their milk to the same industry throughout the year (winter and summer); the second is based on producing cheese solely from milk of transhumant flocks in a dairy situated in the highlands; and the third concerns on-farm cheese production and direct sales to consumers. Here a comparative technical and economic analysis of three case study farms – one from each approach – is presented in order to detect differences in the economic performance, the organization of labor, etc. The findings of this survey will shed light towards the institutional and organizational requirements for more systematic production of dairy products exclusively of transhumant origin.

The impact of climate change on the production and composition of goat milk in Northwest of Croatia

G. Kis[1], D. Mulc[2] and Z. Barac[2]
[1]*University of Zagreb Faculty of Agriculture, Department of Animal Nutrition, Svetosimunska 25, 10000, Croatia,* [2]*Croatian Agricultural Agency, Ilica 101, 10000, Croatia*

For agricultural production environments-climate has a very significant impact. A direct impact on crop production causes a reduction or complete loss of yield in long-term drought or flood periods and changes in chemical composition of plant material. Beside of direct impact on crop production weather conditions have an indirect impact on animal production through the quality of consumed feed materials. Extreme weather (droughts and floods) affects the growth of plants in such conditions and change their nutritional value and utilization. The aim of this paper is to investigate the impact of climate change on the production and composition of goat milk as a result of changes in the nutritional value of feeds in Northwest of Croatia. We investigate goat population in two the most populated Counties in the Northwest of Croatia. During the period of three years we monitored the production and composition of milk of more than 2000 goats in two counties. Beside the parameters of milk production during the investigated years there was analyzed chemical composition of consumed feeding materials. In the years 2012 to 2014 there have been significant changes ($P<0.05$) in the chemical composition of goat's milk reduction of fat, proteins and total milk solids. That can be caused by qualitative changes of feeds and as the effect of rising total goat's milk production. Furthermore changes in chemical composition of feed materials that goats ate are not affected in the proportions as we expected for milk production and composition. It can be conclude that most farmers kept goats in appropriate conditions with adapted meals to new requirements.

Digestibility, nitrogen balance and ruminal fermentation of Korat goats fed different oils

P. Paengkoum[1], S. Thongpea[2] and S. Paengkoum[3]

[1]Suranaree University of Technology, School of Animal Production Technology, Institute of Agricultural Technology, Muang, Nakhon Ratchasima, 30000, Thailand, [2]Suranaree University of Technology, School of Animal Production Technology, Institute of Agricultural Technology, Muang, Nakhon Ratchasima, 30000, Thailand, [3]Nakhon Ratchasima Rajabhat University, Faculty of Science and Technology, Muang, Nakhon Ratchasima, 30000, Thailand; pramote@sut.ac.th

The objectives of this experiment were to check the effects of additional soybean oil and sunflower oil on growth, ruminal metabolism and plasma fatty acid profiles particularly conjugated linoleic acid (CLA) in growing goats fed corn silage, compared and selected either the soybean oil or the sunflower oil for further study. Thirty growing crossbred Korat (Anglo-Nubian × Shami × Kalahari) goats that weighed 18.5±2.5 kg, aged about 7 months, were purchased and allocated to 5 treatments according to Randomized Complete Block Design with 6 goats in each treatment. The results presented as significant increase (P<0.05) of average daily gain (ADG). In addition, presence of soybean oil tended to increase (P<0.05) digestibility of dry matter (DM), organic matter (OM), and neutral detergent fibre (NDF). Ruminal NH_3-N were significantly reduced (soybean oil: P<0.01; sunflower oil: P<0.05). However, TVFA and butyric proportion (P>0.05) were not impacted by additions of soybean oil and sunflower oil, but the acetic proportion (P<0.05) and C2:C3 ratio (P<0.05) significantly increased. Regarding to the N balance, supplementation of sunflower oil resulted in significantly subtraction in dietary N intake, faecal and total N excretion (P<0.05); however, both of soybean oil and sunflower oil supplementations increased N absorption and retention. About the plasma fatty acid profiles, the total saturated fatty acids (P>0.05) composition tended to be increased, CLA content (P>0.05) significantly enhanced, the very long chain fatty acids (P<0.05) significantly reduced and DFA also ratio of n6:n3 (P<0.05) significantly increased owing to supplementations of soybean oil and sunflower oil.

Change in live weigh during pregnancy influences lamb performance – preliminary results

R. Perez-Clariget[1], J. Ithurralde[1], F. Corrales[1], M.B. López-Pérez[2], M.J. Abud[1], A. López-Pérez[1], M.J. Marichal[1] and A. Bielli[2]

[1]Facultad de Agronomia, Animal Production and Pasture, Garzon 780, 12900, Uruguay, [2]Facultad de Veterinaria, Morphology and Development, Lasplaces 1620, 11600, Uruguay; raquelperezclariget@gmail.com

Maternal nutrition can influence the future performance of the offspring. In order to evaluate the effect of the live weight (LW) change during pregnancy in ewes on the performance of their lambs, 36 Corriedale multiparous ewes carrying a single fetus were used in a complete randomized design. After the pregnancy diagnoses, ewes were randomly assigned to one of two different incremental natural pasture allowances (14-20 kg or 5-10 kg DM/100 kg LW) since day 30 of gestation until day 143, and supplemented with 300 g rice bran/animal/day (14% CP, 9% ADF and 24% NDF) since day 100. After day 143, the ewes were managed as a single group. The lambs were weaned at 90 days of age, transferred to a barn, located in individual pens, and fed with a balanced ration (16% of protein) and alfalfa hay *ad libitum*. They were classified according to the LW (corrected by fleece and fetus weight) change of their mothers in: Group A (n=9 females, 9 males): Their mother lost less than 15% of their LW since days 30 to 100 of gestation, and increased less than 15% since days 101 to 143 of gestation; Group B (n=11 females, 7 males): Their mothers lost 15% or more and increased 15% or more of their LW in the same periods. Lambs were weighted at birth, days 45 and 90, and fortnightly until 155 days of age. Data were analyzed with a repeated measures analysis (proc mixed, SAS); the model included the treatment, sex, date and their interactions, and the date as a repeated factor. Lambs of Group A tended to be heavier than Group B since birth until weaning (Group A: 12.9±0.6 kg vs Group B 11.5±0.6 kg; P=0.08) and during the fattening period (Group A: 24.8±0.1 kg vs Group B: 22.4±0.1 kg; P=0.09). The voluntary intake, daily weight gain and efficiency of food conversion were not affected by treatment. No effect of sex nor interactions were found. It is concluded that the LW change of the mother during pregnancy affects the weight of lambs since birth to 155 days of age.

Reproductive response of goats joined to either well or under-fed bucks treated with testosterone

M.A. De Santiago, G. Machado, A. Lopez, V. Contreras, L. Gaytan, G. Veliz and M. Mellado
UAAAN, Posgrado en Produccion Agropecuaria, Periferico y Carr a Sta Fe S/N, Torreon 27054, Mexico;
adesantiago867@gmail.com

To determine the reproductive response of 116 multiparous anovulatory goats (2.0 body condition score; BCS scale: 1-4) in the arid north of Mexico (26° N) were exposed to bucks with different BCS and treated with testosterone, on March 17th, 2015 does received a single dose of 10 mg IM progesterone in order to avoid short estrus cycles and were assigned to 4 groups named according male treatments: 1T (complete diet + testosterone; CC=2.5; n=31), 1/2T (low diet + testosterone; CC=1.7; n=27), 1C (complete diet without testosterone; CC= 2.5; n=30) and 1/2C (low diet without testosterone; CC=1.7; n=28). Eight males were used, (2/group). Four: 1T and 1/2T groups received 25 mg of testosterone IM/3rd day/3 weeks previous to the mating period, the remaining 4 received saline solution. On March 18th males and females were joined for 6 days, and estrus were detected by rotation of males AM/PM, ovulation and pregancy were scaned by ultrasonography on days 10, 48 and 93. Latency and duration of estrus were analyzed with a t test and proportions of does in estrus, with ovulation, pregnant and that gave birth were compared with a X^2 test using SYSTAT 12. None of the does exposed to not-testosterone treated bucks showed sexual response. Number of does in estrus 72 h after mating, was different ($P<0.05$) between 1T and 1/2T groups (22/31, 12/27). However, latency to first estrus (1T; 96±5 h, 1/2T; 112±6 h), duration of estrus (24 h in both groups), does that ovulated (1T:29/31, 1/2T: 22/27), pregnancy rate at d 48 (1T: 20/31, 1/2T: 14/27) and d 93 (1T: 18/31, 1/2T: 10/27), fertility rate (1T:14/31, 1/2T: 9/27), litter size (1T: 2, 1/2T: 1.8) and kidding rate (28 vs 17) were not different ($P>0.05$) between groups. We reaffirm that an adequate nutrition does not bring about sexual drive at 26°N. Furthermore, joining of goats with lean bucks treated with testosterone was as effective as stimulation with well-fed ones, in terms of estrus, ovulatory response, pregnancy and kidding rate. However, a reduction of litter size is expected in does joined to underfed bucks.

Ultrasound measurements of LD muscle properties and subcutaneous fat thickness in Merino lambs

E. Ghita, C. Lazar, R. Pelmus, C. Rotar, M. Gras, T. Mihalcea and M. Ropota
National Research Development Institute for Animal Biology and Nutrition, Animal Biology, 1 Calea Bucuresti, 077015, Balotesti, Ilfov, Romania; elena.ghita@ibna.ro

The stock of sheep belonging to the group of Merino breeds decreased strongly in Romania with the sinking demand and price of wool. Currently, the Merino sheep account for 5.86% of the total stock of sheep. The farmers who preserved this breed earn in comes only from selling the lambs or the growing Merion sheep for meat. The main breeds belonging to the Merino group, reared in Romania, are the Transylvania Merino and the Palas Merino. The purpose of the work was the *in vivo* evaluation of Transylvania Merino (TM) lamb carcass quality. Ultrasound technology was used to measure Longissimus dorsii muscle properties (depth, area and perimeter) and subcutaneous fat thickness on live animals. These measurements were performed on a total of 67 lambs (51 males and 16 females), aged 130 days and weighing 30.46 kg. The ultrasound measurements were performed with an Echo blaster 64 using LV 7.5 65/64 probe, supplied by TELEMED ultrasound medical systems. The first measurement point was 5 cm from the spine, at the 12th rib; the second measuring point was between 3rd and 4th lumbar vertebrae. The average values of the subcutaneous fat layer, LD muscle depth, area and perimeter were: 2.18 mm; 22.60 mm; 9.90 cm^2 and 127.49 mm. All phenotypic correlations between the ultrasound measurements on LD muscle and body weight were significant, the correlation indices ranging between 0.45-0.85. These results were compared with the performance of lambs, of two local breeds Palas Merino (PM) and Teleorman Black Head (TBH), having the same body weight. The following conclusions emerged after the Fisher and Tukey tests were applied: (1) TM lambs scored the best results on LD muscle depth, muscle eye area and perimeter, with significant ($P<0.05$) differences from the other two breeds; (2) TBH lambs scored the highest value of the subcutaneous fat thickness (undesired trait); (3) PM lambs displayed lower LD muscle traits than TM lambs. The expectations for higher incomes from selling lambs with higher carcass traits is a motivation for the producers to develop a strategy for the improvement of carcass value by selecting the animals with higher genetic value.

Milk composition variation of Saanen goats, depending on the rearing system in Romania

S. Zamfirescu[1], N. Dobrin[2], D. Jitariu[2], A. Cismileanu[3], A. Sava[3], D. Voicu[3] and I. Voicu[3]
[1]Ovidius University, Constanta, 900527, Romania, [2]CAPRIROM (National Association of Goat Breeders), Constanta, 900316, Romania, [3]National Research Development Institute for Animal Biology and Nutrition (IBNA), Balotesti, 077015, Romania; voicu.ilie@yahoo.com

The purpose of the work was to determine the response of the imported Saanen goat breed to the rearing systems in Romania namely, indoor rearing, and on pasture, using an optimised diet. The 90 days feeding trial was conducted in a goat farm from Constanta County, on 2 groups of 25 goats each, in their second lactation, with an average body weight of 55 kg. The first group (G1) received 1.2 kg alfalfa hay and 0.6 kg concentrates mix (corn, wheat, barley) plus 8 h of grazing each day. The second group (G2) received 1.4 kg alfalfa hay, 0.5 kg wheat straws and 1.2 kg concentrates mix (corn, wheat, barley and sunflower meal). Both diets had the same energy and protein level: 2,822 g DM, 2.53 milk FU, 207 g IDP, 18.71 g calcium and 8.25 g phosphorus. Milk samples were assayed for protein and fat using Gerber milk analyser, and for fatty acids composition by gas chromatography. The average daily milk yield was 3.66 ± 0.23 litres/goat for G1 and 3.14 ± 0.18 litres/goat for G2 during trial. The average milk fat percentage was 3.42 ± 0.31 in group G1 and 3.81 ± 0.14 in group G2; while protein percentage was 3.41 ± 0.13 in group G 1 and 3.29 ± 0.15 in group G2. The representative milk fatty acids (% of FAME) in group G1 were as follows: SFA 52.58 ± 2.86, MUFA 40.65 ± 1.2, PUFA 6.10 ± 0.44, UFA 46.75 ± 2.25 and CLA 0.74 ± 0.03, $\Omega6{:}\Omega3$ ratio of 5.52 ± 0.71; the corresponding values for group G2 were: SFA 72.35 ± 4.05, MUFA 21.78 ± 0.82, PUFA 5.49 ± 0.38, UFA 27.27 ± 1.22, and CLA 0.72 ± 0.02, $\Omega6{:}\Omega3$ ratio of 10.05 ± 1.02. The conclusion of the feeding trial was that goat rearing during the aestival season on pastures, with a supplement of hay and concentrates, increased the daily milk yield by 17% and improved the fatty acids composition of the milk. This study was supported by Romanian Ministry of National Education – project PN-II-PT-PCCA-2013-4-1828.

Effects of synbiotic use in creep feeding on performance of goat kids

B.V. Raksasiri[1], S. Paengkoum[2], K. Poonsuk[3], P. Thitisak[4] and P. Paengkoum[5]
[1]Institute of Agricultural Technology, Suranaree University of Technology, 111 Muang, Nakhon Ratchasima, 30000, Thailand, [2]Faculty of Science and Technology, Nakhon Ratchasima Rajabhat University, 340 Muang, Nakhon Ratchasima, 30000, Thailand, [3]K.M.P. BIOTECH CO.,LTD, Muang, Chonburi, 20000, Thailand, [4]K.M.P. BIOTECH CO.,LTD, Muang, Chonburi, 20000, Thailand, [5]Institute of Agricultural Technology, Suranaree University of Technology, 111 Muang, Nakhon Ratchasima, 30000, Thailand; took_sw@yahoo.com

This research investigated synbiotic supplementation effects on productive performance of goat kids. Twenty goat kids were grouped with regard to sex and were fed colostrum for 5 days before start the experiment. During the experimental period, the goats were given only milk on days 1-30, thereafter, days 31-75, they were fed with concentrate and roughage, and decreased amount of milk until weaned. Jerusalem artichoke is prebiotic source and with probiotic (BACTOSAC-P). There were five dietary treatments groups; control diet (T1), synbiotic supplemented 0.01% (T2), 0.02% (T3), 0.03% (T4) and 0.04% (T5) of goat kid, respectively. The five treatments were arranged in a randomized complete block design. The results showed that final body weight, average daily gains and feed conversion ratio were significantly different ($P<0.05$), in all the groups are synbiotic supplementation. Moreover, fecal score (14 day) and Lactic acid bacteria (14, 56 and 84 day) were significantly different ($P<0.05$) among dietary treatments. The objectives for this study were to determine an optimal dose of synbiotic supplementation to enhance future studies investigating the effects of synbiotic on parameters associated with performance, immune modulation, or health status in goat kids and other adolescent ruminants.

Evaluation of the performance of milk goats in Romania

I. Voicu[1], A. Anghel[2,3], A. Cismileanu[1], A. Sava[1], D. Nadolu[3] and D. Voicu[1]
[1]*National Research Development Institute for Animal Biology and Nutrition (IBNA), Ruminant Physiology, Calea Bucuresti, Nr.1, Ilfov, Balotesti, 077015, Romania,* [2]*ICDCOC Palas, Constanţa, 900316, Romania,* [3]*CAPRIROM (National Association of Goat Breeders), Constanţa, 900316, Romania; voicu.ilie@yahoo.com*

The aim of the research was to improve the quantity and quality of the milk in the local breed of goats, Carpatina, crossed with specialised milk breeds imported from Western Europe. The feeding trial used three groups of 30 goats each, at their second lactation: one group consisted of Romanian Carpatina breed (C), the second group consisted of Carpatina × French Alpine F1 hybrids (E1) and the third group of Carpatina × Saanen F1 hybrids (E2). The diets consisted in a supplement of alfalfa hay, corn and barley, given before and after the 8 h of grazing (08 to 16 h), every day. The structure and state of the pasture provided 8 kg fresh forages, every day. The energy (FU_{milk}) and protein (IDP) level of the diets was in agreement with the feeding norms for the weight and performance of the animals. Milk samples were collected during each of the 5 months (April-August) of lactation, and assayed for protein and fat using Lactostar milk analyser, and for fatty acids composition by gas chromatography. The average daily milk yield was 1,386±42 ml/goat in group C, 2,106±76 ml/goat in group E1 and 2,172±83 ml/goat in group E2. The average monthly milk fat contents were 3.65±0.12% (C), 3.82±0.33% (E1) and 3.73±0.14% (E2), P=0.849 and respectively, the protein content were 3.15±0.11% (C), 3.43±0.17% (E1) and 3.47±0.14% (E2), P=0.187. The average monthly fatty acids ratio Ω6: Ω3 was also quite similar in all groups: 4.98±1.27 (C), 5.24±0.78 (E1) and 5.54±0.83 (E2), P=0.921. The level of CLA (% of total fatty acids) was also not different between groups, with values of 0.59±0.09 (C), 0.62±0.08 (E1) and 0.6±0.08 (E2), P=0.956. The data show that the cross of the local breed with specialised milk breeds increased significantly the total milk yield in groups E1 and E2, compared with the control group (C) (P<0.001). However, no significant changes were noticed in milk composition. This study was supported by Romanian Ministry of National Education – project PN-II-PT-PCCA-2013-4-1828.

Evaluation of out-of-season estrus induction protocols in progesterone-primed mix-breed dairy goats

V. Contreras-Villarreal[1], C.A. Meza-Herrera[2], G. Arellano-Rodriguez[1], R.A. Delgado-Gonzalez[1], L.R. Gaytan-Aleman[1], O. Angel-Garcia[1] and F.G. Veliz-Deras[1]
[1]*Universidad Autónoma Agraria Antonio Narro, Periférico y carretera a Santa Fe, 27054, Torreón, Coahuila, Mexico,* [2]*Universidad Autónoma Chapingo, Unidad Regional de Zonas Aridas, Bermejillo, 35230, Durango, Mexico; dra.viridianac@gmail.com*

The aim of the present study was to evaluate different out-of-season estrus induction protocols using a single progesterone (P4) priming 24 h prior administration of eCG, GnRH or E2. Thirty-two adult anestrus mix-breed dairy goats were P4-primed (20 mg IM.) and randomly assigned (n=8) to: (1). E2: 1 mg IM estradiol+0.2 ml cloprostenol 24 h later; (2). E2+GnRH: E2+2.5 ml IM GnRH 24 h later; (3). E2+eCG, E2+100 IU IM eCG 24 h later; and (4). eCG, 250 IU IM eCG. Response variables considered the percentages of estrus (ES%, detected by sexually active males), ovulation (OV%), and pregnancy (PR%, diagnosed 45 d after estrus detection). All treatments achieved high ES% (>87%) and OV% (>50%), yet, E2+GnRH reached 88% of OV% and PR%. E2 group had the lowest PR% (P<0.05; 12.5%) while the largest PR% (P<0.05; 81.5%, average) were observed in E2+GnRH and E2+eCG. Results indicate the feasibility of a simple synchronization protocol based in a single i.m. progesterone-priming plus E2+GnRH.

Ruminal digestibility of Leucaena and Mulberry leaves using in sacco technique in goats

P. Paengkoum[1], S. Thongpea[2] and S. Paengkoum[3]
[1]Suranaree University of Technology, School of Animal Production Technology, Institute of Agricultural Technology, Muang, Nakhon Ratchasima, 30000, Thailand, [2]Suranaree University of Technology, School of Animal Production Technology, Institute of Agricultural Technology, Muang, Nakhon Ratchasima, 30000, Thailand, [3]Nakhon Ratchasima Rajabhat University, Faculty of Science and Technology, Muang, Nakhon Ratchasima, 30000, Thailand; pramote@sut.ac.th

The objective of this study was to determine the digestibility of Leucaena (*Leucaena leucocephala*) and Mulberry (*Morus alba*) leaves using nylon bag technique. Three goats with an average 27.5±1.5 kg each fitted with a permanent rumen were used. Ruminal DM and CP disappearances increased with rumen incubation time for all feedstuffs. Dry matter degradation rates of Mulberry leaf was significantly higher (P<0.05) than that of Leucaena leaf. The loss of DM by washing of Mulberry leaf was higher than that of Leucaena leaf. Similarly, the loss of CP by washing of Leucaena leaf was higher than Mulberry leaf, and degradability of water insoluble fraction of Mulberry leaf was higher than Leucaena leaf. The potential degradation and the effective degradability of DM and CP for Mulberry leaf were higher than that of Leucaena leaf. The results of the present study suggested that the total tract digestibility of Mulberry leaf was significantly higher (P<0.05) than Leucaena leaf, especially after 4 h of incubation time. The potential degradation and the effective degradability of DM and CP for Mulberry leaf was high than Leucaena foliage. The digestibilities of the leaves proteins examined in this study were respectably high and could serve as useful protein supplements for ruminant producing in tropical countries.

The future of indigenous cattle breeds of pastoralists in Benin

C. Tamou, I.J.M. De Boer, R. Ripoll-Bosch and S.J. Oosting
Wageningen University, Animal Production Systems group, P.O. Box 338, 6700 AH Wageningen, the Netherlands; raimon.ripollbosch@wur.nl

Pastoralism is under pressure and consequently cattle breed diversity in the pastoralists' herds is at the risk of being affected, as well as the traditional ecological knowledge associated with the use of indigenous cattle breeds. The objectives of this study were to: (1) inventorise pastoralists' indigenous breeds of cattle and their performance in selected traits; (2) analyse pastoralists' preferences for specific breeds and reasons for this preference; and (3) determine whether the knowledge about breeds and their traits was passed from the one generation to the other. Data were collected from focus groups discussions and from a semi-structured questionnaire with 72 pastoralists. Interviewees belonged to three generations and three agro-ecological zones in the periphery of the W Biosphere Reserve in Benin. From the focus groups discussions we identified the most common cattle breeds in the region (i.e. Keteeji, Jaliji, Bodeji, Tchiwali and Gudali) and the most relevant traits (i.e. milk production, meat production, endurance and tolerance to trypanosomiasis). Individual interviewees scored the performance of cattle breeds in the four main traits. Keteeji was valued for its endurance and tolerance to trypanosomiasis, Bodeeji was highly valued for endurance and Gudali was perceived of high value for meat and milk production, but of low value for endurance. Keteeji was the preferred breed by nearly 50% of the pastoralists, because it withstands hunger. Gudali was the least preferred breed; it was preferred by 11% of pastoralists. Hence, to deal with the changing and unfavourable conditions of their environment, which will only aggravate in the future, pastoralists prefer breeds performing well with regard to non-productive traits. Knowledge about breeds did not differ between generations implying that it is still being passed from the one generation to the other.

Native cattle breeds of Norway: genetic- and protein polymorphism and milk coagulation properties

R.A. Inglingstad[1], T.A. Mestawet[1], T.G. Devold[1], G.E. Vegarud[1], E.K. Ulleberg[1], P. Berg[2], A. Rehnberg[3] and N. Sæther[3]
[1]*Norwegian University of Life Sciences, Faculty of Chemistry, Biochemisty and Food Science, Christian Magnus Falsens Vei 1, 1432 Ås, Norway,* [2]*NordGen-The Nordic Genetic Resource Center, Postboks 115, 1432 Ås, Norway,* [3]*Norwegian Institute of Bioeconomy Research, Norwegian Genetic Resource Centre, Postboks 115, 1432 Ås, Norway; nina.sather@nibio.no*

Norwegian Red cattle (NRF) has been the main cattle breed in Norway since 1960. Before the introduction of NRF, local native breeds were kept for milk and meat production. Today six native breeds are endangered: Telemark cattle (TF), Eastern and Western Red Polled (ØR and VR), Western fjord cattle (VFF), Døla cattle (DF), Blacksided Trønder and Nordland cattle (STN). The average milk yield for NRF is 8,000 kg/year, while the old native breeds yields 4,000 kg/year. Lien *et al.* showed a higher frequency of the B-allele of CSN3 (encoding κ-casein) and the A1 –allele of CSN2 (encoding β-casein), which correlates with favorable milk coagulation properties, but lower milk yield. However, milk components and technological properties has not been characterized. Thus, we have investigated the genetic variation, milk protein polymorphism and milk coagulation properties in the seven native Norwegian breeds including NRF. Milk and blood samples from 200 individual cows were collected from 30 different farms. Milk coagulation properties (MCP) of acid- and rennet coagulated milk was assayed by Formagraph and protein profiles were analyzed by capillary zone electrophoresis (CZE). CSN2 A2 were predominant in NRF while A1 were predominant in TF, and the other breeds displayed a more equal distribution of A1 and A2. Higher frequency of CSN3 B was found in TF, VR and VFF compared to STN, DF and NRF, which had higher frequency of CSN3 A. The old native breeds had better MCP than NRF, which had the longest coagulation time and the weakest coagulum. Moreover, the protein profiles of the old native breeds showed a large variation among individuals, with regards to genetic variants and post-translational modifications compared to NRF where less variation among individuals were observed. Other links between the old native breeds and milk characteristics are still under investigation.

Mitogenome contribution of local Aurochs to the modern cattle breeds

V. Cubric-Curik[1], D. Novosel[1], V. Brajkovic[1], S. Krebs[2], J. Soelkner[3], D. Šalamon[1], S. Ristov[1], B. Berger[3], S. Triviziaki[1], I. Bizelis[2], M. Ferenčaković[1], S. Rothammer[2], E. Kunz[2], M. Simčič[1], P. Dovč[1], G. Bunevski[2], H. Bytyqi[2], B. Marković[2], I. Curik[1] and I. Medugorac[2]
[1]*Faculty of Agriculturae, University of Zagreb, Animal Science, Svetošimunska 25, 10000 Zagreb, Croatia,* [2]*LMU Munich, Chair of Animal Genetics and Husbandry, Veterinaerstr. 13, 80539 Munich, Germany,* [3]*University of Natural Sciences and Life Sciences, Augasse 2-6, 1090 Vienna, Austria; vcubric@agr.hr*

Contribution of local Aurochs (*Bos primigenius*) to the development of modern cattle breeds is still an open issue. Our understanding of this issue has progressed recently as NGS enabled more efficient sequencing of the complete mitogenome and today there are around 300 complete cattle mitogenome sequences in the GeneBank. Using Illumina NGS we have additionally sequenced around 200 individuals from over 50 cattle breeds sampled largely in Central and South East Europe. Sequenced haplotypes were classified within P, Q, T1'2'3, T1, T2, T3, T4, T5 and T6 haplogroups. Additional molecular (mtDNA D-loop) and pedigree analyses indicated the presence of the whole maternal lineage with 101 individuals having 'Aurochs' P classified haplotype. The found sequences strongly contributed to the diversification within 'macro-haplogroup T3', resulting with three deeply rooted branches and prompting for the reclassification. Overall, results of this study provide evidence of the contribution of local Aurochs to the formation of modern cattle breeds and the complexity of domestication process.

Genome wide association study using individual milk mid-infrared (MIR) wavenumbers

Q. Wang and H. Bovenhuis
Wageningen University and Research, Animal Breeding and Genomics, Droevendaalsesteeg 1, 6708 PB Wageningen,
the Netherlands; qiuyu.wang@wur.nl

It has been shown that substantial part of the phenotypic variation in bovine milk mid infrared (MIR) wavenumbers is due to genetic differences. The objective of this study was to identify genomic regions associated with individual milk MIR wavenumbers. For this purpose genome wide association studies (GWAS) were performed for 50 individual MIR wavenumbers measured on 1,748 Dutch Holstein milk samples and using a 50K single nucleotide polymorphism (SNP) array. These 50 individual MIR wavenumbers were selected to represent more than 95% of the variability in the full spectra. In total 24 significant regions distributed over 16 bovine chromosomes were identified. Chromosomes 1, 5, 6, 14, 19, and 20 were the main chromosomes with genomic regions associated with multiple MIR wavenumbers. Several of the identified regions have been reported by other studies for their associations with fat % and protein % (e.g. regions containing the DGAT1 and the caseins genes). This study also identified a region on chromosome 6, containing the ABCG2 gene, which showed a strong association with lactose %. Another interesting region, located on chromosome 1, was significantly associated with wavenumbers around 2,587 cm^{-1}. This region harbours the gene solute carrier family 37 member 1 (*SLC37A1*), which functions as glucose-6-phosphate and phosphorus transporter. This gene has been related to milk phosphorus content and therefore this finding suggests that the milk MIR spectra contains direct information on milk phosphorus content. This study revealed the genomic regions affecting milk MIR wavenumbers and identified candidate genes responsible for the variation in these wavenumbers. These findings contribute to a better understanding of the relationships between milk MIR wavenumbers and milk components.

Polymorphism of IGF1 gene and its association with milk production traits in Holstein dairy cattle

M. Hosseinpour Mashhadi[1], F. Golak[1] and N. Tabasi[2]
[1]Department of Animal Science, Mashhad Branch, Islamic Azad University, Mashhad, Iran, [2]Immunology Research
Center, BuAli Research Institute, Mashhad Medical University, Mashhad, Iran; mojtaba_h_m@yahoo.com

To evaluate the IGF1 gene polymorphisms and its association with milk production traits in Holstein cattle, the whole blood samples of 140 dairy cattle were used for DNA extraction and PCR testing. The DNA from whole blood samples extracted by Genet Bio kit (Korea product) and Cinnapure DNA kit (Sinaclon product). Fragment of 249 bp from promoter gene amplified by standard PCR method. The amplified fragment was digested by restriction enzyme SnaBI by PCR-RFLP method. The PCR product visualized on 2% agarose gel and ethidium bromide staining were used. The Popgene 1.31 software was used for calculate genotypic and allelic frequencies. The GLM procedure in SAS software was used for association between IGF1 genotypes with milk production traits. Duncan test was used for comparison between means. Genotype frequencies of AA, AB and BB were 0.26, 0.44 and 0.30 respectively. The frequency of allele A and B were 0.48 and 0.52. The Chi-square Test showed that the population was in Hardy Weinberg equilibrium (P<0/05), so the selection or other systematic or dispersive factors did not had effect on disequilibrium of population for this gene. The heterozygosis and homozygosis in the study population was 0.44 and 0.56. The Nei (expected heterozygosity) and Shannon index were 0.49 and 0.69 respectively. The studied traits were day milk, fat and protein percentage, milk and fat yield of year and breeding value of milk and fat. The mean of AB genotype for protein percentage was more than AA genotype (P?0.05), for other studied traits the value of AB genotype were more than mean of AA and BB genotypes but The differences were not significant (P?0.05). It can be concluded that the cows carrying AB genotype had more production and the heterozygosity in studied population was close to the maximum.

Improving longevity of crossbred sows in a genomic selection breeding scheme

E.F. Mouresan[1], E. Grindflek[2], W.F. Fikse[1] and L. Rydhmer[1]
[1]*Swedish University of Agricultural Sciences, Animal breeding and Genetics, Ulls väg 26, Box 7023, 75007 Uppsala, Sweden,* [2]*Topigs Norsvin, Storhamargata 44, 2317 Hamar, Norway; elena.flavia.mouresan@slu.se*

In a standard pig breeding program selection is performed in purebred lines with the goal to improve crossbred performance. This structure has some limitations as the pure lines are usually kept in a more favourable environment than their crossbred relatives. Moreover, there are traits of high interest for the crossbred population, such as longevity, that are difficult to record in nucleus herds due to a short generation interval. It has been proposed that improving feet and leg traits in purebred lines will lead to improvements in longevity of the crossbred sows. This study aims to investigate the potential of genomic selection in improving longevity, looking into the design of reference populations, evaluation models and methods. Using stochastic simulation, an evolutionary process of 2,000 generations is created by gene-dropping method where the population size is reduced gradually from 2,000 to 200 individuals in order to create LD. Two pure breed lines of 100 individuals each are separated and they diverge for 100 generations. 50.000 biallelic markers are simulated, evenly distributed across 18 chromosomes of varying size in order to resemble the real genome. 1% of these markers are randomly selected to create causative mutations for 2 correlated traits with heritabilities of 0.1 and 0.4 respectively. Genotype-by-environment interactions and dominance effects are simulated for each trait. The preliminary results showed higher accuracies for within pure line estimations, ranging between 0.53 and 0.34 depending on the trait's heritability. The accuracies obtained from across line predictions were very low (~0.10-0.08), while the crossbred reference population yielded accuracies between 0.20 and 0.10. Next, the genetic gain across several generations of selection under different selection scenarios will be evaluated. Additionally, real genotypic data from 2 pig breeds will be available and they will be incorporated into the simulation with the aim of capturing the LD structure of these populations and thus giving more realistic results.

Successful selection of rainbow trout on their ability to grow with a diet devoid of marine products

T. Callet[1,2], F. Médale[1], A. Surget[1], L. Larroquet[1], P. Aguirre[1], T. Kerneis[3], L. Labbé[3], I. Geurden[1], E. Quillet[2], S. Skiba[1] and M. Dupont-Nivet[2]
[1]*INRA, NuMéa, D918, 64310 St-Pée-sur-Nivelle, France,* [2]*INRA, GABI, Allée de Vilvert, 78352 Jouy-en-Josas, France,* [3]*INRA, PEIMA, Barrage du Drennec, 29450 Sizun, France; therese.callet@inra.fr*

Aquaculture significantly expanded since 1950 and provides now half of all fish for human food. This exponential growth, along with the plateauing of the production of fish oil (FO) and fishmeal (FM), the traditional components used in aquaculture, imposed feedstuffs evolution. These last years, scientific research has addressed replacement of FO and FM, and vegetable oils and meals has been increasingly incorporated in fishfeed. In the case of rainbow trout (RBT), it seemed, however, that a threshold has been reached because fish performance (growth and survival) are negatively impacted when more than 80% of FO and FM are replaced by plant ingredients. In this context, a line of RBT was selected for its ability to survive and grow when fed with a 100% plant-based diet from the first feeding. The aims of the present study were to assess the efficiency of the selection after 3 generations and its effects on different nutritional traits. Fish from the third generation and a control line were fed since the first feeding and for 7 months, with either a control marine-based diet (M diet) or a 100% plant-based diet (V diet). Both growth and survival of selected fish were significantly improved in comparison to the control line, showing the success of the selection. The major finding of our study is that after only 3 generations of selection, selected fish fed the V diet reached a final body weight similar as the control fish fed the M diet, while the control fish fed the V diet were 37% smaller than the ones fed the M diet. The higher performance of the selected line was explained by a higher feed intake, rather than an improved feed efficiency. Our results highlight that plant-based diet acceptance is one of the major step impeding the complete replacement of FO and FM, but confirmed that it can be improved by selection. Further investigations are however needed to understand the indirect effects of the selection on starch digestibility, protein retention and fish content in EPA/DHA, two major fatty acids in human nutrition.

Efficiency of different BLUP indexes accounting for polyandry in bee genetic evaluation
B. Gravellier and F. Phocas
INRA, Animal genetic – GABI, Domaine de vilvert, 78350 Jouy-en-Josas, France; bastien.gravellier@labogena.fr

The best linear unbiased prediction (BLUP) applied to an animal model is the standard method for genetic evaluation of domestic livestock. It requires modelling the relationship matrix between all the animals registered in the pedigree of candidates for selection. For diploid species the derivation of relationship coefficients are well established. But the honey bee (Apis mellifera) is a haplodiploid species in which several (10 to 20) haploid drones fertilize a diploid queen. In consequence, the father of a female progeny is in most cases unknown (with the exception of a queen insemination with the semen of a single drone). The bee genetic (male haploïdy) and reproductive (polyandry) specificities have therefore to be accounted for in the derivation of the relationship matrix to model relevant paternal gene flows and uncertain paternity of the workers and queen of the colony. The objective of the study was to compare different approaches for deriving the relationship matrix to estimate the bee breeding values by BLUP indexes. The simplest approach is to ignore the paths of paternal gene transmission in a similar way that in diploid species paths of maternal gene transmission have classically been ignored under BLUP – sire model considering a relationship matrix build from triplets 'animal – sire – maternal grandsire' in the pedigree file. In a symmetric approach for bee a dam model can be considered with a pedigree file composed of triplets 'animal – dam – paternal grand dam'. Such a simplified model was compared to different approaches proposed in the literature to model bee relationship matrix under an animal model. In these last strategies, the group of drone-producing queens is considered as a dummy father, the drones being the male gametes. A stochastic simulation was used to produce a bee population of 5,000 queens with known phenotypes and true genetic values for a trait of heritability varying from 0.1 to 0.5. The different BLUP indexes were computed assuming various degree of polyandry. Correlations between true and estimated breeding values were derived in order to evaluate the impacts of the parameters describing the paternal origins. The number of drone producing queens appeared to have the highest impact on the estimates.

Potential of pure- and crossbred Les Bleues broiler for meat production for small-scale farmers
C. Lambertz, K. Wuthijaree and M. Gauly
Free University of Bozen-Bolzano, Faculty of Science and Technology, Piazza Università 5, 39100, Italy;
christian.lambertz@unibz.it

In dual-purpose chicken production systems both sexes might be raised together during the first weeks, before males are separated during their final fattening stage until slaughter, while females are raised for an entire laying period. Such a system seems to be suitable for small-scale mountain farmers in South Tyrol, Northern Italy to satisfy the increasing demand for high quality and regionally produced products. The performance of male broilers of two genotypes in such a system was compared. One day-old male purebred (P) Les Bleues (n=150) and crossbred (C) New Hampshire × Les Bleues (n=150) chicken were raised together with their female counterparts for 12 weeks in a floor husbandry system. Genotypes were kept in two separate compartments. Thereafter, males were moved to a mobile chicken house with free-range access. Until week 6, a starter (12.9 MJ ME, 22% XP) and until slaughter a broiler diet (12.3 MJ ME, 18% XP) was fed. Animals were slaughtered at weekly intervals from 12th to 23rd week of age. Individual live weight was measured weekly and carcass quality assessed. Feed consumption in the mobile house could only be recorded for both genotypes together due to technical equipment. Results were analysed by one-way ANOVA using a t-test as post-hoc test. At 12 weeks of age, live weight of P (2,075 g) was about 200 g higher than that of C (1,865 g; P<0.05). This equals to a daily weight gain of 24.1 g for P and 21.7 g for C. Until week 16, both genotypes reached a live weight of more than 2,500 g (P>0.05). Dressing percentage was about 1% higher in C than in P when animals were slaughtered with 2,500 to 3,000 g live weight (P>0.05). Proportion of legs, breast and wings was 34.3, 16.0 and 11.0% in P and 34.7, 15.5 and 12.1% in C (P>0.05). Under the specific conditions of the region, production costs per kg of slaughter weight were 5.13 € for P and 7.14 € for C, mainly due to higher animal costs for C. Prices per kg of slaughter weight were 13 € at direct marketing. In conclusion, both genotypes showed an appropriate growth performance. However, production costs clearly demonstrated that production is only economically meaningful if animals are marketed as high-value premium products.

Genetic parameters of medullation percentage in alpacas

A. Cruz[1], R. Morante[1], I. Cervantes[2], A. Burgos[1] and J.P. Gutiérrez[2]
[1]INCA TOPS S.A, Miguel Forga 348, Arequipa, Peru, [2]Universidad Complutense de Madrid, Departamento de Producción Animal, Avda Puerta de Hierro s/n, 28040 Madrid, Spain; gutgar@vet.ucm.es

Alpaca fiber is one of the main natural textiles produced in Peru. The mean fiber diameter goes from 20 to 36 µm, being highly appreciated some categories such as baby alpaca (<23 µm), superbaby alpaca (<20 µm) and royal alpaca (<18 µm). However medullation of alpaca fiber is a usual disappointing feature that prevents being greatly estimated by the textile industry and the final consumer, given that it has been related to prickling. Therefore the present work aimed to estimate the genetic parameters of the percentage of medullation and its genetic correlations with other interesting traits. Data from 2000 to 2016 were obtained from PacoPro v5.4 software belonging to Pacomarca experimental farm, totaling 1,833 records of medullation percentage corresponding to 1,482 Huacaya (HU) and 351 Suri (SU) ecotypes. Fiber samples were taken from the mid side, and analyzed in an OFDA100 device, quantifying the percentage of continuously medullated fibers (PM). Genetic correlation was assessed between PM and other traits. Fiber diameter, standard deviation, coefficient of variation were analyzed concerning fiber traits; density, crimp in HU or lock structure in SU, head conformation, fiber coverage and general aspect were studied as morphological traits, and weaning weight (WW), age at first calving and calving interval as functional and secondary traits. PM ranged from 0.3 to 100% with an average of 28.46±28.98%. PM heritabilities resulted high (0.454) or very high (0.801) for HU and SU respectively. Genetic correlations were found to be moderate to high (0.269 to 0.584) between PM and other fiber traits in both ecotypes, being lower than 0.376 in absolute value between PM and the rest of traits, with only those with WW being unfavorable 0.232 for HU and 0.096 for SU ecotypes. The PM repeatabilities were found 0.746 and 0.840 for HU and SU ecotypes respectively. The genetic parameters and its economical importance make this trait valuable as selection criterion in the breeding program of the population. The high magnitude of the heritabilities suggest the possible existence of a segregating major gene.

CSN2 and CSN3 gene polymorphisms and their effects of milk production in Latvian local breed cows

D. Jonkus, S. Petrovska, L. Cielava and D. Smiltina
Latvia University of Agriculture, Liela 2, Jelgava, 3001, Latvia; daina.jonkus@llu.lv

Latvian Brown (LBGR) and Latvian Blue (LZ) cow breeds that are named local in Latvia. The local breeds characterizes with lower milk yield than Holstein Black and White cows, but their milk has better composition. To preserve the valuable traits of local breeds we need to find out their genetic suitability for production of high quality milk. The purpose of our study was to investigate the genetic diversity of beta (*CSN2*) and kappa (*CSN3*) casein proteins in local Latvian dairy cattle breeds and their association with milk yield and composition traits. Genotypes of milk protein genes, *CSN2* and *CSN3* were detected in 102 cows of LBGR breed, and 86 cows of LZ breed. In the LBGR breed the allelic frequencies for *CSN2* were A1=0.578 and A2=0.422, while for *CSN3*, they were A=0.750, B=0.235 and E=0.015. For LZ breed the allelic frequencies for *CSN2* were A1=0.605 and A2=0.395, while for *CSN3*, they were A=0.512, B=0.482 and E=0.006. Analyzing the frequency of genotypes after *CSN2* gene found that in LBGR and LZ breeds a higher it was for cows with heterozygous genotype A1A2, respectively 0.549 and 0.558, but the lowest – for A2A2 genotype (0.147 and 0.116). For both analyzed breeds by *CSN3* gene were observed four genotypes. Largest amount of LBGR breed cows had AA genotype, but LZ breed cows – AB genotype (respectively 0.539 and 0.558). Unfortunately in both analyzed breeds there were few cows with the optimal genotype for milk processing – BB which for LBGR breed cows had 0.039 frequency, but for LZ breed cows – 0.198. Both LBGR and LZ breeds had a few amount of cows whose genotype had unwanted E allele. Analyzed LBGR breed cows didn't have preferred gene combination (A2A2BB), but in LZ breed group with this combination there was only three cows (3.50%). First lactation LBGR breed cows with genotype A2A2AB had the highest milk yield in closed standard lactation (5,278.3±466.25 kg) and highest protein content in milk (3.43±0.09%), but for LZ breed cows it was higher obtained from cows with genotype A2A2BB (5,028±82.40 kg and 3.38±0.15%). The research is supported by State Research Program AgroBioRes 3 project (LIVESTOCK).

A combined approach for the estimation of recombination rate and linkage disequilibrium in half-sibs

A. Hampel, F. Teuscher and D. Wittenburg
Leibniz Institute for Farm Animal Biology, Institute of Genetics and Biometry, Wilhelm-Stahl-Allee 2, 18196 Dummerstorf,
Germany; wittenburg@fbn-dummerstorf.de

Population genetic parameters, such as linkage disequilibrium and recombination rate (θ), are parameters for characterising livestock populations. The population structure has an influence on these parameters. In half-sib families, which are a typical family structure in dairy cattle, the paternal θ and the linkage disequilibrium of maternal gametes (D) are parameters to be estimated. Previous studies demonstrated a likelihood approach for the estimation of θ and D which is based on an expectation maximisation (EM) algorithm. The special characteristic of the corresponding likelihood function (LF) is its bimodality and thus the possibility of having two maxima. The magnitudes of the maxima depend on the maternal allele frequencies at two loci. EM converges into a local maximum – which one depends on the start values. For this reason, a unique estimation of D and θ is actually not possible. Therefore, the objective of this study was to obtain suitable start values for a differentiation of the bimodality and to develop criteria for determining the maximum likelihood solution. We propose a method of moments (MM) to determine the start values of which the LF is closer to a maximum. The MM contains a quadratic equation, thus yielding two possible start values. The method takes parameters such as maternal allele frequency and empirical covariances between genotype codes for additive and dominance effects, which were obtained from progeny genotypes, at two loci into account. Both start values were used for the EM algorithm. After that, a decision process (DP) selects the most probable final solution of D and θ. The combined approach comprising MM, EM and DP was verified with a simulation study and compared with the outcome if start values of D and θ were fixed at zero. For maternal allele frequencies beyond 0.5, the LF is either unimodal or the two modes have different heights. The DP improved estimation of θ and D if the simulated θ was high (>0.3). For smaller rates, choosing fixed start values for θ and D yielded least biased estimates. If both maternal allele frequencies equal 0.5, additional information is needed to select the most appropriate solution.

Effects of including founder genomic relationship on genetic parameters from a mouse intercross

J. Meng, M. Mayer, E. Wytrwat and N. Reinsch
Leibniz Institute for Farm Animal Biology (FBN) Dummerstorf Germany, Institute of Genetics and Biometry, Wilhelm-
Stahl-Allee 2, 18196, Dummerstorf, Germany; meng@fbn-dummerstorf.de

High informative genotype data such as the SNP array data and the whole genome sequence data are increasingly applied in the animal breeding and QTL identification analyses. Combining pedigree and highly informative genotype data has been extensively researched in the animal breeding literature, while more questions and discussions about the combination of genotype and pedigree in a relationship matrix are raised, for instance, the feasibility of computing the pedigree-genotype combined genomic relationship matrix for a crossbred population. In the article, we demonstrate the protocol of extracting relationships between meta-founders and founders, between founders themselves as well as between founders and their descendants from eight founders' genotype data for the crossbred population which is bred from the high fertility mouse line FL1 and an inbred control line, constructing a relationship matrix using these three types of the relationship and test the effect of this integrated relationship matrix on the estimates of genetic parameters in the crossbred population. Both autosomal and X-chromosomal relationships were considered. The population comprises founders plus additional 29 generations with a total pedigree size of about 19,000. Estimates of genetic variances and heritabilities for six different traits, related to litter size and growth, were computed by including genomic founder relationships. For the sake of comparison in terms of the resulting parameters and estimated genetic trends all analyses were repeated by assuming founders to be unrelated. The subject is of general interest for the genetic analysis of data from synthetic populations in model organisms, livestock and plants.

Improving efficiency of parallel computing in solving very large mixed model equations
M. Taskinen, T. Pitkänen and I. Strandén
Natural Resources Institute Finland (Luke), Myllytie 1, 31600 Jokioinen, Finland; matti.taskinen@luke.fi

Estimation of breeding values (EBV) in dairy cattle is often computationally challenging. EBV are computed by solving mixed model equations (MME) using an iterative solving method such as preconditioned conjugate gradient (PCG). In PCG, the most time consuming step is the coefficient matrix of MME times a vector product. In distributed memory parallel computing, this task is divided to many processors. Parallel computing requires additional work due to communication between processors when the same vector value is needed by many processors. Our original approach for Nordic test-day models implemented in MiX99 was based on herd blocks where each herd block has effects within herd such as herd-year and animal effects. These effects are expected to have only some non-zero coefficients between different herd blocks. Communication is needed only due to a fraction of the effects and these are stored in a separate data structure, a linked list. Suboptimal ordering of the blocks leads to: (1) more communication; (2) unequal work load to processors; and (3) slower memory access of the linked list. Two methods were used in this study to enhance the speed of the parallel computing. First, a faster algorithm can be used to access the linked list. This direct access approach uses more memory instead of searching through the linked list. Secondly, the amount of communication can be reduced by reordering the blocks. Ordering can be improved by taking into account the amount of needed communication between blocks. Thus, the closely related blocks, that need considerable communication between each other, are placed in the same processor. We used Metis program for reordering and partition of the blocks. We tested the two approaches in the joint Nordic production evaluation of Holstein cattle. There were c. 380 million effects. Using 16 processors the first approach decreased the computing time 48% from 31.5 to 16.5 s per iteration. The second approach using Metis was able to decrease communication between processors from 6.3 to 2.1% of total number of effects. With 16 processors the computing time decreased 63% from 31.5 to 11.7 s per iteration. Reordering by Metis was able to balance the work more equally for the processors with increased number of processors.

False positives structural deletions in livestock
M. Mielczarek, M. Frąszczak and J. Szyda
Wroclaw University of Environmental and Life Sciences, Biostatistics group, Department of Genetics, Kozuchowska 7, 51-631, Poland; magda.mielczarek@upwr.edu.pl

The next generation sequencing represents a high throughput technology and thus is very sensitive to technical errors. In this study we showed that, in a context of structural variation, False Positives structural deletions result from gaps in the reference genome. Two datasets used in this study consisted of bovine and equine Copy Number Variant (CNV) type deletions and were obtained from publicly available Database of Genomic Variants archive (www.ebi.ac.uk/dgva). The data was filtered to include CNV deletions detected based on the next generation sequencing data longer than 50 bp. One study of the equine and three studies related to the bovine genome fulfilled the selection criteria. Coordinates of each CNV deletion reported in those studies were matched to the equine EquCab2.0 and bovine UMD3.1 reference genomes and the unknown bases content was calculated for the genomic region equivalent to the CNV deletion reported in each study. For the Split Read CNV detection method 56 CNV deletions corresponding to reference genome regions with at least 90% of unknown base pairs were identified for the bovine genome, while for the Read Depth CNV detection method none was reported. Since 53 of those CNV overlap between investigated individuals it might be hypothesized that they do not represent true structural variants, but are the result of unfavourable features of the reference genome which form a template for the analysis. 1.30% of the total genome length of the equine genome and 0.75% of the total genome length in bovine are represented by unknown base pairs, marked as N in the reference files. The mean length of such regions equals to 726 bp and 289 bp in EquCab2.0 and UMD3.1 respectively. Our analysis indicates that: (1) the presences of such regions causes false positive calls of structural deletions; and (2) there is a need for validation of such deletions either by using multiple software varying in the implemented detection methodology or (better) by a laboratory validation by methods considered as more reliable (e.g. aCGH, qPCR).

Polymorphism of ABCG2 (A/C) gene by PCR-SSCP and sequencing in Iranian Holstein dairy cattle

M. Hosseinpour Mashhadi

Department of Animal Science, Mashhad Branch, Islamic Azad University, Mashhad, Iran; mojtaba_h_m@yahoo.com

The aim of this study was to investigate the single nucleotide polymorphism (A/C) in exon 14 of ABCG2 gene in Holstein dairy cattle by PCR-SSCP and sequencing. ABCG2 (ATP binding cassette subfamily G member 2) gene was mapped on chromosome six in cow, this gene have 16 exon and 15 intron. The gene expressed in the apical membrane of alveolar mammary epithelial cells. ABCG2 gene encodes a protein that responsible for transporting various xenobiotics, cytostatic drugs and cholesterol from the plasma membrane to milk. Some SNPs were reported for this gene but a mutation in base 86 of exon 14 that transversion A to C nucleotide causes to change tyrosine (TAT) to serine (TCT) amino acids. This SNP affected on milk production traits. The blood samples of 320 dairy cattle from three farms were used for extracting genomic DNA by Genet Bio kit (Korea product) and Cinnapure DNA kit (Sinaclon product). A 240 bp fragment includes exon 14 of ABCG2 gene was amplified by following primers: F5'- GTATTCACGAGACTGTCAGGG -3' and R 5'-GGCTTTATTCTGGCTGTTTCC-3' in annealing temperature of 50°c by standard PCR. So for recognizing the SNP (A/C) in exon 14, first the PCR-SSCP technique used and based on single-strand conformation, 13 distinct patterns separated by polyacrylamide gels and detected by silver staining, two sample of each pattern was sequenced. The result showed the population for this (A/C) SNP was uniform, the frequency of allele A and C were one and zero respectively so the population was not in Hardy Weinberg equilibrium (P<0.05). Other researchers reported frequency of allele A as one or close to one so concluded that A allele is ancestral.

Genetic effects of VIP gene polymorphisms on reproductive traits of hen turkeys

S. Nikbin[1], L. Hosseinpour[1], N. Hedaiat[1] and G. Eliasi[2]

[1]University of Mohaghegh Ardabili, Animal Science, st. Daneshgah – Ardabil, 5619911367, Iran, [2]West Azarbaijan Agricultural and Natural Resources Research, Animal Science, Tabriz, West Azarbaijan, 5355179854, Iran; snikbin@uma.ac.ir

Vasoactive intestinal peptide (VIP), a neurotransmitter, is widely distributed in the body. Reports show that VIP plays important role in avian productive and reproductive traits. In the present study, polymorphisms of *VIP* and association of the SNPs with egg production, age of first egg, egg weight, laying period and brooding of Iranian indigenous turkeys were investigated. Genomic DNA of 130 turkeys was isolated from whole blood. A 433 bp of exon 6 and a part of 3'UTR of *VIP* gene (*VIPe6*) was amplified by standard PCR, using specific primers. Analysis of the results of PCR product sequencing showed a novel SNP, A5846G, of *VIP* gene. The frequencies of the genotypes were 0.85, 0.11 and 0.04 for AA, AG and GG respectively. The genotypes of *VIPe6* showed significant associations with egg number, total egg weight, and age at first egg. The results showed that the AG genotype increased significantly egg number and total egg weight compared to AA turkeys (P<0.05). The GG turkeys also showed the highest age at first egg (P<0.05). No association between the *VIPe6* variants and egg weight mean, weight of first egg, laying period and brooding was found. Overall, the genotypes of *VIPe6* showed significant association with egg production traits and may be considered a candidate gene in selection programs of Iranian indigenous turkeys to improve egg production traits.

Genetic parameters of fertility and grading traits in Finnish blue fox

R. Kempe and I. Strandén

Natural Resources Institute Finland (Luke), Biometrical genetics, Myllytie 1, 31600 Jokioinen, Finland; riitta.kempe@luke.fi

The primary goals in the Finnish blue fox breeding scheme are to improve fertility traits, fur quality and to maintain large pelt size. Fertility traits are included in the blue fox breeding scheme, but genetic trend in reproductive performance has been undesired due to unfavourable correlations to the other important production traits. Objective of this study was to estimate (co)variance components between the grading and fertility traits for the development of Finnish blue fox's national breeding scheme. Genetic parameters were estimated with multiple-trait animal model simultaneously for 8 traits, which contained 3 fertility traits (pregnancy rate, whelping success, first litter size) and 5 grading traits (grading size, quality, guard hair coverage, colour darkness, colour clarity). Data included observations of 42,462 animals from 9 farms. The pedigree contained 47,177 animas. Heritability estimates for the first litter size, pregnancy rate and whelping success were low (0.05-0.14). The highest heritability was obtained for colour darkness (0.64). The other grading traits had moderate heritability estimates (0.21-0.27). Genetic correlations between animal grading size and fertility traits were unfavourable (-0.15 to -0.53). Grading quality and guard hair coverage also had antagonistic relationship with all the studied fertility traits (-0.21 to -0.54). Colour clarity had unfavourable genetic correlation with the first litter size (-0.32) and with whelping success (-0.36). Fur quality traits (guard hair coverage, clarity and overall quality) had high genetic correlations (0.71-0.72) with each other. Colour darkness had no genetic correlation with the studied fertility or grading traits except for guard hair coverage (0.25). Larger animals with excellent fur quality tend to have worse than average reproductive results. It would be reasonable to take into account genetic correlations between the grading and fertility traits when selection weights are decided. In addition, litter size, pregnancy rate and whelping success evaluations are likely to benefit from the multiple-trait model where grading traits with highest genetic correlations to the fertility traits are included.

Genotyping of dopamine receptor D1 and somatostatin polymorphisms in Hungarian Yellow hens

K. Tempfli, K. Szalai, E. Lencses-Varga, K. Kovacsne Gaal and A. Bali Papp

Szechenyi Istvan University, Department of Animal Sciences, Mosonmagyarovar, 9200, Var 4., Hungary; tempfli.karoly@sze.hu

The Hungarian Yellow is a dual-purpose indigenous chicken breed developed in the early 20th century. In the present study our objectives were to determine G123A dopamine receptor D1 (DRD1) and A370G somatostatin (SST) allele and genotype frequencies in the population, and to analyse potential associations between genotypes and important production traits, such as egg production intensity or body weight. DRD1 can affect egg production via the regulation of prolactin secretion, whereas SST is an important growth-regulating hormone via the inhibition of growth hormone secretion. Using trap-nests and wing-tagging allowed the individual collection of egg production data. Body weight was measured biweekly from hatching to 14 weeks of age, and at 40 and 45 weeks of age. Genotyping was done by PCR-RFLP method using BsrSI and BsrBI enzymes for the digestion of DRD1 and SST PCR-products, respectively. The analysed DRD1 locus was polymorphic with genotype frequencies as follow: 0.17, 0.49, and 0.34 for AA, AG, and GG, respectively. Chi-square test indicated Hardy-Weinberg equilibrium in the population (P>0.05). Association analyses revealed DRD1 genotype effects on egg production intensity (P<0.05), and body weight at 45 weeks of age (P<0.05). The SST was not polymorphic in the Hungarian Yellow population, and A allele frequency was 1.0. The DRD1 polymorphism is suggested as a marker for egg production, whereas further studies are needed to assess SST frequencies in different chicken breeds.

Genetic evaluation of replacement heifers and monitoring of their growth

N. Bakri, A. Hamrouni, M. Khalifa, A. Arjoun, A. Najjar and M. Djemali
National Institute of Agriculture of Tunisia, Animal Science, 43 Avenue Charles Nicolle, Tunis 1082, 1082, Tunisia;
abirturki@yahoo.fr

Replacement heifers are the source of the new genetic potential in dairy herds. Livestock farmers in Tunisia tend to choose heifers for renewal only according to their conformation by neglecting their genetic value of production. The main objective of this work was to make three evaluations according to conformation, production and a combined conformation and production. The study involved 124 heifers from 33 sires and 45 maternal grandsires belonging to a pilot farm Frétissa. Three Selection Indexes were calculated using indexes of sires and maternal grandsires. A rank correlation between the different indexes was computed. The main results obtained showed that a strong correlation (0.99) was found between heifers' production indexes and their combined genetic values (0.7 × Production index + 0.3 × Conformation Index). The correlation between the conformation index and the production index was -0.05, showing that dairymen in Tunisia should use a combined index if they want to have efficient heifers in production with good conformation traits.

Genetic evaluation for lifetime production and longevity traits on Egyptian buffaloes

A. Khattab[1], S. El- Komey[1], K. Mourad[2] and W. Abdel Baray[1]
[1]Faulty of Agriculture, Tanta Univ., Animal production Department, Faculty of Agric., Tanta, Egypt, [2]Animal production Research insitute, Ministry of Agric., Cairo, Dokki, 002 02 34456, Egypt; Adelkhattab@yahoo.com

A total of 1,621 normal lactation records of Egyptian buffaloes, kept at Mehalet Mousa farm, belonging to the animal production Research Institute, Ministry of Agriculture, Dokki, Cairo, Egypt, during the period from 1991 to 2010 were used. Data were analyzed by using Multi Trait Animal Model, to study the fixed effects of month and year of birth and random effects of animals and errors on lifetime milk yield (LTMY), total lactation period (TLP), age at disposal (AGDS) and number of lactation completed (NLC). Means were 10,552 kg, 1,173 d, 125 mo., and 5.97 for LTMY, TLP, AGDS and NLC, respectively, Estimates of heritability were 0.27, 0.21, 0.12 and 0.06 for above traits studied. Estimates of genetic correlations among traits studied ranged from 0.56 to 0.80 and phenotypic correlations ranged from 0.67 to 0.84.

Family-specific analysis of whole-genome regression models in half-sibs

J. Klosa and D. Wittenburg

Leibniz Institute for Farm Animal Biology, Institute of Genetics and Biometry, Wilhelm-Stahl-Allee 2, 18196 Dummerstorf, Germany; wittenburg@fbn-dummerstorf.de

Identifying positions along the genome which affect a performance trait is of vast interest in quantitative genetics. Bayesian methods are among the most popular and successful mathematical approaches to estimate genetic effects captured by markers and to select the relevant loci, but their results strongly rely on proper choices of prior distributions for model parameters. Dependencies between markers due to linkage and linkage disequilibrium can be expressed in terms of covariance matrices, which can be included as prior information. They depend on population-specific parameters. Hence we focus on a population with half-sib family structure (e.g. dairy cattle). As a sample typically consists of multiple families, two different approaches are possible. Either family-specific effects or population (i.e. average) effects can be estimated for each marker. We expect that considering family-specific effects leads to less precise estimates of genetic effects due to the higher model dimension. On the other hand, more relevant loci shall be detected because the varying linkage phases of sires are respected unlike in population analyses where marker effects may cancel out. Results of these two approaches are compared with a method which does not explicitly account for dependencies between markers in a simulation study. The evaluation criteria are the frequency of detecting significant chromosome segments and the mean squared error (MSE) of the estimated genetic and residual variance component. In population analyses without covariances, we observed a high chance to correctly identify chromosome regions which harbour relevant loci with high effect size and a decent change for medium effects. In family-specific analyses, we obtained a clearer signal of positions with large effect size, the rate of false positive detections was reduced but medium effects were hardly identified. The use of covariance matrices seems advantageous when several causative variants are present in the neighbourhood of each other. Family-specific analyses can help to identify causative genomic regions at the expense of increased MSE, but large family sizes (e.g. n=1000) are required.

Correlated response in environmental genetic variability in two selected rabbit lines

I. Cervantes[1], M. Piles[2], N. Formoso-Rafferty[1], J.P. Sánchez[2] and J.P. Gutiérrez[1]

[1]Departamento de Producción Animal, Universidad Complutense de Madrid, Avda. Puerta de Hierro s/n, 28040, Madrid, Spain, [2]Institut de Recerca i Tecnologia Agroalimentàries (IRTA), Torre Marimon, 08140, Caldes de Montbui, Barcelona, Spain; icervantes@vet.ucm.es

In this research, the correlated effect of selection to increase prolificacy or growth on environmental genetic variability of growth traits at fattening was assessed. Heterogeneous residual variance models were fitted to weaning weight (WW), slaughter weight (SW) and average daily gain (DG) records from two rabbit lines selected for litter size at weaning (Prat line) and post-weaning (aprox. 35-65 d) daily weight gain at fattening (Caldes line). Individual growth records were considered to be repeated records from the mother. Data sets comprised 129,670 records from Caldes line and 155,315 from Prat line. An heteroscedastic model including sex, batch, and litter size as systematic effects, and litter and genetic as random effects, both affecting the trait and its environmental variability was fitted to each set of data. For all traits, the magnitude of genetic component of the environmental residual variance was in the low range found in the literature. The correlations between both genetic effects was almost null for WW and moderate to high and negative for SW and DG. The heritabilities ranged between 0.05 and 0.16. Note that since the data was assigned to the mother, this parameter includes the maternal genetic component plus a quarter of the individual genetic component of each record. In Caldes line, the genetic trend on the mean of the trait did not follow a clear pattern for WW but increased for SW and DG (1.6 g and 0.1 g/d per year) as expected. The genetic trend for the environmental variability of WW and DG tended to decrease (around 0.2% and 0.4% per year) but it tended to increase for SW (0.3% per year). On the contrary, in Prat line, the genetic trend for WW and SW clearly decreased (around 12 and 8 g/year, respectively) as it happened for the environmental variability of WW (around 2%), which has an important maternal component. However, no clear pattern was observed for the genetic component of DG or environmental variability of SW and DG in this line. This study is funded by a MEC-INIA Project (RTA2014-00015-C02-01/02).

New breeding objectives and index development for the Australian dairy industry

J.E. Pryce[1,2] and M. Shaffer[3]
[1]La Trobe University, Applied Systems Biology, 5 Ring Road, 3083 Bundoora, Australia, [2]Agriculture Victoria, 5 Ring Road, 3083 Bundoora, Australia, [3]DataGene, 5 Ring Road, 3083 Bundoora, Australia; mshaffer@datagene.com.au

Genomic selection has opened up opportunities for developing new breeding values that rely on phenotypes that use dedicated reference populations of genotyped cows. There are also opportunities to advance phenotype collection through automation and identifying predictor traits that can be measured cost-effectively in a sufficient sized population for genomic selection. Advances in the use of sequence data, in addition to gene expression studies, can lead to improved persistence of genomic breeding values across generations and potentially greater reliabilities for some traits. The challenge is harnessing optimal genomic prediction methods with developments in phenotype collection. We think that the most effective way of realizing rapid advances in dairy genetics and genomics is development of a highly integrated data collection, research and implementation platform. In Australia this is the foundation to delivering new methodologies and breeding values. For example, we have recently delivered the Feed Saved breeding values to industry and are soon to deliver genomic breeding values for Heat Tolerance. Identifying traits to include in the national objective will be the focus of future breeding value research, such as expanding the number of health traits breeding values available. However, industry, market and social drivers may see the emergence of new breeding values, such as cow level methane emissions, gestation length or niche milk products. Selection index methodology is still needed to ensure that the weights on each trait in the index are appropriate, although the weights can be (and are) subtly altered to respond to industry and consumer requirements. So far nationwide indices remain standard practice, but this may change in the future, especially as tools to deliver information back to farmers become more sophisticated. Already bull selection tools and personalised genetic trends are available, but the capture of economic and farm data will see the emergence of even more tools. Increasing the rate of genetic gain in the genomics era remains a challenge in Australia, so industry engagement is paramount.

The efficient dairy genome project: environmental stewardship through genomic selection

F. Miglior[1,2], C.B. Baes[2], A. Canovas[2], M. Coffey[3], E.E. Connor[4], M. De Pauw[5], B. Gredler[6], E. Goddard[5], G. Hailu[2], J. Lassen[7], P. Amer[8], V. Osborne[2], J. Pryce[9], M. Sargolzaei[2], F.S. Schenkel[2], E. Wall[3], Z. Wang[5], T. Wright[2,5] and P. Stothard[5]
[1]Canadian Dairy Network, Guelph, ON, N1K !E5, Canada, [2]University of Guelph, Guelph, ON, N1G 2W1, Canada, [3]Scottish Rural College, Edinburgh, Scotland, EH9 3FH, United Kingdom, [4]Animal Genomics and Improvement Laboratory, USDA, Beltsville, MD 20705, USA, [5]University of Alberta, Edmonton, AB, T6G 2C8, Canada, [6]Qualitas, Zug, 6300, Switzerland, [7]Aarhus University, Tjele, 8830, Denmark, [8]AbacusBio, Dunedin, 9016, New Zealand, [9]Dept of Econ Dev, Jobs, Transport and Resources, Melbourne, Vic, 3001, Australia; miglior@cdn.ca

The Efficient Dairy Genome Project is an international initiative that brings together research and industry experts to develop genomic selection tools for improving feed efficiency (FE) and decreasing methane emissions (ME) in dairy cattle. Data is being combined from different countries on feed intake, methane emissions, milk MIR spectra and pedigree information. In Canada, data is being collected from the University of Alberta, the University of Guelph, and a large commercial dairy farm. All data will flow to the project database, which is currently being developed for data storage and data exchange between the project and its international partners. A novel method for analysis of ratio traits, such as FE and ME, has been developed so that these traits can be integrated into routine genomic evaluations. In addition to phenotypes related to FE and ME, whole genome sequence information will be used to identify novel markers, which will be incorporated into genomic predictions. Up to 50 animals are currently being sequenced. A new method to select animals for sequencing was developed using an algorithm that optimizes the number of rare haplotypes and increases the overall genetic diversity in the sequenced population. Economic weights for FE and ME have been derived and various selection strategies for these traits are being evaluated. Additionally, research is being conducted at both the consumer and producer levels to determine the social acceptance and understanding of genomics and its use in selecting for FE and ME.

Addressing feed efficiency traits in the absence of feed intake measurements

M.H. Lidauer, E. Negussie, P. Mäntysaari and E.A. Mäntysaari

Natural Resources Institute Finland (Luke), Green technology, Myllytie 1, 31600 Jokioinen, Finland; martin.lidauer@luke.fi

Increasing importance of resource efficient milk production calls for incorporation of a new trait complex, i.e. efficiency, into genetic improvement programmes of dairy cattle. Residual feed (or energy) intake describes the overall efficiency of a cow and has been studied widely. Ratio traits like energy conversion efficiency (ECE) and energy conversion ratio (ECR) describe the efficiency of a cow in transforming energy of the diet into milk energy and therefore add another efficiency aspect to the efficiency trait complex. If desired, such ratio traits can be transferred into a linear index via Taylor series expansion. However, feed intake is not widely available to be used in breeding. Therefore, aim of this study was to estimate the genetic correlations between feed efficiency component traits and simple ratios. Daily feed efficiency records, collected from 539 primiparous Nordic Red dairy cows at Luke's experimental farm in Jokioinen, were used. Traits were lactation records of metabolizable energy intake (MEI), energy corrected milk (ECM), metabolic body weight (MBW), ECE=ECM/MEI and ECR=MEI/ECM. Because so far MEI is not available for cows under conventional recording, analyses included also indicator traits, where MEI was substituted either by its expectation or by the expectation of MEI for maintenance resulting traits EE=ECM/E[MEI], EE_M=ECM/E[MEI_M] and ER=E[MEI]/ECM, ER_M=E[MEI_M]/ECM, respectively. Bivariate analyses were carried out with an animal model including a calving-year-season as a fixed effect. Estimated genetic correlations between ECE and EE, ECE and EE_M, ECR and ER or ECR and ER_M were, 0.70, 0.73, 0.73 or 0.89, respectively. ER_M can be described as a cow's maintenance costs per kg ECM produced. The genetic correlation of ER_M with body weight (BW), ECM and MEI was 0.58, -0.91 and -0.18, respectively. In contrast, genetic correlation of ECR with corresponding traits was 0.82, -0.62 and 0.39, respectively. The high genetic correlation of ER_M with ECE (-0.86) and ECR (0.89) and its reasonable correlations with BW, ECM and MEI, makes ER_M a candidate trait to be considered for inclusion into dairy cows' efficiency index.

Genetic parameters and SNP marker effects for metabolic diseases and stress indicators

S.-L. Klein, K. Brügemann, S. Naderi and S. König

Institute of Animal Breeding and Genetics, Justus-Liebig-University Gießen, Ludwigstraße 21B, 35390 Gießen, Germany; sarah.klein@agrar.uni-giessen.de

Enormous energy demand for milk production in early lactation exceeds energy intake, especially in high yielding Holstein cows. Energy deficiency causes metabolic stress symptoms, e.g. increasing susceptibility to metabolic diseases. In addition to several blood parameters, the fat to protein ratio (FPR) is considered as a valuable indicator for the energy status postpartum. A FPR>1.5 refers to high lipolysis, indicating the metabolic disease ketosis (KET). Aim of this study was to analyse associations between FPR and KET on phenotypic, quantitative-genetic and genomic scales. A total of 9,662 first lactation Holstein cows were phenotyped for KET according to a veterinarian diagnosis key, and for FPR considering the first testday. One entry for KET in the first six weeks after calving implied a score = 1 (diseased); otherwise, score = 0 (healthy). All cows were genotyped with the Illumina BovineSNP50 BeadChip. At the phenotypic scale, increasing KET incidences were associated with significantly higher FPR (P<0.0001), and vice versa. Using generalized linear mixed models with a logit link and an identity link function, heritability estimates were 0.16 for FPR, and 0.17 for KET, respectively. Using bivariate linear-linear animal models, the genetic correlation between KET and FPR was 0.50. Genome-wide association studies identified SNP associated with KET and FPR. The specific SNP ARS-BFGL-NGS-85355 (Chr5, Pos: 115,456,438 bp) contributed to KET, and a gene in close chromosomal distance was associated with non-alcoholic fatty liver disease in humans. The most important SNP marker contributing to FPR was located within the DGAT1 gene. In contrast to quantitative-genetic and phenotypic associations, only weak relationships were identified between FPR and KET at genomic scales. The average correlation coefficient between KET and FPR considering SNP effects from all chromosomes was close to zero (0.06). Nevertheless, from a herd management perspective, routinely measured FPR can be used as metabolic disease indicator, and genetic (co)variance components justify inclusion of FPR into overall breeding goals.

A closer look at longevity in dairy cows: different stages – different genes

J. Heise[1], S. Kipp[2], N.-T. Ha[1], F. Reinhardt[2] and H. Simianer[1]

[1]*University of Goettingen, Animal Breeding and Genetics, Albrecht-Thaer-Weg 3, 37075 Goettingen, Germany,* [2]*IT Solutions for Animal Production (vit), Heinrich-Schroeder-Weg 1, 27283 Verden, Germany; johannes.heise@uni-goettingen.de*

Longevity is an economically important trait in dairy cattle and can be interpreted as consecutive survival of different periods. The genetic background for survival of different periods is known to be different. For the development of new models for routine conventional genetic and genomic evaluations, knowledge about the relationships between the genome and the phenotype is highly desirable. Therefore, we conducted a single-marker as well as a gene-based genome-wide association study, using deregressed estimated breeding values (EBV) from a prototype of the new conventional (pedigree-based) genetic evaluation of longevity in German Holsteins. The model treats survival of 9 different periods (0-49 d, 50-249 d, and 250 d up to the consecutive calving in the first three lactations) as genetically distinct, but correlated traits. In total, data of 4,849 bulls were used for which imputed high-density SNP-genotypes were available. Minimum reliability for the EBV for the first period was 0.7. Different regions on chromosomes 5 (89 Mbp), 6 (89 Mbp), 7 (93 Mbp) and 18 (45, 51, 57, and 63 Mbp) were found to have a significant association to at least one survival trait. These regions were also reported in previous studies on longevity, but our results add knowledge about the differentiated contribution of those regions to survival of different periods. For example, the regions at 45 and 51 Mbp on chromosome 18 were much stronger associated to survival from day 250 up to the consecutive calving than up to day 250 in the same lactation. This period primarily reflects reproduction success. The same pattern was observed for the region on chromosome 7. Conversely, the other regions on chromosome 18 as well as the ones on chromosome 5 and 6 showed highest association to survival up to 250 days from calving. One of these regions on chromosome 18 was previously reported as a candidate for calving traits, the region on chromosome 6 for mastitis resistance. The region on chromosome 14 showed higher significance for survival up to 250 days from calving in second and third lactations, but not to survival of the first.

Genetic evaluation of claw health traits accounting for potential preselection of cows to be trimmed

I. Croué[1,2], K. Johansson[3], E. Carlén[3], G. Thomas[4], H. Leclerc[1], F. Fikse[3] and V. Ducrocq[2]

[1]*Institut de l'Elevage, UMR1313 GABI, 78352 Jouy-en-Josas Cedex, France,* [2]*INRA, UMR1313 GABI, 78352 Jouy-en-Josas Cedex, France,* [3]*Växa Sverige, Institutionen för Husdjursgenetik, SLU, Ulls väg 26, 756 51 Uppsala, Sweden,* [4]*Institut de l'Elevage, Chambre régionale d'agriculture de Picardie, 19 bis rue Alexandre Dumas, 80096 Amiens Cedex 3, France; iola.croue@inra.fr*

Claw lesions are one of the most important health issues in dairy cattle. Although their frequencies greatly rely on herd management, they can be lowered through genetic selection. A genetic evaluation can be developed based on trimming records collected by hoof trimmers. However, not all cows present in a herd are scored by trimmers. The objectives of this study were to investigate the importance of preselection of cows for trimming, ways to account for this preselection and to estimate genetic parameters of claw health traits. The data set contained 25,511 trimming records of French Holstein cows. Analyzed claw lesion traits were digital dermatitis, heel horn erosion, interdigital hyperplasia, sole hemorrhage circumscribed, sole hemorrhage diffused, sole ulcer and white line fissure. All traits were analyzed as binary traits in a multitrait linear animal model, including only trimmed cows in a seven-trait model (scenario 1), or trimmed cows and contemporary cows not trimmed but present at the time of a visit (i.e. considering that non-trimmed cows were healthy) in a seven-trait model (scenario 2), or trimmed cows and contemporary cows not trimmed but present at the time of a visit, in an eight-trait model, including a 0/1 trimming status trait (scenario 3). For scenario 3, heritability estimates ranged from 0.02 to 0.09 on the observed scale. Genetic correlations clearly revealed two groups of traits (digital dermatitis, heel horn erosion and interdigital hyperplasia on the one hand, and sole hemorrhage circumscribed, sole hemorrhage diffused, sole ulcer and white line fissure on the other hand). Heritabilities on the underlying scale did not vary much depending on the scenario. However including untrimmed cows as healthy caused bias in the estimation of genetic correlations. Scenario 3 seemed to be the best way of dealing with cow preselection, as it accounted for all available information without causing bias.

Health and survival of calves in Holstein: a genetic analysis
M. Neubert, R. Schafberg and H.H. Swalve
Martin Luther University Halle-Wittenberg, Institute of Agricultural and Nutritional Sciences, Theodor-Lieser-Str. 11,
06120 Halle, Germany; marie.neubert@landw.uni-halle.de

Health and survival of young stock play an important role for the profitability of a dairy farm. While recording health traits in recent years has become an important tool as an aid in improving the management as well as for genetic analyses and estimation of breeding values, little data and knowledge so far is available on young stock reared as replacements for the dairy herd. In contrast, it is well known that diseases in early life play an important role for later health and performance of the dairy cow. In the project YHealth, health data is collected on commercial dairy farms in the German state Saxony-Anhalt. Data was collected for calves born between 1st January 2013 and 16th June 2016 on 17 large dairy farms. For a genetic-statistical analyses 10,796 female Holstein with 39,574 plausible health records during the first 50 days of life were included. The objectives were to estimate the genetic parameters for diarrhea (DIA) and pneumonia (PNE). Both traits were coded as linear traits and analyzed using a univariate linear animal model. The model included fixed effects of herd and year, and animal as a random effect. Heritabilities, reflecting the genetic pre-disposition for the diseases, were estimated as 0.12 to 0.13 for DIA and 0.12 to 0.14 for PNE, with ranges denoting different definitions of the phenotype. Estimates for genetic correlations between DIA and PNE were 0.59 to 0.64, indicating that the two traits are not identical.

The increase in body size of Holstein cows in Switzerland
H. Joerg, E.A. Boillat, A. Wenger and A. Burren
Bern University of Applied Sciences, School of Agricultural, Forest and Food Sciences, Länggasse 85, 3052 Zollikofen,
Switzerland; hannes.joerg@bfh.ch

The stature continuously increases in Swiss dairy cattle. In the last 10 years, the average height at sacrum of the Holstein cows increased by 30 mm. To analyse the trend of increase estimated breeding values (EBVs) of 10,842 Holstein Friesian (HF) bulls and 17,983 Red Holstein (RH) bulls born between 1985 and 2015 were analysed. The correlations between the EBVs of stature and other traits were calculated to estimate the indirect selection due to the increase of stature. The increase in EBVs of stature between 1985 and 2015 was 30 and 25 index points for HF and RH, respectively. This is more than two times the genetic standard deviation of 12. The increase is not linear and acceleration was shown in the EBVs of bulls born after 2000. The EBV for conformation was strongly correlated with the stature with 0.89 and 0.70 for HF and RH, respectively. One part of the correlation is due to autocorrelation, since the stature is part of the conformation and the correlations between further traits of the frame and the udder were not compensated. The milk yield, the longevity, the somatic cells and the fitness value were correlated moderately with the stature with values between 0.63 and 0.12. The persistency was weakly correlated to stature 0.05 for HF and -0.06 for RH. Due to the trend in the EBV of the Holstein bulls the height at sacrum of the Holstein cows will increase approximately 60 mm in the next 10 years. Swiss dairy cows are well known for their excellent conformation traits and the stature is the most important trait of the conformation, so there is no sign that the statures of Holstein cows will no longer increase in Switzerland.

Transport of delight: 'getting it right' in livestock transport now and in the future?
M.A. Mitchell
SRUC, Animal and Veterinary Science, The Roslin Institute Building, Easter Bush, Midlothian, EH25 9RG Edinburgh,
United Kingdom; malcolm.mitchell@sruc.ac.uk

The reality of about transportation is that it is future orientated. If we're planning for what we have we're behind the curve! Anthony Foxx, American Politician. Animal transport is a key part of the animal production cycle with livestock being transported for purposes of breeding, re-location and slaughter. Estimates of the number of animal journeys per annum vary but it may be safely assumed that in excess of 60 billion animals are transported to slaughter each year with a large proportion being poultry. Despite the recognition that transportation and particularly the conditions encountered in transit may be responsible for animal welfare problems, health issues, reduced production efficiency, increased mortality, poor product quality and economic losses there are still opportunities to develop novel strategies to minimize transport stress and to prevent or reduce these problems. As indicated above (Foxx), it is now essential to plan what measures might be adopted in the long term to improve vehicles, procedures, equipment and technologies, training and guides and legislation to improve the health and welfare of animals in transit. These actions must be undertaken in the face of current and new challenges created by climate change and changes in economic and trade patterns and alterations in cultural and societal perceptions and perspectives relating to animal production and transportation. We must develop strategies now to solve today's problems but also to future-proof animal transport systems to ensure long term benefits in animal health and welfare. These objectives can only be achieved by detailed analysis and understanding of the mechanisms that underlie the problems in animal transportation. This may depend upon the use of predictive modelling to characterize and quantify current and future risks to animal health and welfare in transit which will further underpin the adoption and application of new and developing technologies to improve all aspects of the animal transport process. New disruptive and ground breaking technologies in other fields may inform the development of novel monitoring and control systems for vehicle and animals in the future.

Animal transport guides: a tool box for the transport industry to comply with 1/2005
H.A.M. Spoolder[1], K. De Roest[2], E. Nalon[3], M. Billiet[4] and N. De Briyne[5]
[1]Wageningen Livestock Research, Animal Welfare, P.O. Box 338, 6700 AH Wageningen, the Netherlands, [2]CRPA, Viale Timava 43/2, Reggio Emilia, Italy, [3]Eurogroup for Animals, Rue des Patriotes 6, 1000 Brussels, Belgium, [4]International Roadtransport Union, 32-34 Ave de Tervueren, bte 17, 1040 Brussels, Belgium, [5]Federation of Veterinarians of Europe, 12 Avenue de Tervuren, 1040 Brussels, Belgium; hans.spoolder@wur.nl

The European Parliament has indicated that practices regarding the transportation of 40 million farmed animals differ considerably across Member States. These differences affect the overall level of enforcement across the EU of the legislation. For example, EU Regulation (EC) No. 1/2005 requires that animals are checked for fitness before they are transported on long journeys, but there are differences in the evaluation of the status of the animals and concerns about potential conflicts between the different actors. DG SANTE therefore commissioned a project on the dissemination of best practices for international and domestic animal transport. 'Animal Transport Guides' focusses on the main farm species and key stakeholders involved. It collated and analysed international literature, followed by an iterative stakeholder process to generate Guides to Good Practice for cattle, horse, pig, sheep and poultry transport. They follow the main activities from journey planning, via the actual transportation through to journey evaluation, and include 'Good' as well as 'Better' practices. Recommendations are included on important welfare issues such as animal handling, space allowances, dealing with heat stress, milking lactating animals and providing food and water during the journey. In addition to the Guides (which are aimed at transport organisers and competent authorities), several fact sheets and short video clips for farmers and drivers were developed. These materials are currently presented and discussed with stakeholders of eight EU countries during 'Road Shows' held at major livestock farming and transport events across Europe.

Fitness for transport

M.S. Cockram

University of Prince Edward Island, Sir James Dunn Animal Welfare Centre, Dept. of Health Management, AVC, Charlottetown, PEI, C1A 4P3, Canada; mcockram@upei.ca

IIn the EU and Canada, regulations specify conditions that are considered to make an animal unfit for transport and guidance has been developed on how to assess fitness for transport. There are differences in approach between Canada and the EU, and different stakeholders have different criteria for determining fitness for transport. Examination of enforcement action, condemnation statistics and surveys of animals sent to slaughter show that some animals sent for slaughter have identifiable conditions that were present before transportation. Consideration of the pathophysiology of pre-existing conditions in relation to the physical and physiological challenges of transportation indicates how these conditions can affect an animal's ability to cope with transportation and increase the risk of suffering. Transportation is likely to exaggerate pre-transport health issues. Animals sent for slaughter with pre-existing conditions are more likely to die in transit, become non-ambulatory or be euthanized on arrival than those that are healthy. Many animals with disease or injury are most likely experiencing pain and feeling ill before they are transported. Transportation will almost certainly aggravate any painful conditions e.g. lameness and result in additional suffering. Some conditions affect physiological function, e.g. pneumonia can reduce exercise tolerance and capacity to deal with heat. Emaciated or weak animals may have reduced ability to obtain food and water, are more susceptible to the combined effects of fasting and cold exposure and are less able to respond to events affecting their stability. In Canada, issues that can arise when animals are not fit for their intended journey are reported when appeals are made against enforcement action. Between 2001 and 2016, out of 62 appeal cases, 45% involved lame pigs (journey duration (h) median 1.5, Q_1 0.5, Q_3 3.5 h), 16% pigs with a hernia (2.0, 0.6, 4.6 h), 6% pigs with prolapses or disease (2.2, 1.7, 10.5 h), 15% lame cattle (2.5, 1.8, 7.6 h), 7% recumbent cull cows (1.7, 0.7, 8.9 h), 5% cattle with infections or recently given birth (4.2, 0.1, 4.3 h), 5% slaughter horses (median 16.5 h) and 1% sheep (median 1.7 h). Fitness for transportation is a key factor affecting how well animals cope with transportation.

Statistics and welfare outcomes of livestock transported for slaughter in South of Italy

D. Tullio[1], A. Carone[2] and B. Padalino[3,4]

[1]ASL Ba, Local Health Unit. Veterinary Service, Lungomare Starita 6, 70100 Bari, Italy, [2]Veterinary Clinic Marichyta, Via Principe Amedeo 142/A, 70100 Bari, Italy, [3]University of Bari, Department of Veterinary Medicine, Casamassima St Km 3, 70010 Valenzano (Ba), Italy, [4]The University of Sydney, Faculty of Veterinary Science, 425 Werombi road, 2560 Camden, Australia; barbara.padalino@sydney.edu.au

Millions of animals are moved daily worldwide and transportation may lead to disease and poor welfare. However, statistics on animal transport movements and their implications for health and welfare are limited. To comply with the EC Animal Transport Regulation 1/2005 (Annex 1, Chapter 5), several livestock transported for slaughter from North Europe to Greece stop in a control post in the South of Italy for compulsory rest stop. This study documents the animal transport movements transited across the control post and their welfare status as measured by the incidence of death, injuries and diseases. Surveillance reports issued by the Veterinary Service of ASL Puglia 1 (Italy) from the 1st of August 2005 to the 31st of July 2006 were analysed. Over that period, a total of 69,745 animals in 540 separate shipments transited. In particular, a total of 49,571 (71%) sheep and goats, 16,731 beef cattle (24%), and 3,443 pigs (5%) traveled in 139 (25.7%), 382 (70.7%) and 19 (3.6%) trucks, respectively. While road transportation of beef cattle was constant over the year, sheep and goats transportation peaked in April and May. Animals came from Belgium, Denmark, France, Germany, Holland, Ireland, Luxembourg, Spain and Poland. France was the most prevalent provenience for cattle and pigs, while small ruminants came mainly from Spain. The mortality rate was of 0.035% for beef cattle, 0.030% for sheep and goats, and 0% for pigs. Morbidity (animals presenting injuries, physiological weakness or pathological processes) was 0.897% and 0.062% for cattle and sheep/goats, respectively. Injuries were rarely related to the loading/unloading procedures. Overall, this data highlight the utility of rest stops and surveillance practice during long journey travels. Further studies are needed to identify risk factors during livestock transportation using a larger dataset composed by surveillance reports and the fines issued after the application of the Italian Decree 151/2007.

Investigations on behavioural and thermoregulatory needs of heifers during long transport in Europe

M. Marahrens and K. Von Deylen
Friedrich-Loeffler-Institute, Institute for Animal Welfare and Animal Husbandry, Doernbergstr. 25+27, 29223 Celle, Germany; karin.vondeylen@fli.de

The transport of animals per se is associated with animal welfare concerns. In addition to the transport duration, loading density and compartment height are important aspects when assessing animal welfare on long-term road transports of cattle. The presented study investigated how the combination of different loading densities [low density (LD): 1.5 m²/animal, high density (HD): 1.3 m²/animal] and headspaces (small versus enlarged headspace: 10 resp. 20 cm above the withers of the tallest animal) affect the ventilation conditions in the transport vehicle as well as the behavior and the thermoregulation of breeding heifers during long transports under commercial conditions. A total of eight long road transports of breeding heifers with a distance of more than 1000 km were accompanied across Europe from assembly center to farm. On each transport, sixteen heifers were randomly selected for the monitoring of body temperature and the examination of selected blood parameters. In addition to the video-supported behavioral observation, the climatic conditions within the vehicle (temperature, humidity, air velocity) were recorded. The airflow above the animals showed in some cases results above the recommended values for stable keeping. The ethological assessment, however, revealed no avoidance behavior or signs of undercooling. The loading density had an impact on the resting behavior of the animals during transport regardless of the headspace variants, LD heifers displayed more resting behavior than HD heifers. Blood parameters were not significantly affected by loading density. Results indicate that heifers are more capable of adapting to long transports when loaded at lower densities. At higher loading densities, a larger ventilation area above the animals (headspace) shows a tendency to compensate for the potential greater heat production in the vehicle.

Welfare assessment of cattle at control posts

M. Nardoia[1], S. Messori[1], W. Ouweltjes[2], P. Dalla Villa[1], C. Pedernera[3], B. Mounaix[4] and A. Velarde[3]
[1]Istituto Zooprofilattico dell'Abruzzo e del Molise, Campo Boario, 64100 Teramo, Italy, [2]Wageningen UR, Livestock Research, P.O. Box 338, 6700 AH Wageningen, the Netherlands, [3]IRTA, Veinat de Sies, s/n. Monells, 17121, Spain, [4]Institut de l'Elevage, 149 rue de Bercy, 75595 Paris, France; m.nardoia@izs.it

Long distance transport of live animals is associated with a wide variety of stressors, often resulting in poor animal welfare (AW). Stopping at control posts (CP) after reaching a threshold (Regulation EC No 1/2005) to accommodate animals and to ensure them good welfare conditions while maintaining their health status is mandatory. Despite their importance, CPs are missing in certain locations and sometimes have poor quality standards. This study is part of the European project 'Renovation to promote High Quality CPs in the European Union' aiming to renovate CPs and to set up a reference for a higher AW standard. A scientifically sound tool consisting of three protocols for the welfare assessments of calves on arrival, in the resting pens and at departure from CP was developed. The protocols, based on the Welfare Quality® ones, covered 12 welfare criteria, grouped into four main principles (good feeding, good housing, good health and appropriate behaviour). The protocols were field-tested on 16 calves' transports which stopped at one CP located in Northern Italy. A descriptive analysis was carried out to summarise the main characteristics of the assessments, which were performed during unloading, in the resting pen (1 h after arrival and 1 h before loading), and during loading. Averagely, 20% of calves vocalised at unloading and 5% at loading; at the CP, 35 and 50% vocalised after arrival and before departure, respectively. Less animals slipping (11.7 vs 7.4%) and falling down (2 vs 0.9%) were recorded during unloading and loading. The percentage of calves lying while staying at CP, before departure, was higher after renovation (29.7%) than before (16.0%). Nevertheless, the presented results might not be ascribed to CP renovation only, since other factors might have influenced the assessed parameters. This tool represents a scientific basis for setting up a certification scheme for CPs and provides a useful self-assessment method for CP owners and/or competent authorities to give advice about how to improve animal welfare.

Horse meat production: commercial transport and slaughter procedure

R.C. Roy

Prairie Swine Centre Inc, Box 21057, 2105 – 8th Street East, Saskatoon, SK, S7H 5N9, Canada; rroy@upei.ca

Approximately 785,185 tonnes of horse meat was produced worldwide in 2013, which is a steady increase from 742,695 tonnes in 2009. China and North America (Canada, Mexico and United States) are leading producers of horse meat. The population of horses is also high in China and North America. A utilitarian perspective could suggest that horse meat production could be one of the viable economic options if welfare concerns regarding transport and slaughter are addressed. People holding an animal rights perspective, those who consider horses as pet species and those who follow some religions and cultures may disagree with this view point. Unfortunately, horse meat production is commonly associated with long distances of transport and bad conditions of transport. In reality, there are countries were horses are transported only short distances for slaughter. For example, horses are transported a maximum of 3 h in Iceland and around 8 h if transported within Canada for slaughter. In North America, transport durations are longer when cross-border transportation occurs, for example, from the USA to Canada and the USA to Mexico. Research on commercial transport of horses has found that horses can be transported up to 20 h with minimum effect on their health and welfare if proper preventative management care is provided. Even though mortality has been steadily declining in commercial transport of horses, the prevalence of injuries are still high. Fitness of horses and absence of individual separation for horses during commercial transport leading to non-ambulatory condition is also a significant welfare issue. Slaughter procedure in horses consists of stunning and sticking. There are two techniques used to stun horses. Captive bolt or a rifle. The captive bolt has more safety features for people involved. There are no comparative studies to evaluate the effectiveness of the two techniques for stunnning. Stunning efficiency assessment for captive bolt in Iceland and Chile ranged from 86 to 98%. Grandin suggested that an effective stunning practice for a slaughter plant was to have a minimum standard of 99% horses stunned by one shot. There is a need to undertake research to evaluate the factors that affect efficiency of stunning horses.

Australian consumer concerns about sheep and beef cattle transportation and slaughter

E.A. Buddle[1,2], H.J. Bray[2], W.S. Pitchford[1] and R.A. Ankeny[2]

[1]University of Adelaide, Department of Animal and Veterinary Sciences, Roseworthy, 5371 South Australia, Australia, [2]University of Adelaide, Department of Humanities, North Terrace, Adelaide, 5000 South Australia, Australia; emily.buddle@adelaide.edu.au

Meat consumers are increasingly interested in livestock production practices. There is tension between the desire to consume meat and wanting to make ethical food choices associated with environmental and animal welfare concerns. Concern about the transportation and slaughter of sheep and beef cattle were highlighted within results of a larger research project which attempts to explore how Australian meat consumers conceptualise animal welfare. Sixty-six meat consumers from across Australia participated in focus groups and interviews in 2015 and 2016, consistent with qualitative methods. Interviews were structured according to a series of discussion topics such as on-farm welfare and meat purchasing decisions. Qualitative methods demonstrate significance by continuing to conduct interviews until no new themes emerge, known as saturation. These interviews were digitally recorded, transcribed fully and coded using methods similar to 'open-coding'. Concern about transport extended from road transport in trucks through to the shipping conditions of the live export trade. While participants accepted that slaughter was necessary for the consumption of meat, several did not want to think about the domestic slaughter process. However, many participants held strong opinions about the conditions which Australian livestock are slaughtered overseas, particularly associated with religious festivals and beliefs. Participants also had some understanding about the impact of pre-slaughter stress on meat quality. This research highlights areas of concern amongst Australian meat consumers about the transportation and slaughter of livestock which can be used to generate productive dialogue between industry and the community and potential policy changes around current livestock production practices.

Determination of effects of pre-slaughter resting periods on carcass and meat quality by DIA

Y. Bozkurt[1], M. Saatci[2] and Ö. Elmaz[2]
[1]Suleyman Demirel University, Faculty of Agriculture, Department of Animal Science, Cunur, 32260, Isparta, Turkey,
[2]Mehmet Akif Ersoy University, Faculty of Veterinary Medicine, Department of Animal Science, MAKU kampus, 15030,
Burdur, Turkey; yalcinbozkurt@sdu.edu.tr

In this study, effects of different pre-slaughter resting periods on the carcass and meat quality of slaughtered beef cattle by Digital Image Analysis and traditional methods (colorimeter) was determined. There were 4 resting periods, for those slaughtered without resting (RP0); 3 h resting (RP3), 6 h (RP6) and 12 h (RP12) and formed as four different groups of animals and 375 samples in total from 10 animal carcasses from each group were collected. The samples of 2.5 cm thick *Longissmus dorsi muscle* area (LMA) between the 12 and 13th ribs were taken and pH measurements were obtained from the lumbar region. LMAs were measured using a planimeter and colour values with colorimeter (L*, a* and b*) were determined. Digital images of the LMA samples taken from the carcass were subjected to image analysis. Both LMA and the colour distribution of red, green and blue (RGB) values and their density were determined by DIA. A 10 cm reference card was placed next to each object in order to eliminate the effects of shooting distance between the object and a digital camera. The measurements of pH of each group, L*, a* and b* values and the RGB values obtained by the image analysis were subjected to ANOVA analysis by using GLM (General Linear Model) procedure and Tukey test was applied for pairwise mean comparisons. All RGB colour values of the meat of the animals slaughtered without resting period (RP0) were significantly higher (P=0.01) than those obtained in other resting periods. Similarly, L*, a* and b* values of the meat of the animals slaughtered without resting period (RP0) were significantly higher (P=0.01) than those in other resting periods. It was also found that there were no any statistical differences (P=0.230) in the colour values of the meats of the animals subjected to resting time 6 and 12 h in both methods. Therefore, these results suggested that it would be appropriate for the animals to be rested for at least 6 h before slaughter. It can also be concluded that DIA may open up new dimensions to determine some other meat quality parameters.

Electrolyte and glycerol solution supply to feed deprived bulls before slaughter

I.P.C. Carvalho and J. Martín-Tereso
Trouw Nutrition, Research and Development, Veerstraat 38, 5830 AE Boxmeer, the Netherlands;
isabela.carvalho@trouwnutrition.com

Stress and feed deprivation results in transient endocrine responses leading to dehydration, glycogen depletion, protein catabolism and ultimately to a loss of carcass weight. The present study aimed to evaluate the effects of supplementing a potassium-based electrolyte solution in combination with glycerol to feed deprived Holstein bulls prior to slaughter. Forty-eight Holstein bulls were used following a randomized block design with 6 blocks of 4 pens and 2 animals per pen. Animals were assigned to pens and blocked based on initial body weight (BW). Pens were considered the experimental unit and within blocks were randomised to 2 treatments: CONTROL (water) and SUPPLEMENT (water solution of a Potassium-based electrolyte mixture + glycerol; 27.6 g/l). Animals were feed deprived for 48 h while they had constant access to the liquid treatments. Data was analysed with PROC MIXED in SAS. Solution intake was higher (P<0.01; SEM=2.76) for SUPPLEMENT animals (47.5 vs 25.2 l/animal/d). When expressed as % of BW, SUPPLEMENT bulls drank twice as much solution (14.0 vs7.5%BW; P<0.01; SEM=0.81) and their drop in body temperature was lower (P=0.02; SEM= 0.072) (0.40± vs 0.65 °C after 48 h) indicating less dehydration. Initial BW was equal (P=0.21; SEM=0.82) between CONTROL and SUPPLEMENT groups (333 vs 335 kg). However, after 48 h of feed deprivation, SUPPLEMENT bulls presented higher BW (321 vs 312 kg; P=0.01; SEM=2.22). SUPPLEMENT animals lost on average 13.8 kg or 4.1% BW while CONTROL animals lost 20.9 kg or 6.3% BW during the feed deprivation period (P=0.02; SEM=1.92 and P=0.02; SEM=0.60, respectively). Hot carcass (HC) weight and cold carcass (CC) weight tended to be higher (P=0.06; SEM=0.92 and P=0.05; SEM=1.17, respectively) for SUPPLEMENT. Nevertheless, HC yield, CC yield and cooler shrink, were not different (P=0.17; SEM=0.29, P=0.11; SEM=0.33 and P=0.16; SEM=0.05, respectively) between CONTROL and SUPPLEMENT (50.4%, 49.5%, 1.4% vs 51.0%, 50.3%, 1.3%, respectively). When supplied to animals subjected to a prolonged period of feed deprivation, a potassium-based hypotonic electrolyte solution containing also glycerol prevented animal dehydration prior to slaughter and reduced carcass losses.

Evaluation of pig abattoir-data for animal health assessments in Germany

M. Gertz and J. Krieter

Christian-Albrechts-University, Institute of Animal Breeding and Husbandry, Olshausenstr. 40, 24098 Kiel, Germany; mgertz@tierzucht.uni-kiel.de

During the last years, animal welfare has gained much attention in the public, resulting in a high demand for animal welfare surveillance. Recent endeavors of the German administration aim the inclusion of pig abattoir data into a farm-specific evaluation of health status. Abattoir data is known for having a high validity but low reliability, which makes inference of farm-specific information often difficult. The aim of our study was therefore to evaluate the potential and pitfalls of using abattoir-data as basis for an animal-health assessment in Germany. For this purpose we evaluated Intraclass Correlation (ICC) of findings within and between abattoirs, as well as the suitability of different Boxplot Methods for identification of high prevalences. Therefore, ca. 19 million observations of 12 German abattoirs during 2015-2016 were evaluated on a monthly basis. Considered organ lesions and observed prevalences were pleurisy (9.5%), pneumonia (15.3%), milkspots (9.5%) and pericarditis (3.2%). ICC-Agreement within abattoir was lowest for pneumonia (0.09-0.56) and highest for milkspots (0.57-0.86). In contrast, ICC displayed values of 0.02 for pneumonia and 0.37 for milkspots between the different abattoirs only. Using the Boxplot-Method for identification of high prevalences resulted in unexpected high rates of outliers (expected: 0.35%, observed: 3.2-5.0%). Available data was thereby either log-normal or gamma-distributed, exhibiting a high skewness (2.50-1.35) and kurtosis (14.59-4.16). Subsequently, the *Medcouple-Boxplot* and *General-Boxplot* Method were applied, which are capable of taking distribution skewness in the calculation of outliers into account. Thus, the rate of outliers was reduced to 0.0-1.5% for the Medcouple-Boxplot and 0.0-0.9% for the General-Boxplot, which indicates a strong effect of the data distribution on the classification. Our results showed that a farm-specific evaluation seems reasonable within abattoirs, but remains challenging on an inter-abattoir basis. Results indicate that surveillance of prevalence-levels should incorporate distribution skewness to avoid biased results. A comprehensive animal-health assessment therefore requires further research in standardization and evaluation methods.

European practices to enhance the welfare of transported pigs

P. Ferrari[1] and P. Chevillon[2]

[1]Research Centre for Animal Production, Economics and agricultural engineering, Viale Timavo 43/2, 42121 Reggio Emilia, Italy, [2]IFIP – Institut du porc, La Motte au Vicomte, BP 35104, 35651 Le Rheu Cedex, France; p.ferrari@crpa.it

Implementation of good practices for pig transport can result in the reduction or elimination of stressful experiences and therefore better welfare. Good animal handling generally results in easier and more effective pig movement which means less frustration for the attendant. Other benefits to the industry include a decrease in transport losses, reduced time to load and unload pigs, reduced weight loss, less injuries and wounds and better meat quality. A guideline for pig transport has been developed, including a list of practices which have been identified, selected according to scientific and technical knowledge and literature and discussed and agreed by stakeholders of the pig supply chain, including representatives of the competent authorities, pig producers and processors, transporters and NGOs for animal protection. The initial design of the guide and the basic content has been proposed and agreed by people who will ultimately have to work with it every day. The main categories of risk factors which are addressed by this guide are: lack of competence of staff in charge of pigs, lack of planning before transport, lack of preparation of pigs before loading, prolonged restriction of water before transport, poor handling at loading and unloading at departure and arrival, poor transport conditions, poor management at control post or livestock market or assembly centre, poor management at the arrival. A European stakeholder group and two national stakeholder groups in Italy and France have been formed and involved in the process of proposal, discussion and agreement of transport practices through the organization of five meetings and six email consultation rounds. The guide has been developed within the Animal Transport Guidelines Project promoted and funded by European Commission, DG Sante.

Developing guides for good and best practices for sheep transport
L.T. Cziszter[1], X. Moles Caselles[2], D. Gavojdian[1] and A. Dalmau[2]
[1]USAMVB Regele Mihai I al României din Timişoara, Calea Aradului 119, Timişoara, 300645, Romania, [2]IRTA, Veinat de Sies, s/n. Monells, 17121, Spain; gavojdian_dinu@animalsci-tm.ro

Within a project financed by DG SANTE (Pilot project on best practices for animal transport – SANCO/2015/G3/SI2.701422) a series of guides containing good and best practices for transport of cattle, horse, sheep, pig, and poultry were aimed to be developed. For sheep, two institutions were designed to coordinate the work: IRTA Spain and USAMVBT Romania. To build the sheep guide data was collected from: (1) publications (from national guides, recommendations, scientific literature, manuals, etc) in EU or third countries (UK, ES, RO, NL, AT, IT, FR, DE, IR, AUS, CAN, Chile, Mexico), and (2) stakeholders through survey (face to face or online) and discussions (Delphi rounds). For sheep, over 50 stakeholders responded to the surveys. Based on these, the draft guide was produced. Then, two meetings with stakeholders took place (focus groups) and the final guide was produced. The recommendations are divided into two main categories: good practices, that comprise actions to be taken according to the current legislation and best practices that are actions to be taken above the required legislation in order to better comply with the animal needs. The final sheep guide comprises six chapters, each having a different number of good (GP) and best (BP) practices: (1) Journey planning and preparation (55 GP, 44 BP); (2) Handling and loading sheep (21 GP, 12 BP); (3) Competence and monitoring (14 GP, 9 BP); (4) Travelling (38 GP, 27 BP); (5) Sheep accommodation at control posts, markets and assembly centers (72 GP, 13 BP); and (6). Unloading animals (32 GP, 24 BP). A number of three factsheets will be developed to be handy to those transporting sheep: sheep and lamb fitness for transport; avoiding heat and cold stress: density, head room and partitions, bedding, humidity and ventilation; long distance transport recommendations: inspection, water and food provision and resting times. For dissemination of the guides and factsheets roadshows will be organised in eight EU countries: UK, FR, ES, DE, IT, PL, RO, GR targeting: vets, farmers, transporters, drivers, etc. Finally, the impact of these materials will be assessed.

Nutrition and immunometabolic adaptations to lactation in dairy cows: role in performance and health
J. Loor, F. Batistel, M. Vailati-Riboni and Z. Zhou
University of Illinois, Animal Sciences, 1207 West Gregory Drive, 61801, USA; jloor@illinois.edu

Immunometabolism represents the interface between immunology and metabolism, and is an emerging field of investigation in livestock biosciences. In human medicine, immunometabolism recognizes the link between obesity and the immune system, explicitly acknowledging that obesity-induced inflammation promotes onset of chronic disorders. More importantly, at the core of this concept is the recognition of 'multilevel interactions between metabolic and immune systems', implying 'cross-talk' or 'communication' among key cells and organs, which are orchestrated by unique mechanisms and their effectors. Such mechanisms correlate closely with health or disease status. The field of immunometabolism as it pertains to periparturient dairy cows is in its infancy. Classic studies prior to the 21st century defined the key metabolic, endocrine, and immune adaptations characterizing the transition into lactation. The advent of high-throughput technologies in the past decade further allowed the exploration of interrelationships among these systems. As a result of discoveries from that research, ongoing focus is on the bi-directional communication between immune and metabolic cells with signaling molecules derived intrinsically or as a result of intermediary metabolism or immune responses in tissues such as liver, adipose, and skeletal muscle. Both, macronutrients and micronutrients can be important effectors in the regulation of the immunometabolic response of the cow with effects often broad in nature. These responses are of great importance during the adaptation phase to lactation, as they seem to be key determinants of feed intake and milk production, reproduction, and health status. Management of total dietary energy supply and certain essential nutrients are examples of promising tools that can help regulate and enhance immunometabolic adaptations during the periparturient period and early lactation. A better understanding of the multilevel interactions among the various components of the metabolic and immune systems during the periparturient period has already led to identification of pathogenic mechanisms that underlie certain complications afflicting cows. Future research in this area should lead to promising therapeutic approaches.

The consistency of feed efficiency ranking and efficiency variation among growing calves

A. Asher[1], A. Sabtay[2], M. Cohen-Zinder[2], M. Miron[2], I. Halachmi[3] and A. Brosh[2]
[1]*Northern Research and Development (Northern R&D), Animal science, Migal Building, Kiryat Shemona, P.O. Box 831 Kiryat Shemona 11016, Israel,* [2]*Agricultural Research Organization, Volcani Center, Animal science, Ramat Yishay 30950, P.O. Box 1021, Israel,* [3]*Agricultural Research Organization, Volcani Center, Agricultural Engineering, Bet-Dagan 50250, P.O. Box 6, Israel; avivasher78@gmail.com*

This study investigated the possible mechanism for explaining individual efficiency variation in intact male Holstein calves. We examined whether calves (n=26) changed their feed efficiency (FE) ranking when fed a different diet during their growing periods. These calves were tested during three periods (P1, P2, P3). During P1 and P3, calves were fed a high-quality diet (ME=11.8 MJ/kgDM) and a low quality diet (ME=7.69 MJ/kgDM) during P2. The experimental recording period lasted for: 84, 119, 127 d, respectively. The calves' initial age in P1, P2, and P3 was 7, 11, and 15 months, respectively, and their average initial BW was 245, 367, and 458 kg, respectively. Individual DMI, ADG, diet digestibility, and heat production (HP) were measured in all periods. The measured FE indexes were: residual feed intake (RFI), the gain-to-feed ratio (G:F), residual gain (RG), residual gain and intake (RIG), the ratio of HP-to-ME intake (HP/MEI). In each period the RFI values were ranked and categorized into high-, medium-, and low-RFI groups (H-, M-, L-RFI). In all periods, the L-RFI group had lowest DMI, MEI, HP, retained energy (RE) and RE/ADG. The muscle fat of the L-RFI group had a higher percentage of protein and a lower percentage of fat compare to the H-RFI group. We suggested that the main mechanism separating H- from L-RFI calves is the protein-to-fat ratio in the deposited tissues. The ranking classification of calves to groups according to their RFI efficiency was repeatable throughout the growing periods and when tested in negative energy balance conditions. This novel repeatability of RFI ranking throughout different growing periods and also when tested under close to maintenance conditions highlight the RFI index as the most stable candidate for marker-assist selection for feed efficiency.

Neonatal nutrient supply affects pancreatic development in Holstein calves

L.N. Leal[1,2], G.J. Hooiveld[1], F. Soberon[3], H. Berends[2], M.V. Boekschoten[1], M.A. Steele[4], M.E. Van Amburgh[5] and J. Martín-Tereso[2]
[1]*Wageningen University, Droevendaalsesteeg 4, 6708 PB Wageningen, the Netherlands,* [2]*Trouw Nutrition Research and Development, Veerstraat 38, 5831 JN Boxmeer, the Netherlands,* [3]*Shur-Gain USA, Nutreco Canada Inc., Guelph, N1G 4T2 ON, Canada,* [4]*University of Alberta, 116 St. and 85 Ave., Edmonton, AB, T6G 2P5, Canada,* [5]*Cornell University, Ithaca, NY 14853, USA; leonel.leal@trouwnutrition.com*

The neonatal period is recognized to be of greatest relevance for pancreatic development and more specifically for β-cell expansion and turnover. The present study aimed to investigate the gene expression profiles and major transcription factors that regulate pancreatic development in preweaned calves offered two different levels of energy supply. To this end, 12 Holstein-Frisian calves were fed two distinct daily amounts of the same milk replacer (28% CP, 15% fat), receiving either a fixed rate of 11.7 MJ (LOW, n=6) or 1.26 MJ (HIGH, n=6) of metabolizable energy per kilogram of metabolic body weight ($BW^{0.75}$). At 54 days of age, pancreas tissue samples were taken. Transcriptome analysis revealed that upon HIGH feeding, cell cycle (mitosis), extra-cellular matrix interactions, haemolysis and metabolism of proteins (RNA translation) were induced. Whereas, clusters containing gene sets related with cell cycle (apoptosis), TCA cycle and signal transduction were suppressed. An upstream regulator analysis was also performed to better understand the underlying mechanisms by which HIGH diets modulate gene expression alterations in pancreas. These results suggest that the activation of the transcription factors HNF1A and SMARCA4 orchestrate the effects of HIGH diet on pancreatic gene expression. Overall, these data indicate that pre-weaning energy supply could stimulate the rate of replication and regeneration of β-cells by influencing the balance between β-cell proliferation and β-cell apoptosis.

Modifications in liver transcriptomic profile of fattening lambs by early suckled milk intake level

A. Santos[1], F.J. Giráldez[1], M.A.M. Groenen[2], O. Madsen[2], J. Frutos[1], C. Valdés[1] and S. Andrés[1]
[1]*Instituto de Ganadería de Montaña (CSIC-Universidad de León), Nutrición y Producción de Herbívoros, Finca Marzanas s/n, 24346 Grulleros, León, Spain,* [2]*Wageningen University, Animal Breeding & Genomics Centre, P.O. Box 338, 6700 AH Wageningen, the Netherlands; alba.santos@igm.csic.es*

The increment of world population increases the need of improving feed efficiency traits of livestock. However, the molecular mechanisms behind different feed efficiency traits and their regulation by nutrition remain poorly understood. Nowadays, Next Generation Sequencing methods allow understanding differences in gene expression and identifying functional candidate genes and pathways that control target traits in order to design strategies to increase feed efficiency. The present study was designed to identify (RNA-seq) differentially expressed (DE) genes in the liver tissues of fattening merino lambs caused by milk restriction during the suckling period. Forty male lambs were assigned randomly to two intake levels (n=20 per group) during the suckling period, namely *ad libitum* (ADL) and restricted (RES) groups. When they reached 15 kg of live body weight (LBW), all the animals were offered the same complete pelleted diet at a restricted level (40 g/kg) to ensure no selection of ingredients and no differences in dry matter intake during the fattening period. All the lambs were slaughtered with 27 kg of LBW and four animals from each group were selected for RNA-seq. Thirty-eight DE annotated genes were identified, with 23 DE genes being down-regulated and 15 up-regulated in the RES relative to the ADL group. RES lambs showed over-expression of lipid and xenobiotic metabolism pathways. Moreover, those genes involved in protein synthesis or protease inhibitors were down-regulated in the RES group, whereas those related to proteolytic degradation were up-regulated, thus suggesting a higher catabolism of proteins in these lambs. In conclusion, a restricted milk intake level during the suckling period of merino lambs promoted long term effects on hepatic transcriptomic profile which might have modified fatty acids metabolism and increased catabolism of proteins and detoxification of xenobiotics during the fattening period.

Mineral composition and deposition rates in the empty body of Large White pigs

I. Ruiz-Ascacibar[1,2], P. Schlegel[1], P. Stoll[1] and G. Bee[1]
[1]*Agroscope, Tioleyre 4, 1725 Posieux, Switzerland,* [2]*ETHZ, Universitätstrasse 2, 8092 Zurich, Switzerland; patrick.schlegel@agroscope.admin.ch*

The goal of this study was to model the macro- and trace-mineral composition and deposition rates in the empty body (EB) of grower- finisher Swiss Large White pigs. Diets contained recommended (Control) or 20% reduced levels of crude protein (R). Sixty-six entire male ($\male$), 58 castrated (δ) and 66 female ($\female$) pigs were offered these diets *ad libitum* from 20 kg body weight (BW) on. The EB mineral composition was determined at birth on 8 $\male$ and 8 $\female$, at 10 kg BW on 2 $\male$, 2 δ and 2 $\female$; at 20 kg BW on 8 pigs per sex, and at 20 kg intervals, from 40 to 140 kg BW on 4 pigs per sex and dietary treatment. Organs and empty intestinal tract from each animal were homogenized together and frozen carcasses were grinded. To determine the EB mineral content, the blood, hair, offal and carcass were analyzed for Ca, P, Mg, K, Na, Fe and Zn. While orthogonal contrasts revealed that Ca, P, Mg and Zn content in the EB were independent of gender or diet, $\male$ had greater (P<0.05) Na, K and Fe (P<0.01) content compared to $\female$ and δ. Control-pigs had greater (P<0.05) K content than R-pigs. Each mineral content was fitted to an allometric function (Y=a×EB^{b}+c, where Y is the total mineral content in the EB, a and c are constants and b, the allometric coefficient). All the equations had R^2 greater than 0.97. With increasing EB, mineral deposition rate (g/kg EB, Y'=a×b×EB^{b-1}) declined (b coefficient <1) except for the deposition rate of Zn, which slightly increased. With the exception of Ca and P, information on the dynamics of macro- and trace-mineral deposition rate is scarce. Hence, a constant dietary supply is often provided in practice, regardless of the dynamics of requirements. The developed equations contribute to accurately adjust mineral supply depending on the changing requirements.

Reduction of the amino acids supplied in excess to the growing pig using a precision feeding device

N. Quiniou, L. Ouine and M. Marcon
IFIP, BP 35104, 35651 Le Rheu cedex, France; nathalie.quiniou@ifip.asso.fr

When the body weight (BW) increases, the pigs' requirement in net energy (NE) increases faster than the requirement in amino acids, such as digestible lysine (LYSd). Although the minimum required LYSd/NE ratio decreases with BW, in practice it remains stable over extended periods, depending on the number of diets used by the farmer. During a project co-funded by ADEME, a precision feeding device was developed by IFIP and a French manufacturer (ASSERVA) that allows for individual and dynamic feeding of 96 restrictively fed growing pigs. Each pig wears a RFID ear tag in order to be recognized by the precision feeding device, which consists of a weighing – sorting station and five feeding stations. Each time the pig gets in the weighing station, its BW is recorded but the access to the feeding stations is only possible as long as the cumulated daily feed intake has not reached the daily amount allowed yet. The pig is identified by the electronic feeder and receives doses of 100 g, prepared from two diets formulated at 9.75 MJ NE/kg and 1.0 or 0.5 g LYSd/MJ NE, respectively, and mixed in proportion that depends on the feeding strategy. A 2-phase (2P) was compared to a multiphase (MP) strategy. When all pigs of the 2P group weighed 65 kg on average, the LYSd/NE ratio switched from 0.9 to 0.7 g/MJ. For the MP group, the daily LYSd requirement was assessed per pig according to a factorial approach, the requirement for growth being estimated from the slope of the regression between BW and age over the last 30 days. The daily feed allowance was 4% of initial BW on the first day and then increased by 27 g/d up to 2.4 kg/d for gilts, and 2.7 kg/d for barrows. Over the 40 to 114 kg BW range, the average LYSd/NE ratio was reduced by 11% for the MP group (0.67 vs 0.75 g/MJ for 2P, P=0.001). Despite a tendency toward a reduced feed efficiency with the MP strategy (0.34 vs 0.35 for 2P, P=0.08), the margin on feed cost remained stable (77.0 vs 76.3 €/pig, P=0.05) and the coefficient of N retention was improved by 1.7 point (P=0.05), which reduced the N output by 8% (P=0.001). A more relevant evaluation of the growing rate from the automatic measurement of the BW has to be implemented in order to improve the feed efficiency in addition to the improvement of the nitrogen efficiency.

Determination quali-quantitative of feed contaminants in complementary feedingstuffs for horses

A. Pellegrini[1], M.L. Fracchiolla[1], M. Fidani[1], V. Chiofalo[1], A. Agazzi[2] and V. Dell'orto[2]
[1]Unirelab s.r.l, via Gramsci 70, 20019 Settimo Milanese. Milano, Italy, [2]Università degli Studi di Milano, Dipartimento di Scienze Veterinarie per la Salute, la Produzione Animale e la Sicurezza Alimentare, via G. Celoria 10, 20133. Milano, Italy; alessandro.agazzi@unimi.it

Caffeine, theobromine, morphine, theophylline, atropine, scopolamine, codeine, bufotenine, dimethyltryptamine and tebaine, are widespread natural environmental substances. These compounds, regarded as doping agents by equestrian and racing regulators, can naturally contaminate horse feed affecting the performance of the animal and producing positive results in the anti-doping controls of urine and blood. They are infact well absorbed when administered by oral intake. For this purpose an LC-MS method for the simultaneous determination of all these substances has been developed and validated to analyze the complementary feedingstuffs for horses. A simple and fast liquid-liquid extraction on 0,5 mg of product is used. Chromatographic separation, using caffeine C13 like Internal Standard, was developed on a C18 Synergie Polar® column, with Methanol and 0.1% and Formic acid in water as mobile phases. Triple quadrupole mass spectrometer, equipped with an ESI source was operated in positive ion mode. All the analytes were determined by a Multiple Reaction Monitoring (MRM). For the validation of the method the following parameters were evaluated: limit of detection (LOD), limit of quantification (LOQ), intra-inter assay accuracy and precision. The same method was also validated like quantitative method for caffeine and codeine due to their frequent presence found in different complementary feeds. The developed method is routinely used to control the complementary feedingstuffs for horses.

Severe and short feed quantitative restriction affect on performance of broiler chickens

A.R. Seidavi[1], A. Sefaty[1], M. Bouyeh[1], S. Mohana Devi[2], I.H. Kim[2], V. Laudadio[3] and V. Tufarelli[3]
[1]*Rasht Branch, Islamic Azad University, Pole-Taleshan, 41335-3516, Rasht, Iran,* [2]*Dankook University Dandae-ro 119, Cheonan, 330-714, Korea, South,* [3]*Section of Veterinary Science and Animal Production, University of Bari 'Aldo Moro', Valenzano, Department of Emergency and Organ Transplantation, Italy; alirezaseidavi@iaurasht.ac.ir*

The aim of this study was to determine the effect of an early period of severe quantitative feed restriction followed by re-feeding energy and protein dense diets on growth performance of broiler chickens. Two hundred and seventy Ross 308 day-old chicks were assigned to treatments in a 4×2 factorial design to determine the effect of two feed restriction periods (day 8 to 12 and day 8 to 17), and four levels of quantitative feed restriction (22.5, 25.0, 27.5, and 30% less than the standard guide of Ross 308). Each treatment was considered of three replications. A control treatment without any feed restriction was maintained at *ad libitum* intake (T9; control treatment). Differences among treatments were separated by Duncan's multiple range tests. Probability values of less than 0.05 were considered as statistically significant. Performance parameters including feed intake, body weight gain, feed efficiency, energy intake, energy efficiency, protein intake, and protein efficiency in two different durations and different intensities at 5 days (22.5, 25, 27.5 and 30%), 10 days (22.5, 25, 27.5 and 30%), and control group showed significant differences. The mean were analyzed with significance level of P<0.05. The present study results shows severe and short feed quantitative restriction showed significant effect on performance in broiler chickens.

Effect of dietary composition over food preferences in cats

J. Figueroa[1,2], S. Guzmán-Pino[1] and V. Sotomayor[1]
[1]*Universidad de Chile, Fomento de la Producción Animal, Av. Santa Rosa 11735, La Pintana, Santiago, Chile, 8820000., Chile,* [2]*Pontificia Universidad Católica de Chile, Ciencias Animales, Av. Vicuña Mackena 4860, Santiago, Chile, 6904411, Chile; figueroa.jaime@uc.cl*

Besides the sensorial characteristics of diets, nutrient composition may influence food preferences. The aim of this study was to analyse the relationship between the nutritional compositions of cat's diets and their associated preferences. A database of preference tests from 10 years (2005-2015) was obtained from the Research Centre of Pet Feeding Behaviour (Universidad de Chile). Preferences consisted on the placement of two simultaneous feeders in the front of cat's cannels during 20 h. Food was weighted at the beginning and the end of each test to calculate cat's food intake. The nutritional composition of both diets was analysed, and the difference between the nutrient components of the most preferred diet (A) and the other diet (B) was used for statistical analysis. To evaluate how nutritional components or group of components may explain food preferences, data was analysed with a principal component (PC) analysis by using the statistical software SAS. A linear regression was performed between the PC obtained and cats preferences. Later, a Spearman correlation was performed between the nutritional components of the PC that represented the greatest variability and showed a significance in the linear regression. The first three PC presented eigenvalues closed to 1 that explained the 34.9, 29.7 and 13.8% of variability. The first (Ca, P, crude fibre and ash) and second (crude protein, lipid content, ether extract and metabolizable energy) PC tended to have an effect over cats preferences (P=0.065 and P=0.060, respectively), showing both negative b values (-1.42 and -1.57). The contents of ash, Ca and crude fibre showed negative correlations with preference (r=-0.269, P=0.031; r=-0.241, P=0.054 and r=-0.338, P=0.006, respectively), and P tended to show a negative correlation with preference (r=-0.208, P=0.098). Nutritional components may affect food preferences of domestic cats. The mineral and fibre contents of diets, followed by protein and fat fractions, seem to have the highest repercussion on cat's food preferences.

Evaluation of lentil (*Lens culinaris* cv. Eston or cv. Anicia) as a protein source for broiler diets

G. Ciurescu[1], M. Habeanu[1], C.O. Pana[2] and C. Dragomir[1]
[1]*National Research & Development Institute for Animal Biology and Nutrition, Calea Bucuresti no.1, 077015, Balotesti, Ilfov, Romania,* [2]*University of Agronomic Science & Veterinary Medicine, 59 Marasti Street, 011464, sector 1, Bucharest, Romania; ciurescu@ibna.ro*

Price and availability of soybean meal (SBM) on global markets may change rapidly, thereby stimulating interest in maximizing the use of locally produced protein sources, like seed legumes. This study evaluated the effects of dietary replacement of SBM with two cultivars of lentils (*Lens culinaris*), cv. Eston (green-seeded), and cv. Anicia (green marbled-seeded), on growth performance, digestive organ sizes and pH of the cecal digesta. Two hundred and fifty, one-d-old, broiler chicks (Cobb 500) were randomly allocated to the following five treatments: (1) a diet based on corn and SBM as control; (2) 200 g/kg of raw lentil seeds cv. Eston (LE); (3) 400 g/kg of LE; (4) 200 g/kg of raw lentil seeds cv. Anicia (LA); (5) 400 g/kg of LA. Each diet was fed to 2 replicated groups of 25 chicks each from 0 to 28 d of age. All diets were formulated to be isocaloric, isonitrogenous and with similar content of total lysine, TSAA, calcium and av. phosphorous. Statistical analysis was carried out using the GLM procedure of SPSS. At the end of the experimental period (28 d), the broilers fed raw lentils (*Lens culinaris* cv. Eston or cv. Anicia) had comparable body weights (BW), average daily feed intake (ADFI) and FCR to the control group. However, the replacement of SBM with ra w lentils (cv. Eston) at higher inclusion levels (400 g/kg of diet) did not affect ADFI and FCR, but decreased BW ($P<0.05$) by 4.1% compared to the control. The digestive organ sizes (i.e. gizzard, heart, liver, pancreas, spleen, small intestine, caecum and the small intestine length) and pH of the cecal digesta were not affected ($P>0.05$) by treatments. Nevertheless, only the broilers in the LE400 group had the highest weights for the livers ($P<0.05$). Moreover, carcass, breast and thigh weight and dressing percentage were not affected ($P>0.05$) by treatments. Thus, it was concluded that raw lentils (cv. Anicia) can be used as an alternative protein source to replace SBM in broiler chicken diets, at inclusion levels up to 400 g/kg and raw lentils (cv Eston), at levels up to 200 g/kg.

The effects of starving-refeeding diet on oxidative stress in sturgeons under aquaculture conditions

I.E. Florescu (gune)[1], M. Balas[1], A. Dudu[1], L. Dediu[2], A. Dinischiotu[1], M. Costache[1] and S.E. Georgescu[1]
[1]*University of Bucharest, Faculty of Biology, Department of Biochemistry and Molecular Biology, Splaiul Independentei 91-95, 050095, Bucharest, Romania,* [2]*Lower Danube University of Galaţi, Faculty of Environmental Science and Biotechnology, Department of Aquaculture, Environmental Sciences and Cadastre, Domneasca Street 111, 800201, Galati, Romania; georgescu_se@yahoo.com*

Aquaculture is focused on optimizing the feeding regime of individuals in order to increase the profitability of fish farms without harming the welfare and health of fishes. This study aimed to determine if stellate sturgeon (*Acipenser stellatus*, Pallas 1771), bred intensively mainly for caviar production, can adapt to a starving-refeeding regime by assessing the effects of this diet on oxidative stress biomarkers. During the experiment, 48 juveniles were subjected to different feeding programs in aquaculture conditions: Control group: fed constantly; Group 1: starved 7 days and refed 21 days; Group 2: starved 14 days and refed 21 days. For each individual liver homogenates were obtained and the oxidative stress was evaluated by the level of protein carbonyl groups (PCG) and advanced oxidation protein products (AOPP) which revealed the status of protein oxidation and also of MDA which indicates the degree of lipid peroxidation. The data were statistically analyzed using ANOVA method implemented in GraphPad Prism3. In the liver of the juveniles from Group 1, the PCG level decreased whereas the AOPP one slightly increased compared to the control. In contrast, both AOPP and PCG concentrations increased in the liver of juveniles from Group 2. The level of MDA decreased during the 7/21 days starvation-refeeding regime, while it increased during 14/21 days starvation-refeeding regime. These results suggest that the 7/21 days starvation-refeeding diet reduced the lipid peroxidation and induced minor oxidative changes of the proteins, in contrast to 14/21 days starvation-refeeding regime that induced oxidative changes in both lipids and proteins. In conclusion, stellate sturgeon could adapt to a starving and refeeding regime, the recommended diet being 7/21 days starvation-refeeding. This diet could enhance the profitability of fish farms without affecting the welfare of the individuals.

Effect of calcium und phosphorus on growth performance and minerals status in growing-finishing pigs

P. Schlegel and A. Gutzwiller

Agroscope, Tioleyre 4, 1725 Posieux, Switzerland; patrick.schlegel@agroscope.admin.ch

Dietary calcium (Ca) can be an influencing factor when aiming for an efficient and sustainable use of dietary phosphorus (P) in pig production to minimize P excretion and the use of mineral phosphates, representing a progressively limiting and non-renewable source of P. The 90-day study evaluated the effect of Ca level (2.2, 2.5 and 2.8 g Ca/g digestible P) in a phytase (equivalency of 0.16 g digestible P and 0.0 g Ca/100 FTU) containing diet with recommended (P+) or marginal (P-) levels of digestible P on growth performance and minerals status of 84 growing-finishing pigs. The grower and finisher P+ diets contained 3.0 and 2.4 g and P- diets 2.5 et 1.7 g digestible P/kg, respectively. Measured parameters were weekly feed intake and growth performance, urinary, serum and metacarpal mineral concentration, ulna cartilage scoring, metacarpal bone density and foot bone mineral density by dual-energy X-ray absorptiometry. Mean BW gain was 846 and 951 g/d for the grower and finisher period, respectively and mean final BW was 103.1 kg. Growth performance parameters, serum mineral content at the end of the growing and finishing periods and ulna cartilage scoring were not affected by the dietary treatments, except that 2.2 g Ca/g digestible P deteriorated by 4% the feed conversion ratio during the finishing period. This lowest Ca level also increased urinary P concentration by 36% and as did 2.5 g Ca/g digestible P, reduced metacarpal ash, bone density and foot bone mineral density between 2 and 5%. Diet P- reduced metacarpal ash, bone density and foot bone mineral density between 3 and 4%. No interaction was observed between Ca and digestible P level. Finally, the P- diet enabled a complete removal of mineral phosphates in the finisher diet and reduced by 65% the total phosphate use without any negative effect on growth performance. In a diet containing phytase, a Ca level of 2.2 g/g digestible P was insufficient as bone mineralisation was deteriorated indicating that P could not be sufficiently mineralised and thus was increasingly excreted via urine.

Efficacy of a novel bacterial 6-phytase in grower-finisher pigs

D. Torrallardona[1] and P. Ader[2]

[1]*IRTA, Centre Mas de Bover, 43120 Constanti, Spain,* [2]*BASF SE, G-ENL/MT, F31, 68623 Lampertheim, Germany; david.torrallardona@irta.cat*

Four trials were conducted to investigate the efficacy of a newly developed bacterial 6-phytase (6-Phy) on performance, Ca and P ATTD, bone mineralisation (BM), and P balance in pigs (26 kg initial BW) fed grower-1/grower-2/finisher diets between 0-5/5-10/10-15 wk of trial, respectively. For trial-1, 144 pigs in 48 pens, were offered 4 treatments (TR) consisting of a basal negative control (NC; with 3.8/3.5/3.4 gP/kg at 0-5/5-10/10-15 wk, respectively; Ca:P ratio of 1.4) supplemented with 0, 100, 200 or 300 FTU/kg feed of 6-Phy. In trial-2, 72 pigs in 72 pens, were offered 6 TR: a NC-1 (3.8/3.5/3.4 gP/kg; Ca:P of 1.3) with 0, 100, 200 or 300 FTU/kg 6-Phy and a NC-2 (3.8/3.5/3.5 gP/kg; Ca:P of 1.3/1.1/0.9) with 0 or 300 FTU 6-Phy. In trial-3 (72 male pigs; 18 pens) and trial-4 (72 female pigs; 18 pens) 2 TR were offered: NC (3.8/3.6/3.7 gP/kg; Ca:P of 1.4) with 0 or 100 FTU/kg of 6-Phy. Weight gain (ADG) and feed intake (ADFI) were controlled at 5, 10 and 15 wk in all trials. For trials-1&2, faeces were sampled at the end of each phase to estimate Ca and P ATTD using TiO_2 as indigestible marker, and at the end of the trial, the hoof of the pigs was collected in the slaughterhouse to measure DM and ash of *Os metacarpale III*. P balance (during 4 d) was also measured for the grower-1 and grower-2 diets in trial-3, using 12 male pigs (for each feeding phase) kept in metabolism cages to quantitatively collect faeces and urine. Linear responses ($P<0.05$) to 6-Phy dose were observed for ADG, Ca and P ATTD, and BM (trials-1&2; $P<0.05$), and for ADFI (trial-1) and feed efficiency (G:F; trial-2). In trial-2, 6-Phy (300 FTU) also improved ADG, ADFI, G:F, Ca & P ATTD and BM in the NC-2 diet ($P<0.05$). Significant responses to 6-Phy (100 FTU) were also observed for ADG and ADFI (trials-3&4), for G:F (trial-3) and for P-retention (grower-1 and grower-2 diets). On average, supplementation of feeds with 100, 200 and 300 FTU/kg of 6-Phy increased their digestible P content by 0.42, 0.66 and 0.84 g/kg feed, respectively. It is concluded that the novel bacterial 6-phytase (up to 300 FTU/kg) improves Ca & P ATTD, P retention, bone mineralisation and performance of G/F pigs in a linear dose response manner.

Residual feed intake of lactating Nellore beef cows

R.C. Canesin, T.J.T. Ricci, L.L. Souza, M.F. Zorzetto, J.N.S.G. Cyrillo and M.E.Z. Mercadante
Instituto de Zootecnia, Sertãozinho, SP, 14174-000, Brazil

In order to assess the feed efficiency of beef cows, 27 Nellore first calving cows (37.6±0.9 months of age, 505±32.1 kg of body weight) were evaluated in a performance test (GrowSafe Systems Ltd) started 21 days postpartum, during 81 days (13 days of adaptation and 68 days for dry matter intake-DMI recording). The forage-based diet (dry matter basis) consisted of 90% corn silage, 8.5% soybean meal and 1.5% mineral salt plus urea. Average daily gain (ADG) was obtained as a linear regression of four body weight records on the days in test, and the metabolic body weight ($BW^{0.75}$) was obtained as intercept + (ADG × 0.5 × days in test)$^{0.75}$. Ultrasonic fat thickness was evaluated on the first day of the test in five anatomic sites, and an average of fat thickness (FT) was obtained. Cows were milked 63 days after calving by machine milking, after receiving 20 IU of oxytocin intravenously, quantifying a 6 h interval of production to estimate 24 h milk production. Fat and protein percentage in milk (5.02±1.01% and 3.80±0.288%) were used to calculate energy-corrected milk production in 24 h (ECMP). DMI prediction equation was developed using stepwise regression, by STEPWISE and SLENTRY = 0.25 and SLSTAY = 0.05 options using PROC REG for model selection. DMI, ADG, $BW^{0.75}$, FT and ECMP means were 12.6±1.28 kg/day, 0.430±0.231 kg/day, 109±4.76 kg, 6.61±1.58 mm and 9.73±3.12 l, respectively. The full model of DMI estimation included ADG, $BW^{0.75}$, FT and ECMP and explained 38% of DMI variation. The selected model included ADG (partial R^2=18.34%) and ECMP (partial R^2=12.83%): DMI = 10(±0.894) + 2.78(±0.954) × ADG + 0.150(±0.071) × ECMP + error (i.e. residual feed intake-RFI). RFI of lactating cows was obtained as the difference between DMI and DMI predicted by the selected model. RFI mean was 0±1.06 kg/day (from -2.56 to 2.58 kg/day) and the coefficient of variation, using DMI mean, was 8.42%. Classifying cows in negative (RFI<0) or positive (RFI>0) RFI, significant differences in DMI (11.8 kg/day *versus* 13.3 kg/day) and DMI expressed as percentage of body weight (2.31 vs 2.52%) were observed, without differences in average body weight and FT.

Can artificial lactation program the rumen methanogenic microbiota?

A. Belanche[1,2], J. Cook[3] and C.J. Newbold[2]
[1]EEZ-CSIC, Granada, 18100, Spain, [2]IBERS, Aberystwyth University, SY23 3DA, United Kingdom, [3]Volac International Ltd, Cambridgeshire, SG8 5QX, United Kingdom; a.belanche@csic.es

Microbial fermentation plays a central role in the ability of ruminants to utilize fibrous substrates; however rumen fermentation also has potential deleterious environmental consequences in terms of methane emissions. Early-life interventions in young ruminants could represent an opportunity for rumen microbial programing to decrease such emissions. This experiment studied the effect of using colostrum alternatives (CA) and milk replacer (MR) compared to natural ewe-rearing. A total of 24 pregnant ewes carrying triplets were used and lambs were randomly allocated to the experimental treatments: (1) NN: ewe's colostrum and ewe's milk. (2) NA: ewe's colostrum and MR (Lamlac Instant, Volac) and (3) AA: CA (Lamb Volostrum) and MR. Each treatment was placed in independent pens with free access to hay and creep feed. At 45 d of age all lambs were sampled, weaned, reunited in the same ryegrass pasture and resampled during fattening at 5 months of age. The rumen methanogen population was studied based on quantitative PCR and 16S rDNA amplicon sequencing. At weaning, differences were observed in the structure of the methanogen community among the three treatments, with NN lambs having the more complex methanogen community in terms of diversity (22 vs 16 OTUs) and greater abundance of Methanomasiliicoccaceae Group 9 (+84%), possibly as a result of the presence of rumen protozoa. At fattening, an increase in the total methanogens (+0.24 log) and their relative abundance (+57%) was observed, moreover Methanomassiliicoccaceae Group 12 was replace by increased levels of Group 9, 10 and 11 methanogens (representing 25%, 30% and 9%, respectively) across all treatments. However, after weaning differences between treatments vanished and all animals reached a similar methanogen community structure and diversity as well as similar protozoal levels. Although, all treatments showed similar methane emissions at fattening using *in vitro* incubations, NN lambs showed a faster live weight gain which may indicate more efficient energy utilization. In conclusion, artificial lactation promoted a delay in the development of the rumen methanogen community; however those effects disappeared after weaning.

Effect of *Saccharomyces cerevisiae* boulardii on weanling piglets performance in reduced medication

A. Agazzi[1], V. Bontempo[1], G. Savoini[1] and F. Bravo De Laguna[2]
[1]Università degli Studi di Milano, Dipartimento di Scienze Veterinarie per la Salute, la Produzione Animale e la Sicurezza Alimentare, via G. Celoria, 10. Milano, 20133, Italy, [2]Lallemand SAS, rue Briquetiers, 19. Blagnac Cedex, 31702, France; alessandro.agazzi@unimi.it

The wide use of antibiotics in both human and animal medicine has accelerated the appearance and propagation of resistant microorganisms. In addition, new results show that some of the antibiotic resistance genes increase in the intestinal tract of weaned piglets after two weeks of zinc oxide supplementation. *Saccharomyces cerevisiae* var. *boulardii* CNCM I-1079 (SB) is a probiotic live yeast with proven positive effects on microbiota regulation, intestinal structure and natural defense. The objective of the study was to evaluate the effect of SB on post weaning piglets performance with or without removing antibiotics medication treatments and zinc oxide (ZnO). Two hundred and eighty eight piglets (21 d old, 6.52 kg±0.98) were allotted to 3 treatments and randomly allocated in groups of 9, with a total of 12 replicates per treatment. They followed a 3-phase feeding program: prestarter 0-11 d, starter 1 12-33 d, starter 2, 34-50 d. Treatments were: Control (C), basal diet with antibiotic and ZnO inclusion in the first two phases, and only antibiotic inclusion in the third phase; T1: C + 10^9 cfu/kg of SB; T2: basal diet with antibiotic and ZnO inclusion only in prestarter + 10^9 cfu/kg of SB during all the experimental period. Data were analyzed by a MIXED procedure of SAS for repeated measurements on pen basis. SB increased average daily gain (ADG) in T1 compared to C and T2 (P<0.01). When comparing the same diet (C vs T1), an overall improvement of ADG (P<0.01) and a trend to improved feed conversion ratio (FCR; P<0.10) were observed in T1 (+5.59% and -4.97%, respectively). C vs T2 comparison showed no significant differences in ADG and FCR, despite the removal of the medication for the last 39 days of experiment. The present study shows that *S. cerevisiae* var. *boulardii* CNCM I-1079 can face the post weaning challenges, enhancing performance when added on top or playing a significant role in the nutritional strategies focused on medicated feeds reduction. Furthermore, SB is compatible with antibiotics and ZnO.

Bromide intake via water and urinary excretion rates in a ruminant model

N.H. Casey, B. Reijnders and J.G. Myburgh
University of Pretoria, Animal and Wildlife Sciences and Para-Clinical Sciences, Lynnwood Road, Pretoria, 0002, South Africa; caseynh@icloud.com

Bromide (Br⁻) is endocrine disrupting and suppresses livestock production. While water quality guidelines for Br⁻ are set at 0.01 mg/l, natural exposure varies and can range up to 10 mg/l. Sheep on range lands are exposed to varying levels of Br⁻ by drinking groundwater with high Br⁻ content. The question was the extent to which sheep would excrete ingested Br⁻ under these conditions. The aim was to quantify the excretion of Br⁻ in urine in a sheep model exposed to a range of Br⁻ levels. Five Merino wethers were assigned to four treatments of Br⁻ (dissolved NaBr) administered in a range of concentrations of 0.01, 0.1, 0, 1.0 mg/l as a single dose. Feeding was individual *ad-lib* teff and lucern hay (50:50) for a 72 h-period with *ad-lib* water from a municipal source. Background Br⁻ levels in the water source were added to treatment values, which became the intake values. Water intake was measured twice daily. Base funnel containers collected individual urine output continuously, which was measured after a 12 h period. Standard- and semi-quantitative analyses were done. The mean intake levels of Br⁻ over 24 h were $T_{0.01}$: 0.337, $T_{0.1}$: 0.896, T_0: 1.815, $T_{1.0}$: 2.705 (mg/l) and excretion levels were $T_{0.01}$: 0.043, $T_{0.1}$: 0.029, T_0: 0.029, $T_{1.0}$: 0.021 (R^2=0.7790). Over 48 h, the mean intake levels were $T_{0.01}$: 0.495, $T_{0.1}$: 0.165, T_0: 0.272, $T_{1.0}$: 0.481 (mg/l) and excretion levels were $T_{0.01}$: 0.047, $T_{0.1}$: 0.018, T_0: 0.065, $T_{1.0}$: 0.016 (R^2=0.0001). Over 72 h, the mean intake levels were $T_{0.01}$: 0.0004, $T_{0.1}$: 0.207, $T_{1.0}$: 0.666 (mg/l) and excretion levels were $T_{0.01}$: 0.029, $T_{0.1}$: 0.018, $T_{1.0}$: 0.044 (R^2=0.5621). The mean percentages excreted over all treatment levels and sheep over 24, 48 and 72 h periods were 4.3, 18.7 and 31.6 respectively. Excretion levels for higher intakes were similar to those of the lower intake values, implying a constant level of excretion over time. As the amount of Br⁻ ingested increased over treatments, the proportion of total Br⁻ excreted decreased, indicating that at higher concentrations, Br⁻ was being retained. This could be potentially problematic as long periods of exposure to high concentrations may result in Br⁻ accumulating in the animals' bodies.

SiMMinTM: on line software tool to simulate copper balance in feeding programs of pigs

A. Romeo[1], S. Durosoy[1] and J.-Y. Dourmad[2]
[1]*Animine, 335 Chemin du Noyer, 74330 Sillingy, France,* [2]*INRA-Agrocampus Ouest, UMR Pegase, Domaine de la Prise, 35590 Saint-Gilles, France; aromeo@animine.eu*

At pharmacological dosage, copper (Cu) can improve growth performance of pigs. Consequently, it is commonly supplied in excess in piglets diets and Cu levels in animal wastes may exceed maximal authorized Cu value when manure is used as organic fertiliser. Some scientific methods to estimate copper balance in pig farms have been proposed by INRA and can be used by the pig industry. Cu retention in growing pig is calculated based on the difference in Cu body content between the beginning and the end of a defined period. In order to calculate easily the Cu excretion, the software siMMin™ Cu has been developed with the support of INRA, with the following variables: feeding programs on the farm, growth performance and Cu concentrations in each feed. It focuses on the pig growing life, from the weaning to the slaughter. The software siMMin™ Cu enables to simulate changes in each variable compared to the existing situation, and to measure the rate of improvement in the total reduction of Cu excretion in the life of the growing pig. It is intuitive, user-friendly and available on line since December 2016 at www.animine.eu/simmin for all stakeholders involved in pig production. Depending on the level of interest expressed locally, this software should be later available in national languages for major pig producing countries, like it has been realized in German and in Chinese for the Zn application.

Dietary SOD-rich melon on hepatic gene expression of antioxidant proteins in piglets and broilers

A. Agazzi[1], G. Invernizzi[1], A.S.M.L. Ahasan[2], G. Farina[1], R. Longo[3], F. Barbé[4], M. Crestani[3], V. Dell'orto[1] and G. Savoini[1]
[1]*University of Milan, Department of Health, Animal Science and Food Safety, Via Celoria 10, 20133 Milan, Italy,* [2]*Chittagong Veterinary and Animal Sciences University, Department of Anatomy and Histology, Khulshi, 4225 Chittagong, Bangladesh,* [3]*University of Milan, Department of Pharmacological and Biomolecular Sciences, Via Balzaretti 9, 20133 Milan, Italy,* [4]*Lallemand SAS, 19 rue des Briquetiers, 31702 Blagnac, France; alessandro.agazzi@unimi.it*

The aim of the study was to assess the effects of a SOD-rich feed supplement derived from melon (Melofeed®, Lallemand, France), on hepatic gene expression of antioxidant proteins in piglets and broilers. Two trials were set up: trial 1 involved 6 piglets per treatment that were selected at the end of a larger trial after the administration of: (1) basal diet (C); (2) basal diet plus Melofeed®(MPC; 30 g/ton of complete feed) for 29 days after weaning, and subjected or not to an intramuscular LPS challenge to mimic a chronic inflammation. Trial 2 considered 12 chickens per treatment at the end of the grower phase (day 24) selected from 4 experimental groups receiving a corn-soybean meal based diets supplemented with 0 g/ton (C), 30 g/ton (MPC1), 15 g/ton (MPC2), 15 g/ton (MPC3) of Melofeed® during the starter phase (0-10 d) and the same basal diet for C, MPC1 and MPC2 during the grower phase (11-24 d) or supplemented with 15 g/ton (MPC3) of Melofeed®. Liver tissue samples were collected immediately after euthanization and analysed for GPX1, CAT and SOD1 expression. The relative expression levels were assessed using a standard curve and normalised to internal control genes. *n*-fold change relative to control were then analysed by MIXED procedure of SAS. The dietary treatment did not significantly affect liver gene expression of GPX1 and SOD1 in piglets, while MPC tended to enhance expression of CAT ($P=0.08$). CAT was upregulated also in MPC3 chicken's liver ($P=0.02$) and SOD1 tended to be more expressed than C ($P=0.06$). These results could support an improvement of the antioxidant effect, since SODs and catalase are transcriptionally regulated by the same transcription factors in a coordinated antioxidant effort. The research was funded by Lallemand SAS and Piano Sviluppo Ricerca 2015 Linea 2 Azione B UNIMI.

Animal and run variation for rumen degradation determined with the *in situ* method

S. Lacombe[1,2], J. Dijkstra[1], M. Jacobs[2], J. Doorenbos[2] and H. Van Laar[2]
[1]*Wageningen University & Research, Animal Nutrition Group, P.O. Box 338, 6700 AH Wageningen, the Netherlands,*
[2]*Trouw Nutrition R&D, P.O. Box 220, 5830 AE Boxmeer, the Netherlands; harmen.van.laar@trouwnutrition.com*

Reproducibility of the *in situ* technique to determine rumen effective degradability (ED) has been previously reported in a series of independent *in situ* nylon bag incubations of standard samples. This data was used for analysis of incubation run, animal and residual variation. Standard samples of two grass silages (GS; GS1, n=6; GS2, n=8), two maize silages (MS; MS1, n=4; MS2, n=9) and a compound feed (CF; n=13) were incubated in rumen fistulated cows in 40 incubation runs. In every run, 2 to 3 replicates of each sample were incubated in the rumen of three lactating dairy cows fed *ad libitum*, at 8 time points (0 to 336 h). Fractions of washable (W), degradable (D), and undegradable (U) DM, and DM fractional degradation rate (k_d), were estimated using a first-order exponential model and ED was calculated using a passage rate (/h) of 0.060 (CF) and 0.045 (GS and MS). Mixed procedure of SAS was used to estimate the fixed effect of sample and random effects of run, animal and residual error. For U, D, W, k_d and ED, variation was composed of run and residual variation, the latter including variation caused by animals. For individual incubation time points, variation was composed of run, animal and residual variation, the latter including variation caused by bags. For GS, MS and CF, run represented 84, 92 and 59% (fraction W), 2, 26 and 38% (fraction U), and 59, 84 and 3% (ED) of total variation, respectively. For DM residue at individual time points, in general variation at early incubation time points was lowest, then peaked between 24 and 56 h of incubation, after which variation decreased again for longer incubation times. Variation at early incubation time points was mostly composed of run and residual, with a relative increase for contribution of animal variation with time. To improve reproducibility of the *in situ* technique, reduction of variation due to run (e.g. by using substrate-specific incubation procedures), and due to animal (e.g. by standardizing lactation stage and diet), needs to be considered.

Preservation and nutritive value of mixed silages of fodder beets and by-products

J.L. De Boever[1], E. Dupon[2], S. De Campeneere[1] and J. Latré[2]
[1]*ILVO, Animal Sciences Unit, Scheldeweg 68, 9090 Melle, Belgium, [2]University College Ghent, Experimental Farm Bottelare, Diepestraat 1, 9820 Merelbeke, Belgium; johan.deboever@ilvo.vlaanderen.be*

Due to the reform of the EU's Common Agricultural Policy fodder beets (FB) are regaining importance on dairy farms as third crop besides grass and maize. FB are well-known as a productive and energy-rich feed, but they lost interest at the end of the 20th century because of the labour-intensive cultivation, harvesting and feeding, rhizoctonia disease and limited storability. Nowadays, resistant varieties and adapted harvest-machines open new perspectives and longer preservation is possible by ensiling FB mixed in whole plant maize. However, this combination is not ideal, because FB have to be harvested earlier than maize resulting in lower yield and dry matter (DM) content. Therefore, we studied the potential of mixing and ensiling FB with ready available by-products. Chopped FB with high DM content (22.0%) were mixed with either dry wheat gluten feed (DWG; 90.4% DM), dry maize gluten feed (DMG; 88.7% DM), dry chicory pulp (DCP; 89.2% DM), dried beet pulp (DBP; 89.5% DM) or pressed beet pulp (PBP; 25.1% DM) in a ratio of 50/50 on DM-basis. Each mixture was ensiled in two IBC containers (volume: 1 m³), one during 2 and the other during 12 months. The DM-losses and fermentation quality of the 5 mixed silages were determined. Chemical composition and digestibility of the silages as well as of the fresh FB were determined with sheep to derive net energy lactation (NEl), whereas the content of protein digestible in the intestines (DVE) was obtained with rumen-fistulated cows. In all silages fermentation continued during the whole storage period. After 1 year, DM-losses of the mixtures with DWG, DMG, DCP, DBP and PBP amounted to 26, 26, 10, 13 and 19%. The production of lactate, acetate and ethanol was high and their sum ranged from 138 for DBP to 244 g/ kg DM for DMG, whereas NH_3-fraction was moderate (4.7 to 7.6%). The NEl of DWG, DMG, DCP, DPB and PBP amounted to 7.30, 7.25, 7.46, 7.26 and 7.39 MJ/kg DM respectively, which was slightly higher than for fresh FB (7.23 MJ/kg DM). The DVE of the mixed silages amounted to 72, 79, 87, 95 and 89 g/kg DM respectively as compared with 81 g/kg DM for FB.

Effect of additional oat feeding on the ewe body condition and production in organic farms

P. Piirsalu, T. Kaart, J. Samarütel, S. Tölp and I. Nutt
Estonian University of Life Sciences, Institute of Veterinary Medicine and Animal Science, Kreutzwaldi 62, Tartu, 51014, Estonia; peep.piirsalu@emu.ee

Organic sheep farmers in Baltics and in Scandinavia are feeding their animals with grass based feeds (pasture grass, silage, hay) with minimum use of cereals. During the second half of gestation and suckling periods the ewes cannot cover their nutritional requirements from consumed feed alone, but have to use their body reserves to support their own and the lamb nutritional demand. The aim of this study was to estimate feeding strategies of pregnant and lactating ewes fed only silage/hay or supplemented with concentrates on the ewe body condition score (BCS) and production (lambs born alive (LBA), probability of lambs born dead (LBD), lamb birth weight (LBW) and lamb 100 day weight (L100DW). The test period began with the onset of the mating ewes in October 2013 and was completed at the end of the suckling period with weaning of lambs in July 2014. The Estonian White Face and Texel ewes (107 ewes) were divided into two similar groups: the experimental group (59 ewes) and control group (48 ewes). The experimental ewes were fed in addition to grass feeds (silage, hay or pasture grass) for 45-60 days before lambing (from 90th-105th day of gestation) 0.22 kg of organic oat per day (totally 10 kg in gestation period) and 0.3 kg of oat during suckling period (totally 13.5 kg per ewe). Control group of ewes were fed only with grass based rations (silage, hay, pasture grass) without cereals. Feed samples were analysed regularly and rations were prepared to study if they meet nutritional requirements. Ewe BCS on a scale of 1 (emaciated) to 5 (obese), with 0.5 point graduations, was assessed before mating, at 130-140 day of pregnancy, after lambing (following 7th-10th day), at the 2. suckling months (45 th-60 th suckling day) and at weaning of lambs by two observers during the whole study. It was concluded that additional feeding of minimal amount of oats increased the ewe BCS in experimental group at 130-140 day of pregnancy (BCS was respectively 3.36 and 2.97, P<0.001) and at the 45.-60 suckling day (BCS respectively 2.84 and 2.52, P<0.001) compared with control ewes but doesn't affected BCS of ewes at the end of experiment (at weaning of lambs) and all production traits (LBA, LBW, L100DW).

The effect of *Fusarium* mycotoxins on performance and metabolism of early lactation dairy cows

Z.C. McKay[1], O. Averkieva[2], R. Borutova[2], H.Z. Taweel[2], J. Dunne[2], G. Rajauria[1] and K.M. Pierce[1]
[1]UCD Lyons Research Farm, Lyons Estate, Celbridge, Naas, Co Kildare, W23 ENY2, Ireland, [2]Nutriad International, NV Schietstandlaan 2, 2300 Turnhout, Belgium; zoe.mc-kay@ucdconnect.ie

Fusarium mycotoxins (FM) contaminate cereals and maize corn and may negatively impact dairy cow performance by affecting dry matter intake (DMI) and performance. Therefore, the objective of this research was to investigate the effects of FM in dairy cow diets on performance and metabolism. Twenty-six Holstein Friesian cows were allocated to one of two dietary treatments (n=13) in a randomised complete block design. Diets were offered as a total mixed ration (TMR) [concentrate (maize corn naturally contaminated with low or high FM), straw, grass and maize silage]. T1 (control) maize corn contained low FM levels [deoxynivalenon (DON) 163 µg DON/kg TMR DM and 19 µg ZON/kg TMR DM] and T2 maize corn contained high FM levels (1,966 µg DON/kg TMR DM and 366 µg ZON/kg DM). Cows were blocked on calving date and balanced for parity, predicted milk yield and body condition score. Cows started the trial 28 days post calving. There were 2 dietary periods within each treatment: P1= acclimatization (10 d); P2=low vs high FM TMR (28 d). Cows on the contaminated ration consumed 0.5 kg DM less and produced 0.75 kg milk less, but there were no statistical significant differences in DMI (T1=20.97 kg DM vs T2=20.52 kg DM, P=0.64), milk yield (T1=37.49 kg vs T2=36.75 kg, P=0.38), milk fat and protein kgs (T1=2.61 kg vs T2=2.59 kg, P=0.47). Similarly, there was no difference in rumen pH (T1=6.60 vs T2=6.67, P=0.31), rumen ammonia (T1=2.20 vs T2=2.22 mmol/l, P=0.97), total volatile fatty acids (T1=72.74 vs T2=76.77 mmol/l, P=0.66), alanine aminotransferase (T1=24.82 vs T2=24.13 u/l, P=0.46) or aspartate aminotransferase (T1=34.62 vs T2=29.23 u/l, P=0.28) between treatments. In conclusion, at the level of contamination reported in this study, adding FM contaminated corn to the TMR, had no significant effect on performance, rumen function or metabolism of early lactation autumn calving dairy cows. However, the 0.75 l/cow/d decline in milk yield due to feeding a DON contaminated ration, can represent a significant reduction in farm gate income.

Effects of mixed cassava pulp and Napier Pakchong grass as alternative feedstuff in laying hen diets

S. Khempaka, T. Omura, M. Sirisopapong, A. Khimkem, P. Pasri, S. Chaiyasit, S. Okrathok and W. Molee
Institute of Agricultural Technology, Suranaree University of Technology, School of Animal Production Technology, 111
Suranaree Avenue, Meaung, Nakhon Ratchasima, 30000, Thailand; khampaka@sut.ac.th

Cassava pulp contains 53.55% starch, 2.83% ash, 1.98% crude protein, 13.59% crude fiber and 0.13% ether extract on a dry matter basis. Since cassava pulp still remains high in starch content, therefore the pulp has widely used as an alternative energy source in animal diets. However, the fresh pulp also contains high moisture content approximately 70-80% which can cause a problem of drying in raining season. The combination of fresh cassava pulp and dried Napier Pakchong grass may help to resolve this problem. Therefore, the objective of this study was to evaluate the effects of mixed cassava pulp and Napier Packhong grass (MCN) on productive performance and egg quality in laying hens. A total of 180 ISA Brown, 67-week-old, were randomly distributed to 5 dietary treatments: one control and 4 MCN diets at levels of 5, 10, 15 and 20% with four replications of nine birds each (three birds/cage) using Completely Randomized Design (CRD). The experimental diets were formulated to include a similar level of ME and CP, respectively. Fibrolytic enzymes were added into all MCN diets to prevent the negative effect of high fiber content. Feed and water were provided *ad libitum* throughout the experimental period (8 weeks). Overall, 5-20% MCN diets did not show problematic effects on productive performance and egg quality of laying hens compared to control, except for the egg yolk color reduction of egg yolk color in 20% MCN diet. Thus, MCN can be incorporated in laying hen diets up to 15% without showing detrimental effects on productive performance and egg quality.

Digestibility of different sources of plant protein in growing pigs

J.M. Uriarte, H.R. Guemez, J.A. Romo, R. Barajas, J.M. Romo and N.A. Lopez
Universidad Autonoma de Sinaloa, Facultad de Medicina Veterinaria y Zootecnia, Blvd. Miguel Tamayo Espinoza de Los Monteros, 2358, Desarrollo Urbano 3 rios, 80020, Culiacan, Sinaloa, Mexico; jumanul@uas.edu.mx

The objective of this investigation was to determine the influence of different protein source on apparent digestibility of nutrients in growing pigs; six crossbred pigs (BW=42.14±1.82) were used in a replicated Latin Square Design. Pigs were assigned to consume one of three diets: (1) Diet with 17.80% CP and 3.27 Mcal ME/kg, containing sorghum 69.5%, soybean meal 28%, and premix 2.5% (CONT); (2) Diet with 17.72% CP and 3.28 Mcal ME/kg with sorghum 42.5%, cull chickpeas 40%, soybean meal 12.0%, vegetable oil 3%, and premix 2.5% (CHP), and (3) Diet with 17.88% CP and 3.26 Mcal ME/kg with sorghum 51.4%, cull chickpeas 30%, peanut meal 14%, vegetable oil 2%, and premix 2.5% (CHPN). Pigs were individually placed in metabolic crates (0.6×1.2 m). The adaptation period was 6 days and sample collection period was 4 days. From each diet and period, one kg of diet was taken as a sample and the total fecal production was collected. Apparent digestibility of DM with values of 82.04, 82.89 and 83.36%, for CONT, CHP, and CHPN, was affected among treatments (P<0.05). Apparent digestibility of crude protein was not altered (P=0.77) by CHP and CHPN inclusion (78.35, 78.47 and 79.24%), and apparent digestibility of OM was not altered (P=0.35) by CHP and CHPN inclusion (84.89, 84.45 and 85.36%). It's concluded that cull chickpeas and cull chickpeas-peanut meal can be used in growing pig improving nutrient digestibility, by comparing soybean-soybean meal diets.

Expression of recombinant delta 6 desaturase from Nile tilapia in *Saccharomyces cerevisiae*

S. Boonanuntanasarn and A. Jaengprai
Suranaree University of Technology, Institute of Agricultural Technology, 111 University Avenue, Muang, Nakhon Ratchasima, 30000, Thailand; surinton@sut.ac.th

The enzyme delta 6 desaturase ($\Delta 6$) catalyzes the formation of a double bond between carbon position 6 and 7 (numbered from carbohyl carbon). This study was aimed to produce the recombinant $\Delta 6$ from Nile tilapia (*Oreochromis niloticus*) (Oni-$\Delta 6$) using *Saccharomyces cerevisiae*. Actin promoter (Act) which is the ubiquitous promoter was cloned from genomic DNA of *S. cerevisiae* by PCR. The expression vector (pAct) was constructed by joining Act promoter to CYC 1 poly (A) signal. Recombinant yeast carrying pAct (RY- pAct) was produced by transformation of pAct into *S. cerevisiae*. The Oni-$\Delta 6$ cDNA was cloned from the liver of Nile tilapia. The plasmid vector which expressed Oni-$\Delta 6$ driven by Act was constructed and designated as pAct-Oni-$\Delta 6$. Recombinant *S. cerevisiae* carrying pAct-Oni-$\Delta 6$ (RY- pAct-Oni-$\Delta 6$) was generated. The expression of RY- pAct-Oni-$\Delta 6$ was determined by RT-PCR. In order to test the activity of recombinant Oni-$\Delta 6$, we cultured non-transformed *S. cerevisiae* (NT), RY-pAct and RY-pAct-Oni-$\Delta 6$ in yeast extract peptone-dextrose medium (YPD) in the presence of C18:2n6 substrate at 30 °C for 18 h. Subsequently, the NT, RY-pAct and RY-pAct-Oni-$\Delta 6$ were determined their fatty acid contents. RY-pAct-Oni-$\Delta 6$ contained high level of C18:3n6, comparing with that of NT and RY-ACT, demonstrating that recombinant Oni-$\Delta 6$ could have the $\Delta 6$ activity to convert C18:2n6 into C18:3n6 in *S. cerevisiae*.

Working out a lactic acid bacterium Lactobacillus plantarum TAK 59 NCIMB 42150 based silage additive

K. Märs[1], A. Olt[2], M. Ots[1,2] and E. Songisepp[1]
[1]Tervisliku Piima Biotehnoloogiate Arenduskeskus OÜ (BioCC), Kreutwaldi1, 51014 Tartu, Estonia, [2]Estonian University of Life Sciences, Kreutzwaldi 1, 51014 Tartu, Estonia; kristiina.kokk@tptak.ee

In livestock farming and milk production it is important to ensure that animals get high quality feed all year round. Silage has proven to be one of such feed. The objective of the study was to develop a biological silage additive to increase the fermentation quality of feed through inhibiting the impact of undesirable microbes. *Lactobacillus plantarum* TAK 59 was isolated from naturally ensiled fodder legume crop silage. The strain was thoroughly characterised *in vitro*: sensitivity to EFSA required antibiotics was tested, short-chain fatty acid profile was determined by gas chromatography and antimicrobial activity towards pathogens was evaluated by streak line test. Next ensiling experiment was carried out. The strain was added to feed in ratio of 1×10^5 cfu/g of the fermented feed. *L. plantarum* TAK 59 inhibited the decrease of plant-origin clostridia by 19,81% and improved the quality of silage significantly. When ensiling forage that is difficult to ensile (low dry matter and sugar content), *L. plantarum* TAK 59 reduced the pH of the silage to 4.4, whereas the pH of the control silage was 5.1. As a result of the fast and extensive decrease of pH through targeted lactic acid fermentation with *L. plantarum* TAK 59, the impact of enzymes and undesirable proteolytic pathogenic microorganisms, and the production of degradation compounds produced by them, is inhibited. Concentration of ammonia nitrogen and butyric acid in feed were significantly lower compared to the control silage. TAK 59 reduced the concentrations of acetic acid produced by indigenous yeasts, acetic acid bacteria and enterobacteria into the silage. The properties of the strain have been patented by The Tervisliku Piima Biotehnologiate Arenduskesksus OÜ (EP14752554.7). In 2016 the strain *L. plantarum* TAK 59 passed the panel of experts of the European Food Safety Authority (EFSA) and has been confirmed as a silage additive with established characteristics. Thestrain has been registered as a silage additive in the European Union register of Feed Additives and which has been issued by commission implementing regulation (EU) No 2016/2150.

Supplementation of different levels of urea to sheep fed low quality *Eragrostis curvula* hay
H. Mynhardt, L.J. Erasmus, W.A. Van Niekerk and R.J. Coertze
University of Pretoria, Animal and Wildlife Sciences, Lynnwood Road, 0001 Pretoria, South Africa;
lourens.erasmus@up.ac.za

The objective of the study was to determine whether level of urea supplementation would affect roughage intake, digestibility, or microbial N synthesis (MNS) in ruminants fed tropical roughages. A Latin square design trial was conducted using four Dohne Merino rumen cannulated sheep fed low-quality [0.4% nitrogen (N), 83% neutral detergent fibre (NDF)] *Eragrostis curvula* hay plus starch (240 g/day). The levels of urea supplemented were: 10.4 g urea/sheep/day (LU); 18.4 g urea/sheep/day (MU); 26.4 g urea/sheep/day (HU) and 32.4 g urea/sheep/day (EHU). Urea supplementation did not affect roughage intake or digestibility, however, N intake, as expected, increased with increasing levels of urea. The level of urea supplementation affected total N balance in the sheep increasing from less than 2 g N/sheep/day in the LU treatment to over 11 g N/sheep/day in the EHU treatment. In addition, efficiency of N use, calculated as N retained relative to available N intake, increased from 0.25 in the LU treatment to 0.60 in the HU treatment. Efficiency of N use did not differ between HU and EHU treatments. Rumen ammonia N (RAN) differed with RAN increasing from 7 mg RAN/dl rumen fluid to above 20 mg RAN/dl as urea supplementation was increased. Neither MNS nor efficiency of MNS differed between urea treatments. The ratio of MNS: N intake decreased from levels above 2 to less than 1 as the level of urea supplementation was increased from 10.4 g urea/sheep/day (LU treatment) to 32.4 g urea/sheep/day (EHU treatment). Ratios higher than 1 could be indicative of possible N deficiency in the rumen, since more N from recycling could be incorporated into microbial N synthesis. Results suggested that sheep needed to be supplemented with at least 26.4 g urea/sheep/day (HU treatment) to prevent a deficiency in N supply.

Synchronisation of energy and protein supply in sheep fed low quality *Eragrostis curvula* hay
H. Mynhardt, W.A. Van Niekerk, L.J. Erasmus and R.F. Coertze
University of Pretoria, Animal and Wildlife Sciences, Lynnwood Road, 0001 Pretoria, South Africa;
willem.vanniekerk@up.ac.za

The objective of this study was to determine whether the pattern of starch and/or urea supplementation would impact on roughage intake, digestibility, and microbial nitrogen synthesis (MNS) in sheep fed low-quality [0.4% nitrogen (N), 83% neutral detergent fibre (NDF)] *Eragrostis curvula* hay. Six rumen cannulated Dohne Merino sheep were assigned to a 6×6 Latin square design experiment. Six different supplementation patterns were created by supplementing variable amounts of starch and/or urea during the morning and/or afternoon. The sheep were fed ad lib low quality *E. curvula* hay. Total intake of starch (120 g/day) and urea (13.2 g/day) was similar in all treatments except one treatment where starch was substituted with molasses. Supplementation pattern did not affect roughage intake; however, roughage digestibility appeared to be higher in the treatments where starch was supplemented at least partly, during the morning periods. Neither N intake, nor faecal N excretion were affected by treatment. Urinary N excretion was higher in the treatments where starch was supplemented during both the morning and afternoon periods. The most consistent rumen ammonia N (RAN) concentration, with RAN concentrations appearing to be optimal throughout the day (ranging between 10-15 mg/dl rumen fluid), was achieved in the treatment where both starch and urea was supplemented during both the morning and afternoon periods. Microbial N synthesis and efficiency of MNS (EMNS) were higher in the treatments where starch was supplemented during both morning and afternoon periods. It can be concluded that a continuous supply of starch is more important than a continuous supply of urea in terms of improving forage digestibility, MNS and EMNS.

Effect of prepartum dietary cation – anion difference (DCAD) on dairy cows performance

H.M. Metwally[1], S.H.M. El-Mashed[1,2], H.M. Gado[1], A.M. Salama[2] and H. Abd Elhairth[2]
[1]Ain Shams university, Fac of Agriculture, animal production, Shubra elkhima, 11241, Egypt, [2]Animal Production Research Institute, Animal Production, Doki, 11241, Egypt; sh.elmashed@hotmail.com

Twenty Friesian cows were used to study the effect of dietary cation-anion difference on blood and urine parameters. Animals were divided into four groups. All groups were fed a basal diet consisting of Berseem 30 kg, concentrate feed mixture (CFM) 6 kg, rice straw 6 kg & soya bean meal (SBM) 75 g as a carrier for anionic salts/head/day. Anionic salts were used to control DCAD of the ration at the following levels (groups): group (1): Control group was fed the basal diet without anionic salts. group (2): DCAD was 0 mEq/kg DM. group (3): DCAD was – 150 mEq/kg DM. group (4): DCAD was -150 mEq/kg DM by using Anio – Norel (commercial product). Results obtained showed that: (1) Serum calcium concentration increased significantly (P<0.05) with decreasing DCAD. Animals received low DCAD (-150 mEq/kg DM) whether through anionic salts or Anio-Norel had an overall mean of serum calcium concentration of 8.133 & 8.367 mg/dl, respectively compared with 6.944 and 7.589 mg/dl for DCAD +90 and zero, respectively. Urine calcium concentration showed a similar trend where it was increased with decreasing DCAD levels. (2) Urinary PH decreased significantly with decreasing DCAD level, values were 2.093 for group (1) and 1.883 for group (4). (3) Blood progesterone decreased dramatically after parturition in all groups. However, the second group had lower blood progesterone concentration when compared with the other three groups. (4) Beta hydroxy butyric acid (BHBA) concentration in blood was affected significantly and indicated a significant interaction between treatment and time before and after parturition. (5) Serum T4 concentration was slightly but significantly less (3.629 µg/dl) for group (4) than group 2 and 3. (3.881 and 3.898 µg/dl) respectively. It was concluded that prepartum negative DCAD prevented milk fever and ketosis through controlling Ca metabolism.

Variation in feed intake of 6 months bull calves fed various types of feeds and rations

M. Vestergaard and M. Bjerring
Aarhus University, Department of Animal Science, Foulum, 8830 Tjele, Denmark; mogens.vestergaard@anis.au.dk

A larger variation in day to day (D-D) feed intake in fattening feedlot beef cattle has been associated with high-starch, low-fiber diets compared with moderate starch and fiber diets and this could give rise to larger fluctuations in rumen pH and subacute rumen acidosis that might also lead to a higher level of liver abscesses. A large calf to calf (C-C) variation in feed intake in the pen could be expected to lead to less uniform growth rates in the pen and increased LW differences. However, little is known about the two sources of variation in feed intake among rosé veal calves of dairy breeds usually fed high energy diets and slaughtered before 10 months of age. The objective was to analyse how various types of pelleted concentrates, roughage sources, and total mixed rations (TMR) affected D-D and C-C variation in feed intake. Data from 4 experiments with 15 different feeds and rations, and 271 bull calves were included. Data included individual daily feed intakes (Insentec feeders) measured over a 30 day period from 6 to 7 months of age; a period in which these calves typically has an ADG of 1.4-1.8 kg/d. The CV% calculated as the std. dev. divided by the mean of the 30 d of recordings was used to describe the variation in feed intake. Results showed that for pelleted concentrates varying in starch content from 28 to 40%, the D-D variation was 15% (min-max 13 to 18%). Barley straw is usually the only roughage supplied next to the pelleted concentrate and the D-D variation for straw varied from 50 to 100%; showing days with no straw intake at all. For TMR the D-D variation varied from 14 to 22%; with no clear relation to TMR energy density (high vs low). The C-C variation varied from 7 to 15% for the 5 pelleted concentrates, from 13 to 14% for high-energy TMR, from 24 to 27% for low-energy TMR, and from 24 to 36% for straw. The results suggest that a more variable D-D intake for pelleted concentrate based rations compared to TMR is due to variation in straw intake and not in pelleted concentrates *per se*. The C-C variation is similar for pelleted concentrates and the high-energy TMR usually fed to veal calves, but a low-energy TMR is expected to result in larger C-C variation in feed intake.

Random forests prediction of blood metabolic clusters by milk metabolites and enzymes in dairy cows

L. Foldager[1,2] and Genotype Plus Environment Consortium (www.gpluse.eu)[1]
[1]Aarhus University, Dept. of Animal Science, Blichers Allé 20, DK8830 Tjele, Denmark, [2]Aarhus University, Bioinformatics Research Centre, Aarhus, Denmark; leslie@anis.au.dk

Milk biomarkers as predictors of physiological imbalances and subclinical or clinical diseases in dairy cows may be used in management and selection. The objective was to use milk metabolites and enzymes to predict three clusters differentiating the metabolic status of cows and defined by clustering of plasma hormone Insulin-like Growth Factor-1 (IGF-1) and three plasma metabolites: Glucose, beta-hydroxybutyrate (BHB), and non-esterified fatty acids (NEFA). Data between calving and 50 days in milk (DIM) were obtained from 209 Holstein Friesian cows in six research herds: AU (Denmark), UCD (Ireland) AFBI (UK), CRA-W (Belgium), FBN (Germany), CREA (Italy). Some diets were designed to challenge the cow and provoke production diseases but there were too few incidences to establish predictors of clinical and subclinical diseases. Instead, k-means clustering was performed over scaled residuals from linear mixed effects models for each of IGF-1, Glucose, $\log_{10}$(BHB) and $\log_{10}$(NEFA) measured around 14 and 35 DIM. Each milk metabolite (free glucose, glucose-6-phosphate, BHB, isocitrate, uric acid, urea) and enzyme (LDH, NAGase) from 2 milkings/week was averaged within week and matched with blood sampling. Random forests (RF) predictions with milk markers and parity (1, 2 and 3+) was evaluated by leave-one-cow-out (internal) cross-validation. Overall accuracy: 0.52 (95% CI: 0.45-0.59) on DIM14 and 0.44 (0.37-0.51) on DIM35. Sensitivity for each cluster: (0.77, 0.87, 0.61) on DIM14 and (0.65, 0.71, 0.79) on DIM35. Specificity: (0.54, 0.34, 0.61) and (0.46, 0.51, 0.32). We obtained predictions of metabolic clusters from milk biomarkers of fair accuracy. Sample sizes may constraint both estimation of clusters, number of classes, and RF training. External validation in separate herds is warranted. Still, we see potential in using milk biomarkers to support management of dairy cows.

Exploratory classification of multiparous dairy cows based on fertility related phenotypes

M. Salavati and Genotype Plus Environment Consortium (www.gpluse.eu/index.php/project/partners)
Royal Veterinary College, Hatfield, AL9 7TA, United Kingdom; msalavati@rvc.ac.uk

Tissue mobilization due to major changes in endocrine signals around and after calving influences physiological status and subsequent fertility. Liver function is challenged when nutrient requirements for milk production excessively exceeds that of nutrient intake in early lactation, altering metabolic status and e.g. circulating insulin like growth factor (IGF1), a key signalling molecule. Objective: To use principal component analysis (PCA) to provide a novel integrative approach to differentiate between phenotypes affecting fertility. Multiparous Holstein Friesian cows (n=151) were recruited from 6 herds and allocated between up to 3 diets/herd. Phenotype data were collected between calving and 50 days in milk (DIM). Traits included calving ease, body condition score (BCS) at calving, BCS and circulating IGF1 at 14 and 35 DIM, change in BCS and IGF1, total milk yield (TMY) and fertility (days to first service (DFS) and conception categorised as in calf by <100, 100-200 or >200 DIM (ICB)). PCA was performed in R. Residuals were calculated based on REML prediction of phenotypes taking diet nested within herd as a random effect. K-mean clustering was performed over centre scaled phenotypes to produce 3 significantly discriminant clusters. Comparison of traits between clusters used Kruskal-Wallis or ANOVA with appropriate post-hoc testing. The 3 clusters contained 55, 59 and 37 cows, respectively. Cluster 3 had the best fertility with the lowest DFS and ICB scores. These cows had a medium BCS at calving, high IGF1 at 14 DIM, a high BCS gain but the lowest TMY. Cluster 2 with the worst fertility calved with a low BCS which remained low and produced a medium TMY. Cluster 1 had a high BCS at calving, a high TMY and intermediate fertility. PCA is a useful approach to determine which traits consistently influence fertility across a range of different herds and feeding regimes. Future studies will compare genotypes and tissue gene expression data between clusters to gain insight into differing metabolic pathways involved and to help optimise selection policies to promote better fertility whilst retaining good milk production.

Prediction of energy status of dairy cows using MIR milk spectra

*C. Grelet, A. Vanlierde, F. Dehareng, E. Froidmont and the Genotype Plus Environment Consortium (www.gpluse.
eu/index.php/project/partners)*

Walloon Agricultural Research Center (CRA-W), 24 chaussée de Namur, 5030 Gembloux, Belgium; c.grelet@cra.wallonie.be

A key task within the GplusE project is to undertake a genetic evaluation using ten thousand cows to improve
health traits of dairy cows, with energy status (ES) being a trait of major interest. To achieve this, cost effective
and large scale phenotyping methods are required. This study was designed to evaluate the possibility of using
MIR spectra of milk to predict ES of cows. Data was collected from 241 cows, from calving until 50 days in milk
(DIM) in six research herds of the GplusE consortium distributed in Belgium, Denmark, Germany, Ireland, Italy
and UK. Milk MIR spectra were collected twice weekly during this period. ES was 'quantified' by measuring daily
energy balance (EB), residual feed intake (RFI), dry matter intake (DMI), and measuring at 14 and 35 DIM blood
metabolites/hormones (Glucose, BHB, NEFA and IGF-1). K-means clustering was also performed based on these 4
blood components in order to discriminate 2 groups with healthy vs imbalanced cows. Regression models between
each of these variables and MIR milk spectra have been developed using PLS and classification model with SVM
method. The R^2 of cross-validation obtained when predicting EB, RFI, DMI, Glucose, BHB, NEFA and IGF-1 were
respectively 0.43, 0.46, 0.47, 0.31, 0.40, 0.28 and 0.48. Discriminant model based on blood metabolite clusters was
able to differentiate healthy vs imbalanced cows with sensitivity and specificity of 84% and 81%, respectively. These
preliminary results demonstrate that milk MIR spectra have reasonable potential to provide information on ES related
variables. This could allow large scale predictions for both genetic studies and farm management.

Genetic analysis of milk MIR predicted blood and milk biomarkers linked to the physiological status

*H. Hammami[1], F.G. Colinet[1], C. Bastin[2], C. Grelet[3], A. Vanlierde[3], F. Dehareng[3], N. Gengler[1] and Gpluse Consortium
(www.gpluse.eu/index.php/project/partners)[4]*

*[1]ULg-Gembloux Agro-Bio Tech, Animal Science Unit, 5030, Gembloux, Belgium, [2]Walloon Breeding Association,
R&D, 5590, Ciney, Belgium, [3]Walloon Agricultural Research Center, Valorisation of Agricultural Products Department,
5030, Gembloux, Belgium, [4]University College Dublin, Veterinary Science Centre, Belfield, Dublin 4, Ireland;
hedi.hammami@ulg.ac.be*

Based on reference data from 6 GplusE project partner farms, several equations were developed to predict milk/
blood based biomarkers from milk mid-infrared spectra (MIR). Additional existing MIR prediction equations of
milk based biomarkers were included in this study. Data included predicted biomarkers for test-days between the
5th and the 49th DIM in the first 5 lactations of 57,240 Holstein cows. MIR spectra used to predict those biomarkers
were collected since 2012 in 461 Belgian commercial farms enrolled in the official Walloon milk recording. Genetic
parameters for each trait were estimated using single trait multi-lactation animal linear model. Additionnally bivariate
models were used to investigate the genetic associations of MIR predicted milk and blood biomarkers. The lowest
heritabilities estimates of 0.14, 0.15, and 0.17 were observed for milk urea, blood urea, and milk β-hydroxybutyrate
(BHB) respectively. NEFA, BHB, and IGF-1 in blood have moderate heritability estimates (0.20-0.25). The highest
heritabilties (0.31-0.35) concerned milk lactate dehydrogenase (LDH), milk glucose-6-phosphate, and blood glucose.
Genetic correlations between lactations were relatively strong (≥0.74) for all indicators. Correlations between first-
and later-lactations were the lowest (from 0.74 for blood NEFA to 0.90 for blood glucose). Highest correlations were
observed between second- and later lactations (0.86 to 0.97 for milk BHB and milk LDH respectively). Urea and
BHB in milk have strong genetic correlations with urea and BHB in blood (0.87 and 0.84 respectively). Additional
validation of predictions equations in commercial farms and integration of reference data from other populations
were needed, nevertheless first results showed value of these non-invasive biomarkers for routine monitoring and
for breeding.

Use of metabolic clusters predicted by milk metabolites and enzymes to identify potential risk cows

M.T. Sørensen and the Genotype Plus Environment Consortium (www.gpluse.eu)
Aarhus University, Dept. of Animal Science, Blichers Allé 20, 8830 Tjele, Denmark; martint.sorensen@anis.au.dk

Use of metabolic clusters predicted by milk metabolites and enzymes to identify potential risk cows. Data between calving and 50 days in milk (DIM) were obtained from 209 Holstein Friesian cows in six research herds: AU (Denmark), UCD (Ireland) AFBI (UK), CRA-W (Belgium), FBN (Germany), CREA (Italy). Three clusters were defined at around 14 and 35 DIM from plasma IGF-1, glucose, beta-hydroxybutyrate (BHB), and non-esterified fatty acids (NEFA) and predicted with fair accuracy from milk metabolites (free glucose, glucose-6-phosphate, BHB, isocitrate, uric acid, urea) and enzymes (LDH, NAGase) as described by Foldager. Such clusters may be used as a management tool to identify potential risk cows. Despite that some diets were designed to challenge the cow and provoke incidences of production diseases, the cows were amazingly robust and developed very few incidences clinical diseases. Therefore, we cannot analyse the distribution of diseases among clusters. Instead, we have examined the distribution of the cow's energy balance (Ebal) among clusters, since a very low Ebal may be an indicator of physiological imbalances and a potential triggering factor for production diseases. Average Ebal were -16, -30 and -24 MJ around DIM 14 and -13, -24 and -2 MJ around DIM 35 for cluster #1, 2 and 3. The lowest Ebal were found in cluster #2 around both DIM 14 and DIM 35. Cluster #2 for DIM 14 were characterized by relatively low plasma glucose and IGF-1 and relatively high BHB, and cluster #2 for DIM 35 were characterized by relatively low plasma IGF-1 and relatively high BHB and NEFA; these plasma characteristics are indicative of risk conditions. These data indicate that clusters predicted from milk metabolites and enzymes can be used to identify a group of risk cows.

Investigating metabolic phenotypes in multiparous dairy cows by component analysis and clustering

M. Salavati and Genotype Plus Environment Consortium (www.gpluse.eu/index.php/project/partners)
Royal Veterinary College, Hatfield, AL9 7TA, United Kingdom; msalavati@rvc.ac.uk

The abrupt rise in energy demand for milk production postpartum causes an imbalance between nutrient requirements and dry matter intake (DMI). If mobilization of body reserves is excessive, this causes metabolic stress which affects multiple organs including the liver, immune system and reproductive tract. Objective: To use principal component analysis (PCA) to provide an integrative approach to differentiate between metabolic phenotypes which predict overall performance. Multiparous Holstein Friesian cows (n=113) were recruited from 6 herds and allocated between up to 3 diets/herd. Phenotype data, initially including 30 traits, were collected from calving. Traits included body condition score (BCS) at calving, weight, BCS and circulating IGF1, glucose, cholesterol, beta-hydroxybutyrate (BHB) and NEFAs at 14 and 35 days in milk (DIM), their respective changes between 14 and 35 DIM and cumulative DMI and total milk yield (TMY) over the first 50 DIM. PCA was performed in R. Residuals were calculated based on REML prediction of phenotypes taking diet nested within herd as a random effect. Phenotypes were included based on their contribution to 4 dimensions. K-mean clustering was performed to produce 3 significantly discriminant clusters. Comparison of traits between clusters used Kruskal-Wallis or ANOVA with appropriate post-hoc testing. The 3 clusters contained 50, 28 and 35 cows. Cluster 3 cows were characterised by a low BCS at calving, followed by low weight, cholesterol and DMI in early lactation. Cluster 1 cows had a high BCS at 0 and 14 DIM, high BHB and NEFA at 14 and 35 DIM and a low gain in IGF1 between 14 and 35 DIM. They also weighed more at 35 DIM and produced the most milk. Cluster 2 cows had a medium BCS, higher IGF1 and glucose and low BHB associated with the highest DMI. These data confirm that BCS at calving is associated with different metabolic phenotypes. Cluster 1 cows produced more milk but had an adverse metabolic profile likely to compromise health and fertility. This approach offers a novel way to group cows by phenotype in an unbiased fashion and will next be used to investigate underlying genotypes.

Effect of high protein allocation on plasma metabolite concentrations in postpartum cows
M. Larsen, A.L.F. Hellwing, T. Larsen, L. Hymøller and M.R. Weisbjerg
Aarhus University – Foulum, Blichers allé 20, 8830, Denmark; mogens.larsen@anis.au.dk

The effect of increasing the metabolisable protein or energy supply on plasma metabolite concentrations was investigated in 91 postpartum Holstein cows. Cows were housed in a free stall with automatic milking (AMS) and feed intake recording. Within block (parity and calving date), cows were randomly allocated to 3 diets fed in week 1-4 after calving: control (C), protein (P), and barley (E). The C diet was a partially mixed ration (PMR; maize- and grass/clover silage, ensiled sugar beet pulp, NaOH treated wheat, rapeseed meal, soybean meal) fed ad lib. and 3 kg concentrate fed in the AMS. With P, 9% of sugar beet pulp (DM basis) was exchanged with an equal mix of soypass, maize gluten meal, and potato protein. With E, cows were fed the same PMR as C and 2 kg of rolled barley in the AMS in addition to the 3 kg of concentrate. In week 5-8, all cows were fed a standard PMR ad lib. and 3 kg concentrate in the AMS. Blood samples were taken weekly in week 1-4 and 8. In week 1- 4, ECM yield for multiparous cows was 52.3 kg/d for P as compared with 46.4 kg/d for C and 43.7 kg/d with E, whereas diet did not affect ECM yield for primiparous (P<0.01). In week 5-8, ECM yield did not differ among diet groups. Concentrate intake was greater with E, resulting in a 0.5 kg/d greater DMI for E compared with C and P. For multiparous cows, plasma NEFA tended to be greater in week 1 and 2 after calving for P, but decreased similarly among diets in the following weeks, whereas plasma NEFA did not differ among diets for primiparous (P<0.01). Plasma glucose were lower for C as compared with P and E (P<0.01). Compared with primiparous, plasma glucose was lower for multiparous cows and plasma 3-OH-butyrate (BHB) was greater (P<0.01). Plasma BHB was numerically greater for multiparous cows with C diet compared with other diet by parity combinations (P=0.17). Plasma EAA and urea was greater for P compared with C and E in week 1-4 (P<0.01). Overall, the increased NEFA concentrations in multiparous cows fed a diet high in bypass protein indicate greater fat mobilization; however, this did not lead to greater BHB levels. The greater plasma EAA and plasma urea in cows fed a diet high in bypass protein indicate greater catabolic use of AA.

Accelerometer technology as a non-invasive tool to estimate the state of the dairy cow
L. Munskgaard
Aarhus University, Department of Animal Science, Blichers allé 20, 8830 Tjele, Denmark; lene.munskgaard@anis.au.dk

Structural changes in dairy farming towards larger herds in combination with increased demands for documentation of health and welfare means that there is a need for new tools for monitoring and troubleshooting at dairy farms. Accelerometer technology is relatively cheap and easy to implement in use on dairy farms. This technology to record activity has been used for a number of years to detect cows in heat, and there are a rapidly increasing number of studies including accelerometer technology as a tool to estimate different aspects of the state of dairy cows. This include estimates of lying, standing, walking, number of steps, eating and rumination behaviour as well as detection of lame cows, cows in heat and cows with different diseases. However, the accelerometers are typically placed either on the leg, in a more or less fixed position on the neck or in the ear of the cow. When positioned on the leg more accurate information is derived about movement of the leg and related behaviours such as lying down while position on the neck or especially at the ear lead to more accurate estimates of behaviour more closely related to movements of the head such as rumination. The first symptom of a disease is very often a change in behavioural patterns. The most well-known changes in behaviour in response to disease are reduced activity; decreased feed intake and increased time spend lying. Thus changes in behaviour may be a predictor of different aspects of the health status of the cow. However, there is increasing evidence that more than one indicators needs to be included to obtain reliably predictions. Accuracy in automatic recording of different activities of dairy cows will be discussed as well as the value of the automated recordings of varies activities of dairy cows in relation to prediction of the status of the animal.

Emerging sensor technology for quantifying animal physiological conditions

C. Michie, I. Andonovic, C. Tachtatzis and C. Davison
University of Strathclyde, 204 George Street, Glasgow, G1 1XW, United Kingdom; c.michie@strath.ac.uk

The past 15 to 20 years have witnessed a digital revolution within agriculture. The rapid development of low cost sensors enabled with wireless communications has led to an increase in the deployment of measurement systems that assist farmers to operate more efficiently with fewer staff. Within the dairy sector, such technology has become prevalent as commercial pressures have forced a steady decline in the number of milk producers while milk production has remained largely constant. The number of dairy cows in the UK has declined from 2.6 million in 1996, to 1.9 million in 2015. Concurrently, the average herd size has risen as those holdings with smaller herds have left the industry; in 2014, the average number of cows per herd was 133, compared to 97 in 2004. Increasing herd sizes result in additional complexity of maintaining a close understanding of animal welfare. Heat detection systems that optimise herd fertility and hence milk production are now commonplace. As the industry has grown to embrace the role of technology, the opportunities for harnessing the measurements to give greater understanding of underlying physical conditions has become more valued. Currently a range of sensor systems exist that provide information on activity, eating, rumination, standing, lying, walking and location. These behaviours have been validated by the animal science community to provide insights into animal welfare. In addition, other complementary sensors now provide information such as rumen pH value and body temperature. The paper will present an overview of the state of the art of sensors which are integrated into measurement platforms and connected to cloud platforms. At present, the bulk of these systems are standalone and operate in isolation from other measurement data. Therefore, a significant opportunity exists to integrate multiple sensor data streams with other cloud based data storage platforms in order to derive enhanced information, for example milk production data and/or milk constituents or feed composition and distribution. The opportunities to provide greater insight into animal welfare and in turn inform improved production efficiencies are significant.

GplusE policy dialogue meeting

N. Gengler[1] and C. Knight[2]
[1]University of Liege, Passage des Déportés, 2, Gembloux 5030, Belgium, [2]University of Copenhagen, Dyrlaegevej 100, 1870 Frb C, Denmark; nicolas.gengler@ulg.ac.be

This one-hour Policy Dialogue meeting organised by the EU FP7 GplusE Project will be an opportunity for stakeholders from within and outside the Project to discuss through different perspectives how the non-invasive monitoring of physiological and metabolic state can contribute to solve technical, economic, societal and environmental issues associated with dairy animals. There will be particular emphasis on high performance dairy cows and the transition period, and how current research is addressing the needs of the dairy industry. The Policy Dialogue meeting will also seek to identify what future research needs exist and how these might best be accomplished. GplusE cordially invite you to participate in this Policy Dialogue meeting and look forward to seeing you in Tallin.

Natural killer and γδ T cells characterization in young and old *Bubalus bubalis* by flow cytometry

F. Grandoni[1], M.M. Elnaggar[2,3], S.G. Abdellrazeq[2,3], F. Signorelli[1], L.M. Fry[3,4], F. Napolitano[1], V. Hulubei[2], S.A. Khaliel[2], H.A. Torky[2], C. Marchitelli[1] and W.C. Davis[3]
[1]*Centro di Ricerca per la Produzione delle Carni e il Miglioramento genetico (CREA-PCM), Via Salaria 31, 00015 Monterotondo (RM), Italy, [2]Department of Microbiology, Faculty of Veterinary Medicine, Alexandria University, Al Mealah, Markaz Rasheed, El Beheira Governorate, 22758 Alexandria, Egypt, [3]Department of Veterinary Microbiology and Pathology, College of Veterinary Medicine, Pullman, WA 96140, USA, [4]USDA, ARS, Animal Disease Research Unit, Pullman, 96140, USA; cinzia.marchitelli@crea.gov.it*

Water buffalo (*Bubalus bubalis*) is an important species in livestock industry worldwide. Until now few reagents have been available to characterize its immune system and study the immune response to pathogens and vaccines. Identification of a large set of cross reactive monoclonal antibodies (mAbs) now provide an opportunity to compare similarity and differences with the bovine immune system. The aim of this study was: to use 6 parameter flow cytometry to determine the phenotype and frequency of NK cells and γδ T cells in five heifers (9-10 months of age) and five old lactating buffaloes (12-15 years of age). All mAbs, except anti-CD335 (Pierce-Thermo Fisher Scientific) were provided by the Washington State University Monoclonal Antibody Center (WSUMAC). The SAS GLM procedure was used for statistical analysis. We found significant differences in the frequency of NK cells between young and old buffaloes (2.70%±0.78 vs 8.95±1.11; P≤0.05). We also found expression of CD2 on NK cells was increased (55.22%±4.70 vs 71.50%±8.14) and more uniform in older animals. CD8 expression level was similarly low in both groups. In contrast, γδ T cell frequency was high in young animals and low in old animals (33.99%±1.64 vs 5.36%±2.85; P≤0.01). Expression of CD2 was modestly increased (19.85%±3.45 vs 24.52%±5.98) with age of animals. The most important and unexpected result was the expression of CD8 on γδ T cells: 95% were positive in young and 55% in old buffaloes, while in cattle only a small subset of γδ T cells are CD8[+]. In conclusion, the frequency of NK cells and γδ T cells was similar the frequency in cattle but the difference was in the expression of CD2 and CD8 on γδ T cells.

Creatinine in milk as biomarker for muscle protein mobilisation in postpartum cows

M. Larsen, A.L.F. Hellwing, T. Larsen, P. Løvendahl, J. Sehested and M.R. Weisbjerg
Aarhus University – Foulum, Blichers allé 20, 8830, Denmark; mogens.larsen@anis.au.dk

The effect of increasing the metabolisable protein or energy supply on mobilization of muscle protein was investigated by assessing plasma concentrations of amino acids (AA), creatinine and 3-methyl-His (3MH), and milk creatinine concentrations in 91 postpartum Holstein cows housed in a free stall with automatic milking (AMS) and feed intake recording. Within block (parity and calving date), cows were randomly allocated to 3 diets fed in week 1-4 after calving: control (C), protein (P), and barley (E). The C diet was a partially mixed ration (PMR; maize- and grass/clover silage, ensiled sugar beet pulp, NaOH treated wheat, rapeseed meal, soybean meal) fed ad lib. and 3 kg concentrate fed in the AMS. With P, 9% of sugar beet pulp (DM basis) was exchanged with an equal mix of soypass, maize gluten meal, and potato protein. With E, cows were fed the same PMR as C and 2 kg of rolled barley in the AMS in addition to the 3 kg of concentrate. In week 5-8, all cows were fed a standard PMR ad lib. and 3 kg concentrate in the AMS. Blood samples were taken weekly in week 1-4 and 8, and milk samples were obtained weekly. Plasma essential AA were greater with P compared with C and E (P<0.01) indicating a greater availability of AA for milk synthesis and other maternal body functions. In week 1-4, plasma creatinine were lower with P compared with other diets, whereas concentrations did not differ in week 5-8 after cease of treatment diets (P=0.01). Plasma 3MH were lower with P compared with C and E in week 1-4 after calving (P=0.03). Milk creatinine decreased from 351±6 to 248±3 μM during week 1-8 after calving (P<0.01) and tended to be greater for primiparous cows (P=0.06). For primiparous cows, Pearson correlations were 0.41 for milk:plasma creatinine, 0.59 for milk creatinine:plasma 3MH, and 0.71 for plasma creatinine:plasma 3MH. For multiparous cows, Pearson correlations were 0.52 for milk:plasma creatinine, 0.53 for milk creatinine:plasma 3MH, and 0.60 for plasma creatinine:plasma 3MH. Overall, the high by-pass protein diet reduced plasma levels of creatinine and 3MH indicating reduced mobilization of muscle protein. More research is needed, but milk creatinine may be a potential biomarker for extent of muscle protein mobilization.

Pregnancy diagnosis in dairy cattle by circulating nucleic acid

A. Karimi[1] and R. Hosseini[2]
[1]*University of Tabriz, Ahar Faculty of Agriculture and Natural Resources, Tabriz, 5166616471, Iran,* [2]*Orum Gohar Daneh, Miandoab, 5166616471, Iran; pekarimi@tabrizu.ac.ir*

Each day of an extended calving interval results in economic loss. Pregnancy and pregnancy diagnosis are important factors in commercial dairy cattle. Generally, pregnancy diagnosis in dairy cattle is performed by hormonal measurement in day 22 post mating with low accuracy and precise diagnosis is obtained approximately in day 40 post mating; Circulating nucleic acids (CANs) are extracellular DNA and/or RNA Molecules that circulate in the blood stream of healthy and diseased animals and some diagnostic application are developing as a biomarker. In this experiment, a total of 30 dairy cows with first insemination post-parturition were used in blood sampling in day 22 and 40 after insemination and also, animals were examined for pregnancy determination by palpation technic in day 45 post-insemination. Separated serum from blood samples were used for progesterone (P4) concentration (by RIA technic) and CANs (by Real-Time PCR technic) assays. Real time PCR amplification and analysis were performed using the threshold cycle as determined by the second derivative maximum method. The PCR reagents were used with following conditions: 20 ml reaction volume contained $1\times$ PCR buffer, 3.5 mM $MgCl$, 250 mM of each dNTP, 3 U Fast Start Taq Polymerase, $0.5\times$ EvaGreen Fluorescent DNA Stain, 0.125 mM primer (Art2A) and 5 pg of template DNA. PCR protocol was: initial denaturation at 95 ˚C for 10 minutes, followed by 40 cycles of 95 ˚C for 30 seconds, 61˚C for 15 seconds and 72 ˚C for 30 seconds. The fluorescence signal was measured at the end of each extension step. PCR primers used in this study include: Art2A.f: GGGTAAACTCTGGGAGTCGGTGATG, Art2A.r: AGTCAGTTCAGTCGCTCAGTCGTG. After pregnancy test, the number of 17 dairy cows were pregnant and P4 concentration in day 22 and 40 were significantly higher (5.47 and 7.20 ng/ml; respectively) versus nonpregnant dairy cows (0.18 ng/ml), because of repeat breeding in these animals, there were no blood samples in day 40. Comparison of calculated rational CNA for Art2A after RT-PCR showed higher copies in pregnant cows (about 3.1 fold) than non-pregnant animals (6315 and 2037 copies respectively; P<0.05). It seems putting p4 concentration and CNAs assay together, may have beneficial results in pregnancy test.

Measurement of physiological substances in milk for evaluating welfare level of cows

A. Tozawa[1], S. Ogura[1], Y. Maeda[2], Y. Kato[2] and Y. Nakai[1]
[1]*Tohoku University, Graduate School of Agricultural Science, 232-3, Yomogita, Naruko-Onsen, Osaki, Miyagi 989-6711, Japan,* [2]*Ajinomoto General Foods, Inc., 1-46-3 Hatsudai, Shibuya-ku, Tokyo 151-8551, Japan; akitsu.tozawa.c2@tohoku.ac.jp*

Animal welfare level can be assessed by measurement of physiological state such as hormone concentration in blood as well as behavioural characteristics. However, blood collection is invasive and gives stress to the animals. Milk is potentially useful in welfare assessment because it can be collected easily and non-invasively, and contains physiological substances relating to animal welfare. In this study, relationship of hormone concentrations in serum and milk to behaviour was investigated by using milking cows in five dairy farms with different welfare level. This study also examined the relationship between physiological substances in milk and in serum, to develop a non-invasive technique to evaluate welfare level of animals. Sample collections and behavioural observations were done two times in June and August-September, 2016. Serum and milk samples were collected from six cows in each farm during morning milking period, then cortisol, oxytocin and immunoglobulin G (IgG) were measured. Lactoferrin was also measured in milk samples. In serum components, the concentration of cortisol and oxytocin was related to foraging behavior (cortisol: r=-0.30, P=0.02, oxytocin: r=0.33, P=0.05). Serum oxytocin concentration also related to agonistic behaviour (r=0.39, P=0.03). In milk components, cortisol concentration was related to affiliative (r=-0.38, P<0.01) and redirected behaviour (r=-0.31, P=0.02). Consequently, there was no significant relationship between serum and milk in the concentration of cortisol and oxytocin. In contrast, IgG showed positively relationship between serum and milk (r=0.48, P<0.01). Lactoferrin concentration in milk also negatively related to redirected behaviour (r=-0.29, P=0.04) and positively related to IgG in serum and in milk (serum: r=0.32, P=0.02, milk: r=0.37, P<0.01). The results indicates that these physiological substances in milk can be used to evaluate welfare level of cows.

Legislation as framework conditions and challenges for the upcoming insect industry

D.L. Baggesen

The Technical University of Denmark, National Food Institute, Kemitorvet, building 204, 2800 Kgs. Lyngby, Denmark; dlba@food.dtu.dk

Production of insects for feed and food is an upcoming industry in Denmark as well as in other parts of the Western world. This is partly due to the great potential of insects to address some of the major global challenges described in the UN Sustainable Development Goals in 2015. In both EU and globally, initiatives are in place to boost a green transformation of industrial production. As example, the food- and agriculture industry is now having a strong focus on re- and up-cycling of by-products within the food chain. However, the current legislation seems in some ways to be a barrier for new alternative production forms as e.g. insects farmed on by-products and thereby conflicts with the vision of a more green production. Farming of insects for feed and food is in principle similar to other kinds of food animal production and is therefore underlying the same legislation as these. This is both concerning the substrates to be used for feeding the insects and the insects' application as food and feed. Generally, the entire insect production chain, incl. production, processing, marketing and application needs to adhere to the regulations in place for ensuring safe products. Currently, insects can be farmed using approved feed as substrates, giving the opportunity for a more climate- and environmental-friendly animal production form compared to other food animals. However, shall the insect industry realize its full potential as a new sustainable source of feed and food, legislation shall be changed in order to open for a wider use other kinds of organic by-products as substrates. Revision of the legislative framework is therefore an important precondition for a successful insect industry in the coming years. Amendment of legislation is a complex and time-consuming task that shall be based on scientific risk assessments of the specific steps along the production chain. The European Food Safety Authority (EFSA) initiated the process in 2015 by publishing a scientific opinion 'Risk profile related to production and consumption of insects as food and feed'. Still, more research is needed to close the significant knowledge gaps and provide a scientific basis for political decisions on future legislation to ensure safe insect products.

Safety, regulatory affairs and consumer acceptance in insect industry

T. Arsiwalla

Protix, Industriestraat 3, 5106 NC Dongen, the Netherlands; tarique.arsiwalla@protix.eu

Part A will focus on safety and regulatry affairs in the insect industry and Part B will focus on Consumer accetance of insects. Part A. Safety and EU regulatory affairs in the insect industry: The EU insect sector is significantly guided by regulations, both for applications in human food and animal feed. The International Platform of Insects for Food and Feed (IPIFF) is an EU non-profit organisation which represents the interests of the insect production sector towards EU policy makers, European stakeholders & citizens. Composed of more than 30 members, most of which are European insect producing companies, IPIFF promotes the use of insects & insect derived products as top tier source of protein for human consumption & animal feed. Several topics will be covered in this session: How can insect producers guarantee that their products are safe? Are insect proteins currently authorised for use in EU farmed animals feed ? May insects be used for human consumption in Europe? If yes under which conditions? Part B. Consumer acceptance of insects: In the last decade since 2009, the insect industry has developed rapidly. Products launched on the market are becoming more and more successful. Analysis of what has been core to the successful products on the market has led to five key design and communication principles. Successful insect based products on the market adhered to most of these principles: 1. Make a product that matters; the user must be looking for it 2. Build trust, be transparent 3. Product design; address possible ick factor 4. Generate evidence or proof of added value 5. Capture the added value smartly. The above 5 levers will be discussed and concrete examples will be given including do's and dont's.

Global differences in insect regulation: the background and the impacts

A.M. Lähteenmäki-Uutela[1], L. Hénault-Ethier[2], S.B. Marimuthu[3] and V.V. Nemane[3]
[1]Turku School of Economics, Pori unit, Pohjoisranta 11 A, 28101 Pori, Finland, [2]Université Laval, Département des sciences animales, Faculté des sciences de l'agriculture et de l'alimentation, 2325, rue de l'Université, Québec G1V 0A6, Canada, [3]University of New England, Australian Centre for Agriculture and Law, University of New England, Armidale, 2351 New South Wales, Australia; anu.lahteenmaki-uutela@utu.fi

There are global differences in how insects are regulated as food and feed. We compare the regulations of four first-tier developed countries that have a relatively organized regulatory architecture for developing insect regulation and where market potential is large: the EU, United States, Canada and Australia. Our study focuses on (1) the system-level and historical background for differences in regulation and (2) the impacts of the regulatory differences for insect business operators. Legal systems vary in how risks are perceived and how the precautionary principle is applied in practice. There are differences in how fast the regulatory system can respond to new phenomena such as novel food and feed ingredients. Although the cultural tradition of insect use is an important factor explaining the general attitudes towards insects, there are genuine differences in how the ethical issues are argued and perceived. Private regulation and industry standards have a role in complementing mandatory laws, in some countries more so than in others. Regulation impacts product development strategies of marketers: the pay-back time for R&D investment depends partly on the regulatory requirements and on how fast a product can access the market. The article aims to explore if the insect business would benefit from clear, harmonised and predictable rules by comparing the regulation of insect food and feed in the abovementioned jurisdictions. It examines whether global harmonisation of rules would make international marketing strategies easier or whether tailor-made rules and protectionist practices could be the preferred option, so as to assist local marketers.

The environmental impact of insects as food and feed: a review

H.H.E. Van Zanten, A. Parodi Parodi and I.J.M. De Boer
Wageningen University, Animal Production Systems group, De Elst 1 (Building 122), 6708 WD Wageningen, the Netherlands; hannah.vanzanten@wur.nl

Our current food consumption pattern has a severe pressure on the environment, it causes 25% of the global anthropogenic greenhouse gas emissions and uses 40% of earth's terrestrial surface. The majority of this environmental impact (e.g. 60% for GWP) originates from animal source food (ASF) and the main environmental part of ASF relates to the production of feed ingredients for livestock. To strengthening nutrition supply while preserving our planet, insects can play a crucial role to reduce the impact related to food and feed production. Related to food production, insects can replace – based on their high nutritional value – conventional meat sources (e.g. red meat) with a high impact and, therefore, reduce the environmental impact. Related to feed production, insects can replace conventional protein sources (e.g. soybean meal or fishmeal) and, therefore, reduce the environmental impact. In the last 5 years 7 LCA studies about the use of insects as food and feed were published. The aim of this study was to perform a systematic review of the environmental impact of insects as food and feed. Related to the use of food, our preliminary results showed that expressed by kg of protein, mealworms perform better than milk, pork, chicken and beef in terms of global warming potential and land use and, therefore, insect-based meat is considered as a sustainable substitutes from an environmental perspective. Related to the use of feed, our results showed that the environmental impact of the insects highly depends on the source of 'waste' used to feed the insects. In general, nevertheless, our results showed that insects can be a good alternative protein source to reduce the environmental impact. The LCA studies assessed in our review all express the environmental impact per kg of product or per kg of protein. The unique aspect of insects, however, is that they have a high nutritional content – not only high protein content – as food and feed. In our discussion we elaborate on the nutritional contribution insects can have compared to other protein sources and argue that future LCA studies should include different nutritional aspects when assessing the environmental impact.

Key components for a sustainable entomophagy industry

Å. Berggren[1] and A. Jansson[2]
[1]*Swedish University of Agricultural Sciences, Department of Ecology, P.O. Box 7044, 75007 Uppsala, Sweden, [2]Swedish University of Agricultural Sciences, Department of Anatomy, Physiology and Biochemistry, P.O. Box 7011, 75007 Uppsala, Sweden; asa.berggren@slu.se*

Insects are a promising source of nutrients for humans containing high amounts of good quality protein, fatty acids, vitamins and minerals. However, much around insects as a food source we do not know and just a small number of species have been used as livestock. Plenty of new information and understanding is needed if we are to develop food production systems and mass rearing of insects. The promise and also challenge of this food system is to develop it in a sustainable manner that permeates all parts of the production chain. This means that: (1) choice of species; (2) rearing facilities, resource use in terms of; (3) feed; and (4) energy are core components that needs to be evaluated within a sustainable framework. We suggest important key factors within these components that can guide the way for the future development of insect as mini-livestock. These include that insect species chosen should be native so they do not contribute to the increasing invasion problem facing both natural and production systems. The species should also have a potential to utilise plant products that cannot be used for humans as food. The animals thereby do not compete with humans for food resources, as is the case of many current food systems. Additionally, resources not presently utilised (e.g. by-products from different production systems) could be used as feed. Promising insect taxa are the feeding guild leaf chewers, which includes species from the families orthoptera, coleoptera and phasmatodea. These taxa show a wide diet breadth and serve as a good starting point when assessing suitable species – feed combinations. An evaluation of sustainable resource and energy use for rearing facilities and feed production indicate that western countries relying heavily on fossil fuel will have harder to reach goals in these areas of the food system. This will be the case for approximately 4/5 of the countries that use fossil fuel for more than 50% of the country's total energy use.

Campylobacter contamination level in houseflies after exposure to materials containing *Campylobacter*

A.N. Jensen
Technical University of Denmark, National Food Institute, Division of Food Microbiology and Production, Anker Engelunds Vej 1, 2800 Kgs. Lyngby, Denmark; anyj@food.dtu.dk

Houseflies have been found to carry *Campylobacter jejuni* and play a role in infection of poultry flocks. This study aimed to elucidate the quantitative campylobacter level in naturally infected houseflies depending on their previous exposure to campylobacter-contaminated material in terms of campylobacter level, duration and type of material. Houseflies (*Musca domestica*) reared under laboratory conditions were placed in 250-ml cups containing 5 g of chicken faeces or 1 ml liquid and spiked with 3, 4, 5 or 7 Log cfu *C. jejuni*. Sixteen houseflies were added to each cup at a time. After approx. 1 h of exposure, 4 flies were removed from the cup for direct enumeration of campylobacter by plate spreading. Another 4 flies were transferred onto an Abeyta-Hunt-Bark (AHB) agar plate (9 cm) to assess if flies will contaminate surfaces. After 1 h on the plate, each fly was tested for campylobacter. This procedure was repeated after approx. 4 h of exposure for the remaining eight flies. The percentage of houseflies positive for campylobacter as well as the Log_{10} cfu recovered per fly depended on the contamination level and the material exposed to. For faeces, 90.0% (n=80), 48.4% (n=64), 6.3% (n=48) and 0% (n=16) of flies were campylobacter positive when exposed to 7, 5, 4, and 3 Log_{10} cfu with a mean (±SE) of 2.0±0.1, 0.8±0.1, 0.3±0.0 and 0 Log_{10} cfu recovered per campylobacter-positive fly, respectively. For liquid, 95.7% (n=47), 91.4% (n=47), 20.8% (n=48) and 6.3% (n=16) of flies were campylobacter positive when exposed to 7, 5, 4, and 3 Log_{10} cfu with a mean of 3.3±0.2, 2.0±0.1, 0.8±0.2 and 0.3±0.0 Log_{10} cfu recovered per campylobacter-positive fly, respectively. *Campylobacter* seemed to be taken up readily as there was no significant effect of exposure time (1 vs 4 h) and the uptake was higher from contaminated liquid than faeces. The surface of the AHB plates was only contaminated by houseflies previously exposed to either 5 or 7 Log_{10} cfu and most by flies exposed to liquid (45% n=48) opposed to faeces (31% n=72). The results support that houseflies are likely to become contaminated with campylobacter if exposed to material containing >4 Log_{10} cfu.

Transfer of cadmium, lead and arsenic from substrate to *Tenebrio molitor* and *Hermetia illucens*

H.J. Van Der Fels-Klerx[1], L. Camenzuli[1], M.K. Van Der Lee[1], M. Nijkamp[1] and D.G.A.B. Oonincx[2]
[1]RIKILT Wageningen University & Research Centre, Novel Foods and Agrichains, Akkermaalsbos 2, 6708 WB Wageningen, the Netherlands, [2]Wageningen University, Entomology, Droevendaalsesteeg 1, 6708 PB Wageningen, the Netherlands; monique.nijkamp@wur.nl

Insects have potential as a novel source of protein in feed and food production in Europe, provided they can be used safely. To date, limited information is available on the safety of insects, and toxic elements are one of the potential hazards of concern. Therefore, we aimed to investigate the potential accumulation of cadmium, lead and arsenic in larvae of two insect species, *Tenebrio molitor* (yellow mealworm) and *Hermetia illucens* (black soldier fly), which seem to hold potential as a source of food or feed. An experiment was designed with 14 treatments, each in triplicate, per insect species. Twelve treatments used feed that was spiked with cadmium, lead or arsenic at 0.5, 1 and 2 times the respective maximum allowable levels (ML) in complete feed, as established by the European Commission (EC). Two of the 14 treatments consisted of controls, using non-spiked feed. All insects per container (replicate) were harvested when the first larva in that container had completed its larval stage. Development time, survival rates and fresh weights were similar over all treatments, except for development time and total live weight of the half of the maximum limit treatment for cadmium of the black soldier fly. Bioaccumulation (bioaccumulation factor >1) was seen in all treatments (including two controls) for lead and cadmium in black soldier fly larvae, and for the three arsenic treatments in the yellow mealworm larvae. In the three cadmium treatments, concentrations of cadmium in black soldier fly larvae are higher than the current EC maximum limit for feed materials. The same was seen for the 1.0 and 2.0 ML treatments of arsenic in the yellow mealworm larvae. From this study, it can be concluded that if insects are used as feed materials, the maximum limits of these elements in complete feed should be revised per insect species.

Chairs introduction and poster summary

E.F. Knol
Topigs Norsvin Research Center, P.O. Box 43, 6640 AA Beuningen, the Netherlands; egbert.knol@topigsnorsvin.com

The free communication session has 9 presentations and 10 posters. As a chair I hope to give a short introduction on the presentations and a 1 minute pitch per poster as I understand it from the abstracts, the posters and/or the communication with the authors. The posters with authors will be available during the poster sessions for real insight in the results and discussion on them.

Is the inclusion of edge weights meaningful in the network analysis regarding a pig trade network?

K. Büttner and J. Krieter

Institute of Animal Breeding and Husbandry, Christian-Albrechts-University, Olshausenstr. 40, 24098 Kiel, Germany; kbuettner@tierzucht.uni-kiel.de

The analysis of trade networks focuses mainly on pure animal movements between different farms. However, with additional data such as the number of delivered livestock, the distance between farms or the number of contacts between two farms in a specific time period, the informational content of the network analysis can be complemented. Therefore, the aim of the study was to include this additional information in the network analysis as so-called edge weights and to evaluate their impact on the centrality parameters, i.e. the ranking of the farms according to their importance for the network. The data was recorded from a pig trade network in Northern Germany (2013-2014) containing 866 farms which are connected by 1,884 edges. Besides this unweighted network, 5 different weighted network versions were constructed by adding the following edge weights: Number of contacts; Number of animals; Distance; Number of animals/Number of contacts; 1/Distance. In order to evaluate the differences of the centrality parameters for the unweighted and the weighted network versions Spearman Rank Correlation Coefficients were calculated. Out-degree and outgoing closeness showed for all weighted network versions very high correlation coefficients (0.96-0.98), followed by the betweenness which had only slightly lower correlation coefficients (0.87-0.91). Finally, in-degree and ingoing closeness revealed the lowest but still moderate to high correlation coefficients (0.66-0.85) for all weighted network versions. The results showed that within the same centrality parameter the weighted network versions differed only slightly among each other. But, there are clear differences between the specific centrality parameters. Especially for the centrality parameters based on outgoing trade contacts the inclusion of edge weights did only barely change the ranking of the farms and thus, the chosen edge weights are redundant. However, for the centrality parameters based on the ingoing trade contacts the effect was more prominent. The importance of these differences for disease transmission has to be further analysed and also the correlation between the high ranking farms has to be examined.

Effects of weaning conditions on the evolution of piglet plasma concentration of vitamin E

A. Buchet[1,2,3], C. Belloc[1], N. Le Floc'h[3] and E. Merlot[3]

[1]ONIRIS INRA, UMR BIOEPAR, 44307 Nantes, France, [2]Cooperl Arc Atlantique, BP 60238, 22403 Lamballe, France, [3]INRA, UMR 1348 PEGASE, 35590 Saint Gilles, France; arnaud.buchet@inra.fr

Weaning causes oxidative stress in piglets. Endogenous and exogenous vitamin E (vit E) participates to the neutralization of pro-oxidant molecules in the organism. This study aimed to analyze the effect of weaning conditions on the evolution of plasma vitamin E concentration on piglets. In trial A, piglets born from 12 litters were weaned at either 21 (W21) or 28 (W28) days of age in optimal (OC) or deteriorated (DC) conditions (n=16 per group). Piglets from OC group originating from 2 litters were housed by 4 (0.39 m²/pig) in cleaned and disinfected pens. Those from DC group originating from 4 litters were housed by 8 (0.20 m²/pig) in dirty pens after a waiting time of 4 h at 20 °C. From 12 to 61 days of age, blood was collected and pigs were weighed weekly and then at 88, 119 and 147 days of age. In trial B, piglets (n=288 from 144 sows) from 16 commercial farms were weaned at 28 days of age. Blood samples were collected when piglets were 26- (on sows and piglets) and 33-days old (on piglets only). In both trials, weaning led to a dramatic drop of the plasma concentration of vit E (P<0.001). In trial A, this drop was similar whatever the age at weaning or housing conditions (P>0.05). After weaning, vit E concentration remained stable until 147 days of age. In trial B, piglet plasma concentration of vit E at 26 days of age was not associated with lactating sow vit E concentrations (P>0.05) and was different between farms (P<0.001). The higher plasma concentration of vit E before weaning could be explained by the high vit E content in colostrum and milk leading to the saturation of body fat reserves of the piglet. At weaning, the high requirements of vit E for antioxidant purposes as well as low feed intake, lower fat content of the feed compared to milk and lower activity of enzymes involved in vit E absorption could explain the dramatic drop of plasma vit E. The relatively low concentrations of vit E until slaughter age could reflect an incomplete reconstitution of body reserves. To conclude, weaning led to a drop of plasma concentration of vit E whatever the weaning conditions indicating dramatic changes in vit E status and metabolism at weaning.

Investigating the efficacy of a new *Buttiauxella* sp. phytase on P digestibility in weaned piglets

W. Li[1], F. Molist[2] and Y. Dersjant-Li[1]
[1]Danisco Animal Nutrition, DuPont Industrial Biosciences, Marlborough, SN8 1AA, United Kingdom, [2]Schothorst Feed Research B.V., Lelystad, 8200 AM, the Netherlands; wenting.li@dupont.com

Two trials were conducted to determine the efficacy of a *Buttiauxella* sp. phytase (6-phytase, expressed in *Trichoderma Reesei*) on P digestibility in piglets. In both trials, two barley/maize/SBM based basal diets were prepared: (1) positive control (PC), formulated to meet the nutrient requirements of piglet with 0.75% Ca and 0.36% digestible P; (2) negative control (NC) with same nutrient composition as PC but deficient in Ca (0.60%) and digestible P (0.20%). The phytase was added on-top to the NC diet at 0, 250, 500, 750 (1st trial only), and 1000 FTU/kg, resulting in a total of 6 (5 for the 2nd trial) treatments, with 10 and 8 replicates per treatment in trial 1 and 2, respectively. In trial 1, piglets were fed testing diets from 12 to 24 d post-weaning. Feed was increased from 0.4× maintenance on d 12 to 3.5× maintenance on d 19 based on expected growth. Feed amount was fixed after d 19. In trial 2, test diets were provided at 3.2× maintenance from 8 to 22 d post-weaning. Faeces were collected from d 19 to 24 (trial 1) or d 18 to 22 (trial 2), pooled by animal and analyzed for digestibility. Digestibility data from the two trials were pooled to determine the efficacy of phytase. Across the two trials, in the absence of phytase, P digestibility in piglets fed NC diet (37.1%) was reduced by 17.5 percentage points as compared to those fed PC diet (54.6%, P<0.05). P digestibility was significantly improved by adding phytase to the NC diet. With only 250 FTU phytase/kg inclusion, P digestibility (58.1%) was increased by 21 percentage points vs NC (P<0.05) and was similar to those fed PC diet (54.6%; P>0.05). Feeding piglets with increased phytase from 250 to 1000 FTU/kg resulted in linear improvement in P digestibility (P<0.05), which was 58.1, 61.2, 64.4 and 67.1%, respectively, for those fed diets containing 250, 500, 750 and 1000 FTU/kg phytase. Calculated dig P release for *Buttiauxella* sp. phytase expressed as net digestible P improvement vs NC were 1.06, 1.22, 1.31 and 1.46 g/kg for 250, 500, 750 and 1000 FTU/kg. In conclusion, *Buttiauxella* sp. phytase effectively improved the P utilization in piglets fed P deficient diet.

Effects of physicochemical characteristics on *in vitro* and *in vivo* nutrient digestibility in pigs

D.M.D.L. Navarro[1], E.M.A.M. Bruininx[2], L. De Jong[2] and H.H. Stein[1]
[1]University of Illinois, 1207 W Gregory Dr, Urbana, IL 61801, USA, [2]Agrifirm Innovation Center, Royal Dutch Agrifirm, Estate Laan 20, 7325 AW Apeldoorn, the Netherlands; navarro3@illinois.edu

Two experiments determined correlations between physicochemical characteristics, concentration of DE and ME, and *in vitro* apparent total tract digestibility (IVATTD) and *in vivo* apparent total tract digestibility (ATTD) of DM and nutrients in corn, wheat, soybean meal, canola meal, corn distillers dried grains with solubles, corn germ meal, copra expellers, sugar beet pulp, cellulose, and pectin. In Exp. 1, IVATTD of DM was determined. Results indicated that bulk density was negatively correlated with NDF and ADF (P<0.05; r=-0.78 and -0.69). Soluble dietary fiber was positively correlated with viscosity and swelling (P<0.05; r=1.00 and 0.64). Swelling was also positively correlated with water binding capacity (P<0.01; r=0.89). The IVATTD of DM was negatively correlated (P<0.05) with concentrations of total dietary fiber (TDF; r=-0.76) and insoluble dietary fiber (r=-0.92). In Exp. 2, 80 pigs (initial BW: 48.41±1.50 kg) were allotted to a randomized complete block design with 10 diets and 8 replicate pigs per diet. Diets included a corn-based diet, a wheat-based diet, a corn-SBM basal diet, and 7 diets based on a mixture of the corn-SBM basal diet and each of the remaining ingredients. Results indicated that swelling was positively correlated with ATTD of NDF, ADF, IDF, and TDF (P<0.05; r=0.75, 0.80, 0.89, and 0.84). Viscosity was positively correlated with ATTD of NDF, ADF, and IDF (P<0.01; r=0.92, 0.86, and 0.79). Water binding capacity was positively correlated with ATTD of IDF and TDF (P<0.05; r=0.67 and 0.68). Concentration of TDF, but not concentrations of ADF and NDF, was negatively correlated (P<0.01) with ATTD of GE (r=-0.80) and concentration of DE and ME (r=-0.86 and -0.85), which indicates that TDF is a better estimate of DE and ME than NDF and ADF. However, physical characteristics were not correlated with concentration of DE and ME, which indicates that these parameters may influence fiber digestibility but not digestibility of energy *in vivo*.

Determination of threonine requirements for growing rabbits using plasma urea nitrogen level
P.J. Marín-García
Instituto Universitario de Ciencia y Tecnología Animal, Departamento Ciencia Animal, Calle Camino Vera s/n, 46022
Valencia, Spain; pabmarg2@gmail.com

The protein content in growing rabbits' diets has been reduced to minimize digestive disorders. In this context it is necessary to adjust amino acid profile, especially those limiting-amino acids (lysine, methionine and threonine). Of all of them, threonine (tre) is the least studied. This work evaluated the levels of threonine suitable in growing rabbits using plasmatic urea nitrogen level (PUN). Nine experimental diets were formulated. Starting from the current recommendations for lysine (lys) and sulphure amino acids (saa) (medium; MMX), inclusion of these amino acids were increased or reduced by 15% (high; HHX and low; LLX, respectively). For these 3 combinations, tre content was introduced at 6.2 g/kg (XXM) or by 15% more or less (XXH; 7.2 g/kg or XXL; 5.33 g/kg), A total of 297 rabbits (33 per feed) were used. PUN was analyzed from a samples obtained at 08:00 h (under *ad libitum* feeding) and at 21:00 h (3 h after refeeding after a fasting). Average PUN level was the highest with the current recommendations (13.7, 15.2 and 14.5 mg/dl for L, M and H, respectively; P<0.001), being different interaction with the sample harvesting hour (P=0.0001) and level of other two limiting amino acids (P=0.0133). The results of this study indicates that it does not seem exist an absolute level of tre to be recommended for growing rabbit diets, as it depends of the level of inclusion of lys and saa.

Deviations for sex and halothane genotypes subpopulations in the estimation of pork cuts composition
G. Daumas and M. Monziols
IFIP-Institut du Porc, BP 35104, 35601 Le Rheu Cedex, France; gerard.daumas@ifip.asso.fr

Sex and halothane gene, well known for their effects on carcass composition, are not included in the present EU authorised equations for carcass classification. Consequently, the prediction equations of lean meat percentage (LMP) suffer from systematic deviations for these subpopulations. The aim of this work is to quantify these deviations in the estimation by automatic vision of pork cuts composition. A sample of the French pig slaughtering was selected in 3 abattoirs and stratified according to sex (50% castrated males and 50% females). Carcasses were measured online by the classification method CSB Image-Meater® (IM) and cooled. An ear sample was analysed for Halothane gene (Hal). The left sides were cut according to the EU procedure and the four main joints were CT scanned in order to determine LMP. Among the 208 pigs, the proportions of Nn and NN alleles were respectively of 52% and 48%, and well balanced intra sex too. Five dependent variables were studied: the LMP for carcass classification (the four main cuts) and the LMP in each cut. From the 16 raw IM variables was build a pool of 6 potential predictors: 3 fat depths, 2 muscle depths and 1 length. Carcass weight was added as predictor. Regression models were selected by using the PRESS statistic in the GLMSELECT procedure of SAS software. For each equation were calculated the systematic deviations for each modality of the 2 factors sex and Hal by mean difference between predicted and observed values. In all models females and Nn alleles were underestimated while castrates and NN alleles were overestimated. Deviation between sexes was the lowest (0.49) for the ham and the highest (1.34) for the belly. Deviations between Hal genotypes were lower than between sexes, about the half, except for the ham. They ranged from 0.28 for the shoulder to 0.74 for the belly. Deviation between sexes could be removed by at least a different intercept in the prediction equations. Hal genotype being generally unknown online we recommend to analyse it during the body composition experiments. In case of important deviations, when the estimates of the proportions in population are available and differ from the proportions in the sample, the prediction equations could be calculated by weighed regression.

At-line carcass quality. NIRS determination of fat composition at Swiss commercial slaughterhouses

S. Ampuero Kragten[1], M. Scheeder[2], M. Müller[2], P. Stoll[1] and G. Bee[1]
[1]*Agroscope, Animal Production Systems and Animal Health, Rte de la Tioleyre 4, 1725 Posieux, Switzerland, [2]Suisag, Allmend 8, 6204 Sempach, Switzerland; silvia.ampuero@agroscope.admin.ch*

Fat composition is important and even crucial to various aspects of food quality. Because pig fat tissue is one of the main ingredients of meat products, a limited amount of PUFA in the fat will reduce the risk of oxidation and thus improve their shelf life while limiting the development of off-flavors such as rancidity. A limited amount of double bonds in fatty acids in general is also related to the firmness of the fat tissue hence improving its technological quality. The need for quality control of the fat composition at the slaughterhouse may become a necessity with increasingly leaner breeds, the fattening of entire males and the use of feed ingredients with higher PUFA and MUFA content. In Switzerland, a joint project involving farmers, feed producers, meat processors and scientific institutions identified and agreed on upper thresholds for PUFA and IV (iodine value), 15% and 70 respectively, aiming at assuring a high standard of technological quality of fat tissue. NIRS models were developed based on FAME analysis by GC as reference. More than 400 fat samples were grated from carcasses, on-line at the slaughterhouses, with a spoon-like instrument from the outer layer of backfat overlaying the glutaeus muscle. Three cm diameter cups were filled with mixed fat and diffuse reflection spectra (10,000 to 4,000 cm^{-1}) were taken with three FT-NIR spectrometers (NIRFlex N-500, Büchi, Switzerland) at three different locations. PUFA and IV in the reference set ranged from 7.6 to 25.6% and from 55.8 to 85.4 respectively. NIRS models presented R^2 and SEP (standard error of prediction) values of 0.99 and 0.49 for IV and 0.98 and 0.41% for PUFA. Additional variables were quantified by NIRS: SFA (range 30.3 to 45.4%), MUFA (range 40.1 to 54.8%), total fat (range 66 to 99%) and some individual fatty acids, with R^2 and SEP values from 0.92 to 0.97 and 0.2 to 0.6% (2% for total fat). Since 2014, Swiss commercial slaughterhouses use successfully NIRS at line to rate the price of pig carcasses depending on their fat composition.

Breeding objectives of the traits for dam and sire pig breeds in integrated production system

E. Krupa, Z. Krupová, E. Žáková and M. Wolfová
Institute of Animal Scince, Přátelství 815, 10400, Czech Republic; krupa.emil@vuzv.cz

A general bio-economic model (EWPIG ver. 1.1.0) was used to define breeding objectives for integrated pig production system. Marginal economic weights (MEW) of 30 traits (some of them as an alternative to each other) were calculated for Large White (LW) and Pietrain (PN) breed kept in the two-way crossing system. The production system, i.e. management, housing and feeding systems, was assumed to be the same in both breeds. Artificial insemination was supposed in both links; however boars were used for oestrus stimulation. Maximal number of reproductive cycle was 10 and 8 for LW and PN breed, respectively. Maximal number of mattings was 4 for LW sows after the 1st and 2nd farrowing, 3 for gilts and CL sows after 3rd farrowing. The herd structures for both breeds calculated by using of the Markov Chains was based on the sows conception rate, sow mortality rate and sow culling rate for health problems or low litter size. The average costs to produce one kg of slaughtered animal ranged from 1.27 € in PN to 1.34 € in LW link of the system. Profitability was achieved in both breeds when the subsidies were taken into account but even if not. Generally, maternal traits were more important than growth/carcass traits in dam breed than in sire breed. The ration of the MEW for number of piglets born alive and lifetime daily gain of finished animals were 88 : 1 and 5.5 : 1 in LW and PN breeds, respectively. The highest difference between links was observed for conception rate of gilts (ratio 1 : 0.11 for LW and PN, 6.5 € : 0.7 € in absolute values). The MEW of the productive lifetime of sows was 93.2 € and 28.7 € per farrowing for LW and PN. The highest MEW was obtained for feed conversion in finishing (-1,432 € per kg feed/kg weight gain for LW breed and was almost 9 times less than for PN). The ration of MEW for carcass traits (dressing percentage and lean meat content) was 179.1 € : 1,114 € for LW and PN breeds. The ration for complex of breeding objectives as reproduction : growth : functional : carcass : feed conversion traits were 11.6 : 0.1 : 9.9 : 6.1 : 72.3 for LW and 0.5 : 0.1 : 4.3 : 7.4 : 87.7 for PN breed. This study was supported by the project QJ1310109 of the Ministry of Agriculture of the Czech Republic.

Differentiation between domestic pigs and small wild boar based on growth hormone gene polymorphisms

S. Yoshikawa, A. Maekawa and Y. Mizoguchi
Meiji University, Agriculture, 1-1-1, Higashimita, Tama-ku, Kawasaki, Kanagawa, 214-8571, Japan;
ericsaka1009@yahoo.co.jp

Humans started domesticating wild boars for food about 9000 years ago. In order to increase production, they have repeatedly improved pigs' genetic characteristics, such as body length, number of vertebrae, and growth rate. Growth hormones are major factors promoting the growth of body cells in vertebrates. Pig growth hormone is a peptide hormone containing 191 amino acids, and it is secreted from cells in the anterior pituitary gland. The Ryukyu wild boar, is a subspecies of wild boar which only inhabits the Ryukyu archipelago in Japan. It is one of the world's smaller wild boars because it lives on isolated islands. On these islands there is a large difference in size between the small native wild boars and domestic pigs, but there have been no studies that have looked at differences in their growth-related genes. Therefore, in this study we compared the growth hormone gene haplotypes of 34 Landrace and 31 Large White domestic pigs with 68 Ryukyu wild boar. We found 10 mutations containing five amino acid substitutions, and reconstructed 18 Growth hormone gene haplotypes (16 domestic pigs and 8 wild boar) by using the PHASE algorithm. The number of haplotypes in the domestic pigs was twice that in the wild boar. This increase is due to breeding improvement, and we expect that this differentiation in growth hormone genes will increase further in the future. In addition, four of the mutations were only found in domestic pigs, and so there is a possibility that these mutations could become markers to distinguish domestic pigs from small wild boar.

Association between (PRLR) gene polymorphism and reproduction performance traits of Polish swine

A. Terman[1], D. Polasik[1], A. Korpal[1], K. Woźniak[1], K. Prüffer[1], G. Żak[2] and B.D. Lambert[3]
[1]West Pomeranian University of Technology of Szczecin, Al. Piastów 17, 70-311 Szczecin, Poland, [2]National Research Institute of Animal Production, ul. Sarego 2, 31-047, Poland, [3]Tarleton State University, Stephenville, TX 76402, USA; grzegorz.zak@izoo.krakow.pl

The aim of this study was to assess the relationship between PRLR gene polymorphism and litter size in pigs of synthetic line 990, which derives from crossing of six breeds: Polish Large White, Belgian Landrace, Duroc, Hampshire, Polish Landrace of German origin, and Polish Landrace of British origin. It was selected for the meat deposition in carcass and growth rate, but reproduction traits were not considered in the selection program. A group of 374 sows belonging to the synthetic line 990 from the same farm was selected for the experiment. The maintenance conditions were the same for all individuals, which are necessary to obtain reliable results, because environmental conditions have a significant influence on quantitative traits of animals. All the investigated animals were free of the RYR1 T allele. Prolactin receptor gene variants were determined by PCR-RFLP method. Digestion of porcine PRLR gene amplicons (163 bp) by AluI enzyme allowed to determine three genotypes based on the following restriction fragments length – AA: 85, 59, and 19 bp, AB: 104, 85, 59, and 19 bp, and BB: 104 and 59 bp. In the tested sow herd, allele A was present at 0.67 frequency, whereas allele B with a frequency of 0.33. In the case of genotypes, the most common genotype was the AA (0.50), followed by AB (0.33), and the least frequently occurring BB with frequency of 0.17. It was indicated that animals with AA genotype in the first parity produced 0.39 more piglets than sows with BB genotype ($P \leq 0.05$). Similar relationship was observed for NBA and NW, where the differences between AA and BB genotypes were 0.45 and 0.35, respectively ($P \leq 0.05$). Analysis of II and $\geq$III parities showed tendency of AA genotype being also favorable for TNB, NBA, and NW, but differences between individual genotypes were smaller than in the first parity and not confirmed statistically. It was demonstrated that the analyzed polymorphism had significant influence ($P \leq 0.05$) on reproduction traits in the first parity.

Expression of Wnt5a gene in porcine reproductive tissues at different stages of the cycle

A. Mucha, K. Piórkowska, K. Ropka-Molik and M. Szyndler-Nędza
National Research Institute of Animal Production, Sarego 2, 31-047 Krakow, Poland; aurelia.mucha@izoo.krakow.pl

WNTs (wingless-type MMTV integration site family members) are important factors for sex determination, and reproductive development and function in females; they also contribute to endometrial receptivity and embryo implantation. *Wnt5a* is one of the Wnt family genes. Therefore, the aim of the study was to determine *Wnt5a* gene expression before first estrus and during the follicular and luteal phases in reproductive tissues of gilts. The expression level of *Wnt5a* gene was investigated in ovarian and oviductal tissues as well as in the uterine horn and body collected from 40 Polish Large White and Polish Landrace gilts. To determine estrous cycle phases, ovaries were subjected to morphological analysis in which the number and size of ovarian follicles, corpora lutea and corpora albicans were examined macroscopically. On this basis, the animals were divided into 3 groups according to estrous cycle phase: gilts before first estrus as well as during follicular and luteal phases. Two genes, *OAZ1* and *RPL2*, were used as endogenous control. Analysis of the *Wnt5a* gene expression level in different tissues at different phases of the estrous cycle showed the highest transcript levels in the oviduct, uterine horn and body during the follicular phase. The relative amount of the gene was lowest in the ovary and uterine horn before first estrus as well as in the oviduct and uterine body during the luteal phase, compared to the other phases of the cycle. However, no statistically significant differences were observed between the analysed traits. When comparing *Wnt5a* gene expression between tissues at different phases of the cycle, the highest and comparable transcript level was found in the uterine horn and body during the follicular phase (a difference of 0.27). The highest expression of this gene was noted in the uterine body before first estrus, and in the uterine horn during the luteal phase. Transcript level was the lowest in the ovary at all three phases and it differed significantly ($P<0.01$) compared to uterine body expression before first estrus and in the luteal phase, and in comparison with uterine horn expression during the follicular and luteal phase.

Response to selection in a divergently experiment for birth weight environmental variability in mice

N. Formoso-Rafferty[1], I. Cervantes[1], N. Ibáñez-Escriche[2] and J.P. Gutiérrez[1]
[1]Universidad Complutense de Madrid, Producción Animal, Avda. Puerta de Hierro s/n, 28040, Madrid, Spain, [2]University of Edinburgh, The Roslin Institute, Edinburgh, United Kingdom; gutgar@vet.ucm.es

The use of heterogeneous residual variance models for genetically controlling the environmental variability has been shown to be useful in divergent selection experiments in mice, rabbits and pigs. The objective of this research was to compare the expected with the realized response to divergent selection in a successful mice experiment. Data belonged to fourteen generations of selection for environmental variability of birth weight in mice. A total of 18,626 birth weight records from 1,195 females and 1,999 litters in combination with 18,288 pedigree records were used. The fitted model included generation, litter size, sex and parity number as systematic effects, and litter and additive genetic as random effects. Each record of birth weight was assigned to the mother of the pup assuming that the residual variance was heterogeneous and partially under genetic control. The expected and realized response to selection was computed each generation. The evolution of the actual residual variance was addressed from the mean birth weight variance within litters, while the expected residual variance each generation was obtained computing the genetic selection response as the product of the realized selection intensity for each generation, the accuracy between real and predicted breeding value and the standard deviation of the genetic residual variance. This accuracy was obtained in an empirical way by averaging it across all generations and lines, each of one derived from the observed genetic response, the realized selection intensity and the standard genetic deviation of the environmental variability. Observed evolution of residual variance was irregular but comparable to the predicted one. However, the response seemed to fail in the high variability line from the tenth generation.

Effect of feeding restriction strategies on growth performance and meat quality of gilts

X. Luo[1], A. Brun[1], M. Gispert[1], J. Soler[2], E. Esteve-Garcia[3], R. Lizardo[3] and M. Font-I-Furnols[1]
[1]IRTA, Finca Camps i Armet, 17121 Monells, Spain, [2]IRTA, Veïnat de Sies, 17121 Monells, Spain, [3]IRTA, Ctra. Reus-El Morell km 3.8, 43280 Constantí, Spain; xin.luo@irta.cat

This work aims of evaluate the effect of different feeding strategies on the growth performance and meat quality of gilts by CT. Forty-eight Pietrain × (Large White × Landrace) gilts were distributed into 4 dietary treatments: (1) *ad libitum* feeding (AL) during all fattening (AL-AL) period; (2) AL feeding between 30 and 70 kg followed by restriction (84% of AL) until 120 kg (AL-RV); (3) restricted feeding (78% of AL) between 30 and 70 kg followed by AL until 120 kg (RV-AL); (4) low energy diet (-10%) between 30 and 70 kg followed by AL until 120 kg (RE-AL). Pigs were CT scanned at target body weights (TBW) of 30, 70, 100 and 120 kg with a GE HiSpeed Zx/I (140 kW, 145 mA, matrix 512×512, axial, 7 mm or 10 mm thick). Subcutaneous fat thickness and area, loin area and perimeter were determined between the 3rd and 4th *lumbar vertebrae* in CT images. After 7-10 d of the last CT scanning (120 kg TBW), pigs were slaughtered. Meat pH at 45 min and 24 h, colour (L*a*b*) and drip loss (%) were measured at the last rib level *on Longissimus* (LT) muscle. Global average daily gain was higher in AL-AL and RE-AL compared with AL-RV and RV-AL (911 and 947 vs 859 and 844 g/d, respectively) while there were no differences in feed conversion ratio (between 2.66 and 2.76 kg/kg in all treatments). No significant differences were found in meat quality parameters, except for L* value which was significantly higher in AL-RV and RV-AL than in RE-AL treatments. Loin area and perimeter of pigs were similar for all dietary treatments along growth. However, fat thickness and area at 70 kg from RV-AL pigs were lower compared to the other treatments. At 100 kg, fat area was lower in AL-RV and RV-AL than AL-AL but at 120 kg differences in fat thickness and area were not significant anymore. Thus, under conditions of this work, different feeding strategies does not seem to affect lean and fat contents of the final product.

Pork quality traits in Latvian White breed pig population

L. Paura, D. Jonkus, L. Degola and I. Gramatina
Latvia University of Agriculture, Liela str. 2, Jelgava, 3001, Latvia; liga.paura@llu.lv

Since 90th two maternal breeds Landrace and Yorkshire (including Latvian White breed) and two paternal breed Pietrain and Duroc are used in pig breeding programme in Latvia for improving the production and reproduction traits of swine. Big emphasis on improving the lean meat content and backfat thikness to promote meat quality and intramuscular fat content decreasing and as results its decreased pork sensory properties. Latvian Pig Breeding Organisation (CCC) were organized the sensory evaluation of the pork from Landrace, Yorkshire, Pietrain, Latvian White (LB) and these breeds crosses to get answer about the role of the national – genetic resources LB pigs breed in the meat quality improving. For the meat sensory evaluation was used a taste scale from 1 to 5 and were evaluated the roast and boiled pork meat. The lowest visual evaluation had pork from LB breed, because the samples were with bigger backfat thikness. The highest taste evaluation in average 4 points had pork from Landrace and Yorkshire breed and the LB breed meat was evaluated by 3.7 points in average. To increase the interest between pork producers were analysed the local LB pig breed the chemical composition (protein, intramuscular fat, ash content) and indicators of the technological quality (pH$_{48}$, colour lightness L*) of the *longissimus dorsi* muscle. The chemical compositions in pork of LB breed were 24.5±0.23%, 2.7±0.47% and 1.3±0.05% for protein, intramuscular fat and ash content, respectively. The high protein and optimal intramuscular fat content, which is considered 2.5% as optimum, influence meat tastiness and tenderness. In studied samples the pH$_{48}$ values range from 5.69 to 6.43 and was in average 5.9±0.06 and it was the higher than in other authors investigations about local pig breeds. The average colour lightness L* was 54.4±1.10. The consumer to select pork and determine meat quality by the colour. According to the pH$_{48}$ and colour lightness L* values the pork of LB breed can be classified as red, firm, non-exudative. Investigation supported by VPP 2014-2017 AgroBioRes Project No. 3 LIVESTOCK.

The effect of acidic marinades to the quality of the pork

A. Tänavots[1], A. Põldvere[1,2], K. Veri[1], T. Kaart[1], K. Kerner[1] and J. Torp[1]
[1]*Estonian University of Life Sciences, Institute of Veterinary Medicine and Animal Sciences, Kreutzwaldi 62, 51014 Tartu, Estonia, [2]Estonian Pig Breeding Association, Aretuse 2, 61406 Märja, Tartumaa, Estonia; alo.tanavots@emu.ee*

The pork quality parameters of the longissimus muscle subjected to ageing with white wine vinegar (WWV, pH 3.0), apple vinegar (AV, pH 3.1), mustard-honey (MH, pH 3.9) and kefir (K, 4.5) marinades for 7 days were studied in 8 repeats. pH (raw only), electroconductivity, colour, weight loss and shear force of the raw marinated and cooked meat was determined at 1., 3. and 7. days of storage. MH and K marinades retained its initial pH during the ageing period. A considerable lower pH value was in the samples treated with AV and WWV marinades and additional drop occurred on 3. day after treatment (P<0.05, analysis of variance). The electroconductivity differed in a larger scale between marinades after thermal treatment. The acidity in AV and WWV marinades turned raw samples significantly (P<0.05) lighter (L*). The cooked samples treated with K were lighter within the 7 day period (P<0.05) and the MH samples were darker only on the 7. day (67.31). The raw samples treated with MH marinade had lower redness (a*, P<0.05), but larger yellowness (b*) value, whereas cooking increased b* value considerably. The K marinade decreased raw samples yellowness (P<0.05), but cooking increased b* value close to the value of the WWV and AV treated samples. The weight loss of K treated raw samples was not remarkable during the ageing period (0.27-1.35%), compared to the other groups (4.25-8.70%), which lost a significant amount of weight, especially on the 7. day of storage. Thermal treatment had the modest effect to the samples treated with MH (25.43-27.41%), whereas samples treated with K lost weight almost at the same level as in two other groups. The marinade and storage time had no effect on the raw meat samples shear force. However, the thermal treatment brought out the differences between marinades during storing. The samples treated with WWV and AV turned tougher compared to the other two marinades. Obtained data demonstrated that mustard-honey marinade had the ability to retain moisture better, which turned samples softer after cooking. The meat softening effect had also marinade with kefir.

Determination of N, P, K, Cu and Zn in pig manure at spreading in French pig farms

J.Y. Dourmad[1] and P. Levasseur[2]
[1]*INRA, UMR Pegase, 35590 Saint-Gilles, France, [2]IFIP, Institut du Porc, 35651 Le Rheu, France; jean-yves.dourmad@inra.fr*

The objective of this work was to contribute to a proposal for updated French references for N, P, K, Cu and Zn contents in pig manure at spreading. They are calculated according to a balance approach in agreement with international recommendations. Nutrient excretion is calculated as the difference between nutrient intake and retention. Body retention is determined according to prediction equations derived from the scientific literature. For nutrients without gaseous emissions (P, K, Cu and Zn) nutrient amount in manure at spreading is calculated as the amount excreted plus, in case of litter bedding, the amount in litter substrate. For N, the amount in manure is calculated considering gaseous losses from housing and during storage, with specific emission factors for each chain of manure collection and storage. A first set of average reference values is determined according to a Tier2 approach. Two feeding strategies, either conventional or improved (ie reduced in crude protein and total phosphorus) and three types of manure management: either slatted floor with slurry, slatted floor with V-shaped scrapper under the floor and solid-liquid phase separation, or accumulated straw or sawdust bedding, with or without composting of the solids, are considered. Animal performance is calculated according the French pig farm performance database (1,750 farms). The reference values are expressed per animal and per year for sows, including piglets until weaning, and per animal produced for weaners and fattening pigs. Total farm output is then calculated according to the average number of sows in the farm and the number of piglets and fattening pigs produced each year. The second reference is determined according to a simplified modeling approach (BrsPorc), in agreement with the Tier3 approach. This allows a more precise determination of the nutrient flow according to the actual performance, feeding strategy and manure management in the farm. This approach requires the collection of additional information especially on feed intake and feed composition, the number of sows and the number of piglets and fattening pigs produced each year. Calculation tools using this approach are available (http://www.rmtelevagesenvironnement.org)

The global farm platform initiative - towards sustainable livestock systems

M.C. Eisler[1], M.R.F. Lee[1,2], G.B. Martin[3], D. Ananth[4], J.A.J. Dungait[2], E. Goddard[5], H. Greathead[6], J. Liu[7], N. López-Villalobos[8], S. Mathew[4], F. Montossi[9], S. Mwakasungula[10], C.W. Rice[11], J.F. Tarlton[1], V.K. Vinod[4], R. Van Saun[12], M. Winter[13] and I. Wright[14]

[1]School of Veterinary Sciences and Cabot Institute, University of Bristol, Bristol, United Kingdom; [2]Department of Sustainable Agriculture Sciences, Rothamsted Research, North Wyke, United Kingdom; [3]UWA Institute of Agriculture, University of Western Australia, Perth, Australia; [4]Kerala Veterinary & Animal Science University, Kerala, India; [5]Department of Resource Economics and Environmental Sociology, University of Alberta, Edmonton, Canada; [6]Institute of Integrative and Comparative Biology, University of Leeds, Leeds, United Kingdom; [7]College of Animal Sciences at Zhejiang Agricultural University, China; [8]Institute of Veterinary, Animal & Biomedical Sciences, Massey University, Palmerston North, New Zealand; [9]National Institute of Agricultural Research, Montevideo, Uruguay; [10]Small Scale Livestock and Livelihoods Programme, Lilongwe, Malawi; [11]Department of Agronomy, Kansas State University, Kansas, USA; [12]College of Agriculture, Penn State University, University Park, Pennsylvania, USA; [13]Centre for Rural Policy Research, University of Exeter, Exeter, United Kingdom, [14]International Livestock Research Institute, Nairobi, Kenya; michael.lee@rothamsted.ac.uk

The major challenges of the twenty-first century, namely climate change, population growth, urbanisation, pressure on agricultural land, demand for biofuels, environmental pollution, loss of biodiversity and ecosystem services, demand for animal protein in developing nations and availability of food, water and energy, have engendered concerns about the sustainability of livestock keeping and consumption of their products, particularly so for ruminants. In particular, we urgently need to assess the role of ruminant animals in meeting our requirements for food in the broader context of mixed agricultural systems. Careful consideration of the manifold contributions of ruminant livestock to global food security and rural livelihoods led an international, multidisciplinary group of scientists to create a vision for sustainable ruminant production in the face of these challenges using a network of research farms on every continent of the globe.

Health and welfare challenges in small ruminant production systems

G.I. Arsenos

Aristotle University, School of Veterinary Medicine, Thessaloniki, 54124, Greece; arsenosg@vet.auth.gr

Small ruminants are reared in a wide range of husbandry systems from intensive indoor housing to organic and outdoor semi-intensive and extensive low-input systems. Across systems health and welfare has always been important determining productivity and financial sustainability at farm and whole sector levels. The aim is to discuss the evolution of husbandry systems and the ability of small ruminants to adapt to changes without compromising health and welfare. From a veterinarian perspective, we are interested in research as a driver to enhance our understanding of the role of environmental factors, genetics and management practices on animal health and productivity. The latter will be discussed considering appropriate genetic improvement programmes targeting animal performance, resilience and adaptability. The notion is that balanced breeding programmes can decisively contribute to enhanced health and welfare. The practice has shown that selection to maximise total milk yield has compromised quality and processing performance of produced milk. Sheep and goat systems will face revolutionary changes in reproduction, nutrition and animal transport as well as global competition in trade of animals and their products. The latter present both challenges and opportunities; the role of capital availability and technology uptake will be critical. Recent epidemics of emerging diseases, e.g. bluetongue illustrate how health status can have devastating effects on the financial viability of a farm or entire sector. The fear of repercussions of the disease in terms of deaths and reduced fertility are extremely serious, as are the introduction of other exotic diseases. Consumer awareness and legislative norms have constrained the use of antibiotics and anthelmintics for prophylactic or metaphylactic purposes against infectious and opportunistic pathogens associated with economically important diseases (e.g mastitis, parasitism, lameness and diarrhea). Emphasis will be shifted to effective alternatives subject to improved performance and financial returns. Hence, identifying and integrating all necessary information as well as adopting innovations is practically the only way to tackle endemic and exotic diseases and ensure the sustainability of sheep and goat systems.

In vitro and in vivo responses of gastrointestinal sheep nematodes to marine macroalgae

S. Werne[1], F. Heckendorn[1], M.Y. Roleda[2], L. Baumgartner[1], E. Molina-Alcaide[3], A. Drewek[4] and V. Lind[2]
[1]Research Institute of Organic Agriculture (FiBL), PO Box, 5070 Frick, Switzerland, [2]Norwegian Institute of Bioeconomy Research (NIBIO), Pb 115, 1431 As, Norway, [3]Estación Experimental de Zaidin (CSIC), c/Profesor Albareda 1, 18008 Granada, Spain, [4]ETH Zürich, Rämistrasse 101, 8092 Zürich, Switzerland; steffen.werne@fibl.org

High crude protein (CP) content and digestibility suggest macroalgae to be used as an alternative protein source to soya for animal nutrition. Beyond that, the polyphenol content of some macroalgae might have beneficial effects on the degree of gastrointestinal nematode infection in sheep. Therefore, in a first step, *Porphyra* sp. and *Laminaria digitata* were used to produce two acetone-extracts of each algae, one with the polymeric absorbent AmberliteTM XAD4 and one without it. The 4 extracts were tested *in vitro* via a larval exsheatment assay (LEA) using doses referring to a 100% *in vivo* dry matter (DM) ingestion of lambs: Laminaria (L), Laminaria XAD (LX), Porphyra (P) and Porphyra XAD (PX). Tests were run with 3rd stage larvae of the gastrointestinal nematodes *Teladorsagia circumcincta* and *Trichostrongylus colubriformis*. Six replicates per parasite and extract were run at a time. Each run was repeated 3 times. A logistic regression model was fitted to analyse the effect of 'extract' and 'run' on the exsheatment probability of the larvae. For *T. colubriformis*, the exsheatment probability was significantly lower in 2 (L), 1 (LX), 2 (P) and 2 (PX) out of the 3 runs per extract when compared to the respective controls. For *T. circumcincta*, the exsheatment probability was significantly lower for the larvae exposed to extract 'P' in 2 out of 3 runs only. Due to the higher frequency of significant runs, *Porphyra* sp. was chosen for a subsequent *in vivo* trial. Twenty-four naturally infected lambs were therefore kept individually for a period of 13 days and fed silage *ad libitum* in addition to either 200 g (DM) soya pellets or 200 g (DM) *Porphyra* sp. daily. Analysis of the repeated measured faecal egg counts using a linear mixed model revealed no differences between the two feeding groups. We conclude that the given amount of *Porphyra* sp. had no effect on the faecal egg counts in this study.

Endoparasites in sheep and goats: prevalence and control strategies of mountain farms in Northern Italy

C. Lambertz, L. Flach, I. Poulopoulou and M. Gauly
Free University of Bozen-Bolzano, Faculty of Science and Technology, Piazza Università 5, 39100, Italy; LauraFranziska.Flach@natec.unibz.it

One of the major constraints for sheep and goat production are infections with endoparasites, in particular gastrointestinal nematodes (GIN). The intensive use of medical treatments for control increases the risk for the development of resistances. For mountain livestock farming, where small ruminants are mainly kept in small herds, it remains questionable how effectively parasites are controlled. Therefore, the aim of the study was to: (1) assess the prevalence of endoparasitic infections of goats and sheep in South Tyrol, Northern Italy; and (2) evaluate the use of routine control measures. A total of 94 farms were surveyed to collect data on farm management and control strategies. Additionally, more than 3,400 individual fecal samples were collected in three consecutive seasons (autumn 2015, spring and autumn 2016) and fecal egg counts (FEC) were performed using a modified McMaster technique. Based on log-transformed FEC-values, a mixed model was used with the fixed effects of species, breed within species, age group, season and resulting two-way interactions. Results are presented as LS Means of untransformed FEC values. According to the survey, 6% of the farmers do not treat their animals, while 44% do it once and 50% twice per year. Commonly the whole herd is treated (75% of the farms), while 78% of the farms never determines the infection status. The most common active substances used were so far ivermectin (54%), albendazole (24%) and eprinomectin (8%). Overall, 16, 23 and 24% of the samples were FEC-negative in autumn 2015, spring and autumn 2016, respectively. Goats showed an average number of eggs per gram of faeces of 542 and sheep of 424 (P>0.05). Lambs/kids as well as adult males showed higher infections (P<0.05) than adult females. In both species, differences between the breeds were found (P<0.05). The prevalence of tapeworm-positive animals was higher in autumn (9% in 2015 vs 13% in 2016) than in spring (2%). In conclusion, high infections warrant the intensive use of anthelminthics. Nevertheless, through a more sustainable use of treatments using faecal samples to monitor the actual infection status the development of resistances should be reduced.

A comparison of nonlinear models for estimating heritability of tick-infestation on lambs

P. Sae-Lim[1], L. Grøva[2], I. Olesen[1] and L. Varona[3]
[1]*Nofima, Breeding and Genetics, Osloveien 1, 1433 Ås, Norway,* [2]*Norwegian Institute of Bioeconomy Research, Gunnars veg 6, 6630 Tingvoll, Norway,* [3]*University of Zaragoza, Faculty of Veterinary, Zaragoza, 50013, Spain; ingrid.olesen@nofima.no*

Tick-borne fever (TBF) is stated as one of the main disease challenges in Norwegian sheep farming during the grazing season. TBF is caused by the bacterium *Anaplasma phagocytophilum* that is transmitted by the tick *Ixodes ricinus*. A sustainable strategy to control tick-infestation is to breed for genetically robust animals. The risk of being infected by *A. phagocytophilum* is likely to increase with number of ticks infesting sheep. Hence, tick-count on lambs may reflect the susceptibility of *A. phagocytophilum* infection in lambs. In order to use selection to genetically improve traits we need reliable estimates of genetic parameters. The standard procedures for estimating variance components assume a Gaussian distribution of the data. However, tick-count data is a discrete variable and, thus, standard procedures using linear models may not be appropriate. We compared four alternative non-linear sire-dam mixed models: Poisson, negative binomial, zero-inflated Poisson and zero-inflated negative binomial based on their goodness of fit for quantifying genetic variation, as well as, heritability for tick-count. Models were implemented using a Gibbs Sampler Markov Chain Monte Carlo algorithm. The Gibbs sampler was implemented with a single long chain of 1,250,000 after discarding the first 250,000. Based on Log-marginal probability and deviance information criterion, our results showed that zero-inflated Poisson was the most parsimonious model for the analysis of tick count data. The resulting estimates of variance components and high heritability (0.32) led us to conclude that genetic determinism is relevant on tick count. A selection of 5% downward on whole population for tick infestation would imply a reduction of ticks on the observed scale by 12.7% of the trait mean (1.35) per generation. An appropriate breeding scheme could control tick-count and, as a consequence, probably reduce TBF in sheep.

The body condition score: a central tool to enhance sheep productivity

J.M. Gautier[1], L. Sagot[2] and C. Valadier[2]
[1]*Institut de l'Elevage, BP 42118, 31321 Castanet Tolosan, France,* [2]*Institut de l'Elevage, ferme expérimentale du Mourier, 87800 Saint Priest de Ligoure, France; jean-marc.gautier@idele.fr*

The impact of the body condition score (BCS – scale of 5 with 1 for low BCS and 5 for high BCS) and it's variation on sheep productivity has been newly analysed base on a data collection of 15 years (2000-2014) from the experimental farm Le Mourier (87, France). 5,918 mating ewes are included in this study, distributed into 3,321 natural mating (NM – during sexual season – autumn) and 2,597 artificial insemination with oestrus synchronisation (AI – spring). The BCS has been performed at the start of mating (BCS-SM), at the end of mating (BCS-EM) and at lambing (BCS-L). For NM ewes, a BCS-SM<2 is associated with a lowest fertility rate (0.77% vs 0.87%; P<0.05) and prolificacy (1.61 vs 1.85; P<0.05). The highest fertility during the 1st sexual cycle is obtained with a BCS-SM ≥3 (75.5% vs 65.1%; P<0.05). A positive variation of the BCS during mating is always associated with an increase of the fertility rate. This impact is highest with BCS-SM<2 (+18.1% of fertility rate) compare with BCS-SM>2.5 (+1.8% of fertility rate), reinforcing the interest of flushing for ewes with bad condition score. For AI ewes, the fertility rate increase with the level of the BCS-SM, with a fertility rate of 43.9% for a BCS-SM=1.5 and 84% for a BCS-SM=4. BCS-L≥2.5 is associated with lowest lamb mortality rate in their first 10 days of life (10.1% vs 12.3%; P<0.05). Finally, a BCS≥3 during the whole reproduction and gestation allows a highest sheep productivity (NM: >1.34; AI: >0.96). Thus, body condition score is confirmed as a good tool for sheep management. The challenge is now to find easy ways to performed it in farm. The use of new technologies might be a solution to investigate.

Lameness of dairy ewes in practical conditions

Š. Baranovič[1], V. Tančin[1,2], M. Uhrinčat[2] and L. Mačuhová[2]
[1]Slovak Agricultural University Nitra, Tr. A. Hlinku 2, 94976 Nitra, Slovak Republic, [2]NPPC – Research Institute for Animal Production Nitra, Hlohovecká 2, 95141 Lužianky, Slovak Republic; tancin@vuzv.sk

The aim of the study was to assess the lameness of dairy ewes during milking. The study was performed at three dairy farms under practical conditions with different breeds and crossbreds in Slovakia. The first farm kept crossbreeds of Improved Valachian × Lacaune (IV/LC, crossbred; higher part of IV), the second and the third farm had two groups of animals: 2a (purebred LC), 2b (SD, crossbreed; synthetic population of Slovak dairy sheep) and 3a (T, purebred Tsigai), 3b (purebreed LC). Lameness was assessed once a month in the following order: at the first farm in May and July, at the second farm from April to September and at the third farm from April to July. In total, 4,697 ewes were evaluated. Assessment of lameness was carried out by five point locomotion scoring scale: (1) non-lame; (2) slightly lame; (3) strongly lame 1 (affected limb did not bear weight when standing); (4) strongly lame 2 (affected limb did not bear weight when moving); (5) strongly lame 3 (more than one limb affected). The frequency of lameness was assessed within a farm, breed, month, stage and number of lactation and order of entry into the milking parlour. The best feet conditions were on the third farm, with the highest number of non-lame ewes (88.4%). Second farm had the least non-lame ewes (70.6%). LC was the most sensitive to lameness, having significantly higher frequency of lame ewes (farm 2: 35.7%, farm 3: 15%) compared to SD (farm 2: 27.9%) and C (farm 3: 10.9%). The riskiest season for lameness was in April (e.g. up till 50% LC lamed on farm 2 during this time) and the least risk was in July. Most ewes in the first lactation were non-lame, ewes in the fifth, sixth or higher lactation were on the other hand highly susceptible to lameness. The highest frequency of non-lame ewes was among the ewes entering into the milking parlour in the first turn. Slightly lame ewes entered into the milking parlour in different groups, while strongly lame ewes entered into the parlour often in the last turn. In conclusion lameness was most affected by farm which emphasizes the importance of farm management. This study was supported by APVV 15-0072.

Effect of weaning system on milk production of ewes

L. Mačuhová[1], V. Tančin[1,2], M. Uhrinčat[1], J. Mačuhová[3] and M. Margetín[1,2]
[1]NPPC – Research Institute for Animal Production Nitra, Hlohovecká 2, 951 41 Lužianky, Slovak Republic, [2]Slovak Agricultural University in Nitra, Department of Veterinary Science, Tr. A. Hlinku 2, 949 76 Nitra, Slovak Republic, [3]Institute for Agricultural Engineering and Animal Husbandry, Prof. Dürrwaecher Platz 2, 85586 Poing, Germany; macuhova@vuzv.sk

Improved Valachian (n=45) and Tsigai (n=46) crossbred ewes with Lacaune were used to study the effects of three weaning systems on milk production. Prior to parturition, ewes were assigned to one of the following three treatments for the first 38±5 day of lactation: (1) ewes weaned from their lambs at 24 h postpartum and afterwards machine milked twice daily (MTD); or (2) beginning 24 h postpartum, ewes kept during the daytime with their lambs and allowed them to suckle for 12 h, nights separated from their lambs for 12 h and machine milked once daily in the morning (MIX); or (3) ewes exclusively suckled by their lambs (ES). After the treatment period, lambs were weaned from MIX and ES ewes, and all three groups were machine milked twice daily. Then ewes were evaluated according to live- born lambs (with one (n=37) or with two lambs (n=54)). The measurements of milk yield and milk flow were performed on 110±5 day of lactation by the equipment for graduated electronic recording of the milk level in a jar in one-second intervals. No significant differences were observed in the measured values (total milk yield, machine milk yield, latency time, milking time, machine stripping, peak rate, milk yield in 30 and 60 s) among weaning treatments and between ewes with one or two lambs too (P>0.05). In mid-lactation, the different systems of weaning didn't influence the milk yield and milk flow parameters. This publication was written during carrying out of the project APVV-15-0072.

Factors affecting the viability and live weights of Cyprus Chios lambs from birth to weaning

G. Hadjipavlou[1], A. Pelekanou[2] and C. Papachristoforou[2]
[1]*Agricultural Research Institute, P.O. Box 22016, 1516 Lefkosia, Cyprus, [2]Cyprus University of Technology, P.O. Box 50329, 3036 Lemesos, Cyprus; georgiah@ari.gov.cy*

This study investigated the factors affecting the viability of Cyprus Chios lambs at birth and until weaning, and those influencing birth and weaning weight. Data were collected from 3,088 Cyprus Chios lambs, born during 2010-2014, at the Cyprus Agricultural Research Institute Farm. Lambings took place either in autumn or in late winter-spring. Rearing until weaning occurred via natural suckling or milk replacement, either for 35 or for 28 days after birth. Logistic regression was used for survival phenotypes and linear regression for live weights. Viability at birth was significantly affected by gestation length ($P<2.0\times10^{-16}$), birth weight ($P<2.0\times10^{-12}$) and dam parity ($P<0.01$ or lower). Lambs born to multiparous ($\geq$three years) dams were more likely to survive at birth than lambs from primiparous dams (odds ratio 0.20-0.55). Viability after birth and until weaning was significantly affected by birth weight and rearing system ($P<2.0\times10^{-16}$ for both), the chronological order of birth within lambing season and year (termed date of birth; $P<2.4\times10^{-4}$), and lamb sex (2.1×10^{-4}). Female lambs were less likely to survive until weaning (odds ratio=1.75) and had significantly lower birth and weaning weight than males. Live weight at birth and at weaning were mainly influenced by gestation length ($P<2.2\times10^{-16}$, $P<10.0\times10^{-6}$, respectively), sex ($P<2.2\times10^{-16}$, $P<8.3\times10^{-16}$), year ($P<0.0081$, $P<2.2\times10^{-16}$), litter size ($P<2.2\times10^{-16}$, $P<6.4\times10^{-9}$), lambing season ($P<0.019$, $P<2.2\times10^{-16}$) and lambing season by year of birth ($P<1.1\times10^{-11}$, $P<2.2\times10^{-16}$). Lambs from multiparous ewes has higher live weights at birth and at weaning than lambs from primiparous ewes. Twins and triplets had lower birth and weaning weights than singletons. Weaning weight was also affected by the date of birth, live weight at birth, rearing system and age at weaning ($P<2.2\times10^{-16}$ for all). Lambs reared using milk supplement had lower weaning weight than suckling ones. This study identified specific factors associated with lamb losses at birth and until weaning, highlighting the need for improved management practices during gestation, at birth and until weaning, especially for primiparous ewes, in order to reduce lamb mortality rates.

Effects of sire breed on lamb survival under Eastern European low input production systems

D. Gavojdian[1,2] and L.T. Cziszter[1]
[1]*Banat's University of Agricultural Sciences and Veterinary Medicine, Faculty of Animal Science and Biotechnologies, Timisoara, Calea Aradului 119, 300645, Romania, [2]Academy for Agricultural and Forestry Sciences, Research and Development Station for Sheep and Goats, Caransebes, 325400, Romania; gavojdian_dinu@animalsci-tm.ro*

Aim of the current research was to evaluate the effects that Dorper, German Blackheaded and Bluefaced Leicester terminal sire breeds have on lamb survival rates in extensive low-input production system under European temperate conditions. The study was carried out in a commercial farm from Western Romania (45°34'0"N 20°51'48"E), where groups of 25 to 30 purebred Turcana ewes (3 to 4 years of age) were crossed with Dorper, German Blackheaded, Bluefaced Leicester and Turcana rams by natural service in single sire mating groups, using 2 pedigreed rams from each breed. Lambs were managed under identical conditions at the farm and were weaned at 70±5 days of age. In order to assess the effect of the sire genotype (breed) on lambs survival rates the STATISTICA software was used and the Main Effect ANOVA analysis of variance was applied. Survival rates of lambs at weaning were of 88.4±3.30% in F_1 Dorper × Turcana, 91.6±0.04% in F_1 German Blackheaded × Turcana, 94.0±1.84% in F_1 Bluefaced Leicester × Turcana and 91.1±3.58% for the Turcana purebreds (control group). South African Dorper sired lambs had the lowest survival rates ($P\leq0.05$) when compared to the Turcana purebreds, with the prolific Bluefaced Liecester sired lambs having the highest survival rate ($P\leq0.05$). Compared to the controls, when using meat specialized German Blackheaded the lamb survival was not influenced ($P>0.05$). This pilot study was initiated to test the potential for crossbreeding in order to improve the lamb meat outputs under extensive low input production systems using alternative crossing sire breeds, selected for lamb growth (Dorper), carcass traits (German Blackheaded), prolificacy (Bluefaced Leicester) and hardiness (Turcana). Low input extensive environments require careful consideration when introducing new genotypes, given the economic and welfare implications of lamb mortality.

Scrapie prevention in Estonian sheep population
E. Sild, S. Värv and H. Viinalass
Estonian University of Life Sciences, Institute of Veterinary Medicine and Animal Sciences, Kreuzwaldi Str 1, 51014
Tartu, Estonia; erkki.sild@emu.ee

The breeding programme to select for resistance to ovine TSE was introduced in Estonia in 2005. There are ca 90,000 sheep in Estonia and ca 6,000 of them are breeding sheep. The breeding programme focused on genotyping breeding rams and ewes to eliminate deleterious genetic variants of PRNP gene from the breeding flocks. The aim is to increase the frequency of the TSE-resistant haplotype ALRR. The objective of this study was to assess the temporal effect of the ongoing programme of ~10 year duration and compare the breeding and commercial flock structures on the ground of the PRNP gene polymorphisms. Codons 136, 141, 154 and 171 in PRNP were genotyped in 836 animals. Two groups were formed of the breeding sheep: group I (the basis, sheep born in 2005-2006; n=334) and group II (sheep born in 2013-2016; n=314). Randomly selected sheep born in 2010-2016 (n=188; group III) were sampled to estimate the influence on contemporary commercial flock. The genetic differences between groups were compared using chi-square test applying Bonferroni correction. Six different haplotypes and 15 genotypes were determined among the analysed sheep with prevalence of all risk groups to scrapie (R1-R5) in the all three study-groups. VLRQ, a component of the genotypes with the highest risk status (R4 and R5), designating the sheep removal from any flock, had relatively low frequency in group I (0.024) and showed expected decrease (0.014 in group II). The haplotype ALRR (R1) frequency showed rise from 0.412 (group I) to 0.771 (group II) and had a moderate frequency (0.566) in commercial group III. The study showed highly significant differences in the comparison of breeding sheep born at the beginning of the TSE prevention programme and after operating time by haplotype, genotype, and risk group distributions (all comparisons P<0.001). Although, impact to the commercial population deducible from the genetic difference between breeding group I and the randomly sampled commercial sheep was significant (P<0.05), the divergence from the breeding group of the same age (group II) indicates the need to expand genetic monitoring also among commercial flocks. Estonian Ministry of Education and Research (grants IUT8-2).

Optimum selection of core animals in the efficient inversion of the genomic relationship matrix
H.L. Bradford, I. Pocrnic, B.O. Fragomeni, D.A.L. Lourenco and I. Misztal
University of Georgia, Animal and Dairy Science, 425 River Rd, Athens, GA 30602, USA; heather.bradford25@uga.edu

The Algorithm for Proven and Young (APY) has been effective for implementing single-step genomic BLUP for large, genotyped populations. One question is which animals should be used as core animals. The objective was to determine the effect of using core animals from different generations in single-step genomic BLUP with APY. Simulations comprised a moderately heritable trait for 95,010 animals and 50,000 genotypes for animals across 5 generations. Genotypes consisted of 25,500 SNP distributed across 15 chromosomes. Core animals were defined based on individual generations, equal representation across generations, and at random. For a sufficiently large core size, core definitions had the same accuracies (r^2=0.90±0.01) and biases (β_1=1.02±0.01) for young animals, even if the core animals had imperfect genotypes because of imputation. The accuracies had a range of less than 0.01 with different core definitions. Using the youngest generations as core caused an increase in the number of rounds to convergence indicating some numerical instability with these core definitions. When 80% of genotyped animals had unknown parents, accuracy and bias were significantly better (P≤0.05) for random and across-generation core definitions (r^2=0.71±0.01; β_1=0.75±0.01) than for single generation core definitions (r^2=0.61±0.01; β_1=0.53±0.01). This difference could result from improved relationship estimates between animals in different generations, because all generations were represented in the core partition that was directly inverted in APY. Thus, any subset of genotyped animals can be used to approximate the independent chromosome segments when pedigrees are complete, but core animals should represent all generations when pedigrees are incomplete.

Wide-spread adoption of customized genotyping improves European cattle breeding

B. Guldbrandtsen[1], E. Mullaart[2], S. Fritz[3], S. De Roo[4], G.P. Aamand[5], X. David[4], J.A. Jimenez[6], H. Alkhoder[7], Z. Liu[7], C. Schrooten[2] and K. Zukowski[8]
[1]*Aarhus University, Box 50, 8830 Tjele, Denmark,* [2]*CRV, Box 454, 6800 AL Arnhem, the Netherlands,* [3]*INRA, 147 rue de l'université, 75338 Cedex 07, France,* [4]*EuroGenomics, Box 58052, 1040 HB Amsterdam, the Netherlands,* [5]*NAV, Agro Food Park 15, 8200 Aarhus N, Denmark,* [6]*CONAFE, Ctra. De Andalucia, km. 23,600, 28340 Valdemoro, Spain,* [7]*VIT, Heinrich-Schröder-Weg 1, 27283 Verden / Aller, Germany,* [8]*National Research Institute of Animal Production, Krakowska Street, 32-083 Balice, Poland; bernt.guldbrandtsen@mbg.au.dk*

With the Illumina Bovine 54k chip in 2008, genotyping of cattle populations saw widespread adoption in European cattle. In 2009 a number of cattle breeding companies and academic institutions formed the EuroGenomics (EG) collaboration, initially with members from six countries, Germany, France, The Netherlands, Denmark, Finland and Sweden later joined by Spain and Poland. Genotypes were exchanged, each contributing 4,000 proven bulls. In March 2010, the reference population was 16,000 bulls. The EG partners developed an LD chip called EuroG10K. It is based on the Illumina LD chip and optimized for aid imputation in additional breeds including smaller ones. The chip includes polymorphisms known or suspected of causing various phenotypic effects including, recessive lethal alleles, milk protein variants, polymorphisms marking QTL for traits of interest and the ICAR parentage control set. The EuroG10K chip consists of a shared and a private part which can be used by partners, offering a high flexibility compared to other chips. In addition, the shared use results in better prices and a more affordable service to farmers. Results on imputation efficiency and the accuracies of indices will be presented. The chip has seen widespread including outside the Holstein breed, and in the broader cow population with more than 900,000 genotypes to date. This has led to improved imputation and accuracies of predicted breeding values, increased numbers of bull calves evaluated and enhanced management of recessive disease alleles. The chip allows farmers to use genotyping as a management tool. A significant side effect is easy large scale evaluation of candidate polymorphisms.

Can single-step genomic BLUP account for causative variants?

D.A.L. Lourenco[1], B.O. Fragomeni[1], Y. Masuda[1], A. Legarra[2] and I. Misztal[1]
[1]*University of Georgia, 425 River Rd, Athens, GA 30602, USA,* [2]*INRA, 24 chemin de Borde-Rouge, 31326 Castanet Tolosan, France; danilino@uga.edu*

Single-step genomic BLUP (ssGBLUP) is the method of choice for genomic evaluation because of simplicity and ability to combine pedigree, genotypes, and phenotypes. As the availability of sequence data is increasing, we investigated whether ssGBLUP can be useful for genomic analyses when causative Quantitative Trait Nucleotide (QTN) are known. Simulations included 180k animals from 11 generations. Phenotypes (h^2=0.3) were available for all animals in generations 6-10. A total of 24k parents in generations 6-10 and 5k validation animals in generation 11 were genotyped for 60k SNP. The genetic variance was fully accounted for by 100 or 1000 biallelic QTN. Genomic relationship matrices were computed based on: unweighted 60k SNP; unweighted 60k SNP and QTN; 60k SNP and QTN with variance from GWAS; unweighted 60k SNP and QTN with known variance; only QTN. Accuracy of EBV for validation animals were computed by BLUP and ssGBLUP. To ensure full rank, the genomic relationship matrix (G) was blended with 1% or 5% of the numerator relationship matrix. The inverse of G was computed directly or using APY, the algorithm that exploits the limited dimensionality of G for sparse computations. This dimensionality was calculated as the number of largest eigenvalues explaining 98% of the variance of G. Rank of G with 100 QTN was 16,980 for G based on unweighted 60k SNP; 19,112 after 5% blending; 5,093 for G with 60k SNP and QTN with known variance; 98 when only QTN were used to create G. With 1000 QTN, trends were similar, but the rank with only QTN was 930. BLUP accuracy for validation animals was 0.32. Accuracies increased to 0.49, 0.53, 0.63, and 0.89 for ssGBLUP with unweighted 60k SNP, unweighted 60k SNP and QTN, 60k SNP and QTN with variance from GWAS, and unweighted 60k SNP and QTN with known variance, respectively. When G was constructed based only on QTN, the accuracy was 0.95 with 5% blending and 0.99 with 1% blending. Accuracies with 1000 QTN were lower, with a similar trend. Accuracies using APY G^{-1} were equal or higher than those with direct inverse. Single-step genomic BLUP can account for causative variants with nearly optimum accuracy when QTN and their variances are known.

Single versus two step genomic evaluations over many generations

H. Schwarzenbacher

Zuchtdata GmbH, Dresdner Straße 89/19, 1200, Austria; schwarzenbacher@zuchtdata.at

Potential bias due to preselection based on genomic information is a major concern in routine applications because currently most countries use a two-step evaluation system in their genomic breeding program. Single step genomic evaluation is believed to be less affected by bias due to genomic preselection and to be more accurate. A stochastic simulation study was carried out to assess the performance of single trait genomic selection over several generations based on single versus two step genomic evaluations. For this purpose a simulation software was developed that combined the software 'QMSim' with own routines to allow more flexibility in the design. An evolution process was emulated to obtain an LD structure similar to that observed in the Fleckvieh population. After several generations of selection based on conventional ebvs, ten generations of genomic selection using the two different methodologies were applied to assess their performance relative to conventional blup selection. The single step system produced significantly higher reliabilities in all selection groups and consequently a larger genetic gain. Evaluations were less biased because preselection was accounted for. The results suggest that the current genomic evaluation system should be replaced by a single step genomic evaluation as soon as possible.

The impact of genome editing on the introduction of monogenic traits in livestock

J.W.M. Bastiaansen, H. Bovenhuis, M.A.M. Groenen, H.-J. Megens and H.A. Mulder

Wageningen University & Research, Animal Breeding and Genomics, P.O. Box 338, 6700 AH Wageningen, the Netherlands; john.bastiaansen@wur.nl

The introduction of genome editing technologies provides new tools for the genetic improvement of animals and has the potential to become the next big game changer in animal breeding. The aim of our study was to investigate to what extent genome editing in combination with genomic selection could accelerate the introduction of a monogenic trait as compared to a situation with genomic selection alone. A breeding population under genomic selection for a polygenic trait was simulated. After establishing Bulmer equilibrium different selection scenarios, aiming to increase the polygenic trait and the allele frequency of the monogenic trait with or without genome editing, were compared for time to fixation of the desired allele, selection response in the polygenic trait, and level of inbreeding. In addition, the costs in terms of number of genome editing procedures were compared to the benefits of a higher cumulative proportion of the population with the desired phenotype for the monogenic trait and a lower genetic lag for the polygenic trait. Finally, the impact of efficiency of the editing procedure and survival rate of edited embryos were evaluated. Genome editing resulted in up to fourfold faster fixation of the desired allele, and the loss in long term selection response for the polygenic trait was up to threefold smaller than with genomic selection alone. In a population of 20,000 offspring per generation, the total number embryos edited to achieve fixation ranged from 22,610 with no selection to 7,080 with moderate selection and to 3,830 with high selection emphasis on the monogenic trait. With moderate selection emphasis on the monogenic trait, the total number of animals showing the undesired phenotype before fixation of the desired allele was reduced up to fourfold by adding genome editing. Genome editing had hardly any effect on inbreeding. A low editing efficiency (4%) had a major impact by increasing the number of editing procedures (+65%) and increasing the loss in selection response (+254%). Genome editing in commercial livestock breeding needs careful assessment of technical costs and benefits as well as ethical and welfare considerations.

Benefit cost analysis of aquaculture breeding programs

K. Janssen[1], H. Saatkamp[1,2] and H. Komen[1]
[1]Wageningen University, Droevendaalsesteeg 1, 6708 PB Wageningen, the Netherlands, [2]Wageningen University, Business Economics Group, Hollandseweg 1, 6706 KN Wageningen, the Netherlands; kasper.janssen@wur.nl

Breeding programs aim to improve traits relevant to farm profit. In aquaculture companies, breeding programs are often integrated in production. Objective of this study was to identify technical and economic parameters that determine profitability of breeding programs. A benefit cost analysis based on an existing breeding program for gilthead seabream that included a multiplier tier was used as a baseline. The baseline was compared to a breeding program without multiplier tier where the nucleus itself was used to stock production. With multiplier tier the response to selection was €0.219/kg production, whereas it was €0.193/kg production without multiplier tier. Gene flow was used to simulate the increase in genetic level in the nucleus and multiplier tier. Benefits were calculated as the product of the genetic level of fish used in production and a production output of 5,000 tonnes. Costs of the breeding program were estimated from bookkeeping records, and equalled €150,000 per year. Cost and benefits were discounted to their present values. With multiplier tier, the net present value (NPV) was positive after 5 years and reached 3.2 million euro in year 10. Without multiplier tier, NPV was positive after 4 years and reached 5.1 million euro in year 10. A reduced delay in benefits relative to cost in the breeding program without multiplier tier overcompensated the lower selection response compared to the baseline. The time horizon for benefit cost analysis and production output of the company were varied to study their effect on optimum scale of the breeding program. It was assumed that costs were proportional to the number of selection candidates and benefits were proportional to the selection intensity. Results show that a long time horizon and large production output of a company justify higher annual costs than a short time horizon and small production output. Higher costs result in a larger increase in annual benefits. It is concluded that any delay in benefits relative to costs should be minimized, and production output of a company and the time horizon are essential design parameters for breeding programs.

Challenges of implementing optimum contribution selection in two regional cattle breeds

S.P. Kohl[1], H. Hamann[2] and P. Herold[2]
[1]Institute of Animal Science (460), Animal Genetics and Breeding (460g), Garbenstraße 17, 70599 Stuttgart, Germany, [2]State Office for Spatial Information and Land Development Baden-Württemberg, 35, Stuttgarter Straße 161, 70806 Kornwestheim, Germany; spkohl@gmail.com

Optimum contribution selection (OCS) facilitates the maximization of breeding progress while simultaneously restricting inbreeding rates. Although the superiority of OCS compared with Truncation Selection has been proven several times, the implementation in practical breeding programs is still missing. The aim of this study is to investigate advantages of an implementation of OCS in breeding programs of two German regional breeds, Hinterwaelder and Vorderwaelder cattle. Prior to the implementation of an OCS methodology, the quantity of achievable breeding progress and inbreeding rates are analyzed and compared to the current methods. Additionally, necessary adjustments to the breeding programs are considered. For population analyses and statistical computations as well as the computations of optimum contributions, EVA and the R packages optiSel, GENLIB and Pedigree are used. Results show that in the in Hinterwaelder and Vorderwaelder cattle larger breeding progress could have been achieved at the targeted rate of inbreeding. Especially in Hinterwaelder cattle an overrated weight on genetical conservation lead to big losses in breeding progress while achieving inbreeding rates less than expected under natural selection. In Vorderwaelder cattle, which had been graded up with high-yielding breeds, the foreign breed contributions need to be restricted as part of the OCS. The implementation of OCS in Hinterwaelder and Vorderwaelder cattle could lead to gains in breeding progress while raising inbreeding rates to moderate levels. Particularly small breeds face big gaps in performance traits like milk yield and meat production compared to larger breeds. The implementation of an OCS methodology will not close this gap, but it could prevent the gap from further increasing.

Genetic (co)variation of milk flow and milkability in first lactation Finnish Ayrshire cattle

A. Ewaoche[1], J. Pösö[2] and A. Mäki-Tanila[1]
[1] *University of Helsinki, Department of Agricultural Sciences, P.O. Box 28, 00014 University of Helsinki, Finland,* [2] *FABA, Korpikyläntie 77, 15871 Hollola, Finland; anne.ewaoche@gmail.com*

In dairy cattle, milk flow is an important functional trait which impacts production. Milk flow can be measured accurately by electronic milking meters (EMM) and robots. It is necessary to understand the implications of the transition from subjective scores (milkability, from very slow: 1; to very fast: 5) to objective measurements (milk flow, kg/min), as well as the genetic (co)variation of the traits. The records from Finnish Ayrshire primiparous cows were analysed for milkability, milk flow, annual milk yield and somatic cell count (SCC). Milk flow was recorded by Tru-Test (EMM) and the Lely robot milking systems and the records from the two systems were treated as separate traits. A total of 64,696 cows (2,896 herds, years 2010-16) were analysed for milkability, 1,618 cows (54 herds, years 2010-15 for Tru-Test and 2,232 cows (104 herds, years 203-15) for Lely. To estimate variance components, both single and two-trait animal models were fitted and analysed with REML using the DMU software. The pedigree information covered four generations. Heritability of milkability was 0.25 (standard error 0.01). For milk flow, heritability was 0.41 (0.08) and 0.52 (0.08) for Tru-Test and Lely, respectively. The phenotypic correlation with milk yield was 0.07, 0.31 and 0.13 for milkability, Tru-Test and Lely, respectively, and with SCC 0.10, 0.12 and 0.20 in the same order. The respective genetic correlations with milk yield were 0.10 (0.04), 0.43 (0.14) an 0.37 (0.14) and with SCC 0.50 (0.04), 0.42 (0.17) and 0.35 (0.17). The differences in standard error were due to the number of observations. Common sires provided a way to find the genetic correlation between milkability and milk flow and between Tru-Test and Lely. In conclusion, selection for milk flow is more efficacious than for milkability demonstrating the influence of quality and volume of recording on estimating heritability and genetic correlation.

Genetic and genomic selection for oocyte production in Dutch Holsteins

J.J. Vosman, E. Mullaart and G. De Jong
CRV, AEU, P.O. Box 454, 6800 AL Arnhem, the Netherlands; jorien.vosman@crv4all.com

Reproductive technologies like Ovum Pick-Up (OPU) and Multiple Ovulation and Embryo Transfer (MOET) are important factors determining the success of breeding strategies of breeding companies. By using those techniques more offspring from females can be generated compared to the traditional way of breeding. OPU is based on collection of oocytes, of which maturation and fertilization takes place *in vitro*. With MOET ova (oocytes and embryo's) are collected, after maturation and fertilization of oocytes *in vivo*. Oocyte production is heritable, that make selection for more offspring from females possible. To enhance breeding values for oocyte production, variance components were estimated for number of oocytes collected with OPU and for number of ova collected with MOET based on Dutch Holsteins. Since 1993 OPU and MOET sessions are registered. The dataset included 57,760 OPU sessions, based on 3,884 animals, with 8.0 oocytes per session on average. For MOET 46,884 sessions based on 19,523 animals are included in the dataset. The average number of ova collected per session is 9.1. Both datasets do not follow a normal distribution, records were transformed. For the OPU dataset a logarithm transformation was used, while for the MOET dataset a Anscombe transformation was used. Parameters were estimated using a bivariate animal model with repeatability effects, by using the ss-BLUP approach. The total number of animals in the pedigree was 79,531, of which 7,918 individuals were genotyped. Heritability estimates were 0.34 for OPU and 0.22 for MOET, the genetic correlation between OPU and MOET was 0.84. Breeding values were estimated for bulls and cows. The average reliability of genotyped animals without own observations ranged from 40 to 50%. The average reliability of cows with observations ranged from 55 to 80% for OPU and from 25 to 75% for MOET. Reliability of bull EBVs ranged from 40 to 90% for both OPU and MOET. The results showed that selection for oocyte production in Dutch Holsteins is possible by using breeding values for OPU or MOET.

Effects of conjugated linoleic acids supplementation on bovine sperm integrity and functionality

M.S. Liman[1], C.L. Cardoso[1], B. Gasparrini[2], V. Franco[2], V. Longobardi[2] and G. Esposito[1]
[1]University of Pretoria, Faculty of Veterinary Sciences, Production Animal Studies, Soutpan Road, Onderstepoort, 0110, Pretoria, South Africa, [2]University of Naples Federico II, Veterinary Science, via Delpino, Naples, Italy; giulia.esposito@up.ac.za

Feeding rumen-protected isomers of conjugated linoleic acid (CLA) to dairy cows reportedly improves fertility by reducing the postpartum interval to first ovulation and enhancing circulating IGF-I levels. Recent research has focused on their effect on semen quality and freezability mainly investigating the effect of inclusions in semen extenders. The objective of the study was to evaluate the effect of CLA dietary supplementation and their inclusion in the semen extender on bovine semen quality and freezability. Fourteen bulls blocked by, age, BW and BCS were randomly assigned to 2 groups: control (CTL) and CLA. The animals were supplemented for 10 weeks and samples of ejaculates were collected twice a week during weeks -1 and -2 before supplementation, weeks 4 and 5 during supplementation, and 11 and 12 (after the supplementation). Ejaculate from each bull was frozen according to the addition of CLA to the semen extender namely control (CTL), CLA 9,11 50 uM, CLA 9,11 100 uM, CLA 10,12 50 uM, CLA 10,12 100 uM and CLA mix (9,11 50 uM + 10,12 50 uM). Sperm motility was assessed by CASA system; morphology via eosin-negrosin staining; viability, mitochondrial activity and reactive oxidative species (ROS) using the flow cytometer. CLA supplementation increased Beat cross frequency compared to the control ($P<0.05$). The inclusion of the 9,11 50 uM in the extender increased the percentage of live spermatozoa compared to the control and other isomers 'inclusion ($P<0.1$); it reduced ROS in the cryopreserved semen while the inclusion of the 10,12 100 uM isomer increased it. These results were confirmed in both, CTL and CLA dietary supplemented bulls. However, the positive effect of the 9,11 50 uM isomer was more pronounced in the CLA supplemented group than in the control ($P<0.001$ and $P<0.05$ respectively). These preliminary results support the authors' hypothesis of the potential effect of dietary CLA supplementation and its interaction with CLA inclusion in the semen extender, in improving semen quality and freezability.

Improvement of dna integrity in dog semen treated with different level of cholesterol loaded cyclode

M.E. Inanc[1], K. Tekin[2], K.T. Olgac[2], B. Yilmaz[2], E. Durmaz[3], U. Tasdemir[4], P.B. Tuncer[5], S. Buyukleblebici[5] and O. Uysal[2]
[1]Mehmet Akif Ersoy University, Reproduction and Artificial Insemination, Burdur, 15030, Turkey, [2]Ankara University, Reproduction and Artificial Insemination, Ankara, 06110, Turkey, [3]Gazi University, Toxicology, Ankara, 06330, Turkey, [4]Aksaray University, Technical Sciences Vocational School, Aksaray, 68100, Turkey, [5]Mersin University, Technical Sciences Vocational School, Mersin, 33343, Turkey; enesinanc@hotmail.com

The objective of the study was determined sperm DNA damages in Aksaray Shepherd dog semen with different level of cholesterol-loaded cyclodextrin (CLC) by the single cell electrophoresis (COMET) assay. Twentysix animal, from three different age (≤3, group 1 (n:10); 4-6, group 2 (n:9); >7, group 3 (n:7)), were used in this study. As soon as semen samples were collected with digital manipulation, each ejaculate was divided into four equal aliquots and extended with a tris base extender containing 0 mg (control), 0.5 mg; 1.0 mg; 1.5 mg/120×106 cholesterol loaded cyclodextrin were cooled to 5 °C and frozen in french straws and stored in liquid nitrogen. Frozen straws were then thawed individually at 37 °C for 30 s in the water bath for evaluation. Post-thaw DNA damage was analysed using a fluorescence microscope (Zeiss, Germany). The images 200 randomly chosen nuclei were analysed. Nucleotide DNA extends under electrophoresis to form 'comet tails', and relative intensity of DNA in the tail reflects the frequency of DNA breakage. Thus, tail DNA (tail intensity, %), tail length, tail moment (µm/s) was examined using the Comet Assay III image analysis system (Perceptive Instruments, UK). The analysis was performed with one slide reader. After thawing, there were statistically significant differences among groups for tail intensity and tail moment ($P<0.05$). The both DNA integrity parameters, the highest DNA damage were seen in group 3 (>7 ages). Besides, CLC protected the DNA damage after freezed-thawed of dog semen. In conclusion, the evidence suggests that increased dog age is associated with a DNA damage, CLC were protected to DNA integrity both in group 1 and group 2. This DNA integrity may be positively affect to fertilization in frozen thawed semen. Acknowledgements: This study financed under the TUBITAK, 114O636.

Effect of rose pulp silage on sperm motility stored at 4 °C in yearling rams

S. Gungor[1], M.E. Inanc[1], K.E. Bugdayci[2], M.N. Oguz[2], F. Karakas Oguz[2], O. Uysal[3], O. Ozmen[4] and A. Ata[1]
[1]University of Mehmet Akif Ersoy, Reproduction and Artificial Insemination, Burdur, 15030, Turkey, [2]University of Mehmet Akif Ersoy, Animal Nutrition and Nutritional Diseases, Burdur, 15030, Turkey, [3]University of Ankara, Reproduction and Artificial Insemination, Ankara, 06110, Turkey, [4]University of Mehmet Akif Ersoy, Patology, Burdur, 15030, Turkey; onuysal@veterinary.ankara.edu.tr

Flowers of *Rosa gallica* and *Rosa damascena* species of rose have been used for the production of rose oil in Anatolia since ancient time. Turkey is one of the biggest producers of oil rose and meets in the World. After the rose oil extraction process, waste of oil rose (petal, sepal and pistil) has nutritive value. The nutritional value of this waste (oil rose pulp) have an economic importance especially in seasonal roughage shortage. The objectives of the present study were to determine the effect of rose pulp silage feeding on sperm motility stored at 4 °C in yearling rams. A total 15 rams (average age 8 month and 30 kg average body weights) were randomized assigned to two groups: sugar beet pulp silage (group I, n:8) and rose pulp silage (group II, n:7). Different pens separated the groups according to type of diets. Rams were feeding 2 months with two different experimental diets. At the end of feeding regimen, semen samples collected individually by electroejaculator. Collected semen were diluted and stored at 4 °C for evaluating motility. In every experiment groups, one drop of semen re-warmed up to 37 °C and sperm motility was determined for nine days with phase contrast microscope. At 0. hour motility were 85.0±0.0%; 87.5±2.7% in group I and group II respectively (P>0.05). At 216. hour there were significant differences between the groups. Group II motility (23.3±2.5%) was higher than group I motility (2.3±0.5%) (P<0.05). Especially after sixth day, rose pulp silage group motility was maintained higher than sugar beet pulp silage group (P<0.05). In conclusion, it can be said that rose extract is a good alternative to ram feeding ration. We can use rose extract ration on ram feeeding especially in breeding season. Besides, rose extract positively effect motility on ram semen quality stored at 4 °C. Acknowledgements: This study financed under the Mehmet Akif Ersoy University NAP 16, Project no: 344.

Rose pulp silage could be improve on yearling yearling ram semen spermatological parameters

K.E. Bugdayci[1], S. Gungor[2], M.E. Inanc[2], M.N. Oguz[1], F. Karakas Oguz[1], O. Ozmen[3] and A. Ata[2]
[1]University of Mehmet Akif Ersoy, Animal Nutrition and Nutritional Diseases, Burdur, 15030, Turkey, [2]University of Mehmet Akif Ersoy, Reproduction and Artificial Insemination, Burdur, 15030, Turkey, [3]University of Mehmet Akif Ersoy, Department of Patology, Burdur, 15030, Turkey; enesinanc@hotmail.com

Flowers of *Rosa gallica* and *Rosa damascene* as the king of flowers has been the symbol of love, purity, faith and beauty since the ancient times. Turkey is the most oil rose cultivated areas in the world because of the suitable ecological conditions. After the rose oil extraction process, waste of oil rose (petal, sepal and pistil) has nutritive value. The nutritional value of this waste (oil rose pulp) have an economic importance especially in seasonal roughage shortage. The objectives of the present study were to determine the effect of rose pulp silage feeding on yearling ram semen. A total 15 rams (average age 8 month and 30 kg average body weights) were randomized assigned to two groups: rose pulp silage (group I, n:7) and sugar beet pulp silage (group II, n:8). Different pens separated the groups according to type of diets. Rams were feeding 2 months with two different experimental diets. At the end of feeding regimen, semen samples collected individually by electroejaculator. After collected semen mass activity, motility, semen volume and concentration were evaluated. Besides, testes morphometric measurements (circumference, diameter, height and weight) were recorded for the animals after postmortem. In the examination semen motility 80.8±5.5%; 72.5±11.2%, mass activity 2.8±0.5; 2.2±0.7, concentration $1.009±401×10^6$/ml; $563±280×10^6$/ml in group I and group II respectively. Although there were statistically significant differences between the groups on mass activity, motility and concentration (P<0.05), there were no significant differences between the groups on semen volume and testes measurements (P>0.05). As a result, some spermatological parameters were increased by rose pulp silage feeding on yearling rams. It can be said that rose extract is a well option to ram feeding because of influencing positively motility, mass activity and concentration. Acknowledgements: This study financed under the Mehmet Akif Ersoy University NAP 16, Project no: 344.

Effect of antioxidant on frozen semen quality and fertility rate of native chickens (Leung Hang Kao)

S. Ponchunchoovong and S. Takpukdeevichit
Suranaree University, Institute of Agricultural Technology, School of Animal Production Technology, 111 University avenue, Muang Nakhon Ratchasima 30000, Thailand; samorn@sut.ac.th

Imbalance of reactive oxygen species is considered one of the main triggers of cell damage after cryopreservation, because the spermatozoa antioxidant system is decimated during this process, mainly because the natural antioxidants present in seminal plasma diminish when sperm is diluted in extenders. It has been demonstrated that the addition of antioxidants to the extender improves the quality of frozen sperm. Thus, the aim of the present was to evaluate the status of the antioxidant in cryopreserved native chickens (Leung Hang Kao) sperm. Three antioxidants with three concentrations of vitamin E (1.16, 2.32 and 3.48 mM), cysteine at concentration 0.5, 1.0 and 5.0 mM and glutathione at concentration 0.025, 0.05 and 0.1 mM on the cryopreservation of native chicken (Leung Hang Kao) sperm were investigated. Dimethyl formamide (DMF) at 6% concentration and EK was used to cryopreserve sperm of native chicken (Leung Hang Kao). Semen was collected and pooled from 40 males and mixed with extender, which contained each antioxidant. Fresh semen diluted with 0.9% NaCl and 0.2% glucose was used as a control. Sperm samples were placed into 0.5 ml straws and transferred into liquid nitrogen vapor. Samples were frozen 11 cm above the liquid nitrogen in a Styrofoam box for 12 min and 3 cm above liquid nitrogen for 5 min and then plunged into liquid nitrogen. Sperm were then thawed at 5 °C for 5 min, and motility, viability, fertility, and lipid peroxidation were assessed. It was found that vitamin E at a concentration of 2.32 mM yielded a higher fertility rate of 63.02±5.26% (65% of control) and a lower malondialdehyde (MDA) of 2.68±0.04 nmol/50×106 spz, when compared with the other treatments (P<0.05). The highest motility rate of 68.00±3.46%, progressive motility rate of 39.33±2.91%, and viability rate of 54.00±2.65% were also achieved from vitamin E at a concentration of 2.32 mM. The supplement cysteine at 5.0 mM and 0.1 mM of glutathione had a negative impact on sperm quality.

The effect of supplementation protected fish oil on quality of freeze-thawing sperm of Moghani ram

V. Vahedi, F. Moradi and N. Hedayat-Evrigh
Moghan College of Agriculture and Natural Resources, University of Mohaghegh Ardabili, Department of Animal Science, Moghan College of Agriculture and Natural Resources, University of Mohaghegh Ardabili, Ardabil, Iran, 5158953661, Iran; vahediv@uma.ac.ir

The aim of this study was to evaluate the effects of feeding microencapsulated fish oil on post-thawed ram sperm quality. Eight Moghani rams with an average age of 3-4 years and weight of 60-70 kg were divided to 2 groups (n=4) and fed with palmitic acid (control) and microencapsulated fish oil diet. Both lipid supplements were added at 2.5% of the total diet as fed. Semen was collected twice a week. Semen samples were extended with citrate-yolk-based cryoprotective diluent, then cooled to 5 °C and stored in liquid nitrogen. After freezing-thawing sperm motility parameters, viability and plasma membrane integrity were determined using a Computer-aided sperm analysis (CASA) system, eosin-nigrosin staining and hypo-osmotic swelling test, respectively. Data were analyzed by SAS software using the GLM procedure. Fish oil diet significantly increased (P<0.05) the motility, movement parameters and viability compared to the control group. However, there was no difference between diets on plasma membrane integrity of sperm (P>0.05). Overall, dietary supplementation of rams with protected fish oil has positive effects on the quality of liquid stored semen.

The effect of Fennel alcoholic extract on the quality of frozen-thawed Ghezel rams semen

A. Karimi[1] and R. Hosseini[2]
[1]*University of Tabriz, Ahar Faculty of Agriculture and Natural Resources, Tabriz, 5166616471, Iran,* [2]*Orum Gohar Daneh, Miandoab, 59716-56711, Iran; pekarimi@tabrizu.ac.ir*

Sperm cryopreservation is an important procedure in animal improvement. Because of Oxidative stress there are losses in sperm freezing-thawing. Oxidative stress leads to low viability and consequently low fertility. It has been attempted using of some natural or industrial antioxidants as cryoprotectant. Fennel (Foeniculum vulgare) has antioxidant properties due to a-Pinene, Camphene, phenolic, flavonoids compounds. The aim of this study was to investigate the protective effect of different levels of fennel alcoholic extract in ghezel ram sperm cryopservation. In this study, semen samples were collected from four ghezel rams twice a week using artificial vagina and ejaculates were pooled in order to eliminate the individual effects of rams. Different levels of fennel ethanol extract (0, 2, 4, 6, 8, 10, 12 and 14 ml per dl diluent solution) were added to egg yolk-citrate diluents and then packed in 0.25 ml straws, cooled to 4 °C and then exposed to the liquid nitrogen vapor to be frozen then stored in the tank containing liquid nitrogen until evaluation. Motility, viability, membrane integrity and the lipid peroxidation parameters of the samples were evaluated after thawing. Addition of 6 ml extract in dl diluent increase motility After CASA (computer assisted sperm analysis) versus others (P<0.05). Eosin-nigrosin staining declared ascending viability by increased concentration of extract in diluent except in 12 and 14 experimental groups that had more dead sperm(P<0.05). Sperm of experimental groups 4 and 6 ml/dl diluent showed significantly more plasma membrane integrity than others after HOST (hypo-osmotic swelling test). MDA (Malondialdehyde) concentration assay was carried out In order to determine the lipid peroxidation index. In this experiment the lowest MDA concentration was seen after addition of 4 ml extract in diluent, but there was no significant difference with control group(P>0.05). Addition of 14 ml extract per dl diluent lead to higher MDA concentration than control group (P<0.05) that indicating poisonous effects of this concentration. It seems, addition of 6 ml Fennel ethanol extract in egg yolk-citrate diluents may have beneficial effects for semen cryopreservation.

Impact of FCS-free culture media on cloned bovine embryo development

M. Ivask[1,2,3], M. Nõmm[3], P. Pärn[1,3], Ü. Jaakma[1,3] and S. Kõks[1,2,3]
[1]*The Competence Centre on Health Technologies, Tiigi 61b, 50410 Tartu, Estonia,* [2]*University of Tartu, Institute of Biomedicine and Translational Medicine, Ravila 19, 50411 Tartu, Estonia,* [3]*Estonian University of Life Sciences, Institute of Veterinary Medicine and Animal Sciences, Kreutzwaldi 62, 51014 Tartu, Estonia; marilin.ivask@emu.ee*

Somatic cell nuclear transfer (SCNT) has applications in agriculture and biomedicine. SCNT is currently the most used technique to produce genetically modified cattle, although the efficiency because of developmental abnormalities, like large offspring syndrome (LOS), remains low. The fetal calf serum (FCS) is usually added to the culture media to provide growth factors and energy sources. FCS-free culture systems are therefore often preferred due to the lower risk of contamination and prevention of the development of LOS. The aim of this study was to establish whether elimination of FCS from the bovine SCNT system has an effect on blastocyst rates. Three different *in vitro* cultivation (IVC) media systems were used for cloned embryos: (1) synthetic oviductal fluid with amino acids, sodium citrate and myo-inositol (SOFaaci) supplemented with FCS; (2) SOFaaci with bovine serum albumin (BSA); (3) commercial FCS-free media from IVF Bioscience. Five SCNT trials with each culture media variations were done. The number of reconstructed embryos put into IVC with each setting was: (1) 190, (2) 151, (3) 173 respectively. The number of blastocysts and blastocyst rate varied quite a lot: (1) 8 (4.2%), (2) 22 (14.6%), (3) 39 (22.5%) respectively. However, the differences were not statistically significant (P>0.05, 1-way ANOVA with Tukey's posttest) probably due to the small number of trials. IVF Bioscience commercial media had a tendency for better result with less variation between the trials if compared to self-made FCS-free media. These results indicate that it is possible to have a FCS-free bovine SCNT system, yielding even in a higher blastocyst rate. However, more experiments and transfers of cloned embryos produced in FCS-free media are needed to evaluate the impact on development of cloned calves. This study was supported by Enterprise Estonia grant EU30020 and Horizon 2020 Project SEARMET 692299.

Low molecular weight metabolites as possible new tool for selecting bovine *in vitro* produced embryos

M. Nõmm[1], R. Porosk[2], P. Pärn[1,3], U. Soomets[2], Ü. Jaakma[1,3], S. Kõks[1,2,3] and K. Kilk[2]

[1]Estonian University of Life Sciences, Fr. R. Kreutzwaldi 1, Tartu 51014, Estonia, [2]University of Tartu, Ülikooli 18, 50090 TARTU, Estonia, [3]Competence Centre on Health Technologies, Tiigi 61b, Tartu 50410, Estonia; monika.nomm@emu.ee

Selecting high quality embryos for transfer has been the most difficult task when producing embryos *in vitro*. To date the most used non-invasive method is based on visual observation. Developing a non-invasive method for embryo assessment is essential to have a profitable *in vitro* embryo production (IVP) and embryo transfer system. Molecular characterization of embryo growth media has been proposed as an complementary method to visual assessment of embryo morphology. In this study we are demonstrating a novel method, allowing sample collection at different embryo development stages, without compromising embryo quality, to determine potential viability markers for bovine IVP. Single bovine embryos were cultured in 60 μl SOF+0.4% BSA droplets under mineral oil. Twenty μl of culture media was removed at day 2, 5 and 8 post-fertilization. A total of 58 samples were analyzed using liquid chromatography-mass spectrometry (Q-Trap 3200), followed by principal component analysis. Our results indicate that there are significant differences (P<0.00001) in concentrations for proline (m/z=116), inositol (m/z of sodium adduct = 203) and citrate (m/z of sodium adduct = 215) also in the amino acid group of leucine and isoleucine (m/z=132), phenylalanine (m/z=165) and arginine (m/z=211) between the normally developed and retarded embryo culture media. Platelet activating factor (m/z=524) (PAF) was roughly 3 fold increased in day 5 to day 8 embryo culture media. Unfortunately the increase of PAF was not statistically significant between normally developing and retarded embryos. These results demonstrate that it is possible to remove culture media samples from droplets and not significantly affect embryo development. Applying this method for embryo selection provides a possibility to identify well-developing embryos and provides an opportunity for improving the herds genetic value. This study was supported by Project 3.2.0701.12-0036 of Archimedes Foundation, Enterprise Estonia grant EU30020, Institutional research funding IUT 8-1 and Horizon 2020 Project SEARMET 692299.

Effect of male and cooled storage time in the sperm motility rate and kinetics in Boer male goats

S. Sadeghi[1], R. Del Gallego[1], I. Pérez-Baena[2], C. Peris[2], S. Vázquez[3], E.A. Gómez[3] and M.A. Silvestre[1]

[1]Universitat de València, Biología Celular, Biología Funcional y Antropología Física, Burjassot, 46100, Spain, [2]Universitat Politècnica de València, Instituto de Ciencia y Tecnología Animal, Valencia, 46022, Spain, [3]Instituto Valenciano de Investigaciones Agrarias, CITA, Segorbe, 12400, Spain; miguel.silvestre@uv.es

The introduction of a meat breed, as Boer, and its crossbreeding with a native dairy breed with lower growth rate, could be interesting to improve the growth of the kids and the profitability. However, it is not very much known the reproductive adaptation of this breed to the Spanish latitudes (Valencia; 39°28'long and 0°22'lat). In this work, we evaluated sperm characteristics of 3 Boer males and 1 Murciano-Granadina male and the effect of time of preservation at 4 °C for 48 h. Ejaculates were collected by artificial vagina method during breeding season (October-November) and diluted in glucose-skimmed milk extender. Total motility, progressiveness and kinetics (VCL, VAP, LIN and ALH) of the spermatozoa were studied. Data were analysed using a GLM including these factors: replicate, male, preservation time and interaction between male and preservation time (SPSS Statistics). Replicate affected all parameters. Concerning total motility, all factors affected it except for the interaction (P=0.81). Motility was maintained without a significant decrease during first 24 h (75.2 and 62.4% for 0 and 24 h), however it dropped at 48 h (45.0%). Preservation time has also significant influence on progressive motility, even the first 4 h (60.8, 41.7, 40.2 y 25.6 for 0, 4, 24 y 48 h respectively). VCL was only affected by male individual effect. The effect of male and the storage time, but not their interaction, influenced both the VAP and LIN parameters. Neither male nor storage time affected ALH. The decreasing of temperature of storage affected mainly to progressive motility rather than total motility rate. Interaction between male and refrigerated storage were not detected. A more number of males is needed to compare breed effect. This work was supported by INIA (RTA2013-00107-C03-03).

Different patterns of sperm motility of fluorescent sperm subpopulations in bull

J.L. Yániz[1], C. Soler[2] and P. Santolaria[1]
[1]University of Zaragoza, Department of Animal Production and Food Sciences, Ctra. Cuarte s/n, 22071, Spain, [2]University of Valencia, Department of Cellular Biology, Functional Biology and Physical Anthropology, Burjassot, 46100, Spain; jyaniz@unizar.es

This study was designed to study differences in sperm motility between the fluorescence sperm subpopulations after staining with the ISAS®3Fun kit in bull. Cryopreserved semen samples from 10 Holstein bulls were used in the study. Samples were labeled with the ISAS®3Fun kit (Proiser, Paterna, Spain) developed by the TECNOGAM research group. Briefly, 40 µl sample aliquot was pipetted into 0,6 ml Eppendorf centrifuge tubes, 4 µl of the fluorochrome combination provided by the kit was added and samples were incubated for 5 min at 37 °C in a water bath. Sample aliquots (4 µl) were directly placed in a prewarmed slide to evaluate sperm motility of fluorescent sperm subpopulations. The staining allowed a clear discrimination of sperm subpopulations based on membrane and acrosomal integrity, preserving sperm motility. It was also possible to observe a sperm subpopulation with increased fluorescence intensity in head and flagellum. All spermatozoa with damaged plasma membrane were static. Normal spermatozoa, those with intact plasma membrane and acrosome, showed a higher proportion of motile sperm than those with damaged acrosome or increased fluorescence intensity. Within motile spermatozoa, those with the highest fluorescence intensity showed signs of hyperactivation, as they were faster and had more vigorous movement than normal spermatozoa, whereas those with damaged acrosome were slower and exhibit weak movement. The new method opens the possibility of evaluating the different changes associated to sperm capacitation simultaneously cell by cell. The integrative method described in this paper allows us to establish a single threshold combining different parameters of semen quality. This threshold will determine the proportion of spermatozoa with the potential ability to reach and fertilize the oocyte. Supported by the Spanish MINECO (grant AGL2014-52775-P), and the DGA-FSE-A40.

A new multi-parametric fluorescent test of sperm quality in bull

J.L. Yániz[1], C. Soler[2] and P. Santolaria[1]
[1]University of Zaragoza, Department of Animal Production and Food Sciences, Ctra. Cuarte s/n, 22071, Spain, [2]University of Valencia, Department of Cellular Biology, Functional Biology and Physical Anthropology, Burjassot, 46100, Spain; jyaniz@unizar.es

There is a need to develop more seminal quality analysis with the potential to improve fertility predictions. In this work, we present a new multi-parametric fluorescent test able to discriminate different sperm subpopulations based on their labeling pattern and motility characteristics. Cryopreserved semen samples from 10 Holstein bulls were used in the study. Samples were labeled with the ISAS®3Fun kit (Proiser, Paterna, Spain) developed by the TECNOGAM research group. Briefly, 40 µl sample aliquot was pipetted into 0,6 ml Eppendorf centrifuge tubes, 4 µl of the fluorochrome combination provided by the kit was added and samples were incubated for 5 min at 37 °C in a water bath. Labelled samples (4 µl) were placed on a glass slide, covered and assessed with fluorescence microscopy. The staining allowed a clear discrimination of sperm subpopulations based on membrane and acrosomal integrity. A sperm subpopulation with increased fluorescence intensity in head and flagellum was also observed. It was possible to observe the evolution from a living sperm with intact acrosome to another with total acrosomal loss. Sperm at intermediate stages showed increased acrosome and flagellar fluorescence intensity, and progressive fading of the acrosome, maintaining high flagellar fluorescence intensity. It was also possible to follow the evolution from a live sperm with intact structures to a dead sperm without acrosome. At intermediate stages, a progressive loss of acrosome integrity may be appreciated. We concluded that the ISAS®3Fun is an integrated method that represents an advance in sperm quality analysis with the potential to improve fertility predictions. Supported by the Spanish MINECO (grant AGL2014-52775-P), and the DGA-FSE-A40.

Engineering of Gallinacean oviduct cells using electroporation and lipofection

M. Debowska, M. Bednarczyk and K. Stadnicka

UTP University of Science and Technology, Animal Biochemistry and Biotechnology, 28 Mazowiecka Street, 85-804, Poland; katarzyna.stadnicka@utp.edu.pl

Highly secretive avian oviduct cells carry a premise to be efficient bioreactors of therapeutic proteins upon delivery of human exogenes into the engineered germ cells and expressing them under the promoters of egg white proteins. Apart from the chicken oviduct cells, the cells derived from a quail are considered a feasible tool to prove concepts for engineering techniques. In favour of a quail are its effortless handling, stable genetics and easy cultivation of primary oviduct cells. Here we tested different non viral methods to deliver plazmid pL-OG-OVAIFNEn-Egfp (Institute of Biotechnology and Antibiotics, Warsaw) containing human $\lambda IFN2a$ and a reporter *GFP*, into the cultivated quail oviduct epithelial cells (QOEC). The cells were derived from oviducts of laying quails (n=12, 7-10 wks) and cultivated in DMEM/F-12 with addition of epithelia promoting factors and depletion of serum 3 d post seeding. The exogene was delivered to the freshly isolated QOEC or to the QOEC at 70-80% of confluence, using: (1) lipofection (Lipofectamine 2000); (2) electroporation (Multiporator); (3) nucleofection (Lonza); (4) combination of both methods at $1.0-3.0\times10^5$ cells and 1 or 2 µg of pL-OG-OVAIFNEn-Egfp plasmid. The lipofection was tested at 0.5-5 µl ratios of Lipofectamine®. The electroporation was set at 140 V and 1 or 2 pulses lasting for 75 µs. The average efficiency of lipofection of freshly isolated QOEC measured by means of GFP expression pointed for 3.1%, electroporation for 4.25% and the combination of both methods for 2.07%. Nucleofection resulted in efficiency <1% with lost adherence ability. Electroporation shows to be the promising method for further optimization, but combining both lipofection and electroporation is worth developing in terms of similar efficiency and cell survival rate. One must take into account that various gene constructs may operate differently under diverse conditions of delivery. Research was funded by National Science Centre (2011/03/N/NZ9/03814) and The National Centre for Research and Development (PBS 3/A8/30/2015).

Is a world without meat realistic?

J.F. Hocquette[1], P. Mollier[2] and J.L. Peyraud[3]

[1]Inra, UMR Herbivores, 63122 Theix, France, [2]Inra, Communication, 75000 Paris, France, [3]Inra, UMR Pegase, 35590 Saint-Gilles, France; jean-francois.hocquette@inra.fr

Citizens have growing concerns about livestock and meat consumption. Answers cannot be simplistic solutions such as no meat, to avoid killing animals, improve human health and reduce harmful emission to the environment. First, farm animals are today well adapted to living with humans, and cannot simply return to the wild world without suffering. Second, the relationship between food and health has to be considered for the overall eating pattern, not for one part of the diet. Eating some products in excess as meat will cause health disorders. Nutritional recommendations will be met by consuming in a reasonable amount a variety of foods including lean meats because they provide high-quality proteins, micronutrients (iron, vitamins, etc). Eating meat from herbivores is also an efficient way to indirectly make the most of plants and grass. This is of particular importance in marginal areas, where climate and soil do not allow crop production. Therefore, conversion of plant proteins to animal proteins should be calculated for non-edible proteins (and not for total proteins), since 80% of food consumed by livestock is not consumable by humans. By this way, cattle will appear much more efficient in converting natural resources into edible products. Other claims should be carefully weighted: livestock do occupy 70-75% of agricultural land, but mostly non-arable lands; to produce 1 kg of meat, it is claimed that 15,000 liters are required, but only the blue water (river, ground water) is in competition with human needs, therefore 1 kg of beef requires less than 700 liters of blue water. In addition, carbon storage by grasslands corresponds to 30-80% of methane emissions by ruminants. Livestock manure in Europe represents an important supply of nitrogen and phosphorus to the soils. They are also a source of organic matter essential for soil fertility. Finally, livestock and meat consumption are closely associated with our cultural heritage (food gastronomy). Rearing livestock also offers jobs in rural areas and poor countries. In conclusion, livestock and meat production are faced with a range of sustainability challenges, including changing consumer perceptions and improving rearing practices, but no more meat is not realistic.

Genetic enhancement of meat fatty acid profiles in beef cattle

R. Roehe[1], M. Nath[2], C.-A. Duthie[1], D.W. Ross[1], L.W. Coleman[1], M.D. Auffret[1] and R.J. Dewhurst[1]
[1]Future Farming Systems, Roslin Institute Building, EH259RG Edinburgh, United Kingdom, [2]Biomathematics & Statistics Scotland, King's Buildings, EH93JZ, Edinburgh, United Kingdom; marc.auffret@sruc.ac.uk

Beef provides many essential nutrients associated with human health such as high quality proteins, vitamins, minerals, bioactive substances and antioxidants. However, beef also contains high amounts of fatty acids (FA) to be saturated contributing to obesity in humans, a main risk factor for cardiovascular disease. This adverse effect is expected to be alleviated by an increase in polyunsaturated FA (PUFA) such as omega-3 FA and conjugated linoleic acid (CLA) in beef, which showed evidence of a wide range of benefits to human health. Data of 21 FA profiles from 654 beef cattle of different breeds (Aberdeen Angus, Charolais, Limousin, Luing) were analysed using a statistical model including breed, sex, diet, and farm-year as fixed effects, carcass weight as covariable, and slaughter day and the animal genetic effect (including pedigree) as random effects. The PUFA in beef showed large variation among animals, i.e. linolenic acid and CLA ranged from 6 to 89 and 2 to 91 mg/100 g muscle, respectively. There were significant differences in PUFA among breeds, particularly between Aberdeen Angus and Charolais (LS-means: 23 ± 0.8 vs 14 ± 0.9 mg/100 g linolenic acid or 19 ± 0.9 vs 12 ± 1.1 mg/100 g CLA, respectively). Estimates of heritabilities for important PUFA were 0.30 ± 0.15, 0.46 ± 0.17, 0.74 ± 0.18 and 0.28 ± 0.15 for linolenic, CLA, eicosapentaenoic and docosahexaenoic acids, respectively. Considering the large variation among animals within FA and their moderate to high heritabilities, we conclude that selection for substantial improvement of omega-3 FA and CLA is achievable. The availability of near infrared reflectance spectroscopy to accurately determine FA would facilitate efficient data recording in the abattoir needed for genetic evaluation. Alternatively, rumen samples could be taken in the abattoir to determine the abundance of ruminal microbial genes associated with reduced biohydrogenation of diet PUFA and thus enhanced omega-3 FA and CLA in meat. The present study indirectly suggests that microbial biohydrogenation has a host genetic component.

Fatty acids profiles of food resources in the northern rivers determine the quality of salmonids

S.A. Murzina, Z.A. Nefedova, S.N. Pekkoeva and N.N. Nemova
Institute of Biology, Karelian Research Centre of the Russian Academy of Sciences, Pushkinskaya street 11, 185910 Petrozavodsk, Russian Federation; murzina.svetlana@gmail.com

Condition of food resources is one of the key factors influencing the survival rates and recruitment of fish generations. Nutritive value of food largely depends on the specific characteristics and ratio of individual lipid classes, including fatty acids (FA9, which contribute greatly to the productivity of northern ecosystems. The aim of this study was to investigate the FA composition of rheophilic benthic invertebrate communities on which juvenile of the Atlantic salmon and brown trout forage in rivers with different hydrological characteristics in the White Sea, Onego and Ladoga Lake catchments. The FA were analyzed by gas-liquid chromatography. The research was carried out using the facilities of the Shared Equipment Centre IB KarRC RAS. The structure of macrozoobenthos of the studied rivers was considerably stream-specific – dominant species matching the hydrological, trophic and other conditions. Zoobenthos in some watercourses turned out to be quite specific in terms of essential FA levels and ratios. Zoobenthos from all the studied habitats contained the physiologically active and essential 20:5n-3 derived from food. Its highest levels were found in organisms from rivers inhabited by salmon, in line with the elevated levels of its metabolic precursor, 18:3n-3 FA, and a similar 20:5n-3/18:3n-3 ratio. The elevated 20:5n-3 FA content in zoobenthos from these watercourses points to its presence in phytoplankton, mainly diatoms. Eicosapentaenoic, 20:5n-3 acid influences zooplankton fecundity and growth, as well as the development of the vision system and brain tissue in fish. Where this acid is deficient in an organism, abnormal behavioral reactions will be observed. We found that the ratio of essential 18:3n-3/18:2n-6 FA ratios in zoobenthos from salmon rivers (the White Sea basin and Onega Lake basin) were higher than in zoobenthos from trout rivers. Also, zoobenthos from salmon rivers had higher SFA/PUFA ratios than that from trout rivers. The research was supported by the Russian Science Foundation the project no. 14-24-00102.

Dairy ingredients and human health – application in lifestyle nutrition including infant formula

M. Fenelon

Teagasc Food Research Centre, Moorepark, Fermoy, Co. Cork, Ireland; mark.fenelon@teagasc.ie

Dairy ingredients provide a unique platform, through their innate nutrient composition, on which lifestyle foods and beverages can be formulated. Dairy protein ingredients are commonly used as the nutritional base for infant formula, medical/therapeutic beverages and many currently trending 'high protein' foods. In recent years the development of separation technology, e.g. micro-, ultra-filtration and ion exchange technologies, has extended the range of ingredients with different health benefits that can be developed. As an example, a filtration process has been developed at Teagasc to alter the composition of bovine skim milk to produce a whey protein-dominant ingredient with a protein profile closer to human milk for use in infant formula. The process generates a permeate from a cold (<8 °C) microfiltration stream which contains a casein:whey protein ratio of ~35:65 with no αs-casein present. In another study, heat-induced denaturation of the protein β-lactoglobulin was used to form aggregates for increased stability during thermal treatment, thus improving processability of Dairy based beverages. Controlled aggregation has many applications for generating foods for health, including microparticulated whey proteins combined with inulin to produce a fat mimetic with improved sensory characteristics and lower calorific value. In another study, the Dairy proteins, alpha- and β- casein were shown to have different ability to alter viscosity and glucose release of a protein-starch system *in vitro*. This functionality provides an important mechanism for modulation of glucose in complex food systems, which is an important attribute with potential health benefits. Other components in milk currently under investigation include the milk fat globule membrane (MFGM) which contains glycoproteins and oligosaccharides and has potential as an ingredient for infant and adult nutrition. Due to its excellent nutrient profile, coupled with advancement of *in vitro* techniques, biological assays and separation / drying technology, milk will continue to provide many opportunities for development of innovative ingredients with potential health benefits.

Total antioxidant activity of milk from different dairy species

G. Niero, M. De Marchi, M. Penasa and M. Cassandro

University of Padova, Department of Agronomy, Food, Natural resources, Animals and Environment, Viale dell'Università 16, 35020 Legnaro (PD), Italy; giovanni.niero@phd.unipd.it

Food antioxidants have been associated with positive effects on human health, leading to the definition of nutraceuticals. Several studies demonstrated that milk contains a variety of antioxidant molecules such as thiols, tocopherols, carotenoids, ascorbic acid, phenols, lactoferrin as well as biopeptides derived from caseins and whey proteins showing antioxidant properties. Milk total antioxidant activity (TAA) is defined as the sum of each antioxidant contribution, related to the aforementioned molecules. The characterisation of this new phenotypic trait in different dairy species would represent an important step for the valorisation of milk. Therefore, the aim of the present study was to characterise TAA of cow, buffalo and sheep milk. Individual milk samples from 300 Holstein Friesian cows, 110 Italian Mediterranean buffalos and 85 Comisana sheep were collected, added with preservative and stored at 4 °C. Milk TAA was measured with a near infrared spectrophotometric ABTS based method and expressed as μmol/l. Trolox equivalents (TE), and milk quality traits were predicted through mid-infrared spectroscopy. The greatest TAA was measured in sheep milk (7.33 μmol/l TE), followed by buffalo (7.26 μmol/l TE) and cow milk (6.83 μmol/l TE). The coefficient of variation of TAA was greater in cow (17.91%) and sheep (14.95%), and lower in buffalo milk (9.38%). Moderate to low Pearson correlations of TAA with somatic cell score in sheep (0.35; $P<0.01$) and cow milk (0.19; $P<0.05$) were assessed. Regarding buffalo milk, a low positive correlation was observed between TAA and casein content (0.15; $P<0.05$), and a low negative correlation was estimated between TAA and fat content (-0.20; $P<0.05$). Further analyses are being carried out in order to assess the impact of sources of variation such as stage of lactation and parity on the variability of milk TAA.

Mutations affecting milk quality and associations with production traits in Holstein Friesian cattle

M.P. Mullen[1], L. Ratcliffe[1], J. McClure[2], F. Kearney[2] and M. McClure[2]
[1]*Bioscience Research Institute, Athlone Institute of Technology, Athlone, Co. Westmeath, Ireland, [2]Irish Cattle Breeding Federation, Highfield road, Bandon, Co. Cork, Ireland; mmullen@ait.ie*

DNA mutations that affect milk composition are of particular interest not only to producers, i.e. the livestock breeding industry, but also to consumers. In dairy cattle, κ-casein and β-casein genes harbour mutations which have been associated with positive effects on cheese production and human health, respectively. The objective of this study was to estimate the effects of polymorphisms in κ-casein and β-casein on milk, fertility, carcass and health traits (n=16) in dairy cows. Genotypes and phenotypes on 10,707 dairy cows were obtained through the Irish Cattle Breeding Federation (ICBF). Phenotypes were expressed as predicted transmitting abilities (PTAs). Only animals with an adjusted reliability of >10% were included in the analysis retaining n=6,876, 1,198, 264, 4,566, 8,564, 152, 2,280, 3,194, 518, 360, 1,374, 5,747 cows for milk traits (n=5), calving interval, survival, gestation length, calf mortality, maternal calving difficulty, carcass weight, carcass conformation, carcass fat, cull cow weight, and somatic cell score, respectively. The association between each SNP and deregressed PTA was analysed in ASREML using a weighted mixed animal model. No evidence (P>0.05) was found for an association between a validated κ-casein variant (342T>C) and any of the milk, fertility, carcass or health traits analysed in this population of dairy cows. The β-casein A2 variant (A allele) was associated with: increased milk protein percentage (0.007±0.0018, P<0.0001); increased milk fat percentage (0.0074±0.0038, P≤0.05); increased milk fat (0.51±0.24 kg, P<0.05); protein yield (0.43±0.20 kg, P<0.05); decreased carcass fat (-0.17±0.06 kg, P<0.05) and increased somatic cell score (0.18±0.005, P<0.001). No association was identified between the β-casein A2 variant and any of the fertility or other carcass traits analysed. These results suggest the potential for increasing the frequency of desirable alleles in the national herd without significant negative effects in relation to the milk, fertility and carcass traits tested.

Fatty acid composition in milk of local versus main high producing cattle breeds

O. Vangen[1], B.A. Åby[1], I. Olesen[2] and A. Meås[3]
[1]*Norwegian University of Life Sciences, Animal and Aquacultural Science, P.O. Box 5003, 1432 Ås, Norway, [2]NOFIMA, Aquacultural Genetics, P.O. Box 210, 1431 Ås, Norway, [3]STN-Breeding Organisation, Svorkbygda, 7320, Norway; odd.vangen@nmbu.no*

One reason to maintain local cattle breeds is the well documented difference in kasein variants in milk, with a higher proportion of kappa kasein B. There has been some earlier reports on fatty acid (FA) composition differences, however few have been scientifically proven. The present study deals with fatty acid composition in milk from the local breed STN (Sided Troender and Nordland Cattle) versus NRF (Norwegian Red), the predominant dairy breed in Norway, now exported to more than 20 countries as well. Based on 75 cows in herds with both breeds in seven farms, milk samples were collected during the winter freeding period, from morning and evening milking. Animals were chosn to balance lactation number and stage in lacttation between breeds. Milk samples were frozen and later analysed for individual FA or sums of mono-unsaturated, polyunsaturated and saturated fatty acids (23 traits altogether). The statistical model used included breed, herd, lactation number and days from calving. Even if commercial herds have a relatively comparable feeding practice these days, the diets were not fully comparable over herds, but all diets included consentrate and grass silage. Therefore, the effect of herd in the mode will eliminate the possible the differences in diets. Interaction between breed and lactation number was found insignificant and not included in the final model. The model descibed from 26 to 85% of he variation in the FA, where herd and days from calving were the most important effects. Breed differences were observed in 10 traits. The LS-means show a higher proportion of saturated FA in NRF, and a higher proportion of mono-unsaturated FA in STN. Especially, the unsaturated FA C18:1,t11 level was higher in SN. The results show that, based on present knowledge on healthy milk for humans, that the STN breed has a more favourable FA composition.

Fresh forage dietary inclusion effects on dairy cows milk nutritional properties

F. Righi[1], A. Quarantelli[1], M. Renzi[1], A. Revello Chion[2], E. Tabacco[3], G. Battelli[4], D. Giaccone[2] and G. Borreani[3]
[1]*University of Parma, Department of Veterinary Science, via del Taglio 10, 43126 Parma, Italy,* [2]*Regional Breeding Association of Piedmont, via Livorno, 60 c/o -Enviroment Park, 10100 Torino, Italy,* [3]*University of Torino, Department of Agricultural, Forest and Food Sciences, Via Verdi, 8, 10124 Torino, Italy,* [4]*CNR, ISPA, Via Celoria 2, 20133 Milano, Italy; federico.righi@unipr.it*

Even if fresh forage can improve nutritional properties of milk, its use in Italy is still limited in high producing dairy cows diets. The aim of this study was to evaluate the effects of the inclusion of fresh forage in total mixed ration (TMR) of high producing dairy cows on milk nutritional properties, with special focus on fatty acids and vitamin E levels. Two trials were performed, comparing an Italian TMR diet including high levels of maize silage and concentrates (TMR) with one where TMR was inclusive of fresh forage (TMR+F). Fresh forage was utilized in an advanced stage of maturity, and in an early stage of maturity in Trial 1 and 2 respectively. In both trials, during the experimental periods the dry matter intake (DMI) and milk yield were measured, and TMR, fresh forage and milk were sampled. Chemical composition of feeds and milk quality were determined. Statistical analysis was performed using the software SPSS16.0 through the one-way ANOVA procedure ($P<0.05$). In both trials, the dietary inclusion of fresh forage did not affect milk production, and milk fat content. Fresh forage inclusion significantly affected the milk fatty acid (FA) composition, increasing the vaccenic (Trial 1, 0.69 vs 1.03; Trial 2, 0.79 vs 1.32%) and linolenic acids (Trial 1, 0.29 vs 0.34; Trial 2, 0.31 vs 0.39%), conjugated linoleic acid (CLA: Trial 1, 0.38 vs 0.50; Trial 2, 0.37 vs 0.48%), and vitamin E (Trial 1: 1.03 vs 1.25 mg/l, $P<0.10$; Trial 2: 3.35 vs 5.17 mg/l, $P<0.05$) content. Palmitic acid was decreased only in Trial 2 (31.08 vs 27.96%). The inclusion of fresh forage in cow diet can improve the fatty acid composition of milk fat and increase vitamin E levels without affecting the milk yield and composition.

Fatty acid profile of milk from cows fed diets based on fresh-cut white clover

S. Stergiadis[1,2], D.N. Hynes[1], A.L. Thomson[2], K.E. Kliem[2], C.G.B. Berlitz[2,3], M. Günal[1,4] and T. Yan[1]
[1]*Agri-Food and Biosciences Institute, Large Park, BT26 6DR, Hillsborough, United Kingdom,* [2]*University of Reading, School of Agriculture, Policy and Development, Earley Gate, P.O. Box 237, RG6 6AR, Reading, United Kingdom,* [3]*Federal University of Rio Grande do Sul, Av Bento Gonçalves 7712, RS, 91540-000, Porto Alegre, Brazil,* [4]*Süleyman Demirel University, 32200, Isparta, Turkey; s.stergiadis@reading.ac.uk*

An increased interest in white clover (WC) has been demonstrated for pasture-based dairy systems, in response to environmental, economic, sustainability and resilience concerns arising from the intensive production of monoculture grass pastures. WC has been shown to increase milk yield, while sustaining milk fat and protein content, and reduce reliance on imported protein and nitrogen fertiliser. WC-based diets have been found to improve milk fatty acid (FA) profile, but the mechanisms behind this effect have not been investigated in fresh forages. The current study aimed to assess the effect of increasing WC in cow diets, by substituting fresh grass, on milk FA profile and recovery rates of linoleic acid (LA) and a-linolenic acid (ALNA) from feed to milk. Nine Holstein × Swedish Red cows were used in a changeover design using three iso-nitrogenous and iso-energetic diets (0%, 20%, 40% fresh-cut WC on offered DM) at a constant fresh-cut forage:concentrate ratio of 60:40. An adaptation period of 19-d was followed by a 6-day measurement period in individual tie-stalls, when total feed intakes, milk outputs, and the chemical and FA composition of feed and milk were assessed. Data were analysed by ANOVA linear mixed-effect models, using two fixed (diet, experimental period) and two random (cow, day) factors. Higher WC contribution in cow diets significantly increased milk concentrations of PUFA (n-3, omega-6, LA and ALNA) and the recovery rates of LA and ALNA from feed to milk. The latter indicates a lower rumen biohydrogenation of LA and ALNA when WC is fed. Including WC in grazing swards may improve the concentrations of nutritionally desirable n-3 PUFA in milk, without influencing productivity or milk fat and protein composition. This can be considered an additional benefit to the overall sustainability of WC inclusion in grazing swards.

Analysis of intensively fattened Romanov and Dorper lambs meat chemical composition

D. Kairisa and D. Barzdina
Institute of Agrobiotchnology, Latvia Univeersity of Agriculture, Liela street 2, 3001, Latvia; dace.barzdina@llu.lv

In Latvian sheep industry development contributes to the growing consumer interest in safe, healthy food consumption. The increased interest is offered on the local market of the meat content, not only by protein, fat, minerals, but also about cholesterol and unsaturated fatty acid content. The aim of our study was: to explain in Latvia breeding Romanov and Dorper breeds intensively fattened lamb meat chemical composition. Study on intensively fattened lamb chemical composition carried from 2013 to 2016. In the study used pure-bred Romanov and Romanov and Dorper crossing lambs. During the period of the fattening lambs kept together by 3 to 4 in group and feeding unlimited with the fodder and hay. Lambs are slaughtered in certified slaughterhouse, analysis of meat chemical composition in Institute of Food safety, Animal Health and Environment 'BIOR'. Analyzed 18 samples of Quadriceps muscle. In the meat certain the following chemical composition: dry matter, protein, fat, pH, cholesterol and unsaturated fatty acids. Slaughtered purebred Romanov lambs were significantly older (198.9 ± 3.91 days) for the crossbred lamb's, difference 36.3 days ($P \leq 0.001$). Meat chemical analysis results of protein, fat and ash content between the groups did not differ significantly. Cholesterol content in 100 g Romanov purebred lamb meat was 88.6 ± 11.47 mg. Higher C18:1 n9c content ($43.5 \pm 0.71\%$) was in purebred Romanov lamb meat, but the crossbred lambs by 0.6% lower. Crossbred lamb meat samples obtained higher content of C18:2 n6c (5.1%), C16:1 n9c (2.8%) and C18: 1 trans 3.6%. A significant difference obtained in C15: 1 content Romanov purebred lambs meat $0.4 \pm 0.06\%$, but in the crossbred lamb meat $0.8 \pm 0.18\%$ ($P \leq 0.05$).

The effect of supplement $CoSO_4$ feeding of honey bee families on royal jelly composition

R. Balkanska and M. Ignatova
Institute of Animal Science, Spirka pochivka 1, Kostinbrod, 2232, Bulgaria; r.balkanska@gmail.com

Honey bees (*Apis mellifera L.*) require proteins, carbohydrates, lipids, vitamins, minerals, water and these nutrients must be in the diet in a definite qualitative and quantitative ratio for optimum nutrition. All of these components bees receive from the nectar and the bee pollen. The aim of the present study is to investigate the influence of $CoSO_4$, as a supplement in bee feeding on the main components of royal jelly (RJ). RJ is a secretion from the hypopharyngeal and mandibular glands of young worker honey bees. It is well know that $CoSO_4$ is a supplement with positive results in the bee feeding. From biochemical point of view, cobalt (Co) plays an important role in the work of enzymes and synthesis of vitamin B_{12}. The experiment was conducted from May to August 2014. Two groups of bee families were formed. The control group (3 bee families) was fed only with sugar syrup. The experimental group (3 bee families) was fed with sugar syrup and 4 mg/l $CoSO_4$. The syrup was administered in doses of 300 ml per family, 3 times per week. A total of 13 RJ samples are received. Data was analyzed using SPSS Statistical Package. Correlation was established using Pearson's correlation coefficient (r). The results obtained show the following average values for RJ samples harvested from bee colonies fed with sugar syrup: water content ($61.14 \pm 1.17\%$), proteins ($16.61 \pm 1.23\%$), fructose ($5.63 \pm 0.37\%$), glucose ($3.55 \pm 0.58\%$), sucrose (1.94 ± 0.43), total acidity (4.12 ± 0.22 ml 0.1 N NaOH/g), electrical conductivity (173.86 ± 10.45 µS/cm). The respective physicochemical parameters in RJ samples produced by feeding with sugar syrup and $CoSO_4$ were: water content ($61.65 \pm 1.16\%$), proteins ($17.70 \pm 0.68\%$), fructose ($5.34 \pm 0.90\%$), glucose ($3.32 \pm 0.60\%$), sucrose ($2.73 \pm 0.45\%$), total acidity (3.91 ± 0.23 ml 0.1 N NaOH/g), electrical conductivity (190.67 ± 13.56 µS/cm). The pH values for the all samples are identical about 4. As a conclusion, significant differences for electrical conductivity were found between the RJ samples harvested from the control and experimental group bee families ($P < 0.05$). For the Co content no significant differences were found. Significant correlation ($r = 0.560$, $P < 0.05$) was found between the protein content and electrical conductivity in all analyzed RJ samples.

Changes required in milking equipment cleaning practices to ensure low chlorine residues in milk
D. Gleeson, B. O'Brien and L. Paludetti
Teagasc, Animal & Grassland Research and Innovation Centre, Moorepark, Fermoy, cork, Ireland;
lizandra.paludetti@teagasc.ie

Trichloromethane (TCM) is a residue in milk caused by the interaction of chlorine (hypochlorite) and milk. On the majority of farms in Ireland, the products used for cleaning milking equipment contain sodium hydroxide and sodium hypochlorite (detergent/steriliser). It is thought that incorrect use of these products and/or inadequate rinsing after washing may increase milk TCM levels. A detailed knowledge of the daily milking equipment cleaning practices on dairy farms is required in order to give effective advice to farmers on this issue. A survey of milking equipment cleaning procedures was conducted on 112 farms, previously identified as having milk containing TCM residue. The size of the water trough in the dairy was inadequate on 55% of farms and this is a key factor accounting for the insufficient water used for rinsing milking equipment after washing on 65% of farms. A minimum of 14 litres per milking unit is advised. Meanwhile inadequate rinsing of the bulk milk tank (after washing) was also identified on 30% of farms. Higher than the required volume of cleaning product was used for cleaning the milking machine and bulk milk tank on 18% and 26% of farms, respectively. Products with a high chlorine concentration (>3.5%) were used for cleaning bulk milk tanks on 16% of farms. Using cleaning products with >3.5% chlorine increase the likelihood of residues, particularly when rinse water volumes are inadequate. Even though an adequate chlorine concentration was present in all products used, additional chlorine was added to the wash solution on 15% of farms resulting in very high working solutions of chlorine. Reusing the cleaning solution more than once, using chlorine in a pre-milking rinse to sterilize equipment, and dipping clusters in a chlorine solution between individual cow milking's all represent additional individual causes of high milk TCM. These practices were observed on 19%, 5% and 4% of farms, respectively. Overall, there were 14 incorrect cleaning procedures identified on the farms visited with the number of faults per farm ranging from one to seven. All of these incorrect practices must be addressed to ensure minimum residue levels in milk leaving the farm.

Effects of chamomile (*Matricaria chamomilla* L.) on performance blood and micro flora of broilers
M. Bouyeh, M. Akbari Chalaksari and S. Ashrafi Toochahi
Department of Animal Science, Rasht Branch, Islamic Azad University, Rasht, 4193963115, Iran; mbouyeh@gmail.com

In order to evaluate the effects of dietary chamomile flower powder (at levels of 0, 0.3, 0.6, 0.9 and 1.2% of basal diet) on broiler performance intestine microbial flora and some carcass, immune and blood parameters, a completely randomized design with 5 treatments, 4 replicate, and 10 birds per box, so 200 Ross 308 chicks (male and female) were considered. Food and water were available *ad libitum* during 42 days experimental period. SAS software was used to analyze the variance and Duncan test to compare between the means at level of $P<0.05$. The results showed that the experimental treatments had no significant effect on blood cholesterol, triglycerides, LDL, alkaline phosphatase, body weight, feed conversion ratio, European performance efficiency factor (EPEF), abdominal fat pad, breast and femur muscles ($P>0.05$) but there was significant differences between intestinal *Lactobacillus* and Coliforms population, Newcastle and sheep red blood cell (SRBC), and HDL of the chickens in treatment groups fed different levels of powdered chamomile flowers, so that the two highest levels of chamomile (0.9 and 1.2 percent) obtained the highest levels of Lactobacillus, Newcastle and SRBC antibody and HDL, and the lowest in Coliforms ($P<0.05$). It can concluded that using 0.9% chamomile to chicken diets may tend to more healthy products especially in organic poultry farms.

Diet affects the dominance of antimicrobial resistant genes in the rumen microbial community of cows

M.D. Auffret[1], C.A. Duthie[1], J.A. Rooke[1], T.C. Freeman[2], R.J. Dewhurst[1], M. Watson[2] and R. Roehe[1]
[1]SRUC, Roslin Institute Building, EH25 9RG Edinburgh, United Kingdom, [2]University of Edinburgh, Roslin Institute Building, EH25 9RG Edinburgh, United Kingdom; marc.auffret@sruc.ac.uk

Diet may generate stresses which affect the composition and functionality of the rumen microbiome with potential negative consequences for the overall health and production performance of cattle. This unbalanced microbial community is characterized by a 'bloom' of Proteobacteria. Many pathogenic bacteria classified as Proteobacteria possess several antimicrobial resistance (AMR) genes. Therefore, the overall aim of our research was to use metagenomics to understand the diversity of AMR genes in the rumen microbial community of apparently healthy cattle offered 'forage' or 'concentrate' diets (500 vs 900 g/kg concentrate dry matter basis). The total DNA of post-mortem rumen digesta samples from 50 animals balanced for breed type (Aberdeen Angus, Charolais, Luing and Limousin sires) and diet (without antibiotics) was extracted prior to metagenomic analysis. Identification of phylogenetic and functional microbial genes was based on Greengenes and KEGG genes databases, respectively. The Proteobacteria:(Firmicutes+Bacteroidetes) ratio was calculated as a proxy for rumen microbiome imbalance (ratio>0.15). A general linear model was used to determine signifcance of effects (breed or diet) on AMR genes ($P<0.01$; Gen-Stat 16th edition). The total relative abundance (%) of AMR genes was unaffected by diet. Within the 21 AMR genes identfied, those associated with resistance to macrolide and β-lactamase were significantly enriched by the concentrate diet whilst resistance genes to chloramphenicol and microcin were more abundant in forage-fed cattle. The Proteobacteria ratio was higher in animals offered the concentrate diet, positively correlated with the relative abundance of β-lactamase resistant genes and significantly negatively correlated the relative abundance of chloramphenicol resistance genes. Forage diets could be beneficial for animal health by controlling the abundance of Proteobacteria and the dominance of AMR genes in the rumen microbiome.

Fatty acid supplementation in transition goats: a transcriptional study related to inflammation

G. Farina[1], G. Invernizzi[1], V. Perricone[1], A. Agazzi[1], D. Cattaneo[1], J.J. Loor[2] and G. Savoini[1]
[1]Università degli Studi di Milano, Scienze Veterinarie per la salute, la produzione animale e la sicurezza alimentare, via Giovanni Celoria, 10, 20133 Milano, Italy, [2]University of Illinois, Department of Animal Sciences & Division of Nutritional Sciences, 1207 West Gregory Drive, Urbana, IL 61801, USA; greta.farina@unimi.it

The aim of the trial was to study the expression of genes and miRNA related to inflammation in adipose tissue of periparturient dairy goats supplemented with saturated or unsaturated fatty acids. Twenty-three second-parity alpine dairy goats were either fed calcium stearate (ST, n.7) or fish oil (FO, n.8) and compared to a control group (C, n.8). The dietary treatments lasted from one week before (30 g/head/d total fatty acids) to three weeks after kidding (50 g/head/d total fatty acids) of either ST (26% C16:0, 69.4% C18:0) or FO (10.4% EPA, 7.8% DHA). Adipose tissue samples on day -7 and 7 and 21 from kidding were analyzed for CCL2, IL10, IL18, IL1β, IL6R, SAA3, HP, IL8, RXRA, and TLR4 expression and 99a, 155, 143, 145, 221, 26b, and 378 miRNA expression. RNA was extracted and primers were aligned with available databases using BLASTN at NCBI. A quantitative PCR by SYBR green was used and the relative expression was determined by a 6-point standard curve. Data were analyzed by a MIXED repeated model in SAS 9.3 after normalization with 3 internal reference genes. No diet effect was observed, but there was a time effect for IL10, SAA3, HP, IL8, RXRA ($P<0.05$) and miR-155 ($P=0.02$). IL8 and IL10 were upregulated at days 7 and 21 ($P<0.05$), compared to day -7, as well as miR-221, that plays a role in inflammation and lipolysis, tended to have the same pattern. Furthermore, HP and SAA3 had the highest upregulation at 7 DIM compared to prepartum. In addition, miR-155, which is involved in the infiltration of immune cells, had a similar expression pattern. Obtained data support the existence of an acute-phase response in the first week after kidding in adipose tissue, as observed in cows.

Effects of *Enterococcus faecium* NCIMB 10415 on porcine immune cells

S. Kreuzer-Redmer, N. Wöltje, F. Larsberg, K. Hildebrandt and G.A. Brockmann
Humboldt-Universität zu Berlin, Faculty of Life Sciences, Thaer-Institut, Invalidenstrasse 42, 10115 Berlin, Germany; susanne.kreuzer.1@agrar.hu-berlin.de

Feeding of the lactic acid-producing *Enterococcus faecium* NCIMB 10415 (EF), a licensed probiotic for pigs and piglets, has been described to promote growth performance and health in pigs. However, underlying mechanisms of probiotic additives are still elusive as they may directly influence immune cells or influence the intestinal milieu. We established a porcine *in vitro* cell culture model to explore direct interactions of porcine adaptive immune cells and probiotics. We particularly investigate EF as an alternative dietary additive to improve animals' health. To test the direct effects of EF, we conducted cell culture experiments with primary cultured porcine immune cells of three German Landrace pigs in a co-culture with living or dead EF. For validation, immune cells from blood were repeatedly taken from the same three animals with the same genetic and environmental background. 1, 2, 3, 5×10^6 peripheral blood mononuclear cells (PBMCs) were treated with EF in a ratio of 1 : 2, 2 : 1, 5 : 1 or 10 : 1 (PBMCs : EF) for 1, 1.5 or 3 h. There were higher relative cell counts of CD8b+ cytotoxic T-cells (P<0.05) in the treatment group with living EF after 1 and 1.5 h of incubation compared to untreated controls. In order to identify mechanisms of T-cell activation mediated by soluble factors secreted by EF, trans-well assays were performed. Hence, bacteria and immune cells were separated by a membrane enabling the perfusion of soluble factors, but preventing a direct interaction of cells and bacteria. In this experimental set up, we also detected a higher relative cell count of CD8b+ cytotoxic T-cells (P<0.05), if PBMCs were treated with living EF for 1 h. After 1.5 h of incubation of PBMCs with dead (by UV) EF bacteria a higher relative cell count of CD21+ B-cells (P<0.05) was detected, while there was no effect with living EF. These results suggest that B-cells need a stimulus by a cell surface protein, while cytotoxic T-cells were stimulated by secreted substances of EF. This study could provide evidence of a direct immunomodulatory effect of *E. faecium* on adaptive immune cells *in vitro*.

Effect of zinc oxide and chlortetracycline on antibiotic resistance development in piglets

A. Romeo[1], S. Durosoy[1], W. Vahjen[2] and J. Zentek[2]
[1]Animine, 335 Chemin du Noyer, 74330 Sillingy, France, [2]Freie Universität Berlin, Institut für Tierernährung, Königin-Luise-Str. 49, 14195 Berlin, Germany; aromeo@animine.eu

Chlortetracycline (CTC) and zinc oxide (ZnO) at pharmacological dosage are commonly supplemented in piglet diets in order to reduce diarrhea and to improve growth performance. However, there is a risk of bacterial resistance development. In this trial, CTC and ZnO were used to evaluate their effects on antibiotic resistance genes. The experiment was performed with 4 diets: 2,400 ppm of Zn from standard ZnO vs 110 ppm of Zn from a potentiated ZnO source (HiZox®), with or without 300 ppm of CTC. Each treatment consisted of 10 piglets weaned at 25 days. DNA was extracted from feces (0, 2, 4, 7, 14 d), in order to quantify by qPCR the *Escherichia* group, some antibiotic resistance genes and related genes. As expected, both trials groups with high concentrations of ZnO reduced the Escherichia group, however low concentrations of HiZox and CTC showed the same trend. CTC increased numerically the development of various genes (tetA, bacA) after 6 days; without CTC, there were numerical (tetA) or significant (bacA, zinT) differences between the group fed 2,400 ppm of Zn from standard ZnO and the group fed 110 ppm of Zn from potentiated ZnO. Both ZnO at high level and CTC increase the development of antibiotic resistance genes, and may have an additive effect on these genes.

Impact of GOS delivered *in ovo* on microbiota and intestinal gene expression in broiler chickens

A. Slawinska[1,2], M. Radomska[2], J. Lachmanska[2], S. Tavaniello[1] and G. Maiorano[1]
[1]UNIMOL Università degli Studi del Molise, Via F. De Sanctis snc, 86100 Campobasso, Italy, [2]UTP University of Science and Technology, Mazowiecka 28, 85-084 Bydgoszcz, Poland; slawinska@utp.edu.pl

Galactooligosaccharides (GOS) are potent prebiotics that stimulate gut microflora development. *In ovo* delivery of prebiotics allows for modulation of gut microflora in neonates. This study aimed to determine proportion of beneficial microbes and intestinal gene expression in adult broiler chickens following *in ovo* delivery of GOS. Prebiotic eggs (P) were injected *in ovo* with GOS on 12 day of egg incubation (3.5 mg GOS/embryo). Control eggs (C) were injected *in ovo* with physiological saline. Chickens (Ross 308) were reared until 42nd day. At slaughter, intestinal samples (gut mucosa and digesta) were collected from dudoneum, jejunum, ileum and caecum. Relative abundance of Bifidobacterium spp. and *Lactobacillus* spp. was determined in digesta with qPCR. Molecular responses of the host were analyzed in gut mucosa with RT-qPCR using genes responsible for innate immune responses (*IL1B, IL10, IL12p40, AvBD1* and *CATHL2*), short chain fatty acids signaling (*FFAR2* and *FFAR4*), glucose absorption (*GLUT1, GLUT2* and *GLUT5*), mucose production (*MUC6*) and intestinal permeability (*CLDN1* and *TJAP1*). The highest abundances were determined for *Bifidobacterium* spp. in caecum of P and for *Lactobacillus* spp. in ileum of C. *In ovo* delivery of GOS increased intestinal gene expression: *IL1B, IL10, IL12p40, GLUT1, MUC6* and *CLDN1* in jejunum and caecum; *AvBD1* and *CATHL2* in caecum; *FFAR2* and *FFAR4* in jejunum, ileum and caecum. Expression of *GLUT2* and *GLUT5* was decreased in P group. In summary, GOS delivered *in ovo* had bifidogenic effect on chicken intestinal microflora and up-regulated gene expression in chicken guts. *In ovo* method of GOS prebiotic delivery had beneficial and long-lasting effects in broiler chickens. Acknowledgements: grants: OVOBIOTIC (RBSI14WZCL, MUIR, Rome, Italy) and UMO-2013/11/B/NZ9/00783 (NSC, Cracow, Poland)

Transcriptomic modulation of adult broiler chickens after *in ovo* delivery of *Lactobacillus* synbiotic

A. Dunisławska, A. Sławińska, M. Bednarczyk and M. Siwek
UTP University of Science and Technology, Department of Animal Biochemistry and Biotechnology, Mazowiecka 28, 85-084, Poland; aleksandra.dunislawska@utp.edu.pl

Intestinal microflora is a key factor in maintaining good health and production results in chickens. Bioactive substances such as prebiotics, probiotics or their combination (synbiotic) can effectively stimulate intestinal microflora and therefore replace antibiotic growth promoters. It has been proved that intestinal microflora might be stimulated at the early stage of embryo development. The aim of the study was to estimate a long-term changes of chicken transcriptome after a single *in ovo* administration of synbiotic. On day 12 of incubation 5,850 eggs (Cobb500FF) were distributed to experimental groups and injected with S1– *Lactobacillus plantarum* with raffinose family oligosaccharide (RFO) or S2– *Lactobacillus salivarius* with galactooligosaccharide (GOS). A control group received saline. On day 21 post hatch roosters (n=5) were sacrificed and immunological tissues (cecal tonsils, spleen and jejunum) and metabolic tissue (liver) were collected. Total RNA isolated from tissues was subjected to the whole transcriptome analysis by chicken expression microarray (Affymetrix Gene 1.1ST Array Strips). Data analysis was performed using Affymetrix Expression Console and Transcriptome Analysis Console software, Venn diagrams, DAVID and CateGOrizer. The highest number of differentially expressed genes (DEG) in immune tissues was detected in cecal tonsils of S1 (160 DEG), while the lowest number was detected in spleen of S1 (32 DEG). S2 injection activated 72 genes in jejunum, 70 in spleen and 68 genes in cecal tonsils. There were 10 genes in common between spleen and cecal tonsils of S2. The highest number of DEG in liver was detected in S2 group (159 DEG) compared to S1 (48 DEG). Analysis revealed 9 genes in common in both groups. Research showed targeted effect of gene expression based on used synbiotic. S1 activated mostly genes involved in immune processes, while S2 up-regulated expression of genes involved in metabolic pathways. This research was supported by ECO-FCE project funded by the European Union Seventh Framework Programme FP7-KBBE-2012-6-singlestage Monogastrics Feed Efficiency and Polish Ministerial Funds for Science.

Effects of maternal wheat bran supplementation on microbiota and intestinal parameters of piglets

J. Leblois[1], S. Massart[1], J. Wavreille[2], B. Li[1], J. Bindelle[1] and N. Everaert[1]
[1]Gembloux Agro-Bio Tech (ULg), Precision Livestock and Nutrition, Passage des Déportés, 2, 5030 Gembloux, Belgium, [2]Walloon Agricultural Research Centre, Rue de Liroux, 8, 5030 Gembloux, Belgium; julie.leblois@ulg.ac.be

In pigs, new strategies to avoid infections at weaning focus on the impact of the maternal diet on the microbiota and intestinal health of piglets. In this study, the objective was to determine whether maternal dietary supplementation with wheat bran (WB) could modulate their piglets' microbiota and enhance the health status at weaning. Eight sows were fed a high WB diet from day 43 of gestation (240 g/kg DM of WB during gestation; 140 g/kg DM during lactation) until weaning while 7 sows were fed a control (CON) diet devoid of WB. All diets were formulated to meet sows' requirements and to be iso-nitrogenous and iso-energetic. Two days before weaning, the terminal ileum, caecum and colon contents of piglets were collected, the pH recorded and ileal tissues fixed in formaldehyde 4%. An ex vivo challenge was performed on colon tissues with or without lipopolysaccharide (from *Escherichia coli* O111:B4) to observe immune responses. Microbiota of the colonic contents was determined by sequencing (Illumina MiSeq), short-chain fatty acids (SCFA) production by HPLC and histomorphological parameters after embedding in paraffin followed by haematoxylin-eosin coloration. All statistical analyses were performed in SAS, with the Kruskall-Wallis test (sequencing), the MIXED procedure (SCFA and pH) or the Nested procedure (histomorphology). Results showed differences between maternal treatments for microbiota of piglets, i.e. *Colinsella*, unclassified *Clostridiaceae*, *Methanobrevibacter*. For SCFA, piglets from WB sows showed decreased valerate molar ratio (MR, $P<0.05$) in the terminal ileum, caecum and colon and the CON piglets showed higher butyrate MR in the caecum (13% for the CON piglets, 10% for the WB piglets; $P=0.03$). No impact of the maternal treatment was observed on the intestinal pH or histomorphological parameters. Results concerning the ex vivo challenge will also be presented. In conclusion, SCFA and microbiota were affected by the maternal treatment, and ex vivo results will allow to deepen the importance of these changes on the immune status of the piglets.

Pre-weaning nutrient supply affects gene expression profiles in bone marrow and muscle in calves

L.N. Leal[1], G.J. Hooiveld[2], F. Soberon[3], H. Berends[1], M.V. Boekschoten[2], M.A. Steele[4], M.E. Van Amburgh[5] and J. Martin-Tereso[1]
[1]Trouw Nutrition, R&D, P.O. Box 220, 5830 AE Boxmeer, Netherlands Antilles, [2]Wageningen University, Department of Agrotechnology and Food Sciences, P.O. Box 17, 6700 AA Wageningen, Netherlands Antilles, [3]Nutreco, Shurgain USA, 150 Research Lane, Suite 200, Guelph, Canada, [4]University of Alberta, Department of Agricultural, Food, and Nutritional Science, T6G 2P5, Edmonton, Canada, [5]Cornell University, Department of Animal Science, 507 Tower Road, Ithaca, NY, USA; leonel.leal@trouwnutrition.com

Early life nutrient supply and the resulting gain in calves are positively correlated with the future milk production of dairy cows. This study aimed to assess the effect of pre-weaning nutrient supply on gene expression profiles in bone marrow and skeletal muscle tissue in calves. To this end, 12 Holstein-Friesian calves were fed two distinct daily amounts of the same milk replacer (28% CP, 15% fat), receiving either a fixed rate of 11.7 MJ (LOW, n=6) or 1.26 MJ (HIGH, n=6) of metabolizable energy per kilogram of metabolic body weight (BW0.75). At 54 days of age, tissue samples were harvested. Transcriptome analysis revealed that in bone marrow, upon a HIGH diet, cell mitosis, RNA transcription and immune response were supressed when compared with a LOW diet, whereas overall metabolism such as lipid, vitamin and TCA cycle, was up regulated. Moreover, the changes in gene expression observed in the bone marrow seem to be linked to ESR1 activation by the HIGH diet. In line with the substantial observed increase in growth, in muscle tissue, all gene sets were induced by the HIGH diet. In particular, cell cycle related processes such as cell mitosis, DNA repair, haemostasis and integrin cell surface interactions. In addition, lipid and protein metabolism processes, as well as signal transduction processes, were activated by the HIGH diet. In conclusion, results indicate that pre-weaning energy supply in calves can affect expression of immune-related pathways in bone marrow and expression of pathways related to growth and differentiation in skeletal muscle.

Early nutrition of fattening lambs modifies bacterial community and lymphocytes of the ileum

J. Frutos, S. Andrés, J. Benavides, A. Santos, F. Rozada, N. Santos and F.J. Giráldez
Instituto de Ganadería de Montaña (CSIC-Universidad de León), Finca Marzanas, 24346, Grulleros, León, Spain; sonia.andres@eae.csic.es

Feed efficiency and immunity parameters can be deeply impacted by colonization of gastrointestinal (GIT) mucosa during early life. Thus, the aim of the present study was to determine if the composition of the bacterial populations firmly attached to the ileal mucosa (epimural) is modified by the level of milk intake during the suckling phase of lambs, thus promoting long-term effects during the fattening period. Twenty four merino lambs (average LBW 4.77±0.213 kg) were used, twelve of them (*ad libitum*, ADL) being kept permanently with the dams whereas the other group (restricted, RES) was separated periodically from the dams and milk restricted. After weaning all the animals were penned individually, offered the same complete pelleted diet at a restricted level (40 g/kg LBW to ensure no differences of DMI) and slaughtered with 27 kg. Ileal tissue samples were collected for immunohistochemistry, morphometric analysis and T-RFLP analysis (microbial diversity). All the data were analysed using one-way analysis of variance with the milk intake level as the only source of variation, excepting those corresponding to T-RFLP, which were analysed by discriminant analysis. During the fattening period, the RES group showed higher feed:gain ratios (3.05 vs 3.69, P<0.001). In addition, the ileal epimural bacterial community of the RES group showed lower relative height of two peaks compatible with probiotics and *Proteobacteria*. These changes could be associated with the higher infiltration of lymphocytes (T and B) in the ileal lamina propria, a higher presence of M cells and a greater thickness of submucosa layer when compared to the ADL group. In conclusion, the level of nutrition during the suckling period of merino lambs promoted changes in ileal epimural bacterial community and local immune response that could be related to differences in feed efficiency traits during the fattening period.

The effects of using mix oil herbal extracts in broiler diet on growth performance and immune system

A.Y. Abdullah, M. Khalifeh, R.T. Kridli and M.S. Al Kasbi
Jordan University of Science and Technology, Animal Production, 3030 Irbid, 22110, Jordan; abdullah@just.edu.jo

This study was conducted to evaluate the effect of using essential oil (MixOil™) powder in the diets of broiler chickens on growth performance, carcass characteristics and immune system. A total of 272 one day old mixed sex Hubbard commercial broiler chicks were used and randomly distributed into 4 dietary treatments; control (Con), control treatment supplemented with MixOil powder at 250 g/1000 kg (T1), control treatment supplemented with MixOil powder at 500 g/1000 kg (T2), and control treatment supplemented with MixOil powder at 750 g/1000 kg (T3). Each dietary treatment was divided in a similar design into 4 replicates with 17 chicks each. At the ages of 1, 21 and 42 days, blood samples from 32 chicks from all treatment (2 chick\replicate) were collected. No significant effects of MixOil powder supplementation on broiler body weight, body weight gain and average daily gain (ADG) were observed during starter, grower, finisher and overall periods. Average feed intake and feed conversion ratio (FCR) during the starter, grower and overall rearing periods were significantly affected (P<0.05) by MixOil powder in the diets compared with the control group being lower for T1 and T2 and higher for T3. No significant effects of MixOil powder supplementation on carcasses cuts, dressing percentages and meat quality parameters were detected. There were significant differences (P<0.0001) between treatments in meat chemical composition parameters, while blood biochemistry parameters were unaffected (with the exception of HDL measured at age of 21 days and for total protein measured at age of 42 days). The results showed that MixOil powder supplementation had a positive effect on antibody response generated from vaccination against Infectious Bronchitis (IB) but not from New Castle Disease (NDV) virus. In conclusion, supplementation with MixOil powder at 250 or 500 g/1000 kg can increase feed intake and FCR in addition to enhancing the immune system's response to IB vaccination.

Synbiotics as gut health improvement agents against Shiga toxin-producing *E. coli* in piglets

B.R. Kim[1], J.W. Shin[1], S.Y. Choi[1], M.E.S. Gamo[1], J.H. Lee[1], R.B. Guevarra[1], S.H. Hong[1], M.K. Shim[1], J.S. Kang[2], W.T. Cho[2], K.J. Cho[2], J.Y. Kim[1] and H.B. Kim[1]

[1]*Dankook Univ., Animal Resources and Science, Cheonan, 31116, Korea, South,* [2]*Genebiotech Co., Ltd, Sinwonsa-ro 166, Gongju-Si, 32619, Korea, South; th37gn@naver.com*

Edema disease is one of the major health problems in young piglets resulting in poor health and death. It is caused by *Escherichia coli* producing F18 pili and Shiga toxin 2e. Therefore, it is pivotal to reduce colibacillosis in weaned piglets to enhance production performance. In this study, we evaluated synbiotics as gut health improvement agents in the mouse model challenged with Shiga toxin-producing *E. coli* (STEC) isolated from piglets. Prebiotics, lactulose, was formulated with each 5.0×10^6 cfu/ml of *Pediococcus acidilactici* GB-U15, *Lactobacillus plantarum* GB-U17, and *L. plantarum* GB 1-3, to produce 3 types of synbiotics. A total of 40 three weeks old BALB/c mice were randomly assigned to 4 groups (n=10), the control group and 3 synbiotics treated groups. Each synbiotics treated groups were daily administrated with 5.0×10^6 cfu/ml of one synbiotics for the first week, and every 3 days during the second week. Mice were challenged with 8.0×10^8 cfu/ml of STEC 5 days after piglets began to receive synbiotics. The daily weight gain, fecal index, gross legions and histopathological changes of small and large intestines were compared between groups. Statistical analysis was conducted by Tukey's Studentized Range (HSD) test using General Linear Model Procedure of SAS. The experiments were duplicated. Mice treated with synbiotics based on *P. acidilactici* GB-U15 and *L. plantarum* GB-U17 significantly improved daily weight gain compared to mice in other groups. In particular, mice treated with synbiotics based on GB-U15 showed the best growth performance. While mice treated with GB-U15 showed better fecal index, no significant differences were observed among groups. Gross lesion and histopathological evaluations showed that mice treated with GB-U15 moderately improved recovery from STEC infection. In conclusion, our results suggest that the synbiotics formulated with lactulose and *P. acidilactici* GB-U15 has potential benefits to prevent and improve colibacillosis in weaned piglets.

Use of Bovikalc® calcium boluses and their effect on peri-parturient cow health

H. Scholz[1] and A. Ahrens[2]

[1]*Anhalt University of Applied Sciences, LOEL, Strenzfelder Allee 28, 06406 Bernburg, Germany,* [2]*Thuringia Animal Healths Fund, Victor-Goerttler-Straße 4, 07745 Jena, Germany; heiko.scholz@hs-anhalt.de*

Calcium plays an important role in muscular contraction, conduction of nerve impulses and in the immune response. While the cow needs relatively low amounts of calcium (Ca) during the dry period (ca. 10 g/day), her demand rises quickly when lactation sets in (ca. 30-50 g/day). This demand often exceeds the ability of the cow to replenish her blood plasma Ca pool, leading to hypocalcaemia. Clinical hypocalcaemia is found in about 5% of cows, while subclinical hypocalcaemia occurs in ca. 40% of peri-parturient cattle. Even though these cows don't show clinical symptoms, this condition is related to decreased productivity and a higher risk of developing different health conditions, such as mastitis. Therefore, subclinical hypocalcaemia is economically most relevant. The risk of subclinical hypocalcaemia can be reduced and different strategies are available. Some of these are aimed at the mineral ratios fed to dry cows. Another possibility is to provide calcium to the cow around and shortly after parturition. On a dairy farm with ca. 1,100 cows (rate of clinical milk fever: 11%, 2-phased dry cow feed regime aiming at a low calcium rate), 171 dairy cows (2nd lactation and older) were included in a study. Cows were alternately assigned to the control group (no extra prophylaxis) or the bolus group. These cows received 4 calcium-boluses (Bovikalc®, Boehringer Ingelheim, each containing 43 g of calcium), 1 bolus at 24 h before parturition, 1 bolus shortly after, 1 bolus 12 h and 1 bolus 48 h after parturition). Calcium blood-levels were significantly improved in the Bovikalc group during the first 24 h after calving. There were significantly less cows recumbent in the Bovikalc group (5 vs 11%). Cows in the Bovikalc group developed significantly less puerperal diseases (2.3 vs 8.4%). Following this, there was a tendency for a 14 days shorter calving to first service interval in the Bovikalc group.

Effect of the seabuckthorn berry marc extract on the performance of calves with nutritional problems

L. Liepa, E. Zolnere and I. Dūrītis
Latvia University of Agriculture, Faculty of Veterinary Medicine, Kr. Helmana street 8, 3004 Jelgava, Latvia;
laima.liepa@llu.lv

The study is part of the project LCS 672/2014. The aim was to investigate the influence of the seabuckthorn berry marc extract (SME) on the health and performance indices of newborn calves. The experiment was performed in the herd with 280 dairy cows, June-August, 2016. The nutritional problems were: too small doses and delays in colostrum feeding of newborn calves, later – sometimes over- or underfeeding of milk, bad feeding hygiene. The control (C) and experimental (E) group each consisted of 10 calves. Starting from the second time of feeding during 15 days, E calves received 5 to 10 ml of SME per os ones a day in increasing dosage. Clinical examination of all C and E calves was done every day, but weight gain was controlled and blood samples for biochemical, hematological and cell flow cytometry analyses were collected on Day 1 (D1), D8, D15 and D30. The data were analysed using Microsoft Excel. Results. In a 30-days period, the weight gain for E calves was insignificantly higher than in C calves, 413±63 vs 370±88 g/day. In group E, fed with SME, the concentration of serum haptoglobin decreased significantly (P=0.05) on D1-D8 from 1,235±380 to 557±358 ng/ml, but after finishing to receive SME, it had a tendency to increase again. The concentration of TNF-alfa was in a low concentration for all animals in groups E and C (individually, ranging 5-20 pg/ml). Regarding the cell flow cytometry analyses on D1, a significantly (P<0.05) higher concentration of T lymphocytes subpopulations CD4 (helpers), CD8 (killers), ratio of CD4/CD8, B lymphocytes subpopulation CD21 (antibodies) and mother milk monocytes subpopulation CD14 were in E group than in group C. In a 15-days SME feeding period, for E calves amounts of these indices constantly decreased until the group C levels and remained stable to D30. In group C, all above noted lymphocytes subpopulations constantly increased from D1 to D30 as it was physiologically predicted. Conclusions. SME as a feed additive improves absorption of lymphocytes subpopulations CD4, CD8, CD21 and CD14 from colostrum in the first day of life. In a 30-days period, the SME antiinflammatory action slowed the development of the non-specific immunity in calves.

Blood clinical chemistry parameters and growth performance in post-weaning piglets

Q. Hu, R. Faris, J. De Oliveira and D. Melchior
Cargill Animal Nutrition, 10383 165th Avenue NW, Elk River, MN 55330, USA; qiong_hu@cargill.com

The objectives of the present two trials were to identify biomarkers in the blood clinical chemistry profile that could be correlated with growth performance in post-weaning piglets. In the first trial, 240 post-weaning piglets, mixed genders, 22 days of age and 7.5±0.8 kg of body weight (means ± SD) were group housed, and kept until 42 days post-weaning. A randomized complete block design with a 2×2 factorial arrangement including high vs low level of Zn (2,450 ppm by adding ZnO vs <150 ppm) and high vs low complexity diet were used to induce different growth performance. Pigs and feed intake were weighed every week to determine the average daily gain (ADG) and feed intake (ADFI). Blood and ileum samples were collected to measure clinical chemistry and gut integrity. Data was analyzed using a mixed model with initial body weight (IBW) as covariate. The results showed that high complexity diet induced better ADG and ADFI during the first two weeks compared low complexity diet (P<0.01), whereas high level Zn improved ADG and ADFI only during the 2nd week (P<0.05), which was unexpected. High level Zn resulted in higher villus height to crypt depth ratio on day 7 and 14 (P≤0.07). In high level Zn diet fed pigs, blood alanine aminotransferase (ALT), alkaline phosphatase (ALP) on day 7 and 14, total protein, aspartate aminotransferase and lipase on day 14 were elevated compared to low Zn group (P<0.05). ZnO model was tested alone in the second trial to induce growth differences. Similarly, pig ADG and ADFI was not improved by high level Zn, and Zn effect on blood clinical parameters was similar as in the first trial. Results of partial correlation analysis with IBW and ADFI as partial variables indicated that day 7 and 14 blood phosphate level was positively correlated but total protein on day 7, triglyceride, lipase, ALT, ALP on day 14 were negatively correlated with BW and ADG of the 2nd and 3rd week across two trials (P<0.05). These two trials provided evidence of relations of pig growth and phosphate metabolism, lipid metabolism, liver and pancreas function. Conclusively, selected blood biomarkers could be used to better understand the driver of growth and what is in common for piglets that perform well.

Investigation of bacterial diversity in the feces of Hanwoo steers fed different diets

M. Kim, J.Y. Jeong, H.J. Lee and Y.C. Baek
National Institute of Animal Science, Department of Animal Biotechnology and Environment, 1500, Kongjwipatjwi-ro, Iseo-myeon, Wanju 55365, Korea, South; mkim2276@gmail.com

The objective of this study was to investigate the community structure of fecal microbiota in Korean native Hanwoo steers fed different diets. Fecal samples were obtained from 24 Hanwoo steers (average weight = 288 kg) fed one of two growing diets: (1) 12 steers fed the control diet composed of 35% Timothy, and 65% grains containing typical concentrations of CP and TDN, and (2) 12 steers fed the treatment diet composed of 35% timothy, and 65% grains containing high concentrations of CP and TDN. The community structure of fecal microbiota in Hanwoo cattle was investigated using next generation sequencing of 16S rRNA gene amplicons on the Illumina MiSeq platform. In total, 542,911 sequences were obtained from the feces of 24 steers with at least 17,000 sequences. *Firmicutes* and *Bacteroidetes* were the first and the second dominant phyla in all fecal samples and accounted for 51.7% and 31.7% in collective data, respectively. The relative abundance of *Firmicutes* tended to be greater in the control diet group than in the treatment diet group ($P<0.1$). At the genus level, *Bacteroides* was the most dominant and accounted for 6.2% of the total sequences in collective data, but its relative abundance did not differ between the two diet groups. Genera that accounted for more than 1% of the total sequences in collective data were *Clostridium* XIVa, *Succinovibrio*, *Alistipes*, *Clostridium* IV, *Paraprevotella*, *Ruminococcus*, *Barnesiella*, *Treponema* and *Anaerovorax*, where *Alistipes* tended to be more abundant in the control diet group than in the treatment diet group ($P<0.1$). Unclassified *Ruminococcaceae* was the most dominant unclassified group and accounted for 22.7% of the total sequences in collective data, and its relative abundance tended to be greater in the control diet group than in the treatment group ($P<0.1$). This study indicates that the community structure of fecal microbiota in Hanwoo steers was slightly affected by different levels of CP and TDN.

Influence of olive cake introduction to feed on urine and blood pH for Baladi female goats

N. Mehanna[1] and R. El Balaa[2]
[1]Lebanese University, Faculty of Agriculture and Veterinary Medicine, Beirut, Dekwaneh, Lebanon, [2]University of Balamand, Issam Fares Faculty of Technology, Kalhat, 100, Tripoli, Lebanon; rodrigue.elbalaa@balamand.edu.lb

Olive cake's significant nutritive value and low cost allow its use efficiently in animal feed; However, it may lead to metabolic acidosis. The objective of this study is to measure the potential effect of the introduction of olive cake to Baladi goat feed, on blood and urine pH. Twenty nine Baladi goats were distributed into four groups according to age and feeding regime: 9 adults with olive cake diet (AOC), 9 adults with normal diet (ANR), 6 yearlings with olive cake diet (YOC) and 6 yearlings with normal diet (YNR). The standard diet provided to animals included corn (27%), barley (27%), wheat bran (33%), soja (10%), vitamins (1.8%), minerals (0.8%) and bicarbonate (0.4%). In the experimental diet, 50% of wheat bran was replaced with olive cake. Blood samples were collected 4 h after feeding and urine samples were collected from the first urination after feeding, once a week for a month and a half. For every measure, ANOVA tests were carried out using SPSS 20.0. The average blood pH varied between 7.21 and 7.76 with an average of 7.53±0.24 for YNR, 7.24 and 7.74 with an average of 7.50±0.21 for YOC, 7.29 and 7.74 with an average of 7.55±0.21 for ANR and 7.36 and 7.76 with an average of 7.59±0.19 for AOC. No significant difference between the 6 samples per group was detected. The average urine pH varied between 6.99 and 8.72 with an average of 8.01±0.74 for YNR, 7.77 and 8.85 with an average of 8.19±0.34 for YOC, 6.50 and 8.83 with an average of 8.09±0.85 for ANR and 7.73 and 8.83 with an average of 8.32±0.35 for AOC. Significant difference was noted between the second and the third date for YNR (from 8.04 to 6.99) and ANR (from 8.63 to 6.50). The absence of significant difference shown between groups ($P<0.05$) is probably due to the gradual introduction of olive cake to the diet which gave enough time to the digestive microflora and the digestive system to adapt. This study indicates that olive cake can be introduced to 'Baladi' goat feed gradually without leading to metabolic acidosis.

Nutritive value of seaweed species and extracted seaweed in animal diets

P. Bikker[1], P. Van Wikselaar[1], W.J.J. Huijgen[2] and M.M. Van Krimpen[1]
[1]Wageningen University & Research, P.O. Box 338, 6700 AH Wageningen, the Netherlands, [2]Energy research Centre of the Netherlands (ECN), Westerduinweg 3, 1755 LE Petten, the Netherlands; paul.bikker@wur.nl

The development of the world population increases the requirement of biomass for human and animal consumption. At present, seaweed is not used as feed material to any significant extent because of availability, handling properties and inadequate information on the feeding value. Nonetheless, use of seaweed might be of interest because its cultivation does not compete for land-use. Cost effective use may include biorefinery and application of specific components or residues in animal feed. This study evaluated the nutritive value of the brown *Laminaria digitata*, *Saccharina latissima* and *Ascophyllum nodosum*, the red Palmaria palmata and *Chondrus crispus* and the green *Ulva lactuca*, from Scotland, Ireland and France. In addition, the residues of *L. digitata* and *S. latissma* after mannitol extraction and *U. lactuca* after rhamnose extraction were studied. Observations included (1) nutrient contents, minerals and heavy metals, and (2) *in vitro* digestibility of protein and organic matter using a modified Boisen method, with soybean meal (SBM) as reference. Nutrient composition differed substantially between species (e.g. lysine 2-11 g, fat 3-23 g, CF 27-80 g, P 0.6-3.4 g per kg) and to a lesser extent between different locations of origin. *In vitro* ileal N digestibility was 70-85% for most species, lower for *A. nodosum* (25-60%) and higher for SBM (98%). *In vitro* total tract OM digestibility was 80-85% for most brown and red species, lower for *A. nodosum* (70%) and *U. lactuca* (75-80%), and higher for SBM (97%). Washing intact seaweeds to reduce the salt content prior to *in vitro* digestion reduced the *in vitro* digestibility by about 10%, presumably because of loss of soluble and potentially digestible material. After sugar extraction, the *in vitro* N-digestibility of the residues was about 10% (*L. digitata*) to 20% (*U. lactuca* and *S. latissima*) lower than the original material, presumably due to co-extraction of potentially digestible N and an effect of processing conditions on N digestibility. The results indicate that development of biorefinery processes should include optimisation of the nutritive value of the residue.

Methionine absorption in the porcine GIT affected by dietary methionine sources

S. Romanet[1], L. Mastrototaro[1], R. Pieper[2], J. Zentek[2], J.K. Htoo[3], B. Saremi[3] and J.R. Aschenbach[1]
[1]FU Berlin, Inst. Vet. Physiol., Oertzenweg 19b, 14163 Berlin, Germany, [2]FU Berlin, Inst. Anim. Nutr., Königin-Luise-Str.49, 14195 Berlin, Germany, [3]Evonik Nutrition & Care GmbH, Rellinghauser Str. 1-11, 45128 Essen, Germany; stella.romanet@fu-berlin.de

The limiting amino acid methionine (Met) is an important feed additive for pigs. The project aimed to characterize the influence of 3 Met supplements on intestinal Met absorption in growing pigs. 27 pigs aged 10 wk were pre-fed for >10 d with three diets of identical composition but with different Met supplements that equally accounted for 32% of the dietary Met+Cys requirement: DL-Met (diet 1), L-Met (diet 2) or DL-HMTBA (diet 3). To assess Met absorption, mucosal-to-serosal flux rates (Jms) of 14C-labelled D-Met, L-Met or DL-HMTBA (each at 50 µM and 5 mM) were measured in duodenum (DUO), mid-jejunum (JEJ) and ileum (IL) over 60 min in the presence or absence of Na+. Data were compared by two-factor ANOVA and Student-Newman-Keuls' test. In DUO, Jms of D-Met at 50 µM tended to be higher with diet 1 compared to diets 2 and 3 ($P<0.1$). In JEJ and IL, Jms of D-Met at 50 µM was higher with diet 1 compared to diets 2 and/or 3 ($P<0.05$). The presence of mucosal Na+ had no influence in all intestinal segments. When L-Met was used at 50 µM, diet 1 induced or tended to induce higher Jms than diets 2 or 3 in DUO ($P<0.05$ vs diet 2; $P<0.1$ vs diet 3), JEJ ($P<0.01$ vs diets 2 and 3) and IL ($P<0.1$ vs diet 2; $P<0.05$ vs diet 3). Only in JEJ within diet 1, Jms was higher in presence of Na+ ($P<0.01$). At 5 mM of either D-Met or L-Met, no significant effects of diet or Na+ were observed. The absorption of HMTBA was not altered by diet or Na+, except in DUO where Jms was lower in the presence vs absence of Na+ at 5 mM HMTBA ($P<0.01$). Conclusions: Intestinal Met absorption at physiological luminal concentrations (e.g. 50 µM) adapts to dietary Met sources. In specific, a diet containing DL-Met appears to enhance L-Met and D-Met absorption in all the three intestinal segments. The stimulation involves the induction of a Na+-dependent mechanism for L-Met absorption in JEJ. This study was supported by Evonik Nutrition & Care GmbH.

Sainfoin as a replacement of alfalfa: nutritive value and performances in the rabbit

H. Legendre[1], P. Gombault[2], H. Hoste[3], M. Routier[4], C. Bannelier[1] and T. Gidenne[1]
[1]INRA, GenPhySE, UMR1388, 31326 Castanet Tolosan, France, [2]Multifolia, 1bis grande rue, 10380 Viapres Le Petit, France, [3]Veterinary highschool, UMR 1225 IHAP INRA/ENVT, 31076 Toulouse, France, [4]MG2MIX, La basse haie, 35220 Chateaubourg, France; thierry.gidenne@inra.fr

Since the sainfoin (*Onobrychis viciifolia*) contains high ADF and ADL levels associated with a high level of protein, it could be a good alternative to alfalfa for rabbit feeding. Nowadays dehydrated pelleted sainfoin 'DpS' (PERLY variety) is available on the market (Multifolia company). However no informations are available about the nutritive value of DpS when incorporated in a complete balance pelleted diet for the growing rabbit. Thus we studied the effect of substituting 40% of the alfalfa in a control diet (C) with sainfoin (diet S), on digestion and performances of two groups of 16 rabbits housed in metabolic cages (8 cages of 2 per group), and fed freely C or S diets from weaning (28 d) to slaughter (70 d). DpS chemical composition was: crude protein 'CP'=17.3%, ADF=30.4%, ADL=12.0%. Chemical composition C and S pelleted diets were: CP=15.9 & 16.7% resp., ADF=19.9 & 23.4%. The S diet also differed from C diet by its tannin (1.8% vs 1.0% tannic acid equivalent) and its ADL concentration (8.4 vs 4.3%). A five days fecal collection period (60-64 d old) was performed to calculate the digestibility. Growth rate (28-70 d) was 5% lower in S than in C group (38.2 vs 40.2 g/d; P<0.05), while feed intake was 3% higher for S than for C group (121.5 vs 116.5 g/d, P<0.05), as was the feed conversion ratio (3.18 vs 2.90; P<0.01), probably in relation to the higher ADL level for S diet. Protein digestibility was 5 units lower in S compared to C groups (69.6 vs 75.3%; P<0.01), probably associated with the high tannin concentration. Energy digestibility did not differ between S and C diets (mean = 65.4%). Using the substitution method, the digestible energy 'DE' content of dehydrated sainfoin pellets, as a raw material, was calculated at 11.21 MJ/kg (40% higher than alfalfa), and digestible proteins content at 110 g/kg (similar to alfalfa). In conclusion, dehydrated pelleted sainfoin constitutes a good alternative to alfalfa, since it supplies energy and protein as well fibres, and particularly lignins essential for the growing rabbit.

Cardoon meal as a novel alternative feed: effect on performance and ruminal microbiome in lambs

S.A. Salami[1,2], B. Valenti[1], G. Luciano[3], M.N. O'Grady[2], J.P. Kerry[2], C.J. Newbold[4] and A. Priolo[1]
[1]University of Catania, Di3A, Animal Production Science, 95123 Catania, Italy, [2]University College Cork, School of Food and Nutritional Sciences, Cork, Ireland, [3]University of Perugia, Department of Animal Science, 06121 Perugia, Italy, [4]IBERS, Aberystwyth University, SY23 3DA Aberystwyth, United Kingdom; s.salami@umail.ucc.ie

Cardoon meal is a by-product generated after the extraction of oil from the seeds of cardoon plant (*Cynara cardunculus* var. *altilis*). To date, the potential of cardoon meal as an alternative feed in animal nutrition has not been explored. Fifteen lambs (19.58±2.01 kg) were randomly assigned to a commercial concentrate diet (CON, n=8) or CON containing 15% cardoon meal as a substitute for dehydrated alfalfa hay (CM, n=7) for 75 days pre-slaughter. Rumen contents were collected immediately after slaughter for ruminal fermentation and microbial profiling. Lambs fed CM had reduced dry matter intake without affecting performance traits such as final body weight, average daily gain, feed efficiency and carcass weight compared to CON-fed lambs. Dietary treatments did not influence ruminal fermentation traits. However, there was a trend (P=0.086) for higher NH_3-N concentration in CM-fed lambs. The absolute abundance of ruminal bacteria, methanogens, fungi and protozoa were unaffected by the diets. However, amplicon sequencing of 16s rRNA indicated that dietary treatments had a marked effect on the bacterial community structure with a lower population diversity in CM-fed lambs. Lambs fed CM had a higher abundance of phylum *Proteobacteria* and a lower abundance of *Bacteroidetes* and *Fibrobacteres* phyla compared to CON-fed lambs. These microbial changes were evident at the genus level as CM mediated a shift from hemicellulolytic (*Prevotella* and *Alloprevotella*) and fibrolytic (*Fibrobacter*) bacteria to an amylolytic bacterium (*Ruminobacter*). In conclusion, 15% inclusion of cardoon meal could substitute conventional protein-rich roughages, such as alfalfa, in concentrate diets of growing lambs without compromising animal productivity. However, the apparent effect of cardoon meal on the rumen bacterial community demands further investigation with respect to the impact on rumen function.

The effect of using DDGS for different periods on feedlot performance and meat quality of lambs

S. Karaca[1], S. Erdoğan[1], M. Güney[2], C. Çakmakçı[1], M. Sarıbey[1] and A. Kor[1]
[1]Yuzuncu Yıl University, Department of Animal Science, Faculty of Agriculture, Van, 65080, Turkey, [2]Yuzuncu Yıl University, Department of Animal Nutrition, Faculty of Veterinary, Van, 65080, Turkey; cakmakcicihan@gmail.com

The aim of this study was to determine the effect of using dried distillers' grains with solubles from corn (DDGS) for different periods on some physiological indicators, fattening performance and meat quality of lambs. The animal material consisted of 40 male lambs weaned at 3,5-4 months of age. Lambs were divided into 4 groups as follows; C120: fed with no DGGS containing diet for 120 days; D120: fed with DGGS containing diet for 120 days; D75: fed with no DGGS containing diet for 45 days + fed with DGGS containing diet for 75 days; D45: fed with no DGGS containing diet for 75 days + fed with DGGS containing diet for 45 days. Thus, all lambs were fed *ad-libitum* for 120 days. The substitution rate of DDGS in the diet was proposed to be 27.5% in total (comprised 100% protein requirement supplied soybean meal). Blood and rumen fluid samplings were performed on day 0, 30, 60, 90 and 120 at the fattening period. Average daily weight gain and final body weight were not influenced by dietary treatment. However, feed efficiency was better in D120 and C120 lambs than in other groups. Rumen fluid pH tended to higher in D120 lambs compared to C120 lambs ($P<0.10$). Serum BUN were found higher in C120 lambs than D120 lambs ($P<0.001$). Serum LDH linearly increased in all groups during the fattening period ($P<0.001$). Slaughter and carcass characteristics were found similar among groups. Moreover, the effect of treatment had limited effect on pH, color and water holding capacity of meat. On the other hand, total amount of trans fatty acids of intramuscular and depot fats were found higher in D120, D75 and D45 lambs than C120 lambs whereas total polyunsaturated fatty acids and conjugated linoleic acid contents were higher in all lambs fed with DDGS containing diet groups ($P<0.001$). Overall, these data suggest that corn DDGS can be fed in place of soybean meal as a protein source for at least 120 days in fattening period with no negative effects on feedlot performance and meat quality.

Supplementary carbohydrate source alters rumen nitrogen metabolism of beef heifers

S. Kirwan[1], T.M. Boland[1], A.K. Kelly[1], V. Gath[2], G. Rajauria[1] and K.M. Pierce[1]
[1]School of Agriculture and Food Science, University College Dublin, Lyons Research Farm, Celbridge, Naas, Co. Kildare W23 ENY2, Ireland, [2]School of Veterinary Medicine, University College Dublin, Belfield, Dublin 4, D04 V1W8, Ireland; stuart.kirwan.1@ucdconnect.ie

Grass silage (GS) is the predominant conserved forage fed to beef cattle in Ireland. The main fermentable carbohydrate substrates in GS are slowly fermented plant cell wall components, while the nitrogen (N) substrates are mainly soluble and rapidly available. This imbalance between energy and N supply in the rumen is considered an important cause of the low efficiency of N metabolism in the rumen and leads to excess N excretion and subsequent environmental problems. The objective of this study was to evaluate the impact of differing carbohydrate sources on rumen metabolism of beef heifers offered a 40:60 GS: concentrate diet on a DM basis. Six ruminally cannulated Belgian Blue/Holstein Friesian cross heifers (487±29 kg BW) were used in a 3×3 Latin square design. Experimental periods were 25 d (14 d for dietary adaptation and 11 d for sampling) and diets were offered twice daily as a TMR at 08:00 and 16:00 h to meet maintenance requirements. Dietary treatments were as follows: GS (11.6% CP) plus rolled barley (RB), maize meal (MM) or soya hulls (SH) based concentrate. Rumen fluid (250 ml) was collected via cannula at 1, 2, 4, 6, 8 h after feeding (am and pm) for analysis of ruminal pH, ammonia (NH_3) and volatile fatty acids (VFA) over a 48-hour period. Data was analyzed using the mixed procedure of SAS. There was a treatment × time interaction for rumen NH_3 concentrations ($P<0.01$). At 1 h post feeding MM supplemented animals had higher rumen NH_3 concentrations than RB supplemented animals ($P<0.05$), but this response was reversed at 4 and 6 h post feeding ($P<0.01$), while at 6 h post feeding NH_3 concentrations for the RB supplemented animals was higher than SH ($P<0.05$) supplemented animals. Independent of treatment rumen pH increased over time post feeding. In conclusion, while initially elevated ruminal NH_3 concentration was subsequently reduced in response to MM supplementation, while ruminal pH increased over time independent of treatment.

Effects of a high forage diet on production and metabolism in early lactation dairy cows

J. Karlsson, R. Spörndly, M. Patel and K. Holtenius
Swedish University of Agricultural Science, Department of Animal Nutrition and Management, Box 7024, 750 07 Uppsala, Sweden; johanna.karlsson@slu.se

Ruminants have a unique ability to produce milk and meat from fibrous feed and by-products not suitable for human consumption. However, high-yielding cows such as the Swedish generally have a high proportion of human edible concentrates in their rations, especially in early lactation. Most production related disturbances in early lactation occur due to energy deficiency and metabolic imbalance. Therefore, the aim of the present study was to investigate the performance of early lactation cows fed grass/clover silage *ad libitum* combined with a limited amount of concentrate based on by-products compared with cows fed a conventional ration of the same concentrate. It was hypothesized that cows offered a low concentrate ration in early lactation would maintain milk production but the metabolic status would be compromised. Thus, 22 cows were randomly allocated into two equal groups subjected to either a low concentrate ration or a high concentrate ration. Concentrate was gradually increased with 0.4 kg dry matter (DM) per day in both groups, from 2.6 kg DM at calving to 4.4 and 13.0 kg DM per cow and day, respectively. Blood plasma was collected and body condition (BC) scored at 2, 4 and 6 weeks after calving. The data was analysed by SAS 9.4 PROC MIXED except for BC that was analysed with PROC GLM. The results did not show any effects of concentrate ration on plasma concentrations of glucose, insulin, non-esterified fatty acids, beta-hydroxybutyrate or insulin-like growth factor 1, or on milk yield, milk composition or total DM intake. However, cows offered a low concentrate ration had a greater BC loss. In conclusion, cows in early lactation fed a high forage diet maintained milk production without significant negative effects on metabolism even though they had a greater loss of body tissue compared to those offered a high concentrate ration.

Does feeding extruded linseed to dairy cows improve reproductive performance – an observational study

T. Meignan[1,2], A. Madouasse[2], G. Chesneau[1] and N. Bareille[2]
[1]Valorex, La Messayais, 35210 Combourtillé, France, [2]BIOEPAR, INRA, Oniris, La Chantrerie, 44307 Nantes, France; thomas.meignan@oniris-nantes.fr

The impairment of the dairy cow reproductive performance is one of the key concerns for the modern dairy industry as it is closely related to the dairy herd profitability. Experimental approaches suggested that targeting n-3 polyunsaturated fatty acids (PUFA) in reproductive tissues could improve the follicle development and the oocyte and embryo quality and their environment. However, their impact on reproductive performances couldn't be highlighted in experimental designs due to poor statistical power. Thus the objective of this survey was to quantify the effect on reproductive performances of dairy cows in France fed with feeding extruded linseed (EL), a feed rich in α-linolenic acid. A retrospective study was carried out based on data from dairy herds enrolled in the official Milk Recording Scheme between 2008 and 2015 using Artificial Insemination (AI). Fertility was assessed by the indicator of non-return after a first or second AI between 18 and 200 days. The interval from calving to first service (ICS1) and the interval from calving to conception (ICC) were also investigated. Exposition to EL was obtained from all deliveries of commercial products containing EL. An AI was considered exposed when the cow received EL from calving to 17 days after AI. Two reference populations were used: (1) cows in non-exposed herds with comparative calving systems and located in the same areas, (2) cows in exposed herds that did not receive EL. Number of exposed AI was 200,497 in 2,189 herds with an average daily exposition of 401 g/cow (±285). On the entire database, the non return rate was 45.8%. The effects of different levels of EL exposure on fertility, and on ICSI and ICC will be assessed respectively using mixed-logistic regression model and multivariable Cox model, adjusted for main factors likely to influence the performance in the following weeks.

How high is herbage intake of organic rabbits grazing fescue or sainfoin?

H. Legendre[1], J.P. Goby[2], J. Le Stum[2], G. Martin[3] and T. Gidenne[1]
[1]INRA, GenPhySE, Université de Toulouse, 31326 Castanet Tolosan, France, [2]Université de Perpignan, IUT, 66962 Perpignan, France, [3]INRA, AGIR, 31326 Castanet Tolosan, France; heloise.legendre@inra.fr

Little is known about herbage intake of grazing rabbits. We determined herbage intake and pasture utilization of rabbits fattened under organic production standards. During spring 2016, 30 rabbits were raised for 9 weeks in rolling cages (3 rab. per cage with 0.4 m^2 of grazing area per rabbit, the minimum rate specified by the organic standards) from weaning at 43 days of age. 5 cages were moved on a pure stand of tall fescue (Festuca arundinacea, F), and 5 cages on a grassland dominated by sainfoin (*Onobrychis viciifolia*, S). Cages were moved every day, and rabbits were supplemented with 60 g/d/rab. (52.5 g dry matter, 'DM') of a complete pelleted feed (no refusal). Once a week, herbage samples were collected at two locations (0.25 m^2 each) for each cage: on the side of the cage to measure herbage allowance, and on the line of the cage, just after moving it to measure grazing refusals. Herbage intake was estimated as the difference between herbage allowance and refusals, and expressed per kg of metabolic weight (kg0.75). Over the whole grazing period, herbage allowance was higher in S than in F (5.1 ton DM/ha or 120.1 g DM/kg0.75 vs 2.2 ton DM/ha or 59.5 g DM/kg0.75, P<0.01), and herbage intake (59.2 g DM/kg0.75 vs 37.3 g DM/kg0.75, P<0.01) accounted for respectively 47% and 65% of the allowance. During the first grazing week, rabbits displayed limited herbage intake (resp. 15.8 and 7.8 g DM/kg0.75 for S and F, P=0.39) and utilization (16% and 20%). Thereafter, herbage intake increased and the maximum was reached on week 9 for S rabbits (112.3 g DM/kg0.75 or 947.5 g of fresh matter/2.5 kg rab., 70% of the allowance), and on week 5 for F rabbits (51.2 g DM/kg0.75, 60% of the allowance). Herbage allowance in F dropped during the last 4 weeks (46.8 g DM/kg0.75) yielding lower intake (35.3 g DM/kg0.75) but enhanced utilization, i.e. 77% in average, and up to 93% in week 6. In conclusion, rabbits develop a high herbage intake capacity during fattening (up to 40% of their bodyweight). When herbage biomass is too low (i.e. below 85 g DM/kg0.75), stocking rate per cage must be adjusted in order to increase herbage allowance per rabbit.

Consumer expectations regarding the production of hay and pasture-raised milk in South Tyrol

G. Busch, J. Ortner and M. Gauly
Free University of Bolzano, Universitätsplatz 5, 39100 Bozen-Bolzano, Italy; gesa.busch@unibz.it

In South Tyrol (Italy) the typical landscape in the alpine mountains is shaped by small-scale dairy farms and the use of grasslands, meadows and alpine pastures. This traditional landscape is part of the cultural heritage in the Alps and crucial for tourism. Farms and cultivated land run the risk of abandonment due to high input of manual labor and cost disadvantages compared to farms in less challenging locations. Traditional farming, in combination with high market prices for milk are key drivers for keeping mountain grasslands and pasture in production. Alternative marketing concepts for milk that achieve higher market prices can be one solution to remunerate mountain farmers for their efforts. In Austria for example, hay milk is already established in the market whereas in the Netherlands or Northern Germany, pasture-raised milk marketing concepts are implemented. Crucial for the long-term success of such marketing strategies is consumers' appreciation of the marketed aspects as well as their trust in the concepts implemented on farm-level. Against this background, this study analyses the expectations of South Tyrolean consumers regarding hay and pasture-raised milk. Using a standardized questionnaire, 171 residents from South Tyrol were questioned in an online-survey. The results show that the most frequent free associations with hay milk refer to 'barns', 'hay' and 'hay feeding' whereas associations with pasture-raised milk are 'pasture', 'naturalness' and 'free-roaming cows'. The intention to buy such milks is with 77% of participants for hay milk and 82% for pasture-raised milk rather high. Pasture-raised milk offers more advantages regarding animal welfare, sustainable production, health for dairy cows and the healthiness of the product from a participants' point of view (P≤0.05). Regarding feeding management for hay and pasture-raised milk, only 19 and 15% accept the use of concentrates, respectively. Allowing grass or maize silage and soy in both diets is rejected by more than 90% of participants. The results show that both concepts offer opportunities for milk marketing in South Tyrol, whereas positive effects on animals and environment are perceived to be higher for pasture-raised milk compared to hay milk.

Dairy cow milk protein responses to digestible lysine and methionine in the new INRA feeding system

S. Lemosquet[1], L. Delaby[1], D. Sauvant[2] and P. Faverdin[1]

[1]INRA, UMR PEGASE, 35590 Saint-Gilles, France, [2]AgroParisTech, INRA, UMR MoSAR, 75005 Paris, France; sophie.lemosquet@inra.fr

Dairy cow milk yield, protein yield (MPY) and content (MPC), were predicted in response to the supplies of net energy of lactation (NEL), metabolisable protein (PDI) using curvilinear plateau equations. These responses were combined to marginal responses to Lys and Met Digestible in the Intestine (LysDI and MetDI in % PDI). The responses to NEL and PDI were estimated in a dataset with LysDI on average at 6.7% PDI and MetDI at 1.9% PDI. The curvilinear plateau marginal responses of MPC (dMPC) and MPY (dMPY) to LysDI variations around 6.7% PDI were established on 30 experiments (exp) in which NEL, PDI and MetDI did not vary and on 87 treatments (treat). Similarly, the responses to MetDI variations around 1.9% PDI were established on 21 exp and 63 treat. If LysDI = 7% or MetDI = 2.4% of PDI, 95% of their respective dMPC increase was observed. The best model to predict the dMPC response to combined effects of LysDI and MetDI (RMSE: 0.43 g/kg) was to take the minimum between the linear part of the response to LysDI (dMPCLys = 0.62 + 1.75 × [LysDI – 6.7] with LysDI<6.7; 26 exp and 45 treat; RMSE=0.59 g/kg) and the curvilinear plateau response to MetDI (1.22 + 9.14 × [MetDI – 1.9] – 9.14 × 0.193×ln{1=exp[(MetDI – 1.9)/0.193)]}; RMSE=0.58 g/kg), in accordance with the barrel concept. Neither Lys nor Met post rumen infusions modified significantly milk yield (29.8±5.7 kg/d). The response of dMPY to both LysDI and MetDI was calculated as dMPC multiplied by milk yield estimated in response to NEL and PDI (RMSE=23 g/d on Lys and Met post rumen infusions database). The RMSEp of dMPC and dMPY obtained using a 2nd dataset for validation (38 exp of rumen protected Met lower and higher than 1.9 and 89 treat) were 0.36 g/kg and 22 g/d, respectively. If the dMPY response is added to the predicted MPY response to NEL and PDI (on a 3rd dataset using 105 exp on rumen protected Met and on dietary Lys supplies and 218 treat): MPYobserved = 8 (±13) + 1 (±0.01) × MPYpredicted with a slope not significantly different from 1. The RMSEp (42 g/d) was twice lower than if MPY was predicted using only NEL and PDI (RMSEp=87 g/d). Combining the effects of LysDI (% PDI) and MetDI (% PDI) to PDI and NEL improves MPY prediction.

Confounding effect of intake on milk protein yield response to change in NEL and MP supply

J.B. Daniel[1,2], N.C. Friggens[1], H. Van Laar[2] and D. Sauvant[1]

[1]INRA, AgroParisTech, Université Paris-Saclay, 75005 Paris, France, [2]Trouw Nutrition, R&D, P.O. Box 220, 5830 AE Boxmeer, the Netherlands; jean-baptiste.daniel@trouwnutrition.com

The separate effects of metabolizable protein (MP) supply and net energy (NEL) supply on milk protein yield (MPY) response are difficult to dissociate due to the related dry-matter intake (DMI) response, creating a correlation between the independent variables. In a recent meta-analysis, the within-experiment correlation between MP and NEL supply above maintenance was relatively high (r=0.65). The aim of this study was to assess the impact of this correlation on the estimation of MPY response to change in MP and NEL supply, centred on supply for which MP efficiency is 0.67 and NEL efficiency is 1 (centred variables were called ΔMP and ΔNEL, respectively). For that, the dataset used to quantify MPY response was split into three sub-datasets, according to the measured difference between highest and lowest DMI within experiment. Here only the two extreme datasets were compared. The subset 1, with the lowest variation in DMI had on average, a maximum difference between highest and lowest DMI of 0.47±0.24 kg/d (91 experiments, 242 treatments means) whereas the subset 2, with highest variation in DMI had on average, a difference of 2.62±0.99 kg/d (100 experiments, 328 treatments means). As expected, the within-experiment correlation between NEL supply and MP supply was lower in subset 1 than in subset 2 (r=0.36 vs 0.79). Using a mixed model to split between- and within-experiment variation, MPY response, obtained in subset 1 was: MPY = 979.5 (3.2) + 110.8 (19.8) ΔMP – 264.4 (48.3) ΔMP2 + 3.77 (0.49) ΔNEL (RMSE=26.6 g/d). The response obtained in subset 2 was MPY = 1001.4 (3.0) + 316.5 (23.2) ΔMP – 120.4 (36.4) ΔMP2 + 2.30 (0.26) ΔNEL – 0.017 (0.006) ΔNEL2 (RMSE=32.6 g/d). In both subsets, the interaction between ΔNEL and ΔMP was not significant and the quadratic effect of ΔNEL was only significant in subset 2. The effect of ΔMP on MPY was much greater in subset 2 than in subset 1. Conversely, the effect of ΔNEL on MPY was greater in subset 1 than in subset 2. This result demonstrates that the effects of ΔMP and ΔNEL on MPY were sensitive to the correlated DMI response.

Performance of dual-purpose types in comparison to layer hybrids fed a by-product diet

S. Mueller, R.E. Messikommer, M. Kreuzer and I.D.M. Gangnat

Agricultural Sciences/ETH Zurich, Universitaetstrasse 2, 8092 Zurich, Switzerland; sabine.mueller@usys.ethz.ch

A possible solution to avoid culling of newly hatched male layer cockerels is the use of dual-purpose (DP) types, with females used for egg and males for meat production. However, DP is coupled with a lower performance, which is unfavorable from a feed efficiency perspective unless a diet of lesser quality can be fed. Due to the concept of 'feed no food' and environmental concerns in soy production, alternative feed sources are sought. This study tested whether DP types would better tolerate than layers a diet without wheat and soy replaced by broken rice and wheat bran, molasses and protein sources like lupines. Ten individually kept hens (35-42 weeks in laying) each of three DP types (Lohmann Dual, LD; Belgian Malines, BM; Schweizer Huhn, CH) were compared with a layer hybrid (Lohmann Brown plus, LB). In a cross-over design, each animal received for 4 weeks a commercial layer diet (C; 11.5 MJ/kg ME and 168 g/kg CP) or an experimental diet (E; 10.4 MJ/kg ME and 183 g/kg CP). Effects of type, diet and interaction where tested with ANOVA. There were type effects in many performance and egg quality traits. Diet effects were found for hen performance but not for egg quality. Interactions occurred in feed intake and laying performance. Body weights (kg) of BM were highest (3.4) followed by CH (2.6) and were similar in LB (1.8) and LD (1.9). Feed intake was similar for both diets in BM and CH as well as for C in LB and LD and (101 to 121 g/d). A reduced feed intake occurred for LD and LB with E (80 and 65 g/d, respectively). Feed efficiency (g feed/g egg) was better in LB (2.7) and LD (2.8) than in CH (4.6) and BM (5.4) and was clearly more favorable for C (3.4) than for E (4.4). Laying performance was moderate (38.2 to 53.6%) for BM and CH and remained on the same level with E as with C, whereas for LB and LD it was higher with C (92.1 and 69.6%, respectively) than with E where the level declined to that of BM and CH. The average egg weights were similar for all types, but were higher for C (63 g) than for E (59 g). In conclusion, traditional DP types performed less well and thus tolerated the experimental diet, whereas not only the layer hybrids but also the novel DP type (LD) experienced serious constraints with diet E.

Dehydrated pelleted sainfoin for the growing rabbit: first results from intake and growth test

F. Tudela[1], M. Laurent[1], H. Hoste[2], M. Routier[3], P. Gombault[4] and T. Gidenne[1]

[1]INRA, GenPhySE, UMR1388, 31326 Castanet Tolosan, France, [2]Ecole Nationale Vétérinaire de Toulouse, UMR 1225 IHAP INRA/ENVT, 31076 Toulouse, France, [3]MG2MIX, La basse haie, 35220 Chateaubourg, France, [4]Multifolia, 1bis grande rue, 10380 Viâpres Le Petit, France; thierry.gidenne@inra.fr

Sainfoin (*Onobrychis viciifolia*) could be a good alternative to alfalfa for rabbit since it contains high ADF and ADL levels associated with a high level of protein. However no informations are available about its effect on intake and growth of the rabbit. Nowadays dehydrated pelleted sainfoin 'DpS' (PERLY variety) is available on the market (Multifolia company). Thus we designed two assays to evaluate the intake and growth of rabbit fed DpS (6 mm diameter, 15-23 mm length, hardness= 93 Mpa, crude protein=17.2%, ADF=30.4%, ADL=12.0%) as a sole feed (trial 1), or in free choice (trial 2) with commercial pelleted feed (CpF: 4 mm diam., 9-12 mm length, hardness= 6.5 Mpa, crude protein=15.9%, ADF=18.0, ADL=5.5%), and compared to a control group fed only the CpF. In trial 1, 2 groups of 30 rabbits (weaned at 39 d old) were housed in 2×6 cages, and fed freely either the DpS or the CpF, till 73 d old. In trial 2 (39-73 d old) one group of 96 rabbits (16 cages of 6) was fed freely the DpS and the CpF (in two separate feeders) and was compared to a control group (n=48, 8 cages of 6) fed the CpF only. Among the two trials, no health problems were detected on the animals. In trial 1, DpS intake reached 85.7 g/d (39-63 d period) and was similar to the intake of the control group (84.3 g/d; P>0.15). Although the DpS size and hardness was much higher than for CpF, the palatability of DpS was very high, since during the first week after weaning the intake was 78.5 g/d fo DpS, and 69.7 g/d for CpF (P<0.05). Growth of rabbit fed only DpS reached 18.5 g/d (39-73 d) and was logically lower than in control group (31.2 g/d, P<0.001). In trial 2 (39-73 d old) on a free choice basis rabbits ate 30% of DpS and 70% of CpF. The whole feed intake (DpS+CpF) was 23% higher than that of control (116.3 vs 94.2 g/d, P<0.05) and the rabbit growth was higher (34.6 vs 30.6, P<0.05). Thus dehydrated pelleted sainfoin seemed to be a good alternative to alfalfa as a fibre and protein source for the growing rabbit.

Performance of rabbit does provided with a lactation diet based on white lupine (cv. Zulika)

Z. Volek, L. Uhlířová and M. Marounek
Institute of Animal Science, Department of Physiology of Nutrition and Product Quality, Přátelství 815, 10400 Prague, Czech Republic; volek.zdenek@vuzv.cz

The aim of this work was to compare whole white lupine seeds (WL, cv. Zulika) with a commonly used crude protein sources for rabbit's diet. The effect of a WL diet on performance of does and their litters was observed. Two isonitrogenous and isoenergetic lactation diets were formulated. The first diet (SBM diet) contained soybean meal (130 g/kg as-fed basis) and sunflower meal (50 g/kg as-fed basis) as the main CP sources, whereas the second diet (WL diet) was based on WL (250 g/kg as-fed basis, cv. Zulika). No additional fat was added to any of the diets. A total of 34 Hyplus rabbit does (17 animals per treatment; after the 3rd parturition) were fed 1 of the 2 lactation diets during the entire lactation (32 days). Does were housed in modified cages which allowed controlled suckling and separate access of does and their litters to feed. The litters were standardized to 9 kits immediately after the birth. Litters were offered the lactation diets of their mothers from d 17 of lactation. Feed intake, feed conversion ratio and doe's live weight were not affected by dietary treatments. There was a higher daily milk production between d 2 and 21 (by 30 g; P=0.019) and for the entire lactation period (by 24 g; P=0.067) in rabbit does fed the WL diet than in does fed the SBM diet. The milk of does provided with the WL diet contained less SFA (P=0.015) and linoleic acid (P=0.018) and more oleic acid (P=0.001), α-linolenic acid (P=0.001) and eicosapentaenic acid (P=0.001), with a corresponding increase in the total PUFA n-3:arachidonic acid ratio (P=0.049). A higher daily weight gain between d 2 and 21 of lactation (by 1.4 g; P=0.079), as well as a lower solid feed intake (by 5.1 g; P=0.001) and higher milk intake:solid feed intake ratio (P=0.003) between d 17 and 32 of lactation of litters was observed in does fed the WL diet. It can be concluded that white lupine seeds (cv. Zulika) can be used as the main crude protein source for the rabbit's lactation diet, with a beneficial impact on milk production and fatty acid composition and the growth of their progeny. Supported by project NAZV QJ 1510136.

Replacement of soybean meal with sesame meal improves lactating performance of Awassi ewes

B.S. Obeidat and A.E. Aljamal
Jordan University of Science and Technology, Animal Production, Faculty of Agriculture, 22110 Irbid, Jordan; bobeidat@just.edu.jo

The objective of this study was to evaluate the effect of replacing soybean meal with sesame meal (SM) on milk production and composition, ewes' body weight (BW) change of Awassi ewes and growth performance [(weaning BW, total gain and average daily gain (ADG)] of their suckling lambs. Thirty Awassi ewes suckling single lambs were randomly distributed equally into three diets. These diets were no SM (SM0), 7.5% SM (SM7.5) or 15% SM (SM15) of dietary dry matter (DM) of diets fed *ad libitum* for 8 weeks to replace of the soybean meal. Nutrient intakes were determined daily throughout the experimental period. Body weights of ewes and lambs were measured at the commencement of the study and biweekly thereafter. On the same days, milk yield and composition were recorded. With the exception of Ether Extract intake, which was greater (P<0.05) in SM containing diets compared to SM0 diet, DM, crude protein (CP), neutral detergent fiber and acid detergent fiber intake were unaffected by treatment. No significant differences were noticed in the final BW of ewes among diets. Awassi ewes fed the SM0 diet tended to lose less (P=0.09) BW than ewes fed the SM diets. Lambs fed the SM diets had higher (P≤0.02) weaning BW, total gain and ADG than the SM0 diet. Milk production was higher (P=0.05) in ewes fed the SM7.5 and SM15 diets compared to the SM0 diet. No significant differences were observed in milk total solids (TS) and fat content whereas, CP content, daily TS and daily CP yield were greater (P≤0.05) in the SM containing diets than SM0 diet. Cost/kg milk production was the highest (P<0.05) in SM0 vs with SM containing diets. In summary, results of the study demonstrate the possibility of replacing the soybean meal with the sesame feed in diets of lactating Awassi ewes because it helped in reducing the cost of milk production and the cost of feed as well as it improved production performance.

Effect of dried corn distillers grains plus solubles, in late gestation ewes diets

S. Erdoğan[1], S. Karaca[1], M. Güney[2], A. Kor[1], C. Çakmakçı[1] and M. Sarıbey[1]
[1]Yuzuncu Yil University, Faculty of Agriculture, Department of Animal Science, 65080 Van, Turkey, [2]Yuzuncu Yil University, Faculty of Veterinary Science, Animal Nutrition and Nutrition Diseases, 65080 Van, Turkey; cakmakcicihan@gmail.com

This project was prepared to determine the effects of feeding with distilled grains+ solubles (DDGS) and soybean meal (SBM) on some production performance, and some metabolic hormones and metabolites in sheep. Thirty ewes carried single fetus, were determined by ultrasound 50 days after artificial insemination, were selected and equally divided into two treatment groups (DDGS or SBM) according to their age, BCS and live weights. Experimental diets contain DDGS or SBM were formulated as isonitrogenic, isocaloric and were fed from the d 105 of gestation through lambing. Blood samples were collected prepartum on d 105, 112, 119, 126, 133, 140 and 147 of the feeding period and after lambing at h 1 and 24. In the last period of pregnancy, the SBM and DDGS diets did not affect the live weight of dams and lambs at birth, but the weaning weight of the lambs of ewes from DDGS group was higher (P<0.05). There were effects of prepartum dietary treatments (P<0.0001) or dietary treatment × time (P<0.0001) on the concentrations of serum glucose, BUN, triglyceride and beta hydroxybutyric acid (BHBA) at the last 1/3 of the pregnancy stage. Concentrations of serum glucose and BUN were greater (P<0.0001) on d 105 of prepartum and decreased through d 112 of prepartum. Prepartum dietary treatments did not affect overall meanconcentration of insulin, leptin, IGF-1 and prolactin, but prolactin concentrations begin to increase after d 119 of pregnancy and reach the highest level in late pregnancy (P<0.0001). Serum IgG concentrations were elevated in lamb at h 24 after birth while serum IgG concentration was low in lambs at h 1 after birth (P<0.0001). In conclusion, prepartum winter feed source did not have detrimental effects on prepartum ewe performance. These results indicate that DDGS can be used effectively as an alternative protein source, in state of SBM by using 15% in total ration dry matter. TUBITAK P. No:115O657.

Effect of nutritional component and AFNs on *in vitro* gas production assessed by PCA

R. Primi[1], P.P. Danieli[1], P. Bani[2] and B. Ronchi[1]
[1]Università degli Studi della Tuscia, Dipartimento di Scienze Agrarie e Forestali (DAFNE), Via S.C. de Lellis, snc, 01100 Viterbo, Italy, [2]Università Cattolica del Sacro Cuore, Istituto di Zootecnica, Via Emilia Parmense, 84, 29100 Piacenza, Italy; ronchi@unitus.it

Chickpea (*Cicer arietinum*) could be an alternative to soy products, both under nutritional point of view and in terms of integration of livestock farming with sustainable cropping systems. However, since chickpea seeds contain some anti-nutritional factors (AFNs), the amount of seeds ingested by ruminants may affect positively or negatively their production and welfare. Principal component analysis (PCA) was applied to study and understand the main possible relationships among nutritional and AFNs component and the *in vitro* gas production (IVGP). Seeds from chickpea (cv. Pascià and Sultano) cultivated in Central Italy, were analyzed for main nutritive chemical constituents and for AFNs (total and condensed tannins, inositol phosphates, trypsin inhibitors and α-galactosides). IVGP was performed to assess the fermentability and DM degradation in a 48 h trial. PCA analysis indicates that the first two PCs explain the 76.4% of the total variance. The PC 1 is positively correlated with total tannins and trypsin inhibitors, evidencing a positive role in term of total GP, to offset the negative effect of the protein content of the grains. The PC 2 pointed out that IVGP rate is highly dependent on the total amount of carbohydrates rather than the protein/ AFNs ratio that, in turn play a role on the total gas production. A combination of good farming practices, that can be used to control the content of AFNs in seeds, and a modulation of quantity in rations, can make the chickpea a good substitute to soy in feeding ruminants.

Resistance vs tolerance to pathogens: back to basics?

I. Kyriazakis[1], G. Lough[2] and A.B. Doeschl-Wilson[2]
[1]Newcastle University, School of Agriculture Food and Rural Development, King's Road, NE1 7RU, Newcastle upon Tyne, United Kingdom, [2]The University of Edinburgh, Roslin Institute, Easter Bush, EH25 9RG, Midlothian, United Kingdom; ilias.kyriazakis@newcastle.ac.uk

There is an increased interest in the strategies a host may follow when dealing with exposure to pathogens, especially in the 'post-antibiotic' era. In these situations hosts may limit pathogen load (resistance) or may minimise the impact of the pathogen on performance or fitness (tolerance). The combination of these two traits, as their outcome on performance, is frequently referred to as resilience. From the host perspective, the superficial outcome of both resistance and tolerance can be the same: the retention of fitness. However, the two strategies may have very different outcomes at group performance level and with regards to how the pathogen spreads in a population. Understanding the basis of the strategies and hence deciding how to exploit them is currently hampered by several (methodological) issues: e.g. whilst resistance can be estimated at individual level, tolerance cannot; estimating tolerance requires measurements of both performance and pathogen load, and larger sample sizes to accurately estimate reaction norm slopes. In this presentation we explore novel ways of overcoming these challenges, such as utilizing repeated measures of pathogen load and performance in the statistical models, or expressing the two traits in a dynamic manner, so that we can reach conclusions about the relative contribution of resistance and tolerance to health and performance over time, and estimating host genetic effects on these. In the past it has been assumed that the two strategies may be antagonistically related, due for example to competition for resources. On the basis of the novel approaches indicated above, we show that the two strategies of resistance and tolerance do not necessarily have to be antagonistic, as has been assumed previously, and provide the example of tolerance and resistance to the PRRS virus as a case in point.

Genetic background of resistance, tolerance and resilience traits in dairy cattle

S. König
Institute Animal Breeding and Genetics, University of Giessen, Ludwigstr. 21b, 35390 Giessen, Germany; sven.koenig@agrar.uni-giessen.de

On a biological basis, resilience in cattle refers to the ability to adapt successfully to acute stress. Current indicators for resilience, but also for resistance and tolerance, are numerous and differently defined. Most commonly, they are based on reactions of the 'conventional trait' (e.g. productivity, reproduction, survival, subjectively scored disease traits) in dependency of an environmental descriptor (e.g. temperature, humidity). First, this contribution focusses on possibilities of genetic-statistical modelling (random regression or reaction norm models) for 'conventional traits', aiming at the identification of genetically robust and resilient animals. Second, different bio-markers defined as direct indicators of cattle resilience will be introduced. Potential bio-markers include heat shock proteins, hormone levels, serum metabolites, disease resistance via immunological blood factors, feed efficiency on hormone and enzyme levels, and other body fluids such as bull semen characteristics. From a genetics perspective, individual resilience is reflected by alterations of genetic variances, heritabilities and breeding values for physiological traits with environmental changes. A specific practical breeding approach addresses genetic line comparisons and intra-line genetic and genomic selections on endoparasite resistances. Using an established cross-classified research design for different Black and White selection lines, gastrointestinal nematode infections were lowest for selection lines being adapted to grassland systems. Mostly moderate heritabilities for faecal egg counts (e.g. 0.33 for gastrointestinal flukes in an own study) suggest inclusion of endoparasite traits into overall breeding goals, in order to achieve long-term selection response for disease resistance. Genomic marker data were used to identify single SNP or chromosome segments contributing to resistance against endoparasite infections. In this context, a selective genotyping approach will be discussed. A last aspect focusses on the complexity of immune genetics, i.e. studying aspects of cattle breed specific differences in immunoglobulin combinatorial diversity, somatic hypermutations and putative gene conversions on the 'way from genotypes to phenotypes'.

The variance of daily milk production as predictor for health and resilience in dairy cattle

H.A. Mulder[1], G.G. Elgersma[1], R. Van De Linde[2] and G. De Jong[2]
[1]Wageningen University & Research Animal Breeding and Genomics, P.O. Box 338, 6700 AH Wageningen, the Netherlands, [2]CRV BV, Wassenaarweg 20, 6843 NW Arnhem, the Netherlands; han.mulder@wur.nl

Automatic milking systems (AMS) record an enormous amount of data on milk production and the cow itself. This type of big data is expected to contain indicators for health and resilience of cows. In this study, the aim was to define and estimate heritabilities for traits related with fluctuations in daily milk production and to estimate genetic correlations with existing functional traits such as udder health, fertility, claw health, ketosis and longevity. We used daily milk production records of AMS of 212,433 lactations in parities 1, 2 and 3 from 498 herds in the Netherlands. We defined several traits related to the number of drops in milk production using Student t-tests (DROP) as well as the natural logarithm of the within-cow variance of milk yield (LNvar). ASReml was used to estimate heritabilities and breeding values (EBV) for the new traits. Genetic correlations were estimated using correlations between EBV of the new traits and existing EBV for health and functional traits correcting for non-unity reliabilities using the Calo-method. Heritabilities for the DROP traits and LNvar were around 0.1, similar to heritabilities for health and functional traits. Genetic correlations between the new traits and the existing health traits differed a lot, the strongest correlations (0.4-0.6) were between LNvar and udder health, ketosis, persistency and longevity. LNvar was in this study the best trait, based on the combination of the heritability and the genetic correlations with health traits. Selection of the 20 bulls with the lowest EBV for LNvar, i.e. less fluctuations in milk production, showed that the daughters of these bulls have better udder health i.e. less mastitis, less ketosis, better fertility and stay 150 days longer in the farm than average. This study shows that the variance in daily milk production is heritable and can be used to breed healthy and resilient cows.

Estimation of genetic parameters for rectal temperature in Holstein under heat stress in Beijing

W. Xu[1], X. Yan[1], L. Cao[1], H. Zhang[1], G. Dong[2], R. Zom[3], J.B. Van Der Fels[3], X. Li[2] and Y. Wang[1]
[1]China Agricultural University, No.2 Yuanmingyuan W., 100193 Beijing, China, P.R., [2]Beijing Sunlon Livestock Development Co.,Ltd, No.1 Disheng North Str., Daxing Dist., 100176, China, P.R., [3]Wageningen Livestock Research, P.O. Box 338, 6700 AH Wageningen, the Netherlands; wangyachun@cau.edu.cn

This study was aimed at estimating genetic parameters for rectal temperature (RT) in Holstein cow during summer in Beijing. The data included 6,446 lactating cows with 26,347 RT records. Morning RT (mRT), afternoon RT (aRT), test-day (TD) yield traits and somatic cell sore (SCS) were analyzed simultaneously using multiple-trait animal model, including farm×test-year, parity, stage of lactation, before/after milking, and environmental temperature within barn as fixed effects. The average mRT and aRT were 38.78±0.53 °C and 39.18±0.66 °C, respectively. Heritabilities (and SE) of mRT and aRT were estimated at 0.085±0.02 and 0.045±0.01, respectively. The estimated h^2 of aRT was lower than the published results (0.11-0.17). Moderate repeatabilities of RT (0.25-0.26) in current study was in agreement with literature (0.26), and implied the value of multiple measurements of RT for accurate assessment of dairy cow under heat stress. Estimated genetic correlation between mRT and aRT was 0.90±0.08. Positive low to moderate genetic correlations between mRT and yield traits and SCS implied that a balance between production and heat resistance ability can be achieved through careful selection. Totally 100 bulls with reliability of mRT EBV above 0.4 were selected. The range of their mRT EBV was from -0.15 to 0.41 °C, aRT EBV from -0.11 to 0.27°C, and TD milk yield EBV ranged from -2.14 to 6.62 kg. The rank correlation of these 100 bulls between mRT and TD milk yield EBV was 0.478, and that between aRT and TD milk yield EBV was 0.371. From 1983 to 2010, increasing genetic trends of TD milk yield and RT EBVs were observed, with an annual change of 0.029 kg, 0.003°C, and 0.002°C, for TD milk yield, mRT and aRT, respectively. In conclusion, RT belongs to traits with low heritability, however, needs to be considered within balanced breeding to avoid decreased heat resistance ability with increased yield, and mRT can be considered as a major measurement to evaluate the heat resistance ability in dairy cow.

Selection for feed efficiency and its consequences on fertility and health in Austrian Fleckvieh

A. Köck[1], M. Ledinek[2], L. Gruber[3], F. Steininger[1], B. Fuerst-Waltl[2] and C. Egger-Danner[1]
[1]ZuchtData EDV-Dienstleistungen GmbH, Dresdner Str. 89, 1200 Vienna, Austria, [2]University of Natural Resources and Life Sciences, Dep. Sustainable Agricultural Systems, Div. of Livestock Sciences, Gregor-Mendel-Str. 33, 1180 Vienna, Austria, [3]Agricultural Research and Education Centre Raumberg-Gumpenstein, Raumberg 38, 8952 Irdning, Austria; astridkoeck@gmx.net

This study is part of a larger project, the overall objective of which is to evaluate the possibilities for genetic improvement of efficiency in Austrian dairy cattle. In 2014 a one-year data collection was carried out. Data of approximately 5,400 cows (3,100 Fleckvieh (dual purpose Simmental); 1,300 Brown Swiss; and 1000 Holstein) kept on 167 farms were recorded. In addition to routinely recorded data (e.g. milk yield), data on new traits like body weight, body condition score, lameness, claw health, subclinical ketosis and data about feed quality and feed intake were collected. The specific objective of this study was to estimate genetic parameters for efficiency traits and to investigate their relationships with fertility and health in Fleckvieh cows. The following feed efficiency traits were considered: ratio of milk output to metabolic body weight (ECM/BW$^{0.75}$), ratio of milk output to dry matter intake (ECM/DMI) and ratio of milk output to total energy intake (ECM/INEL). Heritabilities of feed efficiency traits were moderate and ranged from 0.11 for ECM/INEL to 0.18 for ECM/DMI. More efficient cows were found to have a higher milk yield, lower body weight, slightly higher dry matter intake and lower body condition score. Cows with a higher efficiency had a longer calving interval and higher frequency of fertility disorders. Higher efficiency was, however, associated with a slightly better claw health due to the lower body weight of the cows and lower culling rate. Overall, cows with a medium efficiency combine both a high milk yield with good fertility and health.

Dairy cow robustness in fluctuating environments via trade-off analysis of life functions

L. Barreto Mendes, J. Rouel, B. Dumont, B. Martin, A. Ferlay and F. Blanc
INRA 1213 UMRH – VetAgro Sup, Centre de Recherche de Auvergne-Rhone-Alpes, 63122, Saint-Genès-Champanelle, France; luciano.mendes@inra.fr

With climate change and its implications on livestock systems, dairy herds reared based on natural grasslands are especially susceptible. Grassland-based systems are exposed to highly unpredictable fluctuations in terms of diet quality and quantity as well as to other environmental challenges such as thermal stress. In this context, we consider that a livestock system transition towards robustness is crucial. We assume that animal robustness is a function of genotype (breed), management and interactions between genotype and environment that generate individual variability. In this study, we describe a process that allows assessing the life functions trade-offs of individuals facing fluctuating environment, and we calculate an overall individual animal robustness factor (R). The proposed methodology was applied to a database comprising 435 dairy cows of three breeds (160 Prim'Holstein, 200 Montbeliarde and 75 Tarentaise), and a total of 1,174 lactations. The database included variables representative of life functions related to animal productivity (milk yield, fat and protein content, dry period duration), health status (body weight & condition, mastitis and lameness) and reproductive performance (success of insemination, calving interval). The statistical algorithm for calculation of overall robustness (R) consisted in the following steps: (1) linear mixed modeling procedures to model life functions with significant variables; (2) scenario analyses, where models are then used to predict deviations in life functions under nutritional or environmental changes as compared to average conditions; (3) trade-off analysis and (4) determination of the appropriate formula to estimate of R. First results indicate that milk yield is the function most negatively affected by nutritional variability, whatever the breed. Conversely, body weight remained stable across lactation cycles for all breeds, except for Tarentaise, which presented a slight increase in body weight when moving to extreme nutritional conditions. Future research will use the R indicator for identifying animal profiles and system-management practices that are the more resilient to various types of perturbations.

Microbial colonization is essential for proper development of the gut immune system in chickens

B. Kaspers[1], S. Lettmann[1], S. Härtle[1], S. Röll[1], B. Schusser[2], C. Schouler[3] and P. Velge[3]
[1]University of Munich, Veterinary Sciences, Veterinärstrasse 13, 80539 Munich, Germany, [2]Technical University of Munich, School of Life Science Weihenstephan, Liesel-Beckmann-Str. 1/III, 85354 Freising, Germany, [3]INRA, Infectiologie et santé publique bat 311, 37380 Nouzilly, France; kaspers@lmu.de

The development of the immune system is controlled by genetic programs and environmental cues. Microbial colonization of the gut after birth is critical for immune system development as shown in mice reared under germ free conditions. While mammals acquire the initial gut flora from their mother this is not the case in modern poultry production. Whether this impacts on the development and functional maturation of the immune system is largely unclear. We therefore set up a series of experiments to compare immune system development in sterile (germ free), mono- and tetra-reconstituted and SPF birds as well as birds receiving a maternal flora. While no significant differences were observed in the development of selected parameters of the innate immune system between the groups, striking differences were observed in the adaptive immune system. Germ free birds had highly reduced numbers of B-lymphocytes in the gut which was paralleled by a complete absence of IgA production in the gut and blood. IgA production was partially restored by mono-reconstitution and further enhanced in tetra-reconstituted and SPF birds. However, development immune system development was greatly retarded in all groups in comparison with birds that acquired a maternal flora. In contrast, neither B-cell maturation in the bursa of Fabricius nor circulating B-cell numbers were affected. These data predict that molecular cues induced by microbial colonization attract circulating B-cells into the mucosal tissue and regulate maturation towards IgA producing cells. Interestingly, these signals do not only activate homing and maturation of the B-cell compartment but also maturation of the epithelial IgA transporter system as poly-Ig receptor (PIGR) expression was very low in germ free birds but induced in response to microbial colonization. Collectively, our data highlight the importance of early microbial colonization of chickens to support immune system development, disease resistance and animal welfare.

The effect of environmental challenge on health and gut microbiota profiling in growing pigs

X. Guan[1], C. Schot[2], P. Van Baarlen[2] and F. Molist[1]
[1]Schothorst Feed Research, Meerkoetenweg 26, 8218 NA Lelystad, the Netherlands, [2]Wageningen UR, Host-Microbe Interactomics Group, De Elst 1, 6708 WD Wageningen, the Netherlands; xguan@schothorst.nl

Stress may negatively affect performance and welfare of pigs, partially via adverse effects on gut microbiota composition and immune status. To investigate this, the effects of environmental stress on health and gut microbiota composition were analyzed in 320 pigs half entire boars and half gilts weighing 23±2.2 kg. Ten pigs were housed in one pen and one room had 8 pens. During a 68-day experiment, 160 pigs were housed in 4 normal rooms as control, and the other 160 pigs housed in 4 different rooms were exposed to three types of challenges (cold environment, regrouping and fasting). On 8 different days, the temperature was dropped by 10 °C. On days 8, 25 and 41, 2 pigs from each pen were exchanged with 2 pigs from another pen in the same room. Fasting was induced on days 15 and 39. On days 42 and 68, from 32 boars from the control and 32 from the challenge group, blood and faecal samples were taken to measure blood urea nitrogen, pig major acute-phase protein, natural antibodies, specific antibodies against Salmonella typhimurium and faecal calprotectin. Additionally, 64 faecal samples on day 42 was selected for microbiota profiling (16S rRNA gene sequencing; Illumina MiSeq). The raw sequences were processed and clustered to Operational Taxonomical Unit (OTU). Data was analysed by one-way ANOVA. Microbiota sequence data was also analysed using LEfSe. Natural IgG antibody titers on day 68 tended to increase in the challenge pigs (titers of 4.3 vs 3.9, P=0.083). Challenge increased the abundance of *Clostridium perfringens* and *Lactobacillus mucosae* (P<0.05). The most abundant genera were *Lactobacillus* (36%), *Streptococcus* (14.0%), *Prevotella* (13.4%), SMB43 (4.7%), *Blautia* (3.9%), *Clostridium* (2.8%), *Coprococcus* (2.2%) and *Ruminococcus* (1.9%), which were not affected by challenge. Challenge decreased the median abundance of *Aerococcus* and *Campylobacter* (P<0.05), but increased for *Peptococcus* (P<0.05). Collectively, the stressors applied in the current model affect microbiota composition in the faeces of growing pigs, but these changes did not correlate with concurrent changes in five immune response parameters.

Betaine improves performance and reduces rectal temperature in broiler chickens during heat stress

M. Shakeri[1], J.J. Cottrell[1], S. Wilkinson[2] and F.R. Dunshea[1]
[1]*The University of Melbourne, Faculty of Veterinary and Agricultural Sciences, Royal Parade, Parkville 3030, Australia,*
[2]*Feedworks Pty Ltd, P.O. Box 369, Romsey 3434, Australia; mshakeri@student.unimelb.edu.au*

Efforts to control thermoregulation during heat stress (HS) divert energy from growth and therefore dietary amelioration strategies may improve the productivity of poultry meat production. The aim of this experiment was to quantify the benefits of supplementing Betaine (B) (1 g/kg), an anti-oxidant (AO) mixture of Selenium (0.5 mg/kg yeast) and Vitamin E (250 IU/kg, VE) and combination of Betaine and anti-oxidants (BAO) vs control (CON) on production in broilers. A total of 288 day-old-male Ross 308 chicks were randomly allocated to two climate rooms at thermoneutral (TN, 33 °C reduced to 25 °C by 21 d) or heat conditions (HS, 33 °C through daytime and 25 °C overnight from 21-42 d). Weight gain (WG), feed intake (FI) and rectal temperature (RT) were recorded weekly, whereas boneless breast (BR) and respiratory rate (RR) were measured at the end of the study. HS reduced final BW (2.78 vs 2.55 kg, P<0.001) and FI over the entire study (4.73 vs 4.52 kg, P<0.001) compared to their TN counterparts. Overall, B and BAO improved final BW compared to CON (P=0.003) with differences in BW becoming significant (P<0.001) at 4 wk. However, FI was not different amongst diets which indicated that B and BAO had improved FCR (1.79[a], 1.67[b], 1.76[ab] and 1.73[ab] for CON, B, AO and BAO, P<0.004, respectively). Under TN conditions B increased BR by 25% vs CON but no differences were observed for AO and BAO (598[a], 745[b], 589[a] and 583[a] g for CON, B, AO and BAO). In HS chickens B and BAO diets increased BR ~17% and 15% over CON respectively (500[a], 586[b], 498[a] and 574[b] g). By contrast, CON reduced BR by 16% under HS, whereas B and BAO were equivalent to TN CON. Importantly, 88% of the additional BW in the B group was accounted for increased BR during TN and 50% under HS conditions. RT and RR were lower for all dietary treatment groups when under HS respectively (RT: 42.04[a], 41.87[b], 41.76[b] and 41.84[b]; RR: 124[a], 107[b], 112[ab], 104[b]) indicating improved tolerance to HS conditions. In summary, HS hindered growth performance in Ross 308 broilers, which can be partially ameliorated by dietary Betaine.

Response of three divergent chicken genotypes to the most common nematode infections

M. Stehr, C.C. Metges and G. Daş
Leibniz Institute for Farm Animal Biology, Institute of Nutritional Physiology 'Oskar Kellner', Wilhelm-Stahl-Allee 2, 18196, Dummerstorf, Germany; stehr@fbn-dummerstorf.de

Responses of host animals to their pathogens may be associated with the direction of production that they have been selected for. The aim of this study was to compare responses of 3 divergent chicken genotypes, developed either for egg or meat production or for both purposes, to the most common nematodes, namely Ascaridia galli and Heterakis gallinarum. In a two-run-study, a total of 429 male chicks of 3 genotypes namely Ross-308 (R), Lohmann Dual (LD) and Lohmann Brown+ (LB+) were used. One-week-old birds were infected with 500 infective eggs of A. galli and H. gallinarum. Starting from 2 weeks post infection (wpi) on, randomly selected birds of each genotype were necropsied at weekly intervals until an age of 10 weeks (i.e. 9 wpi) to determine worm burdens with either nematode. Overall average A. galli burdens per bird decreased (P<0.001) over time linearly roughly from 96 worms at 2 wpi to 3 worms at 9 wpi, indicating the existence of a strong worm expulsion mechanism over time. Overall A. galli counts tended to be lower (P<0.06) in LB+ than in R birds with no interaction with time (P>0.05). Similar to A. galli, overall average burdens with H. gallinarum declined over time until 6 wpi, indicating the existence of an expulsion mechanism with this nematode, too. Starting by 7 wpi H. gallinarum burdens, nevertheless, increased linearly up to 9 wpi, implying involvement of re-infections with H. gallinarum. Although the three genotypes did not differ in Heterakis-burdens originating from the first generation worms (P>0.05), both LB+ and LD had significantly (P<0.05) higher number of larvae resulting from the reinfections than R. Whether H. gallinarum infections were accompanied by a concomitant Histomonas meleagridis infection, a hyper-parasite that can influence H. gallinarum burdens, has to be clarified yet. It is concluded that there is an effective mechanism responsible for expulsion of both A. galli and H. gallinarum in the chicken host. Effects of host genotype on the expulsion of first generation worms seem to be limited, although there is a clear difference among the genotypes in susceptibility to re-infections, possibly due to behavioral differences.

Serum IL10 provides separable resistance and tolerance traits in chicken response to coccidiosis

K. Boulton[1], S. Bush[1], R. Lawal[2], A. Psifidi[1], M.J. Nolan[3], L. Freem[1], Z. Wu[1], L. Eory[1], R. Hawken[4], M. Abrahamsen[4], O. Hanotte[2], F.M. Tomley[3], D.P. Blake[3] and D.A. Hume[1]
[1]The Roslin Institute and Royal (Dick) School of Veterinary Sciences, University of Edinburgh, Easter Bush, Midlothian, EH25 9RG, United Kingdom, [2]The University of Nottingham, School of Life Sciences, University Park, NG7 2RD, United Kingdom, [3]Royal Veterinary College, University of London, Pathobiology and Population Sciences, Hatfield, AL9 7TA, United Kingdom, [4]Cobb-Vantress Inc, P.O. Box 1030 Siloam Springs, AR 72761-1030, USA; kay.boulton@roslin.ed.ac.uk

While breeding for resistance to *Eimeria* spp. the cause of coccidiosis in chickens, offers one route to mitigation of their impacts on production, tolerance to the disease may be less desirable, with potential impacts on disease transmission and welfare. In a large-scale study seeking evidence of selectable variance in response to coccidiosis, correlations were found between IL10 (serum interleukin-10, an anti-inflammatory cytokine) and intestinal damage. Eigen analysis to dissect the relationships between these variables the production trait, weight gain, produced three distinct vectors. Visualisation of the Eigenvectors confirmed the separation of resistant and susceptible chickens, with both producing high levels of IL10. A third, tolerant subpopulation produced little or no IL10. A GWAS uncovered novel suggestive genome-wide significant SNPs for the three measured traits, although the proprietary 62K SNP array (Cobb-Vantress) used did not cover the location of the chicken IL10 gene. However, SNPs relating to IL10 were located in the region of IL10R2 on the chicken genome. Upregulation of this gene is essential for the production of IL10 in innate immune response to infection. To further investigate, WGS data obtained from >20 broilers and layers are being compared with the Red Jungle Fowl genome to locate selective sweeps for this and other IL10 receptors, plus the IL10 gene itself, to identify mutations in these genes that may relate to disease tolerance.

Response to nematode infections in a high and a lower performing layer chicken genotype

M. Stehr[1], C.C. Metges[1], M. Grashorn[2] and G. Daş[1]
[1]Leibniz Institute for Farm Animal Biology, Institute of Nutritional Physiology 'Oskar Kellner', Wilhelm-Stahl-Allee 2, 18196, Dummerstorf, Germany, [2]University of Hohenheim, Institute of Animal Science, Garbenstr. 17, 70593 Stuttgart, Germany; gdas@fbn-dummerstorf.de

Responses of host animals to their pathogens may be associated with their performance levels determined by their genetic makeup. We investigated whether two chicken genotypes varying markedly in their laying performance respond to nematode infections differently around the peak of egg production. In total, 180 laying hens of a high performing (Lohmann Brown plus, LB+) and a lower performing genotype (Lohmann Dual, D) were either infected at an age of 24 weeks with 1000 eggs of two common nematodes (*Ascaridia galli* + *Heterakis gallinarum*) or kept as uninfected controls. Infected hens were necropsied at 2, 4, 6 weeks post infection (wpi) to determine their infection intensities with both nematodes (n=7-12 hens/genotype and wpi). Laying performance, egg weight and feed intake were determined for 6 wpi at pen levels. Basic egg quality traits (e.g. weight, color and thickness of egg shell, weight and color of egg yolks, Haugh unit) have been assessed so far at 2 and 4 wpi. Individual worm counts with either species did not differ significantly between the two genotypes at 2, 4 and 6 wpi (P>0.05). No significant effect of nematode infections could be quantified on egg quality traits up to 4 wpi (P>0.05). Laying performance of the lower performing genotype (D) was not influenced by the infections (P>0.05), whereas, the high performing genotype (LB+) showed an impaired performance (ca. 9%) due to infection, which was apparently mediated through a lower feed intake (ca. 8%). The preliminary results suggest that genotypes designed to perform high are also able to resist their natural pathogens as the lower performing genotypes whilst penalizing their performance, indicating prioritization of fitness over performance at critically important situations (e.g. peak of laying period) with high physiological pressure.

Genome-wide association study of humoral immunity in chicken using Bayesian analysis

H. Motamed, A. Ehsani and R.V. Torshizi
Tarbiat Modares University, Animal science, Jalal-e-al ahmad highway, Nasr Bridge, 14115-336 Tehran, Iran;
alireza.ehsani@modares.ac.ir

Several statistical methods have been used to identify the association between genomic variables and the phenotypes of complex traits. Single marker regression based on maximum likelihood estimation as well as least square analysis are the most well-known methods up to now. Because these methods are based on a normal distribution for the effects of genomic variables on the traits they may not be appropriate. This is more important for traits with low heritability or difficult-to-measure traits such as immunity-related traits. Studies showed that even when the null hypothesis is rejected using classical statistics, there is still a substantial probability that the null is true and Bayesian methodology may be useful as a complement to p-value. In this study, a random regression Bayes C_{pi} methodology was applied to identify important genomic regions associated to several immunity-related traits in chicken. The population was a F_2 descended from a reciprocal cross between a fast growing commercial chicken and an indigenous Iranian fowl. After quality control, 43,821 SNPs were used analyzed using GS3 software. The results were compared with single SNP fixed model analysis and there was a substantial overlap between the two methods, but still differences were found in many regions. Genetic parameters were estimated using DMU software. The value of π (markers with effect) set to be 0.0005 and the significance threshold greater than 20 was applied for Bayes factors. Although there were no significant SNPs in single SNP analysis with p-value threshold of 0.05, 11 genomic regions on chromosomes 1, 2, 4, 7, 12, 13, 21, 22, 24, 26, 28 were identified using Bayes C_{pi}. The results showed a wide dispersion of causal variables for these traits confirming a polygenic inheritance of the observed traits and explaining why the heritability is low. SNP markers at the closest distance to important genes included ISX, EIF3H, ACOX3, SPPL2B, SLC6A6, EBF1, ACTG2, PLA2G2E, NFRKB, IKBKE, GLI2 genes. These genes are responsible for important functions in relation to immunity. Moreover, the results showed that the Bayesian method is more powerful then the classic method.

Effects of creatine and/or betaine on performance of heat-stressed broiler chicks

H. Al-Tamimi, K. Mahmoud and H. Bani-Khalaf
Jordan University of Science and Technology, Animal Science, Faculty of Agriculture, Irbid, Amman Street, 22110, Jordan; hjaltamimi@just.edu.jo

Improving broiler performance is still an essential task in animal production, especially under certain environmental challenge conditions. The present trial was conducted to assess the effects dietary betaine and/or creatine on productive and thermoregulatory responses of broiler chicks. A total of 440 day-old Indian River broiler chicks were reared to market age (5 weeks). Chicks were randomly distributed four treatment groups; a basal diet acted as the control (CONT) group; a second group (BETA) received 1 g betaine/kg feed, a third group received 1.2 g creatine monohydrate/kg feed (CRET), and finally a forth group received a combination or both supplements (COMB). At 31-d of age, 20 chicks from each group were exposed to acute heat stress for three hours (A-HS; ambient temperature = 34.45±0.20 °C, RH%=37.60±0.28), and hemogramic profiles screened before and after. Feed intake, body weight and feed conversion ratio were assessed at weekly intervals. Final carcass analyses (dressing percentage, pH, cooking loss, water holding capacity, and shear force) were measured. Redness of breast was higher (P=0.011) due to BETA and CRET treatments separately, than the CONT group, but not different when used combined (COMB); 26.74, 24.46, 31.57, 31.27, respectively. Compared to the CONT, dietary supplements alleviated hyperthermia responses, with BETA alone being more efficient than CRET or COMB treatments, especially 150 minutes of A-HS, as reflected by significant (P<0.02) core body temperature values (43.74, 42.47, 43.46 and 43.41±0.28 °C, respectively). The mitigation of hyperthermia seemed to be mediated via enhancement of water balance indicators (an increase of 7.1% packed cell volume in the CONT, compared to a drop of pooled 9.3% in the three other groups). Likewise, hemodilution was parallel to the hematocrit responses. Although not efficient in improving growth performance, BETA and/or CRET are promising dietary additives to alleviate heat stress in finishing broiler chicks.

MHC-B variability in the Finnish Landrace chicken conservation program

M.S. Honkatukia[1], M.E. Berres[2], J. Kantanen[1,3] and J.E. Fulton[4]
[1]Natural Resources Institute Finland (Luke), Green Technology, Myllytie 1, 31600 Jokioinen, Finland, [2]University of Wisconsin, Department of Animal Science, 1675 Observatory Drive, Madison, WI 53706, USA, [3]University of Eastern Finland, Department of Environmental and Biological Sciences, Yliopistonranta 1, P.O. Box 1627, Finland, [4]Hy-Line International, 2583 240th St, Dallas Center, IA 50063, USA; mervi.honkatukia@luke.fi

The Finnish Landrace chicken is adapted to free range and modest living conditions in the northern European climate. The native breed was almost completely replaced by the large-scale egg production hybrids in the 1950s. Only remnants of the native chicken survived and several unknown breeds were likely introduced into the ancestral population in the last century. Conservation efforts to protect this endangered breed were initiated by a hobby breeder in the 1960s. However, the official conservation programme was established in 1998. The conservation program aims to insure breed purity of 12 landrace lineages originally obtained from isolated villages in Finland. The program relies on a network of over 400 non-professional breeders. The chicken Major Histocompatibility Complex B region (MHC-B), located on chromosome 16, contains over 45 highly polymorphic immune response genes. Variation within the MHC has been shown to enhance resistance to multiple viral, bacterial and parasitic pathogens. To assess variability in the chicken MHC-B region in the Finnish Landrace breed, a panel of 90 SNPs encompassing 210 kb was utilized. A total of 195 samples from 12 distinct lineages (average of 15 individuals sampled per population) were genotyped. Variation in the MHC-B region in these populations was examined using the MHC SNP panel. A total of 36 haplotypes were found, 16 of them had been previously identified in other breeds and 20 haplotypes were novel. The average number of MHC-B haplotypes within each lineages was 5.9, ranging from 1 to 13. Some of the haplotypes were lineage-specific. This study shows that substantial MHC-B region diversity exists within the Finnish Landrace breed. Abundant variability within the MHC region indicates overall health and robustness of Finnish Landrace chicken lineages. Finally, the results exemplified successful implementation of the conservation program.

Effect of peri-implantational undernutrition on interferon stimulated gene expression in beef cattle

E. Molina[1], B. Serrano-Pérez[1], A. Noya[2], I. López-Helguera[1], A. Sanz[2], I. Casasús[2] and D. Villalba[1]
[1]Universitat de Lleida (UdL), Animal Science, Av. Alcalde Rovira Roure, 191, 25198 Lleida, Spain, [2]Centro de Investigación y Tecnología Agroalimentaria de Aragón (CITA)-IA2, Avda. Montañana, 930, 50059 Zaragoza, Spain; icasasus@cita-aragon.es

For pregnancy to occur, the maintenance of a functional corpus luteum requires a complex embryo-maternal crosstalk. Reduced nutrient intake during early pregnancy could impair the pregnancy recognition mechanisms in cattle. The aim of this study was to analyse the effects of peri-implantational undernutrition on the expression of Interferon tau Stimulated Genes (ISG) during early pregnancy in two beef cattle breeds. Thirty-five Parda de Montaña (PA) and 19 Pirenaica (PI) multiparous cows were synchronized to estrus and artificially inseminated (AI). Pregnancy was confirmed by transrectal ultrasonography on day 37 post-AI. Dams were randomly allocated to a control (CONTROL, n=30) or nutrient-restricted (SUBNUT, n=24) group, which were fed at 100 or 65% of their estimated energy requirements during the first 82 days of pregnancy. Blood samples were drawn on day 18 after AI. Gene expression of ISG15, OAS1, MX1 and MX2 in peripheral blood mononuclear cells was analyzed by real time PCR. The effects of pregnancy, breed of dams and undernutrition and possible interactions of paired factors on ISG expression were assessed by proc MIXED with SAS and JMPro statistical software (SAS Institute Inc., Cary, NC). Pregnant dams had a higher gene expression of ISG15, MX1 and MX2 compared to those non-pregnant (P<0.001). Breed and undernutrition did not affect ISG gene expression. However, interaction between pregnancy, breed and undernutrition showed that mRNA OAS1 expression was significantly downregulated in SUBNUT PI pregnant dams (P=0.0325), suggesting different immune responses to dietary deficiency in these two cattle breeds during the pregnancy. To sum up, undernutrition during the peri-implantation period down-regulated ISG expression in pregnant PI dams but not in PA dams. This study received financial support from the Spanish INIA RTA2013-059-C02.

Effect of heat stress on metabolic and antioxidant responses in periparturient dairy cows

R. Turk[1], N. Rošić[2], S. Vince[1], Z. Flegar-Meštrić[3], S. Perkov[3], L.J. Barbić[1] and M. Samardžija[1]
[1]*Faculty of Veterinary Medicine, University of Zagreb, Heinzelova 55, 10000 Zagreb, Croatia, [2]Veterinary Practice Jastrebarsko, Trešnjevka 61, 10450 Jastrebarsko, Croatia, [3]Merkur University Hospital, Zajčeva 19, 10000 Zagreb, Croatia; rturk@vef.hr*

Global warming greatly affects performance and production of dairy cows having a significant impact on dairy industry. Exposure to heat stress decreases feed intake of cows compromising energy metabolism and immune system which might reduce disease resistance and production potential. The objective of this study was to evaluate metabolic and antioxidant responses to heat stress during summer in periparturient dairy cows. The study was conducted on 24 Simmental dairy cows kept in a barn with favourable air flow condition. The cows were assigned into two groups according to season: summer (S group, n=12) and autumn (A group, n=12). Blood samples were taken on days -21, -7, 8, 16, 24, 32 and 40 relative to parturition. Serum nonesterified fatty acids (NEFA), β-hydroxybutirate (BHB) and macro mineral (Ca, K, Na, Cl) concentrations were assayed spectrometrically by commercial kits (Randox, UK). Serum superoxide-dismutase (SOD) activity was assayed by the quantitative ELISA method while paraoxonase-1 (PON1) activity was measured spectrometrically by the method of hydrolysis of paraoxon. The heart and respiratory rates and rectal temperature were monitored at each sampling time. The average temperature-humidity index (THI) was statistically higher in summer (78.9) than in autumn (58.6) indicating moderate heat stress in the S group. Rectal temperature and the respiratory rate were significantly higher in the S group (39.1 °C and 50, respectively) than in the A group (38.7 °C and 27, respectively) while the heart rate was not affected. Serum SOD was significantly lower in summer (24 U/ml) than in autumn (39 U/ml) while PON1 activity was only slightly lower in summer. Serum NEFA was not affected while BHB was significantly lower in summer. Electrolyte balance was altered as well. The results indicated that dairy cows responded to heat stress by complex metabolic adaptation affecting metabolic homeostasis and antioxidant protection aiming to counteract detrimental effects of reduced nutrient and energy intake.

Effect of breed and pre-weaning diet on the response of beef calves to abrupt weaning

I. Casasús[1], S. Yuste[2], J. Ferrer[1], A. De Vega[2], M. Fondevila[2] and M. Blanco[1]
[1]*CITA Aragón-IA2, Montañana 930, 50059 Zaragoza, Spain, [2]Universidad de Zaragoza-IA2, Miguel Servet 177, 50013 Zaragoza, Spain; icasasus@aragon.es*

The stress response to weaning and start of intensive fattening were studied in thirty female calves of two breeds (15 Parda de Montaña, PM, and 15 Pirenaica, PIR) subjected to three different feeding strategies during lactation (milk only, MO; milk+hay, MH; milk+concentrate, MC; supplements offered *ad libitum*). At 150 d of age, calves were abruptly weaned from their dams, placed in an adjacent barn and fed on *ad libitum* concentrate and barley straw. Blood was collected at -24 (basal), +6, +24 and +168 h from weaning for analysis of cortisol, fibrinogen, creatine-kinase, leukogram, haematocrit, erythrocyte count and haemoglobin content. Data were analysed with mixed linear models, where interactions between diet or breed and sampling time were never significant. A typical stress response was observed, with increasing fibrinogen, creatine-kinase and cortisol concentrations after weaning (highest values at +6 or +24 h), and return to basal values at +168 h (except for fibrinogen). The leukogram showed a transient neutrophilia and lymphopenia observed at +6 h, and monocytosis and erythrocytopenia that had disappeared by +168 h. The pre-weaning diet affected plasmat cortisol and leukocyte populations, MH calves showing the highest values (10.7, 16.1 and 13.3 ng/ml for MO, MH and MC, P=0.02) but the lowest neutrophil:lymphocyte ratio (0.49, 0.33 and 0.59, respectively, P<0.001). Unsupplemented calves during lactation had the lowest haematocrit (32.7, 35.5 and 37.8% P<0.001) and haemoglobin content (11.5, 12.2 and 12.9 g/dl, P<0.001), but clinical anemia was not observed. Breed influenced concentrations of creatine-kinase (231 and 372 IU/ml in PM and PIR, P=0.01) but not those of fibrinogen and cortisol nor the leukogram. Parda calves had lower erythrocyte count (10.5 and 11.3 10^{-6}x cells/μl in PM and PIR, P=0.01) and haemoglobin content (11.8 and 12.5 g/dl, P=0.04), with larger cells but lower corpuscular haemoglobin concentrations. Providing supplements prior to weaning improved overall health but did not affect the pattern of stress response of calves from both breeds.

Milk yield and composition in dairy goats with different serostatus to SRLVs, PTB and CpHV1

C.E. Pilo[1], G. Puggioni[2], M. Bitti[3], M. Pazzola[1], G. Piazza[2], A.P. Murtino[2], G.M. Vacca[1] and N. Ponti[2]
[1]University of Sassari, Veterinary Medicine, via Vienna, 2, 07100 Sassari, Italy, [2]Istituto Zooprofilattico Sperimentale della Sardegna, Sassari, via Duca degli Abruzzi, 8, 07100 Sassari, Italy, [3]Associazione Interprovinciale Allevatori di Nuoro e Ogliastra, via Alghero, 6, 08100 Nuoro, Italy; gmvacca@uniss.it

SRLVs, PTB and CpHV1 are important diseases and can cause financial losses in dairy small ruminants herds. To evaluate the effects of the seropositivity for Caprine SRLVs, PTB and CpHV1 on milk yield and composition, 1,059 lactating goats, in 31 farms including Saanen (SA), Alpine (AL), Maltese (MA), Murciano-Granadine (MG) and Sarda (SR) breeds were studied. Milk and blood samples have been taken and milk yield recorded. Fat, protein, lactose, pH, SCC (Somatic Cells Count) and TMC (Total Microbial Count) were determined. Antibody rates for PTB and CAEV in milk by ELISA and CpHV1 in blood serum, by seroneutralization, were determined. The 18.1% of goats were positives for PTB, with the lowest prevalence in the AL (9.2%) and higher in MA (31.4%). The 53.0% were positive to the SRLVs (min MG=19.5%; max SR=78.8%) and the 16.2% for CpHV1 (min MG=0.5%; max SR=51.9%). The association analysis was performed by a GLM including breed, DIM (Days In Milk), parity and serostatus of the three diseases as fixed effects. Differences between the breeds ($P<0.01$) were evidenced for daily milk yield (min SR=1,02; max SA=2,84 g/d), fat (min SA=3.4; max SR=5.1%), protein (min SA=3.2; max SR=4.1%) and lactose (min SA and MA=4.5; max SR and MG=4.7%). DIM affected ($P<0.01$) protein, lactose and SCC, while parity had an effect on daily milk yield ($P<0.05$), lactose and SCC ($P<0.01$). None of the performance parameters was affected by the seropositivity for PTB, while the goat positive to SRLVs shown higher values ($P<0.05$) of TMC and those positives for CpHV1 a lower ($P<0.01$) fat content (4.29 vs 4.56%). Whereas PTB and CpHV1 prevalence rates are reduced in allochtonous goats, the high percentage of CAEV and CpHV1 positive subjects in the Sarda breed should be urgently considered, in order to implement the sanitation measures needed. Research funded by Regional Sardinian Government (LR7/2007, 2013).

Association analysis between serological response to SRLVs and PTB with milk traits in Sarda ewes

N. Ponti[1], G. Puggioni[1], E. Pira[2], M.L. Dettori[2], P. Paschino[2], G. Pala[1], M. Pazzola[2] and G.M. Vacca[2]
[1]Istituto Zooprofilattico Sperimentale della Sardegna, Sassari, via Duca degli Abruzzi, 8, 07100 Sassari, Italy, [2]University of Sassari, Veterinary Medicine, via Vienna, 2, 07100 Sassari, Italy; gmvacca@uniss.it

SRLVs are a group of viruses affecting sheep and goat, responsible for chronic debilitating disease known as Maedi Visna; ovine PTB is a chronic enteritis caused by *Mycobacterium avium* subsp. *paratuberculosis*. Both diseases can cause important economic losses in infected herds. The aim of this research was to assess the effect of serological response at SRLVs and PTB on milk yield, composition and renneting properties in sheep milk. From 23 herds located in Sardinia (Italy), 1,071 lactating Sarda ewes, in healthy conditions, have been sampled. Daily milk yield has been recorded and milk sampled. Milk has been analysed for fat, protein, casein, lactose, pH, SCC, TMC (log transformed in SCS and TBC) and renneting properties (RCT, k20, a30). In milk, antibody rates for SRLVs and PTB have been evaluated by ELISA. Data were analysed by a GLM including herd, DIM, parity, SRLVs and PTB as fixed effects. The 40.2% of the samples were positives for SRLVs and 10.6% for PTB. Only 1 herd was negative for SRLVs and 3 were negative for PTB. Prevalence within the herds ranged from 0 to 97.3% for the SRLVs and from 0 to 36.11% for PTB. No herd resulted negative to both the diseases. Association analysis shown for ewes positives to SRLVs higher values ($P<0.05$) for SCS (5.0 vs 4.5 units) and TBC (2.6 vs 2.5 units). Samples that were positive for PTB showed higher values ($P<0.01$) for fat (7.1 vs 6.4), protein (5.7 vs 5.4), casein (4.5 vs 4.2) and RCT (10.7 vs 8.8 min) and higher ($P<0.05$) pH values (6.70 vs 6.67 respectively). As a whole, ewes seropositive to SRLVs provided milk with lower health and hygiene characteristics and those seropositive for PTB milk with a slower coagulation time. In the positive ewes we have not highlighted a worsening in milk traits. Althought this, the danger of these diseases must not be underestimated. Research funded by Fondazione di Sardegna.

Gene expression signatures in bone marrow-derived dendritic treated with prebiotic and probiotic

A. Slawinska[1], A. Plowiec[1], A. Dunislawska[1], M. Siwek[1], K. Zukowski[2], N. Derebecka[3], J. Wesoly[3] and H. Bluyssen[3]
[1]*UTP University of Science and Technology, Mazowiecka 28, 85-084 Bydgoszcz, Poland, [2]National Research Institute of Animal Production, Krakowska 1, 32-083 Balice, Poland, [3]Adam Mickiewicz University, Umultowska 89, 61-614 Poznan, Poland; slawinska@utp.edu.pl*

Dendritic cells (DCs) belong to intestinal antigen presenting cells (APCs). The morphology of DCs allows them to penetrate tight junctions of the gut epithelial tissue and sample intestinal antigens. DCs can differentiate between pathogenic and beneficial intestinal antigens and prime either immune activation or tolerance. Prebiotics and prebiotics are dietary supplements that improve content of the intestinal microflora. They can also cause immunomodulation by communicating with intestinal APCs, including DCs. In this study we attempted to determine gene expression signatures in chicken bone marrow-derived DCs (BMDCs) stimulated with a prebiotic and probiotic *in vitro*. Chicken BMDCs were differentiated *in vitro* from bone marrow cells collected from 18-day-old embryos of Green-legged Partridgelike. Cell differentiation into DCs lineage was carried out with two chicken-specific cytokines: IL-4 and GM-SCF. On day 7 of *in vitro* culture, BMDCs were stimulated with either galactooligosaccharide (GOS) prebiotic or *Lactococcus lactis* subsp. *cremoris* probiotic. After 6 h of stimulation, BMDCs were harvested and RNA was isolated. Total RNA sequencing was performed with HiScan (Illumina). Differential gene expression was determined. Gene expression signatures upon BMDCs stimulation with prebiotic and probiotic were determined. Up-regulated genes included nitric oxide synthase (*iNOS*); interleukines *IL-1B* and *IL-4*; chemokines: *IL-8* and *CCL5*; and immunoresponsive gene 1 (*IRG1*). The most distinct gene expression signatures for prebiotic and probiotic stimulation of BMDCs will be validated at protein level. Further, validated proteins will serve as indicators of immunomodulatory role of prebiotics and probiotics *in vivo*. The results will be used in selecting immunomodulatory prebiotics and probiotics for chickens. Acknowledgements: the research was supported by a project UMO-2013/11/B/NZ9/00783 (NSC, Cracow, Poland)

Toll-like receptor signaling by prebiotic and probiotic in chicken macrophage-like cell line

A. Slawinska[1], A. Plowiec[1], A. Dunislawska[1], M. Siwek[1], K. Zukowski[2], N. Derebecka[3], J. Wesoly[3] and H. Bluyssen[3]
[1]*UTP University of Science and Technology, Mazowiecka 28, 85-084, Poland, [2]National Research Institute of Animal Production, Krakowska 1, 32-083 Balice, Poland, [3]Adam Mickiewicz University, Umultowska 89, 61-614 Poznań, Poland; siwek@utp.edu.pl*

Intestinal microbiota is in a constant cross-talk with gut-associated lymphoid tissue (GALT), triggering either tolerance or activation of the immune system. GALT differentiates between beneficial and pathogenic antigens through pattern recognition receptors (PRRs), such as Toll-like receptors (TLRs), which are expressed by antigen presenting cells. TLRs can bind the array of conserved antigens, called pathogen associated molecular patterns (PAMPs). Some prebiotics and probiotics can also exert immunomodulatory properties on the host by priming adaptive responses by antigen presenting cells. This mechanism is also based on TLRs recognition system and downstream signaling. In this study we analyzed immunomodulatory properties of galactooligosaccharide (GOS) prebiotic and two probiotic strains: *Lactococcus lactis* subsp. *cremoris* and *Saccharomyces cerevisie*, using chicken macrophage-like cell line (HD11). Macrophages are part of innate immune system of the GALT. The ultimate goal was to determine TLR signaling pathway that is activated by a given stimuli (prebiotic or probiotic). For this reason we also used TLR ligands, which activate known TLR pathway in chicken: Pam3CSK (TLR1/2), lipopolysaccharide (TLR4), zymosan (TLR21) and CpG ODN (TLR21). The experiment was conducted as follows: HD11 cell line was stimulated with a prebiotic, probiotic or a given TLR ligand. The stimulation (n=3) was carried out in time course manner (3, 6 and 9 h). RNA was isolated from stimulated HD11 cells and subjected to total RNA sequencing (Illumina). Differential gene expression was determined with Bayseq, Deseq2 and Edger. We have identified multiple genes and pathways activated by each stimuli. The results will help to further understand the immunomodulatory properties of prebiotics and probiotics at the molecular level. Identified gene expression signatures can be used to evaluate prebiotics and probiotics *in vitro*. Acknowledgements: the research was supported by a project UMO-2013/11/B/NZ9/00783 (NSC, Cracow, Poland)

Insects as feed

M. Dicke

Wageningen University, Entomology, Droevendaalsesteeg 1, 6708 PB Wageningen, the Netherlands; marcel.dicke@wur.nl

Some of the major challenges currently faced by mankind are food security for a human population that is projected to grow till ca 9-10 billion in 2050, mitigation of climate change, and conserving biodiversity and the environment at large. These challenges are closely connected and, consequently, require drastic changes in primary production as a result of constraints in the availability of resources, water, land, and energy. However, these constraints imply that we need radical changes in production systems. A dominant issue in food security is the production of sufficient high-quality proteins with minimal waste production and greenhouse gas emission. Important feed sources for livestock production currently include fish meal and soy meal, both competing with availability as a food source to humans. In addition, the use of fishmeal poses an increasing threat to the viability of marine ecosystems. Novel feed stuffs are needed and insects can provide an important source that can be a valuable component of a circular economy. Insects used as feed include e.g. the black soldier fly (BSF; *Hermetia illucens*) and the common housefly (*Musca domestica*), which can convert organic waste to high quality protein, and can be used as feed for a variety of animals. Thus, these insects can be important elements of a circular economy. BSF larvae contain high protein levels (from 37 to 63% DM), and other macro- and micronutrients important for animal feed. The available studies on including BSF larvae in feed rations for poultry, pigs and fish suggest that they can replace traditional feedstuff to some extent. To further develop the opportunities, studies are needed on nutrient composition, digestibility and availability for target species and on improved methods to process larvae, among other aspects. The BSF can be produced by smallholder farmers as well as high-technology companies and thereby this insect may become an important element of food production for individual farmers as well as large private enterprises.

Life-cycle assessment of insects in circular economy

A.M.S. Halloran[1], N. Roos[1], Y. Hanboonsong[2] and S. Bruun[3]

[1]University of Copenhagen, Department of Nutrition, Exercise and Sports, Rolighedsvej 26, 1958 Frederiksberg, Denmark, [2]Khon Kaen University, Department of Entomology, 123 Mitapab Rd., 40002 Khon Kaen, Thailand, [3]University of Copenhagen, Department of Plant and Environmental Science, Torvaldsensvej 40, 1958 Frederiksberg, Denmark; aha@nexs.ku.dk

Compared to their vertebrate counterparts in traditional husbandry, insects are extremely efficient at converting organic matter into animal protein and dietary energy. For this reason, insects for food and feed show great potential as an environmentally friendly choice in future food systems. However, to obtain a true assessment of this, more information is needed about the production systems. This presentation presents Life Cycle Assessment (LCA) to evaluate the environmental impacts of insect farming in a circular economy context. Moreover, this presentation will expand on a proposed reference framework that would allow for the selection of standardised settings for LCA applications in insect production systems, taking both the peculiar nature of each system and the latest developments in food LCA into account.

Some views on insects and livestock providing nature based solutions for a circular bio economy
G. Van Duinkerken and L.B.J. Sebek
Wageningen Livestock Research, De Elst 1, 6708 WD, Wageningen, the Netherlands; gert.vanduinkerken@wur.nl

The main challenge of a circular bio economy is to optimise the value of the entire raw material. All components should be used, to minimise waste. Yield gaps in crop production should be decreased, while maintaining soil fertility. Biorefinery is offering increasing opportunities to fractionate raw materials and to extract high value components; or to take out components without biofunctionality, or with an anti-nutritional effect. Biorefinery technology is developing fast and more fibrous by-products from the food and biofuel industry are becoming available as a raw material. Different organisms offer a natural capacity to convert such biomass to a position higher in the value chain. Fungi can convert lignin into better degradable feed materials, micro-organisms can upgrade cellulose, whereas ruminants convert fibrous materials from marginal grasslands and by-products into human edible proteins. Insects and pigs are champions in utilising organic materials like food waste and residues from vegetable and fruit production. Aquaculture and poultry need more high-grade resources containing sugars, starch, digestible proteins and fatty acids for conversion into human edible proteins. Future animal diets will comprise more fibrous by-products, novel protein sources, food waste and crops from marginal lands. By using such resources, livestock production systems are not in competition with food sources for humans. Thus, animal production plays a key role in a sustainable food chain. However, if a more holistic and agro-ecological view on food production is encouraged, animal production systems are likely to decrease the current focus on feed conversion ratio. This ratio is expressing kg feed consumed per kg animal end product, and inclusion of high-grade feed in the animal diet is improving the ratio. This is potentially leading to less sustainable food production systems. It is essential to clearly define conversion factors which are used for comparison of animal production systems. Poultry has a favourable feed conversion ration (kg feed consumed per kg growth), but cattle have a much higher human edible protein production (HEP) per kg of HEP consumed. Insects can be an intermediate biomass convertor to increase HEP efficiency of poultry.

Vegetable by-products bioconversion for protein meals production through black soldier fly larvae
M. Meneguz[1], A. Schiavone[2], F. Gai[3], E. Bressan[1], A. Dama[1], S. Dabbou[2], M. Renna[1], V. Malfatto[1], C. Lussiana[1], I. Zoccarato[1] and L. Gasco[1,3]
[1]*University of Turin, Department of Agricultural, Forest and Food Sciences, Largo P. Braccini 2, 10095 Grugliasco, Italy,*
[2]*University of Turin, Department of Veterinary Sciences, Largo P. Braccini 2, 10095 Grugliasco, Italy,* [3]*National Research Council, Institute of Science of Food Production, Largo P. Braccini 2, 10095 Grugliasco, Italy; marco.meneguz@unito.it*

The aims of this trial were to evaluate the effect of four different substrates on growth performance and chemical composition of black soldier fly (*Hermetia illucens*) larvae. Two trial were performed: in the first, vegetable and fruit waste (VEG) or only fruit waste (FRU) were used as rearing substrates, while the second trial used winery by-products (WIN) and brewery by-products (BRE). Two thousand four hundred larvae were counted, weighed in batch of one hundred and homogenously divided in plastic containers. Six replicate per substrate were performed to evaluate growth performance. The containers were placed in a climatic chamber with a temperature (27 ± 0.5 °C) and relative humidity (70 ± 0.5%). Larvae were fed with an initial amount of one hundred grams of substrate/container, the containers were monitored everyday and 50 g of feed was added as needed. For each trial, to obtain the needed larvae amount for proximate composition analyses, a second set of six replicates (500 larvae per replicate) per rearing substrate was concurrently prepared. Larvae weights and lengths were monitored and data subjected to a two-way Mixed ANOVA, differences of means by Shapiro-Wilk test ($P<0.05$). Final larvae weights, lengths and chemical composition were further subjected to independent samples Student's *t*-tests to assess differences. Growing time showed significant differences in the two trials. VEG reached 30% of prepupae faster than FRU. In the second trial, the fastest growth rate was reported by BRE. As far as weight is concerned, at the end no statistical differences were highlighted between treatments in both trials. Final larvae chemical composition was dramatically influenced by the rearing substrates.

Nitrogen-to-protein conversion factors for three insect species

C.M.M. Lakemond, R.H. Janssen, J.-P. Vincken, L.A.M. Van Den Broek and V. Fogliano
Wageningen University, P.O. Box 17, 6700 AA Wageningen, the Netherlands; catriona.lakemond@wur.nl

Insects are considered as a nutritionally valuable source of alternative protein. The protein content is usually calculated from total nitrogen using the nitrogen-to-protein conversion factor (Kp) of 6.25. This factor overestimates the protein content, due to the presence of non-protein nitrogen in insects. In this paper, a specific Kp of 4.76±0.09 was calculated for larvae from *Tenebrio molitor*, *Alphitobius diaperinus* and *Hermetia illucens*, using amino acid analysis. After protein extraction and purification, a Kp factor of 5.70±0.39 was found for the three insect species studied. We propose to adopt these Kp values for determining protein content of insects to avoid overestimation of the protein content.

Larvae of black soldier fly upcycle pig manure into highly valuable fat and proteins on the farm

V. Van Linden[1], J. De Koker[2], J. De Boever[1] and J. Pieters[2]
[1]ILVO-Flanders research institute for agriculture, fisheries and food, Burg. Van Gansberghelaan 115, 9820 Merelbeke, Belgium, [2]Ghent University, Biosystems Engineering Department, Coupure Links 653, 9000 Gent, Belgium; johan.deboever@ilvo.vlaanderen.be

This study aimed to optimize the rearing process of black soldier fly (BSF) larvae on solid pig manure on the farm. The optimal rearing conditions were first studied in climate rooms at lab scale. Then, pilot scale bioconversion units were constructed and evaluated on the farm. Larvae of the BSF were reared on the solid fraction of fresh pig manure at 21-24-27 °C until fully grown. Before transfer to the substrate, larvae were grown on poultry meal for 5 days. Then, feeding frequency (all-at-once, 1/week, every 2-3 days) and feed dose (10-15-20 mg DM/larvae/d) were varied. Fresh manure, harvested larvae, and residual substrate were analysed for dry matter (DM), nitrogen and fat. The optimal settings were applied to BSF rearing on farm in self-constructed bioconversion units. These were placed in an occupied fattening pig compartment. Larvae were immediately transferred to the substrate. Manure was stirred during the rearing process. The units were ventilated at air change rates of 30-114/h during the growing phase and up to 371/h during substrate drying. Emissions of NH_3, CH_4, N_2O and CO_2 were monitored and mass balances were made. Rearing BSF larvae on pig manure on farm was possible: the indoor air temperature of fattening pigs was sufficiently high. A correct ventilation is crucial to enable automatic harvesting of the full-grown larvae. Harvested larvae had an average DM content of 27-33% which consisted of 45-50% protein and 20-24% fat (lab & pilot scale). Nitrogen from the substrate was transferred into the larvae (17-21%) and into the air as NH_3 (±30%, increasing with higher temperature). Increased levels of NH_3, CO_2 and CH_4 were measured. Bioconversion rates of DM amounted to 13% (max. 15%) at lab scale and to 9-12% at pilot scale. Differences result from higher mortality rates, less favourable rearing conditions, competition with house fly larvae, immediate transfer to the substrate, and harvest losses. Further optimization of insect rearing on farm is necessary and special attention is required for emission control and economic feasibility.

Potential of insects-based diets for Atlantic salmon (*Salmo salar*)

I. Belghit[1], N.F. Pelusio[2], I. Biancarosa[1], N. Liland[1], R. Waagbø[1], Å. Krogdahl[3] and E.-J. Lock[1]
[1]*National Institute of Nutrition and Seafood Research (NIFES), Department of Requirements & Welfare, Strandgaten 229, 5004 Bergen, Norway, [2]Univerita Politecnica delle Marche, Italy, Department of Life and Environmental Science (DISVA), Piazza Roma, 22, 60121 Ancona, Italy, [3]Norwegian University of Life Sciences (NMBU), Universitetstunet 3, 1430 Ås, Norway, Department of basic sciences and aquatic medicine, Universitetstunet 3, 1430 Ås, Norway; ikram.belghit@nifes.no*

The rapid growth in aquaculture has increased the demand for feed ingredients. Insects, which are part of the natural diet of salmonids, may represent a sustainable ingredient for aquaculture feed. In the present study, we aimed to assess the effect of dietary insect meal (IM) and insect oil (IO) on growth performance, body composition and nutrient digestibility of freshwater Atlantic salmon (Salmo salar). The IM and IO were produced from black soldier fly (BSF) larvae (*Hermetia illucens*). The larvae were grown on (1) media containing organic waste streams, or on (2) media partially containing marine seaweed (*Ascophyllum nodosum*). A factorial two way-Anova experimental design with six diet groups was used for 8 weeks. A standard diet for this species was fed to the control group (CTL); protein from fish meal (FM) and soy protein concentrate (SPC) (50:50) and lipid from fish oil (FO) and vegetable oil (33:66). Five experimental diets were formulated, where 85% of the protein was replaced with IM and/or all the vegetable oil was replaced with IO, either produced from the media 1(IO1) or from the media 2 (IO2). The apparent digestibility coefficients (ADC) were significantly lower for protein, amino acids and for fatty acids in IM fed groups compared to the CTL, though no differences were recorded in growth performance, feed intake, feed conversion ratio or serum biochemical parameters between groups, irrespective of protein or lipid source. Whole-body protein content was not affected by protein or lipid source, while body lipid increased in IM fed groups, associated with increased saturated fatty acids composition, specifically lauric acid (C12). In general, this study showed that BSF larvae meal and oil, represent a great potential as a functional feed sources for Atlantic salmon.

Manipulation of *Tenebrio molitor* unsaturated fatty acid content using a variety of feed oils

E.R. Tetlow, T. Parr, I.C.W. Hardy, C. Essex, C. Greenhalgh and A.M. Salter
University of Nottingham, School of Biosciences, North Laboratory, College Road, LE12 5RD, United Kingdom; ellen.tetlow@nottingham.ac.uk

There is increasing interest in using insect-derived material as a production animal feed ingredients. The aim of this study was to determine if the unsaturated fatty acid (UFA) content of *Tenebrio molitor* larvae (yellow mealworm) could be manipulated utilising feeds containing oils rich in different UFA and whether the larvae were capable of elongating and/or further desaturating these fatty acids. 6-week old larvae were assigned to 4 treatment groups with 100 larvae in each replicate per group (n=8). A control group was fed chick crumb feed (crude protein 190 g/kg, crude fat 40 g/kg, fibre 40 g/kg). Treatment groups were fed the same feed with 5% (w/w) replaced with either rapeseed oil (RS, enriched in C18:1), safflower oil (SF enriched in C18:2) or linseed oil (LS, enriched in C18:3). All treatment groups were fed *ad libitum* and had continuous access to carrot, acting as a water source, for 14 days. Total biomass was recorded at day 0, 7, 11 and 14 along with feed and carrot consumption. At day 14, proximate nutrition analysis (crude fat, protein and energy) and fatty acid (FA) composition (% mole/100 mole FA methyl ester) was determined. One-way ANOVA, one-way MANOVA, Kruskal-Wallis and generalised linear regression assuming Poisson distribution, were used to analyse normally distributed, UFA profiles, skewed and count data respectively. Inclusion of oils into feed had no effect on accumulation of biomass or longevity. However, gross energy, crude fat and total feed consumption were significantly increased compared to controls. Protein deposition was significantly reduced in RS fed larvae. Relative to the control the RS group had significantly decreased C14:0, C16:1, C22:6 and increased C18:1 and 18:3 concentrations. Larvae fed SF had significantly decreased C16:1, C18:1, C22:6 and increased C18:2, whilst larvae fed LS had decreased C16:0, C18:2 and increased C18:3 concentrations. Supplementing the diet of larvae with specific UFA increased the deposition of these fatty acids without effecting the overall growth or longevity. However, there was no clear indication that *T. molitor* larvae are able to further elongate or desaturate UFAs.

Partially defatted black soldier fly larvae (*Hermetia illucens*) meal in piglets' nutrition

E. Bressan[1], F. Gai[2], A. Schiavone[2,3], S. Dabbou[3], A. Brugiapaglia[1], G. Perona[4], M. Meneguz[1], A. Dama[1], I. Zoccarato[1] and L. Gasco[1,2]

[1]University of Turin, Department of Agricultural, Forest and Food Sciences, Largo P. Braccini 2, 10095 Grugliasco, Italy, [2]National Research Council, Institute of Science of Food Production, Largo P. Braccini 2, 10095 Grugliasco, Italy, [3]University of Turin, Department of Veterinary Sciences, Largo P. Braccini 2, 10095 Grugliasco, Italy, [4]University of Turin, SDSV-Veterinary Practical Teaching Special Unit, Largo P. Braccini 2, 10095 Grugliasco, Italy; enrico.bressan01@gmail.com

The aim of this trial was to investigate the effects of different inclusion levels of a partially defatted black soldier fly (*Hermetia illucens*, BSF) larvae meal on growth performance and dorsal muscle composition of piglets. Forty-eight newly weaned piglets were individually weighted (initial body weight (IBW): 6.09±0.16 kg) and allocated in 12 different boxes to have a homogeneous initial live weight. Three isonitrogenous and isoenergetic diets were formulated with increasing inclusion levels of BSF (0% (BSF0), 5% (BSF5) and 10% BSF10)) in substitution of conventional protein sources. Each diet was assigned to four replicates. Two phases feeding program (I from day 0 to day 23; II from day 24 to day 61) were studied. Average daily gain (ADG), average daily feed intake (ADFI) and feed conversion ratio (FCR) were calculated for each feeding phase and for the whole trial while the weight gain (WG) was calculated for each period. The Longissimus dorsi analysis was performed on 6 animals per treatment. Data were analysed by one-way ANOVA and differences of means by Duncan's test (P≤0.05). No significant differences were observed for all the parameters neither growth performance or proximate composition. The obtained results showed that a partially defatted BSF larvae meal can be used as feed ingredient in piglet diets without impacting growth performance and dorsal muscle composition.

Highly defatted insect meal in Siberian sturgeon juveniles feeds

L. Gasco[1,2], A. Schiavone[1,2], M. Renna[2], C. Lussiana[2], S. Dabbou[2], M. Meneguz[2], V. Malfatto[2], M. Prearo[3], A. Bonaldo[4], I. Zoccarato[2] and F. Gai[1]

[1]National Research Council, ISPA, Largo P. Braccini 2, 10095 Grugliasco, Italy, [2]University of Turin, DISAFA – DSV, Largo P. Braccini 2, 10095 Grugliasco, Italy, [3]Veterinary Medical Research Institute for Piemonte, Liguria and Valle D'Aosta, Via Bologna 148, 10154 Torino, Italy, [4]University of Bologna, Via Tolara di Sopra, 50, 40064 Ozzano dell'Emilia (Bo), Italy; laura.gasco@unito.it

Recent investigations highlighted that insect protein meals can be a more sustainable alternative to conventional protein used so far in aquaculture. *Hermetia illucens* (HI) is a good candidate due to its valuable nutritional properties. The aim of this research was to evaluate the effects of fishmeal (FM) substitution by a highly defatted HI larvae meal in sturgeon juveniles feeds. Four diets were formulated: a control (70% of FM – CF), two diets where FM was replaced by 25 (HI25) and 50% (HI50) of HI and a vegetable protein based diet without HI (CV). 352 *Acipenser baerii* were distributed in 16 fiberglass tanks. Each diet was assigned to 4 groups of 22 fishes and feed was distributed to apparent satiation. At the end of the trial (118 days) fish growth performances were calculated, and whole body (WBC) proximate and fatty acid (FA) composition were analysed. Data were statistically analysed by one way ANOVA. Significance level was set at P<0.05. Results indicate that the inclusion of HI affected fish performances and WBC. Generally, up to 25% of FM substitution, fish performance was comparable to those of fish fed CF or CV while the 50% substitution induced a worsening of performance parameters and the same trend was observed for WBC. Lauric acid and total saturated FA contents were higher in fish fed HI when compared to CF and CV groups. Monounsaturated, polyunsaturated, and n3 and n6 FA contents showed differences among groups, with lowest values for CF; however, no differences were found in the n3/n6 FA ratio in WB of CF and HI25 groups. From this preliminary investigation, highly defatted HI larvae meal seems to be a promising and valuable alternative to replace up to 25% of FM in sturgeon juveniles feeds.

Hermetia illucens meal in poultry nutrition: effect on productive performance and meat quality

A. Schiavone[1,2], F. Sirri[3], M. Petracci[3], M. Montagnani[2,3], S. Dabbou[2], M. Meneguz[2], I. Zoccarato[2], F. Gai[1] and L. Gasco[1,2]
[1]*National Research Council, ISPA, Largo P. Braccini 2, 10095 Grugliasco, Italy, [2]University of Turin, DSV – DISAFA, Largo P. Braccini 2, 10095 Grugliasco, Italy, [3]University of Bologna, Via del Florio 2, 40064 Ozzano dell'Emilia, Italy; achille.schiavone@unito.it*

Insect meal represent a promising ingredient in poultry nutrition. The aim of the trial was to evaluate the effects of dietary administration of partially defatted *Hermetia illucens* (HI) larvae meal on growth and slaughter performance, and on breast meat quality in broiler chickens. 192 male broiler chickens (Ross 308) were reared from day 1 to day 39 and assigned to 4 dietary treatments (6 replicates/treatment and 8 birds/replicate). HI larvae meal was included at increasing level (0, 5, 10, 15%; HI0, HI5, HI10 and HI15 respectively) in isonitrogenous and isoenergetic diets formulated for 3 periods: starter (1-11 d), growing (11-26 d) and finisher (26-39 d). HI5 showed the lowest final body weight and the worst feed conversion ratio ($P<0.05$), while no differences were observed among groups HI0, HI5 and HI10 for the same parameters. No differences were observed for slaughtering performance, except for breast weight that resulted higher for HI10. Breast meat color was influenced by dietary treatments resulting in a red (a*) and yellow (b*) indexes increased and decreased respectively for increasing HI inclusion rates ($P<0.05$), L* index resulted unaffected by treatments. Chemical composition (dry matter, crude protein, total lipids and ash) was unaffected by diets. Breast meat fatty acid composition was influenced by diets, showing increasing (C 12:0, 14:0, 16:0, 14:1, 16:1, 18:1, 20:1) or decreasing (18:0, 24:0, 18:2, 20:4, 22:5, 22:6) concentration in relation to increasing dietary HI inclusion. Total SFA and n6/n3 ratio were unaffected by treatments while total PUFA, n-3 PUFA and n-6 PUFA showed decreasing value whith increasing HI dietary inclusion. In conclusion, HI larvae meal represent a valuable protein source for broiler chicken up to 10% dietary inclusion without affecting growth and slaughter performance. Even if a total reduction of PUFA in breast meat is observed, the improved oleic acid could be considered beneficial for human health.

Poster presentations insects

T. Veldkamp[1] and J. Eilenberg[2]
[1]*Wageningen University & Research, Livestock Research, De Elst 1, 6708 WD Wageningen, the Netherlands, [2]University of Copenhagen, Faculty of Science, Bülowsvej 17, 1870 Frederiksberg, Denmark; teun.veldkamp@wur.nl*

Beyond theatre presentations also a number of high-quality abstracts have been accepted as poster presentations: (1) *Campylobacter* contamination le in houseflies after exposure to materials containing Campylobacter (2) Nitrogen-to-protein conversion factors for three insect species (3) Growth of Cambodian field crickets (*Teleogryllus testaceus*) fed weeds and agricultural by-products (4) Finishing diets in insect farming, a preliminary study on yellow mealworm and superworm larvae (5) Let's make the future fly (6) Insect value chain in a circular bioeconomy (inVALUABLE) (7) Metagenomics analysis of gut microbiota of herbivorous insects *Bombyx mori* (8) The beginning of a new neolithic revolution? Honeybees for proteins (9) Amino acid contents of field crickets fed chicken feed, cassava tops and a weed Poster presenters will have the opportunity to present their posters in this session briefly (2-3 minutes per poster).

Growth of Cambodian field crickets (*Teleogryllus testaceus*) fed weeds and agricultural by-products

P. Miech[1], Å. Berggren[2], J.E. Lindberg[3], T. Chhay[1], B. Khieu[1] and A. Jansson[4]
[1]Center for Livestock and Agriculture Development-CelAgrid, www.celagrid.org, 855, Cambodia, [2]Swedish University of Agricultural Sciences (SLU), 75007 Uppsala, Department of Ecology, P.O. Box 7044, 306, Sweden, [3]Swedish University of Agricultural Sciences (SLU), 75007 Uppsala, Department of Animal Nutrition and Management, P.O. Box 7024, 306, Sweden, [4]Swedish University of Agricultural Sciences (SLU), 75007 Uppsala, Department of Anatomy, Physiology and Biochemistry, P.O. Box 7011, 306, Sweden; anna.jansson@slu.se

Chicken feed is commonly used as feed in mass rearing systems of crickets. Chicken feed has high nutritive value also for humans and the sustainability of this feed source in cricket rearing can be questioned. In addition, chicken feed may account for up to 60% of the total production cost. This study aimed to evaluate the growth of Cambodia field cricket fed weeds and agricultural and food industry by-products. Wild crickets (male and female, 50:50) were caught using light traps at night. The crickets were nursed in concrete block pens and fed chicken feed and cassava tops until the second generation of nymphs hatched. These nymphs were used in the experiment. Chicken feed was used as control diet and was compared with cassava plant tops, *Cleome rutidosperma*, *Cleome viscosa*, *Synedrela nodiflora*, residue from mungbean sprout production, *Commelina benghalensis*, spent grain and rice bran. The diets were fed for 70 days to 80 nymphs randomly divided into 4 replicates per diet. Water was provided *ad libitum*. The weight of the crickets was recorded weekly and there was no difference between treatments in body weight at start of the experiment (P=0.36). Throughout the study, body weight did not differ between crickets fed chicken feed, cassava tops and *C. rutidosperma* (P>0.05), but body weight was lower in crickets fed *C. Viscosa*, *S. Nodiflora*, residue from mungbean sprout production, *C. Benghalensis* (P=0.05), spent grain, water spinach and rice bran (P=0.01) compared to crickets fed chicken feed. We conclude that cassava plant tops and *C. rutidosperma* have a great potential as cricket feed. However, the other feed types evaluated also showed potential.

Finishing diets in insect farming, a preliminary study on yellow mealworm and superworm larvae

I. Biancarosa[1], K. Zilensk[2], I. Belghit[1] and E.-J. Lock[1]
[1]National Institute of Nutrition and Seafood Research (NIFES), Requirement and Welfare, P.O. Box 2020, Nordnes, 5817 Bergen, Norway, [2]Järva County Vocational Training Centre, Tallinna 46, 72720 Paide linn, Estonia; elo@nifes.no

The yellow mealworm (*Tenebrio molitor*, TM) and superworm (*Zophobas morio*, ZM) are promising insect species for use in feed. The larvae of these species are rich in protein and fat, and low in ash. TM and ZM larvae are omnivorous and can grow on a wide range of substrates. Meals from TM and ZM larvae have been investigated as alternative feed ingredients in the diet of farmed animals, successfully replacing fishmeal in compound aquafeeds. However, low levels of certain nutrients in these insect species (e.g. Ca, P, methionine and omega-3 fatty acids) could limit their use in the diet of certain animals. The goal of this study was to investigate the plasticity of the nutrient profile of TM and ZM by providing the larvae a finishing diet before harvesting. Larvae of TM and ZM were fed substrates containing bread and vegetables and only for the last 8 days before harvesting transferred to a finishing diet based on fishmeal, mackerel, brewery waste, kelp, tunicates, sealice or pizza factory waste. After harvesting, the larvae were analysed for protein, lipid and fatty acid composition, mineral, ash and DM contents. As expected, larvae fed on marine substrates contained higher levels of omega-3 fatty acids compared to larvae fed on terrestrial substrates. When fed on kelp, the larvae had relatively low lipid and high protein contents, which is in line with previous data on insect larvae fed seaweed-enriched media. Larvae fed fishmeal and mackerel had higher dry matter and lipid contents, compared to the other groups. TM and ZM fed tunicates (naturally high in vanadium), readily accumulated this metal. Larvae fed sealice had the highest ash contents and highest levels of most minerals. In particular, iron levels in these larvae were more than 10 times higher than the other feeding groups. Overall, the macro- and micronutrient composition of the insect larvae reflected the nutrient composition of the diet. Especially lipid and mineral compositions were affected. Feeding a finishing diet for one week can help tailoring the nutrient composition of the insect larvae in a more favourable profile for specific applications.

Let's make the future fly

N. De Craene and J. Jacobs
Millibeter, R&D, Dambruggestraat 200, 2060 Antwerpen, Belgium; nouchkadecraene@millibeter.be

Millibeter was founded in 2012 by Johan Jacobs in Antwerp, Belgium. He saw the black soldier fly as the missing link between underused side streams on one hand and the exhaustion of resources on the other hand. At their larval stage, these insects are able to process a wide variety of organic side streams and convert them into their own biomass, largely consisting of lipids, proteins and chitin. Besides developing know-how on breeding and rearing the black soldier fly on a large scale and in a cost-efficient way, we succeeded in developing a unique technology to extract these different components in an inexpensive and straightforward manner. The potential of these products as feed for livestock is enormous. Live larvae can be fed to fish, chickens and pigs, while larvae meal is a great substitute for fishmeal. The extracted lipids, chitin, proteins and protein-isolate are high quality additives for feed. On top of that, this biomass production is extremely compact; a black soldier fly rearing unit with a surface area of about 1 hectare is able to produce 63 tons of larvae per day, all year round. This makes it very easy to place these rearing units wherever it is most convenient, either being close to the source (side streams) or the to the consumer, depending on the circumstances. We are working on a large scale project that is designed to process 50,000 tons of side streams from a palm oil plantation, producing a 23,000 of larval biomass a year, in the first phase. We teamed up with well renowned international partners, who will take this innovative production model to a truly industrial scale, and roll it out globally. In the framework of this consortium, we are elaborating a range of innovations: new insects as bioconvertors, new sidestreams to be bioconverted, new end-products with higher added value, and new business models on combining insect bioconversion with agriculture.

Insect value chain in a circular bioeconomy (inVALUABLE)

J. Eilenberg[1], L.H. Heckmann[2], N. Gianotten[3], P. Hannemann[4], A.N. Jensen[5], J.V. Norgaard[6], N. Roos[1] and L. Bjerrum[2]
[1]University of Copenhagen, Plant and Environmental Sciences, Thorvaldsensvej 40, 1871 Frb C, Denmark, [2]Technological Institute, Life Science, Kongsvang Allé 29, 8000 Aarhus, Denmark, [3]Proti-Farm, R&D, Harderwijkerweg 141B, 3852 AB Ermelo, the Netherlands, [4]Hannemann Engineering, Stødagervej 5, 1 sal, 6400 Sønderborg, Denmark, [5]Technical University of Denmark, National Food Institute, Anker Engelundsvej 1, 2800 Lyngby, Denmark, [6]Aarhus University, Department of Animal Science, Blichers Allé 20, 8830 Tjele, Denmark; jei@plen.ku.dk

inVALUABLE is a major collaboration project involving (mainly Danish) research institutions and companies and was initiated 2017. The project aims to contribute to improvement and development of major focus areas in the insect value chain; insect production and processing, and product application. The project will focus on optimizing reproduction, growth and health of two beetle species, namely lesser mealworm (*Alphitobius diasperinus*) and common mealworm (*Tenebrio molitor*). There will be specific focus on the future rearing facilities and automation of these in order to ensure a competitive end-product. The processing of the biomass will be investigated to find the most viable solution regarding nutritional quality for animal feed and human consumption, including optimization of protein digestibility. Animal feeding trials will be performed and the composition and ingredients in insect products will be evaluated in relation to human consumption. There will be major focus on feed and food safety and different challenges will be addressed to support related legislation. inVALUABLE will touch upon most of the insect-supply chain and produce a range of new data for use in development of insects as future feed and food. inVALUABLE is supported by Innovation Fund Denmark and has a total budget of approx. 3.7 Million EUR.

Metagenomics analysis of gut microbiota of herbivorous insects *Bombyx mori*

J.W. Shin, B.R. Kim, S.Y. Choi, M.E.S. Gamo, J.H. Lee, R.B. Guevarra, S.H. Hong, M.K. Shim, S.O. Jung and H.B. Kim
Dankook University, Department of Animal Resources and Science, Dandae-ro 119, Cheonan, 31116, Korea, South; hbkim@dankook.ac.kr

The gut microbiome of insect has been known to play important roles in host physiology, nutrition and development. The microbial communities inside the gut of silkworm have not been studied much with the use of modern molecular method. In this study, we characterized the bacterial communities of the herbivorous insects *Bombyx mori* (silkworm) (n=10) through 16S rRNA gene and whole metagenome analysis. Total DNA was extracted from fresh feces of silkworm and PCR was conducted targeting the hypervariable regions V5 to V6, and the whole metagenome were sequenced by using the Miseq chemistry. Quality control of raw sequence data was done using Mothur. Alpha and Beta diversity analysis was done using QIIME software. Functional analysis of gut-derived microbiota was conducted by using the Microbial Genomics Module of CLC Genomics Workbench using the GO Term database Amigo2. Results showed that the most abundant phylum was *Firmicutes* (>63.41%) in silkworm. Gut bacterial community was highly diverse in silkworm as revealed by average Chao1 and Shannon diversity indices at 1,058.90 and 1.83, respectively. 75 genera were found in the gut content of silkworm. *Enterococcus*, *Streptophyta* and *Clostridiales* were abundant, and accounted for more than 89.58% of the total bacteria in silkworm at the genus level. The GO term analysis showed that functional properties related with xylan, cellulose, pectin catabolism of the gut-derived microbiota were abundantly present. In summary, this study increases our knowledge of the gut microbiota of silkworm and the potential benefits of cellulolytic bacteria derived from the gut of silkworms. Even though, future studies are required to elucidate the exact functions of insect-associated gut bacteria, our results suggest that gut microbiome of silkworms has potentials to make animal feeds for better nutrient absorption by increasing digestion of fiber in pig feeds thereby increases yield and quality of animal products.

Amino acid contents of field crickets fed chicken feed, cassava tops and a weed

G. Håkansson[1], P. Miech[1,2], J.E. Lindberg[3], Å. Berggren[4], T. Chhay[2], B. Khieu[2] and A. Jansson[1]
[1]Swedish University of Agricultural Sciences, Dept of Anatomy, Physiology and Biochemistry, Box 7011, 75007 Uppsala, Sweden, [2]Center for Livestock and Agriculture Development, P.O. Box 2423, Phnom Penh 3, Cambodia, [3]Swedish University of Agricultural Sciences, Box 7024, 75007 Uppsala, Sweden, [4]Swedish University of Agricultural Sciences, Box 7044, 75007 Uppsala, Sweden; anna.jansson@slu.se

Chicken feed is commonly used as feed in mass rearing systems of crickets. Chicken feed is often based on high quality ingredients such as cereals and soybean meal and the sustainability of this feed source in cricket rearing can be questioned. We have recently shown that the growth of Cambodian field crickets may be equally good on cassava tops and the weed *Cleome Rutidosperma* as on chicken feed. The aim of this study was to compare the content of lysine, methionine and cysteine in field crickets (*Teleogryllus testaceus*) fed chicken feed, cassava tops or *C. Rutidosperma*. The diets were fed to nymphs of the cricket for 70 days (80 nymphs randomly divided into 4 replicates per diet). The crude protein content of the chicken feed, cassava tops and *C. Rutidosperma* was 23.4, 22.2 and 28.6% respectively. At the end of the study, a pooled sample of crickets from each diet was dried and frozen (-20 °C). Before analysis, wings and hind legs (femur, tibia and tarsus) were separated from the body and were analysed separately from the bodies. Amino acid (AA) content was analysed by HPLC according to ISO standard 13903:2005. ANOVA (proc GLM in SAS) and a model including diet and body sample was used to analyse differences between diets. Statistical significance was set to $P<0.05$. Methionine content was higher in crickets fed cassava tops than in crickets fed chicken feed and *C. Rutidosperma* (LSmeans ± SEM: 0.90±0.06, 0.55±0.06 and 0.80±0.06 g/100 g dry matter, respectively, P=0.04). There was a tendency for higher lysine, cysteine and cystine content in crickets fed cassava tops than in crickets fed chicken feed and *C. Rutidosperma* ($P<0.1$). Although this is a small study, it indicates that the AA profile of field crickets can be altered with diet and that cassava tops seems to be a good feed resource with respect to AA quality of the crickets.

Prevelance of health problems associated with transport of horses and associated risk factors

R.C. Roy

Prairie Swine Centre Inc, Box 21057, 2105 – 8th Street East, Saskatoon, S7H 5N9, Canada; rroy@upei.ca

Horses are transported for sporting, recreation, and commercial reasons mostly by road and air. In developing countries, draft horses are transported by foot or by hitching them to a cart. Prevalence of welfare issues and associated risk factors can change according to the type of transport. Prevalence of road transport associated injuries ranges from 2 to 28%. However, there is paucity of information regarding the prevalence of death and injuries of horses transported by air and working horses in developing countries. Road accidents involving horse transportation is a significant welfare issue. There are 60 to 100 trailer accident reported every year in the USA. However, prevalence of mortalities and injuries are not published. The UK reported 315 horse related road traffic accidents in 2015, of which 12 horses died, and 8 had severe injuries. There is a definite need to keep track of all accidents involving horses and assess risk factors with a fresh perspective. Causal analysis should include trailer type, fitness of trailer and driver related factors. Survey and cross sectional studies conducted in Europe, UK, Australia and North America has showed that injuries, dehydration, fatigue and respiratory diseases are significant welfare issues. Some mitigation strategies developed in response to these studies are; changes in regulation regarding duration of journey, water provisions and journey breaks. When transport durations are short, even minimal management provisions seem to work fine. However, as the transport duration increases, provisions for comfort has to increase correspondingly. Horses transported for long distances in North America and Australia (environmental temperature range of -40 to + 50 °C) will need provisions to control the temperature and humidity inside the vehicle along with provisions for watering. Prevalence of mortality and non-ambulatory animals during transport continues to be high (0.25%), particularly when transport durations are long. Apart from journey duration, fitness before transport, stocking density and interaction between horses are identified as risk factor for non-ambulatory condition during journey duration. Effect of transporting horses in group on non-ambulatory condition needs to be studied further.

Oxidative stress, respiratory health and equine transport

B. Padalino[1,2,3], S.L. Raidal[2], P. Celi[3], P. Knight[4], L. Jeffcott[3] and G. Muscatello[3]

[1]University of Bari, Department of Veterinary Medicine, Casamassima St, Km 3, 70010 Valenzano, Italy, [2]Charles Sturt University, School of Animal and Veterinary Sciences, Locked Bag 588, 2678 Wagga, Australia, [3]The University of Sydney, Faculty of Veterinary Science, 425 Werombi Road, 2570 Camden, Australia, [4]The University of Sydney, School of Medical Science, P.O. Box 170, 1825 Lidcombe, Australia; barbara.padalino@sydney.edu.au

Transportation of horses is a matter of welfare concern. It may predispose animals to shipping fever (i.e. transport pneumonia) and induce oxidative stress (OS); however, the role of OS in the pathogenesis of shipping fever has yet to be explored. The aim of this study was to evaluate markers of OS and other systemic and local respiratory indicators of inflammation on horses undergoing an 8 hour road transportation event. Twelve horses were examined prior to transportation (T1), at unloading (T2), one (T3) and five days (T4) post transportation. At each time point, blood samples were collected to evaluate leucogram and OS biomarkers (derivatives of reactive oxygen metabolites (d-ROMs) and total plasma antioxidant status (PTAS)). The respiratory tract was examined by endoscopy and tracheal washes (TW) were collected for differential cell count and bacteriological evaluation. Transportation resulted in an increase in the proportion of neutrophils in blood and TW, tracheal mucus and TW bacterial concentration with evidence of reduced bacterial diversity due to an increase in *Pasteurellacea*. Tracheal bacterial concentration (cfu/ml) was highest at unloading (P<0.0001), particularly in six out of 12 horses which also presented a TW differential cell count with marked neutrophilia persisting until T4. At T3, these horses showed evidence of systemic OS as indicated by elevated d-ROMs concentration and oxidative stress index values (P=0.05 and P=0.02, respectively) compared to horses that did not present TW neutrophilia at T4. Furthermore, PTAS decreased only in horses that presented persistent TW neutrophilia post transport (T4 vs T1; *P*= 0.02). Transportation induces respiratory system inflammation resulting in horses being exposed to OS. As proposed for other species, monitoring OS could be a useful tool to assess stress and disease-susceptibility, and consequently the welfare of transported horses.

Stress level effects on sporting performance in Spanish Trotter horses

S. Negro[1], A. Molina[2], M. Valera[1] and E. Bartolomé[1]
[1]University of Seville, Agro-Forestal Sciences, Ctra. Utrera km 1, 41013, Seville, Spain, [2]University of Córdoba, Genetics, C5, Ctra. Madrid-Córdoba, km 396a, 14072, Córdoba, Spain; z12neras@uco.es

The aim of this study was to measure stress levels in the Spanish Trotter Horse through reliable non-invasive systems in order to include this effect on the performance evaluations and to determine the threshold level of stress of the animal during trotter racing which lead to the best results. 130 animals aged between 2 and 10 years-old and from different trotting competitions held in October of 2013 and 2016 were evaluated, measuring eye temperature (ET) with infrared thermography and heart rate (HR), 2 h before the race (ETB, HRB) and just after the race (ETJA, HRJA), including the estimation of the percentage of eye temperature increase (ΔET) between ETB and ETJA. Statistically significant differences associated with the performance level (evaluated based on the racing time per kilometre, TPK) were found among ΔET (P<0.05), ETB (P<0.001) and the age of the animals (P<0.005), showing all stress and performance variables a decreasing tendency with age. A segmented Regression analysis indicated that when ETB values were higher than ETJA ones (ΔET<0), the animal showed the worst results during the race. Furthermore, from the break point, ΔET=-0.97%, horse performance improved. When comparing all race and eye temperature variables, TPK (81.04 s) was optimum when ETB and ΔET reached values of, 35.04 °C and 3.67%, respectively. These results show that the initial stress levels of the horse influenced its competition results, and that the eye temperature increase during trotting competitions until a certain point is related to an improvement in the performance results in trotting races.

Attitudes towards horsemeat consumption in the UK

R. Roberts-Fitzsimmons and C. Brigden
Myerscough College, Equine, St Michael's Road, PR3 0RY, United Kingdom; cbrigden@myerscough.ac.uk

Consumption of horse meat defies traditional cultural norms in the U.K. The European Meat Adulteration (a.k.a. 'Horsemeat Scandal') of 2013 provoked fresh debates regarding the morality of the use of horses for meat. This study aimed to explore the public's attitudes to horse meat using a self-administered, online questionnaire (n=906). The survey collected demographic data and closed and open answers regarding horsemeat, allowing a combined quantitative (Chi-square test of association) and qualitative (thematic analysis) approach to analysis. Despite participants being predominantly U.K. based, a medium proportion (17.26%) had eaten horsemeat. Previous consumption of horsemeat was positively associated with male gender, being aged over 50 years, being rurally located, having a higher education (bachelors or postgraduate) qualification and being professionally involved with horses (each statistically significant by the Chi squared test of association, P<0.01 or less). Being a horse owner or being non-UK based were not associated with whether or not horsemeat had been consumed. Of the respondents who were horse owners, 26.09% would consider disposing of their horse for human meat consumption. Those more likely to consider this were male, higher education qualified and/or professionally involved with horses (P<0.01). Age, being UK or non-UK based, and rural or urban location were not associated with either response here. Emerging themes regarding the use of horses for meat from the positive perspective included ensuring humane processes and providing outlets for unwanted horses; negative perspectives included cruelty, breeding for meat and cultural beliefs. Self-selection bias likely influenced results; however the study has illustrated that some portions of society perceive the horse as a valid meat product, despite contradicting British cultural traditions. There appear to be links between socio-cultural aspects, such as education and involvement with horses or rural living, and a positive consideration towards horsemeat. These links may be useful in enhancing further education to encourage a more open mind-set towards the positive aspects horsemeat production.

Nutritional status of horses in Finnish horse stables

E. Autio and M.-L. Heiskanen
Equine Information Centre, Neulaniementie 5 E, 70210 Kuopio, Finland; elena.autio@hevostietokeskus.fi

As part of the development of a preventive health care program for Finnish horse stables, the nutritional status of horses was evaluated in different stables and in different horse types: exercised/trained horses, breeding horses and growing horses. The aim of the study was to reveal the major concerns in feeding horses and targeting the counselling of owners. Nine horse stables (3 full liveries for leisure horses, 3 racing stables for trotters, 3 breeding stables) and 83 horses (age: 8.1±5.0 years; breed: 17 Warmblood, 33 Standardbred, 33 Coldblooded) participated the study. Horses were measured for body weight and assessed for body condition score (BCS), and their nutrient intake was calculated and compared to Swedish feeding recommendations for horses (SLU 2013). The average forage-to-concentrate ratio in the horses' diet was 84.16%. Thus, forage and starch intakes were mainly in accordance with the recommendations. The average intake of nutrients was appropriate although commonly imbalanced. The average energy intake in horses was 21±21% above and digestible crude protein intake 6±35% above the recommendations. 12% of the horses studied were thin and 17% obese. BCS was significantly above the target score in riding, Coldblooded and moderately exercised horses (P≤0.001, Wilcoxon signed rank test). Inadequate protein intake, which existed in 34% of the horses, was related to low crude protein content in forages used in the stable. Ca intake and Ca:P ratio were inadequate in about 20% of the horses; however, trace mineral and fat-soluble vitamin intakes were generally below the recommendations: Cu in 55%, Zn in 41%, Mn in 36%, vitamin A in 53%, vitamin D in 63% and vitamin E in 42% of horses. Intakes of K and Fe were generally above the recommendations (429±136% and 184±120%, respectively), most likely due to their high content in forages and the additional Fe supplementation in trotters. The type and severity of nutritional imbalances varied between stables and individual horses. Therefore, imbalances were more stable- and horse-specific than horse type-specific. This study highlights the need for comprehensive forage analysis and counselling horse owners and stable managers on nutrition and feeding.

Prevalence and genetic variation of osteochondrosis in the Finnhorse population

S. Back[1], L. Muilu[2], K. Ertola[3], M. Mäenpää[1] and A. Mäki-Tanila[4]
[1]The Finnish trotting and breeding association, Suomen Hippos ry, Tulkinkuja 3, 02650 Espoo, Finland, [2]Equine clinic Equivet, Vermo racetrack, 00370 Espoo, Finland, [3]Tampere Equine Clinic, Mikkolantie 62, 33420 Tampere, Finland, [4]University of Helsinki, Department of Agricultural Sciences, Koetilantie 5, P.O. Box 28, 00014 University of Helsinki, Finland; susanna.back@hippos.fi

Osteochondrosis is a failure of normal endochondral ossification and is exhibited as altered lesions in leg joints. The undesired changes are widely recognized in young warmblood horses across many breeds with lower prevalence in coldblood breeds, such as Finnhorse. The radiographic examination of young (9 month - 5 year old) Finnhorses (n=654) in the years 2014-2015 revealed increased incidence of the disease with 17% of individuals showing an osteochondral lesion at least in one of the studied joints. Joint-specific heritabilities (h^2) were estimated using linear sire model for categorical data and the analyses were augmented by using non-linear animal model on a normal liability scale. Statistical analyses with pedigree information were made by R program MCMCglmm resorting to Bayesian methodology. The lesions were most common (5.2%) in hock where the h^2 estimate with linear sire model was 0.18 (95% credibility area 0.03-0.35) while the h^2 estimate for hoof joint of front legs was 0.31 (0.12 – 0.55), for stifle joint 0.31 (0.09-0.54) and 0.22 (0.07-0.40) for fetlock including the dorsodistal osteochondral fragments which are originating from metacarpal bone. The analysis of the hock data with threshold (probit) animal model gave an h^2 estimate of 0.12 (0.03 – 0.24). Similar estimate values with linear sire model were found in a larger data set (n=1,829) made of the previous data extended with radiographic examination records on recently used stallions with a high number of progeny. The binary scaled trait, low frequency of the disease, low number of analysed horses and the pedigree structure with highly variable family size were experienced as challenges in genetic analyses. The observed genetic variation allows the use of breeding selection for the reduction of osteochondrosis in the Finnhorse population, which would require systematic radiographic survey of young individuals.

Consumption, cost and workload effect of different horse bedding materials– a case study
S. Airaksinen and M.-L. Heiskanen
Equine Information Centre, Neulaniementie 5 E, 70210 Kuopio, Finland; sanna.airaksinen@hevostietokeskus.fi

Today, horse bedding is a highly developed product and the international trade of different materials has increased extensively. In this study, the consumption and the costs of different horse bedding as well as the time spent on stable cleaning and handling the bedding was recorded for at least 30 days each. Bedding materials studied were: Finnish wood pellets, shavings, reed canary grass pellets and peat. The study was carried out during the indoor feeding season in a horse stable with six boxes. Horses were held outdoors an average of 5-6 h a day. Covered bedding and manure storages were located next to the stable. The time required for the same employees to muck out dung and urine, and add bedding to boxes as necessary, was recorded daily. The amount of bedding needed for six boxes in 30 days were as follows: shavings 1,520 kg, peat 1,600 kg, reed canary grass pellets 1,620 kg and wood pellets 1,940 kg. When compared to wood pellets (500 kg/bag), relative material cost (not including transport costs) for the consumed bedding was 181% with peat (25 kg/bale), 108% with shavings (20 kg/bale) and 101% with reed canary grass pellets (500 kg/bag). Time required to muck out six boxes for 30 days was following: peat 13 h, shavings 14 h, reed canary grass pellets 14 h and wood pellets 15 h. Stable workers reported the shavings used in the study were dust-free; instead, in some boxes, the bedding layer with wood pellets had to be watered regularly to prevent dusting. The bedding material cost was quite similar with wood pellets, reed canary grass pellets and shavings, even though the consumption of those materials was different. The material cost of peat bedding in small bales was the most expensive, but the time required for mucking out the boxes with peat was less than with the other materials studied. There was a difference of 2 h of employee costs between the bedding use of peat and wood pellets for six boxes in 30 days. When highly developed horse bedding products are used, the material choice has an effect on the work and bedding material costs of stable cleaning.

Incidence of extended spectrum b-lactamase producing Enterobacteriaceae colonization in horses
A. Shnaiderman-Torban[1], K. Kondratyeva[2], G. Abells Sutton[1], S. Navon- Venezia[2] and A. Steinman[1]
[1]Hebrew university, Veterinary medicine, Herzel st, Rehovot, Israel, [2]Ariel University, Molecular microbiology, Ariel, 407000, Israel; ashnaiderman@gmail.com

Antibiotic resistance is a global problem with a complex epidemiology. In humans, ESBL-E are considered a marker for multi-drug resistance, associated with increased illness severity, length of hospital stay and costs. We aimed to perform a prospective study, investigating ESBL-E colonizing horses in farms (n=176) and on admission to hospital (n=178). ESBL-E were recovered from enriched rectal swabs, sub-cultured on CHROMagarESBL plates. Species and susceptibility profiles were determined (MALDI-TOF, Vitek2). Medical records and owner questionnaires were screened. Carriage rate of ESBL-E on admission were 19% (n=33/178); including 38 ESBL-E. Main specie was *Escherichia coli* (65%) and major ESBL gene was CTX-M-1 type (54%). Two horses (1%) carried the hyper-virulent *E. coli* ST131 clone. 17% of *E. coli* belonged to virulent phylogroups D and B2. Resistance rates were 80% for tetracyclines, 80% for Trimethoprim-sulfa, 60% for aminoglycosides, 20% for quinolones and no resistance for carbapenems and polymyxins. No risk factors were found. In farms, 9% of horses (16/176) were colonized; including 20 ESBL-E. Main ESBL-E was *E. coli* (60%) and major ESBL gene was CTX-M-1(55%). Resistance rates were 74% for tetracyclines, 84% for Trimethoprim-sulfa, 27% for quinolones and 57% for chloramphenicol. In univariable analysis, originating farm, breed, age, previous antibiotic treatment and hospitalization were associated with colonization (P<0.05). In multivariable analysis, age and antibiotic treatment were significant. Our findings demonstrate the potential reservoir of ESBL-E pathogens in equine community settings. Resistant bacteria should be further investigated to improve antibiotic treatment regimens and equine welfare.

Extended-spectrum b-lactamases-producing Enterobacteriaceae in hospitalized neonatal foals

A. Shnaiderman- Torban[1], Y. Paitan[2], K. Kondratyeva[3], S. Tirosh- Levi[1], G. Abells Sutton[1], S. Navon-Venezia[3] and A. Steinman[1]

[1]Hebrew University, Veterinary Medicine, Herzel, Rehovot, Israel, [2]Meir hospital, Clinical Bacteriology Lab, Tshernihovsky, Kfar Saba, Israel, [3]Ariel University, Molecular Biology, Ariel, 40700, Israel; ashnaiderman@gmail.com

In human neonates, ESBL-E have been identified as a cause of infections outbreaks. In neonatal foals, bacterial infections, leading to sepsis, are major cause of death. Thus, early treatment with appropriate antimicrobials is crucial. We aimed to determine prevalence and risk factors for colonization and infection in foals on admission and during hospitalization. During a prospective study, paired rectal swabs were sampled from mares and their neonatal foals ($\leq$1 month). ESBL-E were recovered from enriched rectal swabs, sub-cultured on CHROMagarESBL plates. Bacterial ID and antibiotic susceptibilities were determined (Vitek2). Medical records were screened for risk factor analysis (SPSS). On admission, 59 foals and 55 mares were sampled, 60% (36/59) of foals and 55% (30/55) of mares were re-sampled. On admission, 34% (20/59) of foals and 16% (9/55) of mares were colonized. During hospitalization, 80% of foals (28/35) and 63% (19/30) of mares were colonized. Foals' colonization was significantly associated with umbilical infection on admission (P=0.026) and not correlated to mares' colonization, clinical signs or outcome. Species diversity increased during hospitalization and resistance rates changed. ESBL-E were recovered from 7% of foals (4/59), from umbilicus and wounds. All foals which had a positive clinical samples were colonized. Our results substantiate the alarming occurrence of ESBL-E in equine neonatal medicine and the possible infection of both mares and foals during hospitalization. These results emphasize the need for better continuous infection control in veterinary hospitals. Future studies will investigate homogenous foal populations and molecular ESBL-E characteristics.

Practical evaluation of live yeast *Saccharomyces cerevisiae* I-1077 on horse diet digestibility

F. Barbe[1], P. Bonhommet[2], M. Hamon[3] and A. Sacy[1]

[1]Lallemand SAS, 19, rue des Briquetiers, 31702 Blagnac, France, [2]Certivet, 48, avenue Pierre Piffault, 72100 Le Mans, France, [3]Clinique vétérinaire équine de Méheudin, 61150 Ecouche, France; fbarbe@lallemand.com

Digestion is a difficult physiological function to investigate routinely in horse: a previous study presented the interest of using double sieving of droppings as a simple and practical tool to evaluate horse digestibility and intestinal health. The objective of the present trial is to measure the effect of live yeast (LY) on ration digestibility. 5 gallopers (4-5 years old) in training received 2 g/horse/day of Levucell SC10ME, providing 20×10^9 cfu of *Saccharomyces cerevisiae* I-1077/horse/day during 2 months (mo). Diet and intake remained constant over the 2 mo period. Fresh droppings were collected before trial and then weekly and were analysed for their physical aspect (colour and firmness), pH and particle size: a sample of 350 g of fresh droppings was sieved through 5 mm and 2 mm-diameter grids and water was used to help the process until clear water is obtained. The two fractions F1 (coarse) and F2 (fine) coming from the 2 sieves respectively were weighted and the ratio F1/F2 was then calculated. For each parameter, results before and after 2 mo of supplementation were analysed by paired T-test. Values of pH were constant during the trial (6.73±0.16). Droppings colour (3.37±0.73) and firmness (4.06±0.76) were indicative of a good intestinal health for these training horses. F2 fraction increased significantly during the trial (before: 58.2±18.4 g; after: 96.5±23.6 g, P<0.05), which augmented also the contribution of F2 fraction in sieved sample after 2 mo with LY (before: 17±5%; after: 28±7%). Ratio F1/F2 was significantly lower at the end of the trial (before: 4.09±1.80; after: 2.01±1.24, P<0.05). These results suggest a better digestibility and improved value of ration with LY. Moreover, body condition score was significantly improved after 2 mo with LY (before: 1.40±0.55; after: 2.20±0.45, P<0.05). In adult horses, confinement and high-starch diets increase the risk of microflora imbalance. These results highlight the beneficial effect of LY on digestive efficacy in this context and reinforce the interest of using droppings sieving as a simple and practical field tool of intestinal health diagnosis, as previously studied.

Changing of the wither height in Hungarian Grey cattle breed in the last 15 years

Á. Maróti-Agóts[1], D. Fürlinger[1], I. Bodó[1], L. Baracskay[2], E. Kaltenecker[2] and A. Gáspárdy[1]
[1]University of Veterinary Science, Budapest, Animal Breeding and Genetics, István u. 2., 1078, Hungary, [2]Hungarian Grey Cattle Breeders Association, Lőportár u. 16., 1134, Hungary; maroti-agots.akos@univet.hu

Currently the population of Hungary's indigenous breed, the Hungarian Grey Cattle, is expanding continuously. Next to gene preservation, the production of goods has also appeared, as a new goal for the breed. In a growing population the chances of a genetic drift are always higher than in the case of a population whose size is stagnating. Therefore, we decided to monitor the body measurements on a population level. In 2001 we measured the body measurements of the cows of the breeding herd (n=1,048), using the Video Aided Measurements optometric system established back then. In 2001 the size of the population was just above the critical level, therefore, the exclusive breeding goal was to preserve genetic diversity and the unchanged appearance of the breed. 15 years later, as the population is growing, a new goal has appeared, and this is meat production. Repeating the population-level optometric body measurements of 2001, in 2016 we have measured the body parameters of cows (n=1,120) in four breeding populations. The changes of wither height are remarkable in the herds of Bugac and Tiszaigar, where this body parameter has increased significantly. In the population of Hortobágy wither height shows a slight increase, while in Sarród a modest decrease is also supported by statistics. In Bugac and Tiszaigar the standard deviations of wither height have decreased significantly. This shows the effect of choosing animals with a higher wither height for breeding. Based on the data of a relatively short period of time, it can be concluded that the wither height of the breed has not changed significantly, although due to the sales objectives a rising trend can already be seen. The decrease in data variance might be the result of selection. In order to maintain the breed unchanged, we need to also keep cows of a smaller body size in the future. It is important to archive the measurement frames created in the system, as this way in the future it will be possible to provide data to be processed by a body assessment system, using a new method.

Lactation curves parameters estimation by season of calving for Holstein dairy cattle in Tunisia

M. Khalifa, A. Hamrouni, A. Arjoune and M. Djemali
National Institute of Agriculture of Tunisia, Animal Science, 43, Street Charles Nicolle Mahrajène 1082, Tunisia, Tunisia; moniakhlifi3h@gmail.com

Holstein dairy cattle have been raised in Tunisia for more than four decades. The objective of this study was to estimate lactation curves parameters of Tunisian Holsteins, by season of calving. The incomplete gamma function was used to adjust milk lactation curves per calving season. A total of 976,188 milk test days of 68,068 cows in 2,911 herds during 2003-2015 was used in this study. A linear model was developed, including herd-day-month-test year, farm ownership, calving month, calving year, and lactation number, age of cow and milk test number. The main results showed that daily milk averages per calving season ranged from 14 ± 6 to 20.31 ± 9 kg/d in season 1 (September to January), 14 ± 6 to 19.94 ± 9 kg/d in season 2 (February to May) and 14 ± 6 to 19.54 ± 8 kg/d in season 3 (June to August).Estimated milk lactation curve parameters by Wood's Incomplete gamma function for the early lactation production (a), ascending phase (b),descending phase (c), Peak production, peak date and persistence were 15.71 kg; 0.083, 0.0012, 20.44 kg; 64 days and 7.20 respectively, in the first season (autumn-winter), 17.85 kg; 0.043; 0.001; 20.13 kg; 42 days and 7.16, respectively in season 2 (February to May), 16.03 kg; 0.066; 0.001; 19.75 kg; 63 days and 7.31 respectively for season 3 (June to August).

Use of mushroom laccase for reducing aflatoxin contamination of maize flour destined to dairy cows

M. Scarpari[1], L. Della Fiora[2], C. Dall'asta[2], A. Angelucci[3], L. Bigonzi[1], A. Parroni[1], C. Fanelli[1], M. Reverberi[1], J. Loncar[4] and S. Zjalic[4]

[1]University Sapienza, DEB, Piazzale Aldo Moro 5, 00185 Roma, Italy, [2]University of Parma, Parco Area dell Scienze 27/A, 43124 Parma, Italy, [3]Istituto zooprofilatico sperimentale della Lombardia e dell'Emilia Romagna, Via Bianchi 7/9, 25124 Brescia, Italy, [4]University of Zadar, Department of Ecology, Agronomy and Aquaculure, Trg kneza Višeslava 9, 23 000 Zadar, Croatia; szjalic@unizd.hr

Aspergillus flavus is a well-known ubiquitous fungus able to contaminate, both in pre- and postharvest period, different feed and food products. During their growth, these fungi can synthesize aflatoxins, secondary metabolites highly hazardous for animal and human health. Aflatoxins represent a serious problem in cattle breeding. They can have negative influence on animal health, like impaired growth or lower immunity and the intake of contaminated feed could result in production of milk and milk products unsuitable for market and human consumption due to high content of aflatoxin M1. Climatic changes make aflatoxins occurring contaminant of maize and other cereals also in EU countries. Even if different strategies for prevention and control of aflatoxins in food and feed are applied, the complete control still has not been achieved. The mitigation of mycotoxin content via enzymatic degradation could be a promising strategy to ensure safer food and feed, and to address the forthcoming issues in view of the global trade and sustainability. Results obtained with mushroom oxidases in *in vitro* and *in vivo* experiments are encouraging, the degradation rate was up to 90%. Nevertheless, the search for active enzymes is still challenging and time-consuming.

Palmitic acid feeding increases ceramide synthesis, milk yield, and insulin resistance in dairy cows

J.E. Rico, A.T. Mathews and J.W. McFadden

West Virginia University, Animal and Nutritional Sciences, 26506 Morgantown, WV, USA; jericonavarrete@mix.wvu.edu

Reduced insulin action facilitates lipolysis and glucose partitioning for milk synthesis during early lactation. Insulin sensitivity increases beyond peak milk yield, while circulating fatty acids (FA) and milk production decline. Palmitic acid (C16:0) promotes insulin resistance in monogastrics through ceramide-dependent mechanisms, and ceramides are elevated during early lactation. We hypothesized that feeding C16:0 would increase ceramide concentrations and restore peripheral insulin resistance during mid-lactation. Twenty multiparous Holstein cows (137±45 DIM) were enrolled in a study consisting of a 5 d covariate and 49 d treatment period. Cows were randomly assigned to a sorghum silage-based diet containing no supplemental fat (control; n=10) or C16:0 at 3.9% of ration DM (PALM; 98% C16:0; n=10). Blood was collected routinely, and liver biopsied at d47 of treatment. Intravenous glucose tolerance tests (GTT) were performed at d -1, 21, and 49 of treatment start. Sphingolipid levels were quantified using mass spectrometry. Expression of ceramide synthesis genes was evaluated using qPCR. Data were analyzed as repeated measures using a mixed model. Pearson correlations were analyzed. Relative to control, plasma ceramide, monohexosyl-, and lactosyl-ceramide decreased as lactation progressed (P<0.01). Plasma total ceramide and C24:0-ceramide were increased by d 8 of treatment in PALM, and remained elevated throughout the 7 wk treatment (80% average; P<0.01). PALM increased C24:0-ceramide and total hepatic ceramide levels by 29 and 20% at wk 7, respectively (P<0.05). PALM decreased hepatic CerS2 and CerS5 mRNA (P<0.05). Plasma total ceramide was positively associated with hepatic total ceramide (r=0.63; P<0.05). Hepatic and plasma C24:0-ceramide were associated with plasma FA (r=0.52; P<0.01) and FA disappearance during GTT (r=-0.57; P<0.01). Plasma C24:0-ceramide was correlated with milk yield (r=0.44; P<0.01), a relationship shared by most detected ceramides. Increasing C16:0 intake to augment ceramide synthesis delayed the decline in plasma ceramide observed with the progression of lactation. Future research should evaluate whether ceramide is intrinsically involved in the homeorhetic adaptation to lactation.

Author index

A

Aad, P.Y.	310, 322	Amills, M.	114
Aamand, G.P.	233, 427	Ammermueller, S.	134
Abdelatty, A.M.	144	Ammon, C.	344
Abdel Baray, W.	374	Amon, B.	344
Abd Elhairth, H.	401	Amon, T.	344
Abdellrazeq, S.G.	407	Ampe, B.	218, 219
Abdullah, A.Y.	448	Ampuero Kragten, S.	416
Abe, H.	260, 348	Amzil, L.	285
Abells Sutton, G.	484, 485	Ananth, D.	421
Abou-Ward, G.A.	338	Andersen, H.M.-L.	284, 291
Abrahamsen, M.	466	Andersen, I.L.	205
Abud, M.J.	360	Anderson, S.T.	117
Åby, B.A.	440	Andersson, G.	351
Adam, A.	285	Andersson, L.	232
Adamski, J.	148	Andonovic, I.	406
Addo, S.	101, 185	Andonov, S.	131
Ader, P.	392	Andrée O'Hara, E.	257
Ådnøy, T.	185	Andrés, S.	388, 448
Adriaens, I.	147, 149	Angel-Garcia, O.	363
Aernouts, B.	147, 149	Angelucci, A.	487
Agastin, A.	324	Anghel, A.	363
Agazzi, A.	389, 394, 395, 444	Ankeny, R.A.	383
Aguirre, P.	367	Annabi, M.	174
Ahasan, A.S.M.L.	395	Anton, I.	308
Ahlskog, K.	154, 163	Anttila, A.M.	177
Ahrens, A.	449	Appel, A.K.	107
Airaksinen, S.	484	Arana, A.	304
Ajmone Marsan, P.	135	Arandas, J.K.G.	286
Ajuwon, K.	305	Arellano-Rodriguez, G.	358, 363
Akbari Chalaksari, M.	443	Argiriou, A.	161
Akhondzade, H.	308	Arija, I.	341
Albrecht, E.	304, 307	Ariko, T.	307
Alcantara-Neto, A.	92	Ariza, J.M.	126
Aldabeeb, S.	357	Arjoun, A.	374
Alfonso, L.	304	Arjoune, A.	486
Ali, A.M.	338	Arranz, J.	288
Alifa, B.	355	Arsenos, G.I.	161, 228, 359, 421
Aljamal, A.E.	459	Arsiwalla, T.	409
Al Kasbi, M.S.	448	Asa, R.	226, 331, 332
Alkhoder, H.	427	Aschenbach, J.R.	452
Allen, P.	327	Asher, A.	387
Alloggio, I.	303	Ashrafi Toochahi, S.	443
Alminana, C.	92	Ask, B.	207
Al-Saiady, M.	357	Astruc, J.-M.	151
Al-Tamimi, H.	467	Ata, A.	432
Aluwé, A.	215	Ataozu, A.	306
Aluwé, M.	111, 214, 216, 218, 219	Auffret, M.D.	438, 444
Alves, A.G.C.	286	Augusto, J.G.	186, 187, 188
Alves Filho, L.M.	120, 277	Autio, E.	483
Amanzougarene, Z.	274, 281	Avdi, M.	209, 357
Ambriz-Vilchis, V.	201	Averbukh, E.	155
Amer, P.R.	79, 154, 181, 255, 376	Averkieva, O.	397

Bhatti, H.S.	267
Biancarosa, I.	475, 478
Bielli, A.	360
Biermann, A.D.M.	182
Biffani, S.	135
Bignon, L.	175
Bigonzi, L.	487
Bijma, P.	100, 190, 207
Bikker, P.	132, 318, 452
Billiet, M.	380
Billon, Y.	240
Bindelle, J.	269, 447
Bindeman, H.	317
Bink, M.C.A.M.	193
Bittante, G.	165
Bitti, M.	470
Bium, B.F.	269, 274
Bizelis, I.	365
Bjerring, M.	401
Bjerrum, L.	479
Black, R.A.	345
Blake, D.P.	466
Blanc, F.	90, 463
Blanc, G.	349
Blanco, A.	102
Blanco, M.	469
Bläuer, A.	296
Blavy, P.	90
Blees, T.	122
Bloch, V.	223
Blokhuis, H.	172
Blunk, I.	191
Bluyssen, H.	471
Bodin, L.	162, 241, 244
Bodó, I.	486
Boekschoten, M.V.	387, 447
Boichard, D.	135
Boillat, E.A.	379
Bojkovski, D.	104
Bokkers, E.A.M.	169, 283
Bolaji, O.J.	341
Boland, T.M.	454
Bolormaa, S.	232
Bonacim, P.M.	186, 187, 188
Bonaldo, A.	476
Bonaudo, T.	211
Bonhommet, P.	485
Bonilha, S.F.M.	328, 331
Bonizzi, L.	303
Bonneau, M.	214, 215, 216
Bonny, S.P.F.	327
Bontempo, V.	264, 394
Boonanuntanasarn, S.	399
Borchersen, S.	195
Borg, R.	175
Borreani, G.	441

Bortoluzzi, C.	207
Borutova, R.	397
Bosi, P.	107, 112
Bou Karroum, S.	310
Boulton, K.	466
Boura, A.	246
Boushaba, N.	87
Boussaha, M.	135
Boussemart, J.P.	325
Bouwman, A.C.	83, 143
Bouyeh, M.	280, 308, 390, 443
Boval, M.	166, 324
Bovenhuis, H.	184, 366, 428
Bovolenta, S.	165
Boyle, A.	259
Boyle, L.	128, 129, 222
Bozkurt, Y.	225, 384
Bozzi, R.	286, 330
Bradford, H.L.	426
Brajkovic, V.	365
Brandt, H.	173
Brandt, H.R.	289
Bravo De Laguna, F.	394
Bray, H.J.	383
Brehmer, L.S.	218
Brenes, A.	341
Bressan, E.	473, 476
Briens, M.	264
Brigden, C.	482
Brinker, T.	193, 207
Broadbent, J.	152
Brochard, M.	195
Brockmann, G.A.	445
Brosh, A.	387
Brossard, L.	244, 267, 319, 321
Brostaux, Y.	282
Brouček, J.	178
Bruckmaier, R.M.	116, 124, 306
Brügemann, K.	173, 377
Brugiapaglia, A.	476
Bruininx, E.M.A.M.	414
Brun, A.	419
Bruschi, P.	286
Bruun, S.	472
Brzáková, M.	249
Buchet, A.	289, 413
Buddle, E.A.	383
Bugdayci, K.E.	432
Buitenhuis, A.J.	199
Bunevski, G.	365
Bunger, L.	180
Burcea, A.	317
Burfeind, O.	129, 290, 291
Burgos, A.	369
Burgstaller, J.	125
Burren, A.	87, 379

Faux, P.	311
Faverdin, P.	167, 253, 457
Fegraeus, K.	352
Fekete, A.A.	146
Fenelon, M.A.	151, 439
Fennessy, P.F.	315
Ferchaud, S.	245
Ferenčaković, M.	310, 365
Ferguson, N.	326
Ferlay, A.	145, 463
Fernandes, S.R.	120, 122, 277, 332
Fernández, A.	102
Fernandez-Álvarez, J.	226
Fernández, J.	102, 194
Fernandez Martin, J.	101
Ferrari, G.	330
Ferrari, P.	385
Ferreira Júnior, R.J.	331
Ferreira, L.M.M.	298
Ferrer, J.	326, 329, 469
Ferris, C.	199, 259
Fidani, M.	389
Fievez, V.	334
Figueiredo-Cardoso, T.	114
Figueiredo, M.	120, 122
Figueroa, J.	293, 390
Fikse, F.	378
Fikse, W.F.	367
Filipe, J.A.N.	321
Finn, A.	323
Finocchiaro, R.	135
Fischer, A.	253
Fischer, D.	108, 316
Flach, L.	422
Flegar-Meštrić, Z.	469
Fleischer, P.	251, 263
Fleming, A.	183, 312
Fleming, H.R.	273
Fletcher, M.T.	117
Florescu, I.E.	317, 391
Flury, C.	87, 141
Flynn, P.	138, 142
Fodor, N.	346
Fogh, A.	195
Fogliano, V.	474
Foldager, L.	402
Fondevila, M.	274, 281, 469
Fonseca, A.	109
Fonseca, A.J.M.	258, 337
Fonseca, I.	337
Fontanesi, L.	215, 216
Font-I-Furnols, M.	214, 215, 216, 419
Fontoura, A.B.P.	118, 119
Formoso-Rafferty Castilla, N.	241, 244, 375, 418
Fortin, F.	221, 227
Fortun-Lamothe, L.	245

Forutan, M.	312
Foskolos, A.	338, 346
Foster, N.	248
Foyer, C.H.	346
Fracchiolla, M.L.	389
Fragomeni, B.O.	426, 427
Franco, V.	431
Frąszczak, M.	233, 371
Frederiksen, J.	218
Fredriksen, B.	215, 216
Freeman, T.C.	444
Freem, L.	466
Freitas, A.P.	188
Friedrich, L.	127
Fries, R.	134, 232
Frieten, D.	303
Friggens, N.C.	90, 92, 118, 457
Frioni, N.	230
Frischknecht, M.	87, 141
Fritz, S.	135, 427
Froidmont, E.	282, 403
Frutos, J.	388, 448
Fry, L.M.	407
Fuerst, C.	125
Fuerst-Waltl, B.	125, 142, 153, 463
Fujiwara, M.	247
Füllner, B.	93
Fulton, J.E.	468
Fürlinger, D.	486
Furre, S.	101, 298

G

Gabriel, I.	238
Gac, A.	213
Gadeyne, F.	334
Gado, H.M.	401
Gagnon, P.	221, 227, 319
Gai, F.	473, 476, 477
Galama, P.J.	88, 345, 346
Galán, E.	343
Galvão, A.	171
Gal, Y.	223
Gameiro, A.H.	186, 211
Gamo, M.E.S.	449, 480
Gangnat, I.D.M.	458
Gaouar, S.B.S.	105
Garcez Neto, A.F.	120, 122, 277, 332
García-Baccino, C.A.	230
García-Ballesteros, S.	350
Garcia-Launay, F.	112, 167, 213
Garcia, M.	144
Gardner, G.E.	327
Gariepy, C.	227
Garnett, T.	164
Garreau, H.	239
Garrick, D.	87

Gasco, L.	473, 476, 477	Goldshtein, S.	343
Gáspárdy, A.	486	Gombault, P.	453, 458
Gasparrini, B.	431	Gomes, A.C.M.	258
Gath, V.	454	Gómez, E.A.	435
Gaudré, D.	112, 167, 212, 213	Gondret, F.	118, 239, 265, 268, 302
Gauly, M.	128, 246, 250, 258, 368, 422, 456	Gonzalez-Mejia, A.	338
Gautier, J.M.	228, 423	González, O.	237
Gautier, M.	311	González-Rodríguez, O.	114
Gavojdian, D.	386, 425	Goodwin, K.	117
Gaytán, L.R.	358, 361, 363	Gootwine, E.	155
Gebregiwergis, G.T.	314	Gorgulu, M.	356
Geertsema, H.	82	Gori, A.-S.	311
Gelasakis, A.	246	Goto, T.	226, 331, 332
Gele, M.	339	Götz, K.-U.	85, 234, 236
Genever, E.	328	Graczyk, M.	97
Gengler, N.	125, 169, 199, 403, 406	Gramatina, I.	419
Genreith, A.	269	Grandl, F.	199
Georgescu, S.E.	317, 391	Grandoni, F.	407
Georges, M.	136	Granier, R.	292
Georgiadou, M.	356	Gras, A.M.	235, 361
Gerber, V.	124	Grashorn, M.	466
Germain, K.	245	Gravellier, B.	368
Gerster, E.	123	Greathead, H.	421
Gertz, M.	385	Greco, V.	303
Getya, A.	220	Gredler-Grandl, B.	87, 139, 141, 153, 183, 376
Geurden, I.	367	Greenhalgh, C.	475
Gharbi, B.	301	Gregorini, P.	91, 98
Gheorghe, A.	272	Grelet, C.	96, 403
Ghita, E.	272, 361	Gress, L.	240
Giaccone, D.	441	Greydanus, J.F.	293
Giannico, R.	233	Grimberg-Henrici, C.G.E.	290
Gianotten, N.	479	Grindflek, E.	193, 367
Gibbons, J.	338	Grinshpun, J.	223
Gicquel, E.	101	Grobler, J.P.	317
Gidenne, T.	333, 453, 456, 458	Grodkowski, G.	172
Giersberg, M.F.	208	Groenen, M.A.M.	388, 428
Giger-Reverdin, S.	354	Grohmann, D.	98
Gilbert, H.	195, 238, 239, 240	Grohs, C.	135
Giles, T.	248	Große-Brinkhaus, C.	107
Gilkinson, S.	259	Gross, J.J.	116, 124, 306
Giráldez, F.J.	388, 448	Grosu, H.	235
Girard, M.	292	Grotta, L.	146
Giromini, C.	146	Grøva, L.	423
Gispert, M.	419	Gruber, L.	142, 463
Givens, D.I.	146	Guan, X.	264, 464
Gjerlaug-Enger, E.	193	Guarino, M.	129, 176
Gjøen, H.M.	190	Guatteo, R.	126, 175
Glasser, T.	223	Guemez, H.R.	398
Gleeson, D.	147, 443	Guevarra, R.B.	449, 480
Gobi, J.P.	111	Guillon-Kroon, C.	175
Goby, J.P.	456	Guimarães, A.L.	331
Goddard, E.	376, 421	Guimarães, S.E.F.	108
Goddard, M.E.	232	Guivernau, M.	237
Godinho, R.M.	108	Guldbrandtsen, B.	82, 427
Godo, Y.	223	Gumilar, R.	299
Golak, F.	366	Günal, M.	441

McClure, M.	140, 142, 440	Michon, G.	285
McClure, M.C.	138	Michot, P.	135
McCoard, S.A.	265	Miech, P.	478, 480
McConnell, D.	259, 328	Mielczarek, M.	233, 371
McDermott, A.	151	Miglior, F.	183, 376
McFadden, J.W.	487	Mignon-Grasteau, S.	238
McGee, M.	201, 326, 330	Mihalcea, T.	272, 361
McHugh, N.	182	Mikko, S.	351
McKay, Z.C.	397	Milanesi, M.	135
McKenzie, C.M.	265	Milerski, M.	162
McKilligan, D.	295	Millán, A.	102
McParland, S.	151	Miller, G.A.	223
Mead, A.	212	Miller, S.P.	119, 181, 327
Meadows, J.	352	Millet, S.	111, 132, 218, 219
Meås, A.	440	Miltiadou, D.	358
Méda, B.	112, 165, 167, 213, 319, 336	Minery, S.	195
Médale, F.	367	Minozzi, G.	233
Medugorac, I.	104, 365	Mirkena, T.	160
Megens, H.-J.	428	Miron, M.	387
Mehanna, N.	451	Misztal, I.	83, 131, 426, 427
Mehinagic, T.	139	Mitchell, M.A.	380
Meier, H.P.	349	Mitka, I.	113
Meignan, T.	333, 455	Mizoguchi, Y.	417
Melchior, D.	450	Moate, P.	201
Mellado, M.	358, 361	Moazami-Goudarzi, K.	87
Melo, M.P.	274	Mock, T.	139
Melzer, N.	131, 312	Mogensen, L.	203
Mendizabal, J.A.	304	Mohamed, F.F.	144
Mendonça, G.G.	186, 187, 188	Mohana Devi, S.	390
Mendoza, J.	358	Mohr, U.	123
Meneguz, M.	473, 476, 477	Moik, M.	93
Meng, J.	370	Moioli, B.	159
Mercadante, M.E.Z.	86, 328, 331, 393	Molee, W.	398
Mercat, M.J.	109, 195	Moles Caselles, X.	386
Mercier, Y.	264, 268	Molina, A.	350, 482
Merlot, E.	289, 305, 309, 413	Molina-Alcaide, E.	422
Mermillod, P.	92	Molina, E.	326, 468
Mero, H.	163	Molist, F.	264, 414, 464
Messikommer, R.E.	458	Mollier, P.	437
Messori, S.	245, 382	Moll, J.	153
Mestawet, T.A.	365	Moloney, A.P.	326, 330
Mészáros, G.	80, 103	Montagnani, M.	477
Metges, C.C.	465, 466	Montanholi, Y.R.	118, 119
Metuki, E.	223	Monteiro, P.	171
Metwally, H.M.	401	Montossi, F.	158, 421
Meunier, B.	172	Monziols, M.	415
Meuwissen, T.H.E.	82, 157, 193, 228, 229, 314	Mooney, M.H.	326
Meyer-Aurich, A.	91	Moorby, J.M.	338, 346
Meyermans, R.	162	Mootse, H.	150
Meylan, M.	139	Moradi, F.	433
Meza-Herrera, C.A.	363	Moraine, M.	213
Mialon, M.M.	172	Morales, A.	91, 98
Miari, S.	152	Morante, R.	369
Michailidou, S.	161	Moravčíková, N.	80, 297, 352
Michel, L.	175	Morbidini, L.	98
Michie, C.	406	Moreno, C.	102

Printed in the United States
by Baker & Taylor Publisher Services